Advances in Measurement Systems

Advances in Measurement Systems

Edited by
Milind Kr Sharma

Advances in Measurement Systems

Edited by Milind Kr Sharma

CBS, Edition **2016**

Published by InTech

Janeza Trdine 9, 51000 Rijeka, Croatia

Technical Editor Teodora Smiljanic

Cover Designer Dino Smrekar

Additional hard copies can be obtained from orders@intechopen.com

Advances in Measurement Systems, Edited by Milind Kr Sharma
p. cm.
ISBN 978-953-307-061-2

Contents

In-field measurement of soil nitrate using an ion-selective electrode

Kevin J. Sibley, Gordon R. Brewster, Tessema Astatkie, John F. Adsett
Nova Scotia Agricultural College
Canada

Paul C. Struik
Wageningen University
the Netherlands

1. Introduction

Standard laboratory methods for measurement of soil nitrate (NO_3-N) use various procedures and instruments to analyze soil samples taken from the field and transported to the laboratory. Concerns with these procedures range from delays in measurement time, the high cost of soil sampling and analysis, high labour requirements, and the need to aggregate samples. With recent advances in using the ion-selective electrode, as presented in this chapter, soil NO_3-N can now be measured directly, rapidly, accurately, at low cost, at a fine scale, and in real-time right in the field. This chapter describes the methodologies and procedures for how this can be done and provides experimental data and results from data analyses that validate measurements of soil NO_3-N obtained with a prototype soil nitrate mapping system (SNMS) developed at the Nova Scotia Agricultural College, Truro, Nova Scotia, Canada. These advances in the in-field use of the nitrate ion-selective electrode (NO_3^--ISE) provide the ability for (i) assessing soil nitrate variation, (ii) linking soil nitrate variation to crop growth, (iii) developing site-specific crop management practices, and (iv) environmental monitoring of soil nitrate.

This chapter will begin with a discussion of the concerns with nitrate in the soil and environment, precision agriculture and site-specific crop management, variation in soil nitrate and its links to crop growth and yield, and issues with assessing soil nitrate variation in a field. Next will be a discussion of ion-selective electrode theory and application for measuring soil nitrate, followed by a presentation and discussion of early experiments conducted for determining electrode operating parameters to enable the electrode to be used in a soil slurry. The development and testing of the mechanical system used for soil nitrate extraction and measurement along with a description of the control sub-unit, measurement methodology, and operation of the nitrate extraction and measurement sub-unit (NEMS) for using the NO_3^--ISE in the field will be presented. And the results of experiments used to validate in-field measurements of soil NO_3-N obtained with the ion-selective electrode will be presented and discussed. There will be a discussion of what is significant about the new measurement advances presented along with some results of experiments conducted using the SNMS in wheat and carrot production systems. Finally, conclusions and recommendations for future research in this area will be made.

1.1 Soil nitrate is an environmental issue

In addition to the fertility needs of farmers, it is important to deal with environmental issues associated with the use of nitrogen fertilizers. As agriculture continues its best efforts to provide the world's rising population with high-quality, safe, and nutritious food, water sources contamination and associated socio-economic costs indicate a great need for precise soil fertility management practices—using the right form of fertilizer, applied at the right time and place, in the right amount, and in the right way (Power & Schepers, 1989; Dinnes et al., 2002).

The seriousness and extent of NO_3^- contamination of water sources and its effect on drinking water quality has been documented and discussed by many researchers in Canada, the United States, and the European Community (USEPA, 1990; Reynolds et al., 1995; Oenema et al., 1998; Henkens & Van Keulen, 2001). As a result, policy makers are revising laws to ensure the safety of public water supplies. These include amendments to the Water Pollution Control Acts in Canada and the United States, the European Community Nitrate Directive, and the Mineral Policy in the Netherlands.

Nitrate leaching from soil into groundwater has been attributed to poor soil nitrogen management practices involving inorganic and manure fertilizer inputs (Geron et al., 1993; Campbell et al., 1994; Patni et al., 1998; Koroluk et al., 2000; Astatkie et al., 2001; Randall & Mulla, 2001; Dinnes et al., 2002). As such, better soil nitrogen management practices, including more accurate fertilizer recommendations and placement, could help minimize the contribution by agriculture to the NO_3^- pollution problem.

1.2 Precision agriculture and site-specific crop management

The profitability of farmed crops can be severely affected if poor nitrogen management practices are used. Precision agriculture technology offers farmers the potential to more intensely and precisely analyze variations in numerous field conditions throughout the growing season, in association with environmental and crop response data in order to make the most sound, and site- and time- specific, management decisions possible. At the same time the public can be assured those practices are being conducted in the most environmentally friendly way (Adamchuk et al., 2004a; Bongiovanni & Lowenberg-DeBoer, 2004; Bourenanne et al., 2004).

The inability to assess soil and plant data rapidly and inexpensively in the field, however, remains one of the biggest limitations of precision agriculture (Adamchuk et al., 2004b). In particular, the lack of a soil NO_3-N measurement system is a major roadblock (Ehsani et al., 1999). If this roadblock could be overcome, a positive contribution toward improving precision agriculture technology would be made.

1.3 Variation in soil nitrate and its links to crop growth and yield

Soil NO_3-N levels in agricultural fields, as well as other chemical and soil physical properties, exhibit high variation spatially and temporally and at different measurement scales and levels of aggregation (Heuvelink & Pebesma, 1999). Much research has been dedicated to assessing and characterizing this variation to improve our understanding of the effects of soil NO_3-N on crop growth and yield within agro-ecosystems (Almekinders et al., 1995).

Growing plants utilize varying amounts of soil NO_3-N during different phenological (growth) stages and its availability should ideally be in response to the plant's need. In wheat, for example, the level of available soil NO_3-N during early plant growth determines yield for the most part by influencing population density and the degree of stimulation of tiller fertility, spikelet initiation, and floret fertility. Soil NO_3-N uptake is greatly reduced shortly after anthesis, and nitrogen is re-translocated from leaves primarily, and other vegetative organs secondarily, to the ears to meet the need of the filling grains (Simpson et al., 1983). The reduction in soil NO_3-N uptake during grain filling varies with weather conditions, disease pressures, and subsequent management practices (i.e. irrigation or chemical applications) which put stress on the plants. Physiologically, soil NO_3-N and crop yields are linked via nitrate uptake and its conversion into proteins and chlorophylls during plant growth (Engel et al., 1999; Schröder et al., 2000) and photosynthesis buffering against soil nitrogen deficits by an abundance of RuBP carboxylase that serves as a reserve of protein in the leaves during unfavourable weather conditions (Hay & Walker, 1989).

The availability and distribution of NO_3-N in the soil depends on many soil forming, chemical, microbial, plant growth, environmental, and management factors that influence soil crop dynamics (Addiscott, 1983; Wagenet & Rao, 1983; Trangmar et al., 1985). Because the effects of these factors and their interactions are highly variable (Almekinders et al., 1995), they also lead to the characteristic behavior of NO_3-N being highly variable within the soil.

Studying the levels of nitrogen in various plant tissues and organs at the various phenological stages simultaneously with the availability of soil NO_3-N, and on a fine-scale, could provide information to researchers and farmers useful for developing better site-specific nitrogen management (SSCM) practices. Collecting this information at the required sampling intensity, however, has been found to be very tedious and generally cost and time prohibitive using current methods (Engel et al., 1999; Ehsani et al., 2001; Adamchuk et al., 2004a).

1.4 Assessing soil nitrate variation

Geostatistical techniques have been developed to provide practical mathematical tools for assessing spatial and temporal variation, and spatial structure of soil properties including soil NO_3-N (Burgess & Webster, 1980; Webster & Burgess, 1984; Webster & McBratney, 1989; McBratney & Pringle, 1999).

Research applying these tools on a field-scale, such as through SSCM-experimentation (Pringle et al., 2004), has led to the development of a multitude of methods for determining minimum soil sample spacing, sampling grid layout and cell size (Russo, 1984; Han et al., 1994; Van Meirvenne, 2003; Lauzon et al., 2005), optimum number of samples (Webster & Burgess, 1984), sampling schemes and protocols for pre-planning experimental designs (Trangmar et al., 1985; Chang et al., 1999; Ruffo et al., 2005) and sample bulking strategies (Webster & Burgess, 1984).

However, when using these methods for implementing precision agriculture practices related to soil nitrogen management, the "most serious obstacles" are still the need to know the spatial structure in advance and the cost of obtaining this information even though the sampling effort required is much less than for full-scale sampling (Lark, 1997; McBratney & Pringle, 1999; Jung et al., 2006).

1.5 Concept of a soil nitrate mapping system

Development of an SNMS could contribute to the advancement of precision agriculture by providing a way to quickly, accurately, and affordably collect the data necessary to analyze small-scale variation in soil nitrate in time and space while crops are being grown, thus enabling this variation to be linked to crop growth and yield. Ideally, an SNMS would automatically collect a soil sample in the field and directly measure nitrate concentration in real-time. Moreover, global positioning system (GPS) geo-referenced data could be simultaneously recorded at each sampling location to enable a nitrate map to be created for the field. An SNMS, thus, would overcome many of the impediments, roadblocks, and serious obstacles of measuring and assessing soil NO_3-N variation using conventional methods in terms of sample analysis lag time, high labour requirements, and high costs as discussed above. The overall objective of the experimental work described in this chapter was to develop and validate such an advanced soil NO_3-N measurement and mapping system.

2. Attempts by others to develop methods for in-field measurement of soil nitrate

Over the last 20 years or so, attempts to develop a real time soil NO_3-N measurement system by other researchers have been based on three types of sensors: (i) ion-selective field effect transistor (ISFET), (ii) ISE, and (iii) spectrophotometer. The majority of this research work has not progressed past laboratory feasibility studies and testing in soil-bins. A brief review of these works is presented below. Details can be obtained by reviewing the cited papers directly, or the summaries contained in the comprehensive review paper recently published by Adamchuk et al. (2004a) who concluded that "sensor prototypes capable of accomplishing this task are relatively complex and still under development."

2.1 Ion-selective field effect transistor sensor based systems

Loreto & Morgan (1996) developed a prototype real time soil NO_3-N measurement system that consisted of a soil core sampling wheel, indexing and processing table, and a data acquisition and control system. This system was quite similar to that of Adsett & Zoerb (1991); however it used a specially developed prototype ISFET as the NO_3^- analysis instrument. In soil bin tests, correlations between ISFET measurements with a NO_3^--ISE and laboratory colorimetric analysis measurements had an R^2 between 0.65 and 0.43, respectively. The system worked reasonably well as a first attempt, but issues with the ISFET's response characteristics and calibration drift were apparent. Work has continued focusing on the development of ISFET technology and its use in combination with novel soil extraction and flow injection analysis (FIA) systems as a potential method of real-time measurement of NO_3^- in filtered soil extracts (Birrell & Hummel, 1997, 2000, 2001; Price et al., 2003). This work has resulted in the development of a promising combination ISFET/FIA system that gives reasonable results compared to a cadmium reduction method using a Lachat FIA (Slope 1:1, $R^2 = 0.78$) with a measurement time ranging between 3-5 s (Price et al., 2003), but it is still at the laboratory level.

2.2 Ion-selective electrode sensor based systems

A prototype nitrate monitoring system (NMS), was developed by Adsett (1990) and Adsett & Zoerb (1991). It used a specially designed unit for NO_3^- extraction wherein the soil was mixed with de-ionized water and then the liquid fraction was clarified before being presented to the electrode for NO_3^- measurement. Although the system functioned reasonably well as a first attempt, it had major difficulties with collecting a soil sample and obtaining a clear extractant for NO_3^- measurement on a consistent basis. This early work was the starting point from which improvements have been steadily made by Thottan et al. (1994), Thottan (1995), Adsett et al. (1999), Khanna & Adsett (2001), and Sibley (2008) that have advanced the system to the form described below in sections 5 and 6 into a fully functioning and field-validated prototype SNMS.

As part of an investigation into the feasibility of a real time soil K and NO_3–N mapping system, Adamchuk et al. (2002a) performed laboratory tests on four commercially available NO_3^--ISEs to simulate the direct soil measurement technique used in an automated soil pH measurement system developed by Adamchuk et al. (1999, 2002b). In the laboratory, manually remoistened previously air dried soil samples were pressed into contact with the sensing membrane of each NO_3^--ISE to determine NO_3^- concentration (liquid basis of mg L^{-1} reported as ppm). These results were compared to a standard cadmium reduction laboratory analysis technique to give an indication of the accuracy of the NO_3^--ISEs. For individual soil samples, R^2 values ranging 0.38–0.63 were obtained, depending on the ISE, while averaging of three repeated measurements yielded R^2 values ranging 0.57–0.86. It was concluded that it is feasible to use a NO_3^--ISE for measuring soluble nitrate concentration of naturally moist soil samples, but one of the main limitations of the proposed method reported was difficulty in maintaining high quality contact between soil and electrode. It should also be noted that use of the proposed method in the field in combination with the pH measurement system's soil sampling mechanism would not enable the NO_3–N content (mg kg^{-1}) of the sample to be directly computed since the 'weight' (mass) of the soil sample would not be known.

2.3 Spectrophotometer sensor based systems

Laboratory testing and field-based experimentation of a near-infrared (NIR) spectrophotometer conducted by Ehsani et al. (1999) using soils samples spiked with ammonium sulfate, ammonium nitrate, and calcium nitrate (10–100 ppm) revealed that soil NO_3-N could be detected with R^2 ranging 0.76–0.99 using partial least squares regression with each data point being an average of 10 sub-samples. However, the calibration equation must be derived from samples taken from the same location, otherwise the analysis procedure fails. Further laboratory-based research work (Ehsani et al., 2001) using soil samples spiked with ammonium nitrate and calcium nitrate (400–3000 ppm) and a spectrophotometer equipped with a deuterated triglycine sulfate (DTGS) sensor showed that the ratio of area under the nitrate peak to area under the water peak in the mid-infrared (MIR) spectra is proportional to NO_3^- concentration (R^2 = 0.81), and that the analysis technique is not dependent on the time of measurement, soil type, or nitrate source. However, as the authors themselves note, the range of NO_3^- concentration in agricultural soils is usually less than 100 ppm so the practicality of this sensing method is questionable unless a more sensitive mercury cadmium telluride (MCT) type sensor can be used.

Use of a real-time portable spectrophotometer using a multi-spectral approach has been

investigated by Shibusawa et al. (1999, 2003). They reported that NIR reflectance could be used to detect soil NO_3-N with an R^2 of 0.50.
Christy et al. (2003) have conducted preliminary field testing of a prototype soil reflectance mapping unit utilizing a NIR spectrophotometer for simultaneously measuring total N, total carbon, pH, and moisture content. Results from testing in a single field indicated the system could repeatably produce clear definition of patterns in these soil parameters related to spectral reflectance with an R^2 of 0.86, 0.87, 0.72, and 0.82, respectively.

3. Ion-selective electrode theory and application for measuring soil nitrate

The nitrate ion-selective electrode (NO_3^--ISE) (Fig. 1) provides a rapid and reliable method for quantitative analysis of soil nitrate. Nitrate ISEs, which are highly selective to NO_3^- ions in solution, were first used around 1967 as quick and reliable alternatives to chemical-based laboratory methods for nitrate measurement (Dahnke, 1971). The NO_3^--ISE electrochemically generates a voltage across its organophilic membrane that varies with ionic strength (molarity) of the solution according to the Nernst equation (Morf, 1981).

$$E = E_0 + S \log (A) \quad (1)$$

where E is the electrochemical cell potential (mV), E_0 is the standard potential (mV) in a 1M solution, ideally a constant, S is the electrode slope (-mV per decade of concentration), and A is the nitrate activity (effective concentration moles L^{-1}) in the solution.
Through calibration with known standards, the logarithm of solution molarity is related to electrode output voltage to determine a linear calibration curve for determining nitrate concentration (mg L^{-1} or ppm) of subsequent soil samples.
Typically in the laboratory, measurement of nitrate concentration of a soil sample then proceeds by mixing together a known 'weight' (mass) of soil with a known volume of deionized or distilled water (e.g. soil:extractant ratio). After an appropriate extraction time, the extractant in the mixture is decanted from the soil particles and clarified by filtration. Then the molarity of the clarified extractant is measured with the NO_3^--ISE. The resulting electrode voltage output is mathematically converted to concentration via the calibration curve, and subsequently to content (mg kg^{-1}) via the soil:extractant ratio.
Many researchers over the years have studied various aspects of NO_3^--ISE performance (accuracy, repeatability, stability, reliability), the potential for measurement interference by other ions, solution ionic strength, and use of deionized or distilled water as an extractant, for a multitude of use conditions, and in comparison with other chemical-based laboratory methods of soil nitrate determination (Myers & Paul, 1968; Mahendrappa, 1969; Milham et al., 1970; Onken & Sunderman, 1970; Dahnke, 1971; Mack & Sanderson, 1971; Yu, 1985; Sah, 1994).

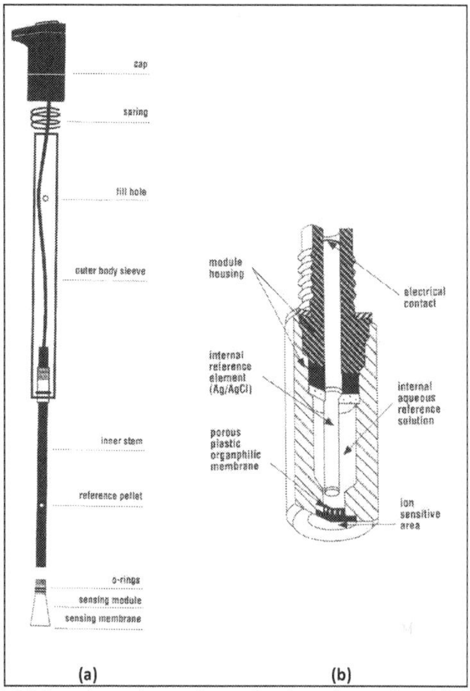

Fig. 1. The Orion 97-07 ionplus nitrate ion-selective electrode. (a) Pictorial diagram, (b) cross-section diagram of the electrode's nitrate sensing module (Orion, 2005).

As a result, NO_3^--ISEs have enjoyed wide acceptability because the results obtained are comparable to other chemical-based methods' results, and they are quick and simple to use. Today, several types of NO_3^--ISEs are manufactured commercially, and they are widely used in laboratories around the world for water quality monitoring and plant tissue sap nitrate measurement in addition to soil nitrate measurement. It is because of their well-defined operating characteristics, reliability, and commercial availability that a NO_3^--ISE was chosen as the analysis instrument for the SNMS to perform direct in-field measurement of NO_3^- in a soil slurry.

4. Experiments conducted for determining electrode operating variable parameters

Laboratory work conducted by Thottan et al. (1994) and Thottan (1995) determined that a NO_3^--ISE could be used in a soil slurry whilst investigating operating variables of soil:extractant ratio, slurry clarity, and electrode response time, repeatability and output signal stability.

Soil samples of sandy loam, silty clay loam, and clay loam were taken from the surface layer (15 cm) of fields in Cumberland and Colchester counties of Nova Scotia, Canada (45^0 N, 63^0 W). The results reported in this chapter relate to Chaswood clay loam, since of the three soils tested it is considered to be more difficult to analyze because of the higher clay content than the coarser textured soils. The Chaswood soil is of the gleysolic order, of the subgroup RegoGleysol. Particle size analysis revealed a composition of 34.0% sand, 37.9% silt, and 28.1% clay. The sampled A horizon was a fine textured alluvial formation which had been deposited above loamy sand.

Testing of the soil:extractant ratio revealed that there was no significant difference ($\alpha = 0.05$) between final NO_3^- concentrations for the three ratios tested. The mean NO_3^- concentrations determined at soil:extractant ratios of 1:15, 1:5 and 1:3 were 18.6, 18.6, and 19.3 ppm, respectively. In terms of mechanical extractor design, these results indicated that any of the three ratios may be used in the field when extracting NO_3^- from soil with equal effectiveness.

Tests to determine the effect of clarity on electrode performance showed that there was no significant difference ($\alpha = 0.05$) between mean final NO_3^- concentration measured in either slurry (34.1 ppm), decanted (32.0 ppm), or filtered (33.8 ppm) soil samples. This result confirmed the hypothesis that the NO_3^--ISE could be used in a soil slurry during in-field use—obviating the need for time consuming filtering of soil extracts required by other nitrate determination methods that would complicate mechanical system design and slow down operation. Using a NO_3^--ISE, Paul & Carlson (1968), Myers & Paul (1968), Dahnke (1971) and Yu (1985) also found that there was no significant difference between nitrate determinations made in a slurry or filtrate.

Fig. 2. shows a typical response curve of the NO_3^--ISE in a soil slurry. The electrode potential drops sharply indicating a rapid release of nitrate into solution. It was found that the electrode detects a large percentage of the nitrate concentration in less than 20 s, but it takes up to two minutes to detect the total nitrate concentration as the electrode signal stabilizes. Electrode signal stability was considered to be achieved when a signal drift of less than 1 mV min^{-1} was obtained. It was also found that the electrode had very consistent response time curves. Therefore, it was hypothesized that it was not necessary to wait until 100% of the NO_3^- in a soil sample is extracted before taking a measurement. This characteristic was utilized to create normalized response curves (Adsett et al., 1999) to speed up the measurement cycle. Accurate and reliable estimates of the sample's total NO_3^- concentration could be made in six seconds, which is within the time required for rapid in-field measurements. A successful mechanical system, however, would depend not only on a properly functioning and calibrated electrode, but also on properly functioning mechanical components, electronics, and controls to enable it to be reliably used in the field.

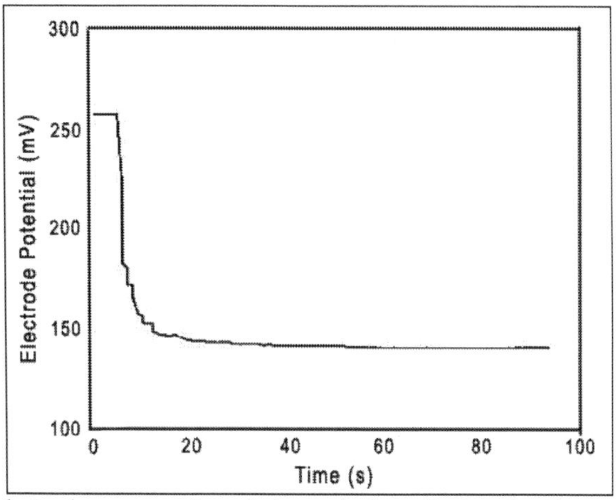

Fig. 2. Typical electrode response in soil slurry during nitrate extraction and measurement (Thottan et al., 2004).

5. Systems developed for in-field measurement and mapping of soil nitrate

In this section, a description of the mechanical systems and their operation for soil nitrate extraction and measurement are presented and discussed. First will be a description of the SNMS, followed by a description of the nitrate extraction and measurement sub-unit (NEMS).

5.1 Soil nitrate mapping system

Sibley (2008) and others (e.g.,Thottan, 1995; Adsett et al., 1999; Khanna & Adsett, 2001) have developed a SNMS (Fig. 3) that uses a nitrate ion-selective electrode (NO_3^--ISE) (Orion Model 9707 ionplus, Thermo Electron Corp., Massachusetts, USA) as the measurement instrument. It is an electro-mechanical machine that automatically collects a soil sample (0–15-cm depth), mixes it with water, and directly analyzes it electrochemically for nitrate concentration in real-time (6 s). Additionally, global positioning system (GPS) geo-referenced position data are simultaneously recorded at each sampling location to enable a nitrate map to be created for the field being sampled.

The SNMS consists of six sub-units: (1) soil sampler, (2) soil metering and conveying, (3) nitrate extraction and measurement, (4) auto-calibration, (5) control, and (6) GPS as indicated in Fig 3.

Fig. 3. Soil nitrate mapping system with six sub-units: (1) soil sampler, (2) soil metering and conveying,(3) nitrate extraction and measurement, (4) auto-calibration, (5) control and (6) global positioning system, with (7) inset showing Orion 97-07 ionplus NO_3^--ISE used for measuring soil nitrate (adapted from Sibley, 2008).

Prior to use, the NO_3^-–ISE is calibrated using pre-prepared reagent-grade NO_3^- standards placed into the calibration cups of the auto-calibration sub-unit. As well, a field (soil condition) calibration is completed to enable rapid measurements of NO_3^- concentration to be taken during system operation. As the tractor moves forward, the SNMS collects a soil sample via the combination of soil sampler and soil metering and conveying sub-units. During sampling, the hydraulic-powered wood-saw blade is lowered into the soil by the carrying frame. Over a travel distance of approximately 0.5 m, the blade cuts a 15-cm deep slot and throws a spray of finely chopped soil onto the head-end area of an automatically positioned flat-belt transfer conveyer. This action creates a sample of uniform bulk density and finely-granulated particles to facilitate the subsequent nitrate extraction process (Sibley et al., 2008). The conveyor belt has an oblong fixed-volume pocket milled into its surface to collect a sample from the soil landing on the conveyor. A specially designed scraper placed above the belt levels the soil sample in the pocket without compaction and removes excess soil from the belt as the belt moves to deliver the soil sample to the NEMS. During delivery, the pocket stretches lengthwise as it passes around the conveyor's tail-end roller to facilitate complete emptying of the pocket.

Just prior to soil sample delivery, water for NO_3^- extraction is pumped into a nitrate extractor to completely submerge the sensing module of the NO_3^-–ISE and the stirrer is activated. The soil sample is received into the extractor where vigorous mixing takes place creating a soil slurry. Nitrate in the soil sample is rapidly extracted into the slurry. The NO_3^- concentration of the mixture is measured by the NO_3^--ISE and stored in the control system's computer memory. Geo-referenced position data are simultaneously recorded by the GPS sub-unit at each sampling location to enable a nitrate map to be subsequently created for the field. All data collected are downloaded to a computer for post-sampling processing via the computer-interface facility built into the control system.

The SNMS can be used to analyze soil samples automatically in real time, or manually while stationary by hand-placing samples into the NEMS. It is envisioned that two configurations of the system will eventually be used in practice—a tractor-mounted version (Fig. 3.) and a 'suitcase' (portable) version. Initial research on developing a 'suitcase version' was completed by Brothers et al., (1997). The prototype developed was capable of measuring NO_3^- and pH with the same mechanical system and control hardware.

5.2 Nitrate extraction and measurement sub-unit

The heart of the SNMS is the NEMS (Fig.4). It consists of an extractor, an impeller and drive motor, a spray nozzle, a gate valve and drive actuator, and the NO_3^--ISE. The electrode and the sample, plus associated electrode circuitry, comprise an electrochemical cell. The extractor was constructed using 9.5 cm ID clear acrylic tubing so that the extraction process could be viewed.

A 7.6 cm ID sliding-knife gate valve was installed to act as the bottom of the extraction chamber, forming the extraction chamber outlet. This arrangement gives a nearly full-diameter chamber pass-through capability for efficient clean-out of each sample and prevents potential jamming by small stones or field debris that might enter the chamber with the soil sample. A 12Vdc linear actuator is used to open and close the valve between samples.

In normal position, the extraction chamber outlet is kept closed by the actuator. When the actuator is powered, it opens the extraction chamber outlet. The extraction chamber was electrically isolated from other components to eliminate any stray voltages that may interfere with the NO_3^--ISE signal.

The added advantage of having the extraction chamber outlet normally closed was that the extraction chamber could be used as a storage unit for the electrode in a dilute NO_3^- standard solution when not being used. To the lower end of the valve, a 3.5-cm diameter polyvinyl chloride (PVC) pipe was connected. The PVC pipe provided structural support and electrical isolation for the extraction chamber, as well as being an extension of the extraction chamber outlet.

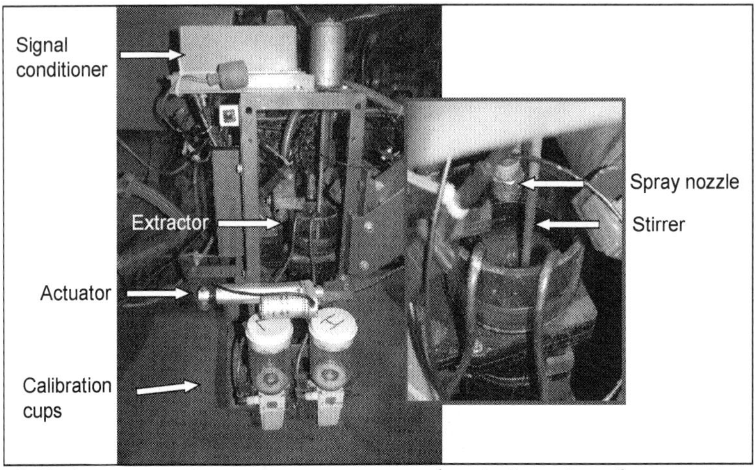

Fig. 4. Soil nitrate extraction and measurement sub-unit. Inset is a close-up view of the extractor showing a measurement being taken in a soil slurry.

A full cone spray nozzle was placed just above the extractor for supplying the water and also for cleaning purposes between successive samples. The nozzle was connected to a supply hose and a pump. A solenoid valve was fitted in between the nozzle and the pump to allow on/off flow control and also to meter in the exact amount of water under computer control by controlling the length of time the valve is turned on.
The mixing mechanism consisted of a fibreglass shaft with an acrylic impeller attached to one end. The fibreglass shaft was used in order to eliminate the possibility of any stray voltage being conducted into the extraction chamber. The shaft was powered using a variable speed 12 Vdc motor and was operated at 300 rpm.

6. Electronic control sub-unit, measurement methodology and operation of the nitrate extraction and measurement sub-unit

Development of the control sub-unit to operate the NO_3^--ISE within the NEMS and the dirty, electrically noisy environment of an agricultural tractor required development of significant advances in instrumentation signal conditioning and processing circuitry. These advances are discussed in this section.

6.1 Electronic control sub-unit

A schematic diagram of the electronic control sub-unit is shown in Fig. 5. The electronic control sub-unit consists of electronic circuitry, relays, a potentiometer, I/O ports, switches, and a key pad all housed in two electrically isolated and waterproof metal boxes. These are the control box shown in Fig. 3 and the signal conditioner box shown in Fig. 4, inside which the functions of signal processing and system control, and signal conditioning occur, respectively. The nitrate electrode sensing module generates a very small signal in mV, which is entered as input to the signal conditioning circuitry. Before leaving the signal conditioner box the signal is converted to mA to prevent degradation as it travels along a cable to the processing circuitry contained in the control box. The heart of the processing circuitry is an EnT board (Advanced Monitoring Technologies Inc., New Brunswick, Canada) with eight analog and eight digital input ports, plus transistor and relay outputs. The BS2SX Stamp main processor and the BS2 Stamp chip are mounted in two on-board 24 pin sockets. The mA inputs received from the conditioning box are converted back to mV and then pass through A/D conversion. The digital signals are then processed by the Stamp BS2SX software program to produce a digital number which is stored in RAM. The digital number is subsequently used after data downloading for calculating the NO_3^- level represented by each electrode reading using a spreadsheet program. Each NO_3^- measurement is accompanied by its geographic position coordinates as determined by a tractor-mounted GPS system (GBX-PRO with GPS/BCN, CSI Wireless Inc., Alberta, Canada). Sample location coordinates are stored in RAM.

The BS2SX Stamp main processor, also through the software program, controls the various mechanical components of the SNMS using 12 Vdc output relays mounted on the EnT board with simple ON/OFF switching. These mechanical components are pumps, solenoid valves, a linear actuator, and electric motors.

6.2 Operation of the nitrate extraction and measurement sub-unit

The operation of the NEMS includes the following procedures:

1. Electrode (NO_3^--ISE) Calibration
2. Field Calibration
3. Soil NO_3^--N analysis

The electrode calibration procedure is important since the reliability of the NO_3^- measurements depends entirely upon proper calibration of the NO_3^--ISE. Two NO_3^- standard solutions of known concentrations (0.0001M and 0.1M) made from reagent grade KNO_3 crystals mixed with de-ionized water are used for calibration. The NEMS software provides an auto-calibration routine which performs a calibration under computer control. The electrode calibration provides the coefficients for the Nernst from which the NO_3^- concentration is calculated. Thottan et al. (1994) and Thottan (1995) describe the calibration theory and process in detail.

Once the electrode calibration is completed, a field calibration is used to speed up the NO_3 analysis time. The calibration determines a scaling factor which allows the prediction of the sample NO_3^- value (which occurs at electrode signal stability and which could take up to two minutes or more) after a short measurement time of six seconds (Thottan, 1995, Adsett et al., 1999).

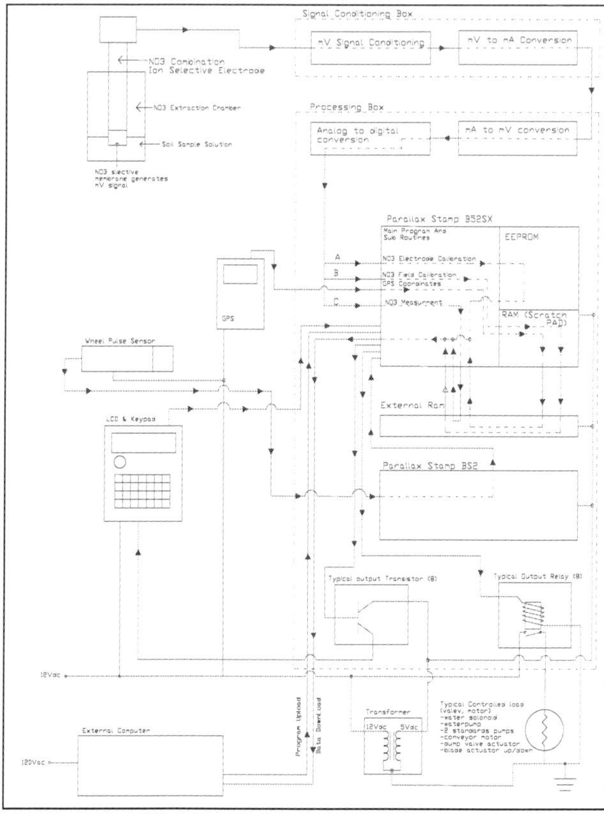

Fig. 5. Schematic diagram of the electronic control unit measurement and control circuitry, and electro-mechanical components.

The scaling factor can differ from field to field according to variations in soil characteristics, thus the name 'field calibration', and thus the necessity of the procedure. Also the field calibration is essential to monitor changes in the speed of response of the electrode's organophilic sensing membrane that have been observed with prolonged usage. A slowing of the response speed indicates it is time to change the electrode's sensing module. The time used to determine when to change the module was when it took longer than two minutes to perform a field calibration measurement. To-date, the NEMS has been used to analyze over 8,000 soil samples. And as many as 2,000 samples have been analyzed before it became necessary to change the sensing module. The membrane of the sensing modules used during this time has not showed any appreciable abrasion wear from soil particle contact.

Early work by Thottan (1994) and Thottan et al. (1999) performed a field calibration by measuring soil NO_3^--N using a sufficient number of samples through to electrode signal stability so as to determine the response characteristics of the NO_3^--ISE for the particular soil type in the field being sampled. Data analysis then used a statistical routine programmed into the control system software to determine the quickest time that resulted in a CV less than 5% with a 95% confidence level. The data corresponding to this time were used to calculate a scaling factor that was subsequently used to predict the final NO_3^- concentration of a sample after measuring for this time. Six seconds always gave reliable results, and even sometimes four seconds were sufficient.

Recent work (Khanna & Adsett, 2001; Sibley, 2008) has advanced the field measurement process even further such that a field calibration can now be based on one soil sample taken at a location considered typical of the soil texture in the field. During field calibration the NO_3^- electrode signal (x) is measured at six seconds, and again at electrode signal stability (y). These x and y values are stored in Scratch Pad RAM and the scaling factor (y/x ratio) is calculated by the software program. The scaling factor is then used for prediction of the final NO_3^- value for all subsequent samples taken in that field. Following field calibration, soil samples can be collected and analyzed for NO_3^- every 40 seconds (6 seconds for NO_3^- analysis and 34 seconds for other mechanical system functions to cycle). Sibley et al. (2009) has found this y/x routine sufficient for field use. After field calibration, soil samples can be collected and analyzed for NO_3^-.

Just prior to a soil sample being deposited into the extractor, water for NO_3^- extraction is pumped into the extractor to completely submerge the electrode's sensing module and the stirrer is activated. The soil sample is deposited into the extractor where the stirrer vigorously mixes the soil slurry into which NO_3^- is rapidly extracted from the soil sample. The NO_3^- concentration of the slurry is measured by the NO_3^--ISE and stored in the control system's computer memory. Geo-referenced position data are simultaneously recorded by the GPS sub-unit at each sampling location to enable a nitrate map to be subsequently created for the field. All data collected are downloaded to a computer for post-sampling processing via the computer-interface facility built into the control system.

7. Experiments conducted to validate in-field measurements of soil nitrate obtained with the ion-selective electrode

Extensive field-scale validation testing of the NEMS in two crop (wheat and carrot) production systems in Nova Scotia, Canada was conducted in 2006 (Sibley et al., 2009). Field conditions used for testing included conventional tillage vs. no tillage, inorganic vs. organic fertilization, four soil groups, and three time points throughout the season. Field moisture content ranged between 12.5–28.5%.

Data was collected with the NEMS and analyzed using four data processing methods: i) integer number (IN), ii) real number (RN), iii) integer number plus moisture content correction (IN+MCC), and iv) real number plus moisture content correction (RN+MCC). Duplicate samples were also analyzed in the laboratory using standard potassium chloride (KCl) extraction and flow injection analysis methods.

The results indicated that the level of agreement between NEMS soil NO_3-N measurements and standard lab soil NO_3-N measurements (Lab), as measured by Root Mean Squared Error (RMSE), Mean Absolute Error (MAE) and Coefficient of Efficiency (CE) was excellent. The results obtained also strongly suggest that the NEMS is quite robust and can be used for measuring soil NO_3-N in both wheat and carrot crops, as well as in different soil groups, fertility levels, tillage conditions, and at any time throughout the season.

Representative graphs comparing NEMS and Lab soil NO_3-N measurements over all field conditions tested are shown in Fig. 6. These graphs illustrate the how well the electrode performed in the field on an individual sample basis, regardless of the field condition from which the sample originated.

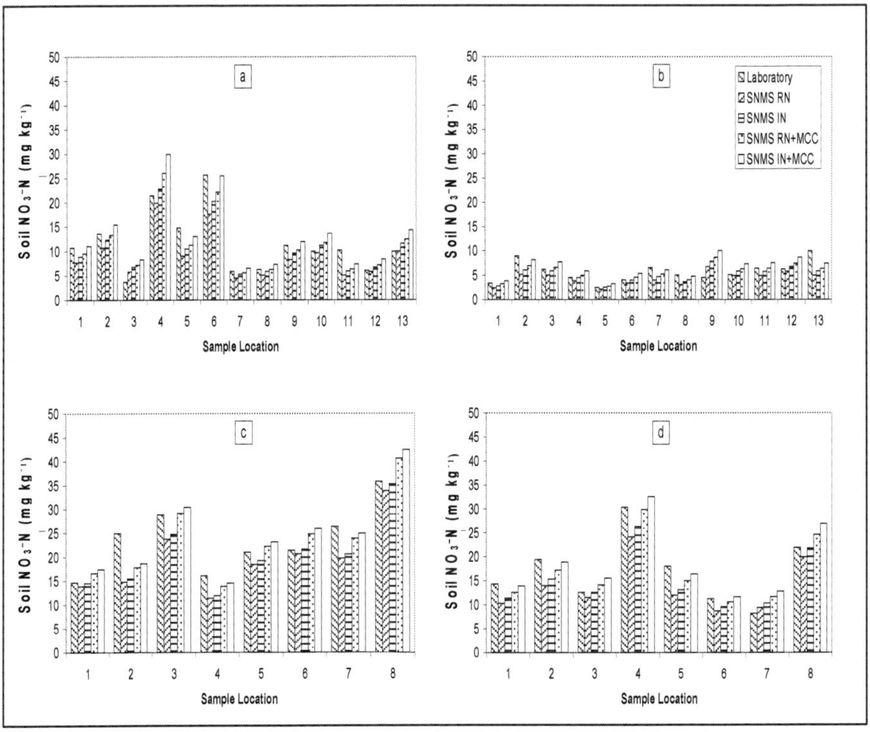

Fig. 6. Comparison of NEMS soil NO_3-N measurements for each data processing method with Lab soil NO_3-N measurements by sample location. (a) Wheat plot 6 (NT, DRT52) on 30 May, (b) wheat plot 8 (CT, PGW52) on 7 Nov., (c) carrot plot 2 (IF, PGW82) on 20 June and (d) carrot plot 3 (LDM, DRT22) on 7 Nov. (Sibley et al., 2009).

As well, they also indicate the responsiveness of the electrode, as the values are displayed by sampling location (x-axis) in the order of measurement. It was found that the electrode responded equally well regardless of whether the NO_3-N level was changing from lower to higher, or higher to lower during measurement.

Nested linear regression analyses revealed that the NEMS had the same level of performance over all field conditions tested regardless of what data processing method was used (Sibley et al., 2009).

It was concluded that any of the regression equations developed for describing the relationship between NEMS measurements and Lab measurements for the four data processing methods tested (Fig. 7) can be used to enable field measurements of soil NO_3-N using the NEMS to be obtained with lab-grade accuracy.

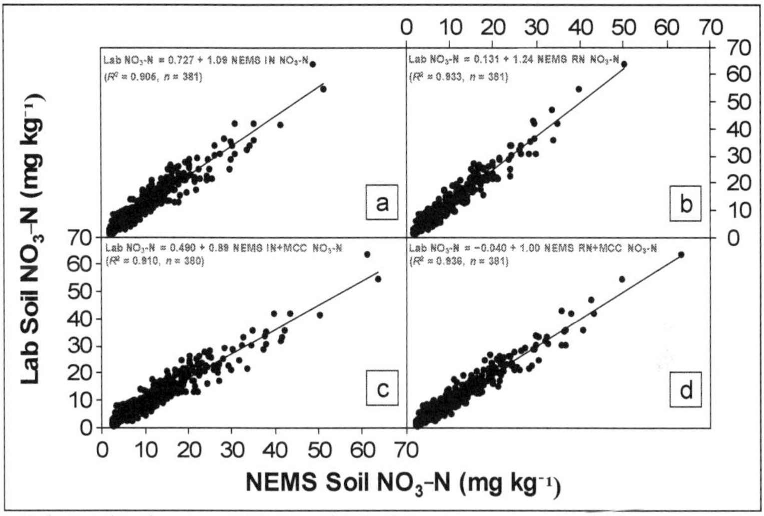

Fig. 7. Relationship between NEMS soil NO_3-N measurements and laboratory soil NO_3-N measurements for each data processing method; all field conditions data combined. (a) Integer number data processing, (b) real number data processing, (c) integer number with moisture content correction data processing and (d) real number with moisture content correction data processing (adapted from Sibley et al., 2009).

8. Significance of the soil nitrate mapping system

The development of the SNMS, including its NEMS, is a significant advancement in real-time field measurement of soil NO_3-N from two main perspectives: (i) linking soil NO_3-N variation to crop growth, and (ii) assessing soil nitrate variation. These perspectives are discussed below along with some results of experiments conducted using the system in wheat and carrot production systems in Nova Scotia, Canada.

8.1 Linking soil nitrate variation to crop growth

Using data collected by the NEMS on small-scale sampling grids (6.0 × 7.5 m) at seven sampling dates before, during, and after crops were being grown, the variation in soil NO_3-N levels in wheat and carrot production systems over time were linked to crop performance (Sibley, 2008). In wheat under organic fertilizer management, the effects of conventional tillage vs. no tillage on soil nitrate, plant nitrogen and yield responses were determined. In carrot under conventional tillage management, the effects of inorganic fertilizer vs. organic fertilizer on soil nitrate, plant nitrogen and yield responses were determined.

In wheat, it was determined that the only significant difference ($\alpha = 0.05$) in mean soil NO_3-N level between the conventional tillage and no tillage treatments occurred early in the growing season shortly after fertilizing, when the level for the conventional tillage treatment was nearly two times higher than for the no tillage treatment (Fig. 8).

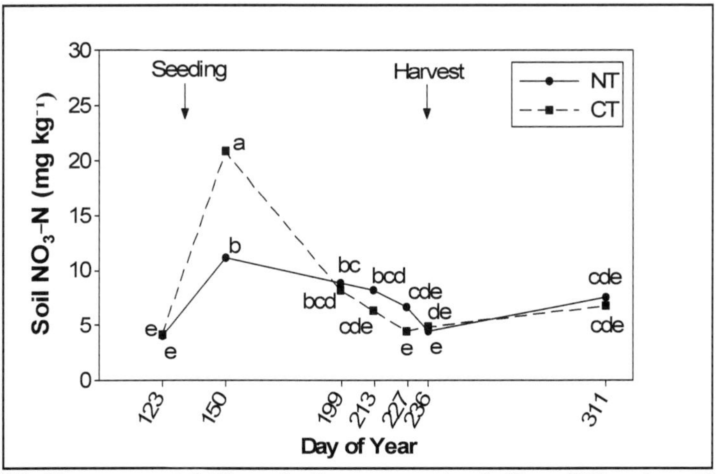

Fig. 8. Soil NO_3-N response at 0–15-cm depth to conventional tillage (CT) and no tillage (NT) treatments in the wheat field over the study period. Means sharing the same letter are not significantly different at the 5% level (Sibley, 2008).

There was no significant difference ($\alpha = 0.05$) in the response of plant tissue Total N to the conventional tillage and no tillage treatments; however a significant Day effect was detected. Significant differences ($\alpha = 0.05$) were found in mean tissue Total N level between grain set, grain filling, and maturity, but no significant difference thereafter. Mean plant tissue-sap NO_3-N, grain yield and grain Total N all showed no significant difference ($\alpha = 0.05$) in response to the conventional tillage and no tillage treatments. These results suggest that the plants responded equally well at producing final grain yield under either the conventional tillage or no tillage management practice and despite there being significant changes in soil NO_3-N level over the growing season.

In carrot, early in the growing season shortly after fertilizing, the soil NO_3-N level for the inorganic fertilizer treatment was nearly three times higher than for the liquid dairy manure treatment, while for the remainder of the growing season it remained in the order of two times higher (Fig. 9). There was a significant ($\alpha = 0.05$) increase in soil NO_3-N for both the inorganic fertilizer and liquid dairy manure treatments after harvest at the end of the study period in late-Fall. It is suspected that these late-Fall increases were due to a 'tillage effect' from mechanical harvesting in combination with a short-term increase in soil temperature typical for the geographic area.

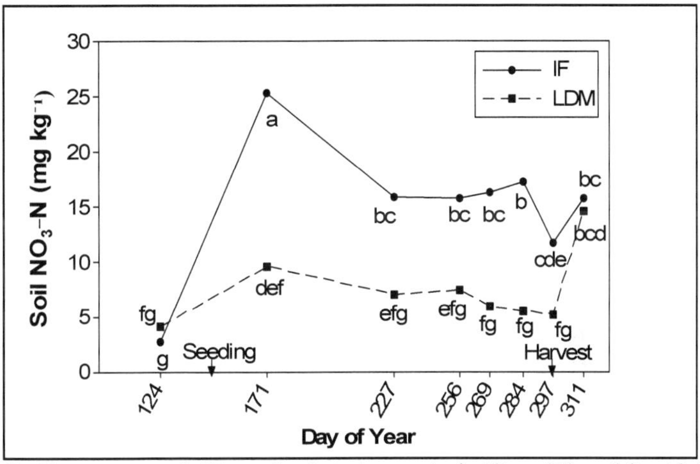

Fig. 9. Soil NO_3-N response at 0–15-cm depth to inorganic fertilizer (IF) and liquid dairy manure (LDM) treatments in the carrot field over the study period. Means sharing the same letter are not significantly different at the 5% level (Sibley, 2008).

There was no significant difference ($\alpha = 0.05$) in mean plant tissue-sap NO_3-N response to the inorganic fertilizer and liquid dairy manure treatments; however a significant Day effect was detected. During mid-growth stage, plant tissue-sap NO_3-N levels were sufficient for maximum top-biomass growth and root yield and then dropped off dramatically during active root bulking until homeostasis was reached. The level remained unchanged between homeostasis and the time the roots were harvested.

Plant tissue Total N level for the inorganic fertilizer treatment dropped significantly during active root bulking and then stabilized for the remainder of the growing season. During this same period for the liquid dairy manure treatment, tissue Total N level also dropped significantly ($\alpha = 0.05$), but unlike for the inorganic fertilizer treatment, continued to drop dramatically instead of stabilizing. This dramatic drop was attributed to a combination of factors including weakness of the liquid dairy manure treated plants at that time resulting in a higher rate of leaf senescence that enabled sufficient N re-translocation to meet the needs of the bulking roots, rainfall-induced nitrification in the soil which was limiting N availability to the plants, and a leaching effect of the heavy rains on the senescing leaves.

There was no significant difference ($\alpha = 0.05$) in fresh root yield or root Total N between the treatments. These results suggest that the plants took up enough N during shoot growth to sustain root bulking, thereby utilizing stored N for root bulking rather than relying on available N in the soil. These results further suggest that although there were significant changes in soil NO_3-N levels over the growing season, and between the inorganic fertilizer and liquid dairy manure treatments, they did not affect very much what was happening in the carrot plants. This finding stresses the environmental implications of managing soil NO_3-N in a carrot production system.

8.2 Assessing soil nitrate variation

Using data collected on a fine-scale sampling grid (6.0 × 7.5 m) and a combination of classical and geostatistical analytical techniques and tools, the spatial and temporal aspects of NO_3-N variation in a wheat production system at seven sampling dates covering pre-seeding, growing season, and post-harvest soil conditions, as well as the intrinsic spatial structure of NO_3-N present in the experimental field, were assessed (Sibley, 2008).

The SNMS was successfully used to measure soil NO_3-N level during the study period and to monitor spatial and temporal variation over time. As well, accurate, high resolution posted values (contour) maps were generated that give excellent visual pictures of the NO_3-N spatial variation that was evident just prior to, at peak nitrogen release, during, and just after the growing season (Fig. 10).

The data generated using the SNMS was also used to assess the spatial structure soil NO_3-N variation in the experimental field. Very strong proportional effect relationships were found between the data sets mean and standard deviation values (R^2 = 0.97) and the squared mean and variance values (R^2 = 0.99). These relationships can be used for predicting high-quality average and proportional variograms, which in turn can be used for determining NO_3-N soil sampling schemes for the experimental field to any desired level of accuracy.

Variogram models of soil NO_3-N spatial structure were developed for each of the sampling dates (Table 1). These variograms were all of the isotropic spherical type and had high R^2 values ranging between 0.90–0.99 combined with very low RSS values. These very high goodness-of-fit measures for all models indicated that they describe the spatial structure of NO_3-N variation very well. Similarities in the spatial structure of soil NO_3-N on the sampling dates were evident as these models had similar slopes, nuggets, ranges, and nugget-to-sill ratios. Spatial dependency was found overall to be moderate. Since the models represent data from time-spaced sampling dates, their similarity indicates a high likelihood that the intrinsic spatial structure of soil NO_3-N in this experimental field exhibited temporal stability over the study period.

A scaled average variogram model that very likely accurately represents the intrinsic spatial structure present in the experimental field was created having a sill of 1.005, a nugget of 0.331, and a range of 44 m. This result indicates that the range of intrinsic spatial structure beyond which NO_3-N values did not have spatial dependency was likely approximately 44 m. This distance, then, would be the minimum sample spacing for conducting soil NO_3-N experiments in this experimental field that require analyses by classical statistical methods.

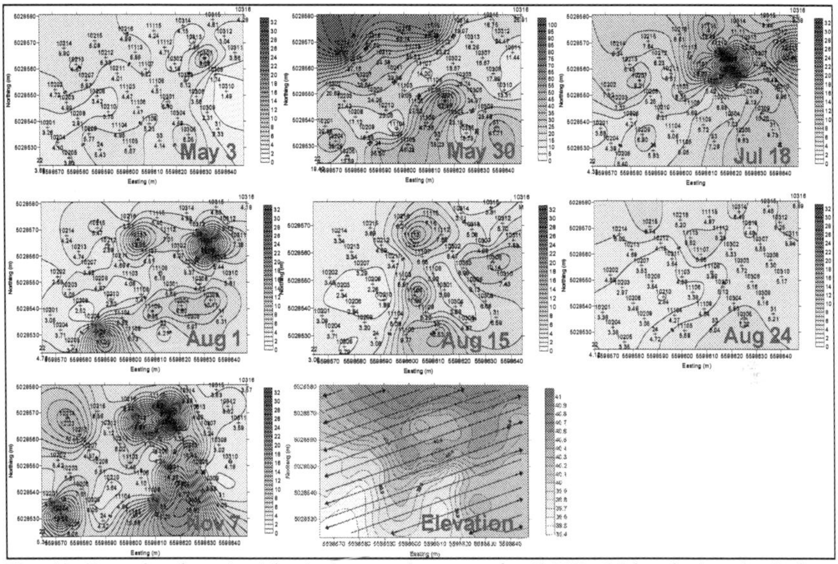

Fig. 10. Posted values isarithmic (contour) maps of soil NO_3-N levels (mg kg^{-1}) for each sampling date and a corresponding elevation map of the soil surface. Darker colour indicates higher amount of soil NO_3-N or higher elevation, respectively. Note the different scale used for May 30 to enable plotting of the higher soil NO_3-N levels resulting from the application of liquid dairy manure two weeks prior. May 3 is prior to field work. Aug 24 is the harvest date. Nov 7 is post-harvest (adapted from Sibley, 2008).

Being able to collect and analyze samples on a fine-scale sampling grid much quicker, as accurately as, and more affordably than conventional laboratory methods means, in practical terms, that the SNMS can provide a long-awaited solution to the problem of conducting soil NO_3-N variation assessment experiments at an affordable cost. Moreover, the SNMS provides a way to collect data so that the spatial structure of the NO_3-N in a field of interest is "known in advance" of the experimental planning. With this knowledge, the experimental sampling scheme and optimal sample size required for statistical analysis reliability can be determined apriori with confidence.

Parameter/ Date	3 May	30 May	18 July	1 Aug.	15 Aug.	24 Aug.	7 Nov.
Nugget, C_o	0.397	0.510	0.291	0.367	0.457	0.304	0.396
Sill, $C_o + C$	1.031	1.162	0.899	1.311	1.296	1.084	1.110
Range†, A_o	41	43	39	62	68	51	27
N:S, $C_o/(C_o + C)$	0.385	0.439	0.324	0.280	0.353	0.280	0.357
R^2	0.99	0.95	0.96	0.90	0.98	0.99	0.93
RSS ($\times 10^{-4}$)	1.10	8.67	6.40	4.36	6.95	3.70	6.52
Lag Pairs‡	33-185	35-192	30-59	47-232	44-263	35-190	30-167
Spatial Class§	M	M	M	M	M	M	M
Model	Spherical	Spherical	Spherical	Spherical	Spherical	Spherical	Spherical

† Range values are in meters.

‡ Range in number of pairs for each lag class interval.

§ S, strong spatial dependency (N:S ≤ 0.25); M, moderate spatial dependency (0.25 < N:S ≤ 0.75); W, weak spatial dependency (N:S > 0.75) (Cambardella et al., 1994).

Table 1. Parameters of final selected scaled variogram models for each sampling date (Sibley, 2008).

9. Conclusions and recommendations for future research

9.1 Conclusions

The SNMS, and its NEMS, have advanced the ability of making in-field measurements of soil NO_3-N by overcoming the impediments, roadblocks, and serious obstacles to measuring and assessing soil NO_3-N variation using conventional methods in terms of sample analysis lag time, high labor requirements, and high costs. It has been demonstrated that soil NO_3-N measurements using the SNMS can be obtained rapidly, on a fine scale, and with lab-grade accuracy. It has been demonstrated that data collected using the SNMS can be used for assessing the spatial and temporal aspects of soil NO_3-N variation, to assess variation in soil NO_3-N levels in space and over time, and link this variation to crop performance.

The SNMS offers the potential to assist farmers and researchers with developing site-specific soil nitrate management practices for agricultural production. It offers farmers the potential to more intensely and precisely analyze variations in soil NO_3-N levels throughout the growing season in association with environmental and crop response data in order to make the most sound and site- and time-specific management decisions possible. As well for farmers, it offers the potential for them to measure and document soil NO_3-N levels in their fields thus improving traceability and their ability to be compliant with any current and future legislation requiring control of nitrogen fertilizers. It offers regulators the potential to conduct environmental monitoring of NO_3^- levels in agricultural fields and water sources. Ultimately as a result of its use, the public may be assured that soil nitrogen management practices in agriculture are being conducted in the most environmentally friendly way.

9.2 Recommendations for further research

Since the beginning of precision agriculture, it has been a goal of many researchers and farmers to develop real-time responding variable rate (VR) fertilizer spreaders. Several types of VR spreaders that use prior-determined prescription maps have been developed and are commercially available. These spreaders are useful to a certain degree, but the missing link to their full effectiveness is still the ability to vary fertilizer application rate in response to precisely what the plant needs in real-time; particularly in response to the plant's need for nitrate, but more generally to several nutrient needs (potassium, calcium, magnesium, etc.) as a 'package' and pH that determine their availability. This was the initial vision for the SNMS. Thus, it has been designed as an expandable platform technology that can be easily modified to incorporate the simultaneous use of several different types of ion-selective electrodes. In the immediate term, further research should be conducted to develop the agronomic-based algorithms linking soil NO_3-N availability to crop performance to enable effective real time control of a fertilizer spreader by the SNMS. In the near term, further research should be conducted in combination with plant-related experiments to begin to build in the capability of the SNMS to measure and respond to the availability of a 'soil nutrient package'.

Currently, the SNMS is a tractor-mounted version. It has been mentioned above that it is envisioned that the system will eventually be used in practice also as a 'suitcase' (portable) version. Further research should be conducted to continue development of this version of the NEMS.

As well, more research should be conducted using the SNMS to further investigate the spatial and temporal variation and spatial structure of soil NO_3-N and the links between soil NO_3-N variation and availability and crop growth and yield under a wide variety of field conditions and crops in different geographic regions around the world.

10. References

Adamchuk, V.I., M.T. Morgan, & D.R. Ess. 1999. An automated sampling system for measuring soil pH. *Trans. ASAE* 42(4):885–891.

Adamchuk, V.I., A. Dobermann, M.T. Morgan, & S.M. Brouder. 2002a. Feasibility of on-the-go mapping of soil nitrate & potassium using ion-selective electrodes. Paper No. 02-1183, *ASABE*, St. Joseph, Michigan, USA.

Adamchuk, V.I., M.T. Morgan, & D.R. Ess. 2002b. System and method for automated measurement of soil pH. *US Patent* No. 6,356,830.

Adamchuk, V.I., J.W. Hummel, M.T. Morgan, & S.K. Upadhyaya. 2004a. On-the-go soil sensors for precision agriculture. *Computers and Electronics in Agriculture* 44:71–91.

Adamchuk, V.I., M.T. Morgan, & J.M. Lowenberg-Deboer. 2004b. A model for agro-economic analysis of soil pH mapping. *Precision Agric.* 5:111–129.

Addiscott, T.M. 1983. Kinetics and temperature relationships of mineralization and nitrification in Rothamsted soils with differing histories. *J. of Soil Sci.* 34:343–353.

Addiscott, T.M., A.P. Whitmore, & D.S. Powlson. 1991. *Farming, Fertilizers and the Nitrate Problem*. CAB International, Wallingford, U.K.

Adsett, J. F. 1990. *Automated field monitoring of soil nitrate levels.* Ph.D. thesis. University of Saskatchewan, Saskatoon, Saskatchewan, Canada.

Adsett, J. F., & G. C. Zoerb. 1991. Automated field monitoring of soil nitrate levels. p. 326-335. In: *Automated Agriculture for the 21st Century*. ASAE Publication No. 1191. Am. Soc. of Agricultural Engineers, St. Joseph, MI., USA.

Adsett, J.F., J.A. Thotton, & K.J. Sibley. 1999. Development of an automated on-the-go soil nitrate monitoring system. *Applied Eng. in Agric.* 15(4):351-356.

Almekinders, C.J.M., L.O. Fresco, & P.C. Struik. 1995. The need to study and manage variation in agro-ecosystems. *NJAS* 43:127-142.

Astatkie, T., A. Madani, R. Gordon, K. Caldwell, & N. Boyd. 2001. Seasonal variation of nitrate-nitrogen in the soil profile in a subsurfaced drained field. *Can. Biosys. Eng.* 43:11-16.

Birrell, S.J., & J.W. Hummell. 1997. p. 459-468. Multi-sensor ISFET system for soil analysis. In: J.V. Stafford (ed.) *Precision Agriculture '97*. Proc. Eur. Conf. on Prec. Agric., 1st. Warwick Univ. Conf. Cent., UK. 7-10 Sept. 1997. BIOS Sci. Pub. Ltd., Oxford, UK.

Birrell, S.J., & J.W. Hummell. 2000. Membrane selection and ISFET configuration evaluation for soil nitrate sensing. *Trans. ASAE* 43(2):197-206.

Birrell, S.J., & J.W. Hummell. 2001. Real-time multi-ISFET/FIA soil analysis system with automatic sample extraction. *Computers and Electronics in Agriculture*: 32(1):45-67.

Bongiovanni, R., & J. Lowenberg-DeBoer. 2004. Precision agriculture and sustainability. *Precision Agric.* 5:359-387.

Bourennane, H., B. Nicoullaud, A. Couturier, & D. King. 2004. Exploring the spatial relationships between some soil properties and wheat yields in two soil types. *Precision Agric.* 5:521-536.

Brothers, J., K. J. Sibley, & J.F. Adsett. 1997. *Automated System for Soil Nitrate Analysis*. Final Project Report, Agri-focus 2000 program, Nova Scotia Department of Agriculture & Marketing. 40pp.

Burgess, T.M., & R. Webster. 1980. Optimal interpolation and isarithmic mapping of soil properties I: the semi-variogram and punctual kriging. 1980. *Journal of Soil Science*. 31:315-331.

Cambardella, C.A., T.B. Moorman, J.M. Novak, T.B. Parkin, D.L. Karlen, R.F. Turco, and A.E. Konopka. 1994. Field-scale variability of soil properties in central Iowa soils. *Soil Sci. Soc. Am. J.* 58:1501-1511.

Campbell, C.A., G.P. Lafond, R.P. Zentner, & Y.W. Jame. 1994. Nitrate leaching in a Udic Haploboroll as influenced by fertilization and legumes. *J. Environ. Qual.* 23:195-201.

Chang, J., D.E. Clay, C.G. Carlson, D. Malo, S.A. Clay, J. Lee, & M. Ellsbury. 1999. Precision farming protocols: Part 1. Grid distance and soil nutrient impact on the reproducibility of spatial variability measurements. *Prec. Agric.* 1:277-289.

Christy, C.D., P.Drummond & D.A. Laird. 2003. An on-the-go spectral reflectance sensor for soil. Paper No. 03-1044, *ASABE*, St. Joseph, Michigan, USA.

Dahnke, W.C. 1971. Use of the nitrate specific ion electrode in soil testing. *Soil Science and Plant Analysis*, 2(2):73-84.

Dinnes, D.L., D.L. Karlen, D.B. Jaynes, T.C. Kaspar, J.L. Hatfield, T.S. Colvin, & C.A. Cambardella. 2002. Nitrogen management strategies to reduce nitrate leaching in tile-drained Midwestern soils. *Agron. J.* 94:153-171.

Ehsani, M.R., S.K. Upadhyaya, D. Slaughter, S. Shafii, & M. Pelletier. 1999. A NIR technique for rapid determination of soil mineral nitrogen. *Precision Agric.* 1:217-234.

Ehsani, M.R., S.K. Upadhyaya, W.R. Fawcett, L.V. Protsailo, & D. Slaughter. 2001. Feasibility of detecting soil nitrate content using a mid-infrared technique. *Trans. ASAE* 44:1931-1940.

Engel, R.E., D.S. Long, G.R. Carlson, & C. Meier. 1999. Method of precision nitrogen management in spring wheat: I fundamental relationships. *Precision Agric.* 1:327-338.

Geron, C. A., Danneberger, T. K., Traina, S. J., Logan, T. J., & J.R. Street. 1993. The effects of establishment methods and fertilization practices on nitrate leaching. *J. Environmental Quality.* 22(1):141-147.

Han, S., J.W. Hummel, C.E. Goering, M.D. Cahn. 1994. Cell size selction for site-specific crop management. *Trans. ASAE* 37(1):19-26.

Hay, R.K.M., & A.J. Walker. 1989. *An introduction to the physiology of crop yield.* Longman Scientific & Technical, Longman Group UK Ltd., Essex, UK.

Henkens, P.L.C.M., & Van Keulen, H. 2001. Mineral policy in the Netherlands and nitrate policy within the European Community. *NJAS* 49:117-134.

Heuvelink, G.B.M, & E.J. Pebesma. 1999. Spatial aggregation and soil process modelling. *Geoderma* 89:47-65.

Jung, W.K., N.R. Kitchen, K.A. Sudduth, & S.H. Anderson. 2006. Spatial characteristics of claypan soil properties in an agricultural field. *Soil Sci. Soc. Am. J.* 70:1387-1397.

Keeney, D.R., & D.W. Nelson. 1982. Nitrogen: Inorganic forms. p. 643-709. In: A.L. Page et al. (ed.) *Methods of soil analysis. Part 2. Chemical and microbiological properties.* ASA, Madison, WI.

Khanna, V., & J.F. Adsett. 2001. *Automated Soil Nitrate Monitoring System and Soil Nitrogen Dynamics.* Project report. Engineering Department, Nova Scotia Agricultural College, Truro, Nova Scotia, Canada. 28pp.

Koroluk, R., D.Culver, A. Lefebvre, & T. McRae. 2000. *Risk of water contamination by nitrogen. In Environmental sustainability of Canadian agriculture: report of the agri-environmental indicator project.* Agriculture and Agri-Food Canada. Publication 2022/E. Ch 5.

Lark, R.M. 1997. Variation in soil conditions and crop performance. p. 127-135. In: J.V. Stafford (ed.) *Precision Agriculture '97.* Proc. Eur. Conf. on Prec. Agric., 1st. Warwick Univ. Conf. Cent., UK. 7-10 Sept. 1997. BIOS Sci. Pub. Ltd., Oxford, UK.

Lauzon, J.D., I.P. O'Halloran, D.J. Fallow, A.P. van Bertoldi, & D. Aspinall. 2005. Spatial variability of soil test phosphorus, potassium, and pH of Ontario soils. *Agron. J.* 97:524-532.

Loreto, A.B., & M.T. Morgan. 1996. Development of an automated system for field measurement of soil nitrate. Paper No. 96-1087, *ASABE*, St. Joseph, Michigan, USA.

Mahendrappa, M.K. 1969. Determination of nitrate nitrogen in soil extracts using a specific ion activity electrode. *Soil Sci.* 108(2):132-136.

Mack, A.R., & R.B. Sanderson. 1971. Sensitivity of the nitrate-ion membrane electrode in various soil extracts. *Can. J. Soil Sci.* 51:95-104.

McBratney, A.B., & M.J. Pringle. 1999. Estimating average and proportional variograms of soil properties and their potential use in precision agriculture. *Precision Agric.* 1:125-152.

Milham, P.J., A.S. Awad, R.E.P., & J.H. Bull. 1970. Analysis of plants, soils and waters for nitrate by using an ion-selective electrode. *Analyst.* 95:751-757.

Morf, W.E. 1981. *The principles of ion-selective electrodes and of membrane transport.* Elsevier Scientific Publishers Co. New York, USA.
Myers, R.J.K, & E.A. Paul. 1968. Nitrate ion electrode method for soil nitrate nitrogen determination. *Can. J. Soil Sci.* 48:369-371.
Oenema, O., P.C.M. Boers, M.M. van Eerdt, B. Fraters, H.G. van der Meer, C.W.J. Roest, J.J. Schroder, & W.J. Willems. 1998. Leaching of nitrate from agriculture to groundwater: the effect of policies and measures in the Netherlands. *Environ. Pollut.* 102:471-478.
Orion Research Inc. 2005. *Nitrate electrode instruction manual.* Laboratory Products Group, Boston, MA.
Onken, A.B., & H.D. Sunderman. 1970. Use of the nitrate electrode for determination of nitrates in soils. *Soil Sci. and Plant Analysis* 1(3):155-161.
Patni, N.K., L. Masse, & P.Y. Jui. 1998. Groundwater quality under conventional and no-tillage: I. Nitrate electrical conductivity and pH. *J. Environ. Qual.* 27:869-877.
Paul, J.L, & R.M. Carlson. 1968. Nitrate determination in plant extracts by the nitrate electrode. *J. Agr. Food Chem.* 16(5):766-768.
Power, J.F., & J.S. Schepers. 1989. Nitrate contamination of ground water in North America. *Agric. Ecosyst. Environ.* 26:165-187.
Price, R.R., J.W. Hummel, S.J. Birrell, & I.S. Ahmad. 2003. Rapid nitrate analysis of soil cores using ISFETs. *Trans. ASAE* 46(3):601-610.
Pringle, M.J., S.E. Cook, & A.B. McBratney. 2004. Field-Scale experiments for site-specific crop management. part I: design considerations. *Precision Agric.* 5 617-624.
Randall, G.W., & D.J. Mulla. 2001. Nitrate nitrogen in surface waters as influenced by climatic conditions and agricultural practices. *J. Environ. Qual.* 30:337-344.
Reynolds, W.D., C.A. Campbell, C. Chang, C.M. Cho, J.H. Ewanek, R.G. Kachanoski, J.A. MacLeod, P.H. Milburn, R.R. Simard, G.R.B. Webster, & B.J. Zebarth. 1995. Agrochemical entry into groundwater. In: *The health of our soils: toward sustainable agriculture in Canada.* Eds. D.F. Acton & L.J. Gregorich. Agriculture and Agri-food Canada. Publication 1906/E. Ch 10.
Ruffo, M.L., G.A. Bollero, R.G. Hoeft, & D.G. Bullock. 2005. Spatial variability of the Illinois soil nitrogen test: Implications for soil sampling. *Agron. J.* 97:1485-1492.
Russo, D. 1984. Design of an optimal sampling network for estimating the variogram. Soil *Sci. Soc. Am. J.* 48:708-716.
Sah, R.N. 1994. Nitrate-nitrogen determination - a critical review. *Commun. in Soil Sci. and Plant Anal.* 25:2841-2869.
Schröder, J.J., J.J. Neeteson, O.Oenema, & P.C. Struik. 2000. Does the crop or the soil indicate how to save nitrogen in maize production? Reviewing the state of the art. *Field Crops Research* 66:151-164.
Shibusawa, S., M.Z. Li, K. Sakai, A. Sasao, H. Sato, S. Hirako, & A. Otomo. 1999. Spectrophotometer for real-time underground soil sensing. Paper No. 99-3030, *ASABE,* St. Joseph, Michigan, USA.
Shibusawa, S., S.W. I Made Anom, C. Hache, A. Sasao, & S. Hirako. 2003. Site -specific crop response to temporal trend of soil variability determined by real-time soil spectrophotometer. p. 639-643. In: J. Stafford, J., and A. Werner (Eds.) *Precision Agriculture,* Wageningen Academic Publishers, Wageningen, the Netherlands.

Sibley, K.J. 2008. *Development and use of an Automated On-the-go Soil Nitrate Mapping System.* Ph.D. Thesis, Wageningen University, Wageningen, the Netherlands. ISBN 978-90-8504-802-2. 172pp.

Sibley, K.J., J.F. Adsett & P.C. Struik. 2008. An On-The-Go Soil Sampler for an Automated Soil Nitrate Mapping System. *Trans. ASABE* 51(6): 1895-1904.

Sibley, K.J., T. Astatkie, G. Brewster, P.C. Struik, J.F. Adsett & K. Pruski. 2009. Field-scale Validation of an Automated Soil Nitrate Extraction and Measurement System. Precision Agric. 10(2): 162-174. DOI : 10.1007/s11119-008-9801-1.

Simpson, R.J., H. Lambers, & M.J. Dalling. 1983. Nitrogen redistribution during grain growth in wheat (*Triticum aestivum* L.). *Plant Physiol.* 71:7–14.

Thottan, J., J.F. Adsett, K.J. Sibley, & C.M. MacLeod. 1994. Laboratory evaluation of the ion selective electrode for use in an automated soil nitrate monitoring system. *Commun. in Soil Sci. and Plant Anal.* 25:3025–3034.

Thottan, J.A. 1995. *Microcomputer Controlled Soil Nitrate Measurement System.* M.A.Sc. Thesis, Technical University of Nova Scotia, Halifax, Nova Scotia, Canada. 182pp.

Trangmar, B.B., R.S. Yost, & G. Uehara. 1985. Application of geostatistics to spatial studies of soil properties. p. 45–94. In: N.C. Brady (ed.) *Advances in agronomy.* Vol. 48. Academic Press, Inc., New York.

USEPA. 1990. *National Survey of Pesticides in Drinking Water Wells.* USEPA, Washington, DC, USA.

Van Meirvenne, M. 2003. Is the soil variability within the small fields of flanders structured enough to allow precision agriculture? *Precision Agric.* 4:193–201.

Wagenet, R.J., & B.K. Rao. 1983. Description of nitrogen movement in the presence of spatially variable soil hydraulic properties. *Agric. Water Management* 6:227–242.

Webster, R., & T.M. Burgess. 1984. Sampling and bulking strategies for estimating soil properties in small regions. *Journal of Soil Science* 35:127–140.

Webster, R., & A.B. McBratney. 1989. On the Akaike Information Criterion for choosing models for variograms of soil properties. *Journal of Soil Science* 40:493–496.

Webster, R., & M. Oliver. 2001. *Geostatistics for environmental scientists.* John Wiley & Sons, Ltd., Chichester, West Sussex, UK. 286pp.

Yu,T.R. 1985. Applications of ion-selective electrodes in soil science. *Ion-Selective Electrode Review* 7:165–202.

2

High-resolution, High-speed 3-D Dynamically Deformable Shape Measurement Using Digital Fringe Projection Techniques

Song Zhang
Iowa State University
USA

1. Introduction

With recent advancements of 3-D geometric shape analysis, acquiring 3-D dynamic scenes in real time becomes increasingly important in enormous fields, such as manufacturing, medical sciences, computer sciences, homeland security, and entertainment, etc.

Over the last decades, 3-D shape measurement techniques have been improving dramatically rapidly. Numerous techniques have been developed including time of flight, stereo vision, structured light, and digital fringe projection. Because of the development of digital display technology, the structured-light-based methods have the potential to be the most important technology for 3-D shape measurement in both scientific research and industrial practices in the long run.

Real-time 3-D shape measurement has recently emerged as a quickly expanding field with the enhancements of computation power of an ordinary personal computer. For real-time 3-D shape measurement, 3-D shapes have to be acquired rapidly, processed quickly, and displayed in real time (Zhang, 2010). In the past few years, a number of technologies have been developed. Among which, the optical method is one of the core techniques. However, due to its fundamentally difficult nature, capturing 3-D dynamic scenes with high quality remains challenging, and even fewer systems are capable of processing 3-D geometry in real time because of the complexity of the problem

Our group has devoted a vast amount of effort in real-time 3-D shape measurement field. Over the past few years, we have been developing technologies using a technique called *digital fringe projection and phase-shifting technique*, a variation of the structured light technique with structured patterns being sinusoidally changing intensity. We have developed a number of algorithms to improve the processing speed, and taken advantage of the most advanced hardware technologies, graphics processing unit (GPU), to increase the speed of 3-D geometry reconstruction and rendering. With these technologies, we have successfully developed a high-resolution, real-time 3-D shape measurement that achieved simultaneous 3-D absolute shape acquisition, reconstruction, and display at a speed of faster than 26 frames / sec with 300K points per frame under an ordinary personal computer (Zhang et al., 2006). Experiments have demonstrated that the system can accurately capture dynamic changing 3-D scenes, such as human facial expressions. The data acquired by such a system have already been applied to

a variety of fields including computer science, medical sciences, and entertainment industry for a variety of purposes.
In this chapter, we will address the most recent advancements in real-time 3-D shape measurement techniques. The principles of different techniques will be explained, and the advantages and limitations of each technology will be discussed. We will especially focus on the high-resolution, real-time 3-D shape measurement techniques using digital fringe projection techniques because of their numerous advantageous features. Finally, we will present some potential applications of real-time 3-D shape measurement, and address some challenges that we are still facing.
We should emphasize that most of the technologies covered in this chapter have already been published in either conference proceedings or journal articles. This chapter, by no means, is an exhaustive survey of real-time 3-D shape measurement technologies. It mainly focuses on the technologies that we have developed over the past few years.

2. Principle of structured light technique

2.1 Introduction

Optical methods to measure 3-D profiles have been extensively used due to their surface noncontact and noninvasive nature. Stereo vision (Dhond & Aggarwal, 1989), which uses two cameras to capture 2-D images from different views, is probably the most well studied of these. This technology simulates human vision, and recovers the depth information via triangulation by knowing the corresponding points between two cameras. Relying on identifying corresponding pairs between two images to recover depth, it is difficult to achieve high accuracy if a surface does not have strong texture variation.
Instead of relying on natural texture, a structured light system uses a projector to actively project predefined structured patterns to assist in correspondence establishment (Salvi et al., 2004). To reach real time, multiple structured patterns are usually switched rapidly so that the number of patterns required to recover one 3-D shape can be captured in a short period of time. Rusinkiewicz et al. (2002) developed a real-time 3-D model acquisition system using a stripe boundary code (Hall-Holt & Rusinkiewicz, 2001). The stripe boundary code is generated by projecting a sequence of binary level structured patterns. Binary-coded structured patterns (only 0's and 1's) are extensively used in a structured light system, because they are (1) simple-it is easy to implement since the coding and decoding algorithms are straightforward, and (2) robust-it is robust to noise because only two intensity levels (0 and 255) are used. Fig. 1(A) shows a binary pattern, and Fig. 1(B) shows its cross section. Here black represents binary level 0, while white represents binary level 1. For a stripe between M and N in Fig. 1(B), all points have the same intensity value, thus cannot be differentiated. Therefore, it is difficult for binary methods to reach pixel-level spatial resolution because the stripe width must be larger than one projector's pixel. Moreover, because it cannot reach pixel level correspondence, it is difficult of this technology to reach very high measurement accuracy.
If more intensity values are used, the spatial resolution will be increased. Coding methods including n-ary (Pan et al., 2004), triangular phase-shifting (Jia et al., 2007), and trapezoidal phase-shifting (Huang et al., 2005) have been developed for higher spatial resolution measurement. As noted by Zhang (2010), all these patterns eventually become sinusoidal if they are blurred properly. The blurring effect often occurs when a camera captures an image out of focus, and all sharp features of an object will be blended together. Therefore, sinusoidal patterns seem to be the natural choice. As shown in Fig. 1(C) and Fig. 1(D), the intensity varies across

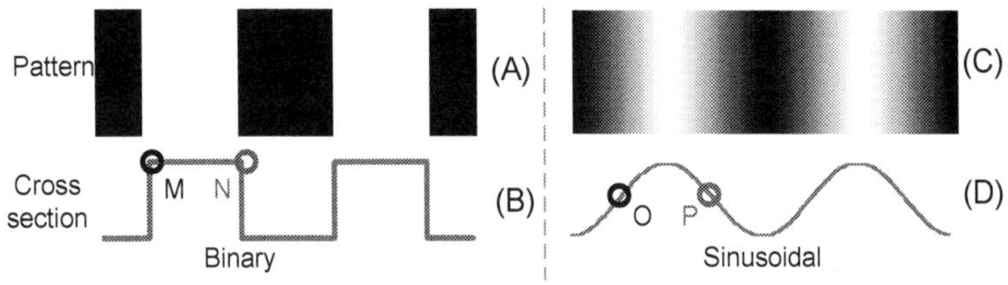

Fig. 1. Comparison between binary and sinusoidal structured pattern. (A) Typical binary structured patterns; (B) Cross section of the binary pattern shown in (A); (C) Typical sinusoidal fringe pattern; (D) Cross section of the sinusoidal patterns as shown in (C).

the image point by point. Therefore, it seems to be feasible to reach pixel level resolution because the intensity values between horizontal neighboring pixels are differentiable. Because the spatial resolution is high, it is possible to achieve high accuracy since the correspondence between the projector and the camera can be determined accurately.

Because using sinusoidal patterns (also known as *fringe patterns*) has the potential to reach pixel-level spatial resolution, it has long been studied in optical metrology where the fringe patterns are generated by laser interference. As noted by Su & Zhang (2009), single B/W fringe image (Guo & Huang, 2009) has been developed to measure dynamic motion of 3-D surfaces using a Fourier method (Takeda & Mutoh, 1983). However, this technique is limited to measuring smooth surfaces. If two fringe patterns are used, the measurement quality is improved (Guo & Huang, 2008). But it is still restricted to measuring surfaces without strong texture and/or geometry variations. To measure generic surfaces, at least three fringe patterns must be used (Schreiber & Bruning, 2007).

3-D shape measurement techniques using color-encoded patterns (Geng, 1996; Harding, 1988; Huang et al., 1999; Pan et al., 2005) have the potential to reach high speed because a structured pattern can be encoded with three fringe patterns. However, the measurement accuracy is affected, to a various degree, by object surface color (Zhang & Huang, 2004). For example, for a blue object, information conveyed by red and green colors will be lost because no signal will be captured. Moreover, the color coupling between red and green, and green and blue will affect the measurement if no filtering is used (Pan et al., 2005).

To avoid the problems induced by color, we developed a 3-D shape measurement system using B/W fringe images (Zhang & Huang, 2006a). By adopting a fast three-step phase-shifting algorithm (Huang & Zhang, 2006), the system achieved simultaneous data acquisition, reconstruction, and display in real time (Zhang et al., 2006). This technique takes advantage of the unique projection mechanism of a single-chip DLP projector: three fringe patterns are encoded into RGB channels of the projector and are switched automatically and naturally. By this means, we achieved 3-D surface measurement at 60 Hz with more than 300K points per frame (Zhang & Yau, 2007). We will discuss the details of this technology next.

2.2 Digital fringe projection system setup

A digital fringe projection system is very simple: it contains a projector to project fringe patterns, a camera to capture the fringe patterns, and the computer software to analyze the fringe images to recover the 3-D shape. Fig. 2 shows the schematic diagram of a 3-D imaging system based on a digital fringe projection technique. A projector projects vertically (varying horizontally) straight fringe stripes (phase lines) onto the object. The object surface distorts the fringe images from straight phase lines to curved ones. A camera images the distorted fringe images from another viewing angle. If the corresponding point between the camera pixel (A) and the projector pixel (C) is identified, depth (z) can be recovered through triangulation (ΔABC).

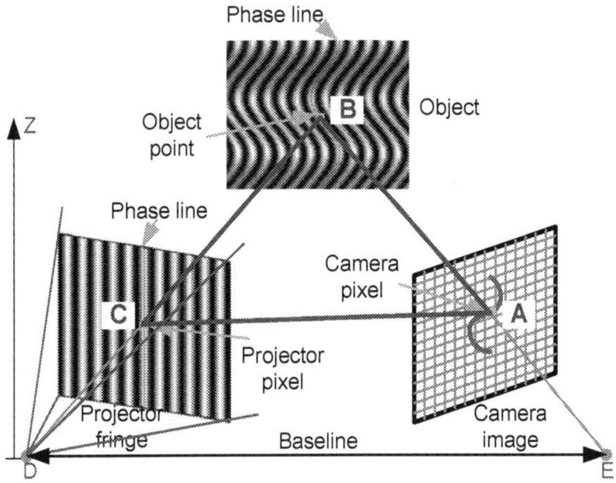

Fig. 2. Schematic diagram of a 3-D shape measurement system with a digital fringe projection technique. A fringe pattern is projected onto an object, the object surface geometry distorts and reflects the fringe pattern, and a camera, viewing from another angle captures the reflected fringe pattern. Because the projector and the camera form a triangular relationship, if the correspondence between the camera and the projector is known, 3-D information can be obtained through triangulation.

2.3 Phase-shifting algorithm

To establish the correspondence between the captured images and the projected images, the structured light system based on binary methods uses *codewords*. The codeword, a unique value, is formed by a sequence of stripes composed of 0's (purely black) or 1's (purely white) and can be determined from captured images. As explained previously, a method using binary structured patterns cannot reach pixel-level resolution spatially, while a technique based on sinusoidal fringe patterns has the potential to do so.

We used a digital fringe projection and phase-shifting method to reach higher spatial and temporal resolution for 3-D shape measurement. For this method, the codeword is determined by the phase that is calculated from the fringe images. To increase measurement speed, the minimum number of fringe images (three) is used. Their intensities can be written as

$$I_1(x,y) = I'(x,y) + I''(x,y)\cos[\phi(x,y) - 2\pi/3], \quad (1)$$

$$I_2(x,y) = I'(x,y) + I''(x,y)\cos[\phi(x,y)], \quad (2)$$

$$I_3(x,y) = I'(x,y) + I''(x,y)\cos[\phi(x,y) + 2\pi/3]. \quad (3)$$

Where $I'(x,y)$ is the average intensity, $I''(x,y)$ the intensity modulation, $\phi(x,y)$ the phase to be solved for. Solving Eqs. (1)–(3) simultaneously, we will obtain the phase

$$\phi(x,y) = \tan^{-1}\left[\frac{\sqrt{3}(I_1 - I_3)}{2I_2 - I_1 - I_3}\right], \tag{4}$$

Equation (4) indicates that the phase ranges from $-\pi$ to $+\pi$ with 2π discontinuities.
From Eqs. (1)–(3), we can also obtain a texture image,

$$\begin{aligned} I_t(x,y) &= I'(x,y) + I''(x,y) \\ &= \frac{(I_1 + I_2 + I_3)}{3} + \frac{\sqrt{3(I_1 - I_3)^2 + (2I_2 - I_1 - I_3)^2}}{3}. \end{aligned} \tag{5}$$

The texture image is the photograph of the object without fringe stripes. Although irrelevant to optical metrology, texture image plays a key role in many fields for shape analysis to industrial inspection, because it provides vital information of the object. In computer graphics, texture image is mostly used for enhancing visual effect by a technique called texture mapping.
In addition, we can also obtain intensity modulation

$$\gamma(x,y) = \frac{I''(x,y)}{I'(x,y)} = \frac{\sqrt{3(I_1 - I_3)^2 + (2I_2 - I_1 - I_3)^2}}{I_1 + I_2 + I_3} \tag{6}$$

Intensity modulation indicates the fringe quality, or the contrast of the fringe images, with 1 being the best. It is usually used to separate the background and the foreground, and sometime used to analyze the phase.

2.4 Phase unwrapping technique

To obtain a continuous phase map from the wrapped phase map determined by Eq. (4), a conventional method is to adopt a phase unwrapping algorithm. The phase unwrapping is essentially to detect 2π discontinuities based on its neighboring phase values and remove them by adding or subtracting multiples of 2π. In other words, the phase unwrapping is to

$$\Phi(x,y) = \phi(x,y) + 2k\pi. \tag{7}$$

Here, k is an integer.
Robust phase unwrapping is usually very time-consuming because it involves a lot computation as well as iterations. The processing time ranges from seconds to minutes, and even to hours (Ghiglia & Pritt, 1998). To improve the robustness of the phase unwrapping step, numerous algorithms have been developed including the branch-cut algorithms (Huntley, 1989), the discontinuity minimization algorithm (Flynn, 1997), the L^p-norm algorithm (Ghiglia & Romero, 1996), and the region growing algorithm (Baldi, 2003). However, they are generally too slow for real-time applications. Some fast phase unwrapping algorithms, such as a flood filling algorithm, can unwrap the phase rapidly (Zhang & Huang, 2006a). However, they are usually very sensitive to the noise of the phase map. To solve the dilemma, we developed a multilevel quality-guided phase unwrapping algorithm (Zhang et al., 2007), which will be discussed in Section 3.2

2.5 Absolute phase retrieval

The phase in Eq. (7), is often regarded as *relative phase*, because it is relative to one point on the phase map that is usually determined from the phase unwrapping stage. *Absolute phase* is the phase value that is relative to the pre-defined phase. In a digital fringe projection and phase-shifting system, the phase obtained from the computer generated fringe images can be regarded as the pre-defined phase values, thus the absolute phase. To obtain absolute phase from the captured fringe images, at least one point within one connected patch must have the known phase value. A marker, such as a point, a line, or any other vivid features can be used as references to convert the relative phase to absolute one. The marker is projected by the projector and captured by the camera. The software algorithm is then used to identify the marker points. Because the phase on the marker point is pre-defined theoretically, the relative phase needs to be shifted to ensure that the marker point phase is the same as that of the pre-defined one. Assume that the reference point on the computer generated fringe images has the absolute phase of 0, and the relative phase ϕ_0 that is calculated from the camera captured images, the relative phase map can be converted to absolute one by the following equation

$$\phi_a(x,y) = \phi(x,y) - \phi_0. \tag{8}$$

2.6 Structured light system calibration

3-D information is carried on by the phase, from which the 3-D coordinates of the object can be recovered. Conventionally, a reference plane is utilized, and any measurement is relative to it (Huang et al., 2002). For smaller range measurement, the depth information (z coordinate) is approximately proportional to the phase difference between the measured phase value and the reference phase value. However, the drawbacks of using a reference plane are:

1. *Approximation.* Accurate coordinates for each point is very difficult to obtain.
2. *Small range measurement.* The depth range must be very small relative to its distance from the hardware system. Otherwise, the approximation error is significant.
3. *Inaccurate x and y coordinates.* This approach in general only obtains the depth z information, while x and y coordinates are ignored.

To circumvent the approximation errors, the system has be to calibrated. One of the key to calibrate the structured light system is to calibrate a camera accurately. The camera calibration is essentially to find the intrinsic parameters (e.g. focal length, principle point), and extrinsic parameter, the transformation between the (x,y,z) coordinates in a world coordinate system (x^w, y^w, z^w), and that in the camera coordinate system. Among the models created, Zhang's calibration model is most widely adopted in computer vision field (Zhang, 2000). In this model, the camera is treated as a pinhole system as shown in Fig. 3, and the intrinsic parameter is represented as,

$$A = \begin{bmatrix} \alpha & \kappa & u_0 \\ 0 & \beta & v_0 \\ 0 & 0 & 1 \end{bmatrix}. \tag{9}$$

Where (u_0, v_0) is the coordinate of the principle point, the intersection between the optical axis and the image sensor, α and β are the focal lengths along u and v axes of the image plane, and κ is the parameter that represents the skewness of uv coordinate system.
And extrinsic parameters can be described as,

$$[R,t] = \begin{bmatrix} r_{11} & r_{12} & r_{13} & t_x \\ r_{21} & r_{22} & r_{23} & t_y \\ r_{31} & r_{32} & r_{33} & t_z \end{bmatrix}. \tag{10}$$

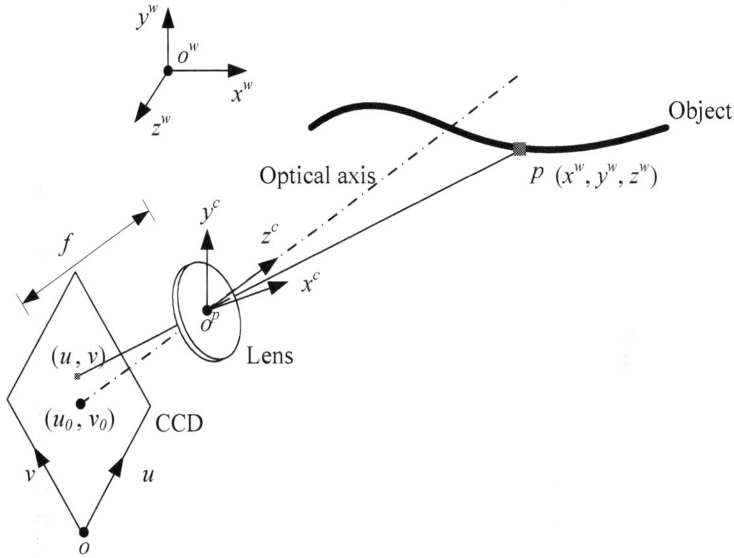

Fig. 3. Pinhole camera model. The camera model is to describe that an arbitrary point in 3-D space under its own coordinate system is transformed to the camera lens coordinate system, and finally 3-D coordinates in lens coordinate system are projected onto a 2-D imaging space.

Here R is the rotation 3×3 matrix, and t is a 3×1 translation vector.
Fig. 3 shows the typical diagram of a pinhole camera model, where p is an arbitrary point with coordinates (x^w, y^w, z^w) in the world coordinate system $(0^w; x^w, y^w, z^w)$, and (x^c, y^c, z^c) in the camera coordinate system $(0^c; x^c, y^c, z^c)$. If p is projected onto the image plane, the relationship between the world coordinates and the image coordinates can be represented as

$$sI = A[\ R, \quad t\ ]X^w, \tag{11}$$

Where $I = [u, v, 1]^T$ is the homogeneous coordinate of the image point in the image plane, $X^w = [x^w, y^w, z^w, 1]^T$ the homogeneous world coordinate for that point, and s a scaling factor. The above equation describes a linear camera model, the nonlinear effects can be compensated for by adopting a nonlinear model.
Unlike a camera-camera (stereo) system, a structured light system calibration involves projector calibration that is usually not simple because it cannot capture images like a regular camera. Numerous techniques have been proposed (Cuevas et al., 1999; Hu et al., 2003). However, most of these techniques are time-consuming and very difficult to achieve high accuracy. Legarda-Sáenz et al. (2004) proposed a method that uses absolute phase to find the marker centers of the calibration board for the projector by projecting a sequence of fringe patterns. Through optimization, this method performed well in terms of accuracy. However, it requires the use of the calibrated camera to calibrate the projector, thus the calibration errors of the camera will bring into the projector calibration, which is not desirable.

Zhang & Huang (2006b) proposed a new structured light system calibration method. In this method, the fringe images are used as a tool to establish the mapping between the camera pixel and the projector pixel so that the projector can "capture" images like a camera. By this means, the structured light system calibration becomes a well studied stereo system calibration. Since the projector and the camera are calibrated independently and simultaneously, the calibration accuracy is significantly improved, and the calibration speed is drastically increased. Fig. 4 shows a typical checkerboard image pair captured by the camera, and the projector image converted by the mapping method. It clearly shows that the projector checkerboard image is well captured. By capturing a number of checkerboard image pairs and applying the software algorithm developed by Bouguet (http://www.vision.caltech.edu/bouguetj/calib_doc), both the camera and the projector are calibrated at the same time.

(a) (b)

Fig. 4. Checkerboard image pair by using the proposed technique by Zhang and Huang (Zhang & Huang, 2006b). (a) The checkerboard image captured by the camera; (b) The mapped checkerboard image for the projector, which is regarded as the checkerboard image captured by the projector.

Following the work by Zhang & Huang (2006b), a number of calibration approaches have been developed (Gao et al., 2008; Huang & Han, 2006; Li et al., 2008; Yang et al., 2008). All these techniques are essentially the same: to establish the one-to-one mapping between the projector and the camera. Our recent work showed that the checker size of the checkerboard plays a key role (Lohry et al., 2009), and a certain range of checker size will give better calibration accuracy. This study provides some guidelines for selecting the checker size for precise system calibration. Once the system is calibrated, the xyz coordinates can be computed from the *absolute* phase, which will be addressed in the next subsection.

2.7 3-D coordinate calculation from the absolute phase

Once the absolute phase map is obtained, the relationship between the camera sensor and projector sensor will be established as a one-to-many mapping, i.e., one point on the camera sensor corresponds to one line on the projector sensor with the same absolute phase value. This relationship provides a constraint for the correspondence of a camera-projector system. If the camera and the projector are calibrated in the same world coordinate system, and the linear calibration model is used for both the camera and the projector, Eq. (11) can be re-written as

$$s^c I^c = A^c [\ R^c, \quad t^c\ ] X^w. \tag{12}$$

Here, s^c is the scaling factor for the camera, I^c the homogeneous camera image coordinates, A^c the intrinsic parameters for the camera, and $[R^c, t^c]$ the extrinsic parameter matrix for the camera.
Similarly, the relationship between the projector image point and the object point in the world coordinate system can be written as

$$s^p I^p = A^p [\ R^p, \quad t^p\ ] X^w. \tag{13}$$

Here s^p is the scaling factor for the projector, I^p the homogeneous projector image coordinates, A^p the intrinsic parameters for the projector, $[R^p, t^p]$ the extrinsic parameter matrix for the projector.
In addition, because absolute phase is known, each point on the camera corresponds to one line with the same absolute phase on the projected fringe image (Zhang & Huang, 2006b). That is, assume the fringe stripe is along v direction, we can establish a relationship between the captured fringe image and the projected fringe image,

$$\phi_a(u^c, v^c) = \phi_a(u^p). \tag{14}$$

In Equations (12)-(14), there are seven unknowns $(x^w, y^w, z^w), s^p, s^p, u^p$, and v^p, and seven equations, the world coordinates (x^w, y^w, z^w) can be uniquely solved for.

2.8 Example of measurement

Fig. 5 shows an example of 3-D shape measurement using a three-step phase-shifting method. Fig. 5(a)-5(c) shows three phase-shifted fringe images with $2\pi/3$ phase shift. Fig. 5(d) shows the phase map after applying Eq. (4) to these fringe images, it clearly shows phase discontinuities. Applying the phase unwrapping algorithm discussed in Reference (Zhang et al., 2007), this wrapped phase map can be unwrapped to get a continuous phase map as shown in Fig. 5(e). The unwrapped phase map is then converted to 3-D shape by applying method introduced in Section 2.7. The 3-D shape can be rendered by OpenGL, as shown in Figs. 5(f)-5(g). At the same time, by averaging these three fringe images, a texture image can be obtained, which can be mapped onto the 3-D shape to for better visual effect, as seen in Fig. 5(h).

3. Real-time 3-D Shape Measurement Techniques

3.1 Hardware implementation of phase-shifting technique for real-time data acquisition

From Section 2, we know that, for a three-step phase-shifting algorithm, only three images are required to reconstruct one 3-D shape. This, therefore, permits the possibility of encoding them into a single color image. As explained in Section 2, using color fringe pattern is not desirable for 3-D shape measurement because of the problems caused by color. To avoid this problem, we developed a real-time 3-D shape measurement system based on a single-chip DLP projection and white light technique (Zhang & Huang, 2006a).
Fig. 6 shows the system layout. Three phase-shifted fringe images are encoded with the RGB color channel of a color fringe image generated by the computer. The color image is then sent to the single-chip DLP projector that switches three-color channels sequentially onto the object; a high-speed CCD camera, synchronized with the projector, is to capture three phase-shifted fringe images at high speed. Any three fringe images can be used to reconstruct one

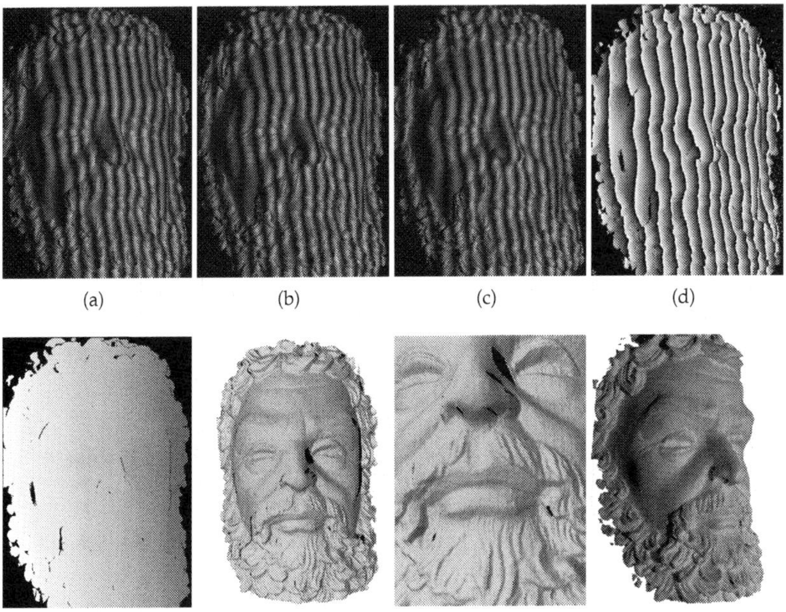

(a) (b) (c) (d) (e) (f) (g) (h)

Fig. 5. . Example of 3-D shape measurement using a three-step phase-shifting method. (a) $I_1(-2\pi/3)$; (b) $I_2(0)$; (c) $I_3(2\pi/3)$; (d) Wrapped phase map; (e) Unwrapped phase map; (f) 3-D shape rendered in shaded mode; (g) Zoom in view; (h) 3-D shape rendered with texture mapping.

3-D shape through phase wrapping and unwrapping. Moreover, by averaging these three fringe images, a texture image (without fringe stripes) can be generated. It can be used for texture mapping purposed to enhance certain view effect.

The projector projects a monochrome fringe image for each of the RGB channels sequentially; the color is a result of a color wheel placed in front of a projection lens. Each "frame" of the projected image is actually three separate images. By removing the color wheel and placing each fringe image in a separate channel, the projector can produce three fringe images at 120 fps (360 individual fps). Therefore, if three fringe images are sufficient to recover one 3-D shape, the 3-D measurement speed is up to 120 Hz. However, due to the speed limit of the camera used, it takes two projection cycles to capture three fringe images, thus the measurement speed is 60 Hz. Fig. 7 shows the timing chart for the real-time 3-D shape measurement system.

3.2 Fast phase-shifting algorithm

The hardware system described in previous subsection can acquire fringe images at 180 Hz. However, the processing speed needs to keep up with the data acquisition for real-time 3-D shape measurement. The first challenge is to increase the processing speed of the phase

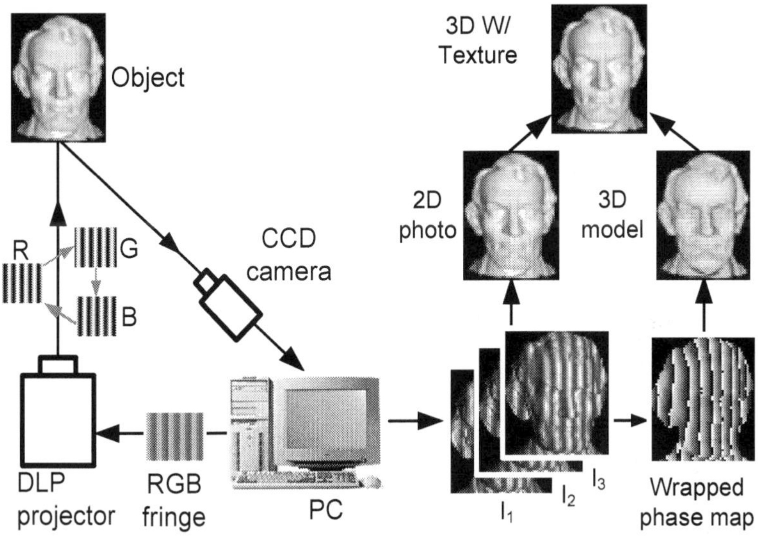

Fig. 6. Real-time 3-D shape measurement system layout. The computer generated color encoded fringe image is sent to a single-chip DLP projector that projects three color channels sequentially and repeatedly in grayscale onto the object. The camera, precisely synchronized with projector, is used to capture three individual channels separately and quickly. By applying the three-step phase-shifting algorithm to three fringe images, the 3-D geometry can be recovered. Averaging three fringe images will result in a texture image that can be further mapped onto 3-D shape to enhance certain visual effect.

wrapping. Experiments found that calculating the phase using Eq. (4) is relatively slow for the purpose of real-time 3-D shape measurement. To improve the processing speed, Huang et al. (2005) developed a new algorithm named trapezoidal phase-shifting algorithm. The advantage of this algorithm is that it processes the phase by intensity ratio instead of arctangent function, thus significantly improves the processing speed (more than 4 times faster). However, the drawback of this algorithm is that the defocusing of the system will introduce error, albeit to a less degree. This is certainly not desirable. Because the sinusoidal fringe patterns are not very sensitive to defocusing problems, we applied the same processing algorithm to sinusoidal fringe, the purpose is to maintain the advantage of processing speed while alleviate the defocusing problem, this new algorithm is called fast three-step phase-shifting algorithm (Huang & Zhang, 2006).

Fig. 8 illustrates this fast three-step phase-shifting algorithm. Instead of calculating phase using an arctangent function, the phase is approximated by intensity ratio

$$r(x,y) = \frac{I_{med}(x,y) - I_{min}(x,y)}{I_{max}(x,y) - I_{min}(x,y)}. \tag{15}$$

Here I_{max}, I_{med}, I_{min} respectively refer to the maximum, median, and minimum intensity value for three fringe images for the same point. The intensity ratio gives values ranging from [0,

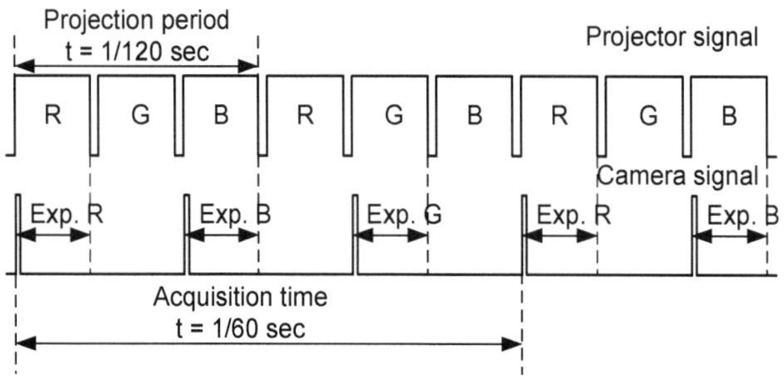

Fig. 7. System timing chart.

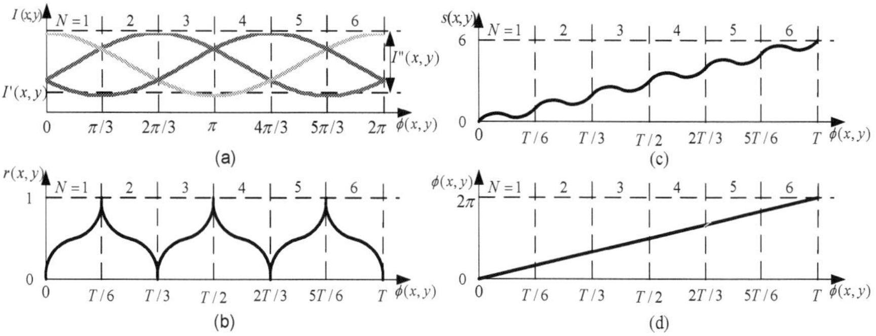

Fig. 8. Schematic illustration for fast three-step phase-shifting algorithm. (a) One period of fringe is uniformly divided into six regions; (b) The intensity ratio for one period of fringe; (c) After slope map after removing the sawtooth shape of the intensity ratio map; (d) The phase after compensate for the approximation error and scaled to its original phase value.

1] periodically within one period of the fringe pattern. Fig. 8(a) shows that one period of the fringe pattern is uniformly divided into six regions. It is interesting to know that the region number N can be uniquely identified by comparing the intensity values of three fringe images point by point. For example, if red is the largest, and blue is the smallest, the point belongs to region $N = 1$. Once the region number is identified, the sawtooth shape intensity ratio in Fig. 8(b) can be converted to its slope shape in Fig. 8(c) by using the following equation

$$s(x,y) = 2 \times Floor\left(\frac{N}{2}\right) + (-1)^{N-1} r(x,y). \tag{16}$$

Here the operator $Floor()$ is used to truncate the floating point data to keep the integer part only. The phase can then be computed by

$$\phi(x,y) = 2\pi \times s(x,y). \tag{17}$$

Because the phase is calculated by a linear approximation, the residual error appears. Since the phase error is fixed in the phase domain, it can be compensated for by using a look-up-table (LUT). After the phase error compensation, the phase will be a linear slope as illustrated in Fig. 8(d). Experiments found that by using this fast three-step phase-shifting algorithm, the 3-D shape measurement speed is approximately 3.4 times faster.

Phase unwrapping step usually is the most timing-consuming part for 3-D shape measurement based on fringe analysis. Therefore, developing an efficient and robust phase unwrapping algorithm is vital to the success of real-time 3-D shape measurement. Traditional phase unwrapping algorithms are either less robust (such as flood-fill methods) or time consuming (such quality-guided methods). We have developed a multi-level quality-guided phase unwrapping algorithm (Zhang et al., 2007). It is a good trade-off between robustness and efficiency: the processing speed of the quality-guided phase unwrapping algorithm is augmented by the robustness of the scanline algorithm. The quality map was generated from the gradient of the phase map, and then quantized into multi-levels. Within each level point, the fast scanline algorithm was applied. For a three-level algorithm, it only takes approximately 18.3 ms for a 640×480 resolution image, and it could correctly reconstruct more than 99% of human facial data.

By adopting the proposed fast three-step phase-shifting algorithm and the rapid phase unwrapping algorithm, the continuous phase map can be reconstructed in a timely manner. In order to do 3-D coordinates calculations, it involves very intensive matrix operations including matrix inversion, it was found impossible to perform all the calculations in real-time with an ordinary dual CPU workstation. To resolve this problem, new computational hardware technology, graphics processing unit (GPU), was explored, which will be introduced in the next subsection.

3.3 Real-time 3-D coordinates calculation and visualization using GPU

Computing 3-D coordinates from the phase is computationally intensive, which is very challenging for a single computer CPU to realize in real-time. However, because the coordinate calculations are point by point matrix operations, this can be performed efficiently by a GPU. A GPU is a dedicated graphics rendering device for a personal computer or game console. Modern GPUs are very efficient at manipulating and displaying computer graphics, and their highly parallel structure makes them more effective than typical CPUs for parallel computation algorithms. Since there are no memory hierarchies or data dependencies in the streaming model, the pipeline maximizes throughput without being stalled. Therefore, whenever the GPU is consistently fed by input data, performance is boosted, leading to an extraordinarily scalable architecture (Ujaldon & Saltz, 2005). By utilizing this streaming processing model, modern GPUs outperform their CPU counterparts in some general-purpose applications, and the difference is expected to increase in the future (Khailany et al., 2003).

Fig. 9 shows the GPU pipeline. CPU sends the vertex data including the vertex position coordinates and vertex normal to GPU which generates the lighting of each vertex, creates the polygons and rasterizes the pixels, then output the rasterized image to the display screen. Modern GPUs allow user specified code to execute within both the vertex and pixel sections of the pipeline which are called vertex shader and pixel shader, respectively. Vertex shaders are applied for each vertex and run on a programmable vertex processor. Vertex shaders takes vertex coordinates, color, and normal information from the CPU.The vertex data is streamed into the GPU where the polygon vertices are processed and assembled based on the order of the incoming data. The GPU handles the transfer of streaming data to parallel computation automatically. Although the clock rate of a GPU might be significantly slower than that of a CPU, it has multiple vertex processors acting in parallel, therefore, the throughput of the GPU can exceed that of the CPU. As GPUs increase in complexity, the number of vertex processors increase, leading to great performance improvements.

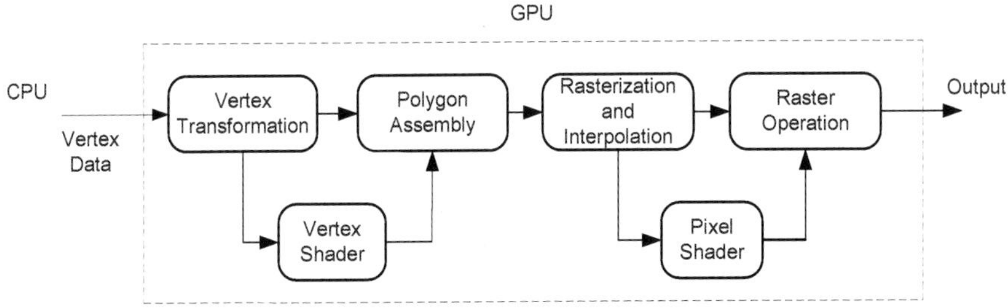

Fig. 9. GPU pipeline. Vertex data including vertex coordinates and vertex normal are sent to the GPU. GPU generates the lighting of each vertex, creates the polygons and rasterizes the pixels, then output the rasterized image to the display screen.

By taking advantage of the processing power of the GPU, 3-D coordinate calculations can be performed in real time with an ordinary personal computer with a decent NVidia graphics card (Zhang et al., 2006). Moreover, because 3-D shape data are already on the graphics card, they can be rendered immediately without any lag. Therefore, by this means, real-time 3-D geometry visualization can also be realized in real time simultaneously. Besides, because only the phase data, instead of 3-D coordinates plus surface normal, are transmitted to graphics card for visualization, this technique reduces the data transmission load on the graphics card significantly, (approximately six times smaller). In short, by utilizing the processing power of GPU for 3-D coordinates calculations, real-time 3-D geometry reconstruction and visualization can be performed rapidly and in real time.

3.4 Experimental results

Fig. 10 shows one of the hardware systems that we developed. The hardware system is composed of a DLP projector (PLUS U5-632h), a high-speed CCD camera (Pulnix TM-6740CL) and a timing generation circuit. The projector has an image resolution of 1024 $\times$ 768, and the focal length of f = 18.4-22.1 mm. The camera resolution is 640 $\times$ 480, and the lens used is a Fujinon HF16HA-1B f = 16 mm lens. The maximum data speed for this camera is 200 frames per second (fps). The maximum data acquisition speed achieved for this 3-D shape measurement system is 60 fps.

With this speed, dynamically deformable 3-D objects, such as human facial expressions, can be effectively captured. Fig. 11 shows some typical measurement results of a human facial expression. The experimental results demonstrate that the details of human facial expression can be effectively captured. At the same time, the motion process of the expression is precisely acquired.

By adopting the fast three-step phase-shifting algorithm introduced in Reference (Huang & Zhang, 2006), the fast phase-unwrapping algorithm explained in Reference (Zhang et al., 2007), and the GPU processing detailed in Reference (Zhang et al., 2006), we achieved simultaneous data acquisition, reconstruction, and display at approximately 26 Hz. The computer used for this test contained Dual Pentium 4 3.2 GHz CPUs, and an Nvidia Quadro FX 3450 GPU. Fig. 12 shows a measurement result. The right shows the real subject and

Fig. 10. Photograph of the real-time 3-D shape measurement system. It comprises a DLP projector, a high-speed CCD camera, and a timing signal generator that is used to synchronize the projector with the camera. The size of the system is approximately 24″ × 14″ × 14″.

the left shows the 3-D model reconstructed and displayed on the computer monitor instantaneously. It clearly shows that the technology we developed can perform high-resolution, real-time 3-D shape measurement. More measurement results and videos are available at `http://www.vrac.iastate.edu/~song`.

4. Potential Applications

Bridging between real-time 3-D shape measurement technology and other fields is essential to driving the technology advancement, and to propelling its deployment. We have made significant effort to explore its potential applications. We have successfully applied this technology to a variety of fields. This section will discuss some applications including those we have explored.

4.1 Medical sciences

Facial paralysis is a common problem in the United States, with an estimated 127,000 persons having this permanent problem annually (Bleicher et al., 1996). High-speed 3-D geometry sensing technology could assist with diagnosis; several researchers have attempted to develop objective measures of facial functions (Frey et al., 1999; Linstrom, 2002; Stewart et al., 1999; Tomat & Manktelow, 2005), but none of which have been adapted for clinical use due to the generally cumbersome, nonautomated modes of recording and analysis (Hadlock et al., 2006). The high-speed 3-D shape measurement technology fills this gap and has the potential to diagnose facial paralysis objectively and automatically (Hadlock & Cheney, 2008). A pilot study has demonstrated its feasibility and its great potential for improving clinical practices (Mehta et al., 2008).

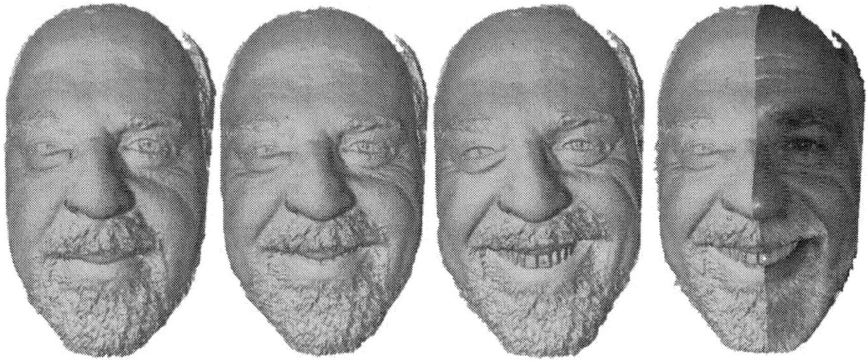

Fig. 11. Measurement result of human facial expressions. The data is acquired at 60 Hz, the camera resolution is 640 × 480.

4.2 3-D computer graphics

3-D computer facial animation, one of the primary areas of 3-D computer graphics, has caused considerable scientific, technological, and artistic interest. As noted by Bowyer et al. (Bowyer et al., 2006), one of the grand challenges in computer analysis of human facial expressions is acquiring natural facial expressions with high fidelity. Due to the difficulty of capturing high-quality 3-D facial expression data, conventional techniques (Blanz et al., 2003; Guenter et al., 1998; Kalberer & Gool, 2002) usually require a considerable amount of manual inputs (Wang et al., 2004). The high-speed 3-D shape measurement technology that we developed benefits this field by providing photorealistic 3-D dynamic facial expression data that allows computer scientists to develop automatic approaches for 3-D facial animation. We have been collaborating with computer scientists in this area and have published several papers (Huang et al., 2004; Wang et al., 2008; 2004).

4.3 Infrastructure health monitoring

Finding the dynamic response of infrastructures under loading/unloading will enhance the understanding of their health conditions. Strain gauges are often used for infrastructure health monitoring and have been found successful. However, because this technique usually measures a point (or small area) per sensor, it is difficult to obtain a large-area response unless a sensor network is used. Area 3-D sensors such as scanning laser vibrometers provide more information (Staszewski, 2007), but because of their low temporal resolution, they are difficult to apply for high-frequency study. Kim et al. (2007) noted that using a kilo Hz sensor is sufficient to monitor high-frequency phenomena. Thus, the high-speed 3-D shape measurement technique may be applied to this field.

4.4 Biometrics for homeland security

3-D facial recognition is a modality of the facial recognition method in which the 3-D shape of a human face is used. It has been demonstrated that 3-D facial recognition methods can achieve significantly better accuracy than their 2-D counterparts, rivaling fingerprint recognition (Bronstein et al., 2005; Heseltine et al., 2008; Kakadiaris et al., 2007; Queirolo et al., 2009). By measuring the geometry of rigid features, 3-D facial recognition avoids such pitfalls of 2-D

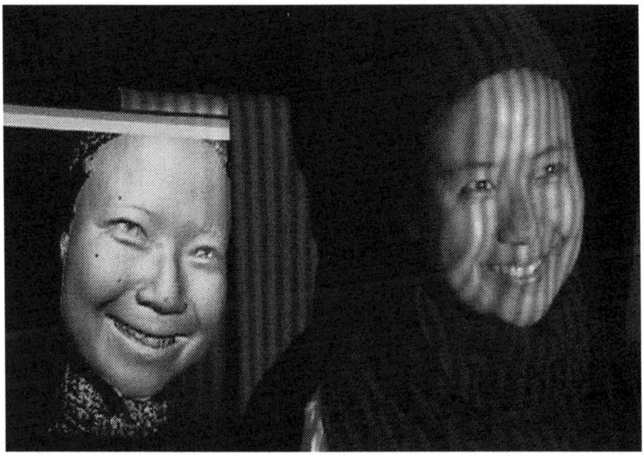

Fig. 12. Simultaneous 3-D data acquisition, reconstruction and display in real-time. The right shows human subject, while the left shows the 3-D reconstructed and displayed results on the computer screen. The data is acquired at 60 Hz and visualized at approximately 26 Hz.

peers as change in lighting, different facial expressions, make-up, and head orientation. Another approach is to use a 3-D model to improve the accuracy of the traditional image-based recognition by transforming the head into a known view. The major technological limitation of 3-D facial recognition methods is the rapid acquisition of 3-D models. With the technology we developed, high-quality 3-D faces can be captured even when the subject is moving. The high-quality scientific data allows for developing software algorithms to reach 100% identification rate.

4.5 Manufacturing and quality control

Measuring the dimensions of mechanical parts on the production line for quality control is one of the goals in the manufacturing industry. Technologies relying on coordinate measuring machines or laser range scanning are usually very slow and thus cannot be performed for all parts. Samples are usually taken and measured to assure the quality of the product. A high-speed dimension measurement device that allows for 100% product quality assurance will significantly benefit this industry.

5. Challenges

High-resolution, real-time 3-D shape measurement has already emerged as an important means for numerous applications. The technology has advanced rapidly recently. However, for the real-time 3-D shape measurement technology that was discussed in this chapter, there some major limitations:

1. *Single object measurement*. The basic assumptions for correct phase unwrapping and 3-D reconstruction require the measurement points to be smoothly connected (Zhang et al., 2007). Thus, it is impossible to measure multiple objects simultaneously.
2. *"Smooth" surfaces measurement*. The success of a phase unwrapping algorithm hinges on the assumption that the phase difference between neighboring pixels is less than

π . Therefore, any step height causing a phase change beyond π cannot be correctly recovered.

3. *Maximum speed of 120 Hz.* Because sinusoidal fringe images are utilized, at least an 8-bit depth is required to produce good contrast fringe images. That is, a 24-bit color image can only encode three fringe images, thus the maximum fringe projection speed is limited by the digital video projector's maximum projection speed (typically 120 Hz).

Fundamentally, the first two limitations are essentially induced by the phase unwrapping of a single-wavelength phase-shifting technique. The phase unwrapping assumes that the phase changes between two neighboring pixel is not beyond π, thus any unknown changes or changes beyond cannot be correctly recovered. This hurtle can be overcome by using multiple-wavelength fringe images. For example, a digital multiple-wavelength technique can be adopted to solve this problem (Zhang, 2009). Using a multiple-wavelength technique will reduce the measurement speed significantly because more fringe images are required to perform one measurement. It has been indicated that at least three wavelength fringe images are required to measure arbitrary 3-D surfaces with arbitrary step heights (Towers et al., 2003). The speed is essentially limited by hardware, and is difficult to overcome for if the traditional method is used, where the grayscale fringe images has to be adopted. The image switching speed is essentially limited by the data sent to the projection system and the generation of the sinusoidal patterns. Recently, Lei & Zhang (2009) proposed a promising technology that realized a sinusoidal phase-shifting algorithm using binary patterns through projector defocusing. This technique may lead a breakthrough in this field because switching binary structured images can be realized in a much faster manner allowed by hardware.

Besides the speed and range challenges of the current real-time 3-D shape measurement techniques, there are a number of more challenging problems to tackle. The major challenges are:

1. *Shiny surfaces measurement.* Shiny surfaces are very common in manufacturing, especially before any surface treatment. How to measure this type of parts using the real-time 3-D shape measurement technique remains challenging. There are some techniques proposed (Chen et al., 2008; Hu et al., 2005; Zhang & Yau, 2009), but none of them are suitable for real-time 3-D measurement cases.

2. *Accuracy improvement.* The accuracy of the current real-time 3-D shape measurement system is not high. Part of the error is caused by motion of the object. This is because that the object is assumed to be motionless when the measurement is performed. However, to measure the object in motion, this assumption might cause problem. To meet the requirement in manufacturing engineering, it is very important to improve its system accuracy. One of the critical issues is the lack of standard for real-time 3-D shape measurement. Therefore, build a higher accuracy real-time 3-D shape measurement as a standard is very essential, but challenging.

3. *High-quality color texture measurement.* Although irrelevant to metrology, it is highly important to simultaneously acquire the high quality color texture, the photograph of the object, for numerous applications including computer graphics, medical sciences, and homeland security. For instance, in medical sciences, the 2-D color texture may convey critical information for diagnosis. We have developed a simultaneous color texture acquisition technique (Zhang & Yau, 2008). However, the object is illuminated by directional light (the projector's light). This is not desirable for many applications that requires very high quality 2-D color textures, where the object must be illuminated

with diffuse light uniformly. How to capture 3-D geometry and the color texture in real time and simultaneously becomes challenging.

6. Summary

We have covered the high-speed 3-D shape measurement techniques, especially focused on the system that was developed by our research group. The technology itself has numerous applications already. We have also addressed the limitations of the technology and the challenging questions we need to answer before this technology can be widely adopted.

7. Acknowledgements

First of all, I would like to thank this book editor, Dr. Vedran Kordic, for his invitation. My thanks also goes to my former advisors, Prof. Peisen Huang at Stony Brook University, and Prof. Shing-Tung Yau at Harvard University for their supervision. Some of the work was conducted under their support. I thank my students, Nikolaus Karpinsky, Shuangyan Lei, William Lohry, Ying Xu, and Victor Emmanuel Villagomez at Iowa State University for their brilliant work. Finally, I would like to thank my wife, Xiaomei Hao, for her consistent encouragement and support.

8. References

Baldi, A. (2003). Phase unwrapping by region growing, *Appl. Opt.* **42**: 2498–2505.

Blanz, V., Basso, C., Poggio, T. & Vetter, T. (2003). Reanimating faces in images and video, *Eurographics*, pp. 641–650.

Bleicher, J. N., Hamiel, S. & Gengler, J. S. (1996). A survey of facial paralysis: etiology and incidence, *Ear Nose Throat J.* **76**(6): 355–57.

Bowyer, K. W., Chang, K. & Flynn, P. J. (2006). A survey of approaches and challenges in 3d and multi-modal 3d+2d face recognition, *Comp. Vis. and Imag. Understand.* **12**: 1–15.

Bronstein, A. M., Bronstein, M. M. & Kimmel, R. (2005). Three-dimensional face recognition, *Intl J. of Comp. Vis. (IJCV)* **64**: 5–30.

Chen, Y., He, Y. & Hu, E. (2008). Phase deviation analysis and phase retrieval for partial intensity saturation in phase-shifting projected fringe profilometry, *Opt. Comm.* **281**(11): 3087–3090.

Cuevas, F. J., Servin, M. & Rodriguez-Vera, R. (1999). Depth object recovery using radial basis functions, *Opt. Commun.* **163**(4): 270–277.

Dhond, U. & Aggarwal, J. (1989). Structure from stereo-a review, *IEEE Trans. Systems, Man, and Cybernetics* **19**: 1489–1510.

Flynn, T. J. (1997). Two-dimensional phase unwrapping with minimum weighted discontinuity, *J. Opt. Soc. Am. A* **14**: 2692–2701.

Frey, M., Giovanolli, P., Gerber, H., Slameczka, M. & Stussi, E. (1999). Three-dimensional video analysis of facial movements: a new method to assess the quantity and quality of the smile, *Plast Reconstr Surg.* **104**: 2032–2039.

Gao, W., Wang, L. & Hu, Z. (2008). Flexible method for structured light system calibration, *Opt. Eng.* **47**(8): 083602.

Geng, Z. J. (1996). Rainbow 3-d camera: New concept of high-speed three vision system, *Opt. Eng.* **35**: 376–383.

Ghiglia, D. C. & Pritt, M. D. (eds) (1998). *Two-Dimensional Phase Unwrapping: Theory, Algorithms, and Software*, John Wiley and Sons, New York.

Ghiglia, D. C. & Romero, L. A. (1996). Minimum l^p-norm two-dimensional phase unwrapping, *J. Opt. Soc. Am. A* **13**: 1–15.

Guenter, B., Grimm, C., Wood, D., Malvar, H. & Pighin, F. (1998). Making faces, *SIGGRAPH*, pp. 55–66.

Guo, H. & Huang, P. (2008). 3-d shape measurement by use of a modified fourier transform method, *Proc. SPIE*, Vol. 7066, p. 70660E.

Guo, H. & Huang, P. S. (2009). Absolute phase retrieval for 3d shape measurement by fourier transform method, *Opt. Eng.* **48**: 043609.

Hadlock, T. A. & Cheney, M. L. (2008). Facial reanimation: an invited review and commentary, *Arch Facial Plast Surg.* **10**: 413–417.

Hadlock, T. A., Greenfield, L. J., Wernick-Robinson, M. & Cheney, M. L. (2006). Multimodality approach to management of the paralyzed face, *Laryngoscope* **116**: 1385–1389.

Hall-Holt, O. & Rusinkiewicz, S. (2001). Stripe boundary codes for real-time structured-light range scanning of moving objects, *The 8th IEEE International Conference on Computer Vision*, pp. II: 359–366.

Harding, K. G. (1988). Color encoded morié contouring, *Proc. SPIE*, Vol. 1005, pp. 169–178.

Heseltine, T., Pears, N. & Austin, J. (2008). Three-dimensional face recognition using combinations of surface feature map subspace components, *Image and Vision Computing (IVC)* **26**: 382–396.

Hu, Q., Harding, K. G., Du, X. & Hamilton, D. (2005). Shiny parts measurement using color separation, *SPIE Proc.*, Vol. 6000, pp. 6000D1–8.

Hu, Q., Huang, P. S., Fu, Q. & Chiang, F. P. (2003). Calibration of a 3-d shape measurement system, *Opt. Eng.* **42**(2): 487–493.

Huang, P. & Han, X. (2006). On improving the accuracy of structured light systems, *Proc. SPIE*, Vol. 6382, p. 63820H.

Huang, P. S., Hu, Q., Jin, F. & Chiang, F. P. (1999). Color-encoded digital fringe projection technique for high-speed three-dimensional surface contouring, *Opt. Eng.* **38**: 1065–1071.

Huang, P. S., Zhang, C. & Chiang, F.-P. (2002). High-speed 3-d shape measurement based on digital fringe projection, *Opt. Eng.* **42**(1): 163–168.

Huang, P. S. & Zhang, S. (2006). Fast three-step phase shifting algorithm, *Appl. Opt.* **45**(21): 5086–5091.

Huang, P. S., Zhang, S. & Chiang, F.-P. (2005). Trapezoidal phase-shifting method for three-dimensional shape measurement, *Opt. Eng.* **44**(12): 123601.

Huang, X., Zhang, S., Wang, Y., Metaxas, D. & Samaras, D. (2004). A hierarchical framework for high resolution facial expression tracking, *IEEE Computer Vision and Pattern Recognition Workshop*, Vol. 01, p. 22.

Huntley, J. M. (1989). Noise-immune phase unwrapping algorithm, *Appl. Opt.* **28**: 3268–3270.

Jia, P., Kofman, J. & English, C. (2007). Two-step triangular-pattern phase-shifting method for three-dimensional object-shape measurement, *Opt. Eng.* **46**(8): 083201.

Kakadiaris, I. A., Passalis, G., Toderici, G., Murtuza, N., Karampatziakis, N. & Theoharis, T. (2007). 3d face recognition in the presence of facial expressions: an annotated deformable model approach, *IEEE Trans. on Patt. Anal. and Mach. Intellig. (PAMI)* **29**: 640–649.

Kalberer, G. A. & Gool, L. V. (2002). Realistic face animation for speech, *Intl Journal of Visualization Computer Animation* **13**(2): 97–106.

Khailany, B., Dally, W., Rixner, S., Kapasi, U., Owens, J. & Towles, B. (2003). Exploring the vlsi scalability of stream processors, *Proc. 9th Symp. on High Perf. Comp. Arch.*, pp. 153–164.

Kim, S., Pakzad, S., Culler, D., Demmel, J., Fenves, G., Glaser, S. & Turon, M. (2007). Health monitoring of civil infrastructurtes using wireless sensor network, *Proc. 6th intl conference on information processing in sensor networks*, pp. 254–263.

Legarda-Sáenz, R., Bothe, T. & Jüptner, W. P. (2004). Accurate procedure for the calibration of a structured light system, *Opt. Eng.* **43**(2): 464Ű–471.

Lei, S. & Zhang, S. (2009). Flexible 3-d shape measurement using projector defocusing, *Opt. Lett.* **34**(20): 3080–3082.

Li, Z., Shi, Y., Wang, C. & Wang, Y. (2008). Accurate calibration method for a structured light system, *Opt. Eng.* **47**(5): 053604.

Linstrom, C. J. (2002). Objective facial motion analysis in patients with facial nerve dysfunction, *Laryngoscope* **112**: 1129–1147.

Lohry, W., Xu, Y. & Zhang, S. (2009). Optimum checkerboard selection for accurate structured light system calibration, *Proc. SPIE*, Vol. 7432, p. 743202.

Mehta, R. P., Zhang, S. & Hadlock, T. A. (2008). Novel 3-d video for quantification of facial movement, *Otolaryngol Head Neck Surg.* **138**(4): 468–472.

Pan, J., Huang, P. S. & Chiang, F.-P. (2005). Accurate calibration method for a structured light system, *Opt. Eng.* **44**(2): 023606.

Pan, J., Huang, P., Zhang, S. & Chiang, F.-P. (2004). Color n-ary gray code for 3-d shape measurement, *12th Intl Conf. on Exp. Mech.*

Queirolo, C. C., Silva, L., Bellon, O. R. & Segundo, M. P. (2009). 3d face recognition using simulated annealing and the surface interpenetration measure, *IEEE Trans. on Patt. Anal. and Mach. Intellig. (PAMI)* . doi:10.1109/TPAMI.2009.14.

Rusinkiewicz, S., Hall-Holt, O. & Levoy, M. (2002). Real-time 3d model acquisition, *ACM Trans. Graph.* **21**(3): 438–446.

Salvi, J., Pages, J. & Batlle, J. (2004). Pattern codification strategies in structured light systems, *Patt. Recogn.* **37**: 827–849.

Schreiber, H. & Bruning, J. H. (2007). *Optical Shop Testing*, 3rd edn, John Wiley & Sons, chapter Phase shifting interferometry, pp. 547–655.

Staszewski, W.J., L. B. C. T. R. (2007). Fatigue crack detection in metallic structures with lamb waves and 3d laser vibrometry, *Meas. Sci. Tech.* **18**: 727–729.

Stewart, B. M., Hager, J. C., Ekman, P. & Sejnowski, T. J. (1999). Measuring facial expressions by computer image analysis, *Psychophysiology* **36**: 253–263.

Su, X. & Zhang, Q. (2009). Dynamic 3-d shape measurement method: A review, *Opt. Laser. Eng* . doi:10.1016/j.optlaseng.2009.03.012.

Takeda, M. & Mutoh, K. (1983). Fourier transform profilometry for the automatic measurement of 3-d object shape, *Appl. Opt.* **22**: 3977–3982.

Tomat, L. R. & Manktelow, R. T. (2005). Evaluation of a new measurement tool for facial paralysis reconstruction, *Plast Reconstr Surg.* **115**: 696–704.

Towers, D. P., Jones, J. D. C. & Towers, C. E. (2003). Optimum frequency selection in multifrequency interferometry, *Opt. Lett.* **28**: 1–3.

Ujaldon, M. & Saltz, J. (2005). Exploiting parallelism on irregular applications using the gpu, *Intl. Conf. on Paral. Comp.*, pp. 13–16.

Wang, Y., Gupta, M., Zhang, S., Wang, S., Gu, X., Samaras, D. & Huang, P. (2008). High resolution tracking of non-rigid 3d motion of densely sampled data using harmonic maps, *Intl J. Comp. Vis.* **76**(3): 283–300.

Wang, Y., Huang, X., Lee, C.-S., Zhang, S., Li, Z., Samaras, D., Metaxas, D., Elgammal, A. & Huang, P. (2004). High-resolution acquisition, learning and transfer dynamic 3d facial expression, *Comp. Graph. Forum* **23**(3): 677 – 686.

Yang, R., Cheng, S. & Chen, Y. (2008). Flexible and accurate implementation of a binocular structured light system, *Opt. Lasers Eng.* **46**(5): 373–379.

Zhang, S. (2009). Digital multiple-wavelength phase-shifting algorithm, *Proc. SPIE*, Vol. 7432, p. 74320N.

Zhang, S. (2010). Recent progresses on real-time 3-d shape measurement using digital fringe projection techniques, *Opt. Laser Eng* **40**: 149–158.

Zhang, S. & Huang, P. (2004). High-resolution, real-time 3-d shape acquisition, *IEEE Comp. Vis. and Patt. Recogn. Workshop*, Vol. 3, Washington DC, MD, pp. 28–37.

Zhang, S. & Huang, P. S. (2006a). High-resolution, real-time three-dimensional shape measurement, *Opt. Eng.* **45**(12): 123601.

Zhang, S. & Huang, P. S. (2006b). Novel method for structured light system calibration, *Opt. Eng.* **45**: 083601.

Zhang, S., Li, X. & Yau, S.-T. (2007). Multilevel quality-guided phase unwrapping algorithm for real-time three-dimensional shape reconstruction, *Appl. Opt.* **46**(1): 50–57. (Selected for February 5, 2007 issue of The *Virtual Journal for Biomedical Optics*).

Zhang, S., Royer, D. & Yau, S.-T. (2006). Gpu-assisted high-resolution, real-time 3-d shape measurement, *Opt. Express* **14**: 9120–9129.

Zhang, S. & Yau, S.-T. (2007). High-speed three-dimensional shape measurement using a modified two-plus-one phase-shifting algorithm, *Opt. Eng.* **46**(11): 113603.

Zhang, S. & Yau, S.-T. (2008). Simultaneous three-dimensional geometry and color texture acquisition using single color camera, *Opt. Eng.* **47**(12): 123604.

Zhang, S. & Yau, S.-T. (2009). High dynamic range scanning technique, *Opt. Eng.* **48**: 033604.

Zhang, Z. (2000). A flexible new technique for camera calibration, *IEEE Trans. Pattern Anal. Mach. Intell.* **22**(11): 1330–1334.

3

High Temperature Superconducting Maglev Measurement System

Jia-Su Wang and Su-Yu Wang
Applied Superconductivity Laboratory of Southwest Jiaotong University
P. R. China

1. Introduction

Melt-textured rare-earth Ba-Cu-O (REBCO, RE=Nd, Sm, Eu, Gd, etc.) bulk samples have high critical current density and high critical magnetic flux, which can produce a strong levitation force and a stable equilibrium. The high temperature superconducting (HTS) REBCO bulk may be cooled using liquid nitrogen instead of liquid helium to reduce the initial construction and running cost in practical application systems. This makes HTS bulks particularly attractive for the applications in magnetic bearings (Moon, 1990), flywheel energy storage devices (Bomemann, 1995), and Maglev vehicle (Wang J. et al., 2002). In order to investigate the magnetic levitation properties (levitation force, guidance force, trapping flux, and so on) of the HTS Maglev vehicle over a permanent magnet (PM) guideway, SCML-01 HTS Maglev measurement system was developed at the Applied Superconductivity Laboratory (ASCLab) of Southwest Jiaotong University in China (Wang J. et al., 2000). The measurement system includes a liquid nitrogen vessel, a permanent magnet guideway (PMG), data collection and processing capabilities, a mechanical drive and an Autocontrol feature. The bottom wall of the vessel has a thickness of 3.0 mm. The PMG has a length of 920 mm, for which its magnetic induction reaches up to 1.2 T. The measuring process is controlled by a computer.

The SCML-01 measurement system is capable of performing real time measurements of Maglev properties through a combination of one or many YBaCuO bulks and one PM or several PMGs. This set up was especially employed on board the HTS Maglev equipment over one and two PMGs. The on board Maglev equipment includes a rectangular-shaped liquid nitrogen vessel containing YBaCuO bulk superconductors.

Based on the original research results (Wang J. & S. Wang, 2005a; Song, 2006) from SCML-01, the first man-loading HTS Maglev test vehicle in the world was successfully developed in 2000 (Wang J. et al., 2002). After 2004, several HTS Maglev vehicle prototypes over a PMG followed in Germany, Russia, Brazil, Japan and Italy (Schultz et al., 2005; Kovalev et al., 2005; Stephan et al., 2004; Okano et al., 2006; D'Ovidio et al., 2008).

Given the lack in measurement functions and measurement precision of the SCML-01, after five years, the HTS Maglev Measurement System (SCML-02) with more functions and higher precision was developed to extensively investigate the Maglev properties of YBaCuO bulks over a PM or PMG (Wang S. et al., 2007). The new features in this measurement system are unique and they include: higher measurement precision, instant measurement upon the movement of the measured HTS sample, automatic measurements of both levitation and guidance forces, dynamic rigidity, three dimensional simultaneous movement of the HTS sample, relaxation measurement of both levitation and guidance forces, and so on.

All these experimental parameters are very helpful to evaluate the load ability of the HTS Maglev vehicle. But the running performance over a PMG cannot be measured by the above mentioned measurement systems.

For the further development of the HTS Maglev vehicle in engineering application, the dynamic Maglev properties should be clearly understood. In order to investigate the dynamic characteristics behavior of the HTS Maglev, an HTS Maglev dynamic measurement system (SCML-03) was designed and successfully developed (Wang J. et al., 2008). The system's main component constitutes of a circular PMG, along with a liquid nitrogen vessel, data acquisition and processing features, mechanical drive, autocontrol, etc. The PMG is fixed along the circumferential direction of a big circular disk with a diameter of 1,500 mm. The maximum linear velocity of the PMG is about 300 km/h when the circular disk rotates around the central axis at 1280 rpm. The liquid nitrogen vessel along with the assembly of HTS bulk samples is placed above the PMG of the dynamic testing. The liquid nitrogen vessel is made to not be rigid along the three principal axes but instead, measurement sensor devices are attached. These sensors can detect weak changes of force along the three-principal directions.

The principles, methods, structure, functions, and specifications of the several HTS Maglev measurement system are discussed in detail in this chapter. These systems were developed at the Applied Superconductivity Laboratory (ASCLab) of Southwest Jiaotong University, P. R. China (Wang J. et al., 2000; Wang S. et al., 2007; Wang J. et al., 2008), and they have unique functions towards the measurement of the HTS Maglev.

2. HTS Maglev measurement system

The potential engineering applications mentioned above are based on high quality HTS bulk samples, and it is especially important to investigate the magnetic levitation properties between the YBCO bulk and the permanent magnet. HTS bulk preparation methods and enhancements are still in progress. The axial symmetry magnetic levitation properties are fully researched, and there are comprehensive review papers elsewhere (Moon, 1994; Hull, 2000; Ma, 2003; Deng, 2008a; 2009a). The Maglev properties between the YBCO bulk and the PMG are discussed by this chapter author (Wang J. & S. Wang, 2005a; Song, 2006; Wang J. et al., 2009).

2.1 Brief History of HTS magnetic levitation

The levitation of a NdFeB permanent magnet 0.7 cm^3 above a piece of 2.5 cm-diameter, 0.6 cm thick disk of YBCO bulk superconductor bathed in liquid nitrogen was observed by Hellman et al. (Hellman, 1988). While Peter et al. (Peter, 1988) had observed the very stable suspension of YBCO samples in the divergent magnetic field, they discovered the suspending phenomenon below the permanent magnet.

An HTS Maglev measurement system was developed (Wang J., 2000; 2001) in order to investigate magnetic levitation properties of the HTS YBCO bulk above a PMG. A series of the properties, for example, levitation force, guidance force, levitation stiffness, etc., of YBCO bulk HTS over a PMG were investigated with this measurement system. The measurement system includes liquid nitrogen vessel (circular and rectangular-shaped), permanent magnet guideway (PMG), data collection and processing, mechanical drive and control system, and scanning of the magnetic flux.

2.2 Permanent magnet guideway (PMG)

Fig. 1 shows the cross-sectional drawing of the PMG. Two construction cross-sectional drawings of the PMG are shown. The PMG is composed of normal permanent magnets and iron plate. The arrows represent magnetic poles where the arrowhead represents north. The length of the PMG is 920 mm, and the concentrating magnetic induction of the PMG is up to 1.2 T at the surface.

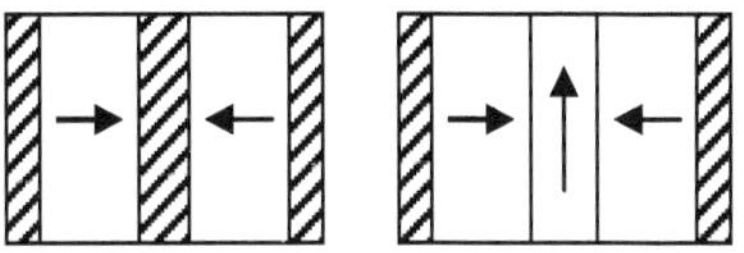

(a) (b)

Fig. 1. Cross sectional view of the PMG
iron → NdFeB

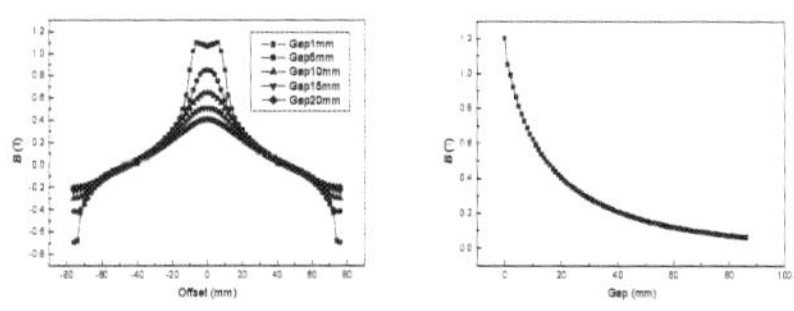

Fig. 2. Measured results of the PMG's magnetic field along the transverse (lift) and vertical (right) directions.

In Fig. 2, the magnetic field of the center of the PMG is stronger than that of any other position, and it decreases rapidly with the increasing of the gap from the surface of the PMG. The surface magnetic field of a single PM is about 0.45 T, while the surface concentrating magnetic flux density of the PMG (a) is up to 1.2 T. The magnetic flux density is 0.4 T at 20 mm above the surface of the PMG which is equivalent to the surface magnetic field of a single PM. The PMG shown in Fig. 2(a) was not solely used for HTS bulk measurements.

2.3 Liquid nitrogen vessel

One of the important technologies developed on board the HTS Maglev vehicle was the thin walled liquid nitrogen vessel. The thickness of said liquid nitrogen vessel wall was generally not considered, and its main feature is its low evaporation rate. Since the superconductors are levitated above the PMG in the HTS Maglev measurement system, a thin bottom wall of the vessel was needed. Only with a thin bottom wall, will a net levitation gap clearance between the outside (bottom) wall of the vessel and the guideway be high. A columnar liquid nitrogen vessel with a thin bottom wall of only 3 mm was developed (Wang S., 2001b) in order to verify the possibility of further developing large sized vessel with thin walls that can be used on the Maglev vehicle. Both the schematic diagram and vapor rate of the liquid nitrogen vessel are shown in Fig. 3.

The outline size of the vessel has an external diameter of 200 mm, an internal diameter of 150 mm, and a height of 250 mm. The liquid nitrogen vessel can operate continuously over 16 hours, and can hold 7 blocks of YBCO samples of 30 mm in diameter. The vessel was used successfully to measure the levitation forces of YBCO bulk over a magnetic guideway. During the experiment, the YBCO is fixed and secured at the bottom of the columnar liquid nitrogen vessel.

According to the experiment results mentioned above, a rectangular-shaped liquid nitrogen vessel on board the HTS Maglev vehicle was developed (Su-Yu Wang, 2003). The wall of the rectangle vessel was made even thinner, and the bottom wall's thickness is only of 3 mm. The schematic diagram of the rectangular-shaped liquid nitrogen vessel is shown in Fig. 4. Its outside outline size is 150 mm × 516 mm. The inside size is 102 mm × 470 mm, and the height is 168 mm. This particular liquid nitrogen vessel can operate continuously for over 6 hours. The rectangular-shaped vessel was used in the measurement of the levitation force of numerous YBCO samples. The vessel was successfully employed on board the HTS bulk Maglev measurement system.

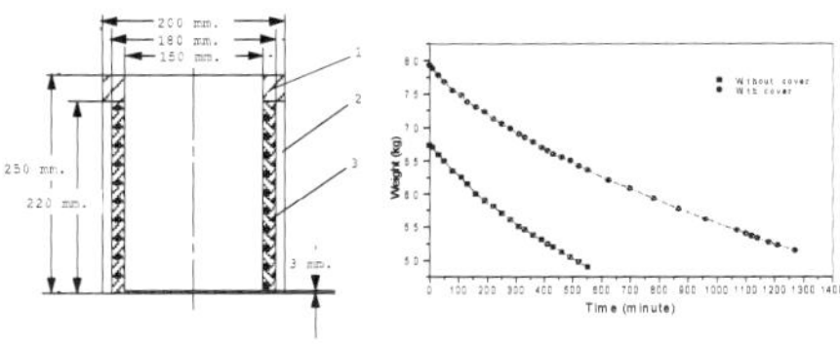

Fig. 3. Schematic diagram and vapor rate of the columnar liquid nitrogen vessel with thin wall

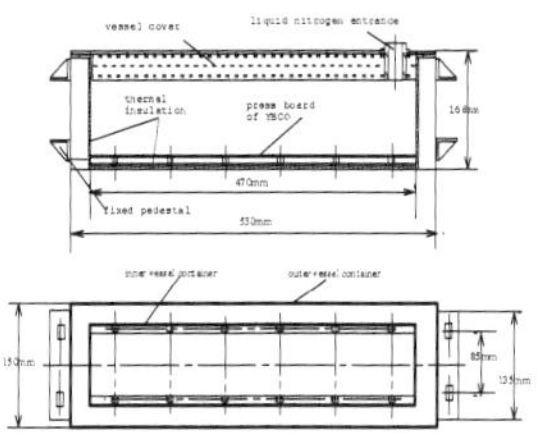

Fig. 4. Schematic diagram of the rectangular-shaped liquid nitrogen vessel

2.4 HTS Maglev measurement system

Fig. 5 shows the schematic diagram of the HTS Maglev measurement system. During the experiment, the YBCO is placed in the columnar liquid nitrogen vessel which is positioned above the PMG. The YBCO is zero field cooled and the vessel is allowed to move up and down at different speeds. The horizontal drive platform is used to measure the guidance force (stable equilibrium force along longitudinal orientation of guideway). The drive device of three dimensions can make scanning measurements of the magnetic field of the PMG and trapped flux inside an HTS.

The specifications of the SCML-01 measurement system are: vertical maximal displacement of 200 mm, ±0.1 mm precision, ±2,000 N vertical maximal support force, 0.2% precision; 100 mm guideway horizontal maximal displacement, ±0.1 mm precision, 1,000 N of horizontal maximal support force, and 0.1% precision. The trapping flux of high T_c superconductors and the magnetic induction of the guideway can be scanned in the range of 100 mm × 100 mm.

In the measurement, the YBCO HTS bulk sample is fixed at the bottom of the thin wall liquid nitrogen vessel and cooled to go into the superconducting state in a zero magnetic field. Secondly, the vessel is fixed at a connecting fixture with a servo electromotor. In order

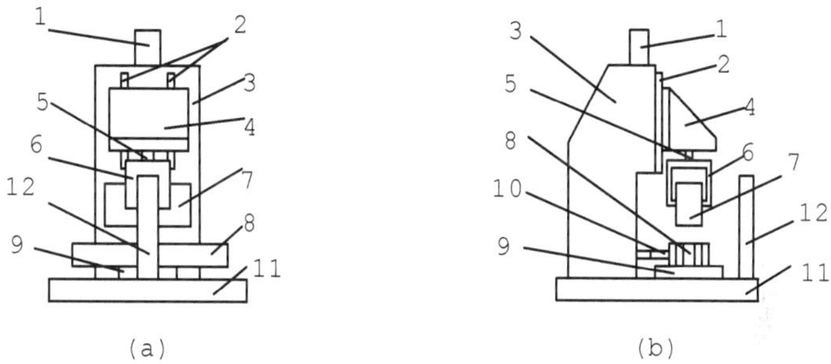

Fig. 5. Scheme of HTS Maglev measurement system.

1) Servo motor , 2) Vertical guided way, 3) Vertical column, 4) Cantlever, 5) Vertical sensor, 6) Fix frame of vessel, 7) Liquid nitrogen vessel, 8) Permanent magnet guideway (PMG), 9) Horizontal drive platform, 10) Horizontal sensor, 11) Base, 12) Drive device of three dimension

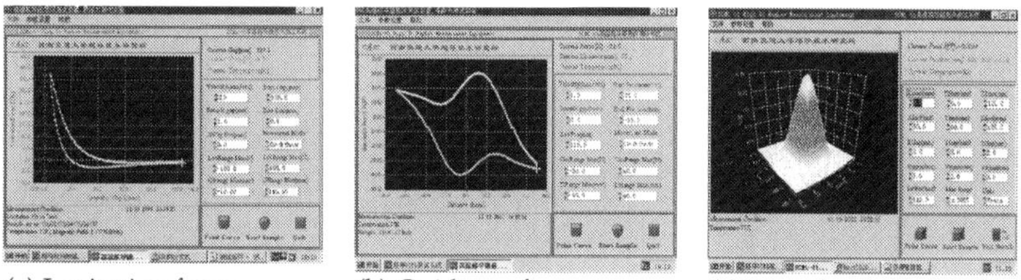

(a) Levitation force (b) Guidance force (c) Scanning trapping flux

Fig. 6. Main interfaces of the measurement results

to avoid collision between the bottom of the vessel and the surface of PMG, there is still a gap of 1.5 mm left in between the bottom of the vessel and the surface of the PMG when the vessel is lowered to the lowest point, so the minimum gap is 5 mm between the bottom of the sample and the surface of PMG. The vessel first moves downward, after reaching the lowest point, then moves upward at a speed of 2 mm/s, and the computer samples a data every 0.5 second. The system can make real time measurements of one or many superconductors. The measurement process is controlled by a computer. The main interfaces of the measurement results of the magletic levitation force, guidance force, and scanning magnetic field of an HTS trapping flux are shown in Fig. 6.

The SCML-01 measurement system is capable to make real time measurement of Maglev properties with one to many YBCO pieces and with a PM or PMGs. This set up was especially employed in on board HTS Maglev equipment over one or two PMGs (Fig. 7). The on board Maglev equipment includes a rectangular-shaped liquid nitrogen vessel and an array of YBCO bulk superconductors. Fig. 7 is a picture of the HTS Maglev measurement system. Three types of liquid nitrogen vessels are shown in Fig. 7, a columnar vessel, a rectangle vessel, and two rectangle vessels.

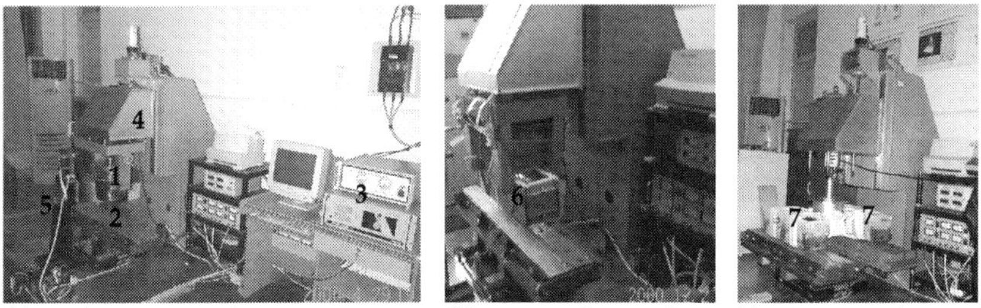

(a) a columnar vessel (b) a rectangle vessel (c) two rectangle vessels

Fig. 7. Photos of HTS Maglev measurement system SCML-01

1 liquid nitrogen vessel, 2 permanent magnet guideway (PMG), 3 data collection and processing, 4 mechanical drive and Autocontrol, 5 scan magnetic flux, 6 a rectangle-shape vessel, 7 two rectangle-shape vessels

Based on the original research results (Wang J. & S. Wang, 2005a; Song, 2006) from SCML-01, the first man-loading HTS Maglev test vehicle in the world was successfully developed in 2000 (Wang et al., 2002). Many of these research results (Ren, 2003; Wang X. R. 2003; Song, 2004; Wang X.Z. 2004; Wang J. & S. Wang, 2005c) were obtained by the SCML-01 HTS Maglev measurement system.

3. Measurement technology of HTS Maglev vehicle

3.1 The first man-loading HTS Maglev test vehicle in the world

High-temperature superconductors are highly attractive because it can operate at liquid nitrogen temperature. Soon after the stable levitation of a PMG above a YBCO bulk superconductor bathed in liquid nitrogen was observed by Hellman et al., (Hellman et al., 1988), people began to consider its applications to superconducting Maglev vehicles. The levitation forces of YBCO bulks over a PMG have been reported (Wang S. et al., 2001a). The onboard HTS Maglev equipment is a key component of the HTS Maglev vehicle that has been developed (Wang S. et al., 2002). The first man-loading HTS Maglev vehicle in the world was tested successfully on December 31, 2000 at the Applied Superconductivity Laboratory of Southwest Jiaotong University, China (Wang J. et al., 2002; 2005b; Wang S. et al., 2001b).

The PMG consists of two parallel PM tracks, whose concentrating magnetic field at a height of 20 mm is about 0.4 T. The total length of the PMG is 15.5 m. The HTS Maglev provides inherent stability along both the vertical and lateral directions, so there is no need to control the vehicle along these two directions. The only control system used are linear motors as driving and breaking devices. The 8 onboard HTS Maglev equipment assemblies (Wang S. et al., 2003) are connected rigidly on the two sides of vehicle body, with 4 Maglev equipment assemblies on each side. The vehicle body (Fig. 8(b)) is 2,268 mm long, 1,038 mm wide and 120 mm high. Both the linear motor of the vehicle and the PMG are under the vehicle body. The vehicle body was lifted by a hydraulic pressure until the gap between the bottom of liquid nitrogen vessels and the surface of the PMG was larger than 75 mm. Then the HTS

(a) (b) (c)

Fig. 8. (a) Photograph of the first pouring of liquid nitrogen; (b) The net levitation gap of the HTS Maglev test vehicle body was more than 20 mm when five people stood on the vehicle; (c) Photograph of the HTS Maglev test vehicle with outer casing.

bulks were cooled with liquid nitrogen. Fig. 8(a) shows the historic moment of the first pouring of liquid nitrogen into the onboard vessels. In Fig. 8(c) we can see the two tracks and the vehicle body.

The net levitation gap of the HTS Maglev test vehicle body is more than 20 mm when five people stood on the vehicle, and the levitation height of the vehicle body was 33 mm when the five people got off the vehicle (Fig. 8(b)). Fig. 8(c) shows the photograph of the HTS Maglev test vehicle with an outer casing. The external outline size of of the vehicle with a shell is 3.5 m long, 1.2 m wide, and 0.8 m high. There are 4 seats in the HTS Maglev vehicle. The total levitation force of the entire Maglev vehicle was measured to be 6,351 N at the net levitation clearance gap of 20 mm, and 7,850 N at the net levitation gap of 15 mm in July 2001. The net levitation gap is the distance between the PMG's upper surface and the liquid nitrogen vessel's bottom.

3.2 Measurement of essential parameters of HTS Maglev test vehicle

The onboard HTS Maglev equipment is of most importance to the HTS Maglev vehicle. There are 8 HTS Maglev equipment assemblies on the vehicle body, each composed of 43 YBCO bulk pieces inside a liquid nitrogen vessel (Wang S. et al., 2001b).

The YBCO bulks are 30 mm in diameter and 17-18 mm in thickness. The YBCO bulks are sealed by a special method in order to preserve the integrity and quality of the superconductors. The melt-textured YBCO bulk superconductors are fixed firmly at the bottom of each of the liquid nitrogen vessels, and cooled by liquid nitrogen. The levitation forces of the 8 HTS maglev equipments were measured and described by Wang et al. (Wang J. et al., 2003a).

Fig. 9(a) shows the levitation force of a single onboard HTS Maglev equipment over the PMG is 1,227 N at the levitation gap of 15 mm, and the levitation force was 1,593 N at the levitation gap of 8 mm. Fig. 9(b) shows the levitation forces of the 8 onboard Maglev equipment assemblies over the PMG are different. There were some differences between each of the equipment assemblies. For example, the levitation force of Maglev equipment No. 7 is 1,493 N at the levitation gap of 10 mm, and 1,227 N of 15 mm. The levitation force of Maglev assembly No. 3 is 1,091 N at the levitation gap of 10 mm, and 902 N of 15 mm. Fig. 9(c) shows the total levitation forces of the 8 onboard Maglev equipment assemblies.

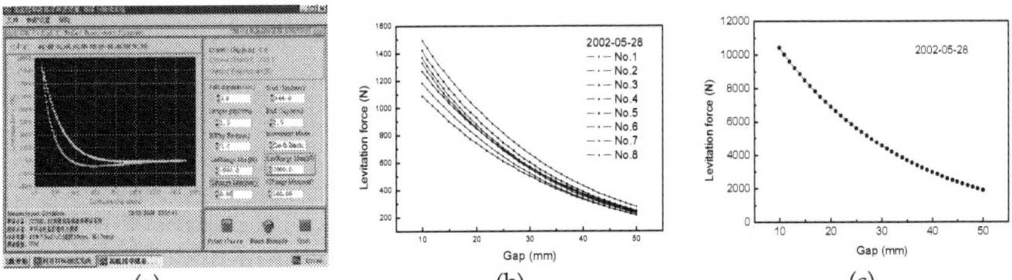

(a) (b) (c)

Fig 9. (a) The measured interface of a single rectangle-shape vessel, (b) Measured results of eight rectangle-shape vessels of on board single HTS magnetic levitation equipment (43 pieces of YBCO bulks), and (c) The total levitation force of 8 onboard HTS Maglev equipments over the PMG.

The total levitation force of 8 onboard Maglev equipments assemblies yielded a force of 10,431 N at the levitation gap of 10 mm, and 8,486 N at 15 mm.

3.3 Guidance forces of the HTS Maglev vehicle(Wang S. et al., 2003b, Wang X.R. et al., 2005)

The guidance force defines the lateral stability of the Maglev vehicle when either standstill or moving. The lateral guidance force is dependent on the trapped flux in the bulk superconductors, so the larger the amount of trapped flux, the stronger the guidance force. This is a distinctive character of the bulk HTS Maglev vehicle. This sort of Maglev vehicle with bulk HTS does not need any lateral stability control systems, which makes it superior to other conventional Maglev vehicle systems. The guidance forces are large and sufficient enough to guide the vehicle when large levitation forces are ensured.

(a) (b) (c)

Fig. 10. Sketch (lift) and Photograph (right) of measuring equipment of guidance force of vehicle

1 Horizontal propulsion system; 2 Vertical propulsion system; 3 Screw of adjusting zero; 4 Force sensor; 5 vehicle body; 6 HTS; 7 Permanent magnetic railway; 8 Linear motor

The measuring equipment of the guidance force of the entire HTS Maglev vehicle is deppicted in Fig. 10(a). The set up includes two probing levers and two sets of force sensors. The force sensors are fixed on the vehicle. Each set of the propulsion system can move in both the horizontal and vertical directions so that they can measure the guidance force of the entire vehicle at different levitation gaps. Two sets of propulsion systems are connected by a chain with a synchronization precision of 0.5 mm. The moving range of the propulsion system along the horizontal direction is 0 to 20 cm and moving precision: 1 mm; and along the vertical direction a range from 0 to 10 cm, vertical moving precision: 1 mm. A photograph of the measuring equipment of guidance force is shown in Fig 10(b).

The experimental results of the guidance force under the maximum lateral distance of 20 mm are shown in Fig 10(c). The data shows that the lateral guidance forces have a large hysteresis effect. The guidance forces rise rapidly when the vehicle leaves its initial position, and the increase rate becomes smaller as the vehicle becomes further away from its original position Fig 10(c). The guidance forces drop rapidly when vehicle moves back to its initial positions from the maximum lateral displacement of 20 mm, and vanishes at about 10 mm. This indicates that the range of effective lateral displacement is smaller than 10 mm. It can be seen from Fig 10(c) that the lateral guidance forces of the entire HTS Maglev vehicle under different field cooling heights (FCH) are sufficiently large enough. The guidance forces with displacements 20 mm are 4,407 N, 2,908 N, and 1,980 N for field cooling heights of 12 mm, 26 mm, and 42 mm, respectively.

The HTS Maglev vehicle can return to its initial position after a lateral displacement from 0 mm to 6 mm, whereas this did not happen when the lateral displacement was from 0 mm to 20 mm. For example, there is a zero guidance force at the displacement 9 mm when FCH is 26mm. The measured guidance pull force, which makes the vehicle return to the initial rest position, was 1,214 N when the FCH was 26 mm. Again, this force is sufficiently large enough to keep the vehicle laterally stable.

3.4 Long-term stability of the HTS Maglev vehicle in 2001~2003(Wang J. et al., 2004)

Fig. 11 shows the total levitation force of eight liquid nitrogen vessels over the PGM at different gaps. In July 2001, the levitation force was found to be 8,940 N at the levitation gap of 15 mm and 7,271 N at the levitation gap of 20 mm. The total levitation force of eight liquid nitrogen vessels over the PMG in March 2003 was 8,000 N at the levitation gap (deduct 3 mm bottom thickness of liquid nitrogen vessel) of 15 mm.

The measurement results are measured by HTS Maglev measurement system SCML-01 (Wang J. et al., 2000) in July 2001, December 2001, May 2002, and March 2003, respectively.

Fig. 11 shows the levitation forces of the entire HTS Maglev vehicle to be 8,486 N at the levitation gap of 15 mm and 6,908 N at the levitation gap of 20 mm in May 2002. At a gap of 30 mm, there was a 46% decrease of the levitation force compared to the gap of 15 mm.

During the 10 month period from July 2001 to May 2002, the levitation force was found to only decrease by 5.0% at the levitation gap of 20 mm.

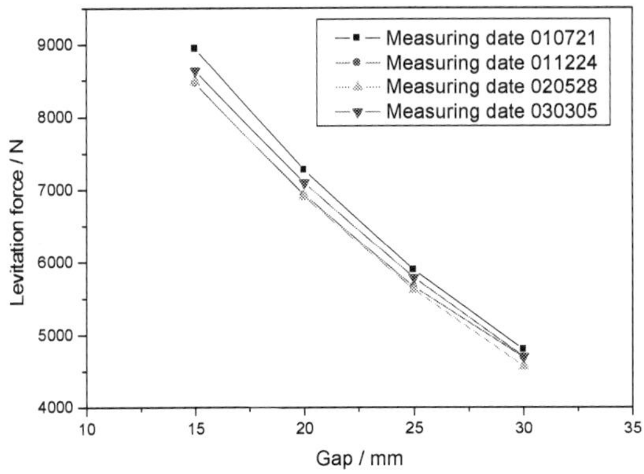

Fig. 11. Change of total levitation force of the entire HTS Maglev vehicle

The data set of levitation forces for the March 2003 was slightly higher than the May 2002 data set was due to one of the PMG was not laid to overlap accurately.
The comparison of total levitation forces of the entire HTS Maglev vehicle in July 2001 with one in May 2002 showed a levitation force decrease of a 5.1%, 5.0%, 4.6%, and 4.8% at the levitation gaps of 15 mm, 20 mm, 25 mm, and 30 mm, respectively. All data are nearly the same, i.e., levitation force decrease is about 5.0% at different levitation gaps.
The levitation forces became small after two years. Up to this day, (September 2009) over 27,000 passengers have taken a ride on the Maglev vehicle, and the total recorded shuttle mileage is about 400 km. Experimental results indicate long-term stability of the HTS Maglev vehicle is better when in the static mode than experiment results mentioned were obtained above when the HTS Maglev vehicle was operating at low speeds. The behavior of an HTS Maglev vehicle is very different at high speeds. Therefore, the investigation of the dynamic properties of the HTS maglev vehicle at high speed operation is extremely important.

3.5 Long-term stability of on-board bulks YBCO in 2001~2009(Liu et al., 2009)

In order to evaluate the long-term operation stability of the HTS bulks for practical application, the levitation performance of bulk YBCO samples are inverstigated by the SCML-02 setup throughout 2001 to 2009. The same bulk YBCO pieces (A5) used in 2000 still remain placed on the man-loading HTS maglev vehicle and have been running for nearly nine years. The measurement results of the levitation force (Fig. 12), hysteresis, and relaxation (Fig. 13) of the bulks YBCO, which were under the liquid nitrogen temperature (77K) were measured by the SCML-02 HTS Maglev measurement system in June, 2009. A batch of YBCO samples were synthesized in 2007 (07-1) and in 2008 (08-1), and some of these samples were loaded in 2007 and others in latter times. This was to analyze the variation of levitation performances of different bulk YBCO samples during different times after fabrication of the YBCO.

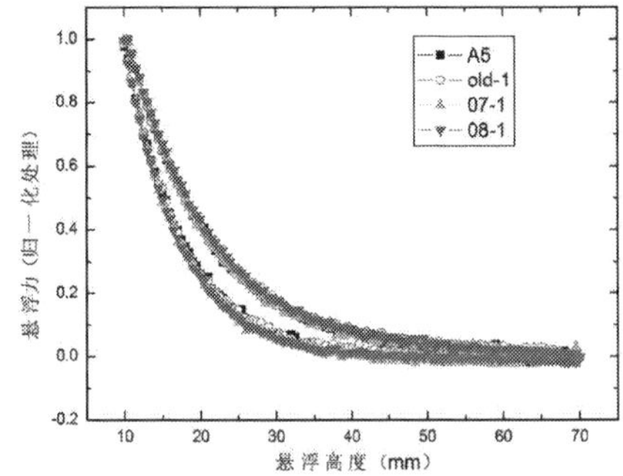

Fig. 12. Measurenment results of levitation force

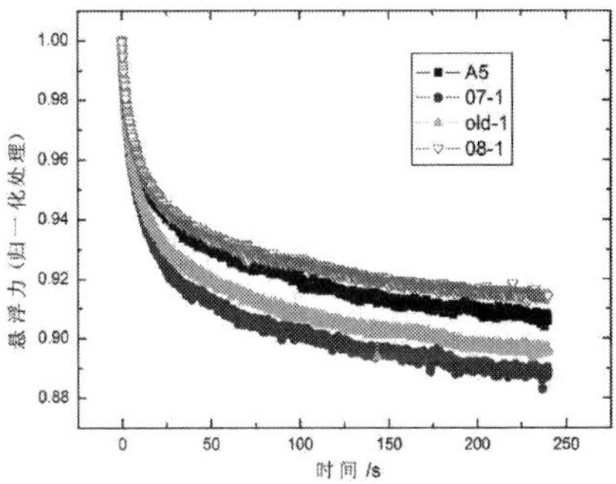

Fig. 13. Measurenment results of levitation force relaxation

(A5) is onboard bulk YBCO in 2000; (Old-1) is bulk YBCO of not loaded in 2000; (07-1) is new in 2007; (08-1) is new in 2008

Fig 12. is the comparison of the levitation force measurement results of different bulks YBCO during 2001~2008, and Fig. 13 is the comparison of the levitation force relaxation measurement results of different bulks YBCO in 2001~2008. A5, old-1, 07-1, and 08-1 in the graphs represent onboard bulk YBCO in 2000, bulk YBCO of not loaded in 2000, new bulk YBCO in 2007, and in 2008, respectively. All measurement results are normalized with respect to the results of A5 all aligned along the centerline of the PMG. In Fig 12 and Fig 13, the measurenment results of the levitation force and relaxation time of the A5, 07-1, 08-1, and old-1 samples remained almost constant.

These experiment results validate the fact that the performances of the bulk YBCO samples above the PMG did not change after nearly nine years. Our research results can be applied to not only the first man-loading HTS maglev vehicle, but also to other superconducting applications such as HTS Maglev bearings and superconducting flywheel energy storage devices.

4. Measurement system with more functions and higher precision

(Wang S. et al., 2007)

Although the first man-loading HTS Maglev vehicle in the world (Wang J. et al., 2002) was developed successfully with the research results from the SCML-01 measurement system (Wang J. et al., 2000), the measurement functions and precision of the SCML-02 platform needed to be improved. Therefore, to make more thorough and careful investigations on the magnetic levitation properties of the HTS Maglev vehicle over a PMG, an SCML-02 HTS Maglev measurement system (Wang S. et al., 2007) with several special functions and high precision was successfully developed.

4.1 System description

The SCML-02 HTS Maglev measurement system includes experiment system, control hardware and control processing software.

The experiment system is shown in Fig. 14. Four vertical support posts 1 are fixed on an optical bedplate. The upside liquid nitrogen vessel 6 with HTSCs is placed above a PM or PMG 7. The underside liquid nitrogen vessel 11 with HTS bulks is placed under a PM. In this way, the experiment system can be fit for different measurement, especially measurement of Maglev properties of superconductor samples on a PMG. The inertial force effect (tension) of movement parts is decreased by four force balance chain wheels 3 and corresponding counterweights. The x-y electromotion seat is fixed onto the optical bedplate. Horizontal directions along the x and y axis can be moved together or separately. Therefore the measured HTS bulk samples can be moved in three dimensions.

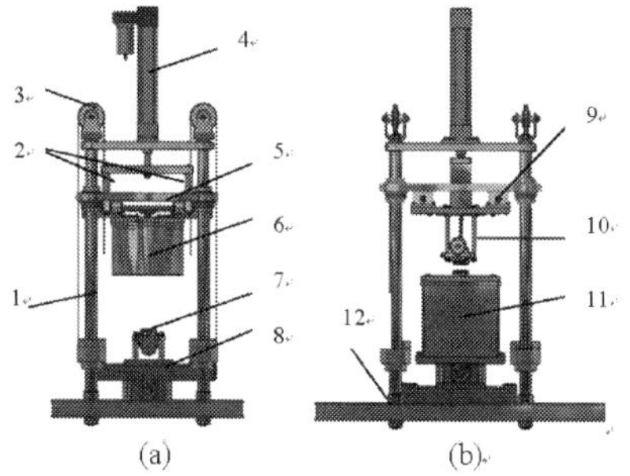

Fig. 14. The design scheme of the SCML-02 HTS Maglev measurement system: (a) Upside liquid nitrogen vessel; (b) Underside liquid nitrogen vessel.

1. Vertical support post; 2. Vertical force sensor; 3. Force balance chain wheel; 4. Servo motor; 5. Slip set; 6. Upside liquid nitrogen vessel; 7. Underside clamp of cylinder PM or PMG; 8. X-Y electro-motion seat; 9. Horizontal force sensor; 10. Upside clamp of cylinder PM or PMG; 11. Underside liquid nitrogen vessel; 12. Optical bedplate.

Fig. 15. Photo of the SCML-02 HTS Maglev measurement system: Levitation force measurement interface (top left corner); Console control desk (under left corner); Experiment system (right).

The vertical levitation force and the horizontal guidance force are measured by four force sensors. The vertical sensor and the horizontal sensor are insulated from each other, and it is possible to avoid interference between the two kinds of force sensors. Therefore SCML-02 Maglev measurement system can measure the levitation force and the guidance force at the same time.
Fig. 15 shows a photograph of the SCML-02 HTS Maglev measurement system.

4.2 The function and technical specification

The system includes a liquid nitrogen vessel, cylindrical PM (or PMG), data collection and processing, mechanical drive and Autocontrol. The special functions of the SCML-02 Maglev measurement system can be summarized as follows:
High precision measurement;
Measurement of both a single or a number of HTS bulk samples;

1. Measured HTS bulk samples and the PM can move along the three dimensions at one time.
2. The location of the HTS bulk and PM can be interchanged;
3. Maglev properties measurement between the measured HTS bulk specimens and a cylindrical PM or a PMG;
4. Simultaneous measurement of the levitation force and guidance force for which the sample can be displaced vertically and horizontally at the same time;
5. Synchronous measurement of not just the levitation and guidance forces, but also the cross stiffness of the levitation force or the guidance force in a real-time independent measurement along the vertical or horizontal;
6. Interaction of dynamic rigidity of levitation force and guidance force synchronous measurement;
7. Relaxation time of levitation force and guidance force measurement along the vertical or horizontal directions;

The SCML-02 Maglev measurement system's main technical specification are as follows:

1. Vertical maximal displacement of 150 mm,
2. Horizontal maximal displacement of 100 mm;
3. Position precision of 0.05 mm;
4. Vertical maximal support force of 1,000N;
5. Vertical force precision of 2‰;
6. Horizontal maximal support force of 500 N;
7. Horizontal force precision of 1‰.

4.3 Measurement precision calibration

In order to calibrate the measurement precision of the SCML-02 HTS Maglev measurement system, two cylindrical PMs were used. The two PMs have the same strength and size dimensions. During the levitation force test, standard weights of 9.8 N, 18.62 N, and 24.95 N, and vertical moving velocities of 0.1 mm/s to 1.0 mm/s (velocity partition 0.1 mm/s) were used. Fig. 16 shows test results of the levitation force at the standard weight of 9.8 N and at different sampling velocities. At the standard weight of 9.8 N, the measurement precision of

the levitation force was 99.23%-99.55%, and the error 0.453%-0.773% with the range of 0.00256-0.02924.
Thus it can be seen from Fig. 17 that the levitation force change is independent of the standard weights of 9.8 N, 18.62 N and 24.95 N, as the total change was between 0.00163-0.0292. The levitation force measurement error was found to be plus or minus 0.5%.
Fig. 18 shows test results of the levitation force at different sampling velocities which justifies the good repeatability of the levitation force measurements of the SCML-02 HTS Maglev measurement system.

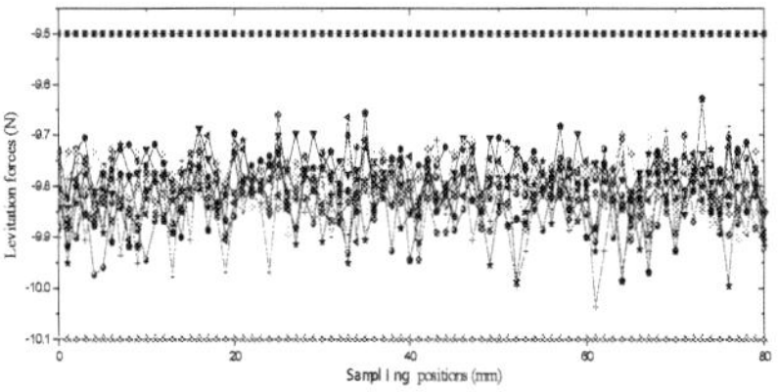

Fig. 16. Test results of the levitation force at the standard weight of 9.8 N at different sampling velocities (the vertical move velocities are 0.1 mm/s to 1.0 mm/s (velocity partition 0.1 mm/s)

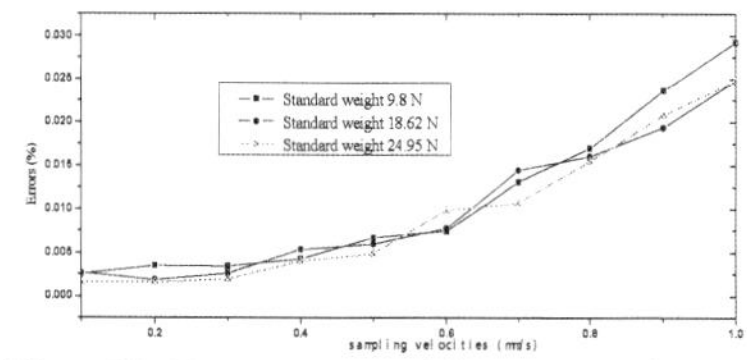

Fig. 17. Test errors of levitation forces at the different standard weight and sampling velocity

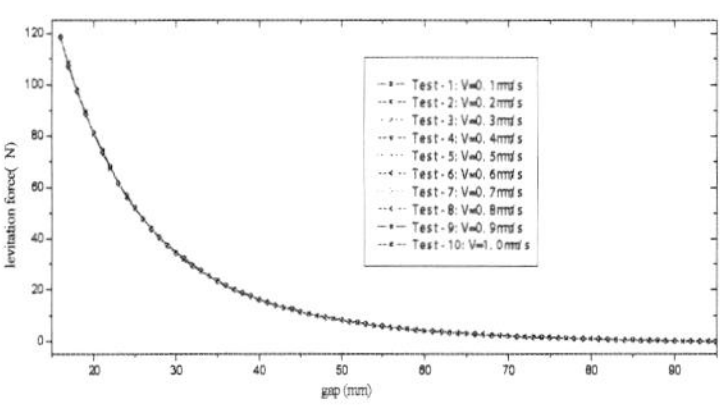

Fig. 18. Test results of levitation forces at different sampling velocities

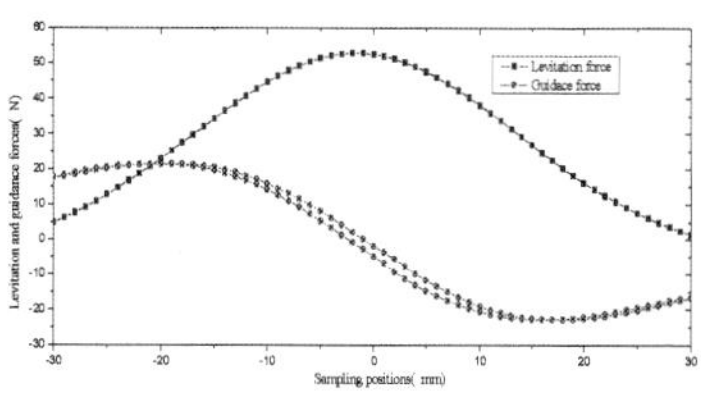

Fig. 19. Synchronous test results of the levitation forces and the guidance forces at sampling velocity of 0.5 mm/s.

Fig. 19. shows the synchronous test results of the levitation force and guidance force at the sampling velocity of 0.5 mm/s. The curves with other sampling velocities of 0.1 mm/s to 1.0 mm/s (velocity partition 0.1 mm/s) are also shown and found to fit well with the profile.
Fig. 20 shows the stiffness test results of the levitation forces at the sampling velocity of 1.0 mm/s. Sampling point gap was 1.0 mm, and 61 sampling points were taken in the measurement range of 95-15 mm.

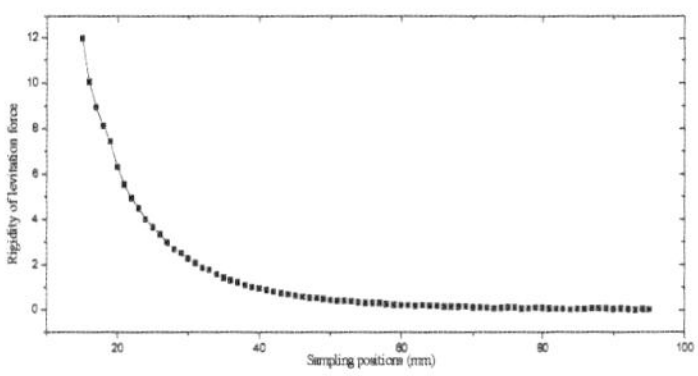

Fig. 20. Stiffness test results of the levitation forces at the sampling velocity of 1.0 mm/s.

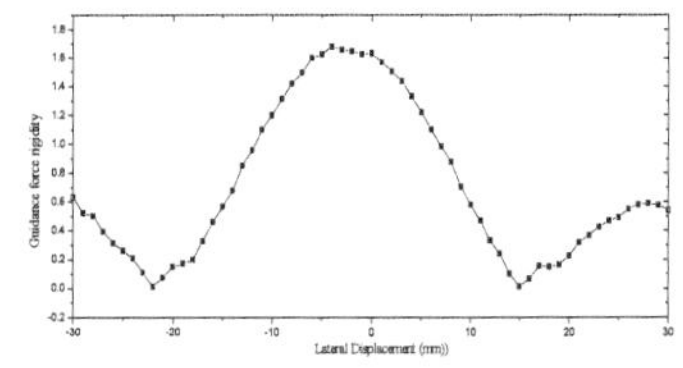

Fig. 21. Test results of levitation force at the standard weight 9.8 N and different sampling velocity (the vertical move velocities are 0.1 mm/s to 1.0 mm/s (velocity partition 0.1 mm/s).

Fig. 21. shows the stiffness test results of the guidance forces with the sampling velocity of 1.0 mm/s. The sampling point gap was 1.0 mm, and 61 sampling points were taken in the measurement range of -30 to 30 mm.

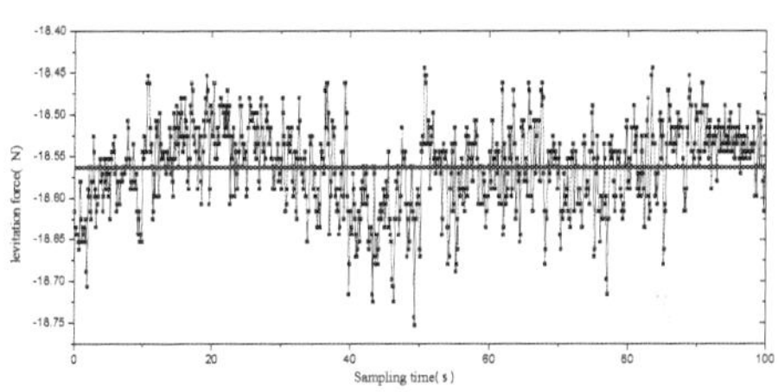

Fig. 22. Test errors of levitation forces at the different standard weight and sampling velocity.

Fig. 22. shows time-dependent levitation forces at the standard weights of 18.62 N. The sampling point interval was 0.1 s, and 1001 sampling points were acquired at the measurement range of 1-100 s.

The above magnetic levitation properties between the PMs were measured by the SCML-02 HTS Maglev measurement system. These measurement results validate the high precision and multi-functionality of the SCML-02.

4.4 Measurement results of HTS Maglev properties

The Maglev properties of YBCO bulk are measured on the basis of the above calibration results of SCML-02 HTS Maglev measurement system. The levitation forces of single YBCO bulk and an array of 7 YBCO bulks above a PMG were measured by the SCML-02 HTS Maglev measurement system. The array of 7 YBCO bulks are concentrically arranged, where 3 bulks share one axis along the axis line of the PMG.

The diameter of the single YBCO bulk was 48 mm, and the diameter of each of the array of 7 YBCO bulks was 30 mm.

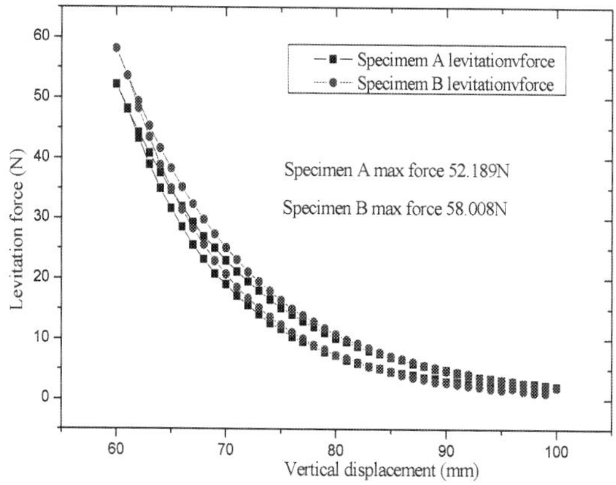

Fig. 23. Levitation forces of two YBCO bulk samples with diameter of 48 mm (FCH of 35 mm, measurement range of 60-100 mm).

Fig. 23. shows the measurement results of the levitation forces of two YBCO bulks with diameter of 48 mm with a field cooling height (FCH) of 35 mm and vertical measurement ranges from 60 to 100 mm and backward.

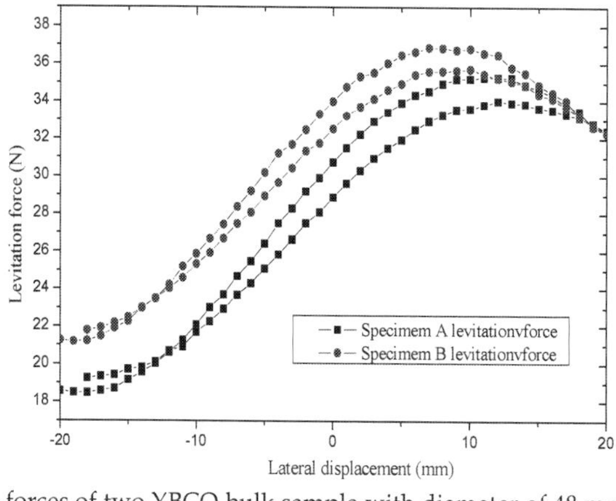

Fig. 24. Guidance forces of two YBCO bulk sample with diameter of 48 mm (FCH of 35 mm, measurement height of 15 mm,).

Fig. 24. shows the measurement results of the guidance forces along the lateral direction of the PMG. The YBCO samples had a diameter of 48 mm, field cooling height (FCH) of 35 mm, measurement height (MH) of 15 mm, and lateral displacement measurement range of -20 to 20 mm.

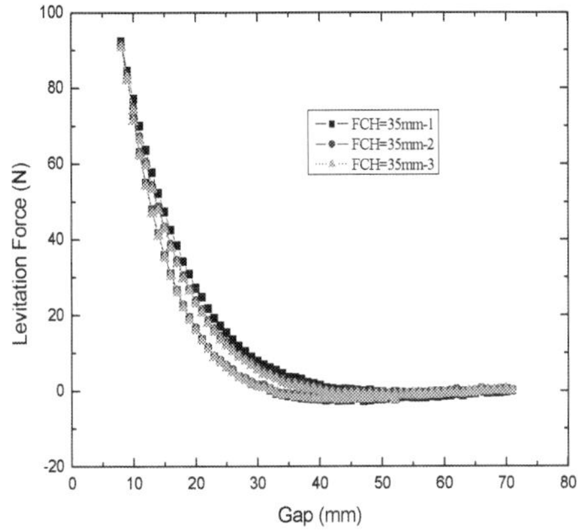

Fig. 25. Levitation forces of an array of 7 YBCO bulk samples with diameter of 30 mm (FCH 35 mm).

Fig. 25. shows the measurement results of the levitation forces of three different arrangements of 7 YBCO samples above the PMG. These 7 YBCO bulk samples had a diameter of 30 mm and FCH of 35 mm.

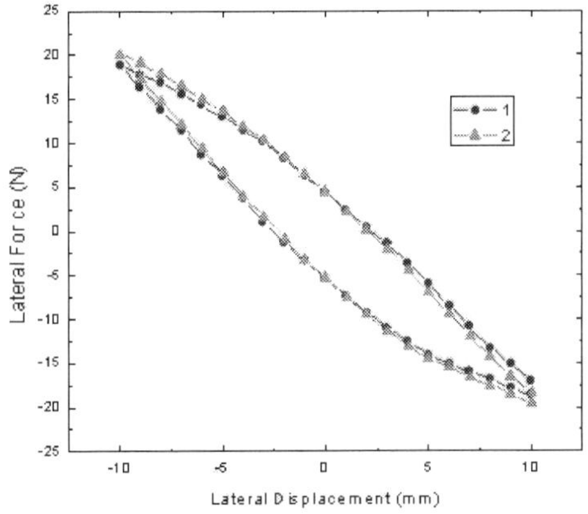

Fig. 26. Guidance forces of an array of 7 YBCO bulk samples of diameter 30 mm (FCH 35 mm, measurement height 10 mm).

Fig. 26. shows the measurement results of the guidance forces of three arrangements of 7 YBCO bulks above a PMG. The MH for this experiment was of 10 mm, and the lateral measurement range from -10 to 10 mm.

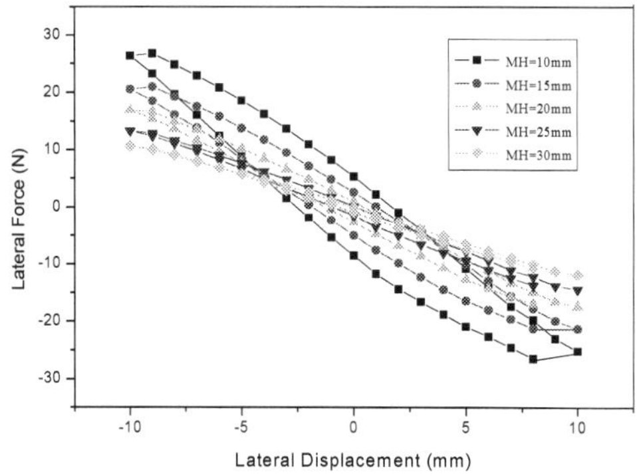

Fig. 27. The guidance forces of 7 YBCO bulks with diameter of 30 mm are measured by SCML-02 (FCH of 20 mm, MH range of 10-30 mm, lateral measurement range of -10 to 10 mm).

Fig. 27. shows the measurement results of the array of 7 YBCO samples above a PMG at different MHs. The FCH was set to 20 mm, the measurement heights ranged from 10 to 30 mm, and the lateral measurement ranged from -10 to 10 mm.

The levitation forces and guidance forces of YBCO bulks are measured by SCML-02 HTS Maglev measurement system. The measurement results of HTS Maglev properties validated the accuracy and reliability of the SCML-02.

The SCML-02 measurement system can make real time measurement and the data acquisition process is controlled by a computer. The main functions and specifications of the system are unique in the world. A lot of precious experimental results were obtained from this system, like the Maglev vehicle's levitation force, guidance force and their stiffness, and cross stiffness, and so on (Deng, 2007; 2008b; 2009b; He, 2008; Jin, 2008; Liu M., 2008; Liu W., 2008; Lu, 2008; Ma, 2008; Qin, 2009; Wang J., 2005a; 2005c; 2007; 2009a; 2009b; Zheng, 2007a; 2007b). All these experimental parameters are very helpful to evaluate the load capability of the HTS Maglev vehicle.

5. HTS Maglev dynamic measurement system (Wang J. et al., 2008)

Although the HTS Maglev measurement system SCML-02 has more functions and higher precision than SCML-01, it can not measure the running performance of bulk YBCO samples above a PMG. For the further engineering application of the HTS Maglev vehicle, the dynamic properties of the Maglev must be clearly understood. From this viewpoint, an HTS Maglev dynamic test system (SCML-03) has been designed and successfully developed in Applied Superconductivity Laboratory (ASCLab) of Southwest Jiaotong University, P. R. China (Wang J. et al., 2008). The test process and results of the HTS Maglev dynamic test system are reported in this section.

5.1 System description

When the HTS Maglev vehicle runs along the PMG, it is difficult to measure its dynamic properties. In SCML-03, the rotational motion of a circular PMG instead of the physical motion of thea YBCO bulk was taken to be the equivalent measure of the dynamic interaction between the superconductor and the PMG. That is, the circular PMG can rotate to different speeds while the onboard HTS Maglev equipment is fixed above the PMG and this will simulate the fact that the superconductor is traveling above a PMG. SCML-03 is composed of a vertical load, horizontal load, three-dimensional measurement systems, liquid nitrogen vessel, circular PMG, drive device, data acquisition and processing, autocontrol, and so on.

The main measurement functions include the dynamic stability of the HTS Maglev equipment (liquid nitrogen vessel including the HTS samples), the levitation force and guidance force of both single and multi HTS bulk samples, the levitation force and guidance force rigidity of both single and multi HTS bulk samples, the levitation force and guidance force change at the levitation gap, and so on.

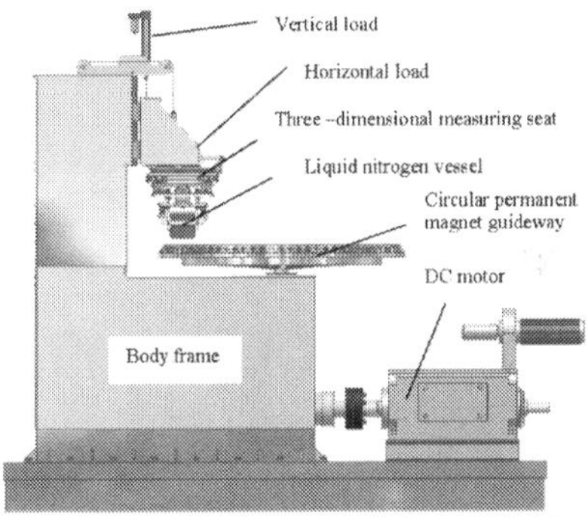

Fig. 28. The design scheme and principal parts of the HTS Maglev dynamic test system SCML-03 (not including power supply and measurement control desk)

Fig. 29. The photo of the airframe of the HTS Maglev dynamic test system SCML-03: 3.3 m long, 2.4 m wide and 3.15 m high With a total weight of 13.95 t

Fig. 30. The measurement scene of the HTS Maglev dynamic test system.
On the right side is the control desk

The main part design scheme of SCML-03 is shown in Fig. 28. The airframe of the HTS Maglev dynamic test system SCML-03 (not including the measurement control desk) is shown in Fig. 29. The total technical parameters of the principal part of the SCML-03 are 3.3 m long, 2.4 m wide and 3.15 m high. The total weight is 13.95 t which includes the circular PMG disk that weights 0.6 t. Fig. 30 shows the measurement scene of the HTS Maglev dynamic test system, the control desk is on the bottom right.

A DC motor is used to drive the circular PMG to rotate and control its rotational speed. The rotational direction of the DC motor is translated into the horizontal rotation direction of the circular PMG by a gear redirection case. The circular PMG is fixed along the circumferential direction of a big circular disk with a diameter of 1,500 mm. The rotating unbalance of the big circular disk is less than 20 gm. The maximum linear velocity of the PMG is about 300 km/h when the circular disk rotates round the central axis at 1,280 rpm. The rotation speed error of the circular disk is less than 3 %.

The three-dimensional measuring seat is fixed on the horizontal load. The seat can be moved along the horizontal direction which is perpendicular to the tangential direction of the circular PMG. Therefore, the HTS Maglev dynamic guidance force can be measured at the same time. Six force sensors are used to measure the vertical, transverse, and longitudinal directional forces of the liquid nitrogen vessel. In order to measure the three-dimensional dynamic response, the liquid nitrogen vessel is connected to the three-dimensional measuring seat by elastomers.

The autocontrol of all components are completed by the total control desk. Both measurement and control software was developed on basis of Labview software by National Instruments. The servo motor movement and the circular PMG speed are controlled by both the software program and relative control card.

In order to calibrate measurement precision of the dynamic measurement system, 11 channel Noise & Vibration Measurement System made by B & K was used. The dynamic vertical levitation force is measured when the circular PMG is rotated by different speed.

5.2 The functions and technical specifications

The SCML-03 can measure the dynamic properties in opposite motion between the vehicle and the PMG. The main measurement functions include:

1. Dynamic stability measurement of the HTS Maglev equipment (liquid nitrogen vessel included HTS bulks) at different velocities;
2. Measurement of levitation force and guidance force of both single and multi HTS bulk samples;
3. Measurement of levitation force stiffness and guidance force stiffness of both single and multi HTS bulk samples;
4. Measurement of levitation force and guidance force change at different levitation gap clearances;

Main technical specifications:

1. Diameter of the circular PMG 1,500 mm;
2. Maximal linear speed of the circular PMG 300 km/h;
3. Rotation speed precision of the circular PMG ±3%;
4. Vertical maximal displacement 200 mm;
5. Horizontal maximal displacement ±50 mm;
6. Position precision ±0.05 mm;
7. Vertical maximal support force 3,350 N;
8. Horizontal maximal support force 500 N;
9. Force sensor precision ±1%.

5.3 Stability of dynamic test system

The most important propertiy of the dynamic test system is its self stability. The rotating unbalance of the PMG circular disk was measured to be less than 20 gm. The rotating unbalance value satisfies the system measurement's needs. The other important parameter of the dynamic test system is the self stability of the liquid nitrogen vessel and the body frame. In order to confirm the self stability, a 11 channel Noise & Vibration Measurement System was used. A 4507-004B accelerometer was fixed onto the body frame, and four 4507-004B accelerometers were fixed on the clamp device of the liquid nitrogen vessel. The vibrations of the perpendicular and horizontal directions on the clamp device were measured by two sensors. A picture of the measurement scene of the SCML-03 is shown in Fig. 30.

The vibration spectrums of the body frame and the clamp device of the liquid nitrogen vessel were measured at the rotation speeds of 50 rpm, 100 rpm, 200 rpm, 300 rpm, and 400 rpm. The vibration spectrum of the clamp device of the liquid nitrogen vessel includes both the perpendicular direction and the horizontal direction. The vibration spectrum of the body frame at 400 rpm is shown in Fig. 31. The vibration spectrums of the perpendicular direction and the horizontal direction on the liquid nitrogen vessel at 400 rpm are shown in Fig. 32 and Fig. 33.

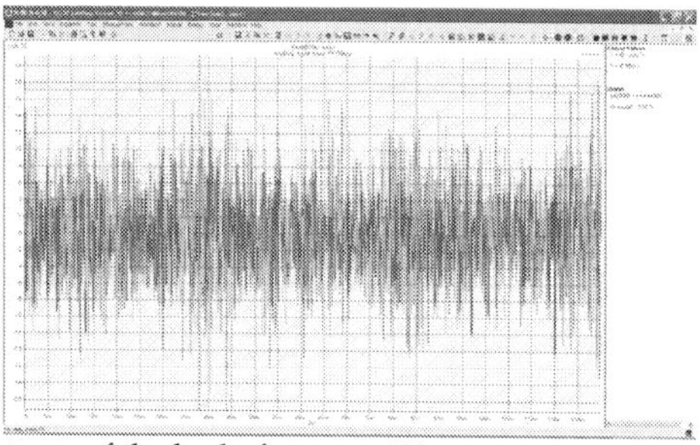

Fig. 31. Vibration spectrum of the body frame at 400 rpm

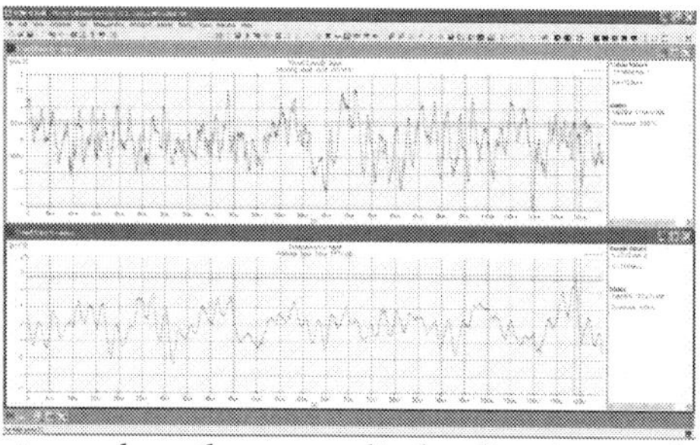

Fig. 32. Vibration spectrum along the perpendicular direction on the clamp device of the liquid nitrogen vessel at 400 rpm.

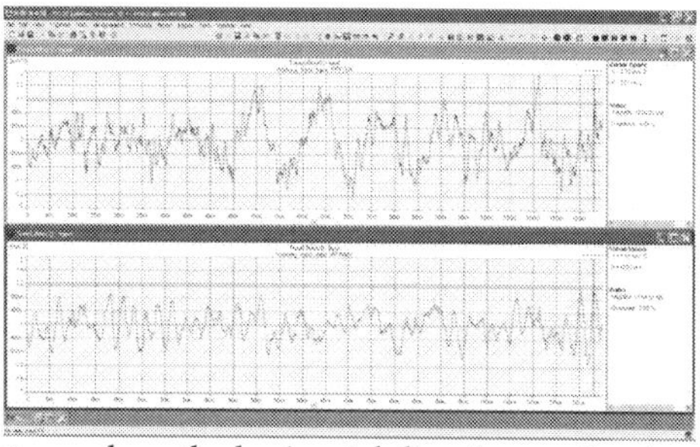

Fig. 33. Vibration spectrum along the horizontal direction on the clamp device of the liquid nitrogen vessel at 400 rpm.

Fig. 34. shows the vibration measuring results of the body frame and the clamp device of the liquid nitrogen vessel (the perpendicular and the horizontal direction) at the rotation speed of 50 rpm, 100 rpm, 200 rpm, 300 rpm, and 400 rpm. The left figure shows the measuring vibration results without the liquid nitrogen vessel on the clamp device. The right figure shows results with the liquid nitrogen vessel attached. After the YBCO superconductors in the vessel are cooled in the zero field, vibration properties are measured at the levitation gap of 20 mm and the lateral displacement of 0 mm. Experimental results show that the vibration of the body frame increases quickly with the increase of rotation speed, and the vibration of the clamp device with or without the liquid nitrogen vessel also increases slowly with the increase of rotation speed. This indicates that the measuring seat has the function of vibration isolation. The vibration of the clamp device with the liquid nitrogen vessel satisfies measurement requirements.

5.3 Measurement of HTS Maglev properties

The main distinguish between the SCML-03 and previous HTS Maglev measurement system (Wang J. et al., 2000; Wang S. et al., 2007) is to measure the Maglev performance at the running case. By SCML-03, both levitation forces and guidance forces are measured experimentally at different velocities.

The experimental results of the levitation forces and the guidance forces of YBCO bulk measured with the SCML-03 are shown in Fig. 35 and Fig. 36, respectively. The levitation forces of YBCO bulk were measured in zero field cooling (ZFC), and the guidance forces of YBCO bulk were measured at 15 mm field cooling height. Those measurement results verified that the SCML-03 is applicable for measurement of the levitation and the guidance forces at movement state.

In order to prove the applicability of the measurement results of the levitation forces and the guidance forces of YBCO bulks, the changes of the measurement results with different

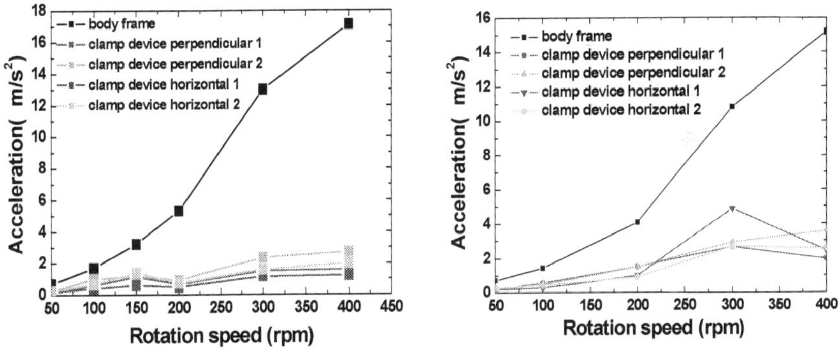

Fig. 34. Vibration measuring results of the body frame and the clamp device of the liquid nitrogen vessel (perpendicular and horizontal direction) at rotation speeds of 50 rpm, 100 rpm, 150 rpm, 200 rpm, 300 rpm, and 400 rpm. Left (right) figure shows the measuring results without (with) the liquid nitrogen vessel on the clamp device.

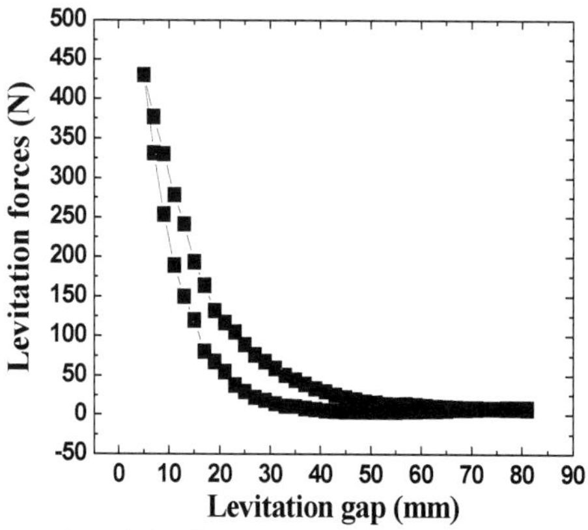

Fig. 35. Measurement results of the levitation forces of YBCO bulk in zero field cooling (ZFL).

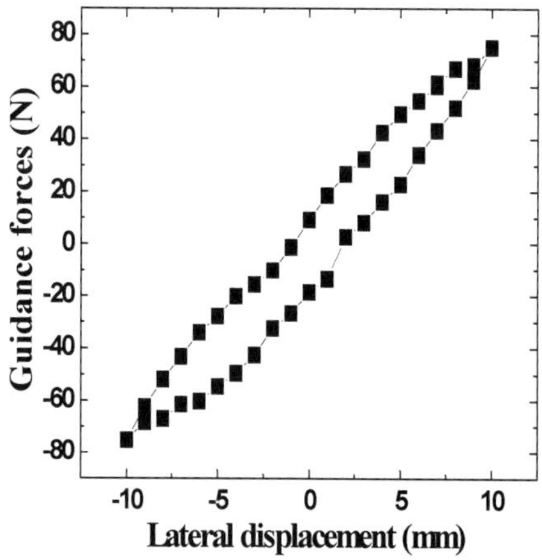

Fig. 36. Measurement results of the guidance forces of YBCO bulk at a 15 mm levitation gap for the field cooling (FL).

measuring time and different rotational speeds were investigated experimentally. Rotational speeds of 50 rpm, 100 rpm, 200 rpm, 300 rpm, and 400 rpm, with time intervals of 12 s were tested and shown in Fig. 37. The experimental results are measured at the levitation gap of 10 mm.

Fig. 37. shows that the changes of the measured levitation forces are small, especially at the 300 rpm and 400 rpm speeds. The higher the PMG rotation speed, the smaller the levitation force changes. At the levitation gap of 10 mm the changes found on the measured levitation forces in a 48 s period were 0.54 N at 300 rpm and 0.10 N at 400 rpm.

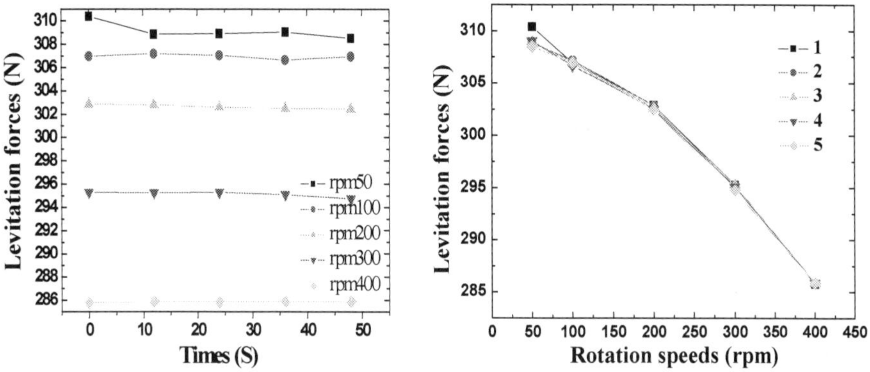

Fig. 37. Experimental results of the levitation force of zero field cooling YBCO superconductor bulks at time intervals of 12 s and special rotation speed (the levitation gap of 10 mm).

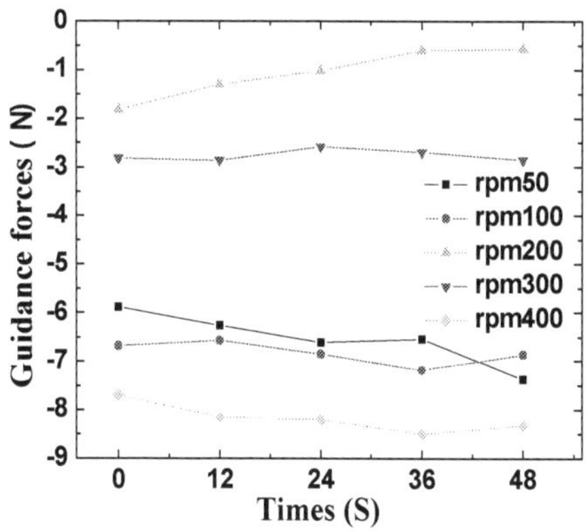

Fig. 38. Experimental results of the guidance force of field cooling YBCO bulks at time intervals of 12 s, levitation gap was of 15 mm, and special rotation speed.

The levitation force decreases with the increase of rotational speed. The decrease in value of the levitation forces is 22.98 N when the rotation speed is increased from 50 rpm to 400 rpm at time intervals of 12 s, which is a decrease of the levitation forces of 7.4% from the original value. The reason for decreasing of the levitation forces can be attributed to some subtle changes in the uniformity of the circular PMG's magnet field along the circumferential direction which at high speeds is equivalent subjecting the HTS to an AC magnetic field.

The experimental results of the guidance force of field cooling YBCO superconductor bulks at the time intervals of 12 s, levitation gap of 15 mm, and special rotation speed are shown in Fig. 38. It indicates that the changes of measured guidance forces at the time intervals of 12 s are small, especially at 300 rpm and 400 rpm. The higher PMG rotation speed, the smaller the guidance force changes. The experimental results of the guidance forces are not better than that of the levitation forces. The most changes of measured guidance forces in the 48 s time frame were 0.28 N at 300 rpm and 0.79 N at 400 rpm).

6. Conclusion

Three high temperature superconducting (HTS) Maglev measurement systems were successfully developed in the Applied Superconductivity Laboratory (ASCLab) of Southwest Jiaotong University, P. R. China. These systems include liquid nitrogen vessel, Permanent Magnet Guideway (PMG), data collection and processing, mechanical drive and Autocontrol features. This chapter described the three different measuring systems along with their theory of operations and workflow.

The SCML-01 HTS Maglev measurement system can make real time measurement of Maglev properties between one or many YBCO bulks and employ a PM or PMG. Also the trapping flux of high T_c superconductors can be measured in the scanning range of 100 mm×100 mm. It was especially employed to develop the on board HTS Maglev equipment which travels over one or two PMGs. The on board Maglev equipment includes a rectangular-shaped liquid nitrogen vessel and YBCO bulk superconductors. Based on the original research results from the SCML-01, the first man-loading HTS Maglev test vehicle in the world was successfully developed in 2000.

In order to make more thorough and careful research investigations, the HTS Maglev, HTS Maglev Measurement System (SCML-02) was subsequently developed with even more function capabilities and a higher precision to extensively investigate the Maglev properties of YBCO bulk samples over a PM or PMG. The new features included: higher measurement precision, instant measurement at movement of the measured HTS sample, automatic measurement of both levitation and guidance forces, dynamic rigidity, ability for the measured HTS sample to be moved along the three principal axes all at once, relaxation measurements of both levitation and guidance forces, and so on. The main specification of the system is: position precision ±0.05 mm. vertical force precision 2 ‰; horizontal force precision 1 ‰; and force measurement precision of 0.02 N.

In order to investigate the dynamic characteristics behavior of the HTS Maglev engineering application, an HTS Maglev dynamic measurement system (SCML-03) was designed and successfully developed. The circular PMG is fixed along the circumferential direction of a big circular disk with a diameter of 1,500 mm. The maximum linear velocity of the PMG is about 300 km/h when the circular disk rotates round the central axis at 1280 rpm. The liquid nitrogen vessel with HTS bulks is placed above the PMG, and the vessel is allowed to move along the three main principal axes so that sensors can detect force variations stemming from the superconductors.

The design, method, accuracy and results have allowed the successful development of these three measurement systems. All systems are calibrated by standard measurement technology, for which its reliability, stability, featured functions, and precision have also been validated through its long-term usage.

7. Acknowledgements

The authors are grateful to Zhongyou Ren, Yiyu Lu, Zigang Deng, Jun Zheng, Fei Yen, Changyan Deng, Youwen Zeng, Haiyu Huang, Xiaorong Wang, Honghai Song, Xingzhi Wang, Longcai Zhang, Hua Jing, Qingyong He, Lu Liu, Guangtong Ma, Wei Liu, Qunxu Lin, Yonggang Huang, Minxian Liu, Yujie Qing, Rongqing Zhao, and Ya Zhang for their contributions towards the abovementioned HTS Maglev measurement systems.

8. References

Bomemann H. J., A. Tonoli, T. Ritter, C. Urban, O. Zaitsev, K. Weber, and H. Rietschel, (1995). Engineering Prototype of a Superconducting Flywheel for Long Term Energy Storage. *IEEE Trans. Appl. Supercond.*, 5(2): 618-621

D' Ovidio G., F. Crisia, G. Lanzara, (2008). A 'V' shaped superconducting levitation module for lift and guidance of a magnetic transportation system. *Physica C*, vol 468, pp. 1036-1040

Deng Zigang, Jun Zheng, Honghai Song, Lu Liu, Lulin Wang, Ya Zhang, Suyu Wang, and Jiasu Wang, (2007). Free Vibration of the High Temperature Supercond. Maglev Vehicle model. *IEEE Trans Appl Supercond*, 17(2): 2071-2074

Deng Zigang, Wang Jiasu, Wang Suyu, Zheng Jun, Ma Guangtong, Lin Qunxu, Zhang Ya, (2008a). Status of High Temperature Supercond. Flywheel Energy Storage System. *Transactions of China Electrotechnical Society*, 23(12): 1-10 (In chinese)

Deng Z., J. Wang, J. Zheng, H. Jing, Y. Lu, G. Ma, L. Liu, W. Liu, Y. Zhang, S. Wang, (2008b). High-efficiency and low-cost permanent magnet guideway consideration for high-Tc superconducting Maglev vehicle practical application. *Superconductor Science and Technology*, 21(11): 115018 (9pp)

Deng Zigang, Wang Jiasu, Wang Suyu, Zheng Jun, Lin Qunxu, Zhang Ya, (2009)a. R & D Status of High-Tc Superconducting Magnetic Bearings. *Transactions of China Electrotechnical Society*, (In chinese)

Deng Z., Q. Lin, J. Wang, J. Zheng, G. Ma, Y. Zhang, S. Wang, (2009b). Basic design and characteristics study of a double-axial superconducting magnetic bearing system. *Cryogenics*, doi: 10.1016/j. *Cryogenics*, 10.006

He Qingyong, Jiasu Wang, Longcai Zhang, Suyu Wang and Siting Pan, (2008). Influence of the Ramp Angle on Levitation Characteristics of HTS Maglev. *Physica C*, 468(1): 12-16

Hellman F., E. M. Gyorgy, D. W. Johnson, Jr., H. M. O'Bryan, and R. C. Sherwood, (1988). Levitation of a magnet over a flat type II superconductor. *J. Appl. Phys.*, vol. 63, pp. 447-450

Honghai Song, Jiasu Wang, Suyu Wang, Zhongyou Ren, Xiaorong Wang, Oliver de Haas, Gunter Fuchs, Ludwig Schultz, (2006). Studies of YBCO Electromagnetic Properties for High-Temperature Superconductor Maglev Technology. In: Barry P. Martins Ed., *New Topics in Superconductivity Research* (Nova Science Publishers, Inc.) pp. 107-156

Hull J. R., (2000). Superconducting bearings. *Supercond. Sci. Technol.* 13(2): R1-R15

Jing Hua, Suyu Wang, Wei Liu, Ming Jiang, Changyan Deng, Jun Zheng, Zigang Deng, Jiasu

Wang, (2008). A high-Tc superconducting Maglev system using T-shaped permanent magnet single-guideway. IEEE Trans Appl Supercond, 18(2): 795-798

Kovalev K. L., S. M.-A. Koneev, V. N. Poltavec, et al., (2005). Magnetically levitated High-speed carriages on the basis of bulk HTS elements," in *Pro. 8th Intern. Symp. Magn. Susp. Technol. (ISMST'8)*, Dresden, Germany, pp. 51-

Liu Minxian, Suyu Wang, Jiasu Wang, Guangtong Ma, (2008). Influence of the Air Gap between Adjacent Permanent Magnets on the Performance of NdFeB Guideway for HTS Maglev System. *Journal of Superconductivity and Novel Magnetism*, 21(7): 431-435

Liu W., J.S. Wang, H. Jing, M. Jiang, J. Zheng, S.Y. Wang, (2008). Levitation performance of high-Tc superconductor in sinusoidal guideway magnetic field. *Physica C*, 468(23): 2345-2350

Liu wei, wang su-yu, lin qun-xu, yang pei, liu hui-fen, ma guang-tong, zheng jun, and wang jia-su, (2009). Research on the long-term operation stability of the bulk YBCO on vehicle. *10th Superdongductivity Conference of China*, beijing, 2009 (In chinese)

Lu Yiyun, Jiasu Wang, Suyu Wang, Jun Zheng, (2008). 3D-Modeling Numerical Solutions of Electromagnetic Behavior of HTSC Bulk above Permanent Magnetic Guideway. *Journal of Superconductivity and Novel Magnetism*, 21(8): 467-472

Ma G T, Q X Lin, J S Wang, S Y Wang, Z G Deng, Y Y Lu, M X Liu, J Zheng, (2008). Method to reduce levitation force decay of the bulk HTSC above the NdFeB guideway due to lateral movement. *Supercond. Sci. Technol.*, 21(6): 065020

Ma K B, Postrekhin Y V, Chu W K, (2003). Superconductor and magnet levitation devices. *Rev. Sci. Instrum.*, 74(12): 4989-5017.

Moon F C. *Superconducting Levitation*. New York: John Wiley and Sons, 1994

Moon F. C., P. -Z. Chang, (1990). High-speed rotation of magnets on high Tc superconducting bearings. *Appl. Phys. Lett.*, 56(4): 397-399

Okano M., T. Iwamoto, M. Furuse, S. Fuchino, I. Ishii, (2006). Running performance of a pinning-type superconducting magnetic levitation guide. *J. Phys. Conf. Ser.*, vol. 43, pp. 999-1002

Peters P. N., R.C.Sisk, E.W.Urban, C.Y.Huang, and M.K.Wu, (1988). □文名. *Appl. Phys. Lett.* 52: 2066

Qin Yujie, Yiyun Lu, Suyu Wang, Jiasu Wang, (2009). Levitation Force Relaxation of HTS Bulk above NdFeB Guideways at Different Approaching Speeds. Journal of Superconductivity and Novel Magnetism, 22(5): 511-516

Ren Zhongyou, Jiasu Wang, Suyu Wang, He Jiang, Min Zhu, Xiaorong Wang, Honghai Song, (2003). Influence of shape and thickness on the levitation force of YBaCuO bulk HTS over a NdFeB guideway. *Physica C*, 384(1-2): 159-162

Schultz L., O de Haas, P. verges, C. Beyer, S. Rohlig, H. Olsen, et al., (2005). Superconductively levitated transport system-the SupraTrans project. *IEEE Trans Appl. Supercond.*, vol. 15, 2301-2305

Song H.H., O. De Haas, Z.Y. Ren, X.R. Wang, et al., (2004). Magnetic interaction between multiple seeded YBCO bulks and the permanent magnet guideway. *Physica C*, 407(1-2): 82-87

Stephan R. M., A. C. Ferreira, R. de Andrade, Jr. M. A. Neves, M. A. Cruzmoreira, M. A. P. Rosario, et al., (2004). A superconducting levitated small scale vehicle with linear synchronous motor. in *2003 IEEE Intern. Sym. Indust. Electro.*, vol. 1, pp. 206-209

Wang Jiasu, Suyu Wang, Guobin Lin, Haiyu Huang, Cuifang Zhang, Youwen Zeng, Shaohua Wang, Changyan Deng, Zhipei Xu, Qixue Tang, Zhongyou Ren, He Jiang, Min Zhu, (2000). High T_c Superconducting Magnetic Levitation Measurement System, *High Technology Letters*, vol. 10 no. 8, pp. 56-58, (In Chinese)

Wang Jiasu, Suyu Wang, Zhongyou Ren, Min Zhu, He Jiang, and Qixue Tang, (2001). *IEEE Trans. on Applied Superconductivity,* 11 1801-1804

Wang Jiasu, Wang Suyu, Zeng Youwen, Huang Haiyu, Luo Fang, Xu Zhipei, Tang Qixue, Lin Guobin, Zhang Cuifang, Ren Zhongyou, Zhao Guomin, Zhu Degui, Wang Shaohua, Jiang He, Zhu Min, Deng Changyan, Hu Pengfei, Li Chaoyong, Liu Fang, Lian Jisan, Wang Xiaorong, Wang Lianghui, Shen Xuming, Dong Xiaogang, (2002). The first man-loading high temperature superconducting maglev test vehicle in the world. *Physica C,* 378-381: 809-814

Wang J.S., S.Y. Wang, Z.Y. Ren, H. Jiang, M. Zhu, X.R. Wang, X.M. Shen, H.H. Song. (2003a). Experiment results of high temperature superconducting maglev vehicle. *Physica C,* 386: 431-437

Wang Jiasu, Suyu Wang, Zhongyou Ren, Xiaorong Wang, Min Zhu, He Jiang, Honghai Song, Xingzhi Wang, Jun Zheng, (2003b). Guidance Forces on High Temperature Superconducting Maglev Test Vehicle. *IEEE Transactions on Applied Superconductivity,* 13 (2): 2154-2156

Wang Jiasu and Wang Suyu, (2005a). Synthesis of Bulk Superconductors and Their Properties on Permanent Magnet Guideway. In: Anant Narlikar Ed., *Frontiers in Superconducting Materials* (Springer Verlag, Germany), pp.885-912

Wang Jiasu, Suyu Wang, Changyan Deng, Youwen Zeng, Honghai Song, Jun Zheng, Xinzhi Wang, Haiyu Huang, and Fu Li, (2005b). Design consideration of a high temperature superconductor Maglev vehicle system. *IEEE Transactions on Applied Superconductivity,* 15(2): 2273-2276

Wang J.S., S.Y. Wang, et al. (2005c). The present status of HTS Maglev vehicle in China. *Supercond. Sci. Technol.,* 18: S215-S218

Wang Jiasu, Suyu Wang, Changyan Deng, Jun Zheng, Honghai Song, Qingyong He, Youwen Zeng, Zigang Deng, et al., (2007). Laboratory-scale high temperature superconducting Maglev launch system. *IEEE Trans Appl. Supercond.,* 17(2): 2091-2094

Wang Jiasu, Suyu Wang, Changyan Deng, Youwen Zeng, Longcai Zhang, Zigang Deng, Jun Zheng, Lu Liu, Yiyun Lu, Minxian Liu, Yaohui Lu, Yonggang Huang, and Ya Zhang, (2008). A High-Temperature Superconducting Maglev Dynamic Measurement System. *IEEE Trans Appl. Supercond.,* 18(2): 791-794

Wang Jiasu, Suyu Wang, Jun Zheng, (2009a). Recent development of high temperature superconducting Maglev system in China. *IEEE Trans Appl. Supercond.,* 2009, 19(3): 2142-2147

Wang Jia-Su, Wang Su-yu, Zheng Jun, Deng Zi-Gang, Lu Yi-Yun, Ma Guang-Tong, Zhang Ya, Yen Fei, (2009b). Recent developments of high temperature superconducting maglev system. *10th Superdongductivity Conference of China,* beijing, 2009 (In chinese)

Wang Suyu, Jiasu Wang, Zhongyou Ren, Min Zhu, He Jiang, Qixue Tang, (2001a). Combination levitation force of multi-block YbaCuO bulk high temperature superconductors, *IEEE Trans. Appl. Superconduct.,* vol.11, no.2, pp. 1808-1811,

Wang Suyu, Qixue Tang, Zhongyou Ren, Min Zhu, He Jing, Jiasu Wang, (2001b). Design of high T_c superconducting Maglev system for vehicles, *Superconductivity and Cryogenics*, vol. 29, no. 1, pp. 14-17 (In Chinese)

Wang S.Y., J.S. Wang, Z.Y. Ren, M. Zhu, H. Jiang, X.R. Wang, X.M. Shen, H.H. Song, (2002). High temperature superconducting maglev equipment on vehicle, *presented at Superconductors for Practical Applications (SPA'2002)*, Xi'an, P. R. China, June 16-20,

Wang S.Y., J.S. Wang, Z.Y. Ren, M. Zhu, H. Jiang, X.R. Wang, X.M. Shen, H.H. Song, (2003). High temperature superconducting maglev equipment on vehicle. *Physica C*, 386: 531-535

Wang Suyu, Jiasu Wang, Changyan Deng, Yiyu Lu, Youwen Zeng, Honghai Song, Haiyu Huang, Hua Jing, Yonggang Huang, Jun Zheng, Xingzhi Wang, and Ya Zhang, (2007). An update High-Temperature Superconducting Maglev Measurement System. *IEEE Trans Appl Supercond*, 17(2): 2067-2070

Wang X.R., H.H. Song, Z.Y. Ren, M. Zhu, J.S. Wang, S.Y. Wang, X.Z. Wang, (2003). Levitation force and guidance force of YBaCuO bulk in applied field. *Physica C*, 386: 536-539

Wang Xiaorong, Zhongyou Ren, Honghai Song, Xingzhi Wang, Jun Zheng, et al., (2005). Guidance force in an infinitely long superconductor and permanent magnetic guideway system. *Supercond. Sci. Technol.*, 18: S99-S104

Zhang Longcai, Jiasu Wang, Suyu Wang, Qingyong He, (2007). Influence of AC external magnetic field perturbation on the guidance force of HTS bulk over a NdFeB guideway. *Physica C*, 459(1-2): 43-46

Zheng J., Z. Deng, S. Wang, J. Wang, Y. Zhang, (2007a). Vibration properties dependence on the trapped flux of bulk high-temperature superconductor. *Physica C*, 463-465: 1356-1360

Zheng Jun, Zigang Deng, Lulin Wang, Lu Liu, Ya Zhang, Suyu Wang, and Jiasu Wang, (2007b). Stability of the Maglev vehicle model using bulk high T_c superconductors at low speed. *IEEE Trans Appl Supercond*, 17(2): 2103-2106

4

Autonomous Measurement System for Localization of Loss-Induced Perturbation Based on Transmission-Reflection Analysis

Vasily V. Spirin
División de Física Aplicada, CICESE,
Apdo. Postal 2732, CP 22860, Ensenada, B.C. México

1. Introduction

The highest state of the art in optical sensing is achieved with optical fiber distributed sensors that allow the measurement of a desired parameter along the test fiber (Hartog, 2000; Byoungho Lee, 2003). The regions where perturbations occur are usually localized by means of optical time-domain reflectometry (OTDR) or frequency domain reflectometry (OFDR) (Tsuji et al., 1995; Pierce et al., 2000; Venkatesh et al., 1990). All these methods utilize time- or frequency-modulated light sources that allow us to localize a number of perturbations along the test fiber simultaneously. Meanwhile, for some applications, it is important to detect and localize a rare but hazardous alarm condition which typically occurs as a single infrequent event, such as a pipe leak, fire or explosion.

For such applications, we proposed a novel simple and inexpensive measurement technique based on so-called transmission-reflection analysis (TRA) (Spirin et al., 2002a). Generally, the TRA method is based on the unique relationships between normalized transmitted and Rayleigh backscattered powers for different locations of the loss-induced disturbance along the sensing fiber. The TRA technique utilizes an unmodulated light source, power detectors and a sensing fiber. Localization of a strong disturbance with a maximum localization error of a few meters along a few km-long single-mode sensing fiber was demonstrated (Spirin et al., 2002b).

The paper presents a systematical review of our works in the TRA-sensing area. In the first parts of the paper we describe general ideas of the TRA, including principle of the operations of TRA-based sensors, localization errors examination, a transmission-reflection analysis for a distributed fiber-optic loss sensor with variable localization accuracy along the test fibre, and theoretical and experimental evidences that the TRA method can be modified for detection and localization of a number of perturbations that appear one after another at different positions along the test fiber.

In the final part we offer completely autonomous measurement system based on transmission-reflection analysis (AMS-TRA). This part includes design of the AMS-TRA system, thermal stability inspection, detailed analysis of experimental accuracy and localization errors, and implementation of the system for gasoline leak detection and localization.

2. Single Perturbation Localization

The basic idea of the TRA method is to localize the perturbation by using the unique relationships between normalized transmitted and Rayleigh backscattered powers of an unmodulated CW light source for different locations of the loss-induced disturbance along the sensing fiber. Indeed, if the bending losses occur at the remote-end of the sensing fiber (see Fig.1), an increase in the load leads to a proportional decrease of the transmitted power. However, it does not change the Rayleigh backscattered power, because all fiber length participate in backscattering and the launched power is the same such as for undisturbed fiber.

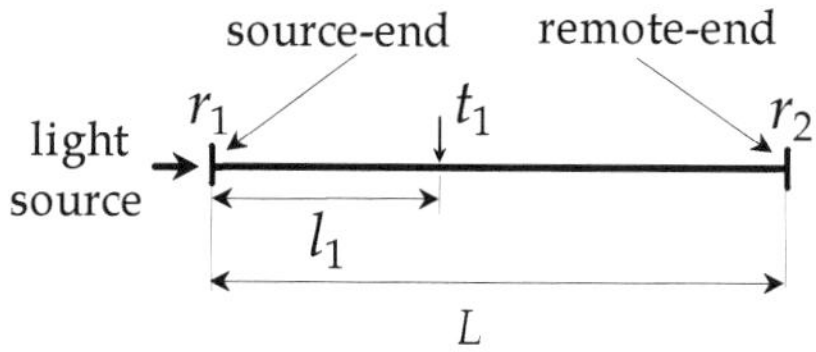

Fig. 1. Test fiber configuration for single perturbation; l_1 - perturbation location, t_1 - transmission of loss-inducing segment, r_1, r_2 - reflections from source- and remote-ends, L- test fiber length.

But if we bend the sensing fiber close to the source-end, the decrease in transmitted power is accompanied by a decrease in the Rayleigh backscattered power. Because in this case the launched into the fiber power is decreased and backscattered power is also decreased due to the induced losses.

Further, if we bend the sensing fiber in the middle, the first half of the fiber, which is closer to the source-end, scatters the light as well as half of undisturbed fiber, but the power scattered from the second half is less due to losses induced in the middle. So, for the identical loss-induced perturbations the value of the decrease in normalized backscattered power depends on the location of the excess loss region.

To find an analytical expression for calculation of the distance from the fiber source-end to the location of the loss region, we will analyze the configuration with two plain fiber sections whose lengths are l_1 and L-l_1 respectively, and a short fiber piece between them affected by a monitored condition (see Fig.1). Plain fiber sections possess Rayleigh scattering and attenuation due to light absorption and short fiber piece induces a losses. The power reflection coefficient of each Rayleigh scattering fiber segment can be calculated as (Gysel & Staubli, 1990; Liaw et al., 2000):

$$R(\delta l_i) = S(\alpha_s / 2\alpha)[1 - \exp(-2\alpha\delta l_i)], \qquad (1)$$

where α_s is the attenuation coefficient due to Rayleigh scattering, α is the total attenuation coefficient of the test fiber, δl_i is the length of the i-th fiber segment, and recapture factor S for the fiber is defined as (Brinkmeyer, 1980):

$$S = b(n_1^2 - n_2^2)/n_1^2, \qquad (2)$$

where b depends on the waveguide property of the fiber and is usually in the range of 0.21 to 0.24 for single-mode step-index fiber (Brinkmeyer, 1980), n_1 and n_2 are the refractive indices of the fiber core and cladding, respectively.

Introducing a parameter $S_\alpha = S(\alpha_s/2\alpha)$, the transmission and backscattering coefficients of plain fiber sections can be written as $T_i = exp\ (-\alpha\delta l_i)$ and $R_i = S_\alpha(1\text{-}exp(\text{-}2\alpha\delta l_i))$, respectively. The short fiber piece is affected by monitored conditions which introduce additional light losses. A transmission of short fiber piece is $t_1 \leq 1$. Let us assume that the scattering is relatively weak and the portion of the scattered light is very small. This allows us to simplify the analysis, neglecting multiple scattering in both directions. The reflections with coefficients r_1 and r_2 from the fiber source- and remote-ends, respectively, have to be taken into account because even a weak reflection can be comparable to the back scattering. However, we can assume that $r_{1,2} << 1$ and neglect multiple reflections as well.

In this case, the transmission T and back-scattering R coefficients of this optical system can be written as:

$$T = T_1 t_1 T_2 = t_1 \mathrm{e}^{-\alpha L}, \tag{3}$$

$$R = r_1 + S_\alpha (1 - e^{-2\alpha l_1}) + T_1^2 t_1^{\ 2} S_\alpha (1 - e^{-2\alpha (L - l_1)}) + T_1^2 t_1^2 T_2^2 r_2. \tag{4}$$

Normalized transmitted T_{norm} and backscattered R_{norm} coefficients are defined as:

$$T_{norm} = \frac{T}{T_{\max}} = t_1, \tag{5}$$

$$R_{norm} = \frac{R}{R_{\max}} = \frac{S_\alpha + r_1 - (S_\alpha - r_2) t_1^2 e^{-2\alpha L} - S_\alpha (1 - t_1^{\ 2}) e^{-2\alpha l_1}}{S_\alpha + r_1 - (S_\alpha - r_2) e^{-2\alpha L}}, \tag{6}$$

where $T_{\max}$ is the maximum transmittance of initially undisturbed sensing fiber when $t_1 = 1$,

$$T_{\max} = \mathrm{e}^{-\alpha L}, \tag{7}$$

and $R_{\max}$ is the maximum back-scattering coefficients of undisturbed optical system

$$R_{\max} = S_\alpha + r_1 - (S_\alpha - r_2) e^{-2\alpha L}. \tag{8}$$

The relationship between the normalized transmitted T_{norm} and Rayleigh backscattered R_{norm} powers for single perturbation can be expressed from (5-6) as:

$$T^2{}_{norm} = \frac{(S_\alpha + r_1)(R_{norm} - 1) - R_{norm}(S_\alpha - r_2) e^{-2\alpha L} + S_\alpha e^{-2\alpha l_1}}{S_\alpha (e^{-2\alpha l_1} - e^{-2\alpha L}) + r_2 e^{-2\alpha L}}. \tag{9}$$

To localize the perturbation with the proposed method, we need to find a parametric curve that passes through the point with coordinates equal to the measured normalized Rayleigh backscattered and transmitted powers. The location of the loss region can also be found directly from Eqn. (9) as:

$$l_1 = -\frac{1}{2\alpha}\ln\frac{(1-R_{norm})(S_\alpha + r_1) + (R_{norm} - T^2_{norm})(S_\alpha - r_2)e^{-2\alpha L}}{S_\alpha(1-T^2_{norm})}. \tag{10}$$

Therefore, the measurement of the normalized transmitted and backscattered powers, as well as the knowledge of the fiber attenuation coefficients α and α_s, provide the calculation of the distance l_1 from the fiber source-end to the fiber section with induced losses.
The slope of dependence of normalized backscattered power R_{norm} versus the square of normalized transmitted power T^2_{norm} can be found from Eqn. (9) as:

$$\frac{\partial R_{norm}}{\partial (T^2{}_{norm})} = \frac{S_\alpha(e^{-2\alpha l_1} - e^{-2\alpha L}) + r_2 e^{-2\alpha L}}{S_\alpha + r_1 - (S_\alpha - r_2)e^{-2\alpha L}}. \tag{11}$$

As we can see this slope uniquely depends on perturbation location l_1. Therefore, the location of the single perturbation can be found from experimentally measured slope as:

$$l_1 = \frac{1}{2\alpha}\ln\left[\frac{S_\alpha}{(S_\alpha + r_1)\dfrac{\partial R_{norm}}{\partial (T^2{}_{norm})} + (r_2 - S_\alpha)e^{-2\alpha L}\left(\dfrac{\partial R_{norm}}{\partial (T^2{}_{norm})} - 1\right)}\right]. \tag{12}$$

The relationship between normalized Rayleigh backscattered power R_{norm} and the square of normalized transmitted power T^2_{norm} is almost linear for a single perturbation which affects the test fiber in any location (see Eqn. (9)). Fig.2 shows the result of the numerical calculation of these relationships when additional losses occur at distances $l_{1,n} = n\Delta l$ from the source-end of the test fiber, where $n = 0,1\ldots10$, and the interval between bending locations Δl = 284.4 meter. Transmitted and backscattered powers were normalized with respect to their initial undisturbed values. A typical value for b equal to 1/4.55 for single-mode fibers (Beller, 1998) was used in the calculations. Reflections from the source-end and the remote-end of the sensing fiber, which are respectively equal to $4.7\text{x}10^{-6}$ and $1.5\ \text{x}10^{-5}$ in our experiment, were also taken into account in the calculations.
For the verification of the proposed method we use firstly a laboratory experimental setup. The schematic diagram of the TRA based fiber-optic sensor is shown in Fig. 3. A continuous wave (CW) light emitted by a amplified spontaneous emission (ASE) optical fiber source operating near 1550 nm wavelength with a linewidth of few nm was launched into a 2.844 km-long standard single mode SMF-28 fiber through 3 dB coupler. The launched optical power was about 1.1 mW, and the attenuation coefficient of the test fiber, which was measured with OTDR, was equal to 0.19 dB/km. An optical isolator was used to cancel back reflections influence on ASE source. An immersion of all fiber ends was employed in order to reduce back reflections. Standard power detectors were used to measure the transmitted

and Rayleigh backscattered powers. We also take into account the ASE power instability by measuring a source power directly (see Fig.3).

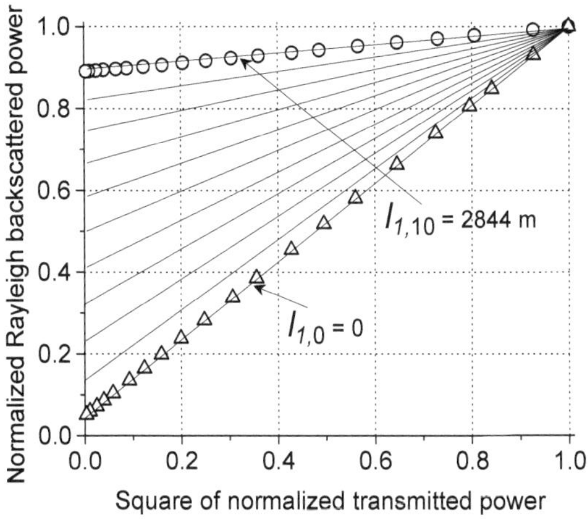

Fig. 2. Relations between normalized Rayleigh backscattered power and the square of normalized transmitted power when additional losses occur at distances $l_{1,n} = n\Delta l$ from the source-end of the test fiber, where $n = 0,1\ldots10$, and the interval between bending locations Δl = 284.4 m. (o, Δ - experimental results, and solid lines – theoretical dependencies).

To induce the bending losses in the sensing fiber, we used bending transducer, which is also shown schematically in Fig.3. By tuning the bending transducer we changed the normalized transmitted power from its initial undisturbed value equal to 1 up to more than -30 dB. The bending losses were induced near source-end and near remote-end of the test fiber.

A good agreement between experimental data and theory was obtained for (α_s/α) = 0.68, that corresponds attenuation coefficient due to Rayleigh scattering α_s equal to 0.13 dB/km that is quite reasonable for the fiber with total attenuation coefficient α = 0.19 dB/km (Beller, 1998).

Experimentally measured slopes for the bending losses, which were induced near remote - and source -ends of the test fiber are equal to 0.109, and 0.95, correspondingly. These values well agree with the values calculated using Eqns. (11), which are equal to 0.108, and 0.96, respectively.

So, for standard telecommunication single mode fibers the TRA method demonstrates concord between calculated and experimentally measured data practically without any fitting parameters. Even using for simulation a typical values of unknown recapture factor S and relation (α_s/α) between Rayleigh-induced and total losses in the fiber guarantee quit reasonable conformity.

Concluding this section we can declare that TRA method provides new opportunity for the localization of loss-induced alarm-like perturbation along few km-length fibers by uncomplicated measuring of transmitted and Rayleigh backscattered powers of an unmodulated CW light.

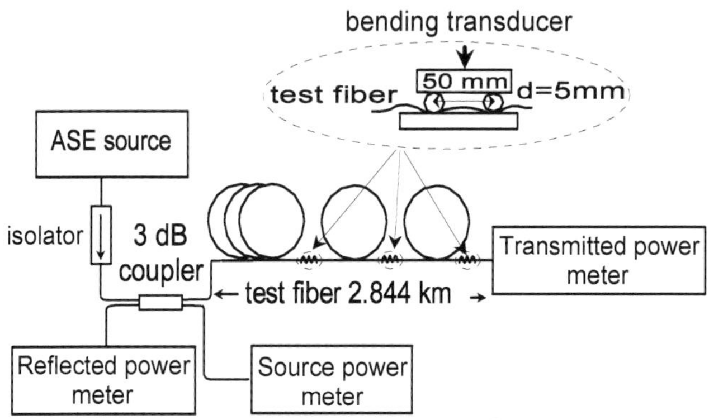

Fig. 3. Schematic diagram of laboratory setup of TRA fiber-optic sensor.

3. Multi-Point Perturbations Localization

Let us now verify that any number of consecutive perturbations can be localized with the TRA method. The proof will be done by mathematical induction. In our analysis we consider the test fiber with the same properties and parameters as in previous case. But now we use a configuration with a number of plain Rayleigh-scattering fiber sections separated by a number of short loss-inducing fiber pieces with transmissions $t_i \leq 1$ (see Fig.4), (Spirin, 2003).

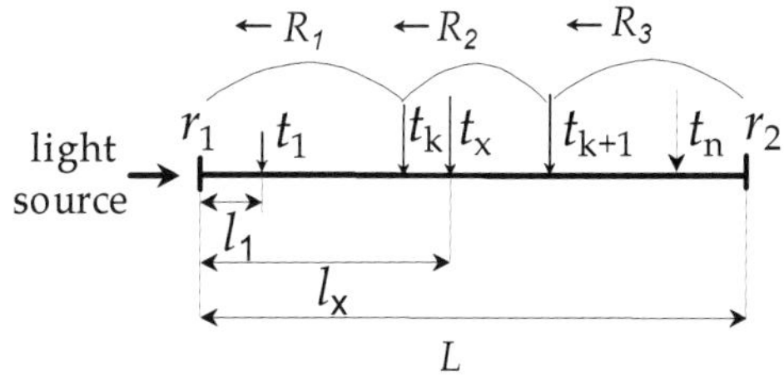

Fig. 4. Test fiber configuration for multi-point perturbations; $t_1 \div t_n$ - transmission of initially disturbed loss-inducing segments, t_x transmission of currently disturbed segment, r_1, r_2 - reflections from source- and remote-ends.

Let us assume that according to the principle of mathematical induction we already determined the values and locations of the first n perturbations and demonstrate that we can find a position of next $(n+1)$-th perturbation without ambiguity. Here we consider only the perturbations that appear one after another at different positions along the test fiber. So, at the current moment all initial n perturbations induce fixed known losses at known locations and only a new $(n+1)$-th perturbation can modify the reflectivity and transmittance of the test fiber.

Because we know the positions and values of all n initial perturbations, we can number these according their positions along the test fiber (see Fig.4). We also can suppose without loss of generality that a new perturbation is located at distance l_x between k-th and (k+1)-th initial perturbations, where k is unknown.
The transmittance of initial n loss-inducing short segments which are located at distances l_j from the source-end is $t_j \leq 1$, where j = 1, n. The transmittance of unknown loss-inducing segment which is located at distance l_x is t_x. Assuming that the scattering and reflections from the fiber ends are relatively weak and neglecting multiple scattering and reflections in both directions, the normalized power reflection coefficient of the optical system can be calculated as:

$$R_{norm} = \frac{R_1 + R_2 + R_3}{R_{\max}}. \tag{13}$$

In Eqn. (13) R_1 corresponds to Rayleigh backscattering from segments placed on the left of unknown perturbation plus Fresnel reflection from the source-end of the test fiber

$$R_1 = r_1 + S_\alpha \sum_{i=0}^{k-1} \{T_{l_i}^2 (1 - e^{-2\alpha(l_{i+1} - l_i)}) \prod_{j=0}^{i} t_j^2\}, \tag{14}$$

R_2 is Rayleigh backscattering coefficient of two segments, which are placed around the unknown perturbation

$$R_2 = S_\alpha \prod_{j=0}^{k} t_j^2 \{T_{l_k}^2 [1 - e^{-2\alpha(l_x - l_k)}] + T_{l_x}^2 [1 - e^{-2\alpha(l_{k+1} - l_x)}] t_x^2\}, \tag{15}$$

and R_3 is power reflection coefficient for segments which are placed on the right of disturbed region plus term related to reflection from the remote-end of the test fiber

$$R_3 = t_x^2 \{S_\alpha \sum_{i=k+1}^{n} T_{l_i}^2 [1 - e^{-2\alpha(l_{i+1} - l_i)}] \prod_{j=0}^{i} t_j + T_L^2 r_2 \prod_{j=0}^{n} t_j^2\}, \tag{16}$$

where $T_{li} = \exp(-\alpha l_i)$ is the transmission coefficient of fiber segment with length l_i (see Fig.4), r_1, r_2 are the reflections coefficients from the fiber source- and remote-ends, respectively.

In the expressions (14 -16) we assign that: $l_0 \equiv 0$, $t_0 \equiv 1$ and, $l_{n+1} \equiv L$. We should emphasize that the condition $t_0 \equiv 1$ does not means that perturbation cannot appear near the source-end of the test fiber. The disturbance can affect the fiber near the source-end but it should be marked as first perturbation with transmittance t_1 at l_1= 0 distance.
The normalized transmitted coefficient of the optical system which is affected by all n+1 perturbations is defined as:

$$T_{norm}^2 = t_x^2 \prod_{j=1}^{n} t_j^2. \tag{17}$$

Note that the changes of first n perturbations are almost finished before a new one is started, so t_j = constant for j =1, n. Therefore, the total normalized transmitted power for all n+1 perturbations T^2_{norm} can change only due to the change of current perturbation, and differential $\partial(T^2_{norm})$ is:

$$\partial(T^2_{norm}) = \partial(t_x^2)\prod_{j=1}^{n} t_j^2. \qquad (18)$$

The derivative of normalized backscattering power with respect to the square of normalized transmitted power is expressed as:

$$\frac{\partial R_{norm}}{\partial(T^2_{norm})} = \frac{}{R_{\max}}[\frac{\partial R_1}{\partial(T^2_{norm})} + \frac{\partial R_2}{\partial(T^2_{norm})} + \frac{\partial R_3}{\partial(T^2_{norm})}] =$$

$$= \frac{}{R_{\max}\prod_{j=0}^{n} t_j^2}\{S_\alpha \prod_{j=0}^{k} t_j^2 [e^{-2\alpha l_x} - e^{-2\alpha l_{k+1}}] + S_\alpha \sum_{i=k+1}^{n} [e^{-2\alpha l_i} - e^{-2\alpha l_{i+1}}]\prod_{j=0}^{i} t_j + T_L^2 r_2 \prod_{j=0}^{n} t_j^2\}. \qquad (19)$$

This derivative (or slope of dependence of normalized backscattering power versus the square of normalized transmitted power) depends only on the backscattering from the plain segments which are located on the right side of unknown perturbation. The plain segments, which are located on the left of the unknown perturbation, do not affect the slope.
Let us introduce for $0 \le k \le n+1$ an auxiliary function F(k, n):

$$F(k,n) = \frac{}{R_{\max}\prod_{j=0}^{n} t_j^2}\{S_\alpha \sum_{i=k}^{n} [e^{-2\alpha l_i)} - e^{-2\alpha l_{i+1}}]\prod_{j=0}^{i} t_j + T_L^2 r_2 \prod_{j=0}^{n} t_j^2\}. \qquad (20)$$

The auxiliary function has a similar structure as the expression for the slope (see Eqn. 19) and it is decreasing with k for any n (see Fig. 5).
The contribution in the slope due to backscattering from segment [l_x, l_{k+1}] which is associated with the term [exp(-2αl_x) - exp(-2αl_{k+1})] in Eqn. (19) is less than the possible contribution due to backscattering from full segment [l_k, l_{k+1}] which is associated with term [exp(-2αl_k) - exp(-2αl_{k+1})] in Eqn. (20). Comparing Eqns. (19) and (20) we can conclude that, if the measured value of the slope for unknown perturbation satisfies the relation:

$$F(k^* + 1, n) < \frac{\partial R_{norm}}{\partial(T^2_{norm})} < F(k^*, n), \qquad (21)$$

the unknown perturbation is located between k^*-th and (k^*+1)-th initial perturbations (see Fig. 5).
Note that, if the measured slope of the unknown perturbation is equal to $F(0,n)$, the new perturbation affects the testing fiber near the source-end of the test fiber. If the slope is equal to $F(n+1,n)$ = exp(-2αL)r_2/R_{max}, the unknown disturbance is located near the remote-end of the test fiber.

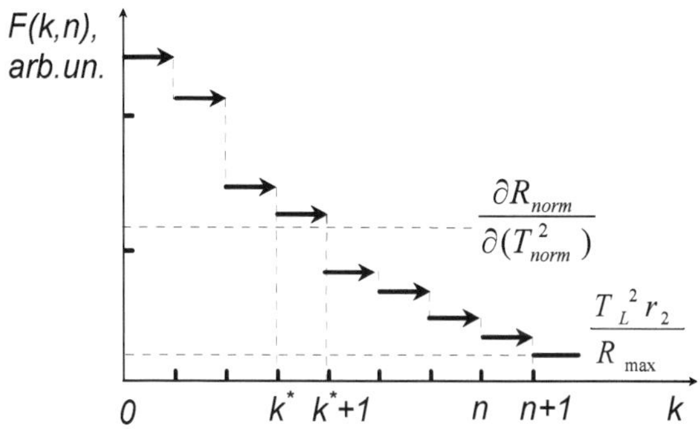

Fig. 5. Preliminary localization of (n+1)-th perturbation with auxiliary function $F(k,n)$.

Finally, the exact location of sought-for short loss segment can be found as:

$$l_x = \frac{1}{2\alpha}\ln\left\{\frac{S_\alpha}{\left[\frac{\partial R_{norm}}{\partial(T^2{}_{norm})} - F(k^*+1,n)\right]R_{\max}\prod_{j=k+1}^{n} t_j^2 + S_\alpha e^{-2\alpha l_{k^*+1}}}\right\}. \qquad (22)$$

In the previous section we have presented the method for single perturbation localization with the TRA technique. Now we have demonstrated the algorithm for localization of a new perturbation when the values and locations of all initial n perturbations are known. Therefore, according to the principle of mathematical induction we have demonstrated that the TRA method can be implemented for the localization of any number of consecutive perturbations.

Fig. 6 shows experimental dependencies of normalized Rayleigh backscattered powers versus the square of normalized transmitted powers for the bending losses consequently induced near the remote- and source-ends of test fiber. Measurements were performed as follows. Initially, the perturbation occurred near the remote-end of test fiber. The increase of the losses leads to decrease of transmitted power (line A in Fig.6). When the square of normalized transmittance decreases to the value equal to 0.241 of its initial undisturbed magnitude, we stop to increase the bending losses. Afterwards, keeping constant losses near the remote-end, we induce additional losses near the source-end of test fiber. This loading continues until the value of the square of normalized transmittance decreases to the 0.061 (line B). Then, keeping the same value of losses near the source-end, we gradually remove the losses near the remote-end of test fiber (line C in Fig. 6). Finally, by eliminating the losses near the source-end, all parameters return to their initial undisturbed values (line D).

All experimental dependencies presented in Fig.6 possess linear behavior. Experimental data show good agreement with theoretical prediction. Namely, experimental dependencies A and D practically coincide with the calculated dependencies using Eqn. 9 (see Fig. 2 and Fig. 6). Experimentally measured slopes for lines A, B, C and D which are equal to 0.109, 3.63, 0.109 and 0.955, correspondingly, agree with the values calculated using Eqns. (11) and (19) which are equal to 0.108, 3.631, 0.108 and 0.957, correspondingly.

Note that the slopes were the same for loading and unloading dependencies. The localization errors that were estimated from the difference between measured and calculated slopes do not exceed 2 meters for any location of perturbation. We should note, however, that in practice, different noise origins and system imperfections such as temporal drifts of fiber and photodetectors parameters, additional uncontrolled losses, etc. may also contribute to the decrease of accuracy. More complete analysis of localization accuracy with the TRA method will be conducted latter.

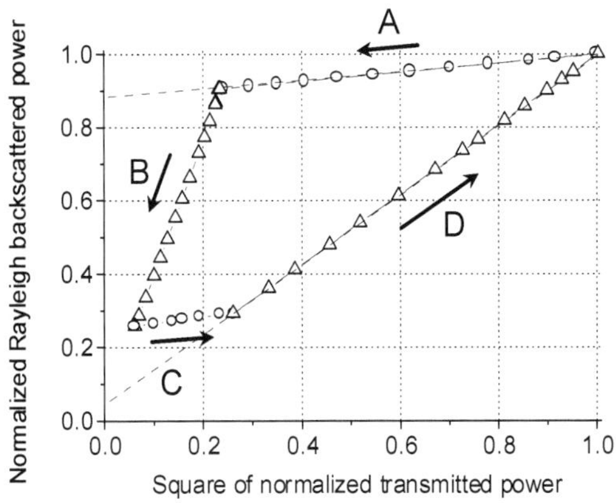

Fig. 6. Relations between normalized Rayleigh backscattered power and the square of normalized transmitted power for the bending losses consequently induced near the remote- and source-ends of test fiber.

Therefore, the TRA method can be implemented for the localization of any number of consecutive perturbations which occur one after another along the test fiber during the monitoring period. In contrast to the OTDR the proposed method cannot be used for the localization of the perturbations in an already installed fiber-optical system that already has many induced losses. The other natural question to be addressed about the TRA method is: what happens if two perturbations affect the test fiber simultaneously? To answer this question, consider calculated dependence of normalized Rayleigh backscattered power versus the square of normalized transmitted power for two equal perturbations which induce the losses near the source- and remote-ends at same time (curve A+D in Fig.7).

The dependence exhibits clear nonlinear behavior. As was shown above for any number of consecutive perturbations this dependence should be linear. Fig. 7 also shows normalized Rayleigh backscattered power versus the square of normalized transmitted power for the perturbations that affect the testing fiber one after another near the remote (line A) and source (line D) ends. Both last dependencies exhibit clear linear behavior. The nonlinear behavior of dependencies of normalized Rayleigh backscattered power versus the square of normalized transmitted power indicates that testing fiber is affected by two or more perturbations simultaneously.

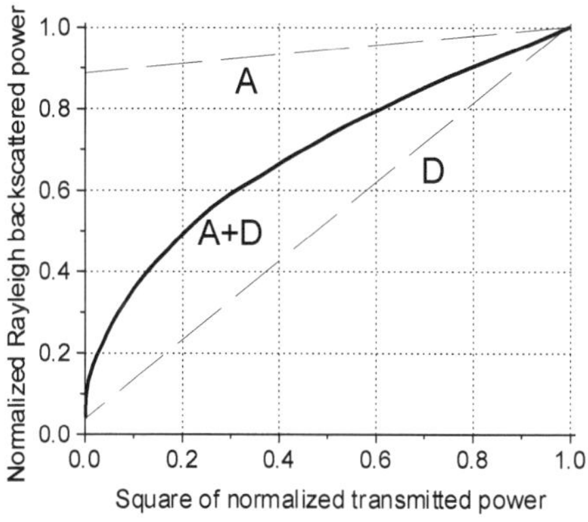

Fig. 7. Relations between normalized Rayleigh backscattered power and the square of normalized transmitted power for two perturbations synchronously (A+D) and independently (A and D) induced near the remote- and source-ends of test fiber.

Using the particular dependence that is shown in Fig. 7 (curve A+D), it is possible to localize at least one perturbation. Indeed, the value of normalized Rayleigh backscattered power at the point when normalized transmitted power is equal to 0 directly shows the location of nearest to the source-end perturbation. Therefore, analyzing the curve A+D, we can conclude that two perturbations affect the test fiber simultaneously and one of the perturbations induces the losses near the source-end. Nevertheless, the complete analysis of different scenarios, even for two synchronous events, is noticeably complicated and is beyond the scope of this paper.

4. Localization Accuracy with TRA Method

The accuracy of excess loss localization with TRA method strongly depends on the value of the induced loss. With TRA method it is easy to localize strong perturbation, but the localization of weak perturbation requires higher accuracy of the transmitted and Rayleigh backscattered powers measurements.

Indeed, normally we measure normalized power transmission and reflection coefficients with some errors. So, the measurements are positioned mainly inside an area: $T_{norm} \pm \sigma_{T_{norm}}$ and $R_{norm} \pm \sigma_{R_{norm}}$, where $\sigma_{R_{norm}}$ and $\sigma_{T_{norm}}$ are deviations of normalized reflected and transmitted powers, respectively. Let us characterize this measurement area in the space T^2_{norm} - R_{norm} by a measurement-rectangular with sides equal to $2\sigma_{T^2_{norm}}$ and 2 $\sigma_{R_{norm}}$, where $\sigma_{T^2_{norm}}$ is deviation of square of normalized transmitted power (see Fig.8). The size of the rectangular indicates the accuracy of the measurements. Smaller measurement-rectangular corresponds to higher measurement accuracy. Fig.8 also repeat dependences already presented in Fig. 2 for the losses induced at different locations with interval Δl = 284.4 m.

Every line which is intersecting the measurement-rectangular corresponds to one of the possible local position of the perturbation. Therefore the localization error depends on the number of the intersecting lines. As we can see for very week losses nearly all lines cross the measurement-rectangular that means that we need significantly higher accuracy of the measurements for the correct localization. For the strong losses the localization error reaches its minimum with the TRA method. In contrast to this, the accuracy of localization of loss with the standard OTDR technique mainly depends on the duration of the optical test pulse and is practically independent on the value of loss.

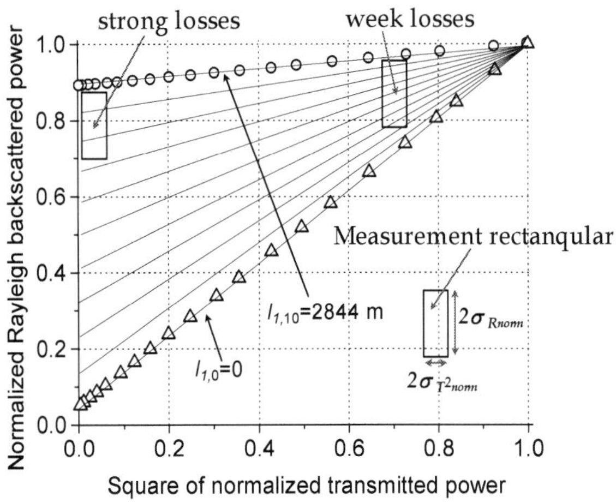

Fig. 8. Localization accuracy with TRA method.

Fig. 9. shows the relative localization error calculated geometrically using the data, which are presented in Fig.8.

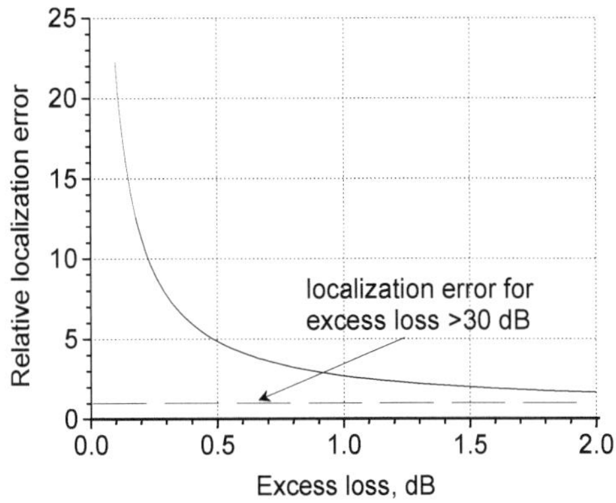

Fig. 9. Relative localization error versus excess loss for the TRA method.

Relative localization error was determined as localization error divided by their value for the loss that decreased the fiber transmission by more than 30 dB. As we can see the localization error for weak loss significantly exceeds the localization error for strong one. If the localization errors for very strong perturbation are equal to ±1meter the localization errors for 1 and 0.1 dB induced loss are equal to ±3 meters and ±22 meters, respectively.
The measurement range and maximum allowable loss for the TRA and OTDR methods are likely to be the same, because both methods measure the Rayleigh backscattering power and the maximum range for both methods is namely restricted by the attenuation of test fiber.
Now, let us examine some special features of the localization accuracy with the TRA method analytically. Firstly let us demonstrate that if the influence of coherent effects is negligibly small, the TRA method can be used for any arbitrary distribution of the reflectivity along the test fiber.
Indeed, let us consider a fiber with total length *L*, initial transmission $T_{max}= T\ (L)$ and an arbitrary distribution of power reflectivity *R(z)* along the fiber. The reflectivity *R(z)* is monotonically non-decreasing function in interval [0, *L*] if we neglect any coherent effects. The reflectivity *R(z)* can be measured as the dependence of the reflected power versus the distance *z* where very strong loss ($t_{norm} \cong 0$) is induced. A derivative $\partial R(z)/\partial z$ can be interpreted as a differential reflectivity that also has an arbitrary distribution along the test fiber. Reflectivity inside the fiber can be induced by Raleigh backscattering or any other way including imprinting of Bragg gratings inside the fiber.
Taking into account the reflection from the ends of the sensing fiber, total power reflection coefficient of the initially undisturbed optical system can be expressed as:

$$R_{\max} = r_1 + R(L) + T^2(L) r_2, \tag{23}$$

where *R(L) is* power reflection coefficient of all fiber.

Let a short fiber piece located at a distance $l_1 \leq L$ induce a loss under influence of a monitored condition. Neglecting coherent effects, multiple scattering and re-reflection in both directions due to its relative weakness, and supposing that no additional reflection comes from loss region the normalized power reflection coefficient R_{norm} of the optical system with a single disturbance located at distance l_1 can be written as:

$$R_{norm} = \frac{r_1 + R(l_1) + [R(L) - R(l_1)] T^2(l_1) T_{norm}^2 + T^2(L) T_{norm}^2 r_2}{r_1 + R(L) + T^2(L) r_2}, \tag{24}$$

where T_{norm} is the normalized power transmission coefficient of the optical system with excess loss, $T(l_1)$ and $R(l_1)$ are transmission and power reflection coefficient of the fiber segment with length l_1, respectively.

Measurands T_{norm} and R_{norm} are normalized by their initial undisturbed values T_{max} and R_{max}, respectively. As follow from Eqn. (24) the dependence between R_{norm} and T^2_{norm} is linear with unique slopes for different locations of the perturbation l_1 along the test fiber.

Therefore, with the TRA method the single loss perturbation can be localized by measuring the slope $\partial R_{norm} / \partial (T^2_{norm})$ for any arbitrary distribution of the reflectivity $R(z)$ along the test fiber.
Neglecting Fresnel reflections from both fiber ends ($r_1 = r_2 = 0$) and losses in undisturbed fiber ($T(l_1)=1$) we can express normalized power reflection coefficient of the segment with length 1_1 from Eqn. (24) as:

$$R_N(l_1) = \frac{R(l_1)}{R(L)} = \frac{R_{norm} - T^2_{norm}}{1 - T^2_{norm}}. \tag{25}$$

As a result, the location of the perturbation can be written as:

$$l_1 = R_N^{-1}\left(\frac{R_{norm} - T^2_{norm}}{1 - T^2_{norm}}\right), \tag{26}$$

where R_N^{-1} is inverse function.

In order to find the loss locacion l_1 from Eqn. (26) we needs to know the function R_N^{-1}, but even without this knovelege we can estimate some features of the localization accuracy. Indeed, for independently measured normalized reflected and transmitted powers, the standard deviation of the disturbance location σ_{l1} can be estimated as:

$$\sigma^2_{l_1} = \left(\frac{\partial R_N^{-1}}{\partial R_{norm}}\right)^2 \sigma^2_{R_{norm}} + \left(\frac{\partial R_N^{-1}}{\partial T_{norm}}\right)^2 \sigma^2_{T_{norm}}, \tag{27}$$

where $\sigma_{R_{norm}}$ and $\sigma_{T_{norm}}$ are standard deviations of normalized reflected and transmitted powers, respectively.

Here we also assume that errors for T_{max} and R_{max} were significantly less than ones for T_{norm} and R_{norm}, because an averaging time for an initial measurements can significantly exceeds the averaging time at normal-mode regime.
Substituting expression (26) in Eqn. (27) we can obtain:

$$\sigma_{l1} = \frac{1}{(1 - T^2_{norm}) \left.\frac{\partial R_N(z)}{\partial z}\right|_{z=l_1}} \sqrt{\sigma^2_{R_{norm}} + \left(\frac{2T_{norm}[R_{norm} - 1]}{1 - T^2_{norm}}\right)^2 \sigma^2_{T_{norm}}}. \tag{28}$$

As it follow from Eqn. (28) the accuracy of localization with the TRA method depends on normalized differential reflectivity $\partial R_N(z)/\partial z$ at the point of measurement l_1. Larger differential reflectivity corresponds to highter accuracy of loss localization.

Therefore, with the TRA method we can achieve variable accuracy distribution along the test fiber with the proper profile of the normalized differential reflectivity.

Enhanced localization accuracy inside several critical intervals can be provided, for example, by imprinting of very week Bragg gratings inside these intervals (Spirin et al., 2004).

Note that the total resolution of the TRA sensor also depends on the uniformity of the waveguide, scattering and loss properties of the test fiber. However, the precise value of the reflectivity $R(z)$ can be directly measured by inducing strong losses at exactly-known locations along the test fiber or, probably, by precise OTDR measurements.

Rayleigh backscattering phenomena meets the condition discussed above because the Rayleigh backscattered power coefficient $R(l_1)$ is independent of the coherent properties of the incident light (Gysel & Staubli, 1990; Liaw et al., 2000):

$$R(l_1) = S_\alpha[1-\exp(-2\alpha l_1)]. \tag{29}$$

Let us estimate the localization errors for the Rayleigh backscattered fiber numerically. Neglecting Fresnel reflections from both fiber ends ($r_1 = r_2 = 0$) we can express the location of the perturbation from Eqn. (10) as:

$$l_1(R_{norm}, T_{norm}) = \frac{1}{2\alpha}\ln\left\{\frac{1-T_{norm}^2}{1-R_{norm}+R_{norm}e^{-2\alpha L}-T_{norm}^2 e^{-2\alpha L}}\right\}. \tag{30}$$

The perturbation location depends on two measurands T_{norm} and R_{norm} which can be defined experimentally with some errors. For independently measured reflected and transmitted powers the error in location can be found directly from Eqn. (30) as:

$$\sigma^2_{l_1} = \left(\frac{\partial l_1}{\partial R_{norm}}\right)^2 \sigma^2_{R_{norm}} + \left(\frac{\partial l_1}{\partial T_{norm}}\right)^2 \sigma^2_{T_{norm}}. \tag{31}$$

Fig.10. shows the calculated with Eqn. (31) localization error versus transmission for two different local positions of the perturbation along 10 km-length test fiber. In the simulation we used standard deviations of the reflected $\sigma_{R_{norm}}$ and transmitted $\sigma_{T_{norm}}$ powers equal to 0.005 and 0.0005, respectively. As we can see the location error depends on the distance where the perturbation takes place and strongly depends on the value of the induced loss. Once again, the localization of a weak perturbation with TRA method requires more accurate measurements of the transmitted and Rayleigh backscattered powers. More precise measurements can be provided by increasing the averaging time, especially for the reflected power because typically the Rayleigh backscattered power is significantly smaller and noisiness than the transmitted one.

Consequently, the TRA method is well adjusted on detection and localization of single strong enough perturbation. However we have also demonstrated that the TRA method provides wide range of localization accuracy for weak and strong loss-inducing perturbations. It was shown that the required accuracy distribution along the test fiber can be achieved with the appropriate profile of the normalized differential reflectivity $\partial R_N(z)/\partial z$.

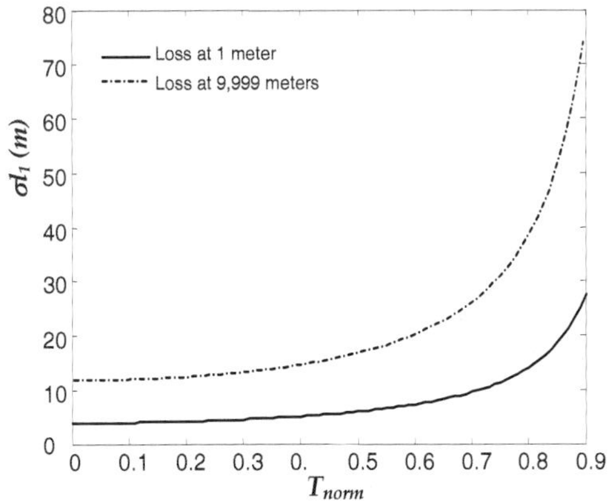

Fig. 10. Localization error versus transmission for 10 km-length test fiber.

5. Autonomous Measurement System

The section describes first completely autonomous measurement system based on transmission-reflection analysis (AMS-TRA). The autonomous system utilizes simple optical scheme with low-cost 2 mW Fabry-Perot (FP) diode laser and original data acquisition and processing system (Spirin et al., 2007). The schematic diagram of AMS-TRA prototype is shown in Fig. 11. CW light emitted by a FP laser operating in few longitudinal modes at 1550 nm with a total linewidth of few nm was launched into a standard single mode SMF-28 test fiber through 3 dB coupler.

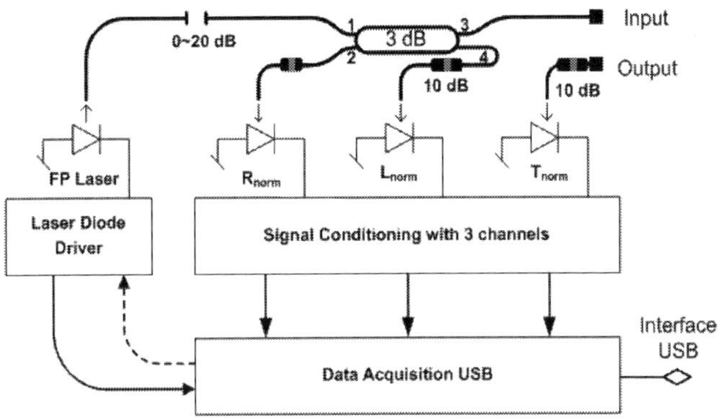

Fig. 11. Schematic diagram of fiber-optic sensor

The launched optical power was about 1 mW, and the measured with OTDR attenuation coefficient of the test fiber was equal to 0.19 dB/km. The angled physical contact (APC) connectors were used to cancel back reflections influence on the FP laser.

Three power detectors with specially designed preamplifiers and filters were used to measure the transmitted (T), Rayleigh backscattered (R), and laser (L) powers.
Fig.12 shows the overview of AMS-TRA system with short peace of sensing cable for hydrocarbon leak detection. After digitization data transferred to PC via USB interface. The data acquisition, signal processing, and the indication of the event location on the map is done by specially designed software.

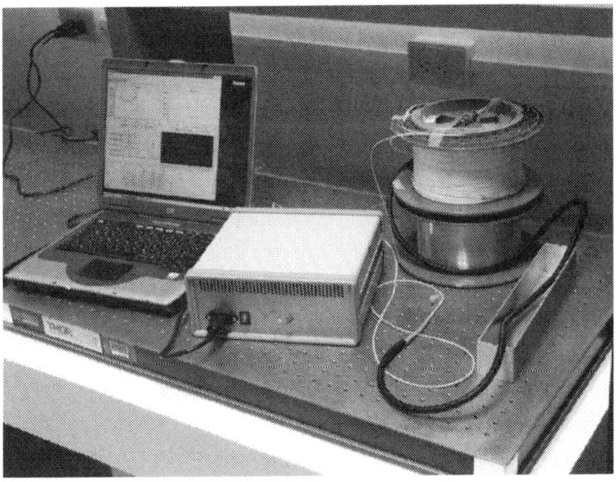

Fig. 12. AMS-TRA prototype.

5.1 Thermal Stability

As we found, considerable problem with our first AMS-TRA prototype version is associated with ambient temperature influence. Fig.13 shows the variation of the measurands T_{norm}, R_{norm} and L_{norm} during 14 hours.

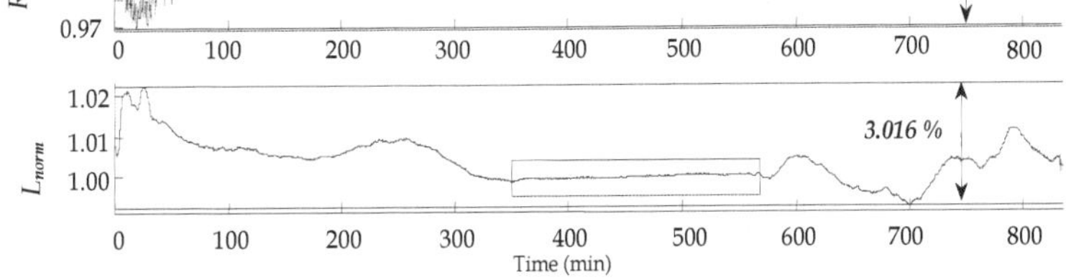

Fig. 13. Measurands variations during 14 hours

Here L_{norm} is directly measured FP laser power (see Fig.11) normalized on its initial value. Smallest variations of the measured values were recorded during the night when the temperature in the room was stable (see Fig. 13).

Significant measurands variations were also recorded every time after AMS-TRA is turn on. Fig.14 shows measurands variations from the moment when an electrical power was switch on. The losses were induced near remote-end of 1.3 km-length test fiber. The measurands became stable only after 1 hour (see Fig.14) when the temperature inside the prototype box stabilized.

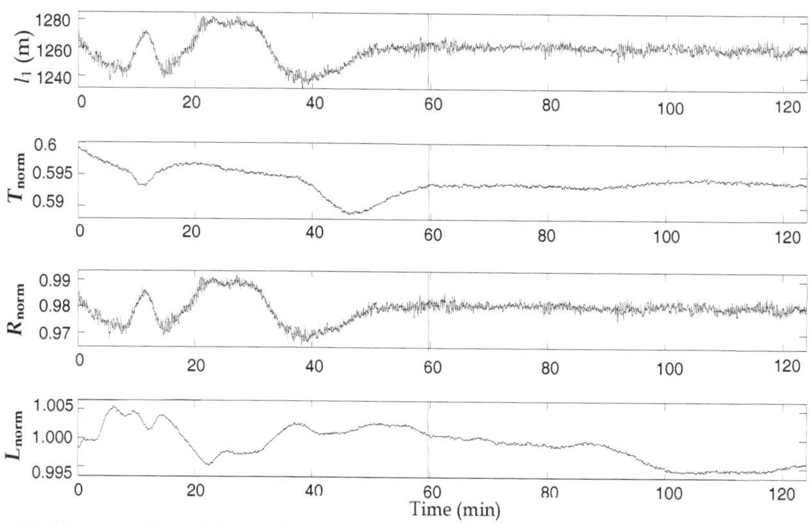

Fig. 14. Measurands variations after power is switch on.

In order to get most temperature-sensitive points we heat different optical and electrical parts of the prototype on 10° C. As we found the most temperature-critical elements are standard optical connectors. However it is possible to eliminate the problem by using special temperature insensitive connectors or simply splicing all fibers.

5.2 Localization Error Measurements

To estimate the accuracy of the perturbation localization with the AMS-TRA prototype, we induced gradually increasing and decreasing perturbations near source- and remote-end of the 1.3 km-length SMF-28 test fiber. Fig.15 shows variation of perturbation location l_1 and other measurands with change the losses near the source-end of the test fiber. For the strong losses that surpass 3dB the variation of l_1 about ±2 meters was recorded.

Fig.16 shows calculated and measured localization errors for 1.3 km-length test fiber versus normalized transmitted power. In the simulation we used experimentally measured standard deviations of the reflected and transmitted powers equal to 0.0015 and 0.0003, respectively. As we can see measured errors for the strong perturbations do not exceed ±5 m for the losses induced near remote-end of the test fiber.

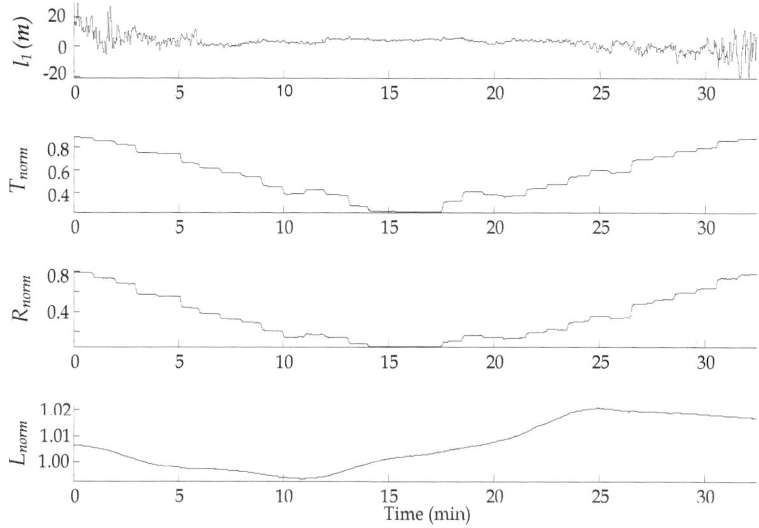

Fig. 15. Measurands variations for losses gradually induced near source-end of 1.3 km-length test fiber.

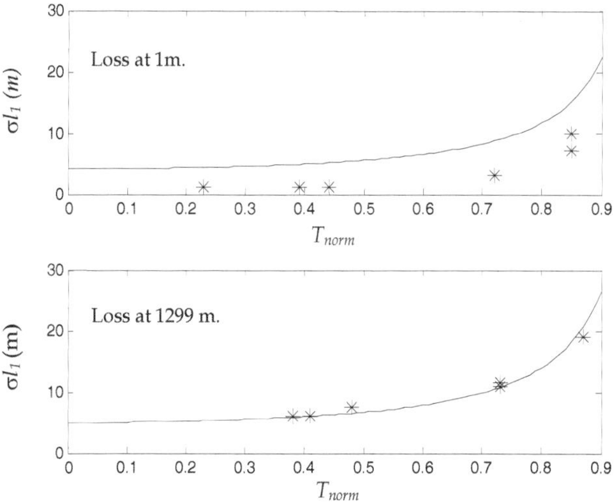

Fig. 16. Calculated (solid lines, $\sigma_{R_{norm}}$ = 0.0015 and $\sigma_{T_{norm}}$ = 0.0003) and measured (stars) localization error for 1.3 km-length test fiber.

Qualitatively the same results were obtained with 5.6 km-length SMF-28 test fiber. Fig.17 shows recorded with AMS-TRA measurands for the gradual change of the loss value. The losses were induced near the remote-end of 5.6 km-length test fiber. Note that in this case normalized Rayleigh backscattered power does not depend on value of losses.

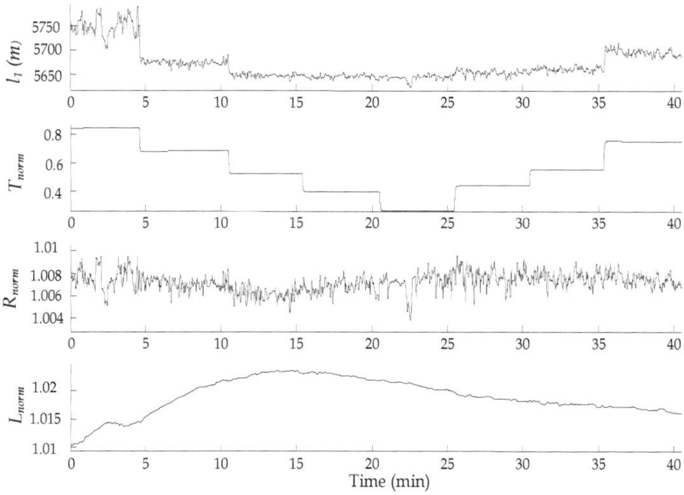

Fig. 17. Localization errors for losses induced near remote-end of 5.6 km-length test fiber.

Fig. 18 shows calculated and measured localization errors for 5.6 km-length test fiber.

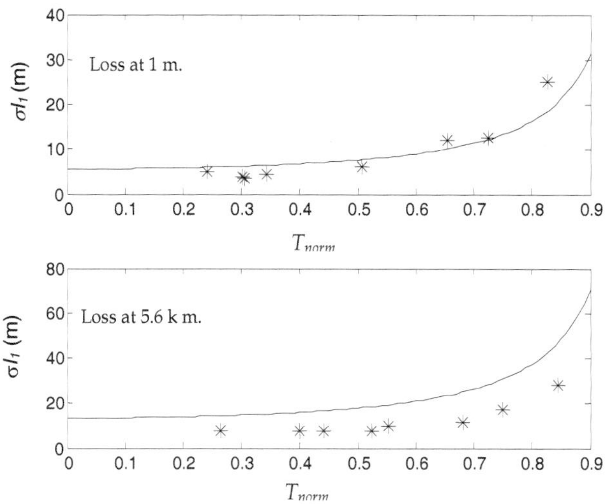

Fig. 18. Calculated (solid lines) and measured (stars) localization error for 5.6 km test fiber.

Variations of the disturbance location l_1 for the strong perturbations do not exceed ±10 meters for the losses induced near remote-end of the test fiber. Experimentally measured standard deviations of the normalized reflected and transmitted powers due to various instability in the prototype were σ_{Rnorm} = 0.0025 and σ_{Tnorm} = 0.0003 for signal bandwidth equal to 1 Hz. These data were used for the numerical estimation of the localization errors.

5.3 Gasoline Leak Detection and Localization

As well known gasoline does not affect optical fiber directly. In order to induce the losses under gasoline influence, special transducer with swellable polymer should be used. The design of the test cable with gasoline- sensitive transducer is shown schematically in Fig.19.

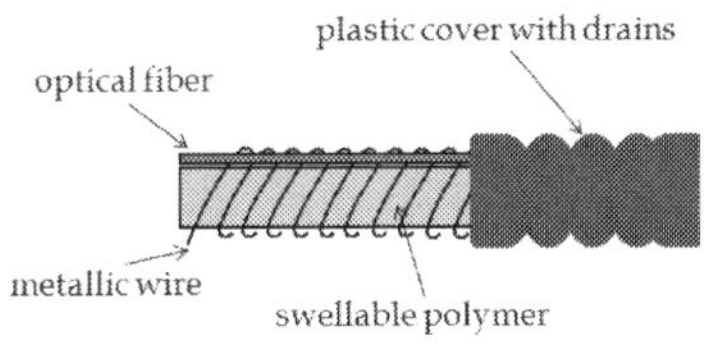

Fig. 19. Fiber optic gasoline-sensitive cable in detail.

The optical fiber and swellable polymer were coupled together mechanically by soft metallic wire winded around them. The fiber was placed inside small groove in the polymer that allows avoid undesirable inoperative losses, but does not significantly increase the response time of the sensor. A white butyl rubber was used as the swellable polymer in the sensitive cable. This material has good aging properties at elevated temperatures, good chemical stability. It also resists weathering, sunlight, ozone, mineral acids, oxygenated solvents (ketones and alcohols), and water absorption (Allen, 1972; Morton, 1987). Butyl rubber absorbs and swells in hydrocarbon media without dissolution and significant change of its mechanical property. The swelling behavior of the polymer produces an increase in the physical dimensions of the material, which can be readily converted to a mechanical response. Our study shows that butyl rubber can increase its volume by more than 2 times under gasoline influence (Spirin et al., 2000; Lopez et al., 2004). Furthermore, its behavior is reversible, i.e. the white butyl rubber can be used over multiple wet-dry cycles. In our experiment we tested the sensor under gasoline influence, but qualitatively the same swelling behavior butyl rubber shows in a range of hydrocarbon fuels (MacLean et al., 2001). The swellable polymer has tube shape with 25 mm diameter and metallic wire was winded with 5 mm spacing [Lόpez, 2002]. Fiber and polymer are protected by an outer plastic cover with drains for gasoline.

To study sensor response, part of sensitive cable located at 1300 meter from the source-end of the test fiber was placed inside a 20-cm length vessel filled with gasoline. Under gasoline influence butyl rubber swells and bends the fiber that induces the losses into the test fiber. Fig. 20. shows the change of transmission of the test fiber during wet-dry cycle. The decrease of transmission indicates gasoline presence and allows estimate an integral value of the disturbance.

Fig. 21. presents AMS-TRA prototype response on gasoline leak at 1300 meter distance. Under gasoline influence the transmission T_{norm} is decreasing and alarm signal is activated when transmission is became less than a threshold level equal to 0.9. With this particular sensitive cable the alarm level was achieved after 19 minutes of gasoline influence. Initial variation of alarm distance l_1 was ±32 meters for 0.5 dB losses and then decreasing to ±13 meters for 2 dB losses. For the losses more than 3 dB the variations not exceeds ±3 meters.

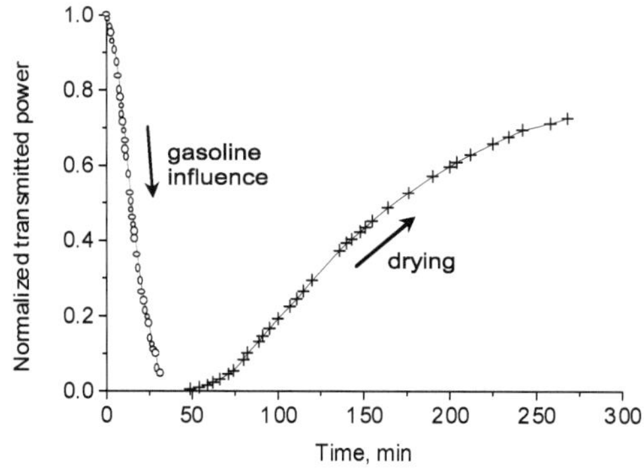

Fig. 20. Normalized transmitted power during wet-dry cycle (o - gasoline influence, + - drying).

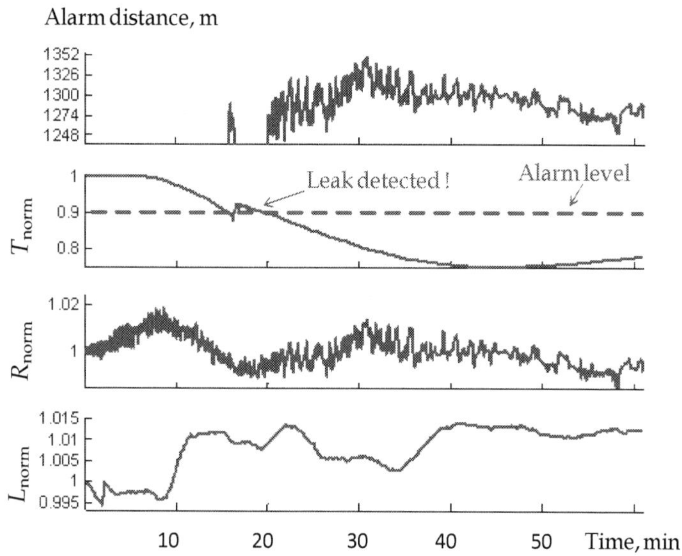

Fig. 21. AMS-TRA response on gasoline leak at 1300 meter.

6. Conclusion

We have demonstrated that the TRA method provides new opportunity for the localization of loss-induced alarm-like perturbation along few km-length test fibers by measuring transmitted and reflected powers of an unmodulated light.

We have designed completely autonomous measurement system based on transmission-reflection analysis. The autonomous system utilizes simple optical scheme with low-cost diode laser and original data acquisition and signal processing system. The AMS-TRA prototype demonstrates localization of loss-inducing perturbation along a 1÷6 km-length standard single mode SMF-28 fiber with few meters error for the losses that surpass 3 dB.

We have employed the AMS-TRA prototype with special sensitive cable for the gasoline leak localization. Alarm signal was activated after 15÷20 minutes of gasoline influence on 20 cm-length piece of sensitive cable which is located at the end of 1.3 km-length test fiber. For the losses more than 3 dB the variations of the alarm distance not exceeds ±3 meters.

We believe the proposed technique will be very attractive for the eventual realization of a compact and inexpensive distributed alarm fiber optic sensor.

7. Acknowledgement

We gratefully acknowledge financial support from the Consejo Nacional de Ciencia y Tecnología de México through Project 23770 SEMARNAT.

8. References

Hartog, A. (2000). Distributed fiber-optic Sensors: Principles and applications, In *Optical Fiber Sensor Technology: Advanced Applications-Bragg Gratings and Distributed Sensors,* edited by K.T.V. Grattan and B.T. Meggitt, pp. 241-302, Kluwer Academic Publishers, ISBN 0-7923-7946-2, Boston

Byoungho, Lee (2003). Review of the present status of optical fiber sensors. *Optical Fiber Technology,* Vol. 9, 2003, pp. 57–79, ISSN 1068-5200

Tsuji, K.; Shimizu, K.; Horiguchi, T. & Koyamada, Y. (1995). Coherent optical frequency domain reflectometry for long single-mode optical fiber using a coherent lightwave source and an external phase modulator, *IEEE Photonics Technol. Letters,* Vol. 7, 1995, pp. 804–806, ISSN 1041-1135

Pierce, S.G.; MacLean, A. & Culshaw, B. (2000). Optical frequency-domain reflectometry for microbend sensor demodulation. *Applied Optics,* Vol. 39, 2000, pp.4569-4581, ISSN 0003-6935

Venkatesh, S. & Dolfi, D.W. (1990). Incoherent frequency modulated cw optical reflectometry with centimeter resolution. *Applied Optics,* Vol. 29, 1990, pp. 1323-1326, ISSN 0003-6935

Spirin, V. V.; Shlyagin, M. G.; Miridonov S. V. & Swart, Pieter L. (2002a). Transmission/reflection analysis for distributed optical fibre loss sensor interrogation. *Electronics Letters,* Vol. 38, No. 3, 2002, pp. 117-118, ISSN 0013-5194

Spirin, V. V.; Shlyagin, M. G.; Miridonov, S. V. & Swart, Pieter L. (2002b). Alarm-condition detection and localization using Rayleigh scattering for a fiber optic bending sensor with an unmodulated light source. *Optics communications,* Vol. 205, No. 1-3, 2002, pp. 37-41, ISSN 0030-4018

Gysel, P. & Staubli, R.H. (1990). Statistical Properties of Rayleigh Backscattering in Single-Mode Fibers", *J. Lightwave Technology* Vol. 8, 1990, pp. 561-567, ISSN: 0733-8724

Liaw, S.-K. ; Tzeng, S.-L. & Hung Y.-J. (2000). Rayleigh backscattering induced power penalty on bidirectional wavelength-reuse fiber systems, *Optics Communications,* Vol. 188, 2000, pp. 63- 67, ISSN 0030-4018

Brinkmeyer, E. (1980). Backscattering in single-mode fibers, *Electronics Letters, Vol.* 16, 1980, pp. 329- 330, ISSN 0013-5194

Beller, J. (1998). OTDRs and Backscatter Measurements, In: *Fiber Optic Test and Measurement,* edited by D. Derickson, pp. 434-474, Prentice Hall PTR, ISBN 0-13-534330-5, New Jersey

Spirin, Vasilii V. (2003) Transmission-Reflection Analysis for Localization of Temporally Successive Multipoint Perturbations in a Distributed Fiber-Optic Loss Sensor Based on Rayleigh Backscattering. *Applied Optics,* Vol. 42, No. 7, 2003, pp. 1175-1181, ISSN 0003-6935

Spirin, V. V.; Mendieta, F.J.; Miridonov, S. V.; Shlyagin, M. G.; Chtcherbakov, Anatoli A. & Swart, Pieter L. (2004) Localization of a loss-inducing perturbation with variable accuracy along a test fiber using transmission-reflection analysis, *IEEE Photonic Technology Letters.,* Vol. 16, 2004, pp. 569-571, ISSN 1041-1135

Spirin, Vasily V.; Miridonov, Serguei V.; Mitrani, Enrique; Morales, Carlos; Shlyagin, Mikhail G.; Castro, Marcial & Manriquez, Manuel. (2007). Autonomous measurement system for localization of loss-induced perturbation based on transmission-reflection analysis, In: Fiber Optic Sensors and Applications V, Proceedings of SPIE Volume: 6770, pp. 6770OH 1-9 ISBN 9780819469304

Allen, P. W. (1972). *Natural Rubber and the Synthetics,* 255 pages, John Wiley & Sons, ISBN: 0470023295, New York

Morton, M. (1999) *Rubber Technology,* 638 pages, Kluwer Academic Publishers, ISBN 0-412-53950-0, Netherlands

Spirin, V.V.; Shlyagin, M.G.; Miridonov, S. V. ; Mendieta, F. J. & López R. M. (2000) Fiber Bragg grating sensor for petroleum hydrocarbon leak detection, *Optics and Lasers in Engineering,* Vol. 32, 2000, pp. 497-503, ISSN 0143-8166

López, R.M. ; Spirin, V. V. ; Shlyagin, M.G. ; Miridonov, S.V. ; Beltrán, G. ;.Kuzin, E.A & Márquez Lucero, A. (2004). Coherent optical frequency domain reflectometry for interrogation of bend-based fiber optic hydrocarbon sensors. *Optical Fiber Technology,* Vol. 28, No. 10, 2004, pp. 79-90, ISSN 1068-5200

MacLean, A.; Moran, C.; Johnstone, W.; Culshaw, B.; Marsh, D. & G. Andrews. (2001). A distributed Fiber Optic Sensor for Liquid Hydrocarbon Detection, In: Smart Structures and Materials 2001: Sensory Phenomena and Measurement Instrumentation for Smart Structures and Materials, Proceedings of SPIE Vol. 4328,2001, ISBN 9780819440143

López, R. M.; Spirin, V.V.; Miridonov, S. V.; Shlyagin, M. G.; Beltrán, G. & Kuzin E.A. (2002). Fiber optic distributed sensor for hydrocarbon leak localization based on transmission/reflection measurement. *Optics&LaserTechnology,* Vol. 34, No. 6, 2002, pp. 465-469, ISSN 0030-3992

5

Radiation Transmission-based Thickness Measurement Systems - Theory and Applications to Flat Rolled Strip Products

Mark E. Zipf
Tenova-I2S (I2S, LLC)
United States of America

1. Introduction

Precise, accurate measurement of strip / sheet thickness is critical in the controlled processing and quality assessment of flat rolled metal products. Through the years, many methods (both contact and non-contact) have been developed, each having specific, relevant applications, and certain characterizable advantages and disadvantages. These systems are provided in a variety of geometries and physical arrangements, and seemingly endless collections of functions and features.

One particular non-contact method employs an understanding of a material's reaction to incident radiation (primarily the photonic / gamma form - although electron / beta radiation can also be considered) in a transmission mode framework. Here, semi-collimated, high energy radiation (a photon beam of a given spectral content) is directed perpendicular to one surface of the flat strip material. Depending on the energy level, the incident radiation interacts with the material's atomic structures and is either passed, absorbed, scattered or involved in high energy pair productions. The resulting transmitted radiation appears as a dispersed beam pattern, having attenuated intensity and modified spectral content. A portion of the exiting radiation is collected by detection instruments which render a signal functionally related to the integral of the received radiation intensity over the detector's spectral bandwidth. Knowledge of the radiation source's intensity and spectral content, the material chemistry and the detector's response characteristics are needed to process the signals and render a thickness measurement.

Typically, the plane of the strip is oriented horizontally with the source and detector mounted above and below the stip. There are a number of different configurations ranging from stationary / physically-fixed source and detector arrangements above and below the strip, to C-Frame / O-Frame mounted configurations that (in some cases) allow for transverse strip thickness profiling, to multi-source / multi-detector arrangements that provide instantaneous measures of the strip profile. Regardless of the physical configuration, the fundamental physics applies.

This chapter is the first of a two-part discussion concerning the nature of radiation transmission-based strip thickness measurement. The intent of this chapter is to examine the underlying physics and methods of this approach, and functions as a tutorial supporting subsequent discussions (Zipf, 2010). Natural and artificial radiation sources are presented and discussed, along with the various means of containing and directing the emitted radiation. The nature of the material's interaction with radiation is analyzed and considered in the presence of possibly complex material chemistries. Detection system sensors and instrumentation are studied and examined with respect to their associated signal processing components and methods of rendering a thickness measurement. Calibration and standardization methods are introduced and combined with the various methods of resolving the material thickness from active measurements. Special functions and features (e.g., transverse profiling, strip quality control, etc.) are discussed and assessed. Classical system architectures and component organizations are presented and considered with respect to typical applications and system implementations.

2. Fundamentals of Non-Contact Radiation Attenuation Gauging

Radiation attenuation gauging involves the measurement of the thickness of a flat sheet of known (or calibrated) material composition, through the assessment of the degree of transmitted attenuation experienced by a beam of high energy ionizing radiation directed perpendicular to the planar surface of the material (I2S, 1992). Figure 2.1 provides a simplified diagram showing the primary concepts and components associated with this approach.

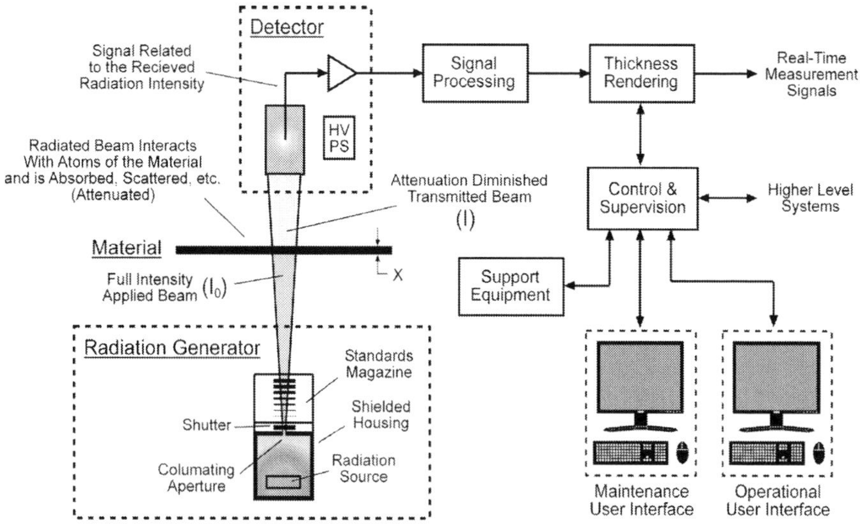

Fig. 2.1 - Diagram illustrating the primary concepts and components associated with transmission-based radiation attenuation gauging.

2.1 General System Objectives and Requirements

Objectives

- Provide a sufficiently accurate, precise, instrumented signal, representative of the measured strip thickness (with accuracy and repeatability < 0.1% of nominal).

Requirements

- Must be "continuous" in nature (not a spot measurement)
- Must measure over a reasonably small surface area (~25mm diameter circle)
- Must provide measurements while the strip is moving (at speed up to 1500 mpm)
- Must be fast responding (5-20msec)
- Must be independent-of or compensate-for alloy variances
- Must compensate for changes in ambient conditions
- Must be highly immune to noise and external interference
- Must not damage the strip surface
- Must provide flexible, multi-facetted interfacing to other systems
- Must provide intuitive, interactive user interfaces (both operational & maintenance)

Desirable Traits

- Measurement of the transverse strip profile
- Insensitive to strip shape / flatness, pass-line height, debris, oil & coolant films, etc.
- Employ Commercial-Off-The-Shelf (COTS) Technologics
- Safe for operational & maintenance personnel
- Physically / mechanically robust
- Real-Time, Interactive, Graphical User Interfaces (GUIs)
- Adaptable, scalable system arrangements & platforms
- Remote Accessibility

2.2 Primary Components

Radiation Generator - This device emits a directed beam of high energy ionizing radiation (of known intensity, I_0 (in photons/sec), and spectral characteristics) and provides radiation containment.

Shielded Housing - This vessel typically consists of a shielded, structural housing containing the associated holders and mounts required to locate and orient the radiation source. The housing may contain dielectric oil immersed components and be supported by an external heat exchanging / cooling system.

Radiation Source - This component generates the radiation that will be applied for measurement. The source may be either natural (radioactive isotope) or artificial (X-Ray tube), and may project a radiation pattern that is sensitive to alignment with the housing aperture.

Collimating Aperture - Radiation is emitted from the housing chamber through a sealed aperture in the form of a beam having a specific, semi-collimated optical geometry needed to support the form and geometry of the application and detector. The aperture is sealed to reduce the infusion of external contaminants and / or the escape of any (possibly pressurized) internally contained dielectric oil.

Shutter - This device provides a means of cutting off the radiation beam, making the radiation generator safe for handling and operations in the proximity.

Standards Magazine - This device contains a group of precision (often NIST traceable) samples that can be introduced into the radiation beam (individually or in groups) to provide a means of measuring the emitted beam's intensity and spectral content for calibration and standardization purposes. (Howard, 1970)

Material Under Measurement - For the purposes of this discussion, we will be considering flat rolled, sheet / strip products, composed of various metals (e.g., steel, aluminum, and copper / brass alloys, etc.) whose width is much larger than the nominal thickness. The strip may be stationary or moving at speeds exceeding 1500 meters/minute.

Detection System - Transmitted / scattered radiation, I (in photons/sec), that results from the incident radiation, I_0, penetrating the strip, is collected and measured by this device, which is typically located above the strip and aligned to the optical axis of the radiated beam. The radiation generator's collimator and detector aperture are sized to provide the detector an optical over-containment of the transmitted beam.

Detector - Collected incident radiation is converted to an electrical signal that is functionally related to the radiation intensity. Ion chambers and scintillation crystal / photomultipliers are often employed (Moore & Coplan, 1983).

High Voltage Power Supply - Detector sensitivity (gain) is related to the applied potential. A high voltage power supply provides the detector potential with sufficient current capacity to provide the necessary charge recovery.

Preamplifier - The feeble detector signal is amplified to usable amplitudes by a high gain, low noise electrometer / transconductance amplifier (Motchenbacher & Fitchen, 1973). To reduce signal noise and interference, it is desirable to place the preamplifier as close as possible to the detector and mounted in a shielded, hermetically sealed enclosure.

Signal Processing - The amplified detector signal requires wide bandwidth signal processing (in both time and amplitude) to render a calibrated measurement of the intensity of the received radiation (i.e., related to material absorption / attenuation). This processing can be provided by discrete electronics and instrumentation, real-time digital signal processors or Field Programmable Gate Arrays (FPGAs).

Thickness Rendering - This subsystem provides the final determination and distribution of the calibrated measurement of strip thickness. Calibration and alloy compensation curves reside in and are supplied by the System Supervisor. The measured thickness is typically transmitted via analog signals or high speed networked numerical data exchanges.

System Supervisor - This subsystem oversees and coordinates the gauging system's control, measurement, calibration and operational activities, along with any operational interfacing to the mill / line control systems.

User and Maintenance Interfaces - Depending on the nature and extent of the system's function, various forms of dedicated operator interfaces may be employed. The user interface can range from simple operator controls and data entry devices, to sophisticated interactive, graphical human machine interfaces (HMIs). The maintenance interface is typically more sophisticated and provides detailed graphical information concerning the status, activity, calibration and performance data, along with trouble shooting and diagnostic assistance.

Interfaces to External Control and Automation Systems - The gauging system must communicate and interact with the mill / line's related control, automation and high level production systems. Measured thickness indications are often transmitted as analog signals or numerically via dedicated network links. Set-up, operational and status data (i.e., nominal gauge sets, alloy / composition, profile / positioning, shutter, etc) are often exchanged via standard network, serial, or even discrete logic (BCD) interconnects.

3. Ionizing Radiation and Radiation Generators

Radiation is a generalized term used to describe a variety of energy forms that are emitted from a body. For the purposes of this discussion, we will focus on ionizing radiation which involves charged particles or electromagnetic waves possessing sufficient energy to dislodge strongly held electrons from atoms or molecules.

3.1 Forms of Radiation

Ionizing radiation comes in three(3) primary forms: (Halliday, 1955), (Kaplan, 1955)

α-Rays - Alpha radiation involves accelerated helium nuclei, composed of 2 protons and 2 neutrons. This particle has a high mass and a double positive charge. Due to its high mass, this form of radiation has low penetrating energy and a limited range. The primary source of formation is during the nuclear transformation process (radioactive decay), where a new element results.

β-Rays - Beta radiation involves accelerated electrons (or positrons). These particles have a low mass and a negative charge (positive for positrons). Beta rays have modest penetrating energy (more than alpha particles), but can be stopped by metals a few centimeters thick. The primary source of formation is during the nuclear transformation process (radioactive decay), where a neutron is converted to a proton (which remains in the nucleus) and an electron and an antineutrino are emitted. Beta radiation can also be formed by an electron gun in the presence of high electric field potentials.

γ-Rays - Gamma rays are high energy photon emissions (electromagnetic waves) (Kraus & Carver, 1973). Gamma radiation has high penetrating energy and is the primary form of radiation employed in strip thickness gauging systems. X-Rays are also a form of electromagnetic (gamma) radiation. Classically, Gamma Rays and X-Rays have been separated by their respective energy levels (with Gamma being of higher energy). However, a more common place distinction involves the means of their generation. We will examine the various aspects of these differences in the next section.

In fact, there are many forms of radiation (when considering the non-ionizing form), which include: neutron or proton emissions, acoustic, low energy electromagnetic radiation (i.e., thermal, black body, light, radio waves), etc. These forms of radiation are not considered within the scope of this discussion.

3.2 Radiation Sources

Radiation sources are components that generate radiation for application to the measurement process. To limit and direct this discussion, we will focus only on sources that produce high energy photons (electromagnetic waves or γ-Rays). Although β-Ray sources are common, a vast majority of the industrial applications employ γ-Ray emissions. As noted previously, it is necessary to draw specific distinctions between the forms of electromagnetic radiation, under consideration, in terms of their origins.

3.2.1 Naturally Occurring Gamma Rays (Isotope Sources)

Naturally occurring gamma rays are specifically produced by radioactive isotopes during the nuclear transformation process, where following the emission of alpha and / or beta radiation, the decaying nucleus releases excess energy (in the form of photons) to obtain an equilibrium (Halliday, 1955), (Kaplan, 1955). These photon emissions form very well defined spectral lines at specific energy levels and relative amplitudes (Halliday, 1955), (Graydon, 1950). Common radioactive isotopes are: Americium 241, Cesium 137, Curium 244. Figure 3.1 shows the spectral characteristics of photonic radiation released by the radioactive isotope Americium 241.

Primary Radiation			
Type	Energy	Unit	Relative Intensity
Alpha	5443	(keV)	84.5
	5485	(keV)	13
Beta	52	(keV)	< 1
Gamma	13.9	(keV)	42
	26.3	(keV)	2.4
	59.5	(keV)	35.9

(a)

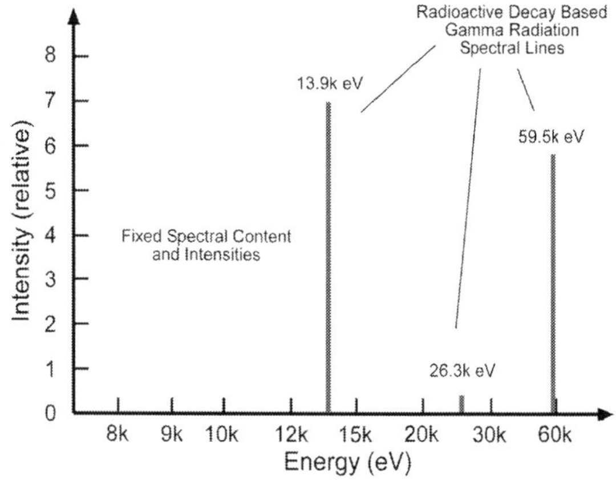

(b)

Fig. 3.1 - Spectral characteristics of the radioactive isotope Americium 241: a) Table defining the form of radiation emitted, the energy level and relative intensity, b) Spectral characteristics of the Gamma radiation components.

3.2.2 Artificially Produced Gamma Rays (X-Ray Sources)

X-Rays (in this context) are specifically generated by high energy, inbound electrons interacting with the inner shell electrons of an atom or the atom's electric fields. These interaction processes, shown in Figure 3.2, produce two distinctly different spectral emissions.

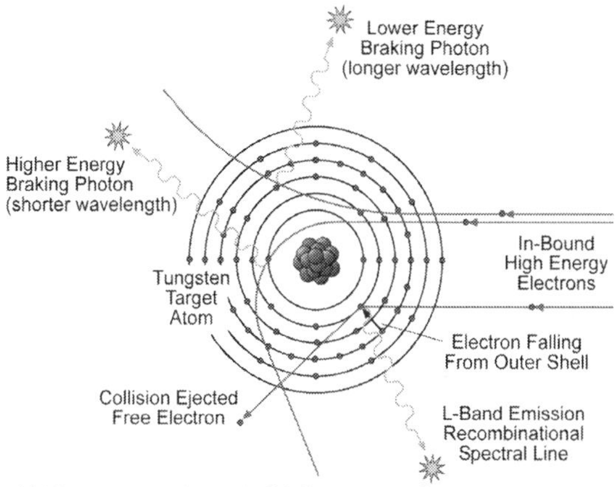

Fig. 3.2 - Nature of X-Ray generation via high energy electron interaction with a Tungsten atom based on a Bohr atomic model of the inbound electron interaction.

Characteristic Spectral Lines - Here an inbound high energy electron has sufficient energy to dislodge an atom's inner shell electron, to the extent of either lifting it to an outer shell (excited state) or removing it from the atomic union (ionized state). The shell's vacated electron position is filled (almost immediately), by a loosely bound electrons from the outer shells, resulting in a release of energy (in the form of a high energy photon), corresponding to the binding energies of the shells involved. The energy released produces discrete, well defined recombinational spectral lines (Mark & Dunn, 1985). The general characteristics of these spectral lines are shown in Figure 3.3 for Tungsten.

Bremsstrahlung Spectra - This spectral content develops when high kinetic energy electrons encounter the electric fields of the atom and are either decelerated or deflected from their previous trajectories (Halliday, 1955). The kinetic energy lost during this deceleration / deflection is emitted as electromagnetic radiation. An electron's inbound kinetic energy can be dissipated as X-Rays either entirely (in a single-stage nucleus encounter) or by several multi-stage encounters, each causing a different radiated energy. When an electron passes-by / interacts-with an atom, the proximity of its trajectory to the nucleus plays a direct role in the amount of energy dissipated. The probability of radiated energy dissipation elevates as the distance from the nucleus increases (i.e., larger distances from the nucleus induce weaker / more frequent radiation events, while shorter distances from the nucleus cause stronger / less frequent radiation events). The spectral content (shown in Figure 3.3) of the Bremsstrahlung component is not a discrete line spectra, but a continuum spanning the initial kinetic energy of the inbound electron (i.e., the maximum spectral energy equals the original kinetic energy of the electron) (Mark & Dunn, 1985).

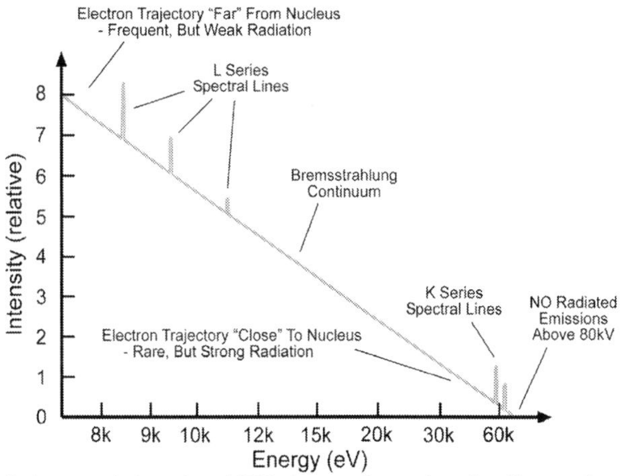

Fig. 3.3 - Spectral characteristics of an 80kV electron beam bombarding and interacting with the atoms of a Tungsten target.

3.3 Radiation Generation

The Radiation Generator emits a directed, collimated beam of high energy ionizing radiation and provides protective radiation containment. When considering isotope source based radiation generators, these devices are very simple (I2S, 1992). They contain only a shielded housing, an isotope source cartridge / pellet, source holder, columating aperture and a shutter. Due to the rather simplistic nature of these generators, we will forgo discussing their associated details.

3.3.1 X-Ray Generators

X-Ray source based radiation generators are far more complex than their isotopic counterparts. In the most classical sense, X-Ray generators are based on the components shown in Figure 3.4 and discussed in the following:

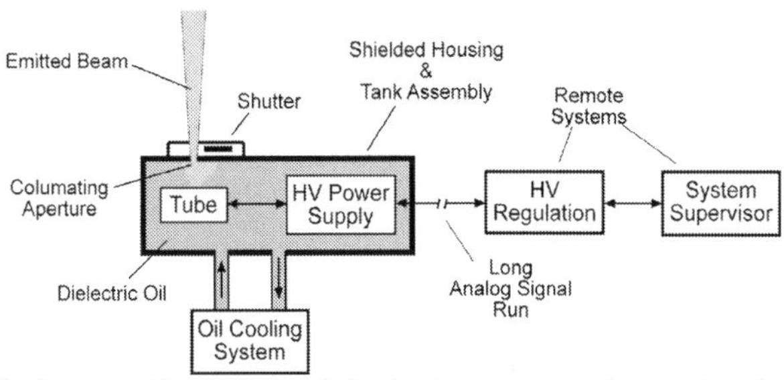

Fig. 3.4 - Block diagram illustration of the basic components associated with an X-Ray Generator.

X-Ray Tube - An X-Ray tube is a vacuum tube that when energized emits a polychromatic gamma ray spectrum (Howard, 1970), (Moore & Coplan, 1983). The spectral range is a direct function of the applied tube potential, and the intensity of the radiation is a direct function of the applied tube current. Figure 3.5 provide a diagram showing the primary components of an X-Ray tube.

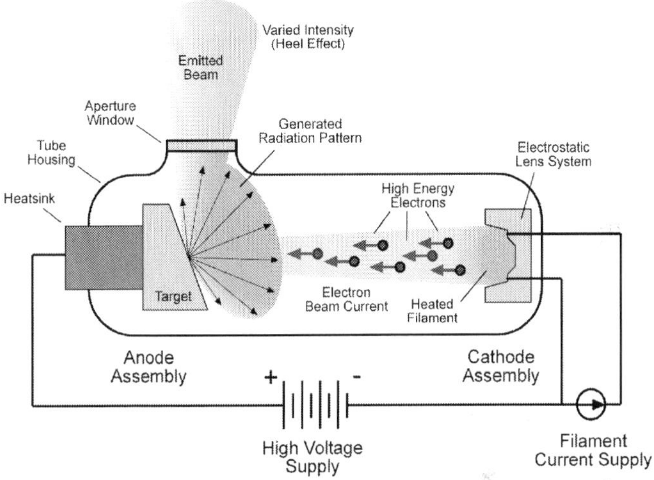

Fig. 3.5 - Simplified diagram showing the basic components of an X-Ray Tube.

Tube Housing - The tube is typically constructed of a sealed, cylindrical glass or ceramic housing and maintains a vacuum. Depending on the aperture material / mounting arrangement and the anode heat sink configuration, the tube geometry may have extensions or added structures, and can be shrouded by a circulating fluid heat exchanger cooling jacket.

Filament - This (typically) Tungsten coil is heated by a constant current source to temperatures that cause sufficient thermal excitation of the valence shell electrons to escape their atomic bonds and form a "cloud" of free electrons.

Target - This (typically) Tungsten plate emits polychromatic X-Rays when its atoms are bombarded by a beam of high kinetic energy electrons. The target's active surface is typically angled to direct the radiation pattern toward the tube's aperture. The angle must be optimized to provide the desired radiation intensity while still maintaining a concentrating projection of the applied electron beam pattern.

High Voltage Power Supply and Tube Potential - A high voltage, direct current (DC) power supply (often 10kV to 200kV) applies a precision regulated, potential between the filament (cathode) and the target (anode), to draw free, thermally excited electrons from the filament and accelerate them to their target impact energy, forming an electron beam. The beam's charge displacement forms a current across the tube (beam current). The power supply's current limits regulate the applied current / tube power. The high voltage electronics /

equipment is often immersed in a dielectric oil bath to provide insulation and allow for a more compact design. The high voltage power supply control and regulation are often provided by external equipment, possibly remotely located.

Electrostatic Lens - The geometric arrangement of this component forms electric field patterns that focus the electron beam to a specific target impact spot geometry (Harting & Read, 1976). Figure 3.6 provides an illustration of the formed electric field lines, and their impact on the electron trajectory.

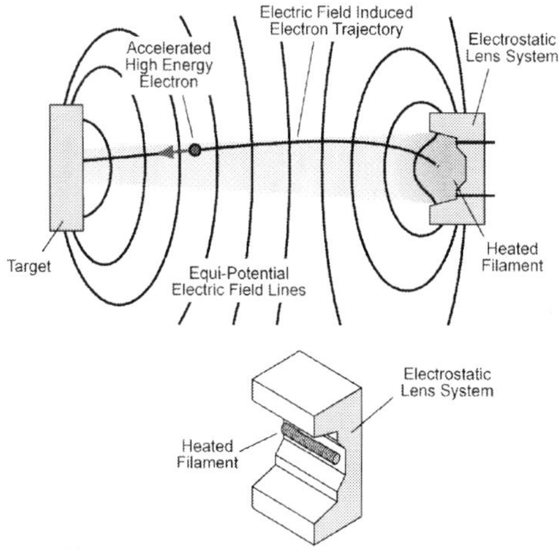

Fig. 3.6 - Block diagram illustration of the X-Ray Tube Electrostatic lens induced electric field lines and associated electron trajectory.

Dielectric Oil - The X-Ray tube and the high voltage power components are often immersed in an oil bath to provide both a high degree of electrical insulation and also a tube heat dissipation capacity.

Thermal Considerations - X-Ray tubes are highly inefficient, with only about 1% of the applied power being converted to X-Ray production. The remainder is converted to heat. Industrial X-Ray tubes are often immersed in an oil bath to dissipate the tube's thermal power. Depending on the tube power and the nature of the generator's housing, the generated heat may exceed the passive dissipation capabilities, thereby requiring an external, oil circulating, cooling / heat exchanging system.

Radiation Pattern and Heel Effect - The target's emitted radiation pattern is dependent on the angular orientation of the target and the spot-size / geometry of the electron beam. Figure 3.7 provides insight into the nature of these radiation patterns and heel effects. The lobed radiation pattern is caused by the angle at which the photons emerge from beneath the surface of the target at the focal point. This causes a differential attenuation of the photons that will ultimately compose a useful, emitted beam. The resulting radiation pattern emitted through the tube's aperture has radial / transverse variations in intensity (often termed "heel effect"), (Mark & Dunn, 1985).

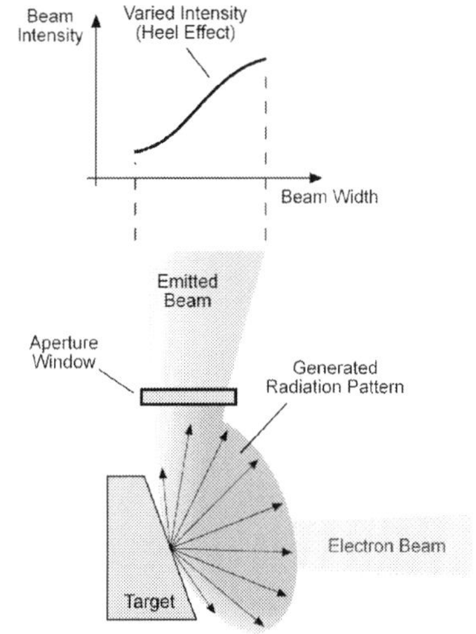

Fig. 3.7 - Illustration showing the varied intensity of the emitted beam associated with the geometry of the target, radiation pattern and location of the tube aperture window.

Tube / Tank Assembly - This component of the generator housing provides a sealed vessel in which the dielectric oil bath is contained and the tube / high voltage power components are mounted. This assembly typically utilizes a lead intermediate liner for radiation shielding and often employs an insulating inner liner (e.g., polypropylene).

Collimating Aperture - The geometry and location of this aperture (with respect to the X-Ray tube mounting and radiation pattern) defines the optical geometry and uniformity of the X-Ray generator's emitted beam. The aperture material will impact the radiated spectrum through energy dependent absorption and scattering processes. Depending on the nature of the tube / tank assembly and shielded housing, the aperture may be required to provide a fluid pressure seal to prevent dielectric oil seepage.

Shutter - A retractable shutter (typically lead) provides the ability to suppress the X-Ray generator's emission, while still allowing the X-Ray tube to be energized (often the tube is kept active to maintain a thermal equilibrium).

System Supervisor - This system component provides the desired references for the tube potential, beam current and filament current, and monitors the tube / tank status, temperature, etc. to control and oversee the generator's operations and performance.

Spectral Characteristics of the Tube / Generator Emitted Radiation - Within the X-Ray tube housing, the polychromatic spectral content of the produced radiation is based on the target material's characteristic and Bremsstrahlung spectra (see Figure 3.3). Beyond the tube's aperture and the aperture of the generator's collimator, the spectral content is modified by the absorption and scattering behavior of the aperture materials (possibly fused silica, calcium fluoride, beryllium, polypropylene, etc.) and any intervening dielectric oil. This causes an attenuation of the lower energy regions of the emitted spectrum (see Section 4.2

concerning the energy dependencies of the Mass Attenuation Coefficient). The resulting spectrum contains a higher energy content, making the beam more penetrating (harder). Figure 3.8 provides example plots of spectral content of Tungsten target radiation attenuated by a glass window aperture for differing applied tube potentials.

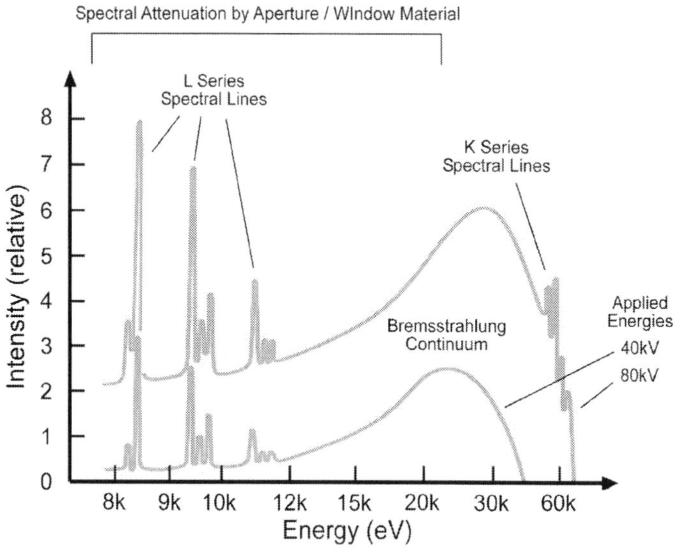

Fig. 3.8 - Graphical representations of the spectral content of the radiation emitted from a Tungsten target X-Ray tube with a glass aperture window, showing the Characteristic and Bremsstrahlung radiation spectrum for 80kV and 40kV tube potentials.

The key feature of this spectrum is the significant attenuation by the aperture window of the lower energy region (as noted by the diminished region of Bremsstrahlung radiation below 20kV). It is important to note that the lower tube potential (40kV) does not provide an electron beam with sufficient kinetic energy to dislodge the target material's K shell electrons (indicated by the lack of K Series recombinational spectral lines).

Controlled Variability of Tube / Generator Emissions - By varying the applied tube potential and beam current, the radiated tube / generator spectral content and intensity can be adjusted to meet the needs of the measurement application. Figures 3.9 and 3.10 show the reactions of the Bremsstrahlung radiation spectra to changes in the tube potential and beam current, respectively.

Beam Hardening - This term traditionally describes the process of increasing the average energy of the emitted spectrum. This causes the resulting beam to have a greater penetrating capability. Beam hardening can be achieved through the used of selected pre-absorbers, whose spectral attenuation characteristics suppress lower energy regions (compare Figures 3.3 and 3.8). This beam hardening effect can also be formed by increasing the applied tube potential. As shown in Figure 3.9, increasing the tube voltage causes the emitted spectrum's peak intensity to shift to higher energies.

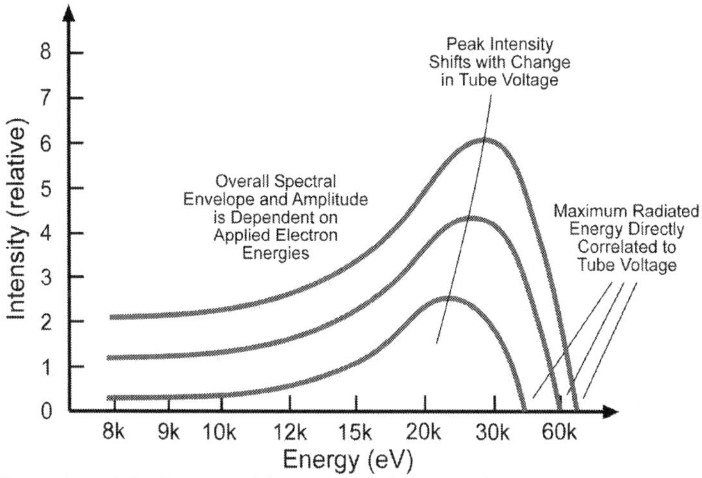

Fig. 3.9 -Illustration of the Bremsstrahlung spectra behavior due to variations in the applied tube potential, while maintaining a constant beam current. This illustrates that an increase in the tube voltage causes a beam hardening effect, by shifting the spectrum's average energy to higher (more penetrating) levels.

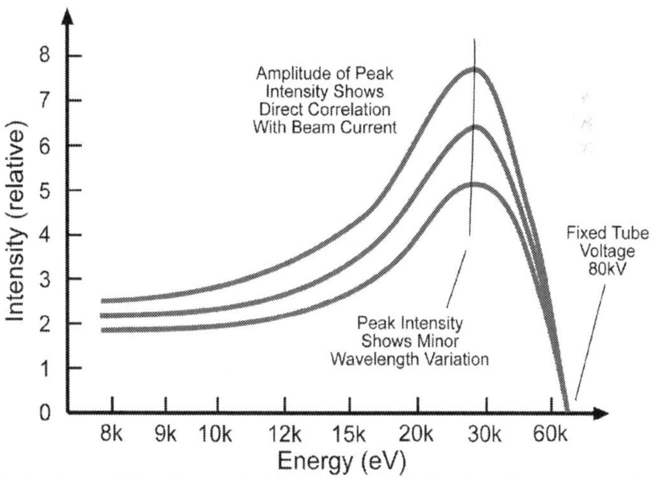

Fig. 3.10 -Illustration of the Bremsstrahlung spectra behavior due to variations in the applied beam current, while maintaining a constant tube potential.

4. Interaction of Radiation with Materials

The collimated beam of radiation emitted by the radiation generator is directed (typically perpendicular) to one surface of the material. The incident radiation interacts with the material's atomic structures and is either passed, absorbed, scattered or involved in high energy pair productions. The nature of this interaction is dependent on the spectral energy content of the applied radiation and the composition of the material. The resulting transmitted radiation appears as a dispersed beam pattern, having attenuated intensity and modified spectral content.

4.1 Attenuation Effects Based on Form of Radiation

The nature of the material interaction is dependent on the form and energy content (wavelength) of the inbound radiation. A number of processes are involved (e.g., collision, photoelectric absorption, scattering, pair production) and their cumulative effect can be characterized as an energy dependent attenuation of the intensity, and a modification of the radiated pattern of the transmitted beam (through scattering processes) (Kaplan, 1955), (Letokhav, 1987).

<u>α-Particles</u> - Due to their dual positive charge and their relatively large mass, Alpha particles interact strongly (through collision processes) with the material's atoms and are easily stopped (Kaplan, 1955).

<u>β-Particles</u> - Due to their physical mass and negative charge, Beta particles also interact through collision / scattering processes. Elastic and inelastic scattering processes are associated with manner in which inbound, high energy electrons interact with the electric fields of the material's atoms (Kaplan, 1955), (Mark & Dunn, 1985).

<u>Inelastic Scattering</u> - A certain amount of the inbound radiation energy is dissipated through an ionization or excitation of the material atoms. Here, the inbound energy is sufficient to dislodge electrons from their shells, forming an ion, or shell electrons are excited to outer shells. Recombinational gamma spectra (electromagnetic) is produced and radiated in all directions, when the excited or ionized electrons fall into the inner shells.

<u>Elastic Scattering</u> - This lesser (secondary) radiation tends to possess lower energy content and is also radiated in all directions. The radiation intensity is an increasing function of the material's atomic number. This attribute is well suited for measuring coating thicknesses on base materials (having different atomic numbers to the coating) via backscattering techniques.

<u>γ-Rays</u> - Gamma rays (electromagnetic energy) are attenuated through reductions in their quanta energies, via the combined processes of photoelectric absorption, scattering and pair production (Hubble & Seltzer, 2004). The experienced attenuation is an exponential function of the inbound radiation energy spectra, and the material composition and thickness. This relationship makes this form of radiation an attractive choice for material thickness measurement via a knowledge of the applied radiation, the material composition and an examination of the resulting transmitted radiation.

4.2 Mass Attenuation Coefficient

The manner in which a composite / alloyed material responds to inbound photonic radiation can be characterized by the composite Mass Attenuation Coefficient (MAC), μ/ρ, of its elemental constituents (typically with units of (cm^2/g)). The MAC is a material density normalization of the Linear Attenuation Coefficient (LAC), μ, where ρ is the density of the material (in g/cm^3), and the MAC is therefore an energy dependent constant that is independent of physical state (solid, liquid, gas). The reciprocal of the LAC, q, is often termed the Mean Free Path. The MAC is typically characterized as an energy cross-section, with the amplitude of attenuation being a function of applied photonic energy, (Hubble & Seltzer, 2004). Figure 4.1 provides a graphical representation of the MAC for the element Iron (Fe, Atomic No.: 26). Radiation attenuation is composed of five(5) primary processes:

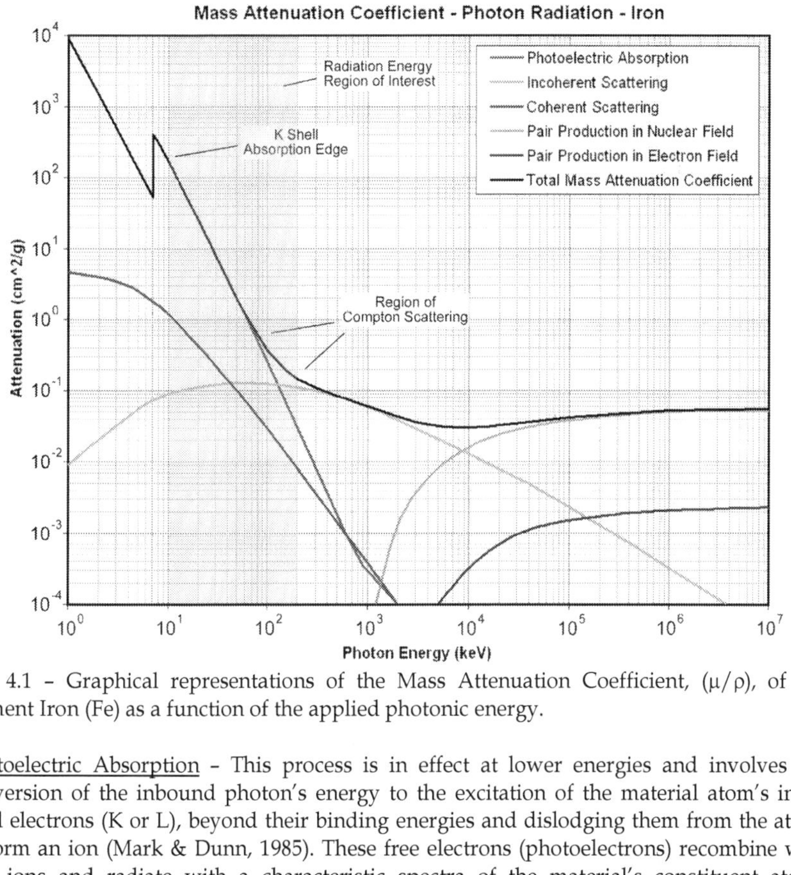

Fig. 4.1 - Graphical representations of the Mass Attenuation Coefficient, (μ/ρ), of the element Iron (Fe) as a function of the applied photonic energy.

Photoelectric Absorption - This process is in effect at lower energies and involves the conversion of the inbound photon's energy to the excitation of the material atom's inner shell electrons (K or L), beyond their binding energies and dislodging them from the atom, to form an ion (Mark & Dunn, 1985). These free electrons (photoelectrons) recombine with free ions and radiate with a characteristic spectra of the material's constituent atoms (recombinational spectral lines). This radiation is emitted in all directions in the form of an X-Ray fluorescence (whose energy increases with atomic number). If the inbound radiation energy is below shell's binding energy, photoelectrons are not formed from that shell and an abrupt decrease in the material's absorption characteristics is noted (see the abrupt, saw-tooth absorption edge in Figure 4.1).

Incoherent Scattering (Compton Scattering) - This absorption process is in effect over a broad range of energies, and involves inelastic scattering interactions between the material atom's electrons and the inbound photonic radiation (Kaplan, 1955). The electrons are transferred part of the inbound radiation energy (causing them to recoil) and a photon containing the remaining energy to be emitted in a different direction from the inbound, higher energy photon. The overall kinetic energy is not conserved (inelastic), but the overall momentum is conserved. If the released photon has sufficient energy, this process may be repeated. The Compton scatter radiation has a directional dependency that results in radiated lobes of having angular intensity dependencies.

Coherent Scattering (Rayleigh Scattering) - This absorption process is in effect in the lower energy regions, and involves the elastic scattering interactions between the inbound photons and physical particles that are much smaller than the wavelength of the photon energy, (Kaplan, 1955).

Pair Production - This absorption process is in effect only at very high energies (greater than twice the rest-energy of an electron (>1.022MeV)), and involves the formation of electron pairs (an electron and a positron), (Halliday, 1955). The electron pair converts any excess energy to kinetic energy, which may induce subsequent absorption / collisions with the material's atoms. This absorption process occurs only at very high energies, and therefore has no practical application in the forms of thickness measurement considered here.

The summation of these components forms the MAC and precision cross-section data is openly published as tabulated lists by the National Institute of Standards and Technology (NIST) (Hubble & Seltzer, 2004), for all the naturally occurring periodic table elements to an atomic number of 92 (Uranium).

It is important to examine the nature of the material absorption characteristics within the region of radiation energy of interest (10keV - 200keV), see Figure 4.1. Here, the attenuation characteristics of the lower energy section is dominated by the Photoelectric absorption. At energies higher than about 100keV, Compton Scattering becomes the primary method of attenuation.

Depending on the nature of a given element's atomic structure and atomic weight, the behavior of the MAC can vary widely. Figure 4.2 provides a comparative plot of four common elements, along with an indication of the energy level associated with the primary spectral line for Americium 241 (59.5keV). The key aspect of this comparison is the extent and energy regions involved in the differences in the attenuation characteristics. Carbon offers very little attenuation and only at low energies, while lead dominates the spectrum, especially at higher energies, illustrating its excellent shielding characteristics. Copper and iron have very similar behavior, and also show K Shell absorption edges at their distinct energies. The differences in attenuation between these metals appear to be relatively small, however, in the region about 60keV, copper has over 30% more attenuation than iron.

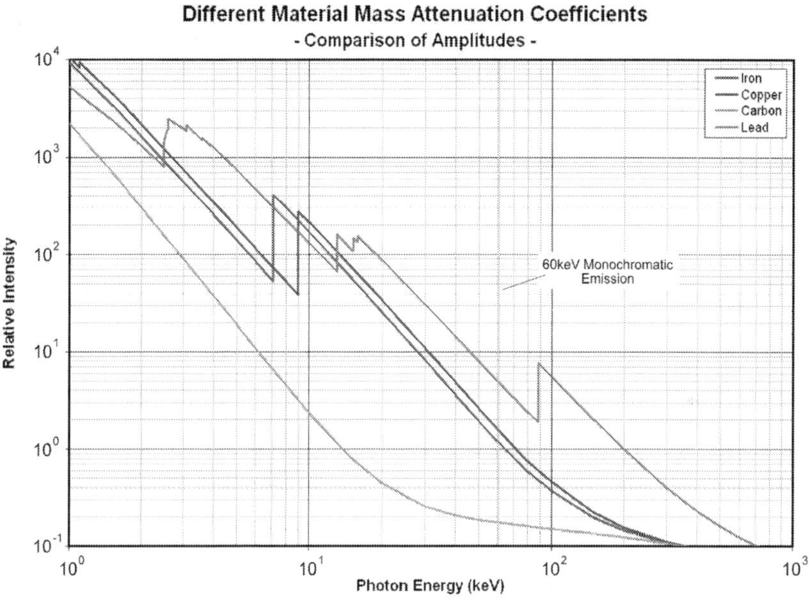

Fig. 4.2 - Graphical comparisons of the energy dependent MACs of differing materials and an indication of the location of 60keV incident radiation.

4.3 Attenuation Characterization

4.3.1 Monochromatic Beer-Lambert Law

When monochromatic radiation of known intensity, I_0, is attenuated by the material, the relationship to the resulting, transmitted radiation, I, is an exponential function of the MAC, the material density and thickness, originating from the differential form:

$$-\frac{dI}{I} = \mu\, dx = \frac{dx}{q} \tag{4.1}$$

where

μ - Linear Absorption Coefficient (LAC - subject to material density variations)
q - Mean Free Path (MFP - subject to density material variations)
x - Material Thickness

Integrating Eq(4.1) results in:

$$I = I_0\, e^{-\mu x} = I_0\, e^{-\frac{x}{q}} \tag{4.2}$$

Expanding Eq(4.2) to employ the MAC, , produces the Beer-Lambert Law (Halliday, 1955), (Kaplan, 1955):

$$I = I_0 e^{-\left(\frac{\mu}{\rho}\right)\rho x} = I_0 e^{-\frac{x}{q}} \tag{4.3}$$

where

$\left(\mu/\rho\right)$ - Mass Attenuation Coefficient (MAC), (cm^2/g)

ρ - Material density (g/cm^3)

Figure 4.3 provides a graphical relations showing the nature of the exponential attenuation characteristics of a monochromatic incident radiation as a function of material thickness in terms of multiples of the material's MFP.

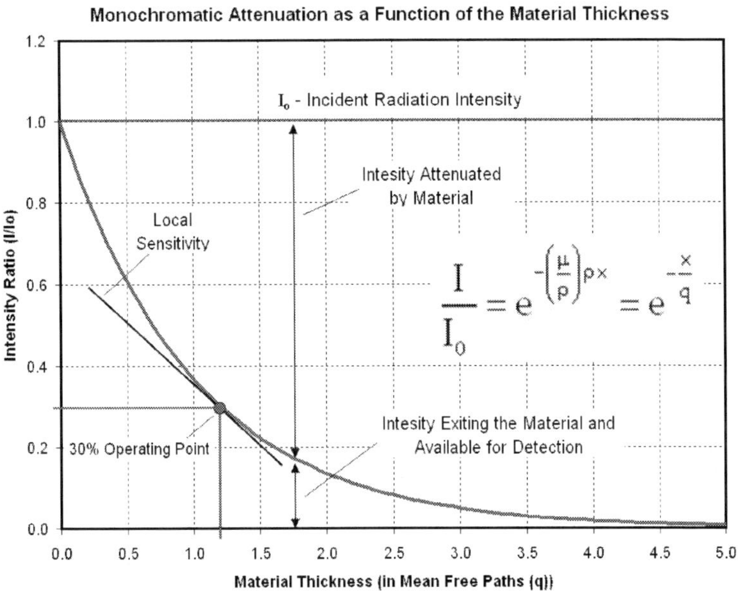

Fig. 4.3 - Monochromatic exponential attenuation as a function of material thickness in terms of multiples of the material's Mean Free Path (q).

4.3.2 Attenuation in Composite Materials

When a material is formed by a combination of constituents (e.g., alloy), the weighted inclusion contributions of the individual components must be taken into account. The composite material's MAC is given by (Hubble & Seltzer, 2004):

$$\left(\frac{\mu}{\rho}\right)\rho = \sum_{i=1}^{N} w_i \left(\frac{\mu}{\rho}\right)_i \rho_i \qquad (4.4a)$$

$$\left(\frac{1}{q}\right) = \sum_{i=1}^{N} \left(\frac{w_i}{q_i}\right) \qquad (4.4b)$$

where

$\left(\mu/\rho\right)_i$ - The MAC of the ith constituent, of N total constituents

w_i - The decimal percentage of inclusion of the ith constituent

$$\sum_{i=1}^{N} w_i = 1.0 \tag{4.5}$$

The single element relationship of Eq(4.3) is therefore extended to the composite material:

$$I = I_0 e^{-\sum_{i=1}^{N}\left(w_i \left(\frac{\mu}{\rho}\right)_i \rho_i\right) x} \tag{4.6}$$

4.3.3 Polychromatic Dependencies of Attenuation

The Beer-Lambert Law of Eq(4.3) (and Eq(4.6)) applies only to monochromatic radiation energy, however, typical radiation sources rarely emit purely singular energies (note the spectral content shown in Figures 3.1b and 3.6a. It is therefore necessary to extend the relationships Eq(4.3) and Eq(4.6) to include the polychromatic spectral content of the applied and transmitted radiation, along with the energy cross-section of the MAC. This is provided through the inclusion of the wavelength (energy) dependency of these components.

$$I(\lambda) = I_0(\lambda) e^{-\left(\frac{\mu}{\rho}(\lambda)\right)\rho x} = I_0(\lambda) e^{-\frac{x}{q(\lambda)}} \tag{4.7}$$

The use of wavelength, as opposed to energy is purely for convenience, and Eq(4.7) can be extended to include the effects of composite materials, Eq(4.4) and Eq(4.6). Figure 4.4 provides graphical examples of how the incident radiation amplitude and polychromatic spectral content is attenuated / modified by its interaction with material. It's interesting to note that manner in which lower energy region attenuating characteristics of the material under measurement causes a beam hardening effect on the radiation available to the detector (note the higher average energy level in Figure 4.4b compared to 4.4a).

5. Radiation Detection / Measurement

Attenuated / scattered, polychromatic radiation, $I(\lambda)$, that results from interaction with the material, is collected and measured by a detector aligned with the optical axis of the generator's radiated beam and has an aperture sized to over-contain the transmitted beam. The detector produces a signal that is functionally related to the total received, polychromatic radiation energy within the spectral bandwidth of the detector's sensitivity.

$$I_D \sim \int D(\lambda)\, I_0(\lambda) e^{-\left(\frac{\mu}{\rho}(\lambda)\right)\rho x} d\lambda = \int D(\lambda)\, I_0(\lambda) e^{-\frac{x}{q(\lambda)}} d\lambda \tag{5.1}$$

where

I_D - The detector's response / measurement signal

$D(\lambda)$ - The detector sensitivity (a function of wavelength / energy)

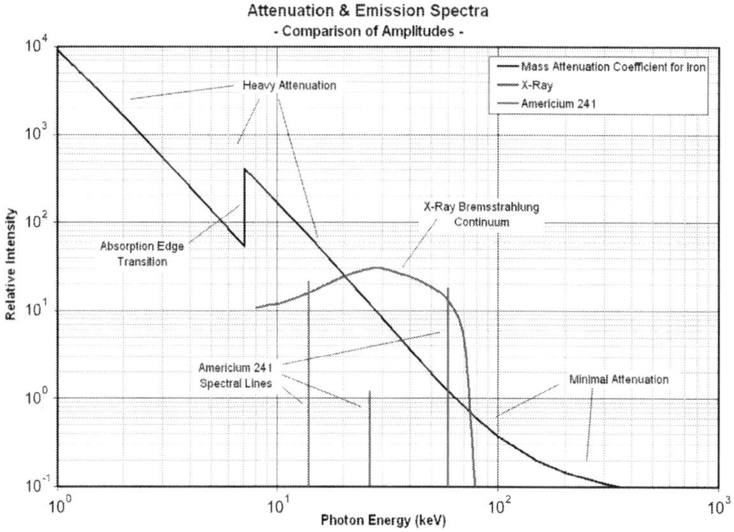

(a)

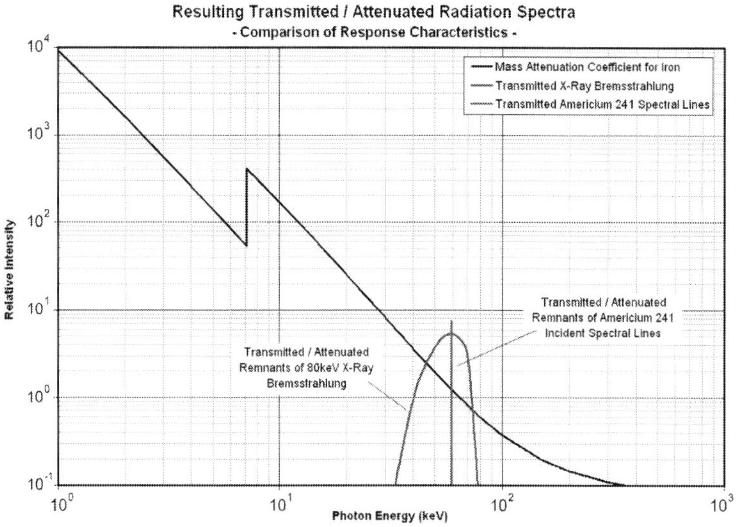

(b)

Fig. 4.4 - Graphical before-and-after comparison of the amplitude and polychromatic spectral modifications of differing sources of incident radiation's interaction with material: a) MAC cross-section of Iron overlaid with the inbound spectral content of both the Americium 241 spectral lines and 80keV X-Ray Bremsstrahlung radiation, b) Transmitted / attenuated spectral content resulting from material interaction.

There are many types of detectors, and we can generally classify them in terms of the nature of their responses to incident radiation.

Ionization Methods - This includes a large class of detectors that respond to incident radiation as a function of the level of ionization occurring within them (Moore & Coplan, 1983). These include: ion chambers, proportional counters, Gieger-Muller counters, cloud chambers, spark chambers, fission chambers and certain semiconductor devices.
Molecular Excitation and Dissociation Methods - This includes detectors that respond to incident radiation as a function of the molecular excitation and dissociation, along with a certain degree of ionization (Moore & Coplan, 1983). These include: scintillation counters, chemical dosimeters and optical properties based systems.

To narrow the focus of this discussion, we will focus on the examinations of ion chamber and scintillation based detectors.

5.1 Ionization Chamber Detectors

Ionization chambers (ion chambers) consist of a media (usually gas) filled chamber containing two(2) charged electrodes, (Halliday, 1955), (Kaplan, 1955), (Moore & Coplan, 1983). The chamber aperture may consist of a sealed window made of a material that either efficiently passes or possibly attenuates the incident radiation, depending on the planned range of radiation intensity. The chamber geometry is typically organized to accommodate the application, generally in the form of a cylindrical arrangement.

5.1.1 Ionization Processes

Depending on the intended radiation form, photons or other charged particles (neutrons, electrons, etc.) enter the chamber aperture and transit through the media, where they interact with the atoms forming the media. Depending on the circumstances, these interactions can strip-away electrons from the outer shells of the media, thereby forming ion / free-electron pairs (ion pairs), via direct or indirect ionization processes. The energy required to form an ion pair is often termed the ionization potential (which is typically on the order of 5-20eV) (Graydon, 1950), (Letokhav, 1987).

Direct Ionization - Charged particles (alpha or beta) passing through the media may either collide with the electrons of media atoms, and impart sufficient kinetic energy eject them from the atom, or they may transfer sufficient energy by their interactions with the atom's electric fields when passing close to a media atom. If these energy transfers do not exceed the electron binding energy (and therefore do not eject the electron), the atom is left in a disturbed / excited state.

Indirect Ionization - Gamma radiation (photons) passing through the media interact with the media atoms and form ion pairs through a photo ionization process, where the photon energy is transferred to the electron's kinetic energy. If this energy exceeds the electron binding energy, the electron is ejected (forming an ion pair). Photons possessing energies not sufficient to form ion pairs are either scattered or absorbed by the atom, leaving it in an excited state.

The number of ion pairs formed within the media is a function of the incident radiation's energy cross-section / spectrum and the nature of the media composition. In most substances, the energy lost in ion pair formation is larger than the ionization potential, which reflects the fact that some energy is lost in excitation.

5.1.2 Gas Filled Ion Chambers

The classical arrangement for an ion chamber is based on a hollow, sealed, gas filled (i.e., Xenon, Argon, etc.), conductive cylinder, typically having an aperture window of a selected material and a conductive filament positioned along the cylindrical axis, insulated from the cylinder's walls (Moore & Coplan, 1983). The dual electrode arrangement is formed by positively charging the filament (anode) and negatively charging the cylinder wall (cathode), through the application of a high voltage / potential (which produces an electric field within the chamber). Figure 5.1 provides a diagram showing the primary components of a gas filled ion chamber and the processes involved in its radiation detection / measurement process.

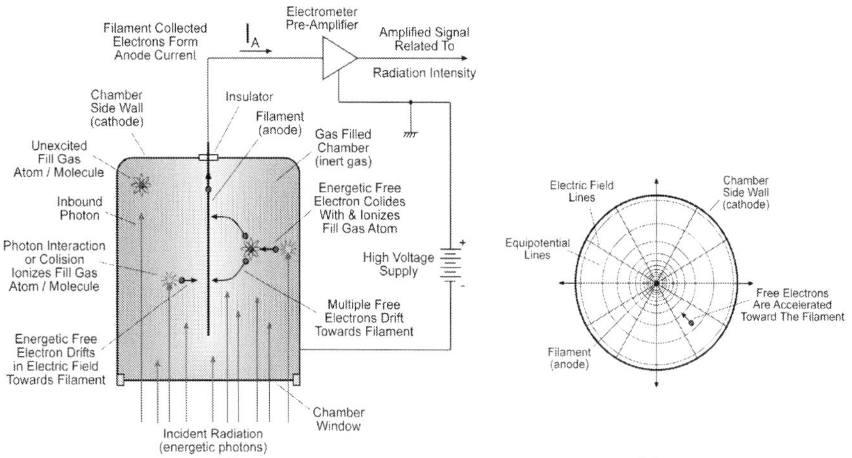

(a) (b)

Fig. 5.1 - Diagram showing the primary components of a gas filled ion chamber: a) Vertical cross-section view of the ion chamber components and processes involved in radiation detection / measurement, b) Cylindrical cross-section view showing the voltage applied electric field lines and associated equi-potential surfaces.

As radiation passes through the ion chamber's gas media, ion pairs are formed (by the processes mentioned above). The ionized gas atoms and free electrons drift and accelerate toward their respective electrodes. The speed (kinetic energy) at which these ion pairs migrate is a function of the chamber's electric field and composition / pressure of the media gas. The low mass of the free electrons causes them to move at much faster speeds toward the central filament.

The electrons (charge) collect on the anode filament, inducing a voltage change / current flow in the external circuitry connected to the anode, resulting in a pulse-like waveform. The amplitude of the pulse is dependent on the number of electrons collected by the filament. Although feeble (as low as tens of femto amperes), these currents can be detected and measured by electrometer class, transconductance amplifiers (Motchenbacher & Fitchen, 1973).

The amplitude of the filament current (for a given intensity of incident radiation) is a function of the applied chamber potential and the composition / pressure of the gas media (Moore & Coplan, 1983). Figure 5.2 provides a graphical description of filament pulse height as a function of chamber voltage.

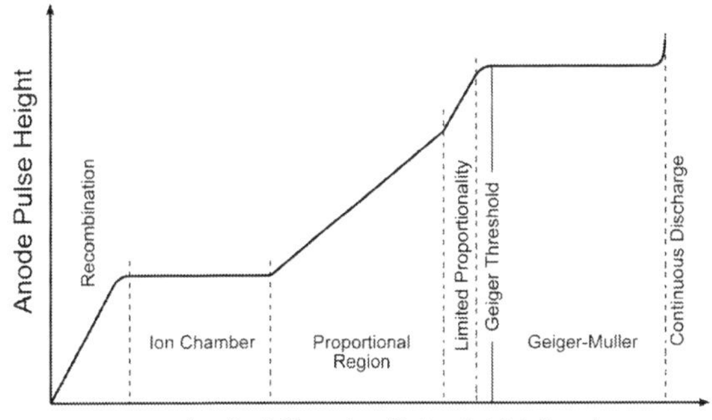

Fig. 5.2 - Behavior of the filament current pulse height as a function of applied potential, for a given constant intensity, incident radiation. This graph also shows the various regions of ion chamber operations.

<u>Recombination Region</u> - The chamber potential is relatively low, and the resulting electric field induced forces on the ion pairs (which draws them to the electrodes), is also low. The speed of electron drift toward the filament (anode) is slow, exposing some free electrons to high probabilities of being captured / neutralized by ions, before reaching the filament. Therefore, not all radiation induced ionization events are evidenced by the filament current. As the voltage is increased, the speed of drift also increases, and the probability of electron / ion neutralization diminishes, causing larger anode currents to be generated.

<u>Ion Chamber Plateau</u> - As the chamber potential is raised further, the rate of ion pair drift begins to reach speeds where the probability of neutralization (through recombination) is negligible. Essentially all of the free electrons formed by incident radiation ionization are collected at the filament. Here, the pulse amplitude levels-off, the anode current reaches a maximum value (for a given incident radiation level) and both no longer vary as functions of the applied voltage. This maximized filament current is often termed the saturation current, whose amplitude is dependent on the amount of received radiation. It is important to note that the formed free electrons gain energy as they drift and are accelerated by the electric field. In this voltage range, the electrons do not gain sufficient energy to induce subsequent ionization processes. If electron kinetic energies were to exceed the binding

energies of the gas atoms, then there would be an increase in the filament current, due to an effective increase in the gas amplification factor.

Proportional Region - With increasing chamber potential, the ion pair's kinetic energies are also increase (primarily noted by the speed of the free electrons). Here, these energies now exceed the gas atoms' binding energies, allowing the primary ions to generate secondary ions. This causes an amplification effect in the filament current. The electric fields are very concentrated local to the filament, and therefore the formation of secondary ions often occurs in the vicinity of the anode, and may induce additional orders of ion formation (essentially an avalanche behavior).

Geiger-Muller Plateau - As the chamber potential is raised further, electron drift speeds (kinetic energies) reach levels that generate photons as part of the secondary ionization processes. These emitted photons induce further ionizations to form throughout the entire chamber media. When operating in this region, it is not uncommon to augment the chamber's media with getter gasses (e.g., halogen) to provide a means of artificially quenching the high energy ion pairs, before secondary ionizations occur. When ion pairs encounter the getter gas molecules, they release certain levels of their energies to the molecules, then proceed with their drift trajectories at kinetic energies below the ionization potential of the base media. The resulting chamber behavior is highly responsive (fast), while also maintaining filament current proportional to the incident radiation intensity without a significant impact of secondary ionization effects.

Continuous Discharge - Beyond the Geiger-Muller region, the chamber potential reaches a level were the radiation induced ionization processes spontaneously erupts to form a sustained, non-dissipating plasma. The filament current can be significant (approaching the current limit of the high voltage power supply) and can damage sensitive external electronics / instrumentation. Under certain conditions, the chamber may experience "arching" discharges between the electrodes due to plasma associated changes in the media's dielectric properties.

5.2 Scintillation Based Detectors

Scintillation based detectors are a family of devices that employ a front-end sensor whose molecules have the property of luminescence (typically in the visible range) when exposed to ionizing radiation (Moore & Coplan, 1983). When incident radiation interacts with the sensor's molecules, the molecules absorb the inbound energy and enter an excited or ionized state. Upon neutralization or relaxation, the molecules emit a characteristics spectra associated with the recombinational spectral lines of the molecules / atoms. These fixed spectral emissions appear as momentary "flashes" of scintillating light (often in the visible range).

If the recombinational emissions occur immediately following the absorption of the inbound radiation energy (< 10 pico seconds), the resulting luminescence process is termed fluorescence. If the recombination / relaxation occurs following a discernable delay, the process is termed phosphorescence or after-glow.

The front-end scintillation sensor is typically composed of a inorganic compounds (e.g., Sodium Iodide, Bismuth Germanate) or an organic fluid, having high quantum efficiencies.

The quantum efficiency is associated with the density of electrons in the compound's molecules / atoms, generally due to high atomic number of the elemental constiuents. Perhaps the most widely used scintillation compound is Sodium Iodide activated with metal ions in Thallium, NaI(Tl).

5.2.1 Scintillation Crystal / Photomultiplier Tube

A classical scintillation based detector involves the pairing of a Sodium Iodide crystal with a photomultiplier tube (PMT) (Moore & Coplan, 1983), (RCA, 1963). Figure 5.3 provides an illustration of this arrangement, the primary components.

Radiation incident to the Sodium Iodide crystal induces a scintillation luminance in the form of characteristic, recombinational spectral emissions, with intensities proportional to the intensity of the inbound radiation. This scintillated light is transmitted-to and strikes the PMTs photocathode, forming free electrons as a consequence of the photoelectric effect.

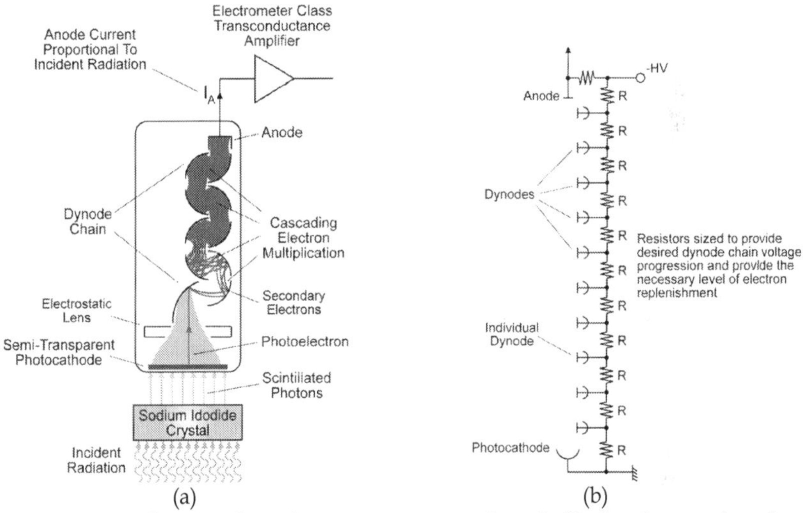

Fig. 5.3 - Diagram showing the primary components of a scintillation detector based on a Sodium Iodide crystal paired with a photomultiplier tube: a) Primary components and illustration of the electron multiplying effect, b) Electrical schematic of the PMTs dinode chain.

The photoelectrons depart the photocathode with a kinetic energy related to the energy of the incoming photon (reduced by the losses of the work function of the photocathode). The PMTs photocathode and electron multiplying dynode chain are charged with progressively more positive voltages, generating an electric field that draws and accelerates the free electrons from the photocathode, to the dynode chain. An electrostatic lens group is often employed to focus / direct the photoelectrons onto the surface of the first dynode.

The accelerated photoelectrons impact the first dynode (at a relatively high energy), and release a group of lower energy electrons (by the process of secondary emission). This group of lower energy electrons are accelerated towards the next dynode, where their increased kinetic energy, releases more electrons (an electron multiplication effect). The organization of the dynode chain causes a cascading / ever-increasing number of electrons to be produced at each stage. The multi-stage / multiplied electrons reach the anode (final stage) with a large charge accumulation, which results in a large current pulse that is directly related to an arrival event of a scintillation photon at the photocathode.

The anode current can be collected and assessed in several ways. For ultra-low radiation intensities (i.e., < 1000 events per second), the individual photon encounter event is directly related to an anode current pulse. Pulse discrimination and counting methods can be applied to measure the broad spectrum radiation intensity. Alternatively, pulse height and height distribution analysis can be employed to measure the radiation cross-section. Typically, the radiation source and measurement system are designed to induce a sufficiently large intensity of inbound radiation, that the resulting anode waveform is a near continuous current. Here, electrometer class, transconductance amplifiers (Motchenbacher & Fitchen, 1973) are applied to create usable signal levels in subsequent signal processing stages.

6. Rendering a Thickness Measurement

The primary function of the systems under consideration, is to render a measurement of the material thickness. Fundamentally, this involves the rearrangement of the complex relationship of Eq(5.1), to isolate "x", the material thickness. This is a non-trivial exercise and is not well suited for this level of discussion. However, it is possible to examine the simplified case for monochromatic incident radiation. Returning to Eq(4.3), and assuming the detector signal, I_D, is directly related to the transmitted radiation intensity, we have:

$$I_D \sim I_0 e^{-\left(\frac{\mu}{\rho}\right)\rho x} = I_0 e^{-\frac{x}{q}} \tag{6.1}$$

Isolating the material thickness, x, and considering the calculated value to be an "estimate" of the thickness, $\hat{x}$, based on the available knowledge of the alloy and generated radiation, I_0, results in:

$$X_M = \hat{x} \sim -\left(\frac{\rho}{\mu}\right)\frac{1}{\rho}\ln\left[\frac{I_D}{I_0}\right] = q\left(\ln\left(I_0\right) - \ln\left(I_D\right)\right) \tag{6.2}$$

The rendered thickness measurement, X_M, is the final indication of the absolute thickness, expressed in a chosen engineering unit of measure. Perhaps the most daunting issue that confronts rendering the thickness measurement, is having sufficient knowledge of the alloy or constituents that form the material, and make-up the Mass Attenuation Coefficient (MAC) energy cross-section. Minor variations in the understood versus the actual MAC can have dramatic effects on the quality / accuracy of the measurement.

6.1 Signal Processing and Data Flow

The digital signal processing sequence that is typically involved in rendering the thickness measurement is shown in Figure 6.1.

Front-End Analog Signals - The detector signal is amplified by an electrometer class pre-amplifier. It is important that this signal set be carefully shielded to prevent interference from external electrostatic noise sources. Often, the detector and pre-amplifier are located within the same shielded housing. The pre-amplifier is band-limited to provide a degree of noise suppression without compromising the temporal dynamics of the material under examination (often the material is in a transport condition with speeds up to 1500 meters / minute, and it is necessary to be able to accurately track thickness variations evolving over 50mm segments (2 millisecond period - 1000Hz minimum Nyquist BW).

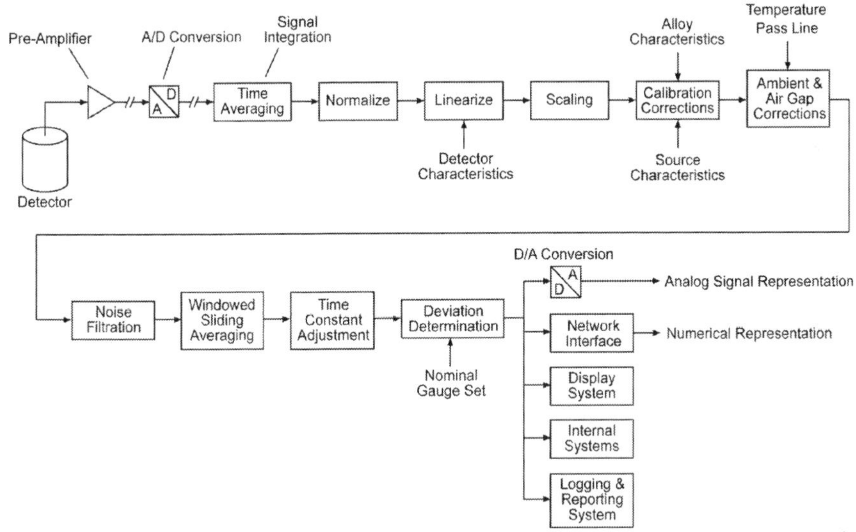

Fig. 6.1 - Simplified block diagram showing a typical digital signal processing sequence for thickness rendering and the transmission of the measurement signal to other systems.

Analog / Digital Conversion (A/D) - The analog pre-amplifier signal is digitized to a numerical form with high resolution (16 bits or greater). It is desirable that this conversion be performed as close as possible to the pre-amplifier to minimize signal runs (length) and the influence of external noise sources on the analog measurement signal. It is not uncommon to locate the A/D converter in close proximity to the detector / pre-amplifier housing (thereby minimizing analog signal runs). The digital / numerical representation of the measurement signal can then be transmitted large distances via noise immune techniques. Typical sampling periods are on the order of 250 microseconds, however slower or faster rates are well within the capabilities of the electronics, and primarily application dependent.

Signal Integration - The numerical measurement signal is integrated over a fixed time interval (typically through simple averaging or low pass filtering methods) to maximize the signal to noise ratio, prior to further signal processing (Bose, 1985). The bandwidth / windowing characteristics of this stage must be selected to not impact the bandwidth of the measurement of the fundamental process (actual material thickness temporal variations).

Normalization - To maximize the signal's dynamic range, the integrated signal can be optionally normalized to a standard signal level framework (often associated with the numerical representation of subsequent signal processing components).

Linearization - Any non-linear aspects of the detector's response characteristics are removed to provide a well defined linear relationship between the measured radiation intensity and the numerical measurement signal.

Scaling - The numerical measurement signal is be converted to engineering / radiometry units from which calibrated adjustments and standardized material / radiation characteristics can be applied.

Calibrated Corrections - This signal processing component often involves the most complex mathematics and numerical methods. The characteristics of the material composition and the radiation source are considered in rendering an initial assessment of the material thickness.

Ambient & Air Gap Corrections - The initial thickness measurement is adjusted to correct for ambient conditions (primarily the air gap temperature and the material temperature) and for known variation in the pass-line. The output signal from this stage is the fundamental thickness measurement. Any subsequent processing of this signal is associated with application specific requirements and compensations.

Noise Filtration - This optional, switchable, programmable filtering stage is applied to situations where expected or unavoidable process related conditions may increase the uncertainties in the measurement accuracy. The nature and operation of this filtration is purely situation dependent and can be as simple as a reduction in the measurement bandwidth, to a sophisticated multi-variable compensation specific to a characterizable disturbance (e.g., uncertainties due to pass-line variations associated with the vertical displacement of vibrating strip), (Bose, 1985).

Windowed Sliding Averaging - This optional signal processing is applied to provide a specific waveform shaping of the measurement signal. This can be employed to assist downstream / related control equipment in compensating for certain process dynamics, (Bose, 1985).

Time Constant Adjustment - To external equipment / observers, the overall measurement system provides an estimate of the material thickness having a programmable, specifiable bandwidth and / or first-order, step function response. Although the fundamental measurement signal may have a relatively wide bandwidth, external equipment may have lower bandwidth input and / or anti-aliasing requirements. This final stage of filtering provides and output signal that corresponds to the bandwith / step function time constant requirement of the external equipment.

Deviation Determination - The rendered signal, X_M, provides an indication of the absolute material thickness, X. External equipment (e.g., Automatic Gauge Control (AGC) systems, Statisical Process Control (SPC) recorders, etc.) may require the measurement system output signal in terms of the Deviation, ΔX_M, about a Nominal Set Thickness, X_{Nom}.

Signal Distribution - The final measurement signal is provided to external equipment via a variety of means. Analog signal representations can be generated to support legacy class systems. Networked interface can provide fully numerical measurement and status data. Display systems ranging from simple metering to sophisticated Graphical User Interface (GUIs) and visualization can be provided to human operators. The measurement system can also provide a variety of internal function based on the final measurement signal (including FFTs, SPC, performance monitoring, status reporting, etc.).

6.2 Characterizing the Measurement Signal

The rendered measurement signal can be transmitted and provided over a number of media and a broad range of formats. At their root, at the completion of the measurement process, the instrument forms a final determination of the material thickness, X_M.

This measurement signal is an indication of absolute thickness, expressed in a chosen engineering unit of measure. The measurement value resides in an internal memory register (possibly fixed or floating point). The value is presented in a number of significant digits, functionally related to the instrument's finest resolution. The signal will operate as a discrete time numeric and be updated at a high frequency.

6.2.1 Signal Formats

There are two(2) primary formats involved in measurement signal transmissions and displays (both graphical and numerical):

Actual Value - This involves the direct transmission of the unipolar material thickness measurement, X_M, in either absolute engineering units or as a percent of Full Scale Range (FSR). When considering analog signals, this format is often expressed in terms of thickness per volt (0.50 mm/volt for a 5mm FSR on a 0-10 volt output range) or percent per volt (0-10 volts equals 0-100% of the FSR). This programmable scaling factor can be adjusted to suit the application of the receiving equipment.

Deviation Value - This involves the transmission of a bipolar deviation signal, ΔX_M, about a nominal thickness value, X_{Nom}, and is provided in either engineering units or a percent of the deviation range. The sum of these signals being the actual thickness value, X_M. The deviation signal is developed as follows:

$$\Delta X_M = X_M - X_{Nom} \tag{6.3}$$

The nominal thickness is typically provided as an operating set-point issued by the end-user / operator or by an overseeing automation system. This arrangement is often preferred when providing thickness readings to thickness control systems (AGC) or quality control / tracking systems (SPC), since the signal accuracy and resolution can be greatly increased. When considering analog signals, this format is often expressed in terms of thickness per volt (25 μm/volt for a +/-0.250mm FSR on a +/-10 volt output range) or percent per volt (+/-10 volts equals +/-100% of the deviation's FSR). This programmable scaling factor can be adjusted to suit the application of the receiving equipment.

6.2.2 Time Response Characterization

The temporal behavior / performance of the measurement signal is classically characterized in terms of a 1st order step response, with the time constant being the key indication of merit. Typical time constants range from 5-200 milliseconds, depending on the accuracy and dynamic response requirements of the application. Most high performance cold rolling and strip processing applications involve time constants on the order of 5-20 milliseconds. Figure 6.2 provides a graphical representation of the time evolution of a classical 1st order step response.

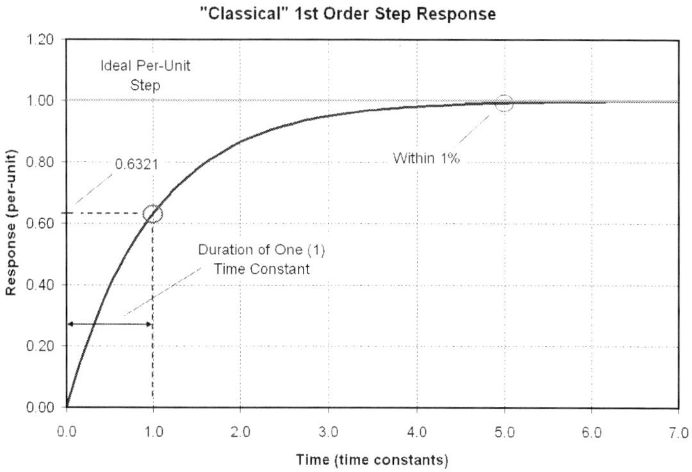

Fig. 6.2 - "Classical" 1st Order Step Response and associated parameters.

The factors (from Figure 6.1) that contribute to the time constant are:

- Pre-Amplifier Band-Limiting Filtration
- Signal Integration / Time Averaging
- Specific Noise Filtration
- Windowed Averaging
- Direct Time Constant Filtration
- Application's Dynamic Response Requirements
- Application's Noise Level and Accuracy Requirements

Realistically, the actual gauging system time response characteristic are more complex and tend to contain higher order dynamics introduced by the signal processing and filtration activities. Specific characterizations of the time response are provided in international standards (IEC, 1996).

Regardless of the signal amplitude resolution, the temporal resolution of the signal's updates plays a key role in the quality and apparent continuity of the provided measurements. Figure 6.3 provides a graphical comparison of the signal fidelity for different signal update rates.

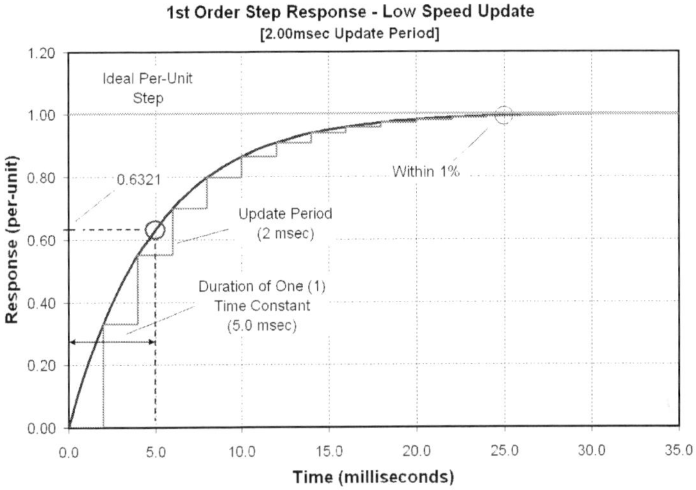

(a)

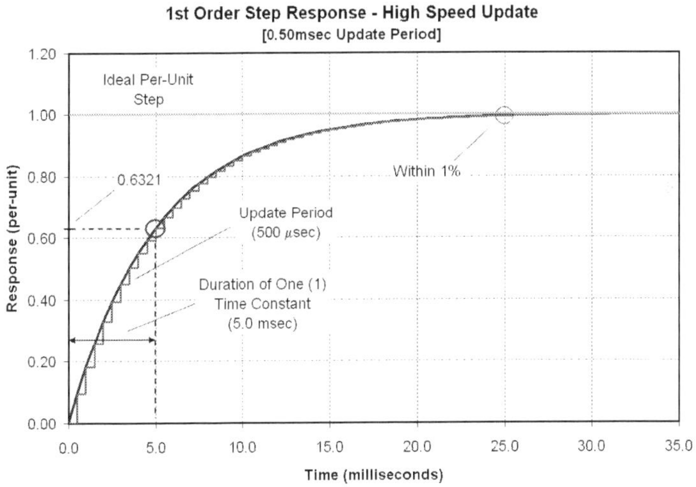

(b)

Fig. 6.3 - Comparison of the influence of update rates on the "fidelity" of the measurement signal having a 5.0 millisecond time constant: a) Time response involving a 2.00 millisecond update period, b) Time response involving a 0.50 millisecond update period.

Beyond the rather crude and abrupt behavior of the slower update signal (see Figure 6.3a), this behavior also introduces a notable phase lag. Downstream systems may require a knowledge of this phase lag to take the necessary compensating actions.

It is possible to further extend the evaluation of the specific time response characteristics to more realistic strip processing through spatial wavefront analysis (I2S, 2004), to account for the passage of abrupt or non-uniform thickness variations through the fast moving strip's beam intercept area (typically circular with a 25mm diameter). However, these subjects are beyond the scope of this discussion.

6.2.3 Measurement Performance Indicators

Accuracy, precision, resolution, repeatability and reproducibility are key terms in specifically describing the fundamental measurement characteristics of the instrument. There are many good sources defining and describing the various aspects of measurement performance (Cooper, 1978) (IEC, 1996) (Nyce, 2004). For the purposes of these discussions, we will concern ourselves with only certain fundamental indicators.

Accuracy - The degree of conformity of a measured or indicated quantity to the actual (true) value. This performance measure is typically expressed as a combination of relative uncertainty and / or absolute uncertainty, in terms of a standardized engineering unit.

Accuracy : +/-0.1% of the nominal thickness or +/-0.50μm, which every is greater, with all alloy compositions resolved and calibrated.

This indication can be based on a statistical sense (2σ - 95.4% of readings) or as a bounding, worst case condition.

Precision - A measure of the degree to which successive measurements differ from one another, over a certain time interval (typically minutes), given the same condition and operating in the same environment. Precision is typically expressed as a combination of relative uncertainty and / or absolute uncertainty, about a nominal operating point, in terms of a standardized engineering unit.

Precision : +/-0.1% of the nominal thickness or +/-0.50μm, which every is greater, in a statistical sense (2σ - 95.4% of readings)

This is often described in terms of a statistical noise envelope of uncertainty associated with any measurement, and the specifications may include an admissible time interval between the measurements and the last standardization.

The definitions of accuracy and precision are often a point of unnecessary confusion and conflict. Figure 6.4 provides an illustration of typical time evolution and amplitude distribution of a thickness deviation signal. Figure 6.5 provides a simple depiction of accuracy and precision when the measurement system's, instantaneous uncertainty characteristics. Figure 6.6 extends these concepts to illustrate the implications of the various combinations of good and bad, accuracy and precision. When formulating or assessing these measures of performance, it is important to grasp the specifics and relevance of their indications with respect to the application requirements.

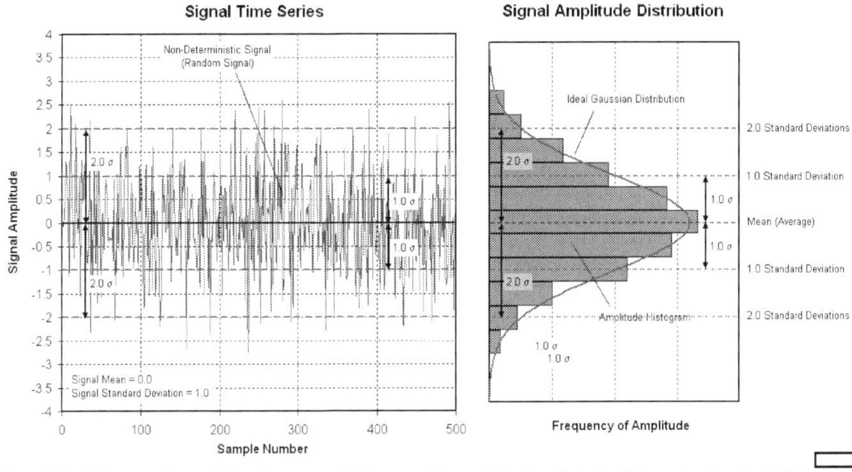

Fig. 6.4 - Illustration of the time evolution and amplitude distribution of a typical thickness measurement deviation signal.

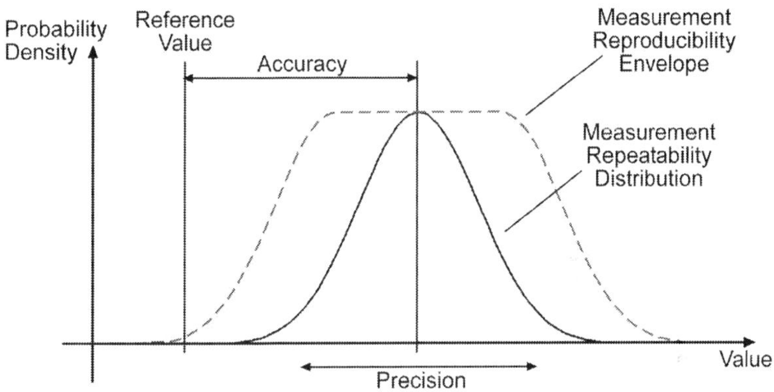

Fig. 6.5 - Fundamental definition and concepts of accuracy and precision, when the measurement uncertainties can be characterized as a white noise having Gaussian amplitude distribution.

Continuing with the definitions:

<u>Repeatability</u> - The closeness of agreement among the results of a number of successive measurements taken for the same value of input under the same operating conditions, over a specific interval of time (often minutes). The standard deviation, σ, of the set of repeatability measurements (often the 2σ indication) is typically used as a measure of precision (see above).

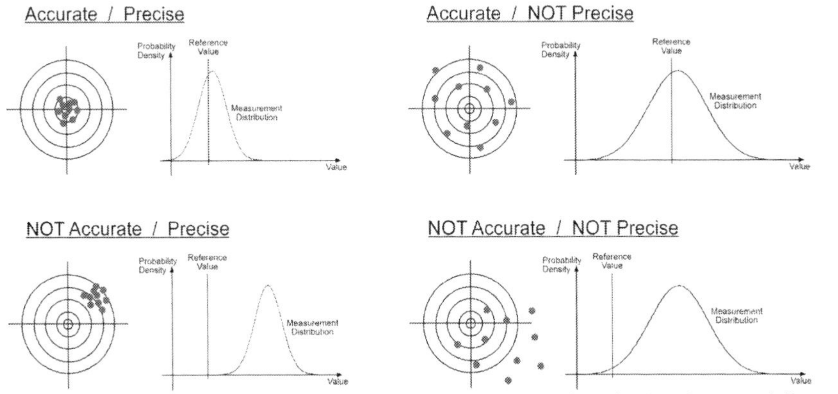

Fig. 6.6 - Comparison of the nature and behavior associated with data having differing levels of accuracy and precision.

Reproducibility - The closeness of agreement among the results of a number of successive measurements taken on the same instrument for the same value of input under the same operating conditions, over an extended interval of time (often hours or days). This means the degree to which a measurement can be replicated independently. This term can be extended to include multiple independent instruments (possibly of the same model or method), to characterize the similarity of measurement s taken (by a product line). Essentially, reproducibility is a bounding envelope, constraining the variability of the accuracy and precision of the instrument, over extended time intervals (see Figure 6.5). Therefore, the instrument's reproducibility is subject to influences of accuracy and precision / noise issues. Reproducibility is typically expressed as a relative or absolute variability bounds and may be based on the standard deviation (often the 2σ indication) of the set of measurements taken over extended periods of time, or a worst case bounding condition.

Reproducibility: +/-0.05% of the nominal thickness or +/-0.25μm, which every is greater, with all alloy compositions resolved and calibrated.

It can also be applied as a permissible range of variability of the mean of the set of repeatability measurement groups taken over long periods of time or on different / independent instruments.

Drift - This describes the variance of the mean of a set of successive groups over a period of time, and serves as a measure of instrument stability. Drift is typically expressed as a relative or absolute variability bound, often as a worst case bounding condition over an interval of time.

Drift : +/-0.1% of the nominal thickness over an 8 hour period

Drift variances are usually determined from an assessment of the mean of repeatability measurements taken at specified intervals of time, and evaluated for long term trending behaviour.

Resolution - The smallest change of the quantity being measured to which the instrument will respond and a change in the measurement signal can be observed or detected. Instrument resolution is related to the precision at which the measurement was made and to the Minimum Detectable Signal (Motchenbacher & Fitchen, 1973) which is typically defined as twice the system noise level (precision). Resolution can also be defined in terms of the quantization or significant digits of the numerical representation (i.e., 16 bit resolution or 6 significant digit display) over a specific operating range. This infers a Least Significant Bit (LSB) / single digit quantization, which may be misleading and have no relationship to the actual instrumentation resolution. Resolution differs from quantization, in that quantization is a mapping process from a continuous range of values to a discrete set of values having a minimum quanta (LSB).

Linearity - This closeness that the calibrated, standardized instrument approximates a straight line response over a specified measurement range. Linearity is typically expressed as an absolute or relative indication of the maximum differences between the instrument's readings and the actual values, over a measurement interval.

Linearity : +/-0.1% of the FSR over a +/-1.0 mm range from the standardized nominal thickness

Some instruments operate in a narrowly defined region of the overall range of measurement capabilities. Here, linearity also provides an indication of the breadth of the localized measurement range, before a distortion appears on the instrument's output.

7. System Sensitivity Characteristics and Relationships

When assessing a measurement produced by these forms of instruments, we must be cognizant of the many influences that can distort or corrupt the accuracy or quality of the measurement.

7.1 General Sensitivity

For a constant incident radiation intensity of, I_0, the general sensitivity is defined as the relative variations in emitted beam intensity, I, corresponding to the relative variations in the material thickness, x.

$$S = \frac{\frac{dI}{I}}{\frac{dx}{x}} \tag{7.1}$$

From Eq(4.1) we have:

$$dx = -\left(\frac{\rho}{\mu}\right)\frac{1}{\rho}\frac{dI}{I} = -\frac{1}{\mu}\frac{dI}{I} = -q\frac{dI}{I} \tag{7.2}$$

Resulting in:

$$S = -\left(\frac{\mu}{\rho}\right)\rho x = -\mu x = -\frac{x}{q} \tag{7.3}$$

In making S as large as possible, we see that there are two(2) possible degrees of freedom.

- Increasing the attenuation by operating at lower energy levels
- Operating at the maximum thickness the instrument can measure.

In practice, the choice of the thickness range and the operating energy is based on the application and instrument characteristics.

7.2 Statistical Noise

The random behavior of the atomic transformation process (fundamentally Poisson statistics) and the random nature of X-Ray photon emissions from electron beam formation and bombardment (primarily a function of micro fluctuations in filament temperature and current, like a shot noise) can be characterized as a zero mean, white noise superimposed on the measured signal. The standard deviation of a monochromatic noise function, when measured over a given integration time, $\Delta\tau$, can be characterized by : (Motchenbacher & Fitchen, 1973)

$$\frac{\sigma(I)}{I} = \frac{K}{\sqrt{\Delta\tau \cdot I}} \tag{7.4a}$$

$$\sigma(I) = K\sqrt{\frac{I}{\Delta\tau}} \tag{7.4b}$$

where K is a constant. When considering the effect of this noise on the measurement signal, $\hat{x}$, can be obtained from combining Eq(7.1) and Eq(7.3)

$$\frac{\partial x}{x} = -\left(\frac{\rho}{\mu}\right)\frac{1}{\rho x}\frac{\partial I}{I} = \frac{-1}{\mu x}\frac{\partial I}{I} = -\frac{q}{x}\frac{\partial I}{I} \tag{7.5}$$

As a consequence, the relative statistical variation in the measured thickness can be provided by:

$$\frac{\Delta x}{x} = -\left(\frac{\rho}{\mu}\right)\frac{1}{\rho x}\frac{K}{\sqrt{\Delta\tau \cdot I}} = \frac{-1}{\mu x}\frac{K}{\sqrt{\Delta\tau \cdot I}} = -\frac{q}{x}\frac{K}{\sqrt{\Delta\tau \cdot I}} \tag{7.6}$$

Combining Eq(6.1)

$$\frac{\Delta x}{x} = -\left(\frac{\rho}{\mu}\right)\frac{1}{\rho x}\frac{K \cdot e^{\frac{1}{2}\left(\frac{\mu}{\rho}\right)\rho x}}{\sqrt{\Delta\tau \cdot I_0}} = \frac{-1}{\mu x}\frac{K \cdot e^{\frac{1}{2}\mu x}}{\sqrt{\Delta\tau \cdot I_0}} = -\frac{q}{x}\frac{K \cdot e^{\frac{1}{2}\frac{x}{q}}}{\sqrt{\Delta\tau \cdot I_0}} \tag{7.7}$$

Figure 7.1 provides a graphical comparison of the relative statistical measurement noise for various integration time intervals. The important aspects of this figure are the integration time independent, noise minimum at the optimal thickness of:

$$x_{Opt} = 2q \tag{7.8}$$

Noise associated with shorter Mean Free Paths have reciprocal behavior, while longer paths have exponential characteristics. Also important is the elevation of the noise function as a function of the reciprocal of the square root of the integration interval.

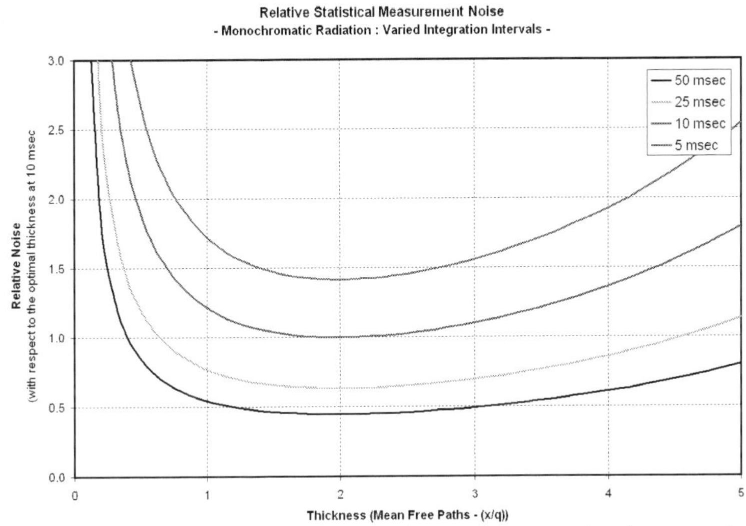

Fig. 7.1 - Comparison of the relative statistical measurement noise for monochromatic radiation as a function of the Mean Free Paths of thickness, for various integration times.

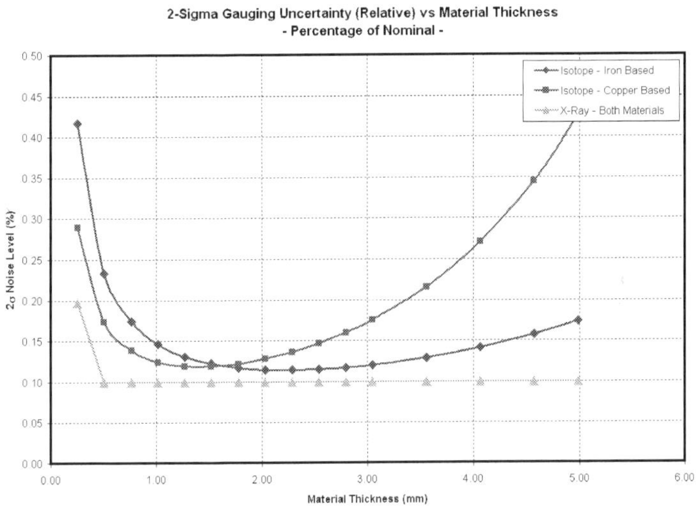

Fig. 7.2 - Comparative plot of the actual noise functions (2σ of the nominal thickness) of a monochromatic 1.0 Ci Americium 241 source applied to NIST Traceable samples of know steel and copper alloys, for a 25 msec integration period. This plot includes the noise function of an X-Ray based instrument (operating with a 10 msec integration period) employing tube potential adjustment to maintain a constant signal-to-noise ratio for thicknesses above 0.5μm.

The relationships of Eqs(7.4) through Eq(7.8) are associated with monochromatic radiation, and therefore are suitable for characterizing isotope sources (e.g., Americium 241). Figure 7.2 provides some instrument performance data for a 1.0 Ci Americium 241 source, applied to NIST Traceable samples of known steel and copper alloys.

This comparative plot shows the variations in the optimal thickness as a function of material composition. This plot also includes the noise function of an X-Ray based instrument employing tube potential adjustment to maintain a constant signal-to-noise ratio. In this case, the choice of applied tube voltage is focused on optimizing the noise level while simultaneously maintain a specific level of signal sensitivity. Further, the X-Ray system is operating with a 10 msec integration period, and achieving better performance. This is primarily because of the degrees of freedom available to the X-Ray gauge's operating point (beam current, tube potential, filament current stability, etc.).

When considering the polychromatic radiation of the X-Ray generator, it's necessary to consider the energy dependencies of the source's applied radiation and the material.

$$\frac{\Delta x}{x}(\lambda) = -\left(\frac{\rho}{\mu}(\lambda)\right)\frac{1}{\rho x}\frac{K\cdot e^{\frac{1}{2}\left(\frac{\mu}{\rho}(\lambda)\right)\rho x}}{\sqrt{\Delta\tau\cdot I_0}} = \frac{-1}{\mu(\lambda)x}\frac{K\cdot e^{\frac{1}{2}\mu(\lambda)x}}{\sqrt{\Delta\tau\cdot I_0}} = -\frac{q(\lambda)}{x}\frac{K\cdot e^{\frac{1}{2}\frac{x}{q(\lambda)}}}{\sqrt{\Delta\tau\cdot I_0}} \tag{7.9}$$

This non-trivial expression is typically not analytically evaluated to determine the optimal noise level, but is often found through empirical characterizations with standardized samples and calibration procedures (Howard, 1970).

It is also possible to express the measurement noise in terms of an absolute thickness variation. Returning to Eq(7.7), statistical noise expressed in terms of absolute thickness is:

$$\Delta x = -\left(\frac{\rho}{\mu}\right)\frac{1}{\rho}\frac{K\cdot e^{\frac{1}{2}\left(\frac{\mu}{\rho}\right)\rho x}}{\sqrt{\Delta\tau\cdot I_0}} = \frac{-1}{\mu}\frac{K\cdot e^{\frac{1}{2}\mu x}}{\sqrt{\Delta\tau\cdot I_0}} = -q\frac{K\cdot e^{\frac{1}{2}\frac{x}{q}}}{\sqrt{\Delta\tau\cdot I_0}} \tag{7.10}$$

This function's behavior is largely exponential with a minimum at zero thickness (i.e., solely a function of the statistical variations of the source's radiation, I_0, and the integration period. Figure 7.3 provides the associated instrument performance data of Figure 7.2, in absolute terms.

7.3 Sensitivity to X-Ray Tube Voltage Variations

As shown in Figure 3.9, the X-Ray Tube's emitted intensity, I_0, varies with the applied potential (tube voltage), V_X. For the range of interest, the relationship between the intensity and applied voltage can be given by:

$$I_0 \sim K_V\left(V_X\right)^3 \tag{7.11}$$

and the MAC behavior (in the 10kV to 100kV range - primarily Photoelectric Absorption) can be described with:

$$\frac{\mu}{\rho} \sim \frac{K_{MAC}}{\rho\cdot\left(V_X\right)^3} \Rightarrow q = K_{MAC}\left(V_X\right)^3 \tag{7.12}$$

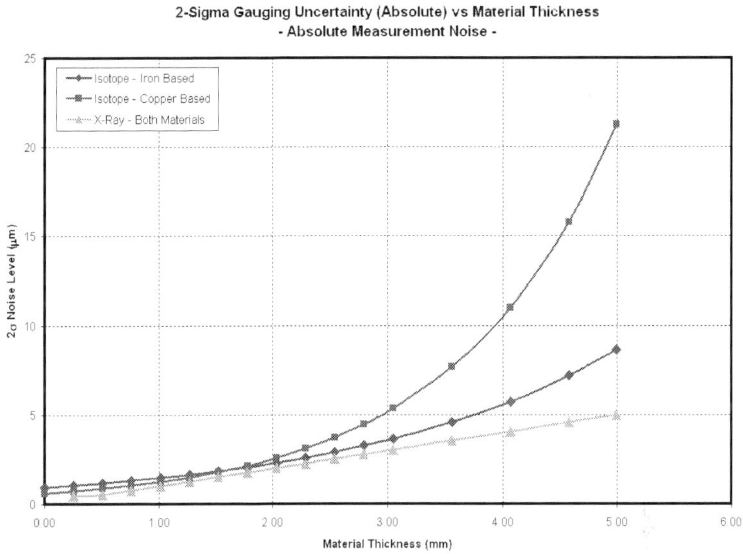

Fig. 7.3 - Comparative plot of the actual noise functions (2σ in absolute measured thickness variations) based on the instruments and situations associated with Figure 7.2.

where K_V and K_{MAC} are instrument dependent constants. From Eq(4.2) we have:

$$\ln(I) = \ln(I_0) - \left(\frac{\mu}{\rho}\right)\rho x = \ln(I_0) - \frac{x}{q} \tag{7.13}$$

Substituting Eqs(7.11, 7.12) results in:

$$\ln(I) = \ln\left(K_V (V_X)^3\right) - \frac{x}{K_{MAC}(V_X)^3} = \ln(K_V) + 3\ln(V_X) - \frac{x}{K_{MAC}}(V_X)^{-3} \tag{7.14}$$

Continuing and differentiating with respect to V_X provides:

$$\frac{\partial(\ln(I))}{\partial V_X} = \frac{1}{I}\frac{\partial I}{\partial V_X} = \frac{3}{V_X} + 3\frac{x}{K_{MAC}}(V_X)^{-4} = \frac{3}{V_X} + 3\frac{x}{K_{MAC}}\frac{(V_X)^{-3}}{V_X} \tag{7.15a}$$

or with Eq(7.12)

$$\frac{\partial I}{I} = 3\frac{\partial V_X}{V_X} + 3\frac{x(V_X)^{-3}}{K_{MAC}}\frac{\partial V_X}{V_X} = 3\left(1 + \left(\frac{\mu}{\rho}\right)\rho x\right)\frac{\partial V_X}{V_X} = 3\left(1 + \frac{x}{q}\right)\frac{\partial V_X}{V_X} \tag{7.15b}$$

Returning to the fundamental sensitivity relationship of Eq(7.2), we have:

$$\frac{\partial I}{I} = -\mu\,\partial x = -\frac{\partial x}{q} \tag{7.16}$$

Introducing this into Eq(7.15b) results in:

$$\partial x = -\frac{3}{\rho}\left(\frac{\rho}{\mu}\right)\left(1+\left(\frac{\mu}{\rho}\right)\rho x\right)\frac{\partial V_X}{V_X} = -3q\left(1+\frac{x}{q}\right)\frac{\partial V_X}{V_X} \tag{7.17}$$

The corresponding relative variations in the measured thickness associated with relative variations in the applied tube voltage are:

$$\frac{\Delta x}{x} = -3\left(\left(\frac{\rho}{\mu}\right)\frac{1}{\rho x}+1\right)\frac{\Delta V_X}{V_X} = -3\left(\frac{q}{x}+1\right)\frac{\Delta V_X}{V_X} \tag{7.18}$$

This is an important relationship, because it defines the impact of the applied voltage regulation quality on the measurement signal. Because of its relative use of the applied voltage, the relationship is a reciprocal function is MAC / MFP, and always acts as an amplification factor (asymptotically approaching its minima of -3). This requires exceedingly good voltage regulation, especially for thicknesses below an MFP of 1.0.

When evaluating the function of Eq(7.18), one must be cognizant that the polychromatic nature of the X-Ray radiation and the tube's energy dependent efficiency are not considered (along with other factors), and therefore the relationship provides only the fundamental guidelines associated with understanding the voltage regulation requirements, and the trends associated with its sensitivities.

7.4 Sensitivity to X-Ray Tube Beam Intensity Variations

For a fixed X-Ray Tube potential, or when considering the Curie (Ci) rating of an Isotope source (μ or q effectively constant), the relative variation in the intensity of the generator / source emitted radiation can be characterized from Eq(7.2).

$$\frac{\partial x}{x} = -\left(\frac{\rho}{\mu}\right)\frac{1}{\rho x}\frac{\partial I}{I} = -\frac{1}{\mu x}\frac{\partial I}{I} = -\frac{q}{x}\frac{\partial I}{I} = -\frac{q}{x}\frac{\partial I_0}{I_0} \tag{7.19a}$$

or

$$\frac{\Delta x}{x} = -\left(\frac{\rho}{\mu}\right)\frac{1}{\rho x}\frac{\Delta I_0}{I_0} = -\frac{q}{x}\frac{\Delta I_0}{I_0} \tag{7.19b}$$

Isotope source variations are dominated by Poisson statistics, and therefore the characteristics of the associated relative measurement noise are also dictated by these statistics. In X-Rays, due to the fundamental physics of the escape of thermally excited filament electrons, the noise characteristics involve a certain amount of Shot Noise. The reciprocal behavior of the MAC / MFP provides an attenuation for larger thicknesses, while thicknesses below an MFP of 1.0 while experience amplified noise levels.

7.5 Sensitivity to Variations in Material Composition

To render an accurate thickness measurement, radiation absorption gauges require a knowledge of material's radiation attenuation characteristics (i.e., (μ/ρ), ρ or q). These parameters can be obtained by direct calibration with samples of known thickness and composition, or through direct entry of the chemical / elemental constituents and their associated percentage inclusions (coupled with tabulated listings of the MACs (Hubble & Seltzer, 2004)). Failure to properly characterize the attenuation parameters of the material under measurement will lead to an improper set-up of the instrument's operating point and inaccurate indications of the material thickness.

To examine the instrument's sensitivity to variances in the alloy of the material, we will examine a case where an alloy of known chemistry and density, is subject to contamination by a low level concentration of a non-accounted constituent.

Constituent #1 - Base alloy (of possibly many components) that is known to the gauge.

Constituent #2 - Contaminating component of *unknown* attenuation characteristics, density and concentration / weighting.

If we assume that the material's constituent attenuation behavior can be characterized as a composition of successive layers of the various elements. The actual thickness, X, is therefore the sum of the individual layer thicknesses.

$$X = x_1 + x_2 \quad \Rightarrow \quad X = \sum_{i=1}^{N} x_i \tag{7.20}$$

The attenuated beam intensity emitted from each layer is given by:

$$I_1 = I_0 e^{-\frac{x_1}{q_1}} \tag{7.21a}$$

$$I_2 = I_1 e^{-\frac{x_2}{q_2}} \tag{7.21b}$$

$$\Downarrow$$

$$I = I_N = I_{N-1} e^{-\frac{x_N}{q_N}} \tag{7.21c}$$

or

$$I = I_0 \left[e^{-\frac{x_1}{q_1}} \cdot e^{-\frac{x_2}{q_2}} \right] \Rightarrow I = I_0 \left[e^{-\frac{x_1}{q_1}} \cdot e^{-\frac{x_2}{q_2}} \cdots\cdots e^{-\frac{x_N}{q_N}} \right] \tag{7.22}$$

The erroneous, measured thickness reading from the gauging system, X_m, is based on the assumption that the material is only composed of the q_1 constituent attenuation.

$$I = I_0 e^{-\frac{X}{q_A}} = I_0 e^{-\frac{X_M}{q_1}} \tag{7.23}$$

where q_A is the composite MFP. Understanding the nature of the differences between Eq(7.23) and Eq(7.22) is key to assessing the sensitivity to alloy variations. The relationship between the actual and measured thickness is based on:

$$I = I_0 \left[e^{-\frac{X}{q_A}} \right] = I_0 \left[e^{-\frac{x_1}{q_1}} \cdot e^{-\frac{x_2}{q_2}} \right] = I_0 \left[e^{-\frac{X_M}{q_1}} \right] \tag{7.24}$$

$$-\ln\left[\frac{I}{I_0}\right] = \frac{X}{q_A} = \frac{x_1}{q_1} + \frac{x_2}{q_2} = \frac{X_M}{q_1} \quad \Rightarrow \quad \frac{x_1}{q_1} + \frac{x_2}{q_2} + \cdots + \frac{x_N}{q_N} = \frac{X_M}{q_1} \tag{7.25}$$

or

$$X_M = q_1\left[\frac{x_1}{q_1} + \frac{x_2}{q_2}\right] \quad \Rightarrow \quad X_M = q_1\left[\frac{x_1}{q_1} + \frac{x_2}{q_2} + \cdots + \frac{x_N}{q_N}\right] \tag{7.26}$$

The relationship between the actual thickness, X, and the measured thickness, X_m, is:

$$X = x_1 + x_2 = K_q X_M \quad \Rightarrow \quad X = \sum_{i=1}^{N} x_i = K_q X_M \tag{7.27}$$

Therefore from Eq(7.25)

$$K_q = \frac{q_A}{q_1} = \frac{x_1 + x_2}{q_1\left[\frac{x_1}{q_1} + \frac{x_2}{q_2}\right]} \quad \Rightarrow \quad K_q = \frac{q_A}{q_1} = \frac{\sum_{i=1}^{N} x_i}{q_1\left[\sum_{i=1}^{N} \frac{x_i}{q_i}\right]} \tag{7.28}$$

which provides the composite MFP, q_A as:

$$q_A = \frac{x_1 + x_2}{\left[\frac{x_1}{q_1} + \frac{x_2}{q_2}\right]} \quad \Rightarrow \quad q_A = \frac{\sum_{i=1}^{N} x_i}{\left[\sum_{i=1}^{N} \frac{x_i}{q_i}\right]} \tag{7.29}$$

Considering our two(2) constituent example, the component weighting follows from Eqs(4.4a,b) and Eq(4.5), and is based on convex function of the alloy inclusion, α.

$$w_1 = (1-\alpha) = \frac{\rho_1 x_1}{\rho_A X} \tag{7.30a}$$

$$w_2 = \alpha = \frac{\rho_2 x_2}{\rho_A X} \tag{7.30b}$$

where ρ_A is the composite density of the total alloyed material and it is assumed that the alloy inclusion is small (i.e., $\alpha \to 0$). Rearranging Eq(7.30a) results in:

$$x_1 = w_1 \frac{\rho_A X}{\rho_1} \quad \Rightarrow \quad x_i = w_i \frac{\rho_A X}{\rho_i} \tag{7.31}$$

Returning to Eq(7.28), we have:

$$K_q = \frac{\frac{w_1}{\rho_1} + \frac{w_2}{\rho_2}}{q_1\left[\frac{w_1}{\rho_1 \cdot q_1} + \frac{w_2}{\rho_2 \cdot q_2}\right]} = \frac{\frac{w_1}{\rho_1} + \frac{w_2}{\rho_2}}{q_1\left[w_1\left(\frac{\mu}{\rho}\right)_1 + w_2\left(\frac{\mu}{\rho}\right)_2\right]} \tag{7.32a}$$

$$\Downarrow$$

$$K_q = \frac{\sum_{i=1}^{N}\frac{w_i}{\rho_i}}{q_1\left[\sum_{i=1}^{N}\frac{w_i}{\rho_i \cdot q_i}\right]} = \frac{\sum_{i=1}^{N}\frac{w_i}{\rho_i}}{q_1\left[\sum_{i=1}^{N} w_i\left(\frac{\mu}{\rho}\right)_i\right]} \tag{7.32b}$$

which provides the composite MFP, q_A as:

$$q_A = \frac{\frac{w_1}{\rho_1} + \frac{w_2}{\rho_2}}{\left[w_1\left(\frac{\mu}{\rho}\right)_1 + w_2\left(\frac{\mu}{\rho}\right)_2\right]} = \frac{\left[\frac{(1-\alpha)}{\rho_1} + \frac{\alpha}{\rho_2}\right]}{\left[(1-\alpha)\left(\frac{\mu}{\rho}\right)_1 + \alpha\left(\frac{\mu}{\rho}\right)_2\right]} \Rightarrow q_A = \frac{\sum_{i=1}^{N}\frac{w_i}{\rho_i}}{\left[\sum_{i=1}^{N} w_i\left(\frac{\mu}{\rho}\right)_i\right]} \tag{7.33}$$

It is important to note that the above formulations assume that the volume of the composite alloy equals the sum of the volume of the constituents (prior to mixing).

The real problem at hand is determining if a change in the transmitted radiation, I, is the result of a change in the actual thickness, X, or a change in composition, which can be examined through the fundamental relationship of Eq(7.25).

$$\frac{X}{q_A} = X\frac{\left[(1-\alpha)\left(\frac{\mu}{\rho}\right)_1 + \alpha\left(\frac{\mu}{\rho}\right)_2\right]}{\left[\frac{(1-\alpha)}{\rho_1} + \frac{\alpha}{\rho_2}\right]} \tag{7.34}$$

We are interested in obtaining the relative error as a function of variations in the alloy inclusion, α. Variations in the received radiation intensity due to variations in actual thickness or to variations in the alloy inclusion are related by:

$$\frac{\partial\left(\frac{X}{q_A}\right)}{\partial X}\partial X = \frac{\partial\left(\frac{X}{q_A}\right)}{\partial \alpha}\partial\alpha \tag{7.35}$$

which results in:

$$\frac{\left[(1-\alpha)\left(\frac{\mu}{\rho}\right)_1+\alpha\left(\frac{\mu}{\rho}\right)_2\right]}{\left[\frac{(1-\alpha)}{\rho_1}+\frac{\alpha}{\rho_2}\right]}\partial X=\cdots$$

$$=\left[\frac{\left[\frac{(1-\alpha)}{\rho_1}+\frac{\alpha}{\rho_2}\right]\left[\left(\frac{\mu}{\rho}\right)_2-\left(\frac{\mu}{\rho}\right)_1\right]-\left[(1-\alpha)\left(\frac{\mu}{\rho}\right)_1+\alpha\left(\frac{\mu}{\rho}\right)_2\right]\left[\frac{1}{\rho_2}-\frac{1}{\rho_1}\right]}{\left[\frac{(1-\alpha)}{\rho_1}+\frac{\alpha}{\rho_2}\right]^2}\right]X\,\partial\alpha \quad (7.36)$$

Since the alloy inclusion, α, is small

$$\frac{\partial X}{X}=\left[\frac{\left[\frac{1}{\rho_1}+\frac{\alpha}{\rho_2}\right]\left[\left(\frac{\mu}{\rho}\right)_2-\left(\frac{\mu}{\rho}\right)_1\right]-\left[\left(\frac{\mu}{\rho}\right)_1+\alpha\left(\frac{\mu}{\rho}\right)_2\right]\left[\frac{1}{\rho_2}-\frac{1}{\rho_1}\right]}{\left[\frac{1}{\rho_1}+\frac{\alpha}{\rho_2}\right]\left[\left(\frac{\mu}{\rho}\right)_1+\alpha\left(\frac{\mu}{\rho}\right)_2\right]}\right]\partial\alpha \quad (7.37)$$

which reduces to :

$$\frac{\partial X}{X}=\left[\frac{\left[\left(\frac{\mu}{\rho}\right)_2-\left(\frac{\mu}{\rho}\right)_1\right]}{\left[\left(\frac{\mu}{\rho}\right)_1+\alpha\left(\frac{\mu}{\rho}\right)_2\right]}+\left[\frac{\rho_2-\rho_1}{\rho_2+\alpha\rho_1}\right]\right]\partial\alpha \quad (7.38a)$$

or

$$\frac{\Delta X}{X}=\left[\frac{\left[\left(\frac{\mu}{\rho}\right)_2-\left(\frac{\mu}{\rho}\right)_1\right]}{\left[\left(\frac{\mu}{\rho}\right)_1+\alpha\left(\frac{\mu}{\rho}\right)_2\right]}+\left[\frac{\rho_2-\rho_1}{\rho_2+\alpha\rho_1}\right]\right]\Delta\alpha \quad (7.38b)$$

A key factor in Eq(7.38) is that the impact of the alloy inclusion is separable in terms of the differences in MACs and in densities. This is important, since the MAC terms are functions of the applied radiation's spectral content, while the density and alloy inclusion remain independent of radiation.

In an X-Ray System, the applied tube potential can be swept across an energy range, while relative sensitivity measurements are made. From these measurements it is possible to identify the differential MAC components, and infer the density difference from projections to a nominal, zero thickness offset. In the bi-alloy case of Eq(7.38b), it is possible to take two(2) measurements at strategically different energies (i.e., tube voltages that yield sufficiently different values of MACs), and resolve two(2) independent equations and two(2) unknowns (the unknowns being the actual thickness, X, and the alloy inclusion, α).

7.6 Sensitivity to Variations in Pass-Line Height

Variations in the vertical location of the material can introduce measurement errors. These primarily stem from the geometries of the radiation lobe patterns associated with coherent and incoherent scattering (see Section 4.2). As shown in Figure 4.1, within the radiation energy region of interest (10keV to 200keV), the MAC is composed of photoelectric absorption, coherent and incoherent scattering. As radiation energy levels approach and exceed 100keV, the nature of the transmitted radiation shifts from a photoelectric absorption to an incoherent scattering form (Compton).

As shown in Figure 7.4a, the generated radiation beam is formed about the optical axis, with conic angles defined by the point-size of the source, and the size, orientation and location of the collimating aperture. The transmitted beam, emerging from the strip, tends to maintain the conic angle relationship and is ultimately over-contained by the detector. A component of the radiation emerging from the strip is associated with scattering. The radiation pattern forms in lobes about the optical axis, largely directed about a primary / dominant scattering angle.

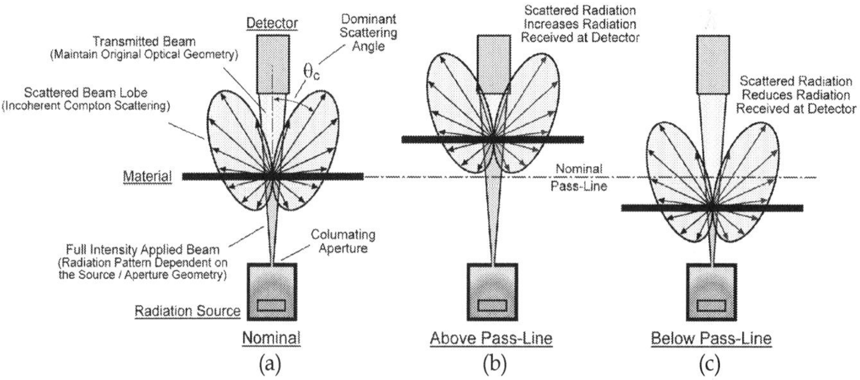

Fig. 7.4 - Illustration of the radiation pattern geometry and impact on measurement accuracy in the presence of pass-line height variations: a) Definition of the radiation patters, including the primary transmitted beam and incoherent scattering (Compton) lobe patterns, b) Exaggerated illustration of the impact of a vertical offset in pass-line height, c) Exaggerated illustration of the impact of a vertical reduction in pass-line height.

At pass-line heights above the nominal (see Figure 7.4b), the geometry of the scattered radiation pattern causes the detector's received radiation intensity to increase, inducing the instrument to report a thinner than actual reading. Measurement errors can approach 0.1% of the nominal thickness for 10mm of pass-line height change when the air gap distances are small and the radiation source collimator projects a wide angled beam.

At pass-line heights below the nominal (see Figure 7.4c), the geometry of the scattered radiation pattern directs too many photons away from the detector. Depending on the extent of the nominal pass-line scattering characteristics, the detector's received radiation intensity may potentially decrease, inducing the instrument to report a thicker than actual reading, or may show no response.

When the air gap distance is relatively large >150mm, and the beam angle is relatively tight, the pass-line height variations are not significant for normal operating conditions on properly aligned equipment. Certain cold mill wiping systems can modify the pass-line height introducing the possibility for measurement errors on small air gap isotope gauges mounted close to the wiping systems.

When X-Ray Systems are operated at higher energies (>100kV), the transmitted radiation is dominated by Compton Scattering, making these systems (and operating regimes) more susceptible to pass-line elevation induced measurement errors.

7.7 Sensitivity to Variations in Pass-Line Inclination

Changes to the angle of the material presented to the radiation beam can result in measurement errors due to the oblique angle of incidence. These errors are associated with the cosine effects leading to a longer pathway through the material. Figure 7.5 illustrates this condition.

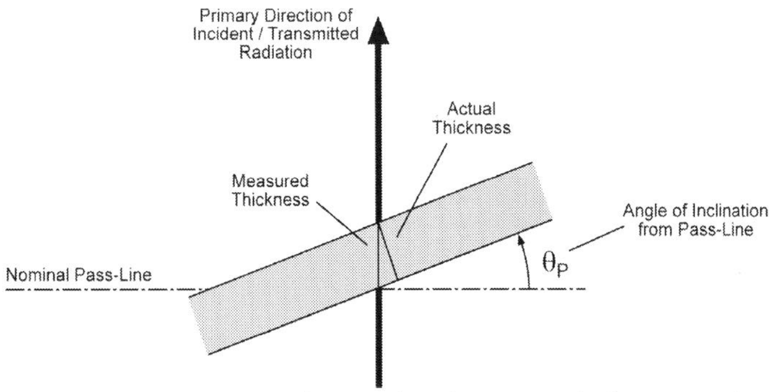

Fig. 7.5 - Illustration of the impact of material inclination, with the increased path-length associated the cosine effect.

The impact on the measured thickness, X_M, for an actual thickness, X, is given by:

$$X_M = \frac{X}{\mathrm{Cos}(\theta_P)} \tag{7.39}$$

where θ_P is the angle of material inclination for the reference pass-line. The longer pathway cause the instrument to report a thicker than actual reading.

It's important to note that combinations of pass-line height and inclination variations can induce complex measurement errors. This is primarily evident when rolling / measuring relatively thick / heavy gauge plate materials (~10-15mm) having center-buckling flatness distortions. These longitudinal waves induce both pass-line height and inclination angle variations, as the strip passes, causing oscillating measurement errors.

7.8 Sensitivity to Variations in Air Gap Temperature and Density

The ambient atmosphere above and below the material, within the air column about the optical axis in the air gap, is also involved in the attenuation of the applied and transmitted / scattered radiation, and therefore the instruments measurements. The composite MAC for air coupled with the instantaneous density of the air column, form the air gap's attenuation characteristics.

Changes in the air gap, air column temperature cause a direct change in the air column density, resulting in a measurement error. This is most apparent when measuring thin materials or materials having reduced absorption characteristics. The general trends are as follows:

- Increases in Air Column Temperature result in a reduction of the air column density, allowing more radiation to pass and be collected by the detector. The instrument will report a thinner than actual reading.
- Decreases in Air Column Temperature result in an increase of the air column density, attenuating more radiation and reducing the radiation collected by the detector. The instrument will report a thicker than actual reading.

Returning to the attenuation equations of Eq(6.2) and Eq(7.25), the actual measurement situation can be more precisely described by:

$$-\ln\left[\frac{I}{I_0}\right]=\frac{X}{q}+\frac{Z}{q_{AG}}=\frac{X_M}{q_S} \tag{7.40}$$

where Z is the air gap dimension, q is the MFP of the material, q_{AG} is the air column MFP and q_s is the effective MFP determined during the last standardization. Variations in the received radiation intensity due to variations in actual thickness or to variations in the air gap attenuation characteristics are related by:

$$\frac{\partial\left(\frac{X_M}{q_S}\right)}{\partial X}\partial X=\frac{\partial\left(\frac{X_M}{q_S}\right)}{\partial q_{AG}}\partial q_{AG} \tag{7.41}$$

which results in:

$$\frac{1}{q}\partial X=-\frac{Z}{\left(q_{AG}\right)^2}\partial q_{AG} \tag{7.42}$$

The variation in thickness which would result from variation in the air gap attenuation is given by:

$$\partial X=-\frac{Z\cdot q}{\left(q_{AG}\right)^2}\partial q_{AG} \tag{7.43}$$

If we consider only monochromatic radiation, the MFP of the air gap, q_{AG}, is related to the average distance between atoms in the air column gasses. For a mass of gas held at the same pressure, we have:

$$\frac{\partial V}{V} = \frac{\partial T_{AG}}{T_{AG}} \tag{7.44}$$

where V is the volume of the gas mass, and T_{AG} is the gas temperature. Since the volume, mass, density and MFP of the air column gas are related, we can use:

$$\frac{\partial q_{AG}}{q_{AG}} = \frac{\partial T_{AG}}{T_{AG}} \tag{7.45}$$

Returning to Eq(7.43), results in:

$$\partial X = -Z \cdot \frac{q}{q_{AG} \cdot T_{AG}} \partial T_{AG} = -Z \cdot \frac{\left(\frac{\rho}{\mu}\right)\frac{1}{\rho}}{\left(\frac{\rho}{\mu}\right)_{AG} \frac{1}{\rho_{AG}\left(T_{AG}\right)} \cdot T_{AG}} \partial T_{AG} \tag{7.46a}$$

or

$$\Delta X = -Z \cdot \frac{q}{q_{AG} \cdot T_{AG}} \Delta T_{AG} = -Z \cdot \frac{\left(\frac{\mu}{\rho}\right)_{AG}}{\left(\frac{\mu}{\rho}\right)} \frac{\rho}{\rho_{AG}\left(T_{AG}\right)} \frac{\Delta T_{AG}}{T_{AG}} \tag{7.46b}$$

It is important to note MACs of the material and air gap column are constant, and not function of density or temperature. The density of the air gap gasses, $\rho_{AG}(T_{AG})$ is a function of temperature, and has a reciprocal relationship within this function, causing an increase in the measurement error as the density is reduced with temperature.

At 20 deg C, an increase in the air gap temperature of 30 deg C at 60% humidity, has a 12.1% reduction in the air density, which can have a dramatic effect on the instrument's behavior. Depending on the radiation energy level and the material chemistry, this level of air density variation can create measurement errors on the order of 0.15% of the nominal thickness.

7.9 Sensitivity to Variations in Material Temperature

Transmission based radiation thickness gauges measure the mass per unit area of the material in the beam. At an elevated temperature, T_1, the material volume increases, causing the unit area of the material engaged with the radiation beam to be reduced (when considering its ambient temperature conditions, T_0). Therefore, the area engaged, A_0, with the beam at the ambient temperature T_0, will expand to the area, A_1, at the material temperature, T_1, in the following manner:

$$A_1 = A_0\left(1 + 2K_T\left(T_1 - T_0\right)\right) \tag{7.47}$$

where K_T is the 1st order thermal expansion coefficient of the material (neglecting higher order terms). Therefore, a measurement taken at a material temperature of T_1, should be adjusted for a final ambient temperature of T_0, in the following manner:

$$X_M\big|_{T_0} = X_M\big|_{T_1} \cdot \left(1 + 2K_T\left(T_1 - T_0\right)\right) \tag{7.48}$$

7.10 Sensitivity to Variations in the Integration Time Period (Time Constants)

As shown in Section 7.2, the system's absolute statistical noise is described by Eq(7.10), where it is noted that the integration time, $\Delta\tau$, appears as a reciprocal of the square root:

$$\Delta x = f\left(\frac{1}{\sqrt{\Delta\tau}}\right) \tag{7.49}$$

The system integration time and response time constant are typically programmable parameters that can be adjusted to improve the system noise levels / performance. Due to the relationships of Eq(7.10) and Eq(7.49), the adjustments to the integration time / time constants can be described by:

$$\Delta x\Big|_{\Delta\tau_2} = \Delta x\Big|_{\Delta\tau_1}\sqrt{\frac{\Delta\tau_1}{\Delta\tau_2}} \tag{7.50}$$

where

$\Delta x\Big|_{\Delta\tau_1}$ & $\Delta x\Big|_{\Delta\tau_2}$ - Absolute statistical noise levels with integration times, $\Delta\tau_1$ & $\Delta\tau_2$.

$\Delta\tau_1$ & $\Delta\tau_2$ - Different integration times

Often, the final time constant / integration time of the system is defined by the application, and therefore noise reduction methods involving this approach may not be available.

8. Architecture of Thickness Measurement Systems

Figure 2.1 provides a base-line overview of the primary components associated with a typical thickness measurement system. Figure 8.1 expands on this representation and partitions the components to their classical physical locations.

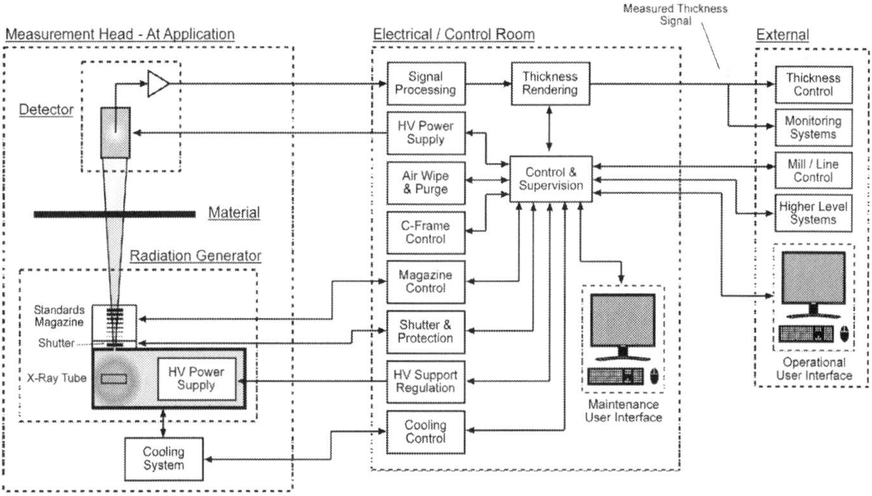

Fig. 8.1 - Typical thickness measurement system architecture and organization.

The key factor in the Figure 8.1 arrangement are the centralization of signal processing and control to a possibly remote location (Electrical Control Room). This suggests long, potentially noise corrupted signal runs, however this arrangement is commonplace in many existing legacy class systems currently in operation.

Modern systems employ highly networked arrangements to support high speed data sets and allow interfacing to a broader array of process related systems (Zipf, et. al., 2007). Figure 8.2 provides a network hierarchical view of a typical system networked interconnect for rolling mill applications.

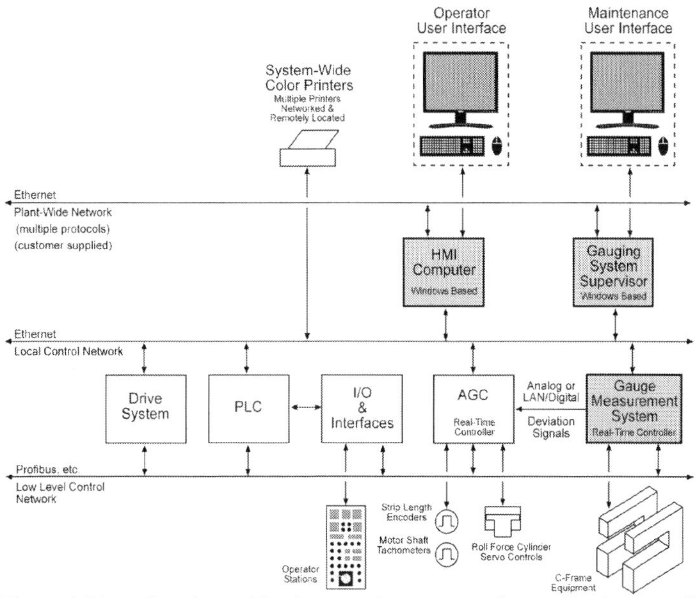

Fig. 8.2 - Network hierarchy of a multi-subsystem interconnect for applications to rolling mills.

9. System Implementations

When considering flat strip / sheet processing, there are a number of thickness measurement interests, including: longitudinal center-line thickness (along the length) as opposed to the transverse thickness profile (along the width). Some implementations may allow the measurement head (detector / radiation generator pair) to be retracted off-line, while others provide only fixed mounting locations. It all depends on the nature of the desired measurement.

9.1 Classical Measuring Head Configurations

A classical measuring head configuration is the C-Frame, shown in Figure 8.1, where the radiation generator is located beneath the strip and the detector above (I2S, 1992). This arrangement has sufficient structural stability to maintain proper alignment of the detector along the optical axis of the radiation generator. It also provides a convenient means of engaging the strip and retracting the unit off-line. The depth of the C-Frame throat can be manufactured to provide full strip width profiling and full extension across the strip for easy maintenance access.

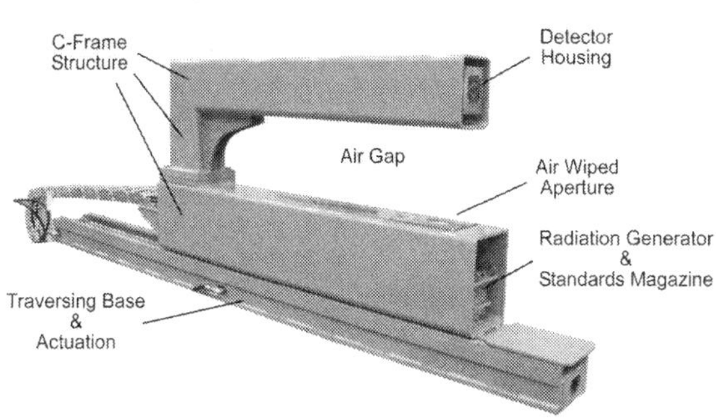

(a)

(b)

Fig. 8.1 - Photo diagrams showing various C-Frame arrangements: a) Primary components, b) C-Frame in flat strip application.

Other common mounting arrangements include O-Frames (that completely encompass the strip width) and Fixed Locations (generator and detector are physically mounted in place).

9.2 Measurement Plans

The application dictates the location and dynamics of the thickness measurement on the strip. Figure 8.2 illustrates some typical measurement plans.

Fixed mounting locations (Figure 8.2a) are uncommon, but necessary when the mechanical structures / surroundings of the application do not provide sufficient space for the mounting of retractable C-Frame equipment and support systems. This arrangement not only has a fixed measurement (typically strip center-line), but also can present difficulties in accessing the components for maintenance and repair. One point of advantage is that this arrangement can afford substantial protective enclosures / guarding, making the equipment essentially impervious to damage from process abnormalities (e.g., strip breaks, mill wrecks, etc.).

Retractable arrangements (Figure 8.2b) are perhaps the most common. These offer the ability to withdraw the C-Frame off-line (for maintenance and repair access), but only provide center-line or rear strip edge measurement capabilities.

Profiling C-Frame arrangements (Figure 8.2c) are the most versatile. They offer the ability to measure at any location across the transverse strip width (including center-line), and can be used to obtain transverse strip profile measurements. The deep C-Frame throat also allows these gauging systems to be traversed fully across the strip, for easy access to the detector and radiation generator equipment.

Dual C-Frame arrangements (Figure 8.2d) are uncommon. They are considered only in applications where simultaneous strip center-line and profile measurements must be provided. Often the center-line measurement is provided to thickness control systems, while the strip profile measurements are provided for quality control monitoring purposes.

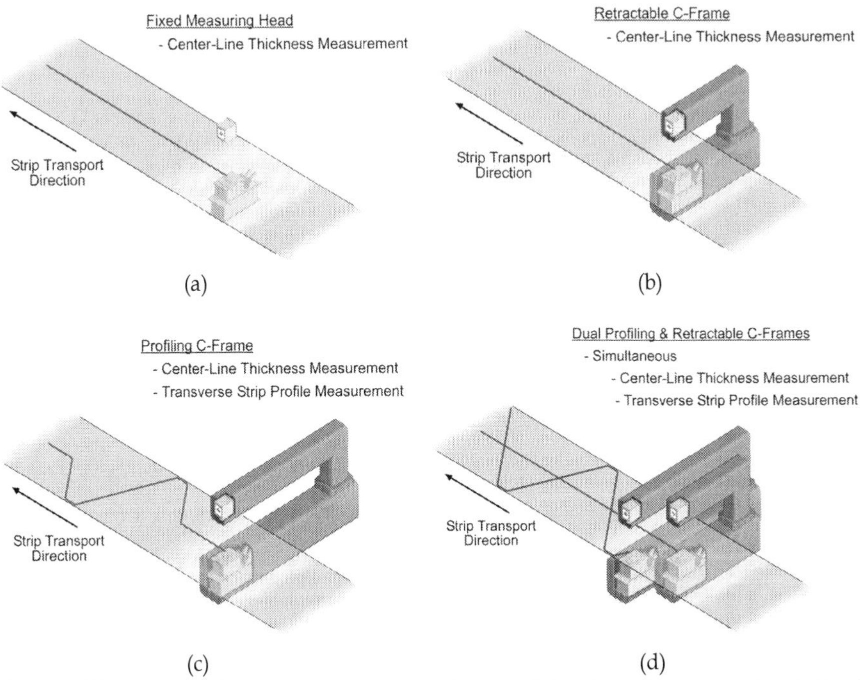

Fig. 8.2 - Illustrations of typical measurement plans on flat strip / sheet products: a) Fixed Measuring Head, b) Retractable C-Frame, c) Profiling C-Frame, d) Dual Profiling & Retractable C-Frames.

10. Conclusion

This chapter has explored the underlying physics and methods associated with radiation based measurements of material thickness. The intent of this chapter has been to provide a tutorial foundation of these principles, to support subsequent discussions (Zipf, 2010). The primary components of this form of measurement system have been presented and discussed. Natural and artificial radiation sources / generators have been examined and considered for application. The nature of the material's interaction with radiation is analyzed and considered in the presence of possibly complex material chemistries have been identified and accommodated in the methods for rendering a measurement. Detection system sensors and instrumentation have been presented and evaluated with respect to their employment in these measurement systems. The typical signal processing flow has been reviewed, along with the manner in which the measurement is resolved and distributed to external systems. A sensitivity analysis has been provided to identify key parameters and relationships that influence the quality, accuracy and precision of the measurement signal. Classical system architectures and implementations have been presented and discussed.

11. References

Bose, N. (1985). *Digital Filters – Theory and Applications*, Elsevier Scientific Publishing Co., ISBN 0-444-00980-9, New York, NY

Cooper, W. (1978). *Electronic Instrumentation and Measurement Techniques* - 2nd Edition, Prentice-Hall, Inc., ISBN 0-13-251710-8, Englewood Cliffs, NJ

Graydon, A. (1950). *Dissociation Energies and Spectra of Diatomic Molecules*, Dover Publishing, Inc., New York, NY

Halliday, D. (1955). *Introductory Nuclear Physics*, John Willey & Sons, Inc., Library of Congress Catalog Card Number : 55-9365, New York, NY

Harting, E. & Read, F. (1976). *Electrostatic Lenses*, Elsevier Scientific Publishing Co., ISBN 0-444-41319-7, Amsterdam, The Netherlands

Howard, R. (1970). Recent Advances in High-Speed X-Ray Thickness Gaging of Steel Strip, *Transactions of the Instrument Society of America*, Vol. 9, No. 4, (1970), pp. 348-354.

Hubble, J. & Seltzer, S. (2004). Tables of X-Ray Mass Attenuation Coefficients and Mass Energy-Absorption Coefficients (version 1.4), National Institute of Standards and Technology, Gaithersburg, MD

I2S – Intergrated Industrial Systems, Inc. (1992), *Gauge Signal Processing Unit Manual*, Intergrated Industrial Systems, Inc., Yalesville, CT

I2S – Intergrated Industrial Systems, Inc. (2004), *X-Ray Gauging System – Time Domain Response Analysis*, Technical Report : 040229 v1.3, Intergrated Industrial Systems, Inc., Yalesville, CT

IEC (1996). Nuclear Instrumentation – Thickness Measurement Systems Utilizing Ionization Radiation : Definitions and Test Methods, IEC-1336 (1996), International Electrotechnical Commission.

Kaplan, I. (1955). *Nuclear Physics*, Addison-Wesley Publishing Co., Library of Congress Catalog Card Number : 54-5732, Reading, MA

Kraus, J. & Carver, K. (1973). *Electromagnetics*, McGraw-Hill Book Co., ISBN 0-07-035396-4, New York, NY

Letokhav, V. (1987). *Laser Photoionization Spectroscopy*, Academic Press, Inc., ISBN 0-12-444320-6, Orlando, FL

Mark, T. & Dunn, G. (1985). *Electron Impact Ionization*, Springer-Verlag, ISBN 3-211-81778-6, Vienna, Austria

Moore, J., Davis, C. & Coplan, M. (1983). *Building Scientific Apparatus*, Addison-Wesley Publishing Co., ISBN 0-201-05532-5, London, England

Motchenbacher, C. & Fitchen, F. (1973). *Low-Noise Electronic Design*, John Willey & Sons, Inc., ISBN 0-471-61950-7, New York, NY

Nyce, D. (2004). *Linear Position Sensors - Theory and Application*, John Wiley & Sons, ISBN 978-0-471-23326-8, New York, NY

RCA – Radio Corporation of America, (1963). *Phototubes and Photocells – Technical Manual PT-60*, Radio Corportation of America, Lancaster, PA

Zipf, M. (2010). Radiation Transmissoin-based Thickness Measurement Systems – Advancement, Innovations and New Technologies, In: *Advances in Measurement Systems*, Chapter 8, Kordic, V. (Ed), In-Tech Publications, Vienna, Austria

Zipf, M., Hummel, C. & Burgess, R. (2007). A New Automation Architecture for Strip Thickness Gauging Systems, *Proceedings of the Associação Brasileira de Metalurgia e Materiais, 44th Rolling Seminar - Processes, Rolled and Coated Products*, Campos do Jordao, SP, Brazil, October 16-19, 2007, Campos do Jordao, SP, Brazil, Associação Brasileira de Metalurgia e Materiais, Sao Paulo, Brazil

6

Design of a Measurement System of End-User Information Competency with a Case Study

Chui Young Yoon
College of Electrical and Computer Engineering
Chungbuk National University
Republic of Korea

1. Introduction

The need for the effective operation of enterprise resources has risen with the intensive competition facing among enterprises. For this reason, enterprises have implemented information systems to improve their competitiveness and performance, and to increase their productivity and business efficiency by using advanced IT (Information Technology). It is important for human resources working on an enterprise information system to have the capability to effectively executing the given tasks by applying their information systems to their business (Marthis and Jackson, 2000; O'Leary, Lindholm, Whitford, and Freeman, 2002). An end-user who directly works his or her business needs the ability to efficiently perform end-user tasks by applying IT and information systems to his or her business. And this directly influences the end-user task performance and the competition edge of the organization.

Therefore, this study presents a measurement system for the end-user information competency, which focuses on the end-user's total capability that an end-user can efficiently use information knowledge, solutions, and information systems for his or her tasks on enterprise information systems.

2. Previous Research

In this study, an end-user is defined as a person who directly interacts with his or her information systems based on previous studies (Rockart and Flannery, 1983; Martin, 1982; Yoon, 2008). Information can explain as an organizational set of data that can be efficiently used for a specific objective (Machlup, 1980; Plotkin, 1994; Freeman, 2001; Yoon, 2008). And, competency is a total set of knowledge, skills, and attitudes as the action characteristics of an organizational member that can do his or her tasks outstandingly and efficiently in an organizational environment (Mirable, 1997; Arthey and Orth, 1999; Rodriguez, Patel, Bright, Gregory, and Gowing, 2002; Yoon, 2008).

By analysis of the major components of competency obtained from the literature, we can extract five major components: Motives, Traits, Self-concepts, Knowledge, and Cognitive and Behavioural Skills (Spencer and Spencer, 1993; Mirable, 1997; Arthey and Orth, 1999;

Rodriguez, Patel, Bright, Gregory, and Gowing, 2002; Yoon, 2008): (1) Motives is a cause of activity leading an end-user to do what he wants to do and what he consistently had in mind to do, and an action which selects and instructs a trigger for a specific activity or an objective. (2) Traits mean a consistent response to physical characteristics and situation or information, and an emotional self-control and careful attitude is 'a consistent response' of a more complicated form. (3) Self-concepts mean attitude, a sense of value, and self-portrait, and a sense of value is an element which reflects on responsible activities in a given situation for a short-period. (4) Knowledge is information that knows for specific department, and only indicates that what a person can do, but does not predict what a person will actually do. (5) Cognitive and Behavioural Skills are the ability to perform specific mental or physical tasks, and mental or cognitive skills include analytical or cognitive thought. In general competency, individual characteristics such as motives, traits, self-concepts and knowledge lead to skills, and the action of a person with skills has an effect on the performance of his or her business in an organizational environment (Spencer and Spencer, 1993). In an information competency, individual characteristics such as motives, traits, self-concepts and knowledge lead to skills, and the action of a person with skills has an effect on the performance of his or her business in an information environment. In other words, information competency can be defined by transforming a general competency into a type of competency based on an information perspective.

Hence, the end-user information competency (EUIC) can be defined as a total set of knowledge, technology, skills and attitudes which function as action characteristics of an organizational member who can do his or her tasks outstandingly and efficiently on an enterprise information system. In other words, EUIC is defined as the total capability that an end-user directly interacts with the information systems to efficiently execute his or her business tasks through using an organizational data for them on information systems. EUIC is the total capability that an end-user can effectively do his or her tasks on information systems.

With these researches, we generated the 24 measurement items that can gauge an end-user information capability in terms of an information competency based on the major components of a general competency such as motives, traits, self-concepts, knowledge, and cognitive and behavioural skills.

3. Methods

3.1 Research Method

In previous literature, the construct validity of the measurement items was studied by many researchers. Kerlinger (1978) presented two methods of model construct validation: (1) correlations between total scores and item scores, and (2) factor analysis. Etezadi-Amoli & Farhoodmand (1996) used factor analysis to verify the validity of the measurement tool construct. Torkzadeh & Doll (1999) and Torkzadeh & Lee (2003) used correlation analysis to verify the validity of the measurement tool construct. This study is likely to verify the validity of the measurement tool construct and the extraction of adequate items by factor analysis and reliability analysis. The ratio of sample size to number of measurement items (11:1) was above the minimum (10:1) ratio suggested for factor analysis by previous literature (Kerlinger, 1978; Rodriguez, Patel, Bright, Gregory, and Gowing, 2002). The items indicating above a criterion value by a factor analysis were selected because they were

closely related to each other, and all the items were thought to be the measures of the same construct. A measurement of criterion-related validity was also examined to identify the items that may not be closely related to the EUIC. Items should present a proper or incongruent element toward the object in question. If the item is ambiguous or indicates a neutral attitude, it should be deleted. A measurement of criterion-related validity was executed to identify items that did not indicate favourable or unfavourable attitudes.
The measurement questionnaire used a five-point Likert-type scale; where, 1: not at all; 2: a little; 3: moderate; 4: good; 5: very good. The questionnaire explains its objectives and contents, and respondents give answers on characteristics such as degree, age, gender, major department, industry and business department, business position level and years of job experience. The survey was gathered data from a variety of industries, business departments, experience, and major educations.

3.2 Sample Characteristics

In this pilot test, a sample of 258 usable responses was obtained from a variety of industries and business departments, and from management levels. All respondents had college or university degrees in: humanities and societies (16.3%), management and economics (20.2%), engineering (51.8%), and science (11.7%). The respondents in terms of business departments were identified as strategy planning (21.1%), development and maintenance (26.8%), business application (38.4%), and administration support (13.7%). The respondents identified themselves as top manager (3.7%), middle manager (44.7%), and worker (51.6%). The respondent had on average of 8.9 years of experience (S.D. =1.118) in their field, their average age was 32.9 years old (S.D. =6.473), and their sex, male (79.8%) and female (20.2%).

3.3 Analysis and Discussion

The analysis of the collected questionnaires was conducted by using SPSS ver.12 software. Items were excluded when their correlation with the collected item-total was < 0.5 or when their correlation with the criterion scales was < 0.6. The correlations with the corrected item-total and the criterion item were significant at $p \leq 0.01$ and similar to those used by others in previous researches (Rifkin, Fineman, and Ruhnke, 1999; McCoy, 2001; Torkzadeh and Lee, 2003; Yoon, 2008). After these analyses, the first 24 measurement items were reduced to 14 items, with 10 items were deleted. The elimination was considered sufficient to ensure that the retained items were adequate measures of EUIC. The validity and reliability of the developed tool were verified through factor analysis and reliability analysis. They were used to identify the underlying factors or components that comprise the EUIC construct. These deletions resulted in a 14-item scale for measuring EUIC. Each of the 14 items had a factor loading > 0.635. The reliability coefficients (Cronbach's alpha) of four potential factors had values > 0.792, above the threshold recommended for exploratory research (Rodriguez, Patel, Bright, Gregory, and Gowing, 2002; Yoon, 2008). The descriptions and loadings for the 14 items are presented (see Table 1 and Table 2).

Variable	Factor Loadings			
	Factor 1	Factor 2	Factor 3	Factor 4
V01	0.754			
V03	0.713			
V06	0.642			
V08		0.839		
V10		0.787		
V11		0.713		
V12		0.702		
V14			0.894	
V16			0.781	
V17			0.719	
V18			0.723	
V21				0.786
V23				0.727
V24				0.635

* Significant at $P \leq 0.01$

Table 1. Factor loadings obtained from factor analysis

In order to research the reliability and validity of the measures, we calculated the corrected item-total correlations between each variable and its corresponding factor. These correlations along with alpha coefficients of each factor are presented (shown in Table 2).

Variable	Corrected item-total correlation	Alpha if item deleted
V01	0.728	0.812
V03	0.689	0.724
V06	0.678	0.627
Coefficient alpha for the above 3 items as a composite measure of Factor =0.792		
V08	0.781	0.847
V10	0.714	0.848
V11	0.817	0.821
V12	0.629	0.723
Coefficient alpha for the above 4 items as a composite measure of Factor =0.884		
V14	0.743	0.852
V16	0.826	0.778
V17	0.634	0.836
V18	0.738	0.798
Coefficient alpha for the above 4 items as a composite measure of Factor =0.901		
V21	0.692	0.724
V23	0.712	0.738
V24	0.624	0.593
Coefficient alpha for the above 3 items as a composite measure of Factor =0.798		

* Significant at $P \leq 0.01$

Table 2. Corrected item-total correlations and coefficient alpha for each factor

This also shows the alpha coefficients for the measurement of factors if a measure was deleted from the scale. These coefficients indicate the relative contribution of a measure to the construction of a scale for measuring a particular factor. They are all in the acceptable range. Most corrected item-total correlations were greater than 0.600, showing that the end-user measures are good indicators of their corresponding factors.

They were grouped by their high factor loading. Each of the 14 items had a corrected item-total correlation > 0.624. The correlation for each of the 14 items was positive and significant ($p = 0.01$ or below). Hence, the measurement items, with a validity and reliability were extracted by carrying two analyses (see Table 1 and Table 2). However, efforts to provide additional evidence of this tool's validity, internal consistency, and stability are encouraged.

4. Measurement Tool

4.1 Structure of Measurement Tool

By factor analysis on the first measurement items, we extracted 14 items to measure EUIC, and the extracted items were classified as 4 factor groups. The 4 factor groups indicate the potential factors that can measure the EUIC and they include 14 measurement items. With investigating the measurement items of each factor, we generated the 4 potential factors as follows: factor 1: information understanding; factor 2: information knowledge; factor 3: information application; and factor 4: information potential. The information understanding means concepts, sense of value, attitude and adaptability related to information, and the information knowledge indicates the knowledge of information solutions and systems. The information application is the skills of information application to efficiently execute his or her tasks on the information systems, and the information potential refers the potential ability to improve information competency in terms of breadth and depth. The 4 potential factors are considered as the major measurement factors of the tool construct. Fig. 1 shows the structure of the measurement tool based on the 4 potential factors and 14 measurement items. Each factor has three or four measurement items, and each item is composed of two or three measurement problems from the measurement problem database.

4.2 Measurement Factor and Item

The information understanding (see Fig.1) is the realm where measures concepts, attitude, and adaptability on information. It includes the measurement items that can identify end-user's concepts on the Internet and information society, understanding of IT progress trend of IT leading countries, and etiquette in using information on enterprise information systems.

The information knowledge (see Fig.1) indicates the knowledge that an end-user has to know to efficiently apply information solutions and systems to his or her works. It comprises the measurement items that can gauge the information knowledge such as knowledge related to hardware, software, networks, and database for operating information systems, solution knowledge related to ERP (Enterprise Resources Planning), SCM (Supply Chain Management), and CRM (Customer Relationship Management), knowledge related to e-Business (B2E, B2C, and B2B), and knowledge related to security measures on enterprise information systems.

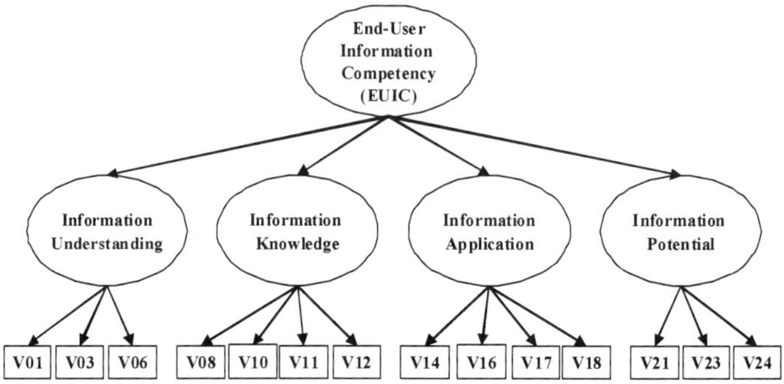

Information Understanding
IU01 Do you understand the Internet and information society ?
IU03 Do you know IT progress trends in IT leading countries ?
IU06 Do you consider an etiquette in using your information systems ?

Information Knowledge
IK08 Do you know hardware, software, networks, and database for your information systems ?
IK10 Do you have solution knowledge related to ERP, SCM, CRM, and e-Commerce ?
IK11 Do you know how to use B2E, B2C, and B2B on your information systems ?
IK12 Do you know how to establish security measures on your information systems ?

Information Application
IA14 Can you use word processing, spreadsheets, and presentation on your information systems ?
IA16 Can you use the solutions such as ERP, SCM, CRM, and E-Commerce ?
IA17 Can you apply your information systems to B2E, B2C, and B2B ?
IA18 Can you establish and manage information security measures on your information systems ?

Information Potential
IP21 How long did you work at IT departments ?
IP23 How many did you participate in oversea or domestic education and training related to IT ?
IP24 How many did you present your articles and ideas for a business improvement on your enterprise's webpage ?

Fig. 1. Structure of the developed measurement tool

The information application (see Fig.1) means the ability that an end-user can effectively apply information knowledge, solutions, and systems to his or her business tasks. It includes OA (Office Automation) ability such as spreadsheet, presentation and word processing, the ability to use business solutions such as ERP, SCM, and CRM, the ability to apply the information systems to an end-user's work such as e-business of the form B to E (Business to Employee), B to C (Business to Customer), and B to B (Business to Business), and the skills related to establish and manage the security system. This factor is a very important department that can mostly influence the performance of an end-user's tasks through applying his or her all abilities of understanding, knowledge, and skills to his or her tasks on an enterprise information system.

The information potential (see Fig.1) refers the potential development probability of the EUIC by job experience, participation of domestic and overseas education and training, and presentation of articles and ideas for a task improvement on the enterprise website.

This is the important factor for the development of information knowledge and ability, and the extension of information competency in terms of the breadth and depth of EUIC.
The tool that has 4 measurement factors and 14 items is an important theoretical construct to measure an end-user's total information ability that can efficiently do his or her tasks on an enterprise information system (see Fig.1).

5. Measurement System

5.1 Framework of Measurement System

The measurement system has a measurement tool and an interpretation tool (shown in Fig. 2). The measurement process of this system is identified as two stages of the measurement process by the measurement tool and the interpretation process by the interpretation tool.

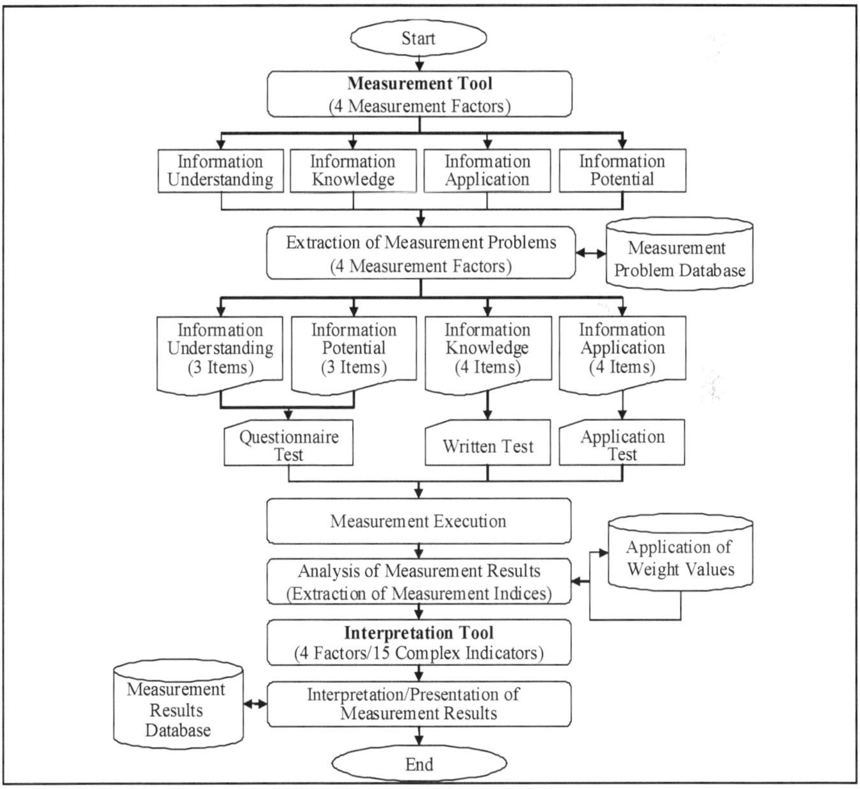

Fig. 2. Framework of the measurement system

The measurement stage is the extraction process of the measurement indices through measuring the EUIC by the measurement tool, and the interpretation stage is the process to explain the measurement results by the interpretation tool. It explains the meanings of the

measurement indices extracted from each measurement factor and the fifteen complex indicators. For measuring a EUIC, the system extracts the measurement problems from the measurement problem database. The form of the measurement problem is different from each measurement factor. The measurement factors such as the information understanding and the information potential are examined by a questionnaire form, and the information knowledge and the information application are examined by a written and application form.
The information understanding and the information potential are tested by a questionnaire form, and the information knowledge and the information application are examined by a written form and an application form.
The measurement index (MI) is calculated by applying each weight value to each measurement value extracted from each measurement factor. The MI is the total value extracted by applying the weight value to the measurement result that the end-user is examined by the measurement items in each measurement factor of the measurement tool. The extracted measurement index is reflected to the interpretation tool.
Finally, the system presents the interpretations of the measurement results of the EUIC based on the measurement factors and the complex indicators. And, the results are stored in the measurement results database and are utilized for presenting the measurement results when an end-user requests them. The total measurement and interpretation results are presented by the measurement report of the EUIC.

5.2 Measurement Method

In this study, we used the weight values for each measurement factor in order to develop an efficient tool considered the relative importance of each factor in measuring the EUIC. The weight values (shown in Table 3) were extracted from the analysis results of the questionnaire survey (AHP) for about 30 experts working in information departments.

Measurement Factor	Weight Value
Information Understanding	0.22
Information Knowledge	0.25
Information Application	0.33
Information Potential	0.20

Table 3. Weight value of each measurement factor

The measurement method first calculates the measurement values of each factor through the analysis of the measurement results that the end-user is tested by the extracted problems based on the measurement items of each factor. It figures out the measurement indices of each factor by multiplying each weight value by the measurement value of each factor. The measurement index (MI) means the value extracted by multiplying the weight value by the measurement value. And, the sum of the measurement indices of each factor becomes the total MI of the end-user. In this way, this tool presents the measurement results of the EUIC based on the total measurement index and the indices of each factor.

Hence, the total MI can be defined as Equation (1):

$$\text{Total MI} = \sum_{i=1}^{4} MV_{MFi} \times WV_{MFi} \quad (1)$$

Where, Total MI: total measurement index (MI) of an end-user

MV_{MFi} : Measurement Value (EV) of the i th Measurement Factor
WV_{MFi} : Weight Value (WV) of the i th Measurement Factor

Here, the sum of the weight values of each factor is 1.00 and i = 1, 2, 3 and 4 indicate the four measurement factors. By equation (1), we calculate the total measurement index of an end-user. In this way, this tool presents and interprets the measurement results of the EUIC based on total measurement index and the measurement indices of each factor.

5.3. Interpretation Tool

The interpretation tool has two kinds of interpretation methods (see Fig. 3). One is to explain the measurement results in each measurement factor, and the other is to present its results on the complex indicators. The interpretation by the measurement factors explains the meanings of the measurement results on four measurement factors.

The interpretation by the complex indicators presents core and general complex indicators based on the measurement results extracted by measurement items of each complex indicator. The complex indicator shows the implicative meanings and states of the EUIC as presenting the significant indicators of his or her measurement results.

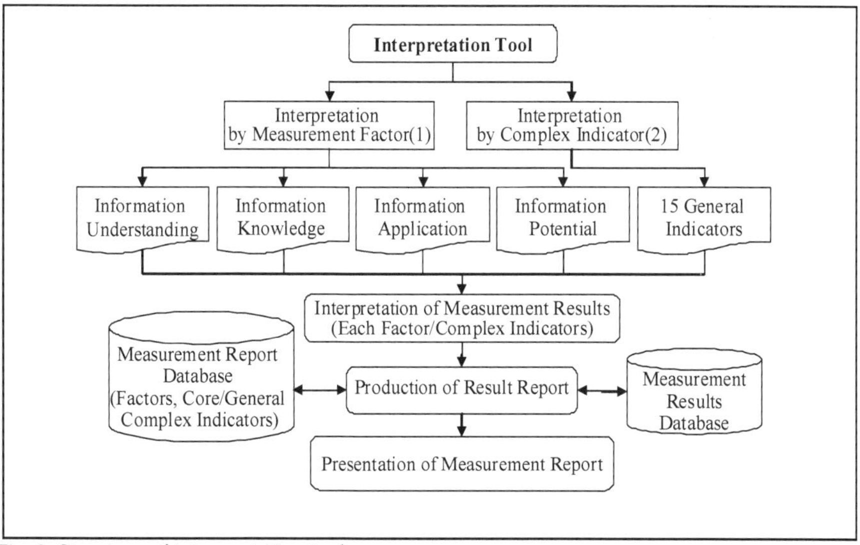

Fig. 3. Structure of interpretation tool

Therefore, the case study presents the measurement results by two methods with the measurement factors and the complex indicators. The final measurement report of the end-user is given by integrating the measurement results and its interpretations.
The complex indicators that present the implicative meanings of the measurement results were identified as 15 general complex indicators. The complex indicators were developed by about 50 experts in IT departments. The core complex indicators are 5 complex indicators (the serial number of 1, 4, 10, 14, and 15 indicator) of general complex indicators. The core complex indicators extracted from the general complex indicators present their meanings more implicative and significant than those of the general complex indicators. If we look at the complex indicators, we can generally know the present state or level of the EUIC. Each complex indicator for the measurement results is yielded by the result values based on the measurement item of it. Table 4 shows the complex indicators and the measurement items belonging to each complex indicator to generate the value of each complex indicator of the EUIC.

General Complex Indicators	Measurement Items
1. Information Understanding Indicator	-Understanding of information plan & implementation and application of business
2. Sense of Value Indicator	-Attitude, acknowledge, etiquette and law & regulation related to information
3. Information Base Knowledge Indicator	-Knowledge of basic information technology
4. Information Knowledge Indicator	-Knowledge of H/W, S/W, N/W, and DB related to information
5. e-Business Knowledge Indicator	-Knowledge of e-Business, e-Commerce and m-Business
6. Information System Knowledge Indicator	-Knowledge of H/W, S/W, N/W, and DB related to operating system
7. Information Security Indicator	-Knowledge of information security, security system, and institution and regulation of information security
8. OA Application Ability Indicator	-Ability using Word processing, Spread sheet, Presentation
9. Internet/Homepage Application Indicator	-Ability related to application of Internet and Intranet
10. Solution Application Indicator	-Ability using ERP,SCM,CRM,KMS, and HRM solutions etc.
11. Information System Application Indicator	-Ability applying information systems to B2E, B2B, B2C etc.
12. Information Management Ability Indicator	-Ability related to utility, security establishment and information management
13. Information Base Ability Indicator	-Degrees, certificates and job experience related to information
14. Information Education & Training Indicator	-Participation of oversea & domestic education and training related to information
15. Information Knowledge Production Indicator	-Presentation in national or international journals, publication of information books, and lectures & education related to information

Table 4. Complex indicators and measurement items

6. Case Study and Discussion

6.1 Sample Characteristics

This case study applied the developed tool to 163 workers working at "B" enterprise in South Korea. The business departments of respondents were identified as follows: strategy plan department (management strategy, plan management, and management plan etc.): 23.1%; development and maintenance department (development, management, and maintenance support etc.): 21.3%; business application department (sale, marketing, customer management, and service etc.): 37.4% and administration support department (personnel, finance, and welfare etc.): 18.2%. The business positions of respondents were classified as follows; top managers (CEO, director etc.): 2.8%; middle managers (chief of department, team manager etc.): 28.4% and workers (working-level person): 68.8%. The respondents had on average 7.7 years of experience (SD = 0.598), and most respondents (69.3%) had college or university degrees.

6.2 Analysis and Discussion based on Measurement Factor

First, application and analysis of each business department: As the analysis of an organizational level, we present the measurement results of each business department of the overall organization. The total measurement index of the overall organization was 61.58, and it was quite high. The strategy plan department and the business application department were 62.12 and 65.78 (shown in Fig. 4).

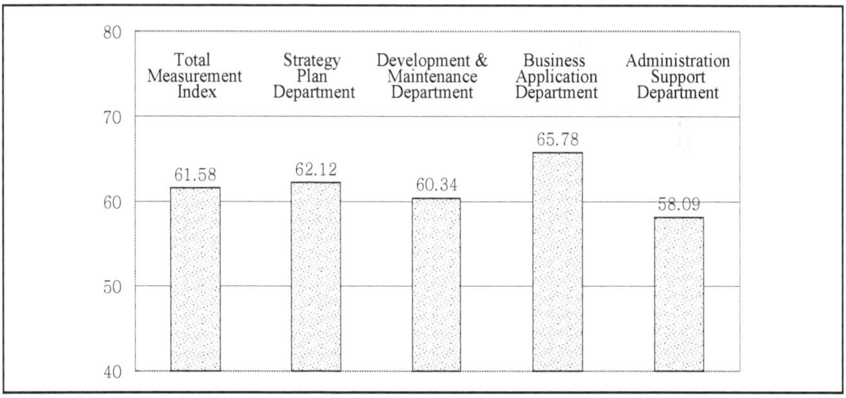

Fig. 4. Measurement indices of each business department

The measurement results of each business department shows that the measurement index of the business application department were higher than those of the other departments. This is due to the ability to effectively accomplish their tasks by frequently applying computing knowledge and computing system to e-Business of the form B to C, B to B and B to E, and the knowledge and abilities to utilize the various solutions such as ERP, SCM, and CRM in order to do their business tasks on an enterprise information system. Especially, the end-users in the administration support department have to make an effort to raise their information competency in general.

Second, application and analysis of a business department: The total measurement index of the strategy plan department (SPD) was 62.12, and it indicates quite high. The measurement indices of the SPD were quite high in the measurement factors of the information understanding, the information knowledge, and the information application, except for the information potential (shown in Fig. 5). But the measurement index of the information potential was 58.96 and it was the lowest level among the measurement factors. Therefore, the end-users of the SPD should make an effort to improve and develop the information departments such as the experience working in information departments, the information education and training, and the presentation of articles and ideas for a task improvement on the organizational website.

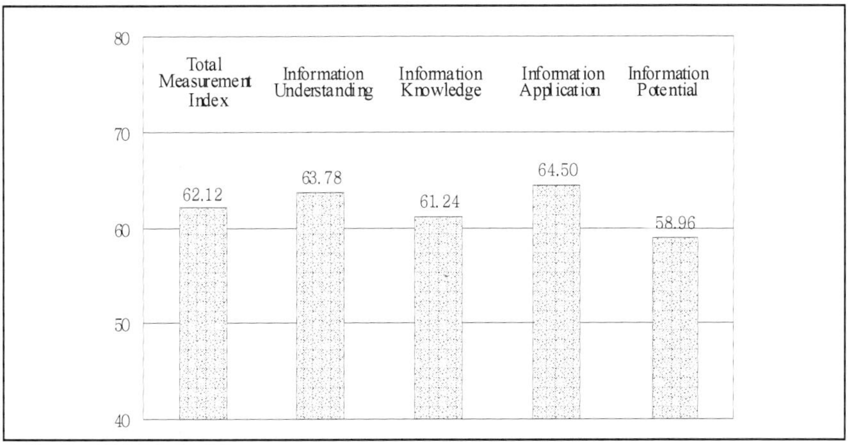

Fig. 5. Measurement indices for each factor of the SPD

Third, application and analysis of an end-user: The measurement results of an end-user working in the administration support department (ASD) were taken as an example. The measurement index of each measurement factor was generated by multiplying each weight value by the measurement value of each factor. The total measurement index is the sum of the measurement indices of each factor (see Table 5).

Division	Information Understanding	Information Knowledge	Information Application	Information Potential	Total Measurement Index
Measurement Indices of Each Factor	61.48	60.12	64.37	56.46	-
Weight Values of Each Factor	0.22	0.25	0.33	0.20	1.00
Calculation of Total Measurement Index	15.53	15.03	21.24	11.29	**63.09**

Table 5. Extraction process of the total measurement index for an end-user

The total measurement index of the end-user computing competency was 63.09 (see Fig. 6), and it was a little high. Especially, the measurement index of the information application was very high. This means the outstanding application ability for applying the information knowledge, solutions, and systems to his or her tasks on an enterprise information system. The measurement indices of the computing understanding, the information knowledge, and the information application were also quite high, except for the information potential.

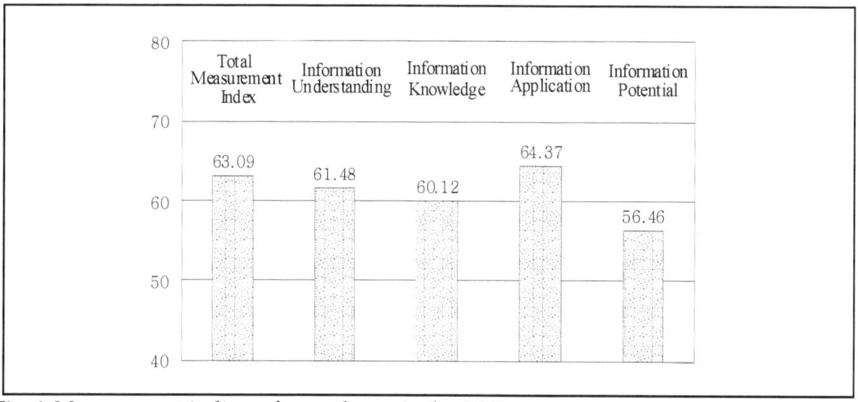

Fig. 6. Measurement indices of an end-user in the ASD

Therefore, this end-user should make an effort to complete information education and training, obtain job experience, and present articles and ideas for a business improvement on the organizational website in order to effectively raise his or her total information competency.

6.3 Analysis and Discussion based on Complex Indicator

The case study based on the interpretation of complex indicators considers an end-user in ASD as a sample. The measurement results were presented based on five core and fifteen general complex indicators (see Fig. 7). In general, the fifteen general indices of the core indicators are low levels, and the indicator of the solution application ability (ERP, SCM, CRM, KMS, and HRM etc.) was a little higher than those of the other core indicators.

The measurement results extracted on fifteen general complex indicators of an end-user working in ASD were also low levels in general (shown in Fig. 7). The complex indicators such as a sense of value, the information system knowledge, information security knowledge, OA application ability, the information system application, and the information base ability were quite a high but the others were low levels. Especially, the measurement results show that five core indicators were lower than the other indicators. Therefore, the end-user has to make a sufficient effort for raising the core indicators to efficiently improve his or her information competency.

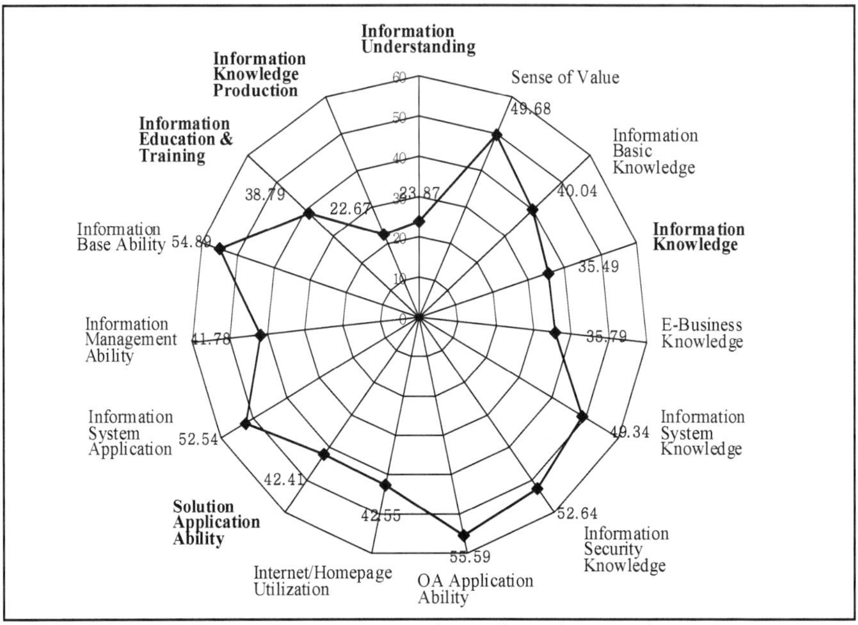

Fig. 7. Measurement indices of core and general complex indicators of an end-user in ASD

7. Conclusion

This study presented a measurement system that can efficiently gauge and interpret an EUIC working on an enterprise information system. The validity and reliability of the developed tool was verified by factor analysis and reliability analysis of the measurement items, and the proper measurement items of each factor were extracted by it. This is an instrument that can measure and interpret an end-user's total information capability based on an information competency. The interpretation tool provides the interpretation indicators that efficiently explain the measurement results of EUIC such as five core and fifteen general complex indicators. The application and utilization of this system was confirmed through a case study.

Therefore, this developed system opens up a new direction and method to develop a measurement system for EUIC since it functions as a measurement system that can entirely gauge and interpret EUIC in respect to the required information ability to efficiently execute an end-user's given tasks on an enterprise information system.

8. Acknowledgements

This work was supported by the second phase of the Brain Korea 21 Project at Chungbuk National University.

9. References

Arthey, Timothy R. & Orth, Michael S. (1999). Emerging Competency Methods for The Future, *Human Resource Management*, Vol. 38, No. 3, pp.215-226.

Etezadi-Amoli, J. & Farhoomand, Ali F. (1996). A structural model of end-user computing satisfaction and user performance, *Information and Management*, 30, pp.65-73.

Freeman, Lee A. (2001). Information Systems Knowledge: Foundation, Definitions, and Application, *Information Systems Frontier*, Vol. 3, No. 2, pp.249-266.

Kerlinger, F. N. (1978). *Foundations of Behavioral Research*, McGraw-Hill, New York.

Machlup, E. (1980). *Knowledge: Its Creation, Distribution, and Economic Significance*, Vol. 1: Knowledge and Knowledge Production, Princeton, NJ: Princeton University Press.

Martin, J. (1982). *Application development without programmers*, Prentice-Hall, Eaglewoods, Cal.

Mathis Robert L. & Jackson, John H. (2000). *Human Resource Management*, South-Western College Publishing.

McCoy, Randall W. (2001). Computer Competencies for The 21st Century Information Systems Educator, *Information Technology, Learning, and Performance Journal*, Vol. 19, No. 2.

Mirable, R. (1997). Everything You wanted to know about Competency Modeling, *Training and Development*, pp. 73-77

Munro, Malcolm C., Huff, Sid L., Marcolin, Barbara L. & Compeau, Deborah R. (1997). Understanding and Measuring User Competence, *Information & Management*, Vol. 33, pp.45-57.

O'Leary, B. S., Lindholm, M. L., Whitford, R. A. & Freeman, S. E. (2002). Selecting the Best and Brightest: Leveraging Human Capital, *Human Resource Management*, Vol. 41, No. 3, pp.325-340.

Plotkin, H. (1994). *Darwin Machines and the Nature of Knowledge*, Cabridge, MA: Harvard University Press.

Rifkin, Kenneth I., Fineman, Michal & Ruhnke, Clara H. (1999). Developing Technical Managers - First You Need a Competency Model, *Research & Technology Management*, pp.53-57.

Rockart, J. & Flannery, L. (1983). The management of end user computing, *Communication of the ACM*, Vol. 26, No. 10, pp.776-784.

Rodriguez, D., Patel, R., Bright, A., Gregory, D. & Gowing, M. K. (2002). Developing Competency Models to Promote Integrated Human Resource Practices, *Human Resource Management*, Vol. 41, No. 3, pp.309-324.

Torkzadeh, G. & Doll, W. J. (1999). The Development of a Tool for Measuring the Perceived Impact of Information Technology on Work, *Omega, International Journal of Measurement Science*, Vol. 27, pp.327-339.

Torkzadeh, Gholamreza & Lee, Jungwoo (2003). Measures of Perceived End-user's Computing Skills, *Information & Management*, 2018, pp.1-9.

Yoon, C. Y. (2008). Measures of End-User Information competency in an Organizational Information Environment, *IEICE TRANS. Fundamentals*, Vol. E91-A, No. 12, PP.3849-3853

7

Radiation Transmission-based Thickness Measurement Systems - Advancements, Innovations and New Technologies

Mark E. Zipf
Tenova-I2S (I2S, LLC)
United States of America

1. Introduction

Radiation transmission-based gauging systems (employing either natural or artificial sources) provide a unique means of non-contact thickness measurement. These systems have experienced a broad acceptance and been employed in a variety of flat rolled sheet product applications and processes, over the last 6+ decades. For a majority of this time, technological and processing speed limitations constrained the avenues and extent of new developments in this field. Systems provided in the 1990's strongly resembled systems from the 1960's, in both equipment and architecture, with the exception of certain computer, signal processing and user interface components.

Recent technologies and system innovations have now made it possible to consider radically different system architectures / partitions, fully digital systems and signal processes, new radiations generator concepts, advanced compensation algorithms, highly networked distributed control arrangements, open interfacing standards, and many other functions / features, beneficial to the performance, operation, maintenance and installation of these systems.

This chapter is the second of a two-part discussion concerning the nature of radiation transmission-based strip thickness measurement systems. The previous chapter (Zipf, 2010) examined the fundamental physics, instruments, signal processes, classical system architectures and implementations. This chapter explores the different paths that recent developmental avenues have considered and efforts taken in the advancement of this form of thickness measurement systems. A comparison of classical and contemporary system architectures is examined, followed by a series of focused discussions on recent evolutions in detectors, radiation generators and signal processing / control systems. New highly networked / highly interface-able / distributed system arrangements will be presented, along with methods of multi-system consolidation to minimize equipment and installation requirements.

2. Modern Departures from Classical Systems Architectures

2.1 Classical System Architecture

Figure 2.1 provides a reiteration of the previous chapter's architectural arrangement (Zipf, 2009) and a layout of a typical installation shown in Figure 2.2.

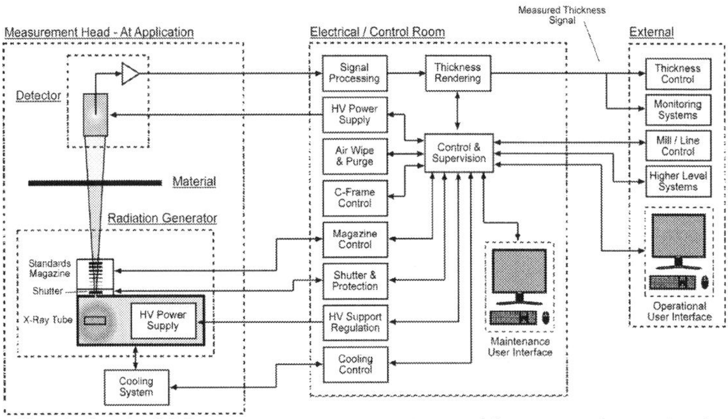

Fig. 2.1 - Classical strip thickness measurement system architecture and organization.

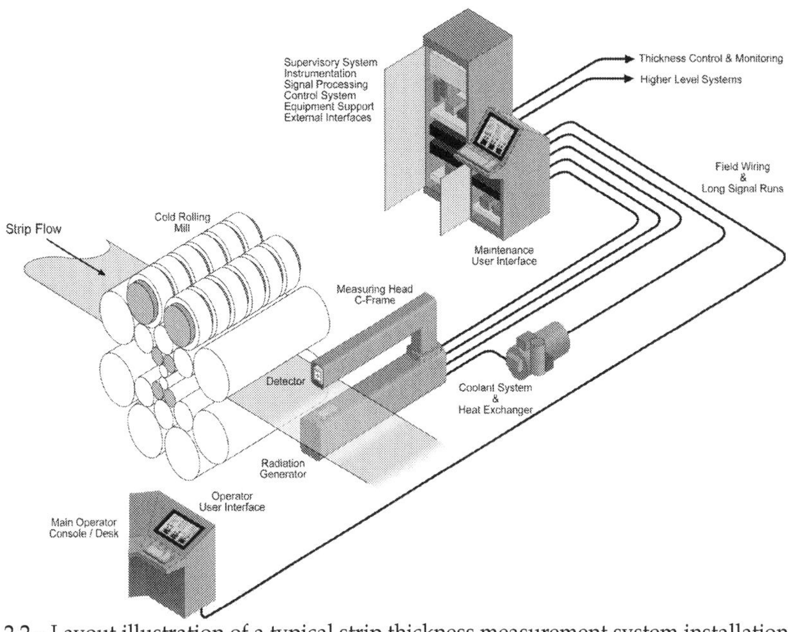

Fig. 2.2 - Layout illustration of a typical strip thickness measurement system installation.

The important architectural factors in the system arrangements of Figures 2.1 and 2.2, are the highly consolidated / highly centralized nature of the system's control, supervision and signal processing components, and its remoteness from the measuring head equipment. Extensive field wiring and long signal runs are an inevitable consequence of this architecture / arrangement in industrial settings. The centralized systems are often very fixed / rigid, and offer no convenient method for expansions of the system capabilities (e.g., additional measuring heads, additional user interfaces, etc.).

This architecture originated in single computer systems, whose processing power was limited and carefully applied to achieve the measurement specifications with fast dynamic responses. Computer equipment technologies were not as robust as we currently enjoy. It was necessary to consolidate / sequester the electronics and computing equipment in environmentally protected / controlled electrical control rooms, distant from the industrial conditions where the measuring head equipment resided. Long field wiring and signal runs were a necessary component of the design and installation. Interfaces to external equipment was typically primitive, employing analog signals (for information bandwidth needs) and low data throughput serial links.

2.2 Traits of Modern System Architectures

The classical system architecture has a number of limitations. Modern electronics, control system equipment and networks have allowed a re-examination of the system organization, and various means of improving the quality of the measurement and transmitted signals. The primary points of interest are:

- Minimization of field wiring and long signal runs
- Maximize noise immunity and information integrity
- Consolidation of critical analog signals and immediate digitization
- High speed, digital signal processing and Field Programmable Gate Array (FPGA) technologies.
- Self-contained, local control of C-Frame equipment
- Numerical thickness measurement rendered at the C-Frame
- Modular organization and distributed control architecture
- Highly networked system interconnects
- Graphical User Interfaces (GUIs) / Human Machine Interfaces (HMIs)
- Fully integrated detector assemblies
- Compact tube / tank assembly
- Passive X-Ray tube thermal dissipation and removal of external cooling systems
- Dry-potted, high voltage power supply (withdrawn from dielectric oil)
- Compact, self-contained X-Ray generators
- Highly interface-able and compliant with network / database standards (OPC, etc.)
- Scalable / expandable (arbitrary number of C-Frames and operator GUIs / HMIs)
- Equipment consolidation through multi-core processors

The basic concepts and objectives of the above list can be achieved by the reorganization of the system architecture of Figure 2.1, and the application of newly available technologies. Figures 2.3 and 2.4 provide illustrations of a contemporary architecture and system organization.

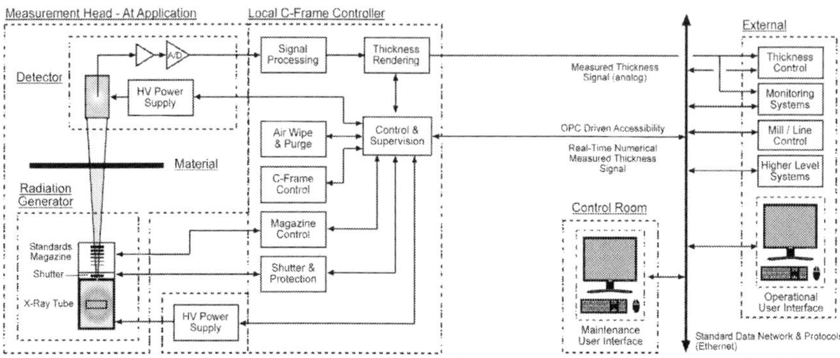

Fig. 2.3 - Desired, contemporary architecture and organization of a modern strip thickness measurement system.

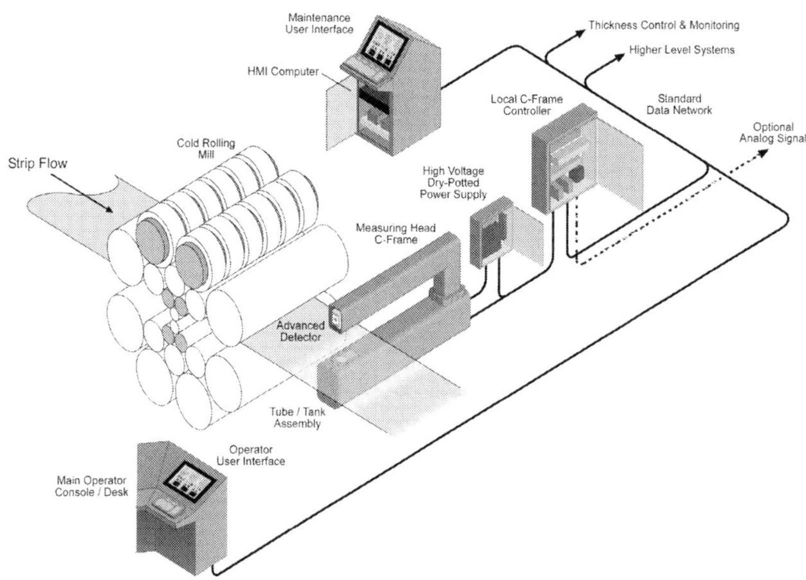

Fig. 2.4 - Layout illustration of a desired, contemporary strip thickness measurement system installation.

The key aspect of the modern system architecture is that what was once a remote, centralized system, is now modular, distributed and localized to the C-Frame / measurement head (Zipf, et. al., 2007a). This consolidation of all control / signal processing (local to the C-Frame / measurement head) allows us to consider the C-Frame equipment as a complete, free-standing system (C-Frame System = C-Frame, HV Power Supply, Local C-Frame Controller). Essentially all other equipment (operator interfaces, GUIs, etc.) are peripheral.

The networked interconnect significantly reduces field wiring, and allows the free-standing C-Frame System to be conveniently integrated into broader industrial applications. The digital / numerical nature of the high speed network data exchanges, eliminates the need for long analog signal runs and provides inherent noise immunity. The distance between system components (e.g., C-Frame System and operator interfaces / GUIs / HMIs) is limited only by the capabilities of the network media (copper wire, fiber optic, etc.). This arrangement is completely scalable / expandable, through direct replication of the C-Frame System and the attachment of additional HMI Computers, as network drops. Figure 2.5 provides a hierarchical view of this form of highly networked system arrangement.

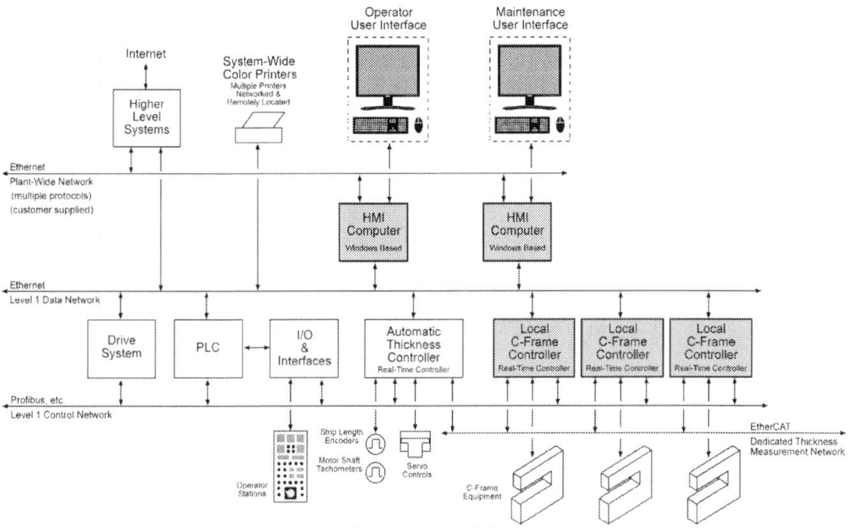

Fig. 2.5 - Hierarchical block diagram illustration of the network topology, components and their interconnections.

An interesting advancement is the Dedicated Thickness Measurement Network (DTMNet). This real-time, deterministic network (EtherCAT, etc.) provides high speed data exchanges (thickness indications and status broadcasts) between the gauging system and the Automatic Thickness / Gauge Control (AGC) system (or other gauge monitoring, quality tracking system). In this way, the separation between equipment is not hampered by noise and interference experienced by long analog signal runs.

The C-Frame Controllers are mounted local to the C-Frame equipment (in industrial / mill duty enclosures) and function as complete stand-alone units, handling and supervising all aspects of the signal processing, control and operation of the measuring head and C-Frame, and rendering a numerical / digital value of the measured thickness local to the C-Frame (Zipf, et. al., 2007a). The controller can interface to surrounding equipment through the Level 1 Control and Data Networks to support commands, status and large data set exchanges. Wide bandwidth thickness measurement data can be transmitted on the deterministic DTMNet, or via analog or serial links (to support legacy systems). The controllers are OPC compliant and can openly publish (allow access to) their entire control / set-up parameter registers for easy integration, remote control and diagnostic support. Advanced Web-based technologies (AJAX, HTML, Java Scripts) allow the controllers to publish real-time graphical and status data to thin client HMI computers, and through directed Internet access (via the higher level systems), remote monitoring capabilities from virtually anywhere, even on mobile phones (iPhone, PDA, etc.) (Zipf, et. al., 2008b).

The maintenance and operator interfaces can be based purely on networked general purpose HMI computers functioning only as thin clients (Internet Explorer, FireFox, Crome, etc.), being supported by / through OPC Servers and the Web-based streaming data from the C-Frame Controllers. Internet-based remote accessibility is possible via network interfaces to the high level systems.

3. Advances in Radiation Detection Systems

Radiation detection systems acquire and measure the intensity of incident radiation, and provide an instrumented signal functionally related to the received radiation intensity. Detectors typically employ either a photomultiplier tube (PMT) in combination with a scintillation crystal (typically Sodium-Iodide) or an ionization chamber (ion chamber) pressurized with appropriate mixtures of ideal and "getter" gases (Zipf, 2009), as the fundamental sensor. Both employ low noise, electrometer class pre-amplifiers mounted local to the sensor, to amplify the detector signal amplitude and provide the necessary line drive to potentially remote signal processing equipment.

In the past, PMT based detectors were favored because of their enhanced speed and sensitivities, unfortunately, this equipment is susceptible to fractures of the scintillation crystal and the PMT dynode chain is inherently delicate and fragile. Historically, ion chambers are of a far more rugged / robust construction and recent advances in ion chamber technologies have allowed them to eclipse PMTs as the preferred radiation detection sensor.

The key recent innovations have come in the following forms:

- Compact, fully integrated detector assemblies
  - Pre-amplifier mounted in detector housing, immediately adjacent to the sensor
  - Programmable high voltage power supply mounted in detector housing
  - Immediate analog-to-digital conversion of pre-amplifier signal
  - Dedicated network interconnect to signal processing system
- Ion chamber arrays
- Solid state / semiconductor sensors

3.1 Compact, Fully Integrated Detector Assemblies

A comparison of Figures 2.1 and 2.3, illustrates the consolidation of the sensor, instrumentation and power supply within the detector housing. The detector is a fully stand-alone unit, capable of supplying high resolution numerical representations of the measured radiation (no analog signal output), requiring only low voltage power, network and air purge connections / cabling.

The sensor (PMT or Ion Chamber) is mounted in a sealed, mill duty, enclosure / housing. The electrometer class pre-amplifier, high voltage power supply, analog-to-digital converter (A/D) and network interface components, are consolidated onto an industrial duty electronics board, mounted within the housing. The entire housing is sealed / submersible and air purged for environmental protection (dust, fluids, grease). Figure 3.1 provides some photos showing a typical detector assembly.

(a) (b) (c)

Fig. 3.1 - Photos of typical detector assemblies: a) Machined monoblock housing containing cylindrical ion chamber sensor, b) Mounted detector assembly with integrated electronics card with sealed cover removed, c) Fully assembled detector.

3.2 Ion Chamber Arrays

There is a fundamental limitation between the extent of the ion chamber's entry aperture and speed of response. Faster responses can be achieved through chamber geometries that minimize the electron drift distances, at the cost of a reduced radiation gathering capability (smaller aperture). The reduction in drift distances reduces the time required for the detector's filament to collect the ionized charge (free electrons), leading to a faster response. The reduction in aperture size can be accommodated by employing multiple, identical, small aperture ion chambers arranged in an array that coincides with the geometry of the received radiation pattern.

The sensor array can be connected in parallel and supported by a single pre-amplifier, or can remain independent, requiring a matching array of pre-amplifiers. Independent pairs of sensors / pre-amplifiers provide the ability to monitor the performance of each sensor, supply downstream signal processing compensation, and evaluate the received beam pattern, all at the cost of added complexity.

Independent pairs also provide the ability to charge each ion chamber differently (different tuned gas mixtures, pressures, voltages and window materials). This offers the ability to detect and compensate for changes in material composition (alloys), and apply the appropriate alloy compensation.

3.3 Solid State / Semiconductor Sensors
Semiconductor sensors (typically in the form of a photodiode), measure the intensity of incident radiation as a function of the number of charge carries (electron / hole pairs) freed within the detector. The electron-hole pair formation energy is relatively low, when compared to the production of ions in gas detectors, making the inherent sensitivity / resolution higher. The time resolution is also very good and cooling systems (typically thermo-electric) can reduce the sensor's thermal noise to improve low radiation level performance. Their small size makes them ideal for high resolution area arrays, which can be used for beam geometry, uniformity and alignment studies.

However, the dynamic range of these typically silicon detectors can be limited and their relatively small size makes it difficult to over-contain the inbound radiation beam. They also suffer from performance degradation from long exposure to radiation.

4. Evolution of Radiation Generators

X-Ray radiation generation techniques have experienced a number of key innovations over the last few years. These developments have allowed for totally new system architectures to be made available, along with improvements in the maintainability of this equipment.

4.1 Classical X-Ray Generators
Classical / legacy X-Ray generation systems employ large, heavy, shielded tank assemblies containing the X-Ray tube and high voltage power supply, all immersed in a bath of dielectric oil. These assemblies are not only large, but also very heavy (often greater than 125 kg due to the large amount of lead shielding required) and difficult to extract from the C-Frame (often requiring several people). The high voltage power supplies are regularly driven, controlled and regulated by remote / external analog electronics. These arrangements often suffer from the limitations of long signal runs and the natural perturbations of discrete analog circuitry, inducing questionable accuracies and drifting characteristics. The oil is often circulated through external cooling systems (outside the C-Frame structure), requiring a mix of flexible and fixed tubing, pumps, filters, heat exchangers, etc. Leaks in these coolant loops could disable or damage the generators, induce uncompensated thermal variations in the produced radiation, or over-pressure the tank / fluid system by the oil's thermal expansion. Figure 4.1a provides a block diagram and photograph of typical internal components associated with this system arrangement.

4.2 Sealed X-Ray Generators and Local Regulation
A developmental step forward involved sealed, fully contained tank assemblies that did not require external cooling systems. This approach utilized high integrity, internal circulation systems and relied on the tank structure's heat dissipation characteristics to provide the necessary thermal relief. Careful design considerations were required to ensure that the necessary heat dissipation was provided, however the removal of the external coolant systems was a welcome advancement. Figure 4.1b provides a block diagram description and photograph of typical internal components associated with this system arrangement.

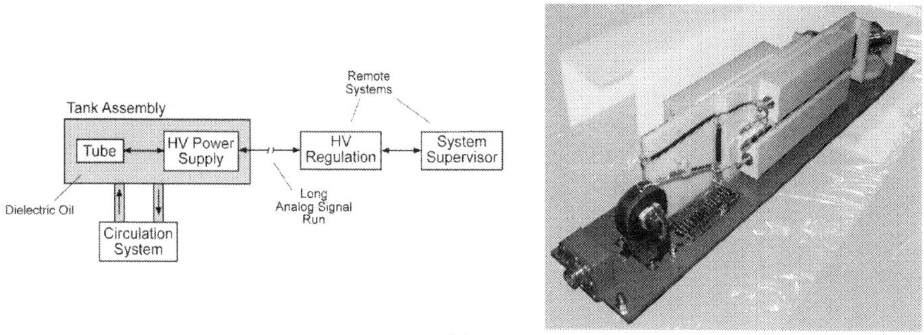

(a)

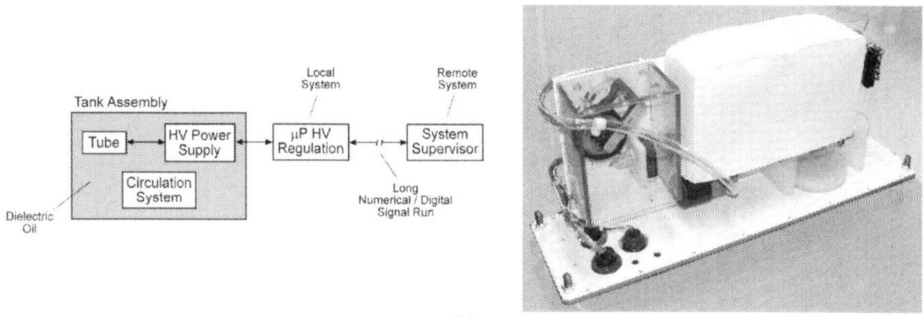

(b)

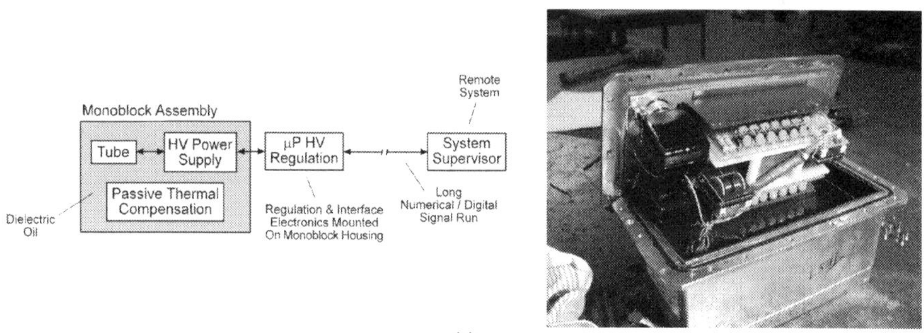

(c)

Fig. 4.1 - Diagrams and photos illustrating various avenues of X-Ray generator development: a) Classical generator arrangement using an external cooling system and the high voltage components immersed in dielectric oil, b) Development of internal circulation and thermal dissipation techniques to provide a more self-contained and reliable operation. This diagram also shows the employment of microprocessor based regulation and interfacing mounted local to the tank assembly and high voltage power supply, c) Ultra-compact monoblock arrangement using passive thermal compensation.

Another advancement involved the development of localized, computer controlled and regulated high voltage power supplies. Here, the high voltage electronics and drive circuitry was located within the oil immersed tank and microprocessor based controllers were mounted in environmentally protected enclosures within the C-Frame. Digital fiber optic communication links were provided to set and monitor the high voltage systems, and to provide inherent isolation to the supervisory and signal processing equipment. This step provided the removal of remote analog electronic drive systems and provided higher system accuracies due to the tighter radiation regulation. However, this arrangement still relied on an internal oil circulation system, and still required a relatively large and heavy tank assembly.

4.3 Compact Monoblock X-Ray Generators

There has been a long standing interest in finding X-Ray based direct replacements for existing isotope sources (Zipf, et. al., 2007b). Isotope sources require special handling and maintenance considerations, and by their fundamental nature, can not be "turned-off". The idea has been to create a highly compact, fully self-contained X-Ray generator, whose physical size matches that of the isotope source housing, provides the necessary radiation intensity and can internally compensate for the X-Ray tube's thermal dissipation needs. Ideally, this compact X-Ray generator would be directly retrofit-able to existing isotope source installations, requiring only minor modifications to support the new equipment, and it would also provide a wide range of interfacing possibilities, to support the broad spectrum of existing isotope based systems and instrumentation.

The development of monoblock systems directly addresses these needs by employing a small format X-Ray tube, compact programmable high voltage power supply, passive thermal dissipation methods and externally / adjacently mounted regulation and interface electronics, all within a sufficiently small physical size that it can reside in the available C-Frame space of existing isotope radiation sources (Zipf, et. al., 2007). Figure 4.1b provides a block diagram and photograph of a typical compact, monoblock assembly.

An important innovation can be seen in the passive thermal compensation technique used in the limited volume of dielectric oil and relatively small cooling surfaces. The idea is to employ internal air bladders that vent to ambient (external to the tank), located within the tank assembly. As the tube's thermal conditions expand the dielectric oil, the bladders volumetrically deflate to absorb / compensate for the oil's expanded volume, and maintain a non-pressurized tank (Zipf, et. al., 2007b).

The monoblock units include a localized microprocessor based controller that oversees, controls and regulates that generator's operations. This controller provides a variety of interfacing possibilities (network, serial, discrete signal, etc.) and the ability to report detailed status and performance information. This allows this unit to be installed in many existing isotope-based systems with only minor instrumentation / interfacing modifications.

These monoblock concepts stem from the existing technologies employed in the medical instrumentation and inspection industries, but have not been considered (until recently) in the harsh environments associated with industrial and mill duty settings. These systems are gaining popularity even in the presence of their non-trivial costs.

4.4 Dry-Potted High Voltage Power Supplies and Compact Tube / Tank Assemblies

An important innovation came with the development of modular, dry, potted high voltage systems mounted external to the tank assembly, but local to the C-Frame, and capable of supporting X-Ray tube potentials ranging from 10kV to 150kV (Zipf, et. al., 2008b). This is a major advancement in that the only component required within the oil immersion tank was the X-Ray tube itself. High voltage is supplied to the tube by protected, flexible high voltage cabling / connectors. The potted high voltage equipment has a compact format, and mounted in an environmentally sealed industrial enclosure located on or near the C-Frame equipment. The drive and regulation systems are also contained within this enclosure and communicate with the supervisory system via a fiber optic link (to provide further isolation and noise immunity). Figure 4.2 provides a block diagram description and photograph showing the internal components associated with this system's arrangement.

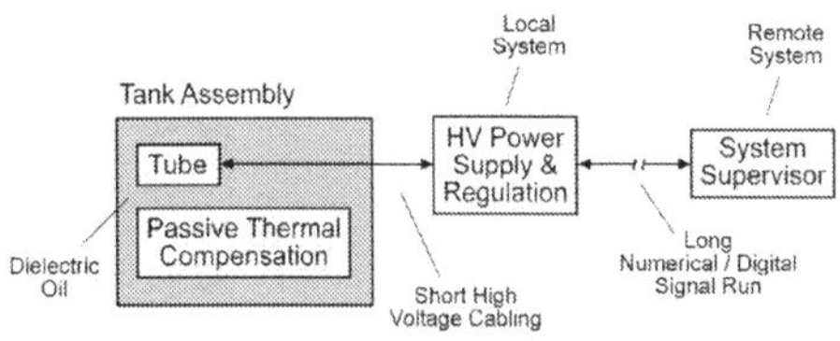

(a)

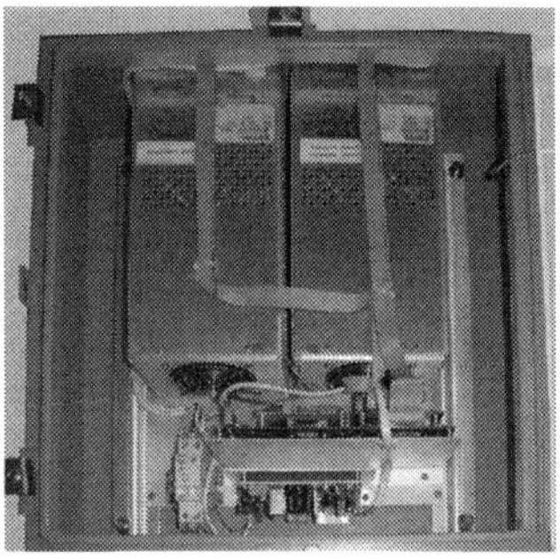

(b)

Fig. 4.2 - Illustrations of X-Ray radiation generation equipment employing a compact tube / tank arrangement and external, modular, dry, potted high voltage system: a) Block diagram of the typical components and the interconnects / relations, b) Photograph of the dry, potted high voltage bipolar power supply and fiber optic interface controller housed in an environmentally secure enclosure.

This created a modular arrangement that was not only easy to maintain, but the tube / tank assembly now only contained the X-Ray tube and its mounting structures. To reduce the tube's thermal loading on the dielectric oil expansion, the beam current can be reduced to levels on the order of 500 micro-amperes, providing lower operating power (temperature) and longer tube life. This reduction in radiation intensity is countered by an increase in detector sensitivity through proper design and manufacture of the sensors, along with lower noise pre-amplifier electronics.

The thermal loading / expansion of the dielectric oil is thereby minimized, and can be accommodated by the passive expansion / dissipation methods (internally mounted air bladders) described in Section 4.3 used in the monoblock generators. The tube / tank assembly was now of a size, form factor and weight that permitted handling by a single person (tube tank assembly weights on the order of 20-30 kg, directly accessible from the front / operator-side of the C-Frame). Figure 4.3 provides photographs of a typical tube / tank assembly.

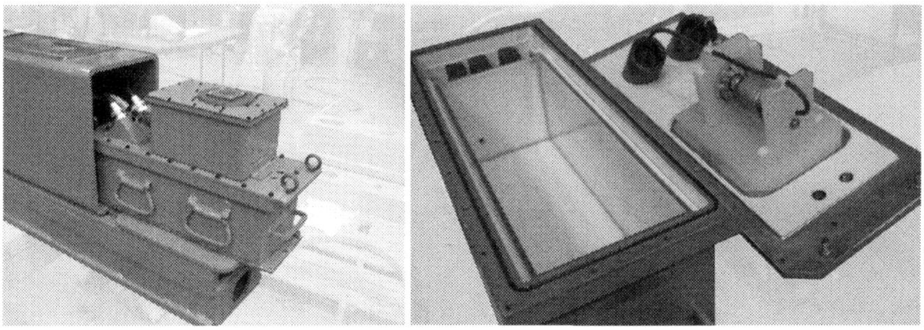

(a) (b)

Fig. 4.3 - Photographs showing a typical tube / tank assembly: a) Tube / tank assembly and standards magazine (top) withdrawn from the base of a C-Frame, b) Internal components of the sealed tube / tank assembly, showing the shielded, insulated dielectric oil tank and air bladders (left), and X-Ray tube mounted to the tank lid (right).

5. Advances in Signal Processing and Control

One of the major system architecture advancements has been to partition, distribute and decentralize the operational responsibilities, by localizing the signal processing and control to the C-Frame / measurement head areas. This work has followed three (3) primary objectives:

- Complete, independent control and supervision of an individual C-Frame's systems with a localized real-time controller
- Calibrated, real-time indications of the measured thickness local to the C-Frame via high speed digital signal processing (DSP) hardware and algorithms
- Fully networked, high speed communications for command and status data exchanges, and distribute deterministic numerical thickness measurements over a dedicated network.

5.1 C-Frame Controller

In the past, locating commercially available control system equipment in harsh, mill duty environments was not an appropriate option. The physical size and required support equipment made them too awkward to consider in this application.

Recent developments in off-the-shelf, industrial controller technologies, not only provide compact physical arrangements, suitable for mill duty industrial applications, but also offer high speed real-time processing and embedded Field Programmable Gate Array (FPGA) circuitry, allowing them to tackle exceedingly fast signal processing and control problems. One such family of devices is the National Instruments Compact Remote I/O (cRIO) Programmable Automation Controllers (PACs), which offer high resolution interfacing, full network compatibility and a flexible framework from which a variety of applications can be accommodated. The PAC contains a programmable CPU running a real-time operating system (RTOS) (Wisti, et. al, 2008) and an underlying programmable FPGA hardware layer

(NI, 2007a,b). Both components are programmed with National Instrument's LabView language and the target code is stored in onboard Flash memory.

A PAC controller and associated support equipment reside in an environmentally protected enclosure, located in direct proximity to an individual C-Frame / measurement head. Figures 5.1 and 5.2 illustrate this system arrangement. The controller's responsibilities are partitioned between the real-time processor and FPGA hardware.

FPGA Hardware - This hardware is programmed to receive the digitized, pre-amplifier measurement of the detector's sensor (from the A/D) and perform the selected digital signals processing (DSP) algorithms (see Figure 6.1 of Zipf, 2009) to immediately render a calibrated thickness measurement. The resulting measurement can be output as a high resolution analog signal (to support legacy systems) or passed to the RTOS for distribution over the DTMNet. The signal processing parameters / calibration coefficients are provided by the RTOS and openly accessible via the OPC interface.

Real-Time CPU / RTOS - This processor is responsible for all C-Frame equipment control and operational activities including: Closed-loop C-Frame motion control, high voltage power supply supervision, air wipe / purge systems, magazine and shutter control, calibration and standardization, safety and protection systems. This processor manages all network interfaces (Level 1 Data Network and DTMNet) and supports the OPC Server interface. All calibration, alloy compensation and signal processing parameters are handled by this processor and distributed on the OPC interface, to maximize interface-ability, remote control, performance monitoring and diagnostic support. This processor also employs advanced Web-based technologies (AJAX, HTML, Java Scripts) to allow publishing of real-time graphical and status data to thin client HMI computers, and through directed Internet access (via the higher level systems), remote monitoring capabilities.

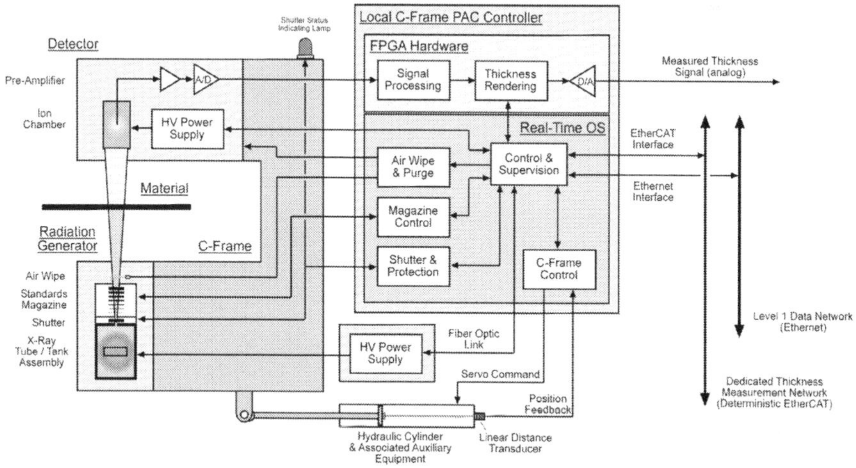

Fig. 5.1 - Block diagram showing the C-Frame Controller, the internal partitions between the RTOS and the FPGA hardware, and the internal and external interconnections.

(a) (b)

Fig. 5.2 - Photographs showing the C-Frame PAC Controller: a) National Instruments cRIO PAC, b) C-Frame Controller and associated support equipment, mounted in a environmental protection enclosure.

5.2 User and External System Interfacing

The highly networked nature of the system architecture (Figure 2.5) and the C-Frame Controllers provide a number of opportunities for interfacing and scalability / expansion. The popularity and commonplace of OPC Server technologies allows extensive availability of the C-Frame Controller's internal parameters, high voltage and detector calibration data, command and control structures, collected data buffers, performance and status data, etc. This broad and selectable exposure of OPC Tags allows the C-Frame Controller to be easily integrated into complex automation systems, and externally controlled / tuned to match the needs of the application and process.

The networked architecture also provides the ability to support sophisticated, fast responding Web-based GUIs (AJAX, HTML, Java Scripts), arbitrarily added as thin client drops via general purpose HMI Computers (Zipf, et. al., 2008b). These GUIs interact through the database provided within the OPC Server and through data streaming methods, to provide real-time graphics and visualizations. Through higher level system interfaces to the Internet, remote control, monitoring and diagnostic assistance can be provided from any Internet access point, world-wide. The extensive OPC interfacing allows for the use of commercially available (off-the-shelf) software packages (e.g., Siemens WinCC, GE Cimplicity, Intellution, Wonder Ware, Interact X, RS View, etc.) to also implement operator and maintenance interface GUIs.

The networked interfacing also provides a substantial reduction in the extent of field wiring required to support the overall thickness measurement system, and has a direct impact on the installation activities. Only low voltage AC power, Ethernet cable, EtherCAT cable and E-Stop wiring are required.

6. System Consolidation Using Multi-Processor Technologies

An interesting side-line of the modern architectural arrangement and implementation (Figures 2.3, 2.4 and 2.5), is the ability to consolidate multiple related systems into a single computer arrangement, through the use of multi-processor technologies. The main driving force behind this interest stems from a desire to minimize and hyper-modularize multiple system equipment, primarily to support surgical modernization, replacement and retrofit efforts of older or obsolete equipment. The idea is to be able to replace older multi-computer systems with a single, modern, multi-processor based system.

6.1 Industrial Case Study

An interesting industrial case study can be found in the thickness measurement and control systems found in cold rolling mills. A classical system arrangement is shown in Figure 6.1.

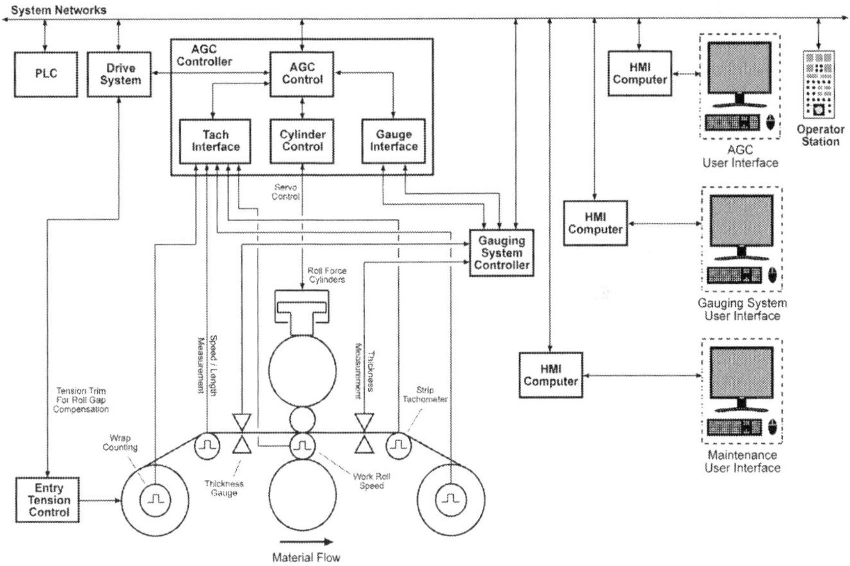

Fig. 6.1 - General arrangement and primary components associated with strip thickness measurement and control, in a cold rolling mill application.

The overall mill systems and operations are controlled / coordinated through the master programmable logic controller (PLC). A motor drive system provides strip transport and tension control. Strip thickness is measured on the entry and exit sides of the mill, and provided to the thickness controller. The thickness controller (AGC) adjusts the compressive roll force and trims the entry tension to provide corrections to the rolled material thickness. The operator controls and indications for the AGC and gauging system are provided through operator station control hardware and metering, while HMI Computers support interactive GUIs. A maintenance / engineering interface is provided to developmental support of the PLC, drive systems, AGC and gauging systems.

This system often consists of a five (5) computer arrangement, integrated into the general mill and drive system controls.

- AGC Real-Time Controller
- AGC HMI Computer
- Gauging System Controller
- Gauging System HMI Computer
- Maintenance Support HMI Computer

Through the use of similar FPGA, RTOS and Multi-Core Processor technologies (Wisti, et. al., 2008) (Zipf, et. al., 2008a), AGC Systems have been reduced to single computer arrangements. Here, using an Intel Dual Core processor, one core (running VxWorks) is dedicated to the real-time activities of the AGC Controller, while the other core (running Windows XP) handles the AGC's HMI tasks and supports the onboard OPC Server.

An extension of this concept involves the integration of the AGC, Gauging System, OPC Server, all HMI's and Maintenance Support into a single computer system. Development is currently underway, with the idea of employing an Intel Quad Core processor and a pair of self-contained C-Frame Controllers to achieve this consolidation. Figure 6.2 provides a block diagram illustration of this system arrangement, with the following core allocation:

- Core 1 : AGC Real-Time Controller running VxWorks
- Core 2 : AGC HMI and OPC Server running Windows XP
- Core 3 : Gauging System HMI running Windows XP
- Core 4 : Maintenance, Support and Development Software running Windows XP

The pair of C-Frame Controllers interface to the AGC Controller via a DTMNet and to the OPC Server and HMI via the Level 1 Data Network.

The first of these highly consolidated, multi-core processor systems is expected to be deployed in the mid-2010 time frame.

7. Conclusion

This chapter has presented a number of avenues of development and evolution in radiation based thickness measurement systems. The classical system architecture was reviewed and compared against various versions of contemporary architectures. It was shown that advancements in network systems have allowed highly decentralized / distributed configurations to be implemented, with a concentration of control / supervision authority local to the C-Frame / measurement head.

Detector developments have centered mainly on the full integration of all components into the sealed detector housing, and the inclusion of immediate digitization with deterministic networked distribution. These compact, self-contained detector assemblies offer convenient maintenance and easy spare / repair actions.

Developments in radiation generators have been interesting and varied. On one side, compact monoblock assemblies using passive thermal compensation and onboard regulation, control and supervision, offer opportunities to replace existing isotope sources.

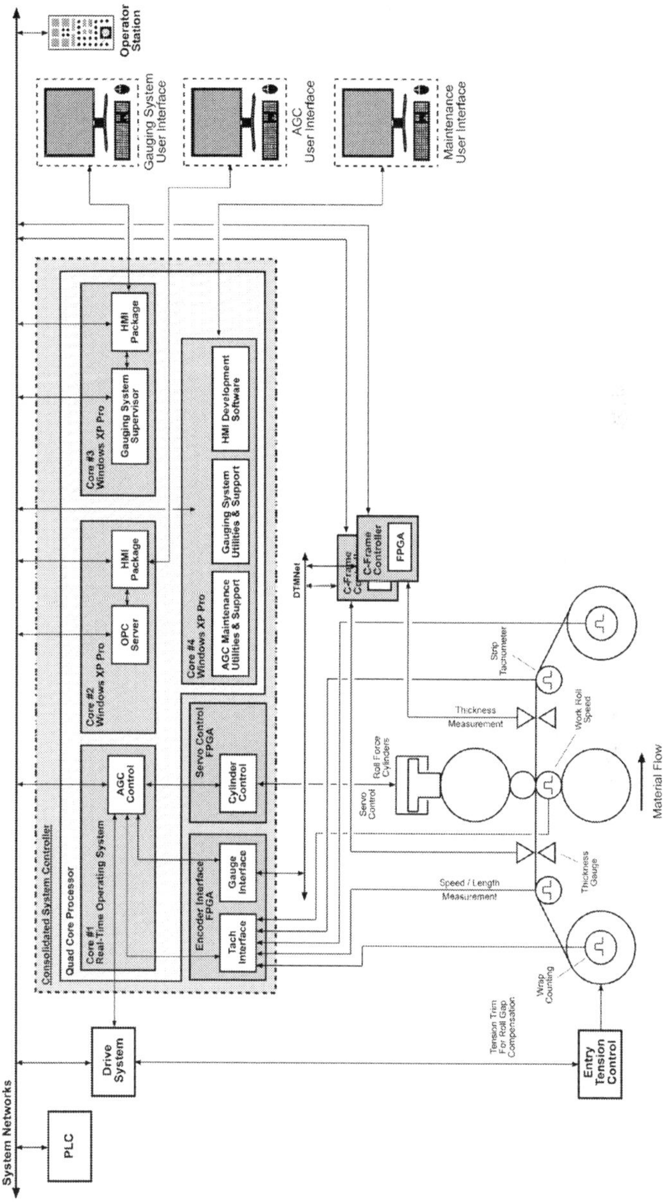

Fig. 6.2 - Block diagram showing a highly consolidated system, employing multi-core technologies, real-time controls, FPGA signal processing and networked interconnects.

On the other side, dry-potted, programmable high voltage power supplies provide an improving degree of modularity and require that only the X-Ray tube be immersed in oil, making it possible to use compact tube / tank assemblies that can be handled by a single person. The compact tube / tank assembly offers convenient maintenance and easy spare / repair actions.

New processor and FPGA technologies have allowed advanced real-time control, signal processing and complex analytic compensation to be performed local to the C-Frame / measurement head. Highly networked, distributed architectures eliminate long analog signal runs, offer broad interface-ability and flexible scaling / expansion. Advanced Web-based technologies coupled with OPC Servers provide the ability to support highly interactive, fast responding GUIs via thin clients executing on general purpose HMI Computers. Further, these methods provide Internet accessibility for remote control, monitoring and diagnostic assistance from any access point, world-wide.

Multi-core processor technologies have opened a new developmental avenue by offering the ability to consolidate multi-computer systems into single computer frameworks. These concepts provide new opportunities for surgical replacement / modernization of aged, obsolete equipment and reductions in installation costs and production outages.

In closing, it's rather interesting to note the various twists and turns in the developmental evolution of these systems. On one side, there is a consuming interest in decentralization, modular independence of subsystems and distribution of control / supervision local to the application. On the other side, new processor technologies draw developers to consolidate multi-computer systems into single computer arrangements. It would appear that the future may embrace the philosophy of decentralized consolidation employing integrated modularity.

8. References

NI - National Instruments (2007), *Operating Instructions and Specifications : CompactRIO cRIO-9012/9014 Intelligent Real-Time Embedded Controller for Compact RIO*, National Instruments Corp., Austin, TX

NI - National Instruments (2007), *Operating Instructions and Specifications : C Series I/O Modules*, National Instruments Corp., Austin, TX

Wisti, D., Krzewki, A., Godwin, C. & Zipf, M. (2008). A New Architecture for Automatic Gauge Control Systems, *Proceedings of the 2008 AISTech Annual Conference*, Pittsburgh, PA, May 5-8, 2008, Association of Iron and Steel Technologies, Pittsburgh, PA.

Zipf, M. (2010). Radiation Transmissoin-based Thickness Measurement Systems - Theory and Applications to Flat Rolled Strip Products, In: *Advances in Measurement Systems*, Chapter 6, Kordic, V. (Ed), In-Tech Publications, Vienna, Austria

Zipf, M., Hummel, C. & Burgess, R. (2007). A New Automation Architecture for Strip Thickness Gauging Systems, *Proceedings of the Associação Brasileira de Metalurgia e Materiais, 44th Rolling Seminar – Processes, Rolled and Coated Products*, Campos do Jordao, SP, Brazil, October 16-19, 2007, Campos do Jordao, SP, Brazil, Associação Brasileira de Metalurgia e Materiais, Sao Paulo, Brazil

Zipf, M., Hummel, C. & Regan, R. (2007). A Compact X-Ray System for Direct Replacement of Existing Isotope Radiation Thickness Measurement Systems, *Proceedings of the 2007 AISTech Annual Conference*, Indianapolis, IN., May 7-10, 2007, Association of Iron and Steel Technologies, Pittsburgh, PA.

Zipf, M., Wisti, D., Godwin, C. & Krzewki (2008). A New FPGA and Multi-Core System Architecture for Automatic Gauge Control, *Proceedings of the Associação Brasileira de Metalurgia e Materiais, 45th Rolling Seminar - Processes, Rolled and Coated Products*, October 21-24, 2008, Ipojuca - Porto de Galinhas, Brazil.

Zipf, M., Wisti, D., Krzewki, A., Carlson, S. & Godwin, C. (2008). A New FPGA and Multi-Core Processor System Architecture for Strip Thickness Gauging Systems, *Proceedings of the 2009 AISTech Annual Conference*, St. Louis, MO, May 4-7, 2008, Association of Iron and Steel Technologies, Pittsburgh, PA.

8

Experimental Radio Indoor Positioning Systems Based on Round-Trip Time Measurement

Alessio De Angelis[1], Antonio Moschitta[1], Peter Händel[2] and Paolo Carbone[1]
[1]Department of Electronic and Information Engineering (DIEI), University of Perugia, Italy
[2]Signal Processing Lab, ACCESS Linnaeus Centre, Royal Institute of Technology, Stockholm, Sweden

1. Introduction

This chapter presents the design issues and performance characterization results of two experimental systems for radio distance measurement and positioning. The main envisioned application area for these systems is the indoor environment, characterized by an insufficient coverage of global navigation satellite systems and by issues related to the indoor radio propagation channel. In particular, such issues include multipath propagation and the absence of line-of-sight. Furthermore, the applications for indoor positioning systems usually require a high degree of accuracy, which can be of the order of a centimetre, in the position estimation. Recently, this research area experienced a considerable growth and the state of the art is characterized by a wide array of solutions.

In this context, the systems described in the present chapter are based on two different approaches: an Ultra-Wideband (UWB) pulse-based solution (De Angelis et al., 2008a), (De Angelis et al., 2009a), (De Angelis et al., 2009c) and a platform developed using commercial devices complying with the ZigBee standard (Santinelli et al., 2009). However, both approaches share the same principle of operation: the measurement of signal Time-Of-Flight. In particular, the systems are capable of measuring the Round-Trip-Time (RTT), eliminating the need for accurate synchronization between transmitters and receivers. The time interval measurement function is performed in both systems by a commercial Time-to-Digital Converter (TDC) with 50 ps rms resolution.

The first approach involves the complete design process of the pulse-UWB radio interface, that is the circuitry necessary to properly generate and detect a sub-nanosecond pulse. These blocks have been designed and built from scratch using off-the-shelf components where possible. Furthermore, other relevant blocks are the timing and control logic, the interface with the TDC and the communication with a PC for data processing. The overall architecture has been developed using a master-slave approach, in which several slaves (pulse repeaters) are placed in fixed and known positions and thus constitute the infrastructure of the system. The master device is capable of measuring its distance with respect to the slaves and of estimating its position by means of triangulation algorithms. A

basic on-off pulse-keying protocol has been implemented in order to enable the communication of the master with multiple slaves.
The second approach uses commercial ZigBee radio interface devices and is based on the RTT measurement of communication packets. The main feature is the introduction of a self-calibration method, aiming at removing the bias introduced by the latencies of the fixed-position nodes.
Detailed descriptions of the architectures as well as experimental test results are presented in this chapter, together with a comparison and discussion of the advantages and disadvantages of each approach.

2. State of the art in the field of indoor positioning

Indoor positioning is the object of a considerable research interest. Accurate geolocation, in fact, represents an enabling technology for indoor location-aware sensor networks (Gezici et al., 2005) (Wymeersch et al., 2009). Some of the real-life applications for research in this field are related to the industry and logistic areas, e.g. for efficient asset tracking (Fontana, 2004). Another application area of interest is the location and tracking of emergency personnel and incident first responders (Rantakokko et al., 2007). In this context, both ad-hoc systems and architectures based on pre-existing communication infrastructures have been developed in the literature. Comprehensive surveys of the fundamental positioning strategies, as well as the main research and commercial systems in the indoor positioning area can be found in (Pahlavan et al., 2002) (Liu et al., 2007).
In the following subsections, some aspects of the state of the art are presented, related to the research activity described in the rest of this chapter.

2.1 UWB indoor positioning

In the specific field of UWB systems for indoor positioning, there is also a large number of research activities. In fact, such systems are considered good candidates for accurate ranging, mainly due to their fine time resolution, which is a consequence of the very large bandwidth (Di Benedetto et al., 2006). Other advantages include robustness to multipath propagation, low-power operation, immunity to communication interception and possible penetration through various types of materials (Fontana, 2004), (Molisch, 2005), (Gezici et al., 2005). In the following subsections, the UWB indoor positioning field is presented by first describing the regulatory context of the technology, then by describing some of the main approaches and systems in the literature.

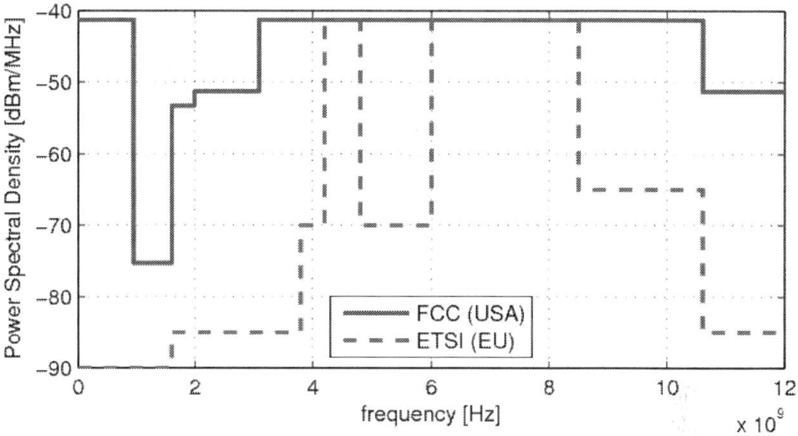

Fig. 1. Power Spectral Density limit mask for UWB.

2.1.1 UWB Regulation and standardization framework

UWB systems are regulated in different ways depending on the geographic areas. According to the most influential and widespread definition, provided by the Federal Communications Commission Regulation (FCC, 2002), an UWB system is defined as any intentional radiator having a fractional bandwidth greater than 20% or an absolute bandwidth greater than 500 MHz. This requirement means that a band-limited signal, with lower frequency f_L and upper frequency f_H, must satisfy at least one of the following conditions:

$$\frac{2(f_H - f_L)}{(f_H + f_L)} > 0.2 \tag{1}$$

$$f_H - f_L > 500\,MHz \tag{2}$$

Furthermore, the regulation defines a limit mask for the equivalent isotropic radiated power in indoor environments. In Europe, the regulation is different, due to the dissimilar usage of the electromagnetic spectrum (ETSI, 2008) (Hirt, 2007). The Equivalent Isotropic Radiated Power masks for the two regulations are shown in Fig. 1.

Within the rather wide regulatory framework, which allows for the development of several different technical approaches, a highly suitable ranging technology is pulse-based UWB, also known as Impulse Radio UWB (Win & Scholtz, 1998) (Dardari et al., 2009). This technology is currently regarded as a viable candidate for low data-rate applications, such as Wireless Sensor Networks, as demonstrated by the introduction of an alternative PHY layer provided by the recent IEEE 802.15.4a-2007 standard. This standard defines the protocol and compatible interconnection for data communication devices using low-data-rate, low-power and low-complexity, short-range radio frequency transmissions in a wireless personal area network (IEEE 802.15.4a, 2007).

However, it is important to mention that another variant of UWB signals, namely Multi-Band Orthogonal Frequency Division Multiplexing, is also currently under development and it is mostly intended for high data-rate short-range wireless communications (Wireless USB, 2007), (WiMedia Alliance, 2009), (Kumar & Buehrer, 2008).

2.1.2 Overview of approaches for UWB indoor geolocation

In the scientific literature, the state of the art in UWB indoor geolocation includes works based on different approaches, methodologies and objectives. In particular, it is possible to find works based on experimental testbeds realized using general-purpose instrumentation (Low et al., 2005). Furthermore, a considerable number of papers deal with the development of integrated circuits. Inside this category, of particular relevance are the System-on-Chip UWB pulse generator implementations (Sim et al., 2009) (Phan et al., 2007), and transceiver prototypes (Terada et al., 2006). Other works are aimed at presenting complete architecture solutions and associated challenges (Chandrakasan et al., 2009) (Mahfouz et al., 2008) (Stoica et al., 2005). Also, the results of additional research activities consist of experimental characterizations of full-featured positioning systems (Oh, 2009), (Fontana, 2004). Moreover, it is possible to mention several commercially-available UWB geolocation systems, mainly intended for industrial and logistics applications (Zebra Enterprise Solutions, 2009), (UbiSense, 2009), (TimeDomain, 2009). These systems provide positioning accuracies in the order of magnitude of 10 cm, with communication ranges of several tens of metres. Such commercial solutions typically include an infrastructure which is flexible and customizable to the user needs. Some of the additional features provided by such solutions include centralized localization algorithms, robust multiple access schemes, alternative channels for network coordination, and cell-based structures for network scalability.

2.2 ZigBee indoor positioning

Apart from UWB ad-hoc positioning solutions, other approaches for geolocation are designed to exploit a pre-existing wireless communication infrastructure. Among these, the systems based on the ZigBee technology (ZigBee Alliance, 2007) represent an emerging research area. ZigBee is considered as an interesting choice for indoor geolocation thanks to several features such as low power consumption, good range, low cost and wide commercial diffusion. Furthermore, by using this technology, the positioning functions can be easily integrated with other communication services in a wireless sensor network.

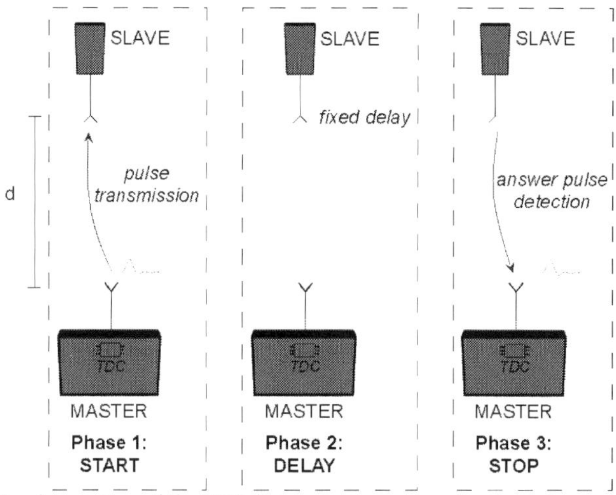

Fig. 2. Principle of operation of the RTT distance-measurement approach.

The most common strategy for geolocation using ZigBee signals is based on the Received Signal Strength (RSS) of communication signals (Blumenthal et al., 2007) (Cho et al., 2007). Using this approach, an estimate of the distance between a transmitter and a receiver is obtained by means of the received power level, given an appropriate model for the power loss of the transmission channel. Such a measure may be provided by some commercial ZigBee hardware as a RSS Indicator (RSSI). However, the achievable accuracy of RSS-based ZigBee positioning systems is usually in the order of magnitude of 1 m. Better results can potentially be obtained with Time-Of-Arrival methods, as demonstrated by some works found in the literature (Santinelli et al., 2009) (Corral et al., 2008) (Schwarzer et al., 2008).

2.3 Indoor Radio Propagation Channel

One of the fundamental issues associated to radio geolocation is represented by the propagation channel. In fact, especially in indoor scenarios, the channel is strongly affected by multipath fading, a phenomenon which is of key relevance in both the approaches considered in this work. In particular, for the UWB case, due to the high bandwidth and fine time resolution of the pulses, the system is intrinsically able to discriminate a large number of paths (Lee & Scholtz, 2002). In this field, channel modeling research activities have been published in the literature, see (Molisch, 2005) (Molisch et al., 2005). Furthermore, the ZigBee solution requires an essentially different approach to the multipath problem, since it can be basically considered as a narrowband system. In particular, as a ZigBee receiver relies on correlation techniques to synchronize itself with an incoming packet, RTT measurements may be affected by increased correlation times when input signal power is reduced by multipath. The goal of research activity in this field is to properly model and compensate this phenomenon, in order to accurately estimate its effect on the measured signal time-of-flight (Hirschler-Marchand & Hatke, 2002).

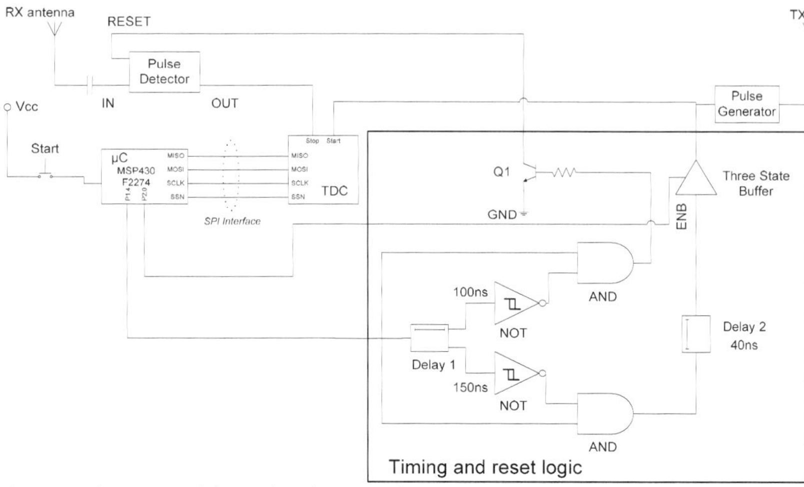

Fig. 3. Architecture of the realized master device prototype (De Angelis et al., 2008a).

3. UWB pulse-based ranging and positioning system

The basic principle of operation of the developed UWB ranging system involves the accurate measurement of the RTT of a pulse propagating between two transceivers, schematized in Fig. 2. In order to implement this accurate measurement in a compact, low-power and portable platform, an approach based on a TDC has been employed. In particular, the commercial integrated TDC-GP2 by Acam Messelectronic Gmbh, has been used, having a rms resolution of 50 ps (Acam GmbH, 2008). This order of magnitude is appropriate for accurate distance measurement in a typical indoor application, since it implies a 0,75 cm theoretical resolution when a RTT approach is used. Apart from the time-interval measuring unit, a significant challenge in the development of the system has been the design from scratch of all the main building blocks, such as the radio front-end (pulse generation and pulse detection), timing, control and results processing.

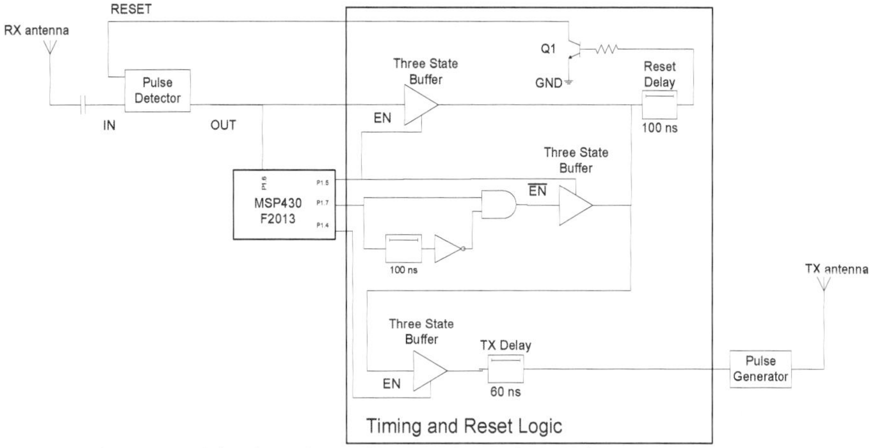

Fig. 4. Architecture of the slave device (De Angelis et al., 2008a).

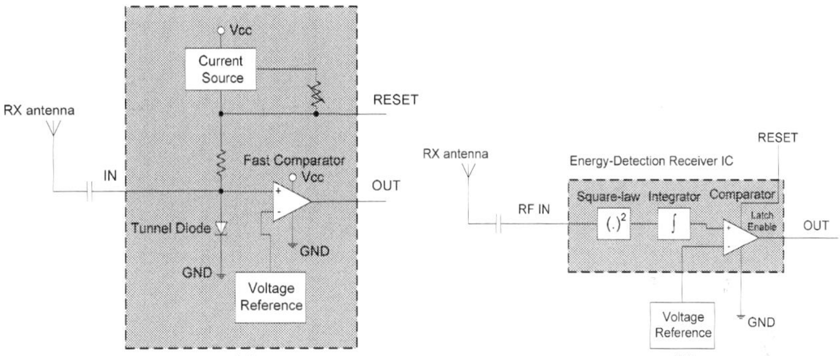

Fig. 5. Block diagram of the two receiver architectures: a) Constant-current biased tunnel-diode threshold detector. b) Energy detection receiver.

3.1 Ranging system architecture

The system is based on a master-slave architecture, where the master is a transceiver with the capability of measuring the time interval between transmission and reception of a pulse. On the other hand, the slave device has a simpler architecture, lacks the time-interval measuring capability and is designed as a pulse repeater. The block diagram of the master device is shown in Fig. 3. As mentioned above, the RTT of a pulse propagating between the master and the slave is measured by the TDC, while the low-power microcontroller (Texas Instruments MSP430) is used to program it and read measurement results through a Serial Peripheral Interface (SPI). The data are also transmitted by the microcontroller to a Personal Computer for subsequent processing, in order to estimate the master-slave distance.

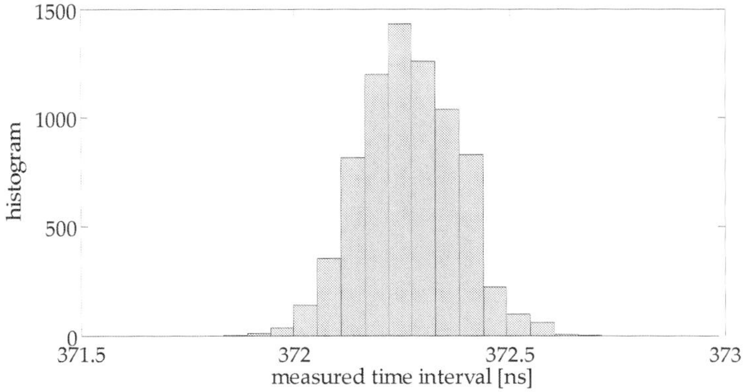

Fig. 6. Experimental results, raw RTT data, obtained using the tunnel-diode threshold detector. Normalized histogram, 7500 RTT measurement results at a distance of 140 cm.

Furthermore, an important block is represented by the timing and reset logic network, entirely realized with commercial off-the-shelf CMOS circuits. Its purpose is to provide proper timing for the reset path of the pulse detector. Using this network, the self-detection condition, that is the detection of a pulse transmitted by the same device, is avoided.
The block diagram of the slave device is shown in Fig. 4. It can be noticed that its architecture is similar to that of the master. The main difference is that this device doesn't implement the time-interval measurement hardware. There are also differences in the timing and reset logic.
A relevant section of the research activity has been dedicated to the comparison of several different UWB pulse generation strategies that exist in literature. The nature of this comparison has been mainly experimental with requirements of architectural simplicity and of the use of commercial off-the-shelf components, where possible. The results of this experimental comparison activity are presented in (De Angelis et al., 2008b). The various strategies investigated were based on logic gates, step recovery diodes and avalanche transistors. At the end of the comparison it was established that the step recovery diode pulse generator approach is the one that provides the best trade-off between transition time, amplitude and power consumption. In particular, it provides a negative pulse with an amplitude of about -8.2 V and a measured fall time of about 690 ps. Furthermore a -10 dB bandwidth of about 500 MHz has been observed, thus satisfying both the requirements in equations (1)-(2). This generator has been chosen for use in the complete UWB geolocation system experimental prototype.
Another fundamental building block for the system is the pulse detector. In this regard two strategies have been analysed: a signal-threshold detector and an energy detector. For details on the design and implementation of these devices, the reader can refer to (De Angelis et al., 2008a). Fig. 5a shows the diagram of the threshold detector, which is based on a constant-current biased tunnel diode.

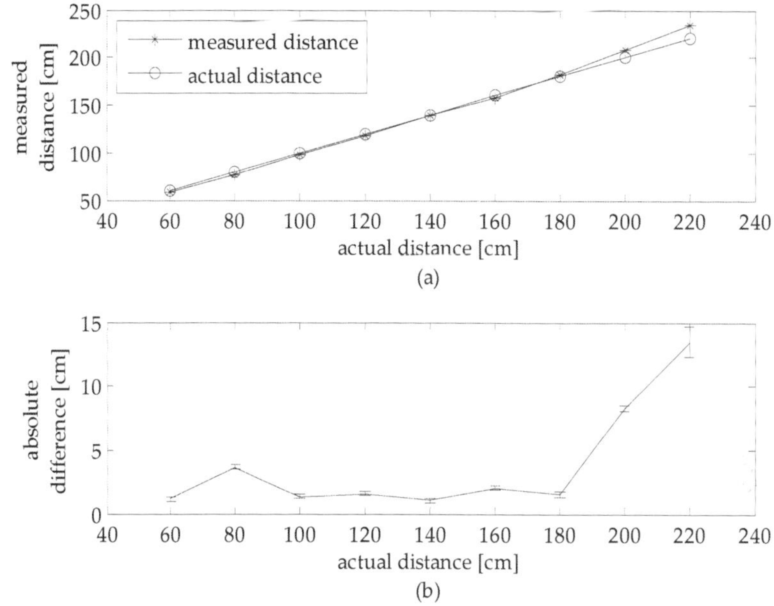

Fig. 7. Experimental results obtained with the tunnel-diode threshold detector, with the application of the calibration factor. (a) Measured distance vs actual distance. (b) Absolute difference between the measured and actual distances. The bars along the absolute difference curve represent the standard deviation in centimetres.

The block diagram of the energy detector, which has been implemented by means of a commercial RF circuit with an integrated comparator (Linear Technologies LTC5536), is shown in Fig. 5b. As evidenced by the experimental results shown in the next section, the energy-detection approach leads to better results in terms of operating range, even with a slight deterioration of the accuracy. Considering this trade-off, however, in the following research activity the energy detection receiver was selected.

Furthermore, the employed antenna consists in a disc-cone with an height of about 12 cm. This antenna design is widely used in wideband applications, in particular in the UWB field, and its introduction can be dated to a 1945 patent (Kandoian, 1945). This design has been chosen for its relative simplicity of construction, wide bandwidth and omnidirectional pattern in the horizontal plane, particularly interesting for positioning applications.

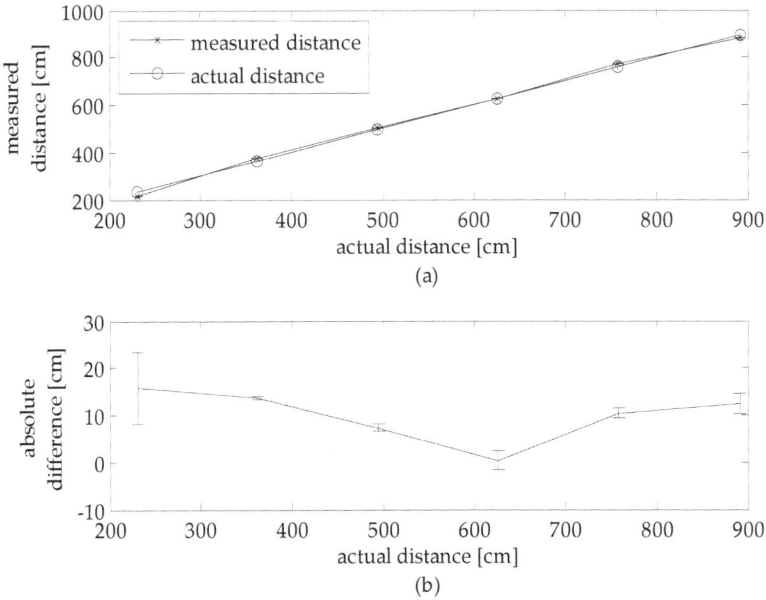

Fig. 8. Experimental results obtained with the energy detector, with application of the calibration factor. (a) Measured distance vs actual distance. (b) Absolute difference between the measured and actual distances. The bars along the absolute difference curve represent the standard deviation in centimetres.

3.2 Distance-measurement experimental results

The two transceivers have been placed at known distances in line-of-sight conditions in order to characterize their performance. For each distance, multiple RTT measurements have been performed. An histogram for a single distance is shown in Fig. 6, and the mean measured distance values with respect to test distance are plotted in Fig. 7 and 8, together with the related standard deviations. This procedure has been executed in a typical laboratory environment, in the presence of multipath fading and interference from other commonly-used electronic and radio systems. Prior to the actual measurement phase, a calibration procedure has been carried out, consisting in the calculation of the linear regression of experimental data. This makes it possible to obtain a calibration factor to compensate for the effects of systematic errors in the RTT measurement procedure.

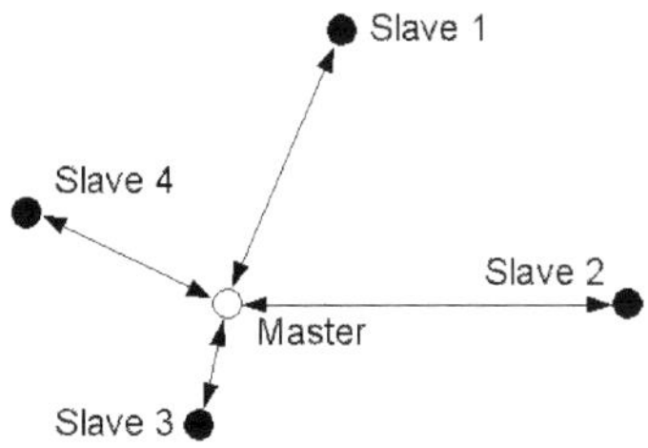

Fig. 9. Star network topology.

For the tunnel-diode threshold detector, Fig. 7 shows the obtained plot for all distances. At each distance, apart from the last one, a standard deviation of about 0.2 ns has been observed, which implies an accuracy of about 3 cm in the distance estimation. The maximum range achievable by the system using this receiver is about 2.5 m.
Moreover, Fig. 8 shows the results obtained with the energy-detection approach. It can be noticed that the measurement accuracy is worse (maximum deviation from the reference distance is about 10 to 30 cm), but the communication range is considerably extended (up to about 9 m). However, by averaging repeated measurement results, assuming that the RTT measurement results are uncorrelated, it could be possible to improve the accuracy when using the energy detector. Therefore, depending on the application, a trade-off between accuracy and processing time can be taken into account.

3.2.1 Metrological Characterization

Following the approach described in the ISO "Guide to the expression of Uncertainty in Measurement" (GUM), it is possible to perform a "Type A" analysis to evaluate the uncertainty associated with the estimated distance (ISO GUM, 1995). Using this approach the uncertainty is estimated by statistical methods. In particular, the sample standard deviation obtained from the experimental data provides an estimate of the standard deviation of the underlying distribution, which is identified with the standard uncertainty. Therefore it can be noticed that it grows with distance. However an upper limit for distances up to 9 m can be quantified at 30 cm, while for distances up to 4 m the standard uncertainty is less than 10 cm. Furthermore, it must be emphasized that these results are relative to a single RTT measurement. As explained above, by averaging multiple results, it is possible to improve accuracy, since we can assume that the measurement results are uncorrelated. Particular considerations could also be done on the systematic deviations. It can be noticed that these effects may be greatly reduced by performing a proper calibration of the system.

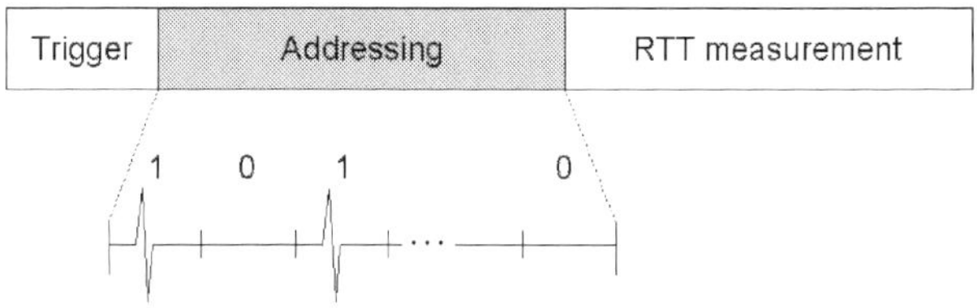

Fig. 10. On-Off Keying addressing scheme using UWB pulses

3.3 Complete positioning system and experimental tests

The distance-measuring scheme described in the previous paragraphs can be exploited to implement a full positioning system. In order to do so, however, a number of design choices and trade-offs have to be considered. This section aims at highlighting some of the main decisions made in the design phase for our positioning project, and how they affected the prototype features. A detailed presentation of the system-level architecture and the prototype specifications can be found in (De Angelis et al., 2009b).

3.3.1 Extension of the UWB ranging system with multiple slaves

Using the terminology associated to the sensor networks field (Zhang et al., 2009), a fundamental aspect of a location system is the network topology. From this point of view the realized system is characterized by a star configuration with a centralized access scheme, as shown in Fig. 9. Following this approach, only one of the nodes (the mobile node) is responsible for the timing and multiple access of the network. The reason for the choice of this topology is that it is suitable for small geographical extension networks, and it reduces the complexity and overhead of channel access. Regarding the access scheme, a simple protocol based on Time-Division Multiple Access (TDMA) was adopted. The devices can only communicate, for distance-measuring purposes, during a predefined time period, which is assigned by the master node. This time slot assignment procedure is performed by means of an addressing phase, during which the master sends a unique slave identifier code, using binary on-off keying with UWB pulses. Basically, this coding scheme consists in a subdivision of the temporal axis in sequential intervals. In each interval, the presence of a UWB pulse represents the "1" bit, while its absence represents the "0" bit, as illustrated in Fig. 10.

If the microcontroller in a particular slave recognizes the transmitted sequence as its own identification sequence, it drives the timing and reset logic in a proper way, thus putting the slave in the measurement mode. In this mode, the slave acts as a pulse repeater for a predefined period of time (measurement period), activating an automatic signal path. Through this path it generates an answer pulse after a known delay following each received pulse from the master. If the transmitted sequence is not recognized as the slave's own identification sequence, the microcontroller disables pulse transmission for the measurement period. The motivation for the use of this technique is that it represents a low-complexity scheme allowing the master device to unequivocally identify the slave with which it is communicating, thus avoiding collisions that could result in incorrect position determination.

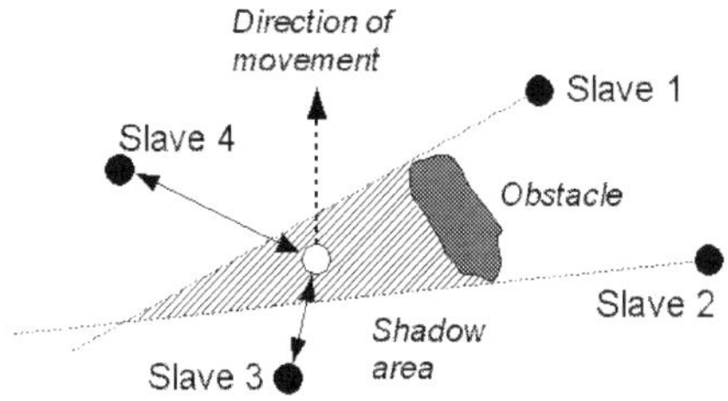

Fig. 11. Example of a situation where the INS supports the positioning system by tracking the position in an obstructed area.

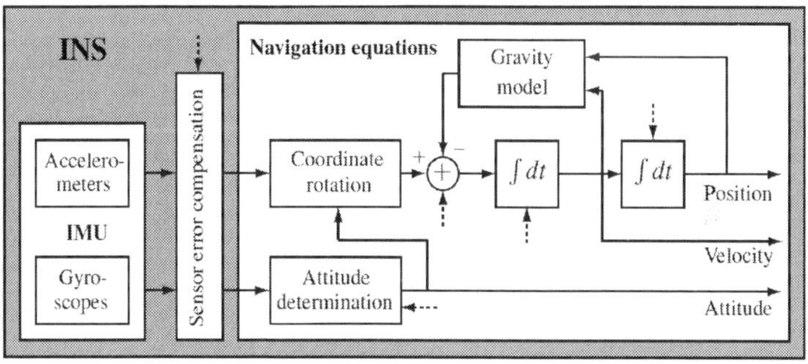

Fig. 12. Inertial Navigation Subsystem architecture (De Angelis et al., 2009b).

3.3.2 Inertial Measurement System

The main motivation for integrating an inertial navigation system (INS) with the UWB ranging and positioning devices described in the above sections is to obtain a complete and robust system. In fact, the INS adds information about the attitude of the mobile device, that is the orientation with respect to a reference frame. Furthermore, it provides the knowledge of the velocity and acceleration, thus enabling a complete characterization of the navigation states. Regarding the contribution of the INS to the system's robustness, it can be useful to support the ranging system in those areas where there is insufficient coverage, e.g. there is line-of-sight connection with less than three slaves. In such situations the INS can track the mobile node's position for a limited period of time, as shown in Fig. 11.

The INS is composed of two sections: the Inertial Measurement Unit (IMU), an hardware device that houses accelerometers and gyroscopes, and the software component which implements the navigation equations. A detailed illustration of the INS architecture is shown in Fig. 12.

	Navigational states	Dynamic range	Stand-alone error
UWB	No	Low	Bounded
INS	Yes	High	Unacceptable Growth

Table 1. Complementary properties of the two subsystems.

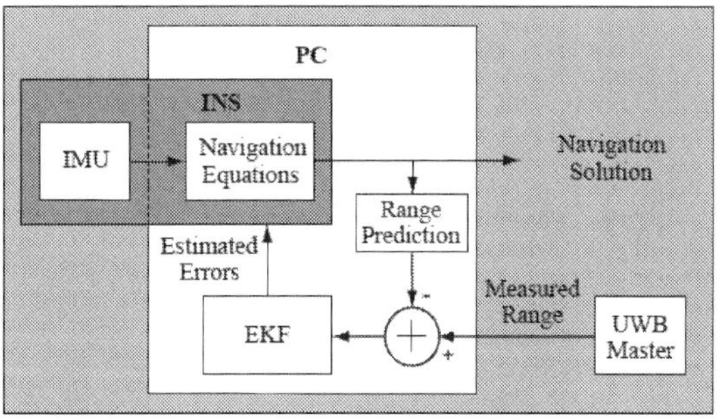

Fig. 13. Information fusion architecture (De Angelis et al., 2009).

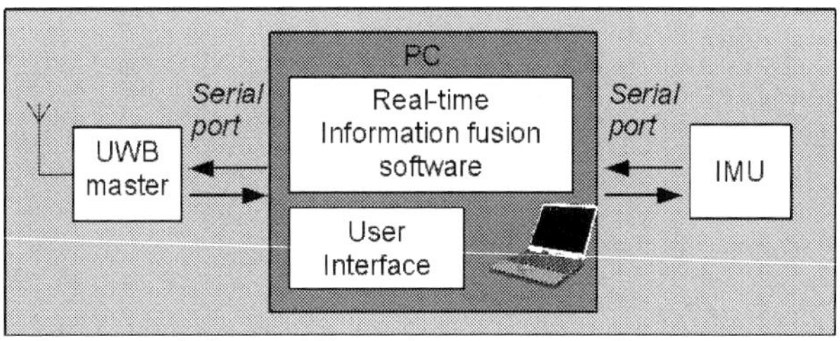

Fig. 14. Block diagram of the mobile node prototype. The sensor fusion platform is implemented by a software running in a PC, connected to the UWB ranging master device and to the IMU by means of serial ports.

3.3.2 Information fusion

The information fusion technique has been designed with the main goal of exploiting the complementary properties of the UWB and INS systems. In particular, the INS readily provides information about the attitude, velocity and acceleration, which represent a complete set of navigational states. Furthermore, the high update rate of the employed INS (100 Hz) supports a high dynamic range, while the UWB ranging system has a relatively slow update rate (2 Hz) when used as a stand-alone system. The INS, however, intrinsically amplifies low-frequency noise and has an unacceptable growth rate with time. On the other hand, the UWB ranging system has a bounded error for the position estimation and therefore may be used to compensate for inertial measurements deviations. A summary of the complementary properties is provided by Table 1.

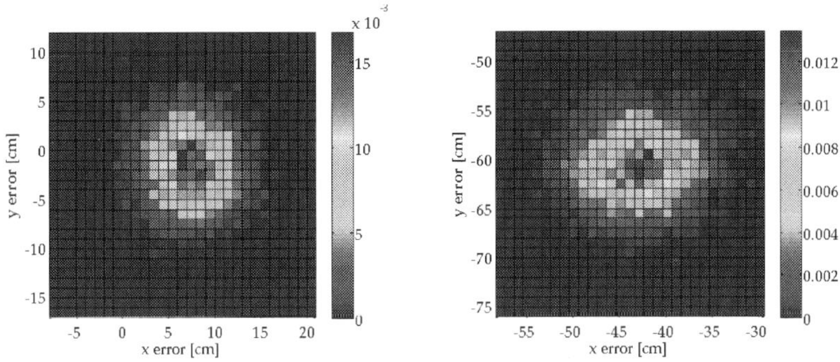

(a) (b)

Fig. 15. Histogram of the estimated positions. (a) Reference position: (150 cm, 150 cm), coinciding with the centre of the measurement area. (b) Reference position: (100 cm, 100 cm).

Fig. 13 shows a block diagram of the information fusion approach used in the system. The basic principle is a UWB-radio range aided INS architecture, which is developed in a similar manner as a widely adopted solution in outdoor GPS-aided INS. Following such an approach, the INS provides the main navigation solution and when a UWB-radio range measurement is available a range prediction is calculated based upon the current position estimate of the INS. The difference between the measured and predicted range is used as an observation for an Extended Kalman Filter (EKF), implementing a model for how the errors in the INS develop with time and how they relate to errors in the range predictions. The errors are estimated and fed back to the INS. For an exhaustive description of the signal processing algorithm, together with the main issues addressed and the solutions implemented in the system, the reader can refer to (Nilsson et al., 2009).

The complete information fusion architecture has been implemented in a prototype, as shown in Fig. 14. In the prototype, the two information sources, namely the UWB ranging system and the INS, are connected to the sensor fusion platform implemented via software in a PC by means of serial ports.

3.3.2 Experimental Characterization

In order to experimentally investigate the behaviour of the system and the achievable performance, three kinds of tests have been performed. All the tests have been executed in an indoor environment with line-of-sight signal propagation conditions. Furthermore, as a preliminary phase, all the slave devices were calibrated by performing a series of RTT measurements at known distances, thus obtaining the calibration parameters. Following the preliminary phase, a static position-measuring procedure has been implemented, in which four slave nodes were placed in fixed and known positions at the corners of a 3×3 meter square and a master node was placed in the centre of the same square. During this test, only the UWB ranging system was used, without the fusion with the INS. Repeated individual master-slave distance measurements were collected and the position was estimated following a least squares approach. The results obtained from this measurement set are shown in the histograms of Fig. 15a and 15b. Each of the figures is relative to a different master position, and it can be noticed that the second image shows a larger histogram

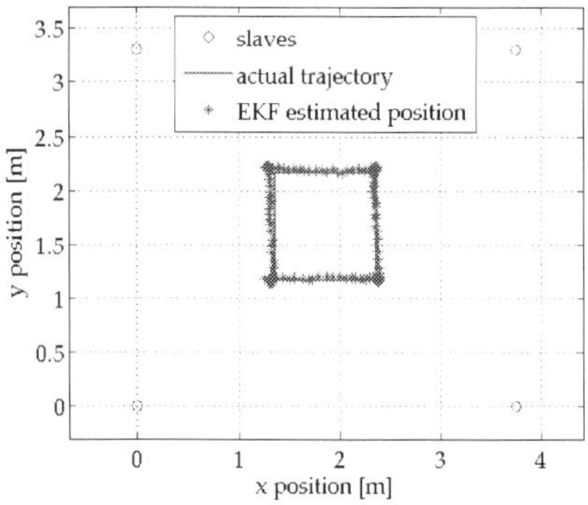

Fig. 16. Low-dynamic test, stand-alone UWB-positioning system. A good agreement between the estimated and the actual trajectory can be observed (De Angelis et al., 2009b).

dispersion. This behaviour is consistent with what was expected, since the positioning accuracy is dependent on the geometrical configuration of the master with respect to the slaves and their relative distance. From this point of view, the geometric centre position is the most favourable.

By analysing these experiments, it is possible to evaluate the uncertainty associated with the measurement result. Such an evaluation, similarly to the one provided in Section 3.2.1, can be performed using statistical methods, according to the GUM (ISO GUM, 1995). The sample mean and the sample covariance of the experimental estimated positions for the geometrical configuration in Fig. 15a are given by:

$$sample\ mean[cm] = \begin{bmatrix} 157 \\ 148 \end{bmatrix}$$

$$sample\ covariance\left[cm^2\right] = \begin{bmatrix} 10.2 & -0.5 \\ -0.5 & 11.2 \end{bmatrix}$$

Therefore, referring to the sample covariance matrix, the “Type A” standard uncertainty can be quantified as 3.3 cm for each spatial coordinate.
Subsequently, in order to investigate the position-tracking capability of the UWB ranging platform as a stand-alone system, a second set of measurements have been performed. In this case the master device was moved slowly, at about 0.1 m/s, along a predefined 1×1 m square trajectory. The tracking function has been implemented by means of an algorithm based on the EKF approach. The results of this experiment are shown in Fig. 16, demonstrating a good tracking capability with a maximum deviation of the measured trajectory from the reference path of about 10 cm.

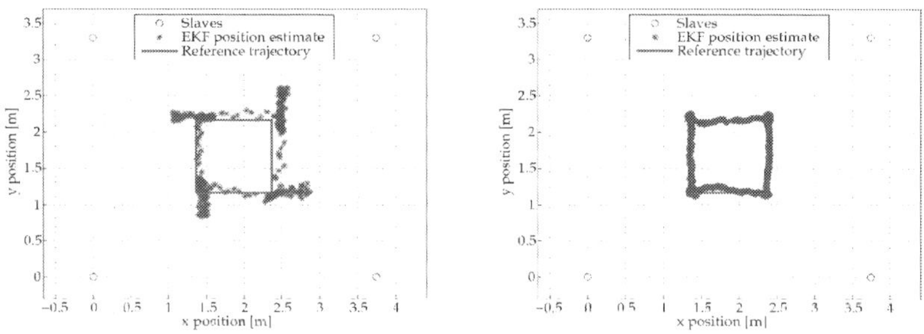

(a) (b)

Fig. 17. High-dynamic tests, (a) UWB-positioning stand-alone: it can be noticed that overshoot is present, and poor tracking performance is observed. (b) information fusion with INS: no overshoot and better tracking performance has been observed (De Angelis et al., 2009b).

The last set of tests consisted in high-dynamic experiments involving the complete system, including the INS and information fusion platform. The experiment was executed on the same trajectory as that of the previous set, but in this case the master was moved quickly, at approximately 1 m/s. Initially the test was performed using the UWB system alone. This test allowed to demonstrate some of the limits of the UWB stand-alone system, mainly caused by the slow update rate, as shown by the overshoot and by the low tracking performance in Fig. 17a. Subsequently, another test using the full system with INS measurement integration was executed in the same experimental conditions. The resulting data is shown in Fig. 17b, demonstrating the considerable performance improvement obtained. In particular, the trajectory overshoot has been eliminated and an overall better tracking behaviour can be observed.

3.4 Future developments

Some of the research activities expected in the future involve the improvement of the hardware ranging platform realized so far. In particular, a modification of the baseband pulse generator is currently being studied, with the goal of realizing a system which is fully compliant with both the current European and USA regulations. The step recovery diode generator described in Section 3.1, in fact, provides an output spectrum roughly from DC to 500 MHz. By means of a pulse amplitude-modulation approach, it is possible to shift this spectrum around the frequency of a carrier signal in the 6 GHz region, therefore operating inside the UWB regulation masks shown in Fig. 1. Furthermore, such a solution could potentially reduce problems associated with the interference from other radio systems, since there is a lower number of narrowband users in the 6 GHz band. Other future developments are aimed at increasing the communication range, coverage and update rate of the system, also considering advances in the information fusion technique and signal processing algorithms.

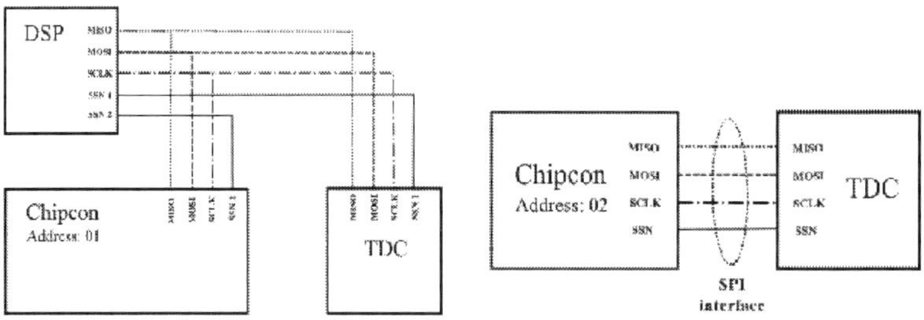

(a) (b)

Fig. 18. Block diagram of the ZigBee positioning system: (a) Mobile node, (b) Calibrator node.

4. ZigBee ranging system

The principle of operation of the ZigBee system is based on the RTT measurement of Direct Sequence Spread Spectrum (DSSS) modulated pulses, propagating between several transceiver devices. The system-level design, presented in detail in (Santinelli et al., 2009), has been focused on a trade-off between accuracy, cost and synchronization hardware. In particular, it is a low-cost solution, realized by applying some modifications to an existing communication network, without the need for external synchronization hardware, but characterized by a good accuracy for the majority of the applications. Moreover, the implemented system is capable of planar positioning and can easily be extended to 3-D positioning or combined with other ZigBee based applications.

A key feature of the realized prototype is its self-calibration capability, designed to remove the bias introduced by the fixed beacons latencies. In fact, in order to properly implement a RTT distance measurement system, it is necessary to estimate and compensate for the responder node latency, as already seen in the previous sections for the UWB device.

4.1 Architecture

The ZigBee transceivers used as a basis for the system prototypes are the CC2431 by Chipcon, while the time-interval measuring hardware is the same TDC, the TDC-GP2 by Acam Messelectronic GmbH, used for the UWB system described in the previous sections. The system is comprised of three kinds of devices: the mobile node, the fixed node and the calibrator node. Diagrams for the mobile and calibrator are shown in Fig. 18a and 18b respectively.

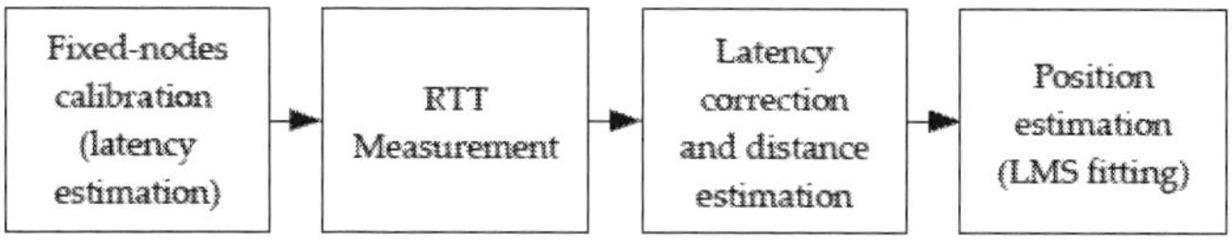

Fig. 19. Flow-chart of the operation of the ZigBee positioning system.

In the mobile node the Chipcon CC2431 device is connected via a Serial Peripheral Interface (SPI) with a DSP (Texas Instrument TMS320C6713) and a TDC. The Chipcon device is used as an RF interface, while the time interval measurement function is performed by the TDC. The DSP is used to program the TDC, to read measurement data through the SPI and to triangulate the mobile node position by properly combining the various RTT measurement results. In particular, the DSP coordinates the communications with slave nodes, by polling them one at a time.

The calibrator node has a simpler architecture, without the use of the DSP. By placing this node in a known position, the bias introduced by the fixed nodes can be evaluated by polling them with the calibration node to perform RTT measurements. Finally, each fixed and known position node is a Chipcon CC2431 device, used as pulse repeater.

The operation of the system is described in the flow-chart of Fig. 19. Initially the mobile node programs and initializes its TDC. Then it transmits a packet to a given fixed node, which is received and retransmitted after a certain delay. The master then receives the retransmitted packet, and the TDC determines the signal RTT by measuring the time interval occurring between the transitions of its Start and Stop input signals. By repeating this procedure for a set of fixed nodes, and using calibration data, the mobile node DSP may determine the unknown position using data fusion techniques.

4.2 Experimental Results

The ZigBee positioning system has been experimental validated by implementing two configurations in a LOS environment scarcely affected by multipath. The first one consisted of a network of 6 fixed nodes (beacons), without the self-calibration capability. The second one, having fewer nodes, consisted of the mobile node, the calibration node, and a cell of 3 beacons. Experimental results obtained using the first setup are reported in Fig. 20. The white square represents the actual node, the 6 black squares are the fixed beacons, and the points represent the estimated positions. Each position measurement has been obtained by averaging 500 RTT measurements for each fixed position node. The latency of a single beacon was manually evaluated, and the obtained result was used to compensate all of the RTT measurements. Subsequently, the triangulation was achieved by applying a Least Mean Square (LMS) fitting approach on a linearised model (Santinelli et al., 2009). By analysing the obtained results, a bias of -0.74 m on the x coordinate and of -0.59 m on the y coordinate have been observed, while the standard deviation is about 0.41 m for both coordinates.

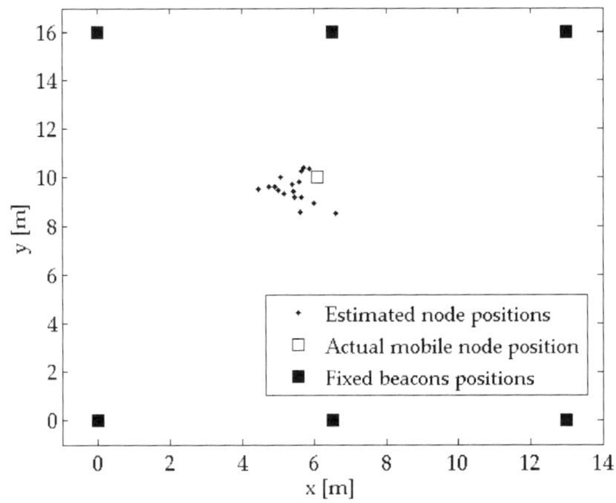

Fig. 20. Experimental results for the first configuration, without self-calibration feature. (Santinelli et al., 2009).

Following this experiment, the second setup was implemented and tested. As a first step, the calibration procedure was performed for each beacon, in order to estimate and compensate for the corresponding latencies. In particular, the average of 500 RTT calibration measurements performed at increasing distances between the calibration node and one of the beacons was calculated. Then, the calibration line was obtained by applying numerical fitting to the measurement results, as shown in Fig. 21a. It can be observed that the RTT increases linearly in the considered range, with a slope of approximately 22.5 ns/m. A possible cause of this behaviour may be the reduced signal-to-noise ratio at the receiver input, inducing increased correlation latencies. Consequently, as the signal propagation contributes to the RTT with a slope of 6.6 ns/m, we may infer that also the detection time varies linearly, with a slope of about 15.9 ns/m. Moreover, the time required to retransmit a detected pulse, which is not dependent on the distance d, is the line intercept point, which is about 888 μs.

Measurement results for the second setup are reported in Fig. 21b, where the positions of the various nodes and estimated positions are shown. Similarly to Fig. 20, the white square represents the mobile node, located in (3,3), the 3 filled squares are the fixed beacons, located in (2,1), (4,1), and (1,6) respectively, the white circle is the calibration node, located in (3,6), and the points are measurement results. Again, each position measurement has been performed by averaging 500 RTT measurements per fixed position node. Then, bias removal and range estimation have been performed, using the calibration results obtained for each beacon. Finally, as only 3 fixed beacons were present, a simplified triangulation algorithm was used, obtained by modifying the algorithm which calculates the intersection of 3 circumferences. The measurement results correspond to a standard deviation of 0.2 m and a bias of 0.14 m for the x coordinate, and a standard deviation of 0.16 m and a bias of 0.01 m for the y coordinate. Consequently the results obtained show that, by using self calibration, a system based on a reduced number of beacons may achieve better location accuracy than a larger system without self-calibration capabilities.

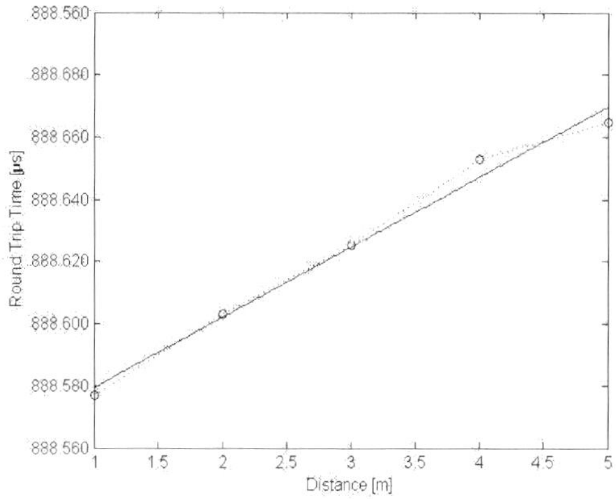

(a)

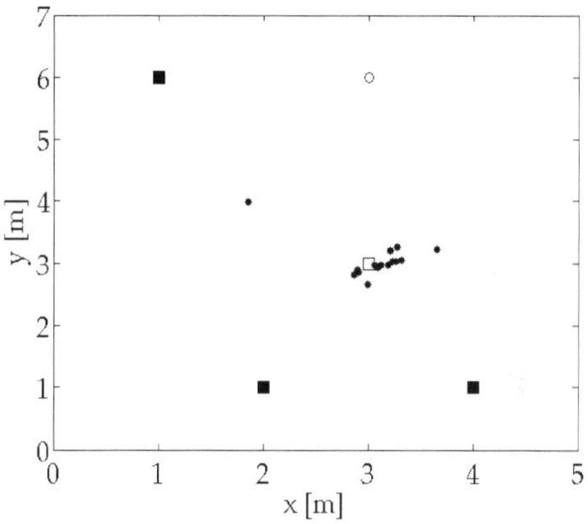

(b)

Fig. 21. Experimental results for the second configuration, with self-calibration. (a) Node calibration line. The circles are measurement results, while the straight line has been obtained by fitting. (b) Positioning results, the white square is the actual node position, the black squares are the reference nodes, the white circle is the calibration node and the points are estimated positions. (Santinelli et al., 2009).

4.3 Future developments

Some possible developments of this research activity involve the theoretical and experimental study of the performance of RTT-based ranging and positioning techniques for ZigBee systems. Of particular interest is the performance evaluation in multipath and non-line-of-sight channels. A further development is represented by the study and implementation of methods aimed at increasing system immunity to multipath, such as those presented in (Schwarzer et al., 2008). An alternative approach involves the information fusion between the

RTT and readily-available signal-strength data, in order to identify measurements affected by strong multipath and weight their contribution to LMS fitting. Additionally, multipath components may possibly be identified by performing ZigBee transmissions at different frequencies and by analysing the variations when measuring both the RSS and the RTT.

5. Conclusion

Two indoor positioning systems based on RTT measurement have been presented in this chapter, focusing on a description of the basic principle of operation, main building blocks and experimental evaluations. Regarding the technical performance aspects, as shown by the experimental test results described in the previous sections, the UWB system has the advantage of a higher positioning accuracy. This system, in fact, is capable of measuring position with a standard uncertainty of about 3 cm under favourable conditions, while the ZigBee platform is characterized by an accuracy of about 20 cm. The latter solution, instead, provides an higher communication range, up to the order of magnitude of 100 m, as opposed to the 10 m order provided by the UWB prototypes.

Apart from the purely performance-related analysis, it is also possible to perform a more general comparison, considering other aspects. In this regard, the ZigBee system has the advantage of exploiting a pre-existent communication platform. This feature could be of relevance from the point of view of a planning strategy, depending on the specific application. Using this feature, in fact, a fast, robust and convenient data transfer in the wireless sensor network could be provided in addition to the position evaluation. On the other hand, the UWB system, being a non-commercial solution, allows for a greater control of the design aspects and specifications. The additional bandwidth-related benefits associated to the UWB approach must also be taken into account, as discussed in the state of the art section of this chapter.

6. Acknowledgements

This research activity has been partially funded by the research grant COTR09SS provided by "Consorzio per lo Sviluppo Universitario della Provincia di Terni", whose support the authors gratefully acknowledge.

7. References

Acam messelectronic GmbH, (2007) TDC-GP2 Universal 2 Channel Time-to-Digital Converter Data Sheet. [Online]. Available: http://www.acam.de/fileadmin/Download/pdf/English/DB_GP2_e.pdf

Blumenthal, J.; Grossmann, R.; Golatowski, F. & Timmermann, D. (2007) Weighted Centroid Localization in ZigBee-based Sensor Networks, *IEEE International Symposium Intelligent Signal Processing, WISP*, Alcala de Henares, Oct.3-5, 2007, ISBN: 978-1-4244-0829-0

Corral, P.; Pena, E.; Garcia, R.; Almenar, V.; de C. Lima, A.C. (2008) Distance Estimation System based on ZigBee, *Proceedings of 11th IEEE International Conference on Computational Science and Engineering Workshops, CSEWORKSHOPS '08*, San Paulo, 16-18 July 2008, pp.405-411, ISBN: 978-0-7695-3257-8

Chandrakasan, A.P.; Lee, F.S.; Wentzloff, D.D.; Sze, V.; Ginsburg, B.P.; Mercier, P.P.; Daly, D.C. & Blazquez, R. (2009) Low-Power Impulse UWB Architectures and Circuits, *Proceedings of the IEEE* , Vol. 97, No. 2 (Feb. 2009), pp. 332–352, ISSN: 0018-9219

Cho, H.; Kang, M.; Park, J.; Park, B. & Kim, H. (2007) Performance Analysis of Location Estimation Algorithm in ZigBee Networks using Received Signal Strength, *Proceeding of 21st IEEE International Conference on Advanced Information Networking and Applications Workshops AINAW'07.* Niagara Falls, Ont., CA. ISBN: 978-0-7695-2847-2

Dardari, D.; Conti, A.; Ferner, U.; Giorgetti, A.; Win, M.Z. (2009) Ranging With Ultrawide Bandwidth Signals in Multipath Environments, *Proceedings of the IEEE* , Vol. 97, No. 2, (Feb. 2009), pp. 404-426, ISSN: 0018-9219

De Angelis, A.; Dionigi, M.; Moschitta, A.; Giglietti, R. & Carbone, P. (2008a) An experimental UWB distance measurement system, *Proceedings of the IEEE International Instrumentation and Measurement Technology Conference, I2MTC,* pp. 1016-1020, ISBN: 1-4244-1541-1, Victoria, BC, Canada, 12-15 May 2008, IEEE, Piscataway, NJ, USA

De Angelis, A.; Dionigi, M.; Moschitta, A.; Giglietti, R. & Carbone, P. (2009a). Characterization and Modeling of an Experimental UWB Pulse-Based Distance Measurement System," *IEEE Transactions on Instrumentation and Measurement,* Vol. 58, No. 5, (May 2009), pp. 1479–1486, ISSN: 0018-9456

De Angelis, A.; Nilsson, J.O.; Skog, I.; Händel, P. & Carbone, P. (2009b) Indoor positioning by ultra wide band radio aided inertial navigation, *Proceedings of the XIX IMEKO World Congress,* Lisbon, Portugal, Sept. 6-11 2009

De Angelis, A.; Dionigi, M.; Moschitta, A. & Carbone, P. (2009c) A Low-Cost Ultra-Wideband Indoor Ranging System, *IEEE Transactions on Instrumentation and Measurement,* Vol. 58, No. 12 (December 2009), pp. 3935-3942, ISSN: 0018-9456.

Di Benedetto, M.G.; Kaiser, T.; Molisch, A.F.; Oppermann, I.; Politano, C. & Porcino, D. (2006) *UWB Communication Systems: A Comprehensive Overview. EURASIP book series on signal processing and communications, Volume 5.* Hindawi Publishing Corporation, ISBN 977-5945-10-0

ETSI, European Telecommunications Standards Institute (2008) Electromagnetic compatibility and radio spectrum matters (ERM); ultra wideband (UWB) technologies for communication purposes; ETSI EN 302 065, Feb. 2008

Federal Communications Commission, FCC (2002) Revision of part 15 of the commission's rules regarding ultra-wideband transmission systems. Report and order, FCC 02 48, Apr. 2002

Fontana, R.J. (2004) Recent system applications of short-pulse ultra-wideband (UWB) technology, *IEEE Transactions on Microwave Theory and Techniques,* vol.52, no.9, (Sept. 2004), pp. 2087-2104, ISSN: 0018-9480

Gezici, S.; Tian, Z.; Giannakis, G.B.; Kobayashi, H.; Molisch, A.F.; Poor, H.V. & Sahinoglu, Z. (2005) Localization via ultra-wideband radios: a look at positioning aspects for future sensor networks, *IEEE Signal Processing Magazine,* Vol. 22, No. 4, (July 2005) pp. 70-84, ISSN: 1053-5888

Hirschler-Marchand, P.R.; Hatke, G.F. (2002) Superresolution techniques in time of arrival estimation for precise geolocation, *Conference Record of the Thirty-Sixth Asilomar Conference on Signals, Systems and Computers*, pp. 1272-1277, ISBN: 0-7803-7576-9, Asilomar, CA, USA, 3-6 Nov. 2002, IEEE, Piscataway, NJ, USA

Hirt, W. (2007) The european UWB radio regulatory and standards framework: Overview and implications, *Proc. IEEE International Conference on Ultra-Wideband (ICUWB)*, pp. 733–738, 24–26 Sept. 2007, Singapore, ISBN: 978-1-4244-0521-3

IEEE 802.15.4a-2007, IEEE Standard for Information Technology - Telecommunications and information exchange between systems - Local and metropolitan area networks - specific requirement Part 15.4: Wireless Medium Access Control (MAC) and Physical Layer (PHY) Specifications for Low-Rate Wireless Personal Area Networks (WPANs), ISBN: 978-0-7381-5538-8

ISO GUM (1995), *Guide to the expression of uncertainty in measurement*, International Organization for Standardization, Genève, Switzerland, 1995.

Kandoian, A. G. (1945) Broad Band Antenna, U.S. Patent n. 2368663, February 1945.

Kumar, N. & Buehrer, R. M. (2008) The ultra wideband WiMedia standard [standards in a nutshell], *IEEE Signal Processing Magazine*, vol. 25, No. 5 (Sept. 2008), pp. 115–119, ISSN: 1053-5888

Lee, J.Y. & Scholtz, R.A. (2002) Ranging in a dense multipath environment using an UWB radio link, *IEEE Journal on Selected Areas in Communications*, vol. 20, No. 5, pp. 1677-1683, ISSN: 0733-8716

Liu, H.; Darabi, H.; Banerjee, P.; Liu, J. (2007) Survey of Wireless Indoor Positioning Techniques and Systems, *IEEE Transactions on Systems, Man, and Cybernetics, Part C: Applications and Reviews*, , Vol. 37, No. 6, (Nov. 2007), pp.1067-1080, ISSN: 1058-6393

Low, Z. N.; Cheong, J. H.; Law, C. L.; Ng, W. T. & Lee, Y. J. (2005) Pulse detection algorithm for line-of-sight (LOS) UWB ranging applications, *IEEE Antennas and Wireless Propagation Letters*, vol. 4, pp. 63–67, ISSN: 1536-1225

Mahfouz, M. R.; Zhang, C.; Merkl, B. C.; Kuhn, M. J. & Fathy, A. E. (2008) Investigation of High-Accuracy Indoor 3-D Positioning Using UWB Technology, *IEEE Transactions on Microwave Theory and Techniques*, vol. 56, No. 6 (June 2008), pp. 1316–1330, ISSN: 0018-9480

Molisch, A.F. (2005) Ultrawideband propagation channels-theory, measurement, and modeling, *IEEE Transactions on Vehicular Technology*, Vol. 54, No. 5, (Sept. 2005), pp. 1528-1545, ISSN: 0018-9545

Molisch, A.; Balakrishnan, K.; Chong, C.; Emami, S.; Fort, A.; Karedal, J.; Kunisch, J.; Schantz, H.; Schuster, U. & Siwiak, K. (2005) IEEE 802.15.4a channel model - final report," tech. rep., IEEE 802.15.04-0662-02-004a.

Nilsson, J.O.; De Angelis, A.; Skog, I.; Carbone, P. & Händel, P. (2009) Signal processing issues in indoor positioning by ultra wide band radio aided inertial navigation, *Proceedings of the 17th European Signal Processing Conference, EUSIPCO*, August 24-28, Glasgow, Scotland

Oh, M.K.; Park, J.H. & Kim, J.Y. (2009) IR-UWB packet-based precise ranging system for u-Home networks, *IEEE Transactions on Consumer Electronics*, vol. 55, no. 1 (Feb. 2009), pp. 119-125, ISSN: 0098-3063

Pahlavan, K.; Li, X. & Makela, J.P. (2002) Indoor geolocation science and technology, *IEEE Communications Magazine*, Vol. 40, No. 2 (Feb. 2002), pp. 112 - 118, ISSN: 0163-6804

Phan, T.A; Lee, J.; Krizhanovskii, V.; Han, S.K. & Lee, S.G. (2007) A 18-pJ/Pulse OOK CMOS Transmitter for Multiband UWB Impulse Radio, *IEEE Microwave and Wireless Components Letters*, Vol. 17, No. 9 (Sept. 2007), pp. 688-690, ISSN: 1531-1309

Rantakokko, J.; Händel, P.; Eklöf, F.; Boberg, B.; Junered, M.; Akos, D.; Skog, I.; Bohlin, H.; Neregård, F.; Hoffmann, F.; Andersson, D.; Jansson, M.; Stenumgaard, P. (2007) Positioning of emergency personnel in rescue operations - possibilities and vulnerabilities with existing techniques and identification of needs for future R&D, Technical Report TRITA-EE 2007:037, Royal Institute of Technology, Stockholm, Sweden, July 2007. [Online]. Available: http://www.ee.kth.se.

Santinelli, G.; Giglietti, R. & Moschitta, A. (2009) Self-calibrating indoor positioning system based on ZigBee Devices, *Proceedings of the IEEE International Instrumentation and Measurement Technology Conference, I2MTC*, pp. 1205 - 1210, ISBN: 978-1-4244-3352-0 Singapore, 5-7 May 2009, IEEE, Piscataway, NJ, USA

Schwarzer, S.; Vossiek, M.; Pichler, M. & Stelzer, A. (2008) Precise distance measurement with IEEE 802.15.4 (ZigBee) devices, *IEEE Radio and Wireless Symposium*, Orlando, FL, 22-24 Jan. 2008, pp. 779 - 782, ISBN: 978-1-4244-1463-5

Sim, S.; Kim, D.W.; Hong, S. (2009) A CMOS UWB Pulse Generator for 6–10 GHz Applications, *IEEE Microwave and Wireless Components Letters*, Vol. 19, No. 2 (Feb. 2009), pp. 83-85, ISSN: 1531-1309

Stoica, L.; Rabbachin, A.; Repo, H.O.; Tiuraniemi, T.S. & Oppermann, I. (2005) An ultrawideband system architecture for tag based wireless sensor networks, *IEEE Transactions on Vehicular Technology*, vol. 54, no. 5 (Sept. 2005), pp. 1632–1645, ISSN: 0018-9545

Terada, T.; Yoshizumi, S.; Muqsith, M.; Sanada, Y. & Kuroda, T. (2006) A CMOS ultra-wideband impulse radio transceiver for 1-mb/s data communications and ±2.5-cm range finding, *IEEE Journal of Solid-State Circuits*, Vol. 41, No. 4 (April 2006), pp. 891-898, ISSN: 0018-9200

Time Domain Corporation (2009) PulsON P220 Reference Design. [Online]. Available: http://www.timedomain.com/products/P220aRD.pdf

UbiSense (2009), UWB location system Fact Sheet. [Online]. Available: http://www.ubisense.net/pdf/fact-sheets/products/software/Precise-Location-EN090624.pdf

WiMedia Alliance (2009) [Online], Available: http://www.wimedia.org

Win, M.Z. & Scholtz, R.A. (1998) Impulse radio: How it works, *IEEE Communication Letters*, Vol. 2, No. 2 (Feb. 1998). IEEE, Piscataway, NJ, USA

Wireless USB (2007) USB Implementers Forum, "Wireless USB Specification Revision 1.0." [Online]. Available: http://www.usb.org/developers/wusb

Wymeersch, H.; Lien, J. & Win, M.Z. (2009) Cooperative Localization in Wireless Networks, *Proceedings of the IEEE*, Vol. 97, No. 2 (Feb. 2009), pp. 427-450, ISSN: 0018-9219

Zebra Enterprise Solutions, Multispectral Solutions, (2009), Sapphire DART Real-Time Location System data sheet [Online], Available: http://zes.zebra.com/pdf/products-datasheets/ds_sapp_tech.pdf

Zhang, J.; Orlik, P.V.; Sahinoglu, Z.; Molisch, A.F. & Kinney, P. (2009) UWB Systems for Wireless Sensor Networks, *Proceedings of the IEEE*, Vol. 97, No. 2 (Feb. 2009), pp. 313-331, ISSN: 0018-9219

ZigBee Alliance (2007) ZigBee Specification r17, October 19th, 2007. [Online]. Available: http://www.zigbee.org

Metrology for non-stationary dynamic measurements

Jan Peter Hessling
Measurement Technology, SP Technical Research Institute of Sweden

1. Introduction

The widely applied framework for calibrating measurement systems is described in the GUM - The Guide to the Expression of Uncertainty in Measurements (ISO GUM, 1993). Despite its claimed generality, it is evident that this guide focuses on measurements of stationary quantities described by any finite set of constant parameters, such as any constant or harmonic signal. Non-stationary measurements are nevertheless ubiquitous in modern science and technology. An expected non-trivial unique time-dependence, as for instance in any type of crash test, is often the *primary* reason to perform a measurement. Many formulations of the guide are indeed difficult to interpret in a dynamic context. For instance, correction and uncertainty are referred to as being *universal constant* quantities for *direct* interpretation. These claims seize to be true for non-stationary dynamic measurements.

A measurement is here defined as stationary if a time-independent parameterization of the quantity of interest is used. The classification is thus relative, and ultimately depends on personal ability and taste. A given measurement may be stationary in one context but not in another. Static and stationary measurements can be analysed similarly (ISO GUM, 1993) since constant parameters are used in both cases.

The term 'dynamic' is frequently used, but with rather different meanings. The use of the term 'dynamic measurement' is often misleading as it normally refers to the mere time-dependence, which itself never requires a dynamic analysis. Instead, the classification into a dynamic or static measurement that will be adopted refers to the relation between the system and the signal: "The key feature that distinguishes a dynamic from a static measurement is the speed of response (bandwidth) of the measurement systems as compared to the speed at which the measured signal is changing" (Esward, 2009, p. 1).

This definition indirectly involves the acceptable accuracy through the concept of response time or bandwidth. In this formulation, a dynamic responds much slower than a static measurement system and therefore needs a dynamic rather than a static analysis. The relativity between the system and the signal is crucial - no signal or system can be 'dynamic' on its own. A static analysis is often sufficient whether it refers to a measurement of a constant, stationary or non-stationary time-dependent quantity. Indirectly and misleadingly the guide (ISO GUM, 1993) indicates that this is always the case. This is apparent from the lack of discussion of e.g. differential or difference model equations, time delay, temporal correlations and distortion. The difference between ensemble and time averages is not mentioned, but is very important for non-stationary non-ergodic measurements.

Even strongly imprecise statements of measurement uncertainty may in practice have limited consequences. It may be exceedingly difficult to even illustrate an incorrect analysis due to neglect or erroneous treatment of dynamic effects. Dynamic artefacts or fundamental physical signals corresponding to fundamental constants like the unit of electric charge do not exist. These aspects are probably the cause to why dynamic analysis still has not penetrated the field of metrology to the same extent as in many other related fields of science and engineering. By limiting the calibration services to only include characterizations and not provide methods or means to translate this information to the more complex targeted measurement, the precious calibration information can often not be utilized at all (!) to assess the quality of the targeted measurement. The framework Dynamic Metrology presented in this chapter is devoted to bridge this gap from a holistic point of view. The discussion will focus on concepts from a broad perspective, rather than details of various applications. Referencing will be sparse. For a comprehensive exploration and list of references the reader is advised to study the original articles (Hessling, 2006; 2008a-b; 2009a), which provide the basis of Dynamic Metrology.

2. Generic aspects of non-stationary dynamic measurements

The allowed measurement uncertainty enters into the classification of dynamic measurements through the definitions of response time of the system and change of rate of possible signals. This is plausible since the choice of tools and analysis (static/stationary or dynamic) is determined by the acceptable accuracy.

As recognized a long time ago by the novel work of Wiener in radar applications (Wiener, 1949), efficient dynamic correction will always involve a subtle balance between reduction of systematic measurement errors and unwanted amplification of measurement noise. How these contributions to the measurement uncertainty combines and changes with the degree of correction is therefore essential. The central and complex role of the dynamic measurement uncertainty *in* the correction contrasts the present stationary treatment.

Interactions are much more complex in non-stationary than in stationary measurements. Therefore, engineering fields such as microwave applications and high speed electronics dedicated to dynamic analysis have taken a genuinely system-oriented approach. As microwave specialists know very well, even the simplest piece of material needs careful attention as it may require a full dynamic specification. This has profound consequences. Calibration of only vital parts (like sensors) might not be feasible as it only describes one ingredient of a complex dynamic 'soup'. Its taste may depend on all ingredients, but also on how it is assembled and served. Testing the soup in the relevant environment may be required. Sometimes the calibration procedures have to leave the lab (*in vitro*) and instead take place under identical conditions to the targeted measurements (*in situ*).

The relativity between signal and system has practical consequences for *every* measurement. Repeated experiments will result in different signals and hence different performance. The dynamic correction will be unique and must be re-calculated for every measured signal. There will thus never be universal dynamic corrections for non-stationary measurements, as can be found for stationary measurements in a calibration laboratory.

The need for repeated in situ evaluation illustrates the pertinent and critical aspect of transferability of the calibration result. This is seldom an issue for stationary calibration methods where a limited set of universal numbers is sufficient to describe the result.

Stationary measurements have minute variations in comparison to the enormous freedom of non-stationary events. Physical generation of *all* possible non-stationary signals in calibrations will always be an insurmountable challenge. Indirect analyses based on uncertain and potentially abstract dynamic models are required. In turn, these models are deduced from limited but nevertheless, for the purpose complete testing against references. This testing is usually referred to as 'dynamic calibration' while the model extraction is denoted 'system identification'. Here calibration will be associated to the *combined* operation of testing and model identification. The testing operation will be called 'characterization'.
The performance of the targeted non-stationary measurements will be assessed with calculations using measured signals and an uncertain *indirect* model of the measurement. Essentially, the intermediate stage of modelling reduces the false appearance of extremely complex measurements due to the high dimensionality of the *signals* (directly observable) to the true much lower complexity of the measurement described by the *systems* (indirectly observable). It is the indirect modelling that makes the non-stationary dynamic analysis of measurements possible.
The measurand is often a function of a signal rather than a measurable time-dependent quantity of any kind. The measurand could be the rise time of oscilloscopes, the vibration dose R for adverse health effects for whole body vibrations, complex quantities such as the error vector magnitude (EVM) of WCDMA signals in mobile telecommunication (Humphreys & Dickerson, 2007), or power quality measures such as 'light flicker' (Hessling, 1999). Usually these indexes depend on time-dependent signals and do not provide complete information. A complete analysis of signals and systems is required to build a traceability chain from which the measurands can be estimated at any stage. As illustrated in Fig. 1, the measurement uncertainty is first propagated from the characterization to the model, from the model to the targeted measurement, and perhaps one step further to the measurand. The final step will not be addressed here since it is application-specific without general procedures. Fortunately, the propagation is straight-forward using the definition of the measurand often described in detail in standards (for the examples mentioned in this paragraph, EA-10/07, ISO 2631-5, 3GPP TS, IEC 61000-4-15).
The analysis of non-stationary dynamic measurements generally requires strongly interacting dynamic models. In situ calibrations of large complex systems as well as repeated in situ evaluations for every measurement may be needed. The uncertainty must be propagated in two or more stages. Realized in full, this requires nothing less than a new paradigm to be introduced in measurement science.

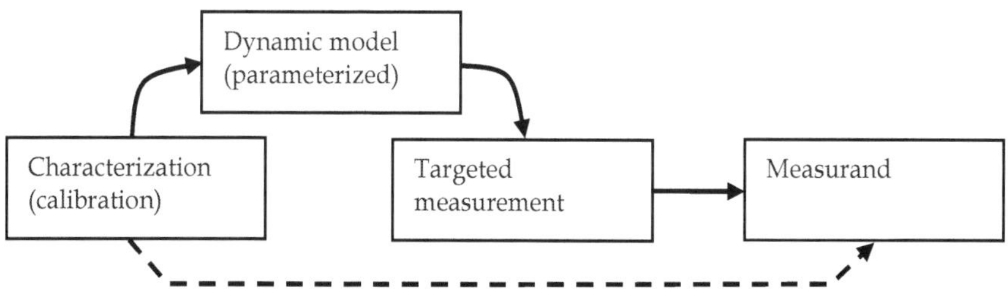

Fig. 1. Propagation (arrows) of non-stationary (full) measurement uncertainty compared to the conventional stationary case (ISO GUM, 1993) (dashed).

3. State-of-the-art dynamic analysis

The level of applications of dynamic analyses varies greatly. Leading manufacturers of measurement equipment are often well ahead of measurement science, as the realization of dynamic operations is facilitated by detailed product knowledge. However, dynamic design usually requires a substantial amount of compromises. Maximal bandwidth or slew-rate may for instance be incompatible with good time domain performance. A neutral evaluation in terms of an extended calibration service is strongly needed. Nevertheless, in many cases the motivation is low, as the de-facto standards of calibration are restricted and simplified. The motivation should primarily originate from end-users. The field of dynamic analysis in the context of calibration, or waveform metrology, is currently emerging at major national metrology institutes in a number of applications. This is the precursor for changing calibration standards and procedures to better account for the experimental reality.

The development of calibration of accelerometers illuminates the progress. One part of a present standard (ISO 16063, 2001) is based on the shock sensitivity of accelerometers. The calibration has suffered from poor repeatability due to large overseen dynamic errors (Hessling, 2006) which depend on unspecified details of the pulse excitation. A complete specification would not solve the problem though, as the pulse would neither be possible to accurately realize in most calibration experiments, nor represent the variation of targeted measurements. Despite this deficiency, manufacturers have provided dynamic correction of accelerometers for many years (Bruel&Kjaer, 2006). The problem with the present shock calibration is now about to be resolved with an indirect analysis based on system identification, which also provides good transferability. The solution parallels the analysis to be proposed here. Beyond this standard, dynamic correction as well as uncertainty evaluation for this system has also recently been proposed (Elster et al., 2007).

Mechanical fatigue testing machines and electrical network analyzers are comparable, as they both have means for generating the excitation. Their principles of dynamic calibration are nonetheless different in almost all ways. For network analyzers, simple daily in situ calibrations with built-in software correction facilities are made by end-users using calibrated calibration kits. Testing machines usually have no built-in correction. Present dynamic calibration procedures (ASTM E 467-98a, 1998) are incomplete, direct and utilize calibration bars which are *not* calibrated. This procedure could be greatly improved if the methods of calibrating network analyzers would be transferred and adapted to mechanical testing machines.

The use of oscilloscopes is rapidly evolving. In the past they were used for simpler measurement tasks, typically detecting but not accurately quantifying events. With the advent of sampling oscilloscopes and modern signal processing the usage has changed dramatically. Modern sampling techniques and large storage capabilities now make it possible to accurately resolve and record various signals. Dynamic correction is sometimes applied by manufacturers of high performance oscilloscopes. Occasionally national laboratories correct for dynamic effects. A dissatisfying example is the standard (EA-10/07, 1997) for calibrating oscilloscopes by evaluating the rise time of their step response. Unfortunately, a scalar treatment of a non-stationary signal is prescribed. This results in the same type of calibration problem as for the shock sensitivity of accelerometers described above. The uncertainty of the rise time cannot even in principle be satisfactorily evaluated, as relevant distortions of the generated step are not taken into account. Further, interaction effects are not addressed, which is critical for an instrument that can be connected to a wide range of different equipments. Correction is often synthesized taking only an approximate amplitude response into account (Hale & Clement, 2008). Neglecting the phase response in this way is equivalent to not knowing if an error should be removed by subtraction or addition! Important efforts are now made to account for such deficiencies (Dienstfrey et al., 2006; Williams et al., 2006). Many issues, such as how to include it in a standardized calibration scheme and transfer the result, remain to be resolved.

A general procedure of dynamic analysis in metrology remains to be formulated, perhaps as a dynamic supplement to the present guide (ISO GUM, 1993). As advanced dynamic modelling is currently *not* a part of present education curriculum in metrology, a substantial amount of user-friendly software needs to be developed. Most likely, the present calibration certificates must evolve into small dedicated computer programs which apply dynamic analysis to each measured signal and are synthesized and optimized according to the results of the calibration. In short, the infra-structure (methods and means) of an extended calibration service for dynamic non-stationary measurements needs to be built.

4. Dynamic Metrology – a framework for non-stationary dynamic analysis

In the context of calibration, the analysis of non-stationary dynamic measurements must be synthesized in a limited time frame without detailed knowledge of the system. In perspective of the vast variation of measurement systems and non-stationary signals, robustness and transferability are central aspects. There are many requirements to consider:

- **Generality**: Vastly different types of systems should be possible to model. These could be mechanical or electrical transducers, amplifiers, filters, signal processing, large and/or complex systems, hybrid systems etc.
- **Interactions**: Models of strong interactions between calibrated subsystems are often required to include all relevant influential effects.
- **Robustness**: There exists no limit regarding the complexity of the system. This requires low sensitivity to modelling and measurement errors etc.
- **In situ calibration and analysis**: Virtually all methods must be possible to transfer to common measurement computers and other types of computational hardware.
- **Transferability**: All results and methods must be formulated to enable almost fool-proof transfer to end-users without virtually any knowledge of dynamic analysis.

A framework (Dynamic Metrology) dedicated to analysis of dynamic measurements was recently proposed (Hessling, 2008b). All methods were based on standard signal processing operations easily packed in software modules. The task of repeated dynamic analysis for every measurement may effectively be distributed to three parties with different chores: Experts on Dynamic Metrology (1) derive general synthesis methods. These are applied by the calibrators (2) to determine dedicated but general software calibration certificates for the targeted measurement, using specific calibration information. The end users (3) apply these certificates to each measurement with a highly standardized and simple ['drag-and-drop'] implementation. Consequently, Dynamic Metrology involves software development on two levels. The calibrators as well as the end users need computational support, the former to *synthesize* (construct and adapt), the latter to *realize* (apply) the methods. For the steps of system identification, mathematical modelling and simulation reliable software packages are available. Such tools can be integrated with confidence into Dynamic Metrology. What

remains is to adapt and combine them into 'toolboxes' or modules for synthesis (calibrator) and realization (end-user), similar to what has been made for system identification (Kollár, 2003). This fairly complex structure is not a choice, but a consequence of; general goals of calibration, the application to non-stationary dynamic measurements and the fact that only the experts on Dynamic Metrology are assumed to have training in dynamic analysis.
Prototype methods for all present steps of analysis contained in Dynamic Metrology will be presented here. Dynamic characterization (section 4.1) provides the fundamental information about the measurement system. Using this information and parametric system identification (section 4.2), a dynamic model (section 4.2.1) with associated uncertainty (section 4.2.2) is obtained. From the dynamic model equation the systematic dynamic error can be estimated (section 4.3). The dynamic correction (section 4.4) is supposed to reduce this error by applying the optimal approximation to the inverse of the dynamic model equation. To evaluate the measurement uncertainty (section 4.5), the expression of measurement uncertainty (section 4.5.1) is derived from the dynamic model equation. For every uncertain parameter, a dynamic equation for its associated sensitivity is obtained. The sensitivities will be signals rather than numbers and can be realized using digital filtering, or any commercially available dynamic simulator (section 4.5.2). The discussion is concluded with an overview of all known limitations of the approach and expected future developments (section 4.6), and a summary (section 5). The versatility of the methods will be illustrated with a wide range of examples (steps of analysis given in parenthesis):

- Material testing machines (identification)
- Force measuring load cells (characterization, identification)
- Transducer systems for measuring force, acceleration or pressure (correction, measurement uncertainty - digital filtering)
- All-pass filters, electrical/digital (dynamic error)
- Oscilloscopes and related generators (characterization, identification, correction)
- Voltage dividers for high voltage (measurement uncertainty - simulations)

4.1 Characterization

The raw information of the measurement system required for the analysis is obtained from the characterization, where the measurement system is experimentally tested against a reference system. In perspective of the targeted measurement, the testing must be *complete*. All relevant properties of the system can then be transformed or derived from the results, but are not explicitly given. Using a *representation* in time, frequency or something else is only a matter of practical convenience (Pintelon & Schoukens, 2001). For instance, the bandwidth can be derived from a step response. The result consists of a numerical presentation with associated measurement uncertainty, strictly limited to the test signal(s). Different parts of the system can be characterized separately, provided the interactions can also be characterized. The simplest alternative is often to characterize whole assembled systems. When the environment affects the performance it is preferable to characterize the system in situ with a portable dynamic reference system. One example is the use of calibration kits for calibrating network analyzers. The accuracy of characterization of any subsystem should always be judged in comparison to the performance of the whole system. There is no point in knowing any link of a chain better than the chain as a whole.
How physical reference *systems* are realized is specific to each application. The test *signals* are however remarkably general. The test signals do not have to be non-stationary to be

used for analysis of non-stationary measurements. On the contrary, stationary signals generally give the highest accuracy. Usually there are important constraints on the relation between amplitude and speed of change/bandwidth of the signals. Harmonic sweeps generally yield the most accurate characterization, but test signals of high amplitude and high frequency may be difficult to generate and the testing procedure can be slow. Various kinds of impulses can be generated when stored mechanic, electric or magnetic energy is released. To control the spectrum of the signal frequency sweeps are superior, while high amplitude and sometimes high speed often requires the use of pulses. The design of excitation signals is important for the quality of modelling, and is thus an integral part of system identification (Pintelon & Schoukens, 2001).

The two mechanical examples will illustrate an in-vitro sensor and a corresponding in-situ system characterization. The non-trivial relation between them will be explored in section 4.2.3. The two electrical examples illustrate the duality of calibrating generators and oscilloscopes, or more generally, transmitters and receivers. The least is then characterized with the best performing instrument. The examples convey comparable dynamic information of the devices under test, but differently, with different accuracy and in different ranges. The often used impulse (h), step (v) and continuous $H(s = i2\pi f)$ or discrete time $(G(z = \exp(i2\pi f T_S))$, T_S sampling time) frequency (f) response characterizations are related via the Laplace- (L) and z-transform (Z), respectively,

$$h(t) = \frac{dv}{dt}, \quad \begin{array}{rcl} G(z) & = & Z(h(t)) \\ H(s) & = & L(h(t)) \end{array}. \tag{1}$$

Stationary sources of uncertainty (e.g. mass, temperature, pressure) are usually rather easy to estimate, but often provide minor contributions. For non-stationary test signals the largest sources of the uncertainty normally relate to the underlying time-dependent dynamic event. Typical examples are imbalances of a moving mechanical element or imperfections of electrical switching. All relevant sources of uncertainty must be estimated and propagated to the characterization result by detailed modelling of the experimental set up. The task of estimating the measurement uncertainty of the characterization is thus highly specialized and consequently not discussed in this general context. The uncertainty of the dynamic characterization is the fundamental 'seed' of unavoidable uncertainty, that later will be propagated through several steps as shown in Fig. 1 and combined with the uncertainty of the respective measurement (sections 4.2.2 and 4.5.1).

4.1.1 Example: Impulse response of load cell

An elementary example of characterization is a recent impulse characterization of a force measuring load cell displayed in Fig. 2 (left) (Hessling, 2008c, Appendix A). This force sensor is used in a material testing machine (Fig. 4, left). The load cell was hung up in a rope and hit with a heavy stiff hammer. As the duration of the pulse was estimated to be much less than the response time of the load cell, its shape could be approximated with an ideal Dirac-delta impulse. The equivalent force amplitude was unknown but of little interest: The static amplification of the load cell is more accurately determined from a static calibration. The resulting oscillating normalized impulse response is shown in Fig. 2 (right).

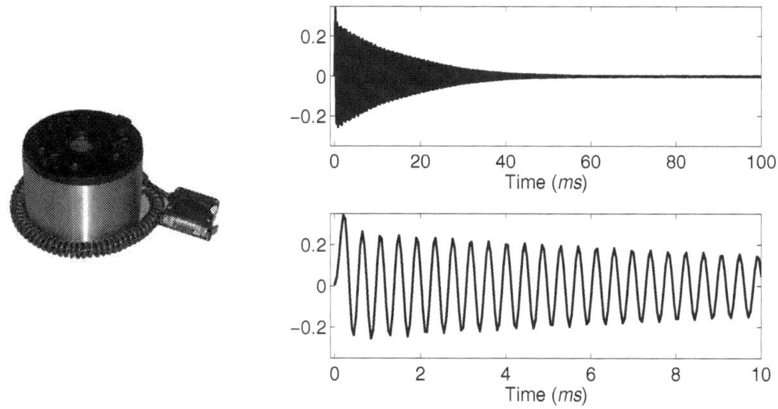

Fig. 2. Measured impulse response (right) of a load cell (left), on different time scales. The sampling rate is 20 kHz and the response is normalized to unit static amplification.

4.1.2 Example: Impulse response of oscilloscope and step response of generator

Oscilloscopes can be characterized with an optoelectronic sampling system (Clement et al., 2006). A short impulse is then typically generated from a 100-fs-long optical pulse of a pulsed laser, and converted to an electrical signal with a photodiode. A reference oscilloscope characterized with a reference optoelectronic sampling system may in turn be used to characterize step generators. Example raw measurements of a photodiode impulse and a step generator are illustrated in Fig. 3. A traceability chain can be built upon such repeated alternating characterizations of generators and oscilloscopes. If the measurand is the rise time it can be estimated from each characterization but not propagated with maintained traceability.

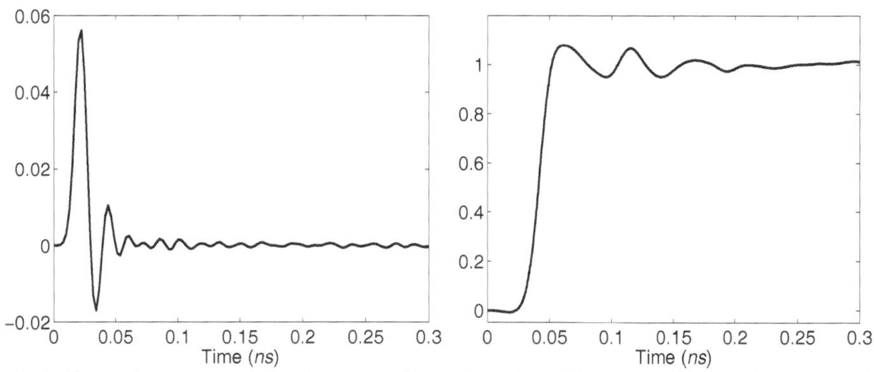

Fig. 3. Example measurement of an optoelectronic pulse with a sampling oscilloscope (left) as described in Clement et al., 2006, and measurement of a voltage step generator with a different calibrated oscilloscope (right).

4.2 System identification

System identification refers here to the estimation of parametric models and their uncertainty (Pintelon & Schoukens, 2001), even though the subject also includes non-parametric methods (Ljung, 1999). To adapt the basic procedure of identifying a model from the experimental characterization (section 4.1) to metrology, follow these steps:

1. Choose a criterion for comparing experimental and modelled characterization.
2. Select a structure for the dynamic model of the measurement. Preferably the choice is based on physical modelling and prior knowledge. General 'black-box' rather than physical models should be utilized for complex systems.
3. Find the numerical values of all model parameters:
    a. Choose method of optimization.
    b. Assign start values to all parameters.
    c. Calculate the hypothetical characterization for the dynamic model.
    d. Compare experimental and modelled characterization using the criterion in step 1.
    e. Adjust the parameters of the dynamic model according to step 3a.
    f. Repeat from c until there is no further improvement in step 3d.
4. Evaluate the performance of the dynamic model by studying the model mismatch.
5. Repeat from step 2 until the performance evaluation in step 4 is acceptable.
6. Propagate the measurement uncertainty of the characterization measurement to the uncertainty of the dynamic model.

There are some important differences in this approach compared to the standard procedure of system identification (Ljung, 1999). Validation of the model is an important step for assessing the correctness of the model, but validation in the conventional sense is here omitted. The reason for this is that it requires at least two characterization experiments to form independent sets of data, one for 'identification' and another for 'validation'. Often only one type of experimental characterization of acceptable accuracy is available. It is then unfortunately impossible to validate the model against data. This serious deficiency is to some extent compensated for by a more detailed concept of measurement uncertainty. The correctness of the model is expressed through the uncertainty of the model, rather than validated by simulations against additional experimental data. All relevant sources of uncertainty should be estimated in the preceding step of characterization, and then propagated to the dynamic model in the last step (6) of identification. Corresponding propagation of uncertainty is indeed discussed in the field of system identification, but perhaps not in the widest sense. The suggested approach is a pragmatic adaptation of well developed procedures of system identification to the concepts of metrology.

The dynamic model is very often non-linear in its parameters (not to be confused with linearity in response!). This is the case for any infinite-impulse response (IIR) pole-zero model. Measurement noise and modelling errors of a large complex model might result in many local optima in the comparison (step 3d). Obtaining convergence of the numerical search may be a challenge, even if the model structure is valid. Assigning good start values to the parameters (step 3b) and limiting the variation in the initial iterations (step 3e), might be crucial. The model may also be identified and extended sequentially (step 2) using the intermediate results as start values for the new model. A sequential approach of this kind is seldom the fastest alternative but often remarkably robust.

A common criterion (step 1) for identifying the parameters is to minimize the weighted square of the mismatch. For frequency response characterization and a symmetric positive definite weighting matrix $W(\omega_k, \omega_l)$, the estimated parameters $\{\hat{q}\}$ are expressed in the residual $\Delta H(q, i\omega_k)$ defined as the difference between modelled and measured response (T represents transposition and $*$ complex conjugation),

$$\{\hat{q}\} = \arg\min_{q}\left[\Delta H(q)^{T*} W \, \Delta H(q)\right]. \tag{2}$$

4.2.1 Modelling dynamic measurements

Dynamic models are never true or false, but more or less useful and reliable for the intended use. The primary goal is to strongly reduce the complexity of the characterization to the much lower complexity of a comparatively small model. The difference between the dimensionality of the characterization and the model determines the confidence of the evaluation in step 4. The maximum allowed complexity of the model for acceptable quality of evaluation can be roughly estimated from the number of measured points of the characterization. For a general linear dynamic measurement the model consists of one or several differential (CT: continuous time) or difference (DT: discrete time) equations relating the (input) quantity $x(t)$ to be measured and the measured (output) signal $y(t)$,

$$\begin{array}{llcl} \mathrm{CT}: & \tilde{a}_0 y + \tilde{a}_1 \partial_t y + \tilde{a}_2 \partial_t^2 y + \ldots \tilde{a}_n \partial_t^n y & = & \tilde{b}_0 x + \tilde{b}_1 \partial_t x + \tilde{b}_2 \partial_t^2 x + \ldots \tilde{b}_m \partial_t^m x \\ \mathrm{DT}: & a_0 y_k + a_1 y_{k-1} + a_2 y_{k-2} + \ldots a_n y_{k-n} & = & b_0 x_k + b_1 x_{k-1} + b_2 x_{k-2} + \ldots b_m x_{k-m} \end{array}, \tag{3}$$

where $x_k = x(kT_S)$, $y_k = y(kT_S)$, $k = 0,1,2,\ldots$, T_S being the sampling time. In both cases it is often convenient to use a state-space formulation (Ljung, 1999), with a system of model equations linear in the differential (CT) or translation (DT) operator. State space equations also allow for multiple input multiple output (MIMO) systems. The related model equation (Eq. 3) can easily and uniquely be derived from any state space formulation. Thus, assuming a model equation of this kind is natural and general. This is very important since it provides a unified treatment of the majority of LTI models used in various applications for describing physical processes, control operations and CT/DT signal processing etc.

An algebraic model equation in the transform variable s or z is obtained by applying the Laplace s-transform to the CT model or the z-transform to the DT model in Eq. 3,

$$\begin{array}{llcl} \mathrm{CT}: & \left(\tilde{a}_0 + \tilde{a}_1 s + \tilde{a}_2 s^2 + \ldots \tilde{a}_n s^n\right) Y(s) & = & \left(\tilde{b}_0 + \tilde{b}_1 s + \tilde{b}_2 s^2 + \ldots \tilde{b}_m s^m\right) X(s) \\ \mathrm{DT}: & \left(a_0 + a_1 z^{-1} + a_2 z^{-2} + \ldots a_n z^{-n}\right) Y(z) & = & \left(b_0 + b_1 z^{-1} + b_2 z^{-2} + \ldots b_m z^{-m}\right) X(z) \end{array}. \tag{4}$$

These relations are often expressed in terms of transfer functions $H(s) = Y(s)/X(s)$ or $G(z) = Y(z)/X(z)$. The polynomials are often factorized into their roots, 'zeros' (numerator) and 'poles' (denominator). Another option is to use physical parameters. In electrical circuits lumped resistances, capacitances and inductances are often preferred, while in mechanical applications the corresponding elements are damping, mass and spring constants. In the examples, the parameterization will be a variable number of poles and zeroes. The fundamental reason for this choice is that it provides not only a very general and effective, but also widely used and understood parameterization. Good initial values of the parameters are in many cases fairly easy to assign by studying the frequency response, and it is straight-forward to extend any model and identify it sequentially.

4.2.2 Uncertainty of dynamic model

The performance of the identified model (step 4) can be explored by studying the properties of the residual $\Delta H(\hat{q}, i\omega_k)$. If the model captures all features of the characterization, the autocorrelation functions of the residual and the measurement noise are similar. For instance, they should both decay rapidly if the measurement noise is uncorrelated (white).

There are many symmetries between the propagation of uncertainty from the characterization to the model (step 6), and from the model to the targeted measurement (section 4.5.1), see Fig. 1. The two propagations will therefore be expressed similarly. The concept of sensitivity is widely used in metrology, and will be utilized in both cases. Just as in section 4.5.1, measured signals must be real-valued, which requires the use of real-valued projections $\varphi(\Delta q, q)$ of the pole and zero deviations Δq (Hessling, 2009a). The deviation in modelled characterization for a slight perturbation φ is given by, $-E^T(\hat{q}, i\omega)\varphi(\Delta q, \hat{q})$ (compare Eq. 12). The matrix $E(\hat{q}, i\omega)$ of sampled sensitivity systems organized in rows is represented in the frequency domain, since the characterization is assumed to consist of frequency response functions. Row n of this matrix is given by $E_n(\hat{q}, i\omega)$ in Eq. 18. The minus sign reflects the fact that the propagation from the characterization to the model is the inverse of the propagation from the model to the correction of the targeted measurement. If the measured characterization deviates by an amount $\Lambda(i\omega)$ from its ensemble mean, the estimated parameters will deviate from their ensemble mean according to $\hat{\varphi}$,

$$\{\hat{\varphi}\} = \arg\min_{\varphi \in R} \left[\left(\Lambda^{T*} + \varphi^T E^* \right) W \left(\Lambda + E^T \varphi \right) \right]. \tag{5}$$

The problem of finding the real-valued projected deviation φ (Eq. 5) closely resembles the estimation of all parameters q of the dynamic model (Eq. 2). The optimization over $\varphi \in R$ is constrained, but much simpler as the model of deviation is linear. Contrary to the problem of non-linear estimation, the linear deviations due to perturbed characterizations can be found explicitly; $\hat{\varphi} = -\mathrm{Re}(\Gamma\Lambda)$, $\Gamma \equiv \left[\mathrm{Re}\left(E^* W E^T\right)\right]^{-1} E^* W$. From this solution, the covariance of all projections is readily found ($\langle\cdot\rangle$ denotes average over an ensemble of measurements and let $\hat{\varphi} \to \varphi$),

$$\left\langle \varphi\varphi^T \right\rangle = \frac{1}{2} \mathrm{Re}\left[\Gamma \left\langle \Lambda\Lambda^T \right\rangle \Gamma^T + \Gamma \left\langle \Lambda\Lambda^{T*} \right\rangle \Gamma^{T*} \right]. \tag{6}$$

This expression propagates the covariance of the characterization experiment $\left\langle \Lambda\Lambda^{T(*)} \right\rangle$ to $\left\langle \varphi\varphi^T \right\rangle$, which is related to the covariance of the estimated dynamic model. Within the field of system identification (Pintelon & Schoukens, 2001) similar expressions are used, but with sensitivities denoted Jacobians. If the model is accurate the residual is unbiased and reflects

the uncertainty of the characterization, $\langle \Lambda\Lambda^{T(*)} \rangle = \langle \Delta H \Delta H^{T(*)} \rangle - \langle \Delta H \rangle \langle \Delta H^{T(*)} \rangle \approx \langle \Delta H \Delta H^{T(*)} \rangle$. Otherwise there are systematic errors of the model and the residual is biased, $\langle \Delta H \rangle \neq 0$. Not only the covariance of the characterization experiment but also the systematic errors of the identified model should be propagated to the uncertainty. The systematic error $\langle \Delta H \rangle$ can however not be expressed in the model or its uncertainty, as that is how the residual ΔH is defined. A separate treatment according to section 4.3 is thus required: A general upper bound of the dynamic error $\langle \Delta H \rangle$ valid for the targeted measurement should be added directly and *linearly* to the final uncertainty of the targeted measurement (Hessling, 2006).
Only one or at most a few realizations of the residual ΔH are known, since the number of available characterizations is limited. Therefore it is difficult to evaluate *ensemble* averages. However, if the residual is 'stationary' (i.e. do not change in a statistical sense) over a frequency interval $\Delta\omega$, the system is ergodic over this interval. The average over an ensemble of experiments can then be exchanged with a restricted mixed average over frequency and just a few $(m \geq 1)$ experimental characterizations,

$$\left\langle \Delta H(i\omega_1) \Delta H^{(*)}(i\omega_2) \right\rangle \approx \frac{1}{2\alpha m} \sum_{k=1}^{m} \int_{-\alpha}^{\alpha} \Delta H^{(k)}(i(\omega_1 + \varpi)) \Delta H^{(k,*)}(i(\omega_2 + \varpi)) d\varpi, \tag{7}$$

where $\alpha = (\Delta\omega - |\omega_1 - \omega_2|)/2 > 0$. If the correlation range is less than the interval of stationary residual, $\left\langle \Delta H(i\omega_1) \Delta H^{(*)}(i\omega_2) \right\rangle \approx 0$ for $\alpha < 0$, all elements can be estimated.
As model identification requires experimental characterizations, some of the previous examples will be revisited. The comparison of the load cell and the material testing machine models will illustrate a non-trivial relation between the behaviour of the system (machine) and the part (load cell) traditionally assumed most relevant for the accuracy of the targeted measurement. The propagation of uncertainty (Eq. 6) will not be discussed due to limited space, and as it is a mere transformation of numerical numbers.

4.2.3 Example: Models of load cell and material testing machine

Load cells measure the force in mechanical testing machines often used for fatigue testing of materials and structures, see Fig. 4 (left). It is straightforward to characterize the whole machine by means of built-in force actuators and calibration bars equipped with strain-gauges for measuring force. The calibration bar and load cell outputs can then be compared in a calibration experiment. The load cell can also be characterized separately as explained in section 4.1.1. A recent frequency response characterization and identification of a machine and a load cell (Hessling, 2008c) is compared in Fig. 4 (right) and shown in full in Fig. 5. There is no simple relation or scaling between the amplitude and phases of the frequency responses of the load cell and the installed testing machine. Accurate dynamic characterization in situ thus appears to be required.

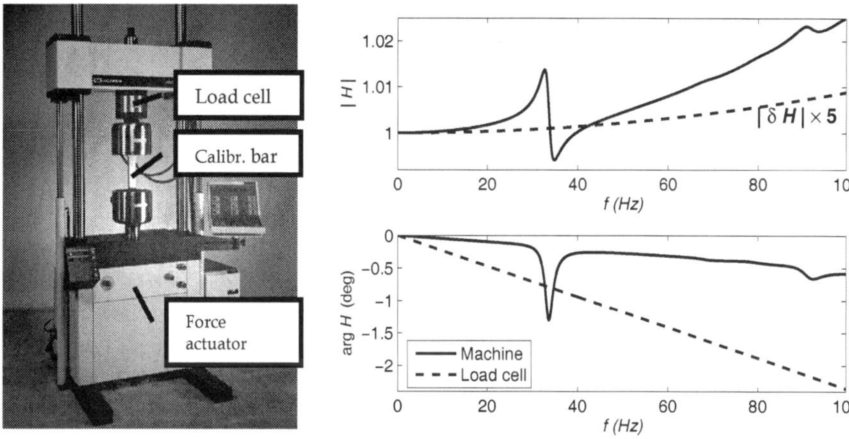

Fig. 4. Magnitude (right, top) and phase (right, bottom) of the frequency response of identified models of a testing machine (left, Table 1: 2*) and its load cell. The legend applies to both figures. The magnitude variation of the load cell is magnified 5 times for clarity.

The autocorrelation of the residual for the load cell (Fig. 6, bottom left) clearly indicates substantial systematic errors. The large residual may be caused by an insufficient dynamic model or a distorted impulse used in the characterization (section 4.1.1). The continuous distribution of mass of the load cell might not be properly accounted for in the adopted lumped model. If the confidence in the model is higher than the random disturbances of the excitation, the residual ΔH may be reduced before propagated to the model uncertainty. Confidence in any model can be formulated as prior knowledge within Bayesian estimation (Pintelon & Schoukens, 2001). The more information, from experiments or prior knowledge, the more accurate and reliable the model will be. However, the fairly complex relation between the machine and the load cell dynamics strongly reduces the need for accurate load cell models. The more important testing machine model is clearly of higher quality (Fig. 6).
Pole-zero models of different orders were identified for the testing machine (Table 1). A vibration analysis of longitudinal vibration modes in a state-space formulation (Hessling, 2008c) provided the basic information to set up and interpret these models in terms of equivalent resonance and base resonances. Model 2* is considered most useful.

	Load cell	Machine				
Model (complexity)	-	0	1	2*	3	4
Equiv. resonance (Hz)	2380	635	607	614	616	617
Base resonances (Hz)	-	-	33.7	33.6	33.6	33.6
				91.4	91.3	91.2
					79.6	79.3
						69.3
Weighted residuals[1]	22e-3	48e-4	9.8e-4	6.4e-4	6.4e-4	6.3e-4

Table 1. Identified load cell and material testing machine models. [1]Defined as the root-mean-square of the residuals, divided by the amplification at resonance (load cell), or zero frequency (machine). Not all parameters are displayed (Hessling, 2008c).

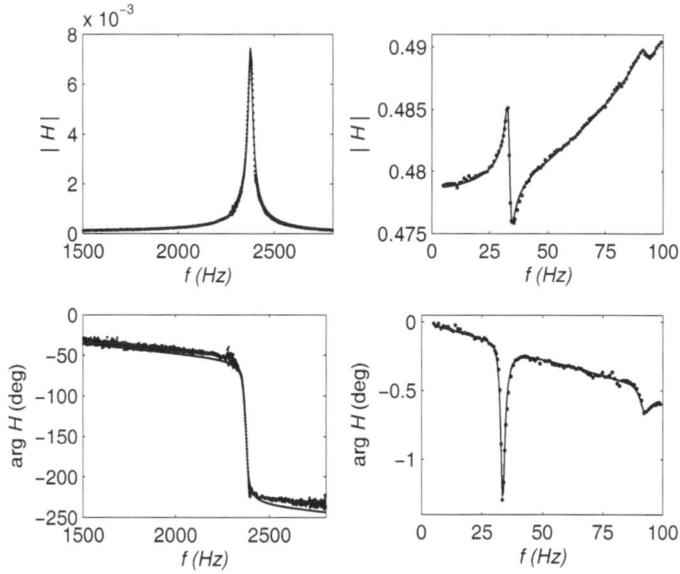

Fig. 5. Model fits for the load cell (left) and the material testing machine (right, Table 1: 2*).

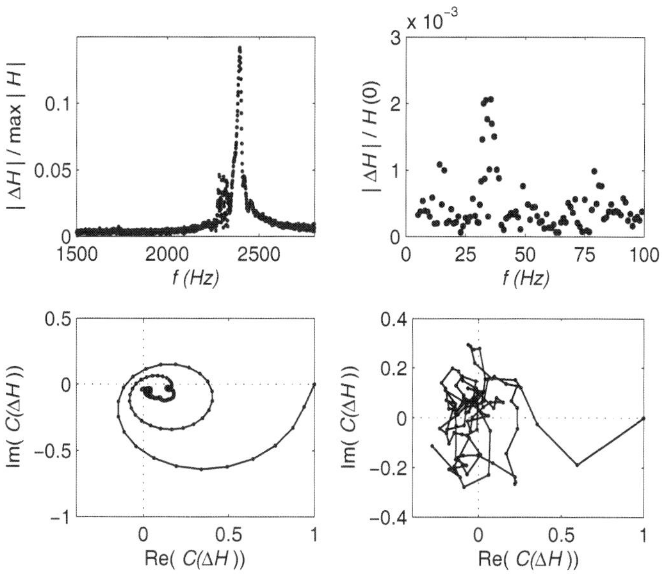

Fig. 6. Magnitude (top) and auto-correlation (bottom) of the residuals ΔH for the load cell (left) and model 2* (Table 1) of the material testing machine (right).

4.2.4 Example: Model of oscilloscope / step generator

The oscilloscope and the generator (section 4.1.2) can be identified in the same manner (Eq. 1). When differentiation is applied, attention has to be paid to noise amplification. It can be mitigated with low-pass filtering. The generator is *modelled* as consisting of an ideal step generator and of a linear time-invariant system which describes all physical limitations. The identified generator model in Fig. 7 may be used to compensate oscilloscope responses for imperfections of the generator. Several corrections for generators and oscilloscopes can be accumulated in a traceability chain and applied in the *final* step of any evaluation, as all operations commute. The model has non-minimum phase (zeros outside the unit circle of the z-plane). This will have consequences for the synthesis of correction (section 4.4.1).
The structure of the generator model could not be derived as it did not correspond to any physical system. Instead, poles and zero were successively added until the residual did not improve significantly. The low resolution of the characterization limited the complexity of possible models. In contrast to the previous example, a discrete rather than continuous time model was utilized for simultaneous identification and discretization in time.

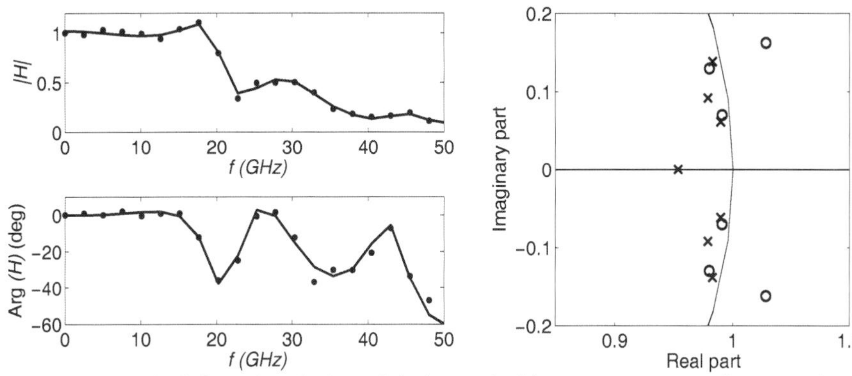

Fig. 7. Magnitude (left, top) and phase (left, bottom) of frequency response functions for an oscilloscope step generator. The characterization (•) is derived from the step response in Fig. 3 (right). The model (line) corresponds to poles and zeros shown in the z-plane (right).

4.3 Systematic error

The dynamic error of a dynamic measurement depends strongly on the variation of the measured signal. The more accurately an error needs to be estimated, the more precise the categorization must be. This manifests a generic problem: To estimate the error with high precision the variation of the physical signal must be well known, but then there would be no need to make a dynamic measurement! Ergo, error estimates are always rather inaccurate. This predicament does however not motivate the common neglect of important error mechanisms (Hessling, 2006). If not taken into account by any means, there is no definite limit to how imprecise the error estimates can be! For substantial correction or precise control though, a low uncertainty of the characterization is an absolute requirement.
The concept of dynamic error is intimately related to the perceived time delay. If the time delay is irrelevant it has to be calculated and compensated for when evaluating the error!

Distortion of the signal caused by non-perfect dynamic response of the measurement system makes the determination of the time delay ambiguous. The interpretation of dynamic error influences the deduced time delay. A *joint* definition of the dynamic error and time delay is thus required. The measured signal can for instance be translated in time (the delay) to minimize the difference (the error signal) to the quantity that is measured. The error signal may be condensed with a norm to form a *scalar* dynamic error. Different norms will result in different dynamic errors, as well as time delays. As the error signal is determined by the measurement system, it can be determined from the characterization (section 4.1) or the identified model (section 4.2), and the measured signal.

The norm for the dynamic error should be governed by the measurand. Often it is most interesting to identify an event of limited duration in time where the signal attains its maximum, changes most rapidly and hence has the largest dynamic error. The largest (L^1 norm) relative deviation in the time domain is then a relevant measure. To achieve unit static amplification, normalize the dynamic response $y(t)$ of the measurement system to the excitation $x(t) \in B$. A time delay τ and a relative dynamic error ε can then be defined jointly as (Hessling, 2006),

$$\begin{aligned} \varepsilon &\equiv \min_{\tau}\left[\max_{x(t)\in B,t}\left(\frac{|y(t)-x(t-\tau)|}{\max_t[|x(t)|]}\right)\right] \cong \min_{\tau}\left(\left\langle\left|\frac{\delta\tilde{H}(i\omega,\tau)}{H(0)}\right|\right\rangle_B\right). \\ \langle f\rangle_B &\equiv \frac{1}{\omega_B}\int_0^{+\infty} f(\omega)B(\omega)d\omega, \qquad \omega_B \equiv \int_0^{+\infty} B(\omega)d\omega \end{aligned} \tag{8}$$

The error signal in the time domain is expressed in terms of an error frequency response function $\delta\tilde{H}(i\omega,\tau) \equiv H(\sigma(\omega))\cdot\exp(i\omega\tau) - H(0)$ related to the transfer function H of the measurement system. The expression applies to both continuous time $(\sigma \to i\omega)$, as well as discrete time systems ($\sigma \to \exp(i\omega T_s)$, T_s being the sampling time interval). It is advanced in time to adjust for the time delay, in order to give the least dynamic error. The average is taken over the approximated magnitude of the input signal spectrum normalized to one, $B(\omega) \leq 1$, which defines the set B. This so-called *spectral* distribution function (SDF) (Hessling, 2006) enters the dynamic error similarly to how the *probability* distribution function (PDF) enters expectation values. The concept of bandwidth ω_B of the system/signal/SDF is generalized to a 'global' measure insensitive to details of $B(\omega)$ and applicable for any measurement. The error estimate is an upper bound over *all* non-linear phase variations of the excitation as only the magnitude is specified with the SDF. The maximum error signal (x_E) has the non-linear phase $-\delta\tilde{H}(i\omega,\tau)$ and reads (time t_0 arbitrary),

$$\frac{x_E(t)}{\max_t |x_E(t)|} \approx \frac{1}{\omega_B}\int_0^{+\infty} B(\omega)\cos\left[\omega(t-t_0) - \arg\left(\delta\tilde{H}(i\omega,\tau)\right)\right]d\omega\,. \tag{9}$$

The close relation between the system and the signal is apparent: The non-linear phase of the *system* is attributed to the maximum error *signal* parameterized in properties of the SDF.

The dynamic error and time delay can be visualized in the complex plane (Fig. 8), where the advanced response function $\tilde{H}(i\omega,\tau) = H(\sigma(\omega))\cdot\exp(i\omega\tau)$ is a phasor 'vibrating' around the positive real axis as function of frequency.

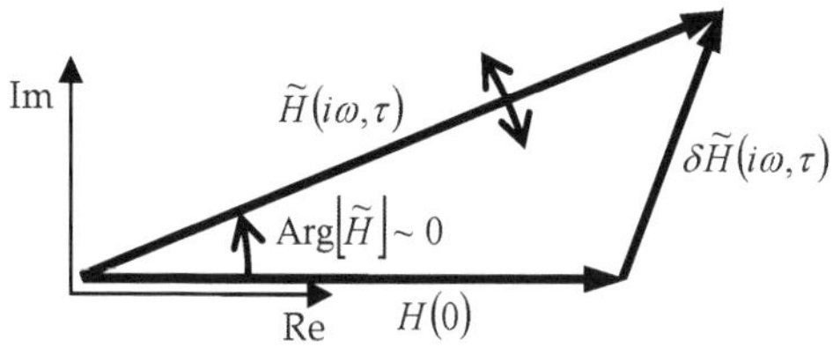

Fig. 8. The dynamic error ε equals the weighted average of $\left|\delta\tilde{H}(i\omega,\tau)\right|$ over ω, which in turn is minimized by varying the time delay parameter τ.

For efficient numerical evaluation of this dynamic error, a change of variable may be required (Hessling, 2006). The dynamic error and the time delay is often conveniently parameterized in the bandwidth ω_B and the roll-off exponent of the SDF $B(\omega)$. This dynamic error has several important features not shared by the conventional error bound, based on the amplitude variation of the frequency response within the signal bandwidth:

- The time delay is presented separately and defined to minimize the error, as is often desired for performance evaluation and synchronization.
- All properties of the *signal* spectrum, as well as the frequency response of the measurement *system* are accounted for:
  - The best (as defined by the error norm) linear phase approximation of the measurement system is made and presented as the time delay.
  - Non-linear contributions to the phase are effectively taken into account by removing the best linear phase approximation.
  - The contribution from the response of the system from *outside* the bandwidth of the signals is properly included (controlled by the roll-off of $B(\omega)$).
- A bandwidth of the system can be uniquely defined by the bandwidth of the SDF for which the allowed dynamic error is reached.

The simple all-pass example is chosen to illustrate perhaps the most significant property of this dynamic error - its ability to correctly account for phase distortion. This example is more general than it may appear. Any incomplete dynamic correction of only the magnitude of the frequency response will result in a complex all-pass behaviour, which can be described with cascaded simple all-pass systems.

4.3.1 Example: All-pass system

The all-pass system shifts the phase of the signal spectrum without changing its magnitude. All-pass systems can be realized with electrical components (Ekstrom, 1972) or digital filters (Chen, 2001). The simplest ideal continuous time all-pass transfer function is given by,

$$H(s) = \frac{1 - s/\omega_0}{1 + s/\omega_0} \to \begin{cases} 1 & s \to i0 \\ -1, & s \to i\infty \end{cases}. \tag{10}$$

The high frequency cut-off that any physical system would have is left out for simplicity. For slowly varying signals there is only a static error, which for this example vanishes (Fig. 9, top left). The dynamic error defined in Eq. 8 becomes substantial when the pulse-width system bandwidth product increases to order one (Fig. 9, top right), and might exceed 50% (!) (Fig. 9, bottom left). For very short pulses, the system simply flips the sign of the signal (Fig. 9, bottom right). In this case the bandwidth of the system is determined by the curvature of the phase related to $f_0 = \omega_0 / 2\pi$. The traditional dynamic error bound based on the magnitude of the frequency response vanishes as it ignores the phase! The dynamic error is solely caused by different delays of different frequency components. This type of signal degradation is indeed well-known (Ekstrom, 1972). In electrical transmission systems, the same dispersion mechanism leads to "smeared out" pulses interfering with each other, limiting the maximum speed/bandwidth of transmission.

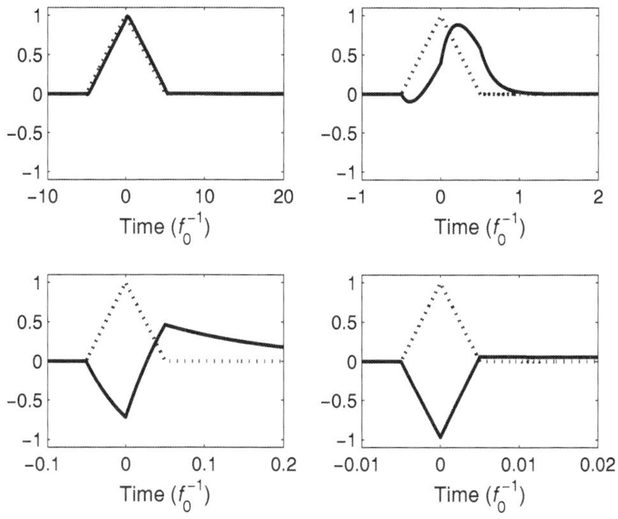

Fig. 9. Simulated measurement (solid) of a triangular pulse (dotted) with the all-pass system (Eq. 10). Time is given in units of the inverse cross-over frequency f_0^{-1} of the system.

Estimated error bounds are compared to calculated dynamic errors for simulations of various signals in Fig. 10. The utilization f_B/f_0 is much higher than would be feasible in practice, but is chosen to correspond to Fig. 9. The SDFs are chosen equal to the magnitude of the Bessel (dotted) and Butterworth (dashed, solid) low-pass filter frequency response functions. Simulations are made for triangular (◊), Gaussian (o), and low-pass Bessel-filtered square pulse signals (+, □). The parameter n refers to both the order of the SDFs as well as the orders of the low-pass Bessel filters applied to the square signal (FiltSqr). The dynamic error bound varies only weakly with the type (Bessel/Butterworth) of the SDFs: the Bessel SDF renders a slightly larger error due to its initially slower decay with frequency. As expected, the influence from the asymptotic roll-off beyond the bandwidths is very strong. The roll-off in the frequency domain is governed by the regularity or differentiability in the time domain. Increasing the order of filtering (n) of the square pulses (FiltSqr) results in a more regular signal, and hence a lower error. All test signals have strictly linear phase as they are symmetric. The simulated dynamic errors will therefore only reflect the non-linearity of the phase of the system while the estimated error bound also accounts for a possible non-linear phase of the signal. For this reason, the differences between the error bounds and the simulations are rather large.

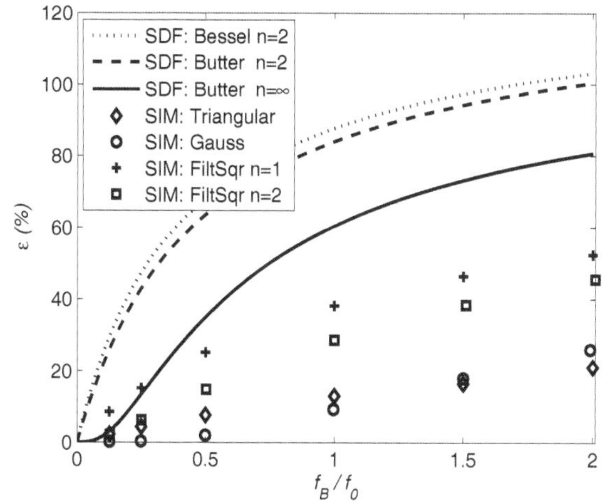

Fig. 10. Estimated dynamic error bounds (lines) for the all-pass system and different SDFs, expressed as functions of bandwidth, compared to simulated dynamic errors (markers).

4.4 Correction

Restoration, de-convolution (Wiener, 1949), estimation (Kailath, 1981; Elster et al., 2007), compensation (Pintelon et al., 1990) and correction (Hessling 2008a) of signals all refer to a more or less optimal dynamic correction of a measured signal, in the frequency or the time domain. In perspective of the large dynamic error of ideal all-pass systems (section 4.3.1), dynamic correction should never even be considered without knowledge of the phase response of the measurement system. In the worst case attempts of dynamic correction result in doubled, rather than eliminated error.

The goals of metrology and control theory are similar, in both fields the difference between the output and the input of the measurement/control system should be as small as possible. The importance of phase is well understood in control theory: The phase margin (Warwick, 1996) expresses how far the system designed for negative feed-back (error reduction - stability) operates from positive feed-back (error amplification - instability). If dynamic correction of any measurement system is included in a control system it is important to account for its delay, as it reduces the phase margin. Real-time correction and control must thus be studied jointly to prevent a potential break-down of the whole system! All internal mode control (IMC)-regulators synthesize dynamic correction. They are the direct equivalents in feed-back control to the type of sequential dynamic correction presented here (Fig. 11).

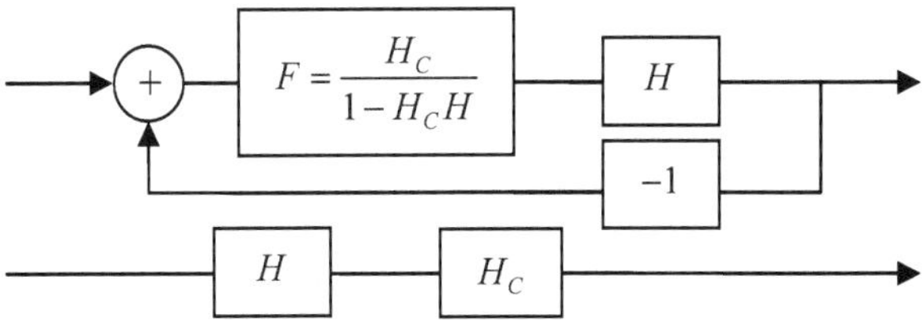

Fig. 11. The IMC-regulator F (top) in a closed loop system is equivalent to the direct sequential correction $H_C \approx H^{-1}$ (bottom) of the [measurement] system H proposed here.

Regularization or noise filtering is required for all types of dynamic correction, H_C must not (metrology) and can not (control) be chosen identical to the inverse H^{-1}. Dynamic corrections must be applied differently in feed-back than in a sequential topology. The sequential correction H_C presented here can be translated to correction within a feed-back loop with the IMC-regulator structure F. Measurements are normally analyzed afterwards (post-processing). That is never an option for control, but provides better and simpler ways of correction in metrology (Hessling 2008a). Causal application should always be judged against potential 'costs' such as increased complexity of correction and distortion due to application of stabilization methods etc.

Dynamic correction will be made in two steps. A digital filter is first synthesized using a model of the targeted measurement. This filter is then applied to all measured signals. Mathematically, measured signals are corrected by propagating them 'backwards' through the modelled measurement system to their physical origin. The synthesis involves inversion of the identified model, taking physical and practical constraints into account to find the optimal level of correction. Not surprisingly, time-reversed filtering in post-processing may be utilized to stabilize the filter. Post-processing gives additional possibilities to reduce the phase distortion, as well as to eliminate the time delay.

The synthesis will be based on the concept of filter 'prototypes' which have the desirable properties but do not always fulfil all constraints. A sequence of approximations makes the prototypes realizable at the cost of increased uncertainty of the correction. For instance, a time-reversed infinite impulse response filter can be seen as a prototype for causal application. One possible approximation is to truncate its impulse response and add a time delay to make it causal. The distortion manifests itself via the truncated tail of the impulse response. The corresponding frequency response can be used to estimate the dynamic error as in section 4.3. This will estimate the error of making a non-causal correction causal. Decreasing the acceptable delay increases the cost. If the acceptable delay exceeds the response time, there is no cost at all as truncation is not needed.

The discretization of a continuous time digital filter prototype can be made in two ways:

1. Minimize the numerical discrepancy between the characterization of a digital filter prototype and a comparable continuous time characterization for
   a. a calibration measurement
   b. an identified model
2. Map parameters of the identified continuous time model to a discrete time model by means of a unique transformation.

Alternative 1 closely resembles system identification and requires no specific methods for correction. In 1b, identification is effectively applied twice which should lead to larger uncertainty. The intermediate modelling reduces disturbances but this can be made more effectively and directly with the choice of filter structure in 1a. As it is generally most efficient in all kinds of 'curve fitting' to limit the number of steps, repeated identification as in 1b is discouraged. Indeed, simultaneous identification and discretization of the system as in 1a is the traditional and best performing method (Pintelon et al., 1990). Using mappings as in 2 (Hessling 2008a) is a very common, robust and simple method to synthesize any type of filter. In contrast to 1, the discretization and modelling errors are disjoint in 2, and can be studied separately. A utilization of the mapping can be defined to express the relation between its bandwidth (defined by the acceptable error) and the Nyquist frequency. The simplicity and robustness of a mapping may in practice override the cost of reduced accuracy caused by the detour of continuous time modelling. Alternative 2 will be pursued here, while for alternative 1a we refer to methods of identification discussed in section 4.2 and the example in section 4.4.1.

As the continuous time prototype transfer function H^{-1} for dynamic correction of H is unphysical (improper, non-causal and ill-conditioned), many conventional mappings fail. The simple exponential pole-zero mapping (Hessling, 2008a) of continuous time $(\tilde{p}_k, \tilde{z}_k)$ to discrete time (p_k, z_k) poles and zeros can however be applied. Switching poles and zeros to obtain the inverse of the transfer function of the original measurement system this transformation reads (T_S the sampling time interval),

$$\begin{aligned} z_k &= \exp(\tilde{p}_k T_S) \\ p_k &= \exp(\tilde{z}_k T_S) \end{aligned} \quad (11)$$

To stabilize and to cancel the phase, the reciprocals of unstable poles and zeros outside the unit circle in the z-plane are first collected in the time-reversed filter, to be applied to the time-reversed signal with exchanged start and end points. The remaining parameters build up the other filter for direct application forward in time. An additional regularizing low-pass noise filter is required to balance the error reduction and the increase of uncertainty (Hessling, 2008a). It will here be applied in both time directions to cancel its phase. For causal noise filtering, a symmetric linear phase FIR noise filter can instead be chosen.

4.4.1 Example: Oscilloscope step generator

From the step response characterization of a generator (Fig. 3, right), a non-minimum phase model was identified in section 4.2.4 (Fig. 7, right). The resulting prototype for correction is unstable, as it has poles outside the unit circle in the z-plane. It can be stabilized by means of time-reversal filtering, as previously described. In Fig. 12, this correction is applied to the original step signal. As expected (EA-10/07), the correction reduces the rise time T about as much as it increases the bandwidth.

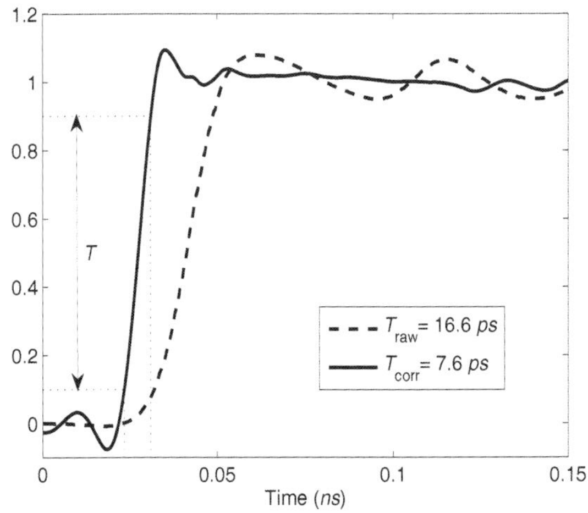

Fig. 12. Original (dashed) and corrected (full) response of the oscilloscope generator (Fig. 3).

Two objections can be made to this result: 1. No expert on system identification would identify the model and validate the correction against the same data. 2. The non-causal oscillations before the step are distinct and appear unphysical as all physical signals must be causal. The answer to both objections is the use of an extended and more detailed concept of measurement uncertainty in metrology, than in system identification: (1) Validation is made through the uncertainty analysis where all relevant sources of uncertainty are combined. (2) The oscillations before the step must therefore be 'swallowed' by any relevant measure of time-dependent measurement uncertainty of the correction.

The oscillations (aberration) are a consequence of the high frequency response of the [corrected] measurement system. The aberration is an important figure of merit controlled by the correction. Any distinct truncation or sharp localization in the frequency domain, as described by the roll-off and bandwidth, must result in oscillations in the time domain. There is a subtle compromise between reduction of rise time and suppression of aberration: Low aberration requires a shallow roll-off and hence low bandwidth, while short rise time can only be achieved with a high bandwidth. It is the combination of bandwidth and roll-off that is essential (section 4.3). A causal correction requires further approximations. Truncation of the impulse response of the time-reversed filter is one option not yet explored.

4.4.2 Example: Transducer system

Force and pressure transducers as well as accelerometers ('T') are often modelled as single resonant systems described by a simple complex-conjugated pole pair in the s-plane. Their usually low relative damping may result in 'ringing' effects (Moghisi, 1980), generally difficult to reduce by other means than using low-pass filters ('A'). For dynamic correction the s-plane poles and zeros of the original measurement system can be mapped according to Eq. 11 to the z-plane shown in Fig. 13. As this particular system has minimum phase (no zeros), no stabilization of the prototype for correction is required. A causal correction is directly obtained if a linear phase noise filter is chosen (Elster et al. 2007). Nevertheless, a standard low-pass noise filter was chosen for application in both directions of time to easily cancel its contribution to the phase response completely.

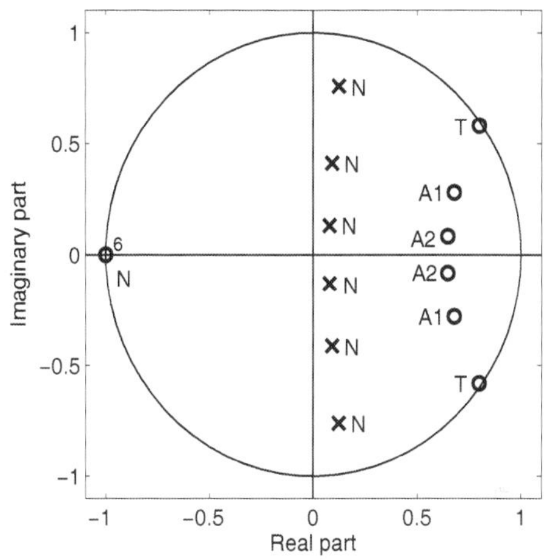

Fig. 13. Poles (x) and zeros (o) of the correction filter: cancellation of the transducer (T) as well as the analogue filter (A), and the noise filter (N).

The system bandwidth after correction was mainly limited by the roll-off of the original system, and the assumed signal-to-noise ratio $(50\,\text{dB})$. In Fig. 14 (top) the frequency response functions up to the noise filter cut-off, and the bandwidths defined by 5% amplification error before (β) and after (α) correction are shown. This bandwidth increased 65%, which is comparable to the REq-X system (Bruel&Kjaer, 2006). The utilization of the maximum 6 dB bandwidth set by the cross-over frequency of the noise filter was as high as $\eta = 93\%$. This ratio approaches 100% as the sampling rate increases further and decreases as the noise level decreases. The noise filter cut-off was chosen $f_N = 2 f_A$, where f_A is the cross-over frequency of the low-pass filter. The performance of the correction filter was verified by a simulation (Matlab), see Fig. 14 (bottom). Upon correction, the residual dynamic error (section 4.3) decreased from 10% to 6%, the erroneous oscillations were effectively suppressed and the time delay was eliminated.

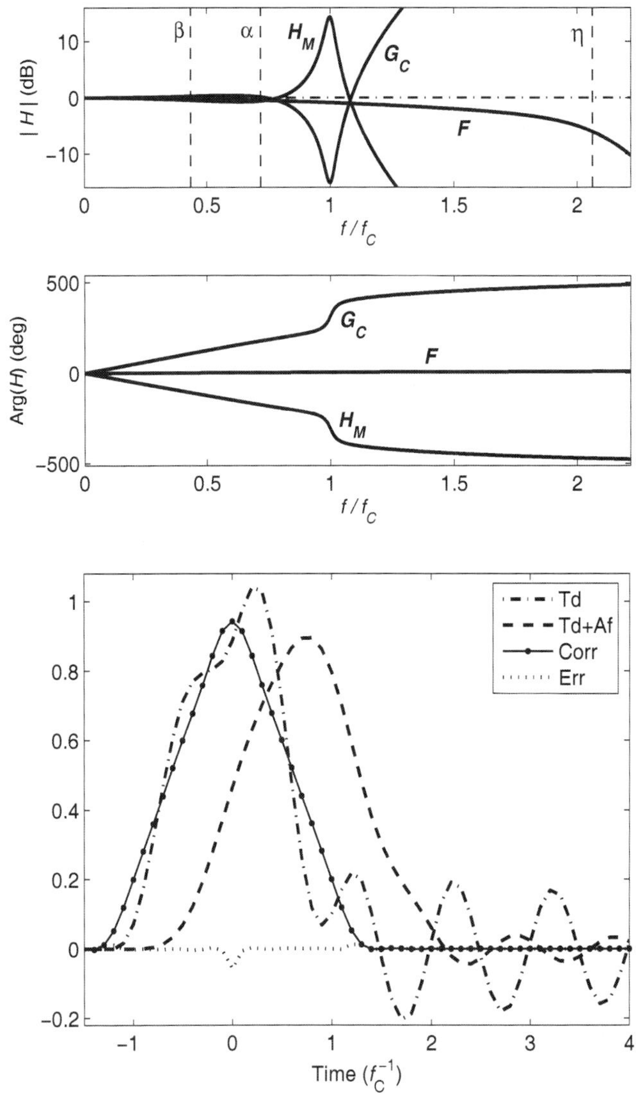

Fig. 14. Magnitude (top) and phase (middle) of frequency response functions for the original measurement system (H_M), the correction filter (G_C) and the total corrected system (F), and simulated correction of a triangular pulse (bottom): corrected signal (Corr), residual error (Err), and transducer signal before (Td) and after (Td+Af) the analogue filter. Time is given in units of the inverse resonance frequency f_C^{-1} of the transducer.

4.5 Measurement uncertainty

Traditionally, the uncertainty given by the calibrator is limited to the calibration experiment. The end users are supposed to transfer this information to measurements of interest by using an uncertainty budget. This budget is usually a simple spreadsheet calculation, which at best depends on a most rudimentary classification of measured signals. In contrast, the measurement uncertainty for non-stationary signals will generally have a strong and complex dependence on details of the measured signal (Elster et al. 2007; Hessling 2009a). The interpretation and meaning of uncertainty is identical for all measurements - the uncertainty of the conditions and the experimental set up (input variables) results in an uncertainty of the estimated quantity (measurand). The unresolved problems of non-stationary uncertainty evaluation are not conceptual but practical. How can the uncertainty of input variables be expressed, estimated and propagated to the uncertainty of the estimated measurand? As time and ensemble averages are different for non-ergodic systems such as non-stationary measurements, it is very important to state whether the uncertainty refers to a constant or time-dependent variable. In the latter case, also temporal correlations must be determined. Noise is a typical example of a fluctuating input variable for which both the distribution and correlation is important. If the model of the system correctly catches the dynamic behaviour, its uncertainty must be related to constant parameters. The lack of repeatability is often used to estimate the stochastic contribution to the measurement uncertainty. The uncertainty of non-stationary measurements can however *never* be found with repeated measurements, as variations due to the uncertainty of the measurement or variations of the measurand cannot even in principle be distinguished.

The uncertainty of applying a dynamic correction might be substantial. The stronger the correction, the larger the associated uncertainty must be. These aspects have been one of the most important issues in signal processing (Wiener, 1949), while it is yet virtually unknown within metrology. The guide (ISO GUM, 1993, section 3.2.4) in fact states that "it is assumed that the result of a measurement has been corrected for all recognized significant systematic effects and that every effort has been made to identify such effects". Interpreted literally, this would by necessity lead to measurement uncertainty without bound. Also, as stated in section 4.4.1 the correction of the oscilloscope generator in Fig. 12 only makes sense (causality) if a relevant uncertainty is associated to it. This context elucidates the pertinent need for reliable measures of non-stationary measurement uncertainty.

The contributions to the measurement uncertainty will here be expressed in generalized time-dependent sensitivity *signals,* which are equivalent to the traditional sensitivity constants. The sensitivity signals are obtained by convolving the generating signals with the virtual *sensitivity systems* for the measurement. The treatment here includes one further step of unification compared to the previous presentation (Hessling, 2009a): The contributions to the uncertainty from measurement noise and model uncertainty are evaluated in the same manner by introducing the concept of generating signals. Digital filters or software simulators will be proposed tools for convolution. Determining the uncertainty of input variables is considered to be a part of system identification (section 4.2.2), assumed to precede the propagation of dynamic measurement uncertainty addressed here.

The measurement uncertainty signal is generally *not* proportional to the measured signal. This typical dynamic effect does not imply that the system is non-linear. Rather, it reveals that the sensitivity systems differ fundamentally from the measurement system.

4.5.1 Expression of measurement uncertainty

To evaluate the measurement uncertainty (ISO GUM, 1993), a model equation is required. For a dynamic measurement it is given by the differential or difference equation introduced in the context of system identification (section 4.2). Also in this case it will be convenient to use the corresponding transformed algebraic equations (Eq. 4), preferably given as transfer functions parameterized in poles and zeros, or physical parameters.

The measurement uncertainty is associated to the quantity of interest contained in the model equation. For measured uncorrected signals, the uncertainty is probably strongly dominated by systematic errors (section 4.3). The model equation for correction is the inverse model equation/transfer function for the direct measurement, adjusted for approximations and modifications required to realize the correction. Generally, a system analysis (Warwick, 1996) of the measurement and all applied operations will provide the required model. For simplicity, this section will only address random contributions to the measurement uncertainty associated to the dynamic correction discussed in section 4.4.

The derivation of the expression of uncertainty in dynamic measurements will be similar for CT and DT, due to the identical use of poles and zeros. Instead of using the inverse Laplace and z-transform, the expressions will be convolved in the time domain with digital filters or dynamic simulators. The propagation of uncertainty from the characterization to the model (section 4.2.2), and from the model to the correction of the targeted measurement discussed here will be evaluated analogously; the model equation or transfer function will be linearized in its parameters and the uncertainty expressed through sensitivity signals. For an efficient model only a few weakly correlated parameters are required. The covariance matrix is in that case not only small but also sparse. As the number of sensitivity signals scales with the size of this matrix, the propagation of uncertainty will be simple and efficient.

The time-dependent deviation χ of the signal of interest from its ensemble mean can be expressed as a matrix product between the deviations φ of all m variables from their ensemble mean, and matrix ξ of all sensitivity signals organized in rows,

$$\chi = \xi^T \varphi, \quad \varphi \equiv \begin{pmatrix} \varphi_1 & \varphi_2 & \cdots & \varphi_m \end{pmatrix}^T, \quad \xi_{nk} = (e_n * \mu_n)_k . \tag{12}$$

The sensitivity signal ξ_{nk} for parameter φ_n, evaluated at time t_k, is calculated as a convolution $(*)$ between the impulse response e_n of the sensitivity system E_n and a generating signal μ_n. Both the response e_n and the signal μ_n are generally unique for every parameter. In contrast to the previous formulation (Hessling, 2009a), the vector φ here represents ***all*** uncertain input variables, noise (δy) as well as static and dynamic model parameters (q). The covariance of the error signal is found directly from this expression by squaring and averaging $(\langle\cdot\rangle)$ over an ensemble of measurements,

$$\left\langle \chi\chi^T \right\rangle = \xi^T \left\langle \varphi\varphi^T \right\rangle \xi . \tag{13}$$

The variance or squared uncertainty at different times are given by the diagonal elements of $\left\langle \chi\chi^T \right\rangle$. The matrix $\left\langle \varphi\varphi^T \right\rangle$ and columns of ξ is the covariance matrix of input variables and sensitivity at a given time often written as (ISO GUM, 1993) $u(x,x)$ and c, respectively.

The combination of Eq. 6 and Eq. 13 propagates the uncertainty of the characterization (Λ) to any time domain measurement (χ) in two steps via the model (Fig. 1), directly (ξ) or indirectly (Γ) via the sensitivity systems E. Physical constraints are fulfilled for all realizations of equivalent measurements (φ), for the parameterization (poles, zeros), and for all representations (frequency and time domain).
The covariance matrix $\left\langle \varphi\varphi^T \right\rangle$ will usually be sparse, since different types of variables (such as noise $\left(u_N^2\right)$ and model parameters $\left(u_D^2\right)$, as well as disjoint subsystems $\left(u_{D1}^2, u_{D2}^2, \ldots u_{Dn}^2\right)$ characterized separately) usually are uncorrelated,

$$\left\langle \varphi\varphi^T \right\rangle = \begin{pmatrix} u_N^2 & 0 \\ 0 & u_D^2 \end{pmatrix}, \quad u_D^2 = \begin{pmatrix} u_{D,1}^2 & 0 & 0 & 0 \\ 0 & u_{D,2}^2 & 0 & 0 \\ 0 & 0 & \ddots & \vdots \\ 0 & 0 & \cdots & u_{D,n}^2 \end{pmatrix}. \tag{14}$$

For each source of uncertainty, the following has to be determined from the model equation:

a. Uncertain parameter φ_n.
b. Sensitivity system $E_n(z)$, or $E_n(s)$.
c. Generating signal for evaluating sensitivity, $\mu_n(t)$.

The presence of measurement noise $\delta y(t)$ is equivalent to having a signal source without control in the transformed model equation (Eq. 4). It is thus trivial that the noise propagates through the dynamic correction $\hat{G}^{-1}(z)$ just like the signal itself, $\delta\hat{X}(z) = \hat{G}^{-1}(z)\delta Y(z)$:

a. The uncertain parameters are the noise levels at different times, $\varphi_n = \delta y_n = \delta y(t_n)$.
b. The sensitivity system is identical to the estimated correction, $E_n(z) = \hat{G}^{-1}(z)$.
c. The sensitivity signal is simply the impulse response of the correction, $\xi_{nk} = \hat{g}_{k-n}^{-1}$. The generating signal[1] is thus a delta function, $\mu_{nk} = \delta_{nk}$.

The contribution due to noise to the covariance of the corrected signal at different times is directly found using Eq. 13,

$$u_N^2 = \left(\hat{g}^{-1}\right)^T \left\langle \delta y \delta y^T \right\rangle \hat{g}^{-1}. \tag{15}$$

The covariance matrix $\left\langle \delta y \delta y^T \right\rangle$ will be band-diagonal with a width set by the correlation time of the noise. This time is usually very short as noise is more or less random. The band of $\left\langle \delta y \delta y^T \right\rangle$ is widened by the impulse response $\hat{g}^{-1}$, as it is propagated to u_N^2. The matrix u_N^2 is thus also band-diagonal, but with a width given by the sum of the correlation times of the noise and the impulse response $\hat{g}^{-1}$ of the correction. Evidently, not only the probability

1 The introduction of generating signals may appear superfluous in this context. Nevertheless, it provides a completely *unified* treatment of noise and model uncertainty which greatly simplifies the general formulation. In addition, the concept of generating signals provides more freedom to propagate any obscure source of uncertainty.

distributions but also the temporal correlations of the noise and the uncertainty of the correction are different.

If the noise is independent of time in a statistical sense, it is stationary. In that case the covariance matrix will only depend on the time difference of the arguments, $\langle \delta y_l \delta y_k \rangle = u_Y^2 \cdot \lambda_{|k-l|}$, and thus has a diagonal structure (lines indicate equal elements),

$$\langle \delta y \delta y^T \rangle = u_Y^2 \cdot \begin{pmatrix} \lambda_0 & \lambda_1 & \lambda_2 & \ddots & \cdots \\ \lambda_1 & \lambda_0 & \lambda_1 & \ddots & \\ \lambda_2 & \lambda_1 & \lambda_0 & \ddots & \\ \vdots & \ddots & \ddots & \ddots & \end{pmatrix}. \tag{16}$$

Further, if the noise is not only stationary but also uncorrelated (white), $\lambda_k = \delta_{k0}$. Only the diagonal will be non-zero. The noise will in this case propagate very simply,

$$u_N^2 = \left(\hat{g}^{-1}\right)^T \hat{g}^{-1} u_Y^2, \quad \operatorname{diag}\left[\left(\hat{g}^{-1}\right)^T \hat{g}^{-1}\right] = \left\|\hat{g}^{-1}\right\|^2 \equiv c_N^2. \tag{17}$$

The variance given by $u_N^2 = c_N^2 u_Y^2$ is as required time-independent since the source is stationary. The sensitivity c_N to stationary uncorrelated measurement noise is simply given by the quadratic norm of the impulse response of the correction.

The propagation of model uncertainty is more complex, because model variations propagate in a fundamentally different manner from noise. Direct linearization will give,

$$\delta H^{-1}(s) H(s) = \sum_n \frac{q_n}{H^{-1}} \frac{\partial H^{-1}}{\partial q_n} \frac{\delta q_n}{q_n} = \sum_n \frac{\partial \ln H^{-1}}{\partial \ln q_n} \frac{\delta q_n}{q_n} \equiv \sum_n E_n(q,s) \frac{\delta q_n}{q_n}. \tag{18}$$

Logarithmic derivatives are used to obtain relative deviations of the parameters and to find simple sensitivity systems $E_n(q,s)$ of low order. Therefore, the generating signals are the corrected rather than the measured signals. This difference can be ignored for a minor correction, as the accuracy of evaluating the uncertainty then is less than the error of calculation.

If the model parameters $\{q_n\}$ are physical:

a. The uncertain parameters can be the relative variations, $\varphi_n = \delta q_n / q_n$.
b. The sensitivity systems are $E_n(q,s)$.
c. The generating signals are all given by the corrected measured signal, $\mu_{nk} = \hat{x}(t_k)$.

For non-physical parameterizations all implicit constraints must be properly accounted for. Poles and zeros are for instance completely correlated in pairs as any measured signal must be real-valued. This correlation could of course be included in the covariance matrix $\langle \varphi \varphi^T \rangle$. A simpler alternative is to remove the correlation by redefining the uncertain parameters. The generating signals $\mu_{nk} = \hat{x}(t_k)$ remain, but the sensitivity systems change accordingly (Hessling, 2009a) ($\bullet$ denotes scalar vector/inner product in the complex s- or z-plane):

a. For complex-valued pairs of poles and zeros, two projections can be used as uncertain parameters, $\rho_r(q_n) = [\delta q_n / |q_n|] \bullet [q_n^r / |q_n|^r]$, $r = 1,2$. For all real-valued poles and zeros q the variations can still be chosen as $\varphi_n = \delta q_n / q_n$.
b. The sensitivity systems can be written as $E_q^{(mn)}(\hat{s}) = \hat{s}^m / (q/|q| - \hat{s})(|q|/q - \hat{s})^{n-1}$, $E_q^{11}(s/|q|)$ for real-valued and $-E_q^{22}(s/|q|)$ and $E_q^{12}(s/|q|)$ for the projections $\rho_1(q)$ and $\rho_2(q)$ of complex-valued pairs of poles and zeros, respectively.

Non-physical parameters require full understanding of implicit requirements but may yield expressions of uncertainty of high generality. Large, complex and different types of measurement systems can be evaluated with rather abstract but structurally simple analyses. Physical parameterizations are highly specific but straight forward to use. The first transducer example uses the general pole-zero parameterization. The second voltage divider example will utilize physical electrical parameters.

The conventional evaluation of the combined uncertainty does not rely upon constant sensitivities. As a matter of fact, the standard quadratic summation of various contributions (ISO GUM, 1993) is already included in the general expression (Eq. 13). The contributions from different sources of uncertainty are added at each instant of time, precisely as prescribed in the GUM for constant sensitivities. The same applies to the proceeding expansion of combined standard uncertainty to any desired level of confidence. In addition, the temporal correlation is of high interest for non-stationary measurement. That is non-trivially inherited from the correlation of the sensitivity signals specific for each measurement, according to the covariance of the uncertain input variables (Eq. 13).

4.5.2 Realization of sensitivity filters

The sensitivity filters are specified completely by the sensitivity systems $E(q,s)$. Filters are generally *synthesized* or constructed from this information to fulfil given constraints. The actual filtering process is *implemented* in hardware or computer programs. The *realization* of sensitivity filters refers to both aspects. Two examples of realization will be suggested and illustrated: digital filtering and dynamic simulations.

The syntheses of digital filters for sensitivity and for dynamic correction described in section 4.4 are closely related. If the sensitivity systems are specified in continuous time, discretization is required. The same exponential mapping of poles and zeros as for correction can be used (Eq. 11). The sensitivity filters for the projections ρ_n will be universal (Hessling, 2009a). Digital filtering will be illustrated in section 4.5.3, for the transducer system corrected in section 4.4.2.

There are many different software packages for dynamic simulations available. Some are very general and each simulation task can be formulated in numerous ways. Graphic programming in networks is often simple and convenient. To implement uncertainty evaluation on-line, access to instruments is required. For post-processing, the possibility to import and read measured files into the simulator model is needed. The risk of making mistakes is reduced if the sensitivity transfer functions are synthesized directly in discrete or continuous time. The Simulink software (Matlab) of Matlab has all these features and will be used in the voltage divider example (Hessling, 2009b) in section 4.5.4.

4.5.3 Example: Transducer system – digital sensitivity filters

The uncertainty of the correction of the electro-mechanical transducer system (section 4.4.2) is determined by the assumed covariance of the model and of the noise given in Table 2.

$u_Y^2 = 50\,\mathrm{dB}$, stationary, uncorrelated ('white')	Measurement noise
$u_M^2 = 10^{-4}\begin{pmatrix} 0.5^2 & 0 & 0 & 0 & 0 & 0 & 0 \\ 0 & 1^2 & 0.1^2 & 0 & 0 & 0 & 0 \\ 0 & 0.1^2 & 0.2^2 & 0 & 0 & 0 & 0 \\ 0 & 0 & 0 & 1^2 & -0.1^2 & 0.05^2 & 0.02^2 \\ 0 & 0 & 0 & -0.1^2 & 0.4^2 & 0.01^2 & -0.03^2 \\ 0 & 0 & 0 & 0.05^2 & 0.01^2 & 1^2 & -0.9^2 \\ 0 & 0 & 0 & 0.02^2 & -0.03^2 & -0.9^2 & 0.8^2 \end{pmatrix}$	Covariance of: static amplification K, transducer (T) and low-pass filter $(A1, A2)$ zero projections ρ_1, ρ_2

Table 2. Covariance of the transducer system. The projections ρ_1, ρ_2 are anti-correlated as the zeros approach the real axis (Hessling, 2009a), see entries (6,7)/(7,6) of u_M^2 and Fig. 13.

The cross-over frequency f_N of the low-pass noise filter of the correction strongly affects the sensitivity to noise, $c_N = 36$ for $f_N = 3f_A$ but only $c_N = 2.6$ for $f_N = 2f_A$ (section 4.4.2), where f_A is the low-pass filter cut-off. In principle, the stronger the correction (high cut-off f_N) the stronger the amplification of noise. The model uncertainty increases rapidly at high frequencies because of bandwidth limitations. The systematic errors caused by imperfect discretization in time are negligible if the utilization is high, $\eta \approx 100\%$ (section 4.4.2). The uncertainty in the high frequency range mainly consists of:

1. Residual uncorrected dynamic errors
2. Measurement noise amplified by the correction
3. Propagated uncertainty of the dynamic model

For optimal correction, the uncorrected errors (1) balance the combination of noise (2) and model uncertainty (3). Even though the correction could be maximized up to the theoretical limit of the Nyquist frequency for sampled signals, it should generally be avoided. Rather conservative estimates of systematic errors are advisable, as a too ambitious dynamic correction might do more harm than good. It should be strongly emphasized that the noise level should refer to the *targeted* measurement, not the calibration! As the optimality depends on the measured signal, it is tempting to synthesize adaptive correction filtering related to causal Kalman filtering (Kailath, 1981). With post-processing and a recursive procedure the adaptation could be further improved. This is another example (besides perfect stabilization) of how post-processing may be utilized to increase the performance beyond what is possible for causal correction.

The sensitivity signals for the model are found by first applying the correction filter $\hat{g}^{-1}$ and then the universal filter bank of realized sensitivity systems $E_n(q,z)$ (Eq. 18) (Hessling, 2009a) (omitted for brevity). Three complex-valued pole pairs with two projections, one for the transducer (T) and two for the filter $(A1, A2,\ \operatorname{Im} z_{A1} > \operatorname{Im} z_{A2})$ results in six unique sensitivity signals. For a triangular signal, some sensitivity signals $(T, A1)$ are displayed in

Fig. 15 (top). The sensitivities for the transducer and filter models are clearly quite different, while for the two filter zero pairs they are similar (sensitivities for $A2$ omitted). The standard measurement uncertainty u_C in Fig. 15 (bottom) combines noise ($u_Y \rightarrow u_N$) and model uncertainty $(u_M \rightarrow u_D)$, see covariance in Table 2. Any non-linear static contribution to the uncertainty has for simplicity been disregarded. To evaluate the expanded measurement uncertainty signal, the distribution of measured values at each instant of time over *repeated measurements* of the *same* triangular signal must be inferred.

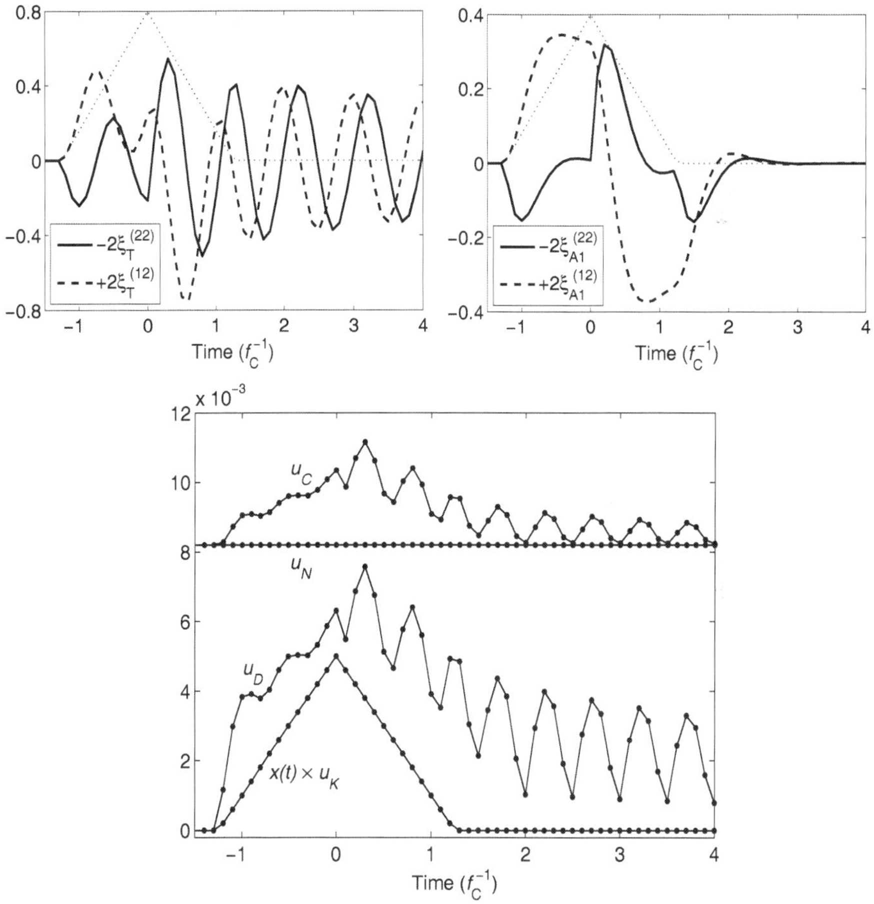

Fig. 15. Measurement uncertainty u_C (bottom) for correction of the electro-mechanical transducer system (Section 4.4.2), and associated sensitivities for the transducer zero pair T projections (top left) and the filter zero pair A1 projections (top right). The measurand (x) (top: dotted, bottom: $u_K \equiv u_M(1,1)$ (Table 2)) is rescaled and included for comparison. Time is given in units of the inverse resonance frequency f_C^{-1} of the transducer.

4.5.4 Example: Voltage divider for high voltage – simulated sensitivities

Voltage dividers in electrical transmission systems are required to reduce the high voltages to levels that are measurable with instruments. Essentially, the voltage divider is a gearbox for voltage, rather than speed of rotation. The equivalent scheme for a capacitive divider is shown in Fig. 16. The transfer function/model equation is found by the well-known principle of voltage division,

$$H(s) = K \frac{1+(RC)_{LV}\,s+(LC)_{LV}\,s^2}{1+(RC)_{HV}\,s+(LC)_{HV}\,s^2}, \quad K = \frac{C_{HV}}{C_{LV}} << 1\,. \tag{19}$$

Linearization of H in $K, R_1, R_2, L_1, L_2, C_1, C_2$ yields seven different sensitivity systems which can be realized directly in Simulink by graphic programming (Fig. 17).

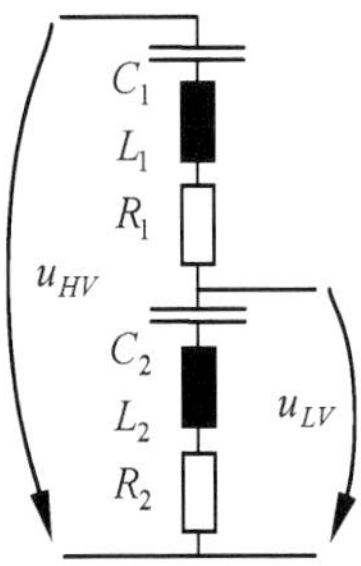

u_{LV} = 40 dB	Signal to noise ratio, measurement noise (LV)
$u_M^2 = 10^{-4}\begin{pmatrix} 2^2 & 0 & 0 & 0 & 0 & 0 & 0 \\ 0 & 2^2 & 1^2 & 0 & 0 & 0 & 0 \\ 0 & 1^2 & 10^2 & 2^2 & 0 & 0 & 0 \\ 0 & 0 & 2^2 & 5^2 & 0 & 0 & 0 \\ 0 & 0 & 0 & 0 & 1^2 & 0.5^2 & 0 \\ 0 & 0 & 0 & 0 & 0.5^2 & 8^2 & 1^2 \\ 0 & 0 & 0 & 0 & 0 & 1^2 & 6^2 \end{pmatrix}$	Relative covariance of $K, R_1, L_1, C_1, R_2 L_2, C_2$

Fig. 16. Electrical model of capacitive voltage divider for high voltage (left) with covariance (right). The high (low) voltage input (output) circuit parameters are labelled HV (LV).

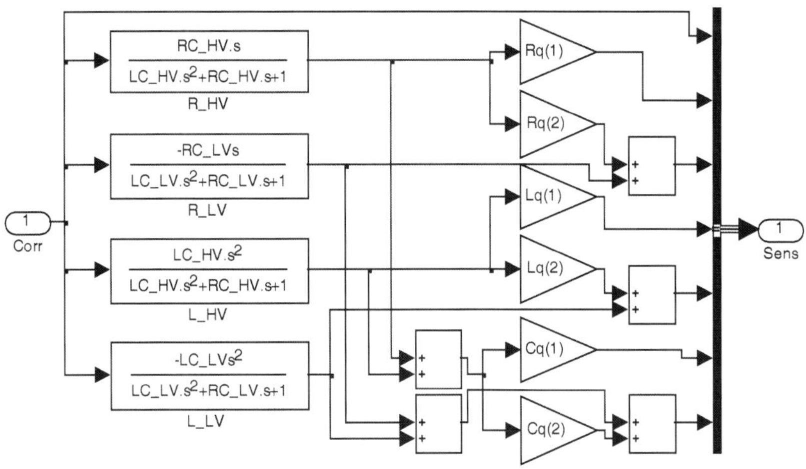

Fig. 17. Simulink model for generating model sensitivity from corrected signals (Corr). Here, $\mathrm{Xq}(\mathrm{n}) \equiv X_n / X_{HV}$, where $X = \{R, L, C\}$, $n = \{1,2\}$ and X_{HV} the total for the HV circuit.

As the physical high-frequency cut-off was not modelled (Eq. 19), no noise filter was required. To calculate the noise sensitivity from the impulse response (Eq. 17), proper and improper parts of the transfer function had to be analyzed separately (Hessling, 2009b). In Fig. 18 the uncertainty of correcting a standard lightning impulse (u_{HV}) is simulated. The signal could equally well have been any corrected voltmeter signal, fed into the model with the data acquisition blocks of Simulink. The R, L, C parameters were derived from resonance frequency $f_C = [2.3 \quad 0.8]$MHz and relative damping $\zeta = [1.2 \quad 0.4]$ of the HV and LV circuits, and nominal ratio of voltage division $K = 1/1000$. The resulting sensitivities are shown in Fig. 18 (left). The measurement uncertainty of the correction (u_C) in Fig. 18 (right) contains contributions from the noise ($u_{LV} \rightarrow u_N$) and the model $(u_M \rightarrow u_D)$.

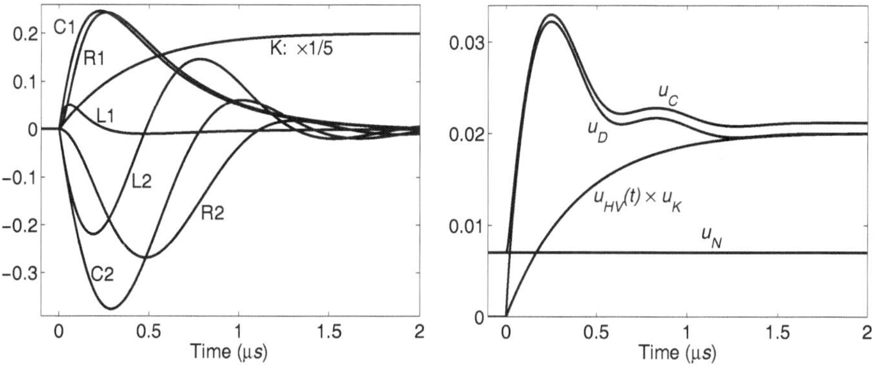

Fig. 18. Model sensitivities ξ (left) for the standard lightning impulse u_{HV} (left: $u_{HV} \propto \xi_K$, right: $u_K = u_M(1,1)$), and measurement uncertainty u_C (right) for dynamic correction.

4.6 Known limitations and further developments

Dynamic Metrology is a framework for further developments rather than a fixed concept. The most important limitation of the proposed methods is that the measurement system must be linear. Linear models are often a good starting point, and the analysis is applicable to all non-linear systems which may be accurately linearized around an operating point.

Even though measurements of non-stationary quantities are considered, the system itself is assumed time-invariant. Most measurement systems have no measurable time-dependence, but the experimental set up is sometimes non-stationary. If the time-dependence originates from outside the measurement system it can be modelled with an additional influential (input) signal.

The propagation of uncertainty has only been discussed in terms of sensitivity. This requires a dynamic model of the measurement, linear in the uncertain parameters. Any obscure correlation between the input variables is however allowed. It is an unquestionable fact that the distributions often are not accurately known. Propagation of uncertainty beyond the concept of sensitivity can thus seldom be utilized, as it requires more knowledge of the distributions than their covariance.

The mappings for synthesis of digital filters for correction and uncertainty evaluation are chosen for convenience and usefulness. The over-all results for mappings and more accurate numerical optimization methods may be indistinguishable. Mappings are very robust, easy to transfer and to illustrate. The utilization ratio of the mapping should be defined according to the noise filter cross-over rather than, as customary, the Nyquist frequency. This fact often makes the mappings much less critical.

The de-facto standard is to evaluate the measurement uncertainty in post-processing mode. Non-causal operations are then allowed and sometimes provide signal processing with superior simplicity and performance. Instead of discussing causality, it is more appropriate to state a maximum allowed time delay. When the ratio of the allowed time delay to the response time of the measurement system is much larger than one, also non-causal operations like time-reversed filtering can be accurately realized in real-time. If the ratio is much less than one, it is difficult to realize any causal operation, irrespectively of whether the prototype is non-causal or not. Finding good approximations to fulfil strong requirements on fast response is nevertheless one topic for future developments.

Finding relevant models of interaction in various systems is a challenge. For analysis of for instance microwave systems this has been studied extensively in terms of scattering matrices. How this can be joined and represented in the adopted transfer function formalism needs to be further studied.

Interpreted in terms of distortion there are many different kinds of uncertainty which need further exploration. The most evident source of distortion is a variable amplification in the frequency domain, which typically smoothes out details. A finite linear phase component is equivalent to a time delay which increases the uncertainty immensely, if not adjusted for. Distortion due to non-linear phase skews or disperses signals. All these effects are presently accounted for. However, a non-linear response of the measurement system gives rise to another type of systematic errors, often quantified in terms of total harmonic distortion (THD). A harmonic signal is then split into several frequency components by the measurement system. This figure of merit is often used e.g. in audio reproduction. Linear distortion biases or colours the sound and reduces space cognition, while non-linear distortion influences 'sound quality'. Non-linear distortion is also discussed extensively in the field of electrical power systems, as it affects 'power quality' and the operation of the equipment connected to the electrical power grid. A concept of non-linear distortion for non-stationary measurements is missing and thus a highly relevant subject for future studies.

5. Summary

In the broadest possible sense *Dynamic Metrology* is devoted to the analysis of dynamic measurements. As an extended calibration service, it contains many novel ingredients currently not included in the standard palette of metrology. Rather, Dynamic Metrology encompasses many operations found in the fields of system identification, digital signal processing and control theory. The analyses are more complex and more ambiguous than conventional uncertainty budgets of today. The important interactions in non-stationary measurements may be exceedingly difficult to both control and to evaluate. In many situations, in situ calibrations are required to yield a relevant result. Providing metrological services in this context will be a true challenge.

Dynamic Metrology is currently divided into four blocks. The calibrator performs the characterization experiment (1) and identifies the model of the measurement (2). The dynamic correction (3) and evaluation of uncertainty (4) are synthesized for *all* measurements by the calibrator, while these steps must be realized by the end user for *every* single measurement. The proposed procedures of uncertainty evaluation for non-stationary quantities closely resemble the present procedure formulated in the Guide to the Expression of Uncertainty in Measurement (ISO GUM, 1993), but its formulation needs to be generalized and exemplified since for instance:

- The sensitivities are generally time-dependent *signals* and not constants.
- The sensitivity is *not* proportional to the measured or corrected signal.
- The uncertainty refers to distributions over *ensembles* and *temporal correlations*.
- The model equation is one or several *differential* or *difference* equation(s).
- The uncertainty, dynamic correction or any other comparable signal is *unique* for every combination of measurement *system* and measured or corrected *signal*.
- Proper estimation of systematic errors requires a robust concept of *time delay*.
- Complete dynamic correction must never be the goal, as noise would be amplified without any definite bound.

6. References

ASTM E 467-98a (1998). *Standard Practice for Verification of Constant Amplitude Dynamic Forces in an Axial Fatigue Testing System*, The American Society for testing and materials

Bruel&Kjaer (2006). Magazine No. 2 / 2006, pp. 4-5, http://www.bksv.com/4279.asp

Chen Chi-Tsong (2001). *Digital Signal Processing*, Oxford University Press, ISBN 0-19-513638-1, New York

Clement, T.S.; Hale, P.D.; Williams, D. F.; Wang, C. M.; Dienstfrey, A.; & Keenan D.A. (2006). Calibration of sampling oscilloscopes with high-speed photodiodes, *IEEE Trans. Microw. Theory Tech.*, Vol. 54, (Aug. 2006), pp. 3173-3181

Dienstfrey, A.; Hale, P.D.; Keenan, D.A.; Clement, T.S. & Williams, D.F. (2006). Minimum phase calibration of oscilloscopes, *IEEE Trans. Microw. Theory. Tech.* 54 No. 8 (Aug. 2006), pp. 481-491

EA-10/07, EAL-G30 (1997) *Calibration of oscilloscopes*, edition 1, European cooperation for accreditation of laboratories

Ekstrom, M.P. (1972). Baseband distortion equalization in the transmission of pulse information, *IEEE Trans. Instrum. Meas.* Vol. 21, No. 4, pp. 510-5

Elster, C.; Link, A. & Bruns, T. (2007). Analysis of dynamic measurements and determination of time-dependent measurement uncertainty using a second-order model, *Meas. Sci. Technol.* Vol. 18, pp. 3682-3687

Esward, T.; Elster, C. & Hessling, J.P. (2009) Analyses of dynamic measurements: new challenges require new solutions, *XIX IMEKO World Congress*, Lisbon, Portugal Sept., 2009

Hale, P. & Clement, T.S. (2008). Practical Measurements with High Speed Oscilloscopes, *72rd ARFTG Microwave Measurement Conference*, Portland, OR, USA, June 2008

Hessling, P. (1999). Propagation and summation of flicker, *Cigre' session 1999*, Johannesburg, South Africa

Hessling, J.P. (2006). A novel method of estimating dynamic measurement errors, *Meas. Sci. Technol.* Vol. 17, pp. 2740-2750

Hessling, J.P. (2008a). A novel method of dynamic correction in the time domain, *Meas. Sci. Technol.* Vol. 19, pp. 075101 (10p)
Hessling, J.P. (2008b). Dynamic Metrology - an approach to dynamic evaluation of linear time-invariant measurement systems, *Meas. Sci. Technol.* Vol. 19, pp. 084008 (7p)
Hessling, J.P. (2008c). Dynamic calibration of uni-axial material testing machines, *Mech. Sys. Sign. Proc.*, Vol. 22, 451-66
Hessling, J.P. (2009a). A novel method of evaluating dynamic measurement uncertainty utilizing digital filters, *Meas. Sci. Technol.* Vol. 20, pp. 055106 (11p)
Hessling, J.P. & Mannikoff, A. (2009b) Dynamic measurement uncertainty of HV voltage dividers, *XIX IMEKO World Congress*, Lisbon, Portugal Sept., 2009
Humphreys, D.A. & Dickerson, R.T. (2007). Traceable measurement of error vector magnitude (EVM) in WCDMA signals, *The international waveform diversity and design conference*, Pisa, Italy, 4-8 June 2007
IEC 61000-4-15 (1997). *Electromagnetic compability (EMC), Part 4: Testing and measurement techniques, Section 15: Flickermeter - Functional and design specifications*, International Electrotechnical commission, Geneva
ISO 2631-5 (2004). *Evaluation of the Human Exposure to Whole-Body Vibration*, International Standard Organization, Geneva
ISO 16063 (2001). *Methods for the calibration of vibration and shock transducers*, International Standard Organization, Geneva
ISO GUM (1993). *Guide to the Expression of Uncertainty in Measurement*, 1st edition, International Standard Organization, ISBN 92-67-10188-9, Geneva
Kailath, T. (1981). *Lectures on Wiener and Kalman filtering*, Springer-Verlag, ISBN 3-211-81664-X, Vienna, Austria
Kollár, I.; Pintelon, R.; Schoukens, J. & Simon, G. (2003). Complicated procedures made easy: Implementing a graphical user interface and automatic procedures for easier identification and modelling, *IEEE Instrum. Meas. Magazine* Vol. 6, No. 3, pp. 19-26
Ljung, L. (1999). *System Identification: Theory for the User*, 2nd Ed, Prentice Hall, ISBN 0-13-656695-2, Upper Saddle River, New Jersey
Matlab with System Identification, Signal Processing Toolbox and Simulink, The Mathworks, Inc.
Moghisi, M.; Squire, P.T. (1980). An absolute impulsive method for the calibration of force transducers, *J. Phys. E.: Sci. Instrum.* Vol. 13, pp. 1090-2
Pintelon, R. & Schoukens, J. (2001). *System Identification: A Frequency Domain Approach*, IEEE Press, ISBN 0-7803-6000-1, Piscataway, New Jersey
Pintelon, R.; Rolain, Y.; Vandeen Bossche, M. & Schoukens, J. (1990) Toward an Ideal Data Acquisition Channel, *IEEE Trans. Instrum. Meas.* Vol. 39, pp. 116-120
3GPP TS (2006). *Tech. Spec. Group Radio Access Network; Base station conformance testing*
Warwick, K. (1996). *An introduction to control systems*, 2nd Ed, World Scientific, ISBN 981-02-2597-0, Singapore
Wiener, N. (1949). *Extrapolation, Interpolation, and Smoothing of Stationary Time Series*, Wiley, ISBN 0-262-73005-7, New York
Williams, D.F.; Clement, T.S.; Hale, P. and Dienstfrey, A. (2006). Terminology for high-speed sampling oscilloscope calibration, *68th ARFTG Microwave Measurement Conference*, Broomfield, CO, USA, Dec. 2006

10

Sensors Characterization and Control of Measurement Systems Based on Thermoresistive Sensors via Feedback Linearization

M. A. Moreira [1], A. Oliveira[2], C.E.T. Dórea[2], P.R. Barros [3] and J.S. da Rocha Neto[3]

[1] *Universidade Estadual de Campinas*
[2] *Universidade Federal da Bahia*
[3] *Universidade Federal de Campina Grande*
Brazil

1. Introduction

In this work the application of feedback linearization to characterize thermoresistive sensors and in feedback measurement systems which uses this kind of sensors, is presented.

Thermoresistive sensors, in the general sense, are resistive sensors which work based on the variation of an electrical quantity (resistance, voltage or current) as a function of a thermal quantity (temperature, thermal radiation or thermal conductance). An important application of thermoresistive sensors is in temperature measurement, used in different areas, such as meteorology, medicine and motoring, with different objectives, such as temperature monitoring, indication, control and compensation (Pallas-Areny & Webster, 2001), (Doebelin, 2004), (Deep et al., 1992).

Temperature measurement is based upon the variation of electrical resistance. In this case heating by thermal radiation or self-heating by Joule effect may be null or very small so that the sensor temperature can be considered almost equal to the temperature one wants to measure (contact surface temperature, surround temperature, the temperature of a liquid, etc.).

A successful method for measurement using thermoresistive sensor uses feedback control to keep the sensor temperature constant (Sarma, 1993), (Lomas, 1986). Then, the value of the measurand is obtained from the variation of the control signal. This method inherits the advantages of feedback control systems such as low sensitivity to changes in the system parameters (Palma et al., 2003).

A phenomenological dynamic mathematical model for thermistors can be obtained through the application of energy balance principle. In such a model, the relationship between the excitation signal (generally electrical voltage or current) and the sensor temperature is nonlinear (Deep et al., 1992), (Lima et al., 1994), (Freire et al.,1994).That results in two major difficulties:

- The static and dynamic characterization of the sensor has to be done through two different experimental tests,
- Performance degradation of the feedback control system as the sensor temperature drifts away from that considered in the controller design.

As an alternative to circumvent such difficulties, the present work proposes the use of feedback linearization. This technique consists in using a first feedback loop to linearize the relationship between a new control input and the system output (Middleton & Goodwin, 1990). Then, a linear controller can be designed which delivers the desired performance over all the operation range of the system. Here, feedback linearization is used in measurement systems based on thermoresistive sensors, allowing for:

- Sensor characterization using a single experimental test,
- Suitable controller performance along the whole range of sensor temperatures.

The Section 2 presents the fundamentals of measurement using thermoresistive sensors. In Section 3, the experimental setup used in this work is described. In Section 4 it is presented the proposed feedback linearization together with the related experimental results. Finally, in Section 5 the control design and the related experimental result are discussed, which attest the effectiveness of the proposed technique.

2. Measurement Using Thermoresistive Sensors

2.1 Mathematical model

The relationship between temperature and electrical resistance of a thermoresistive sensor depends on the kind of sensor, which can be a resistance temperature detector - *RTD* or a thermistor (thermally sensitive resistor). Details of construction and description about the variation in electrical resistance can be found, for example, in (Asc, 1999), (Meijerand & Herwaarden, 1994).

For a *RTD* the variation in electrical resistance can be given by:

$$R_S = R_0(1 + \alpha_1(T_S - T_0) + \alpha_2(T_S - T_0)^2 + \ldots + \alpha_n(T_S - T_0)^n) \tag{1}$$

where R_0 is the resistance at the reference temperature T_0 and α_0 are temperature coefficients.

For a *NTC* (Negative Coefficient Temperature) thermistor the variation in electrical resistance can be given by:

$$R_S = R_0 e^{B(1/T_S + 1/T_0)} \tag{2}$$

where R_0 is the resistance at the reference temperature T_0 , in Kelvin, and B is called the characteristic temperature of the material, in Kelvin.

For some thermoresistive sensor not enclosed in thick protective well, the equation of energies balance (the relationship of incident thermal radiation, electrical energy, the dissipated and stored heat) can be given by:

$$\alpha SH + P_s(t) = G_{th}[T_s(t) - T_a(t)] + C_{th}\frac{dT_s(t)}{dt} \tag{3}$$

where:
α is the sensor transmissivity-absorptivity coefficient,
S is the sensor surface area,
H is the incident radiation,
$P_s(t)$ is the electric power,
G_{th} is the thermal conductance between sensor and ambient,
$T_a(t)$ is the ambient temperature,
C_{th} is the sensor thermal capacity,

At the static equilibrium condition ($dT_s(t)/dt = 0$) the (Eq. 3) reduces to:

$$\alpha SH + P_e = G_{th}(T_S - T_a) \tag{4}$$

In experimental implementations it is common to use voltage or electric current as the excitation signal, as it is not possible to use electric power directly. Using electric current:

$$P_s(t) = R_s(T_s(t))I_s^2(t) \tag{5}$$

and considering the ambient temperature T_a, constant, (Eq. 3) can be rewritten as:

$$\alpha SH + R_s(T_s(t))I_s^2(t) = G_{th}T_\Delta(t) + C_{th}\frac{dT_\Delta(t)}{dt} \tag{6}$$

where $T_\Delta(t)=T_s(t)-T_a$.

Temperature measurement is based upon the variation of electrical resistance (Eq. 1 or Eq.2).

The measurement of thermal radiation (H) or fluid flow velocity (fluid flow velocity related to G_{th} variation) is based on (Eq. 4). With the sensor supplied by a constant electrical current, or kept at constant resistance (and temperature), the measurement is based on the variation in the sensor voltage.

The thermoresistive sensor used in this research is a *NTC* (Negative Coefficient Temperature).

For S, G_{th} and C_{th} constant, (Eq. 3) is a first order linear differential equation with variable $T_s(t)$ and excitations H, $T_a(t)$ and $P_S(t)$. In experimental implementations it is common to use voltage or electric current as excitation signal of the sensor (it is not possible to use electric power directly), and the electrical power is done by $V_S \times I_S$, $R_S \times I_s^2$ or V^2 / R_S. In this case, (Eq. 3) becomes a nonlinear differential equation (Eq. 6) and causes difficulties for the characterization of the sensor (determination of G_{th} and C_{th}) with electrical signal, and for the measurement of thermal radiation.

2.2 Sensor Characterization

Usually the characterization of a thermoresistive sensor is done by combination of three experimental tests. In the first experimental test the parameters of the equation associating electrical resistance and temperature are determined (Eq. 1 or Eq. 2). The other experimental tests are for the determination of G_{th} e C_{th} (Eq. 3).

The parameters of (Eq.1) and (Eq.2) are given from de measurement of the electrical resistance at fixed temperatures, (0.01 °C - triple point of water and 100 °C - boiling water), or with the sensor in an equipment with adjustment and control of temperature. In this experimental test the heating by thermal radiation or self-heating by Joule effect may by null or very small. The result of the first experimental test is employed in the other two, for the determination of sensor temperature.

For the second experimental test, the sensor is immersed in a fluid medium (air, water, etc), where one wants to determine G_{th}. The temperature of this medium is kept constant and the sensor is heated by Joule effect, supplied by a constant electrical current with values I_i (i = 1 to n). In the condition of static thermal equilibrium ($dT_S/dt = 0$) the equivalents voltages, V_i, of the sensor are measured. With the values of I_i e V_i, the values of R_{Si}, T_{Si} e P_{Si} e $(T_{Si} - T_a)$ are determined. The value of G_{th} is then determined by curve fitting from the values of P_{Si} and $(T_{Si} - T_a)$. With low accuracy the value of G_{th} can be determined from only two points.

In the third experimental test the sensor electric resistance is measured during a given time interval. The values of the sensor temperature, during this time interval, are determined from the resistance values, and then the time constant is calculated $\tau = C_{th}/G_{th}$. The variation of the sensor temperature is done by a step of electrical signal or of medium temperature.

By application of a step in the electrical signal, the sensor is maintained in the medium of the previous test and heated by a constant electrical current, reaching a temperature higher then that of the medium. When the electrical current is reduced to zero at instant t_0 (negative step), the temperature of the sensor (T_S) falls exponentially from a temperature T_0 to the temperature T_a of the medium, according to:

$$T_S = T_a + (T_0 - T_a)e^{-(t-t_0)/\tau} \quad (7)$$

Through a step of medium temperature, the sensor is initially maintained in the environment temperature and, at instant t_0, it is suddenly immersed in the fluid of the second experimental test with temperature higher then that of the environment.

The sensor temperature changes exponentially from T_0 (environmental temperature) to T_a (temperature of the fluid), with:

$$T = T_a - (T_a - T_0)e^{-(t-t_0)/\tau} \tag{8}$$

The equations (Eq. 7) and (Eq. 8) are particular solutions of (Eq. 3) with $H = 0$, and the time constant, τ, may be determined by curve fitting or from the falling (rising) time. As shown in the sequel, with feedback linearization G_{th} and C_{th} can be determined from a single experimental test.

2.3 Methods of Measurement

Considering the measuring operations, the measurement methods can be classified as: substitution or deflection methods; difference methods and null methods (Klaassen, 1996), (Tse & Morse, 1989).

The substitution method uses a single transformation from the value of the measurand to the value of the measuring variable, as, e.g. the temperature measurement with clinic thermometer where temperature of a patient is transformed into the expansion of the liquid in the thermometer.

In the difference method the measuring variable is related to the difference between an unknown quantity and a know reference quantity as, e.g. in the measurement of temperature with *RTD* in a Wheatstone bridge.

In the null method, the unknown value of the measurand is compared with a know reference quantity. A well know example is the balance, where an unknown mass is put on one extremity and a known mass is put on the other extremity until the equilibrium of the balance is reached (null). Another example is resistance measurement using Wheatstone bridge. In these two examples, feedback is provided by the operator.

The method employed in feedback measurement system can be considered as a null method, where a quantity of the system is automatically set to a reference value (equilibrium value) by the variation of the measure variable. This method preserves some known proprieties of a feedback system, like the reduction of the time constant (or response time) and of the nonlinear behavior, as well as enlargement of frequency pass band.

In measurement with thermoresistive sensor excited by a constant current, a variation in the value of the measurand (thermal radiation or fluid flow velocity) causes a variation in the sensor voltage. This method is know as constant current method and may be classified as a substitution method.

Another possible method is the so-called constant temperature method. In this case, the electrical resistance of the sensor is kept constant by feedback. A variation of the value of the measurand is compensated by a variation of the sensor voltage or electrical current (and indirectly the electrical power) to keep the electrical resistance in a constant reference value.

3. Experimental Setup

The experimental setup developed for tests, characterization and control of the sensor temperature is shown in Fig. 1. The block *ambient with sensor* refers to a closed housing where the sensor is kept. Hence, the incident radiation on the sensor surface is null ($H=0$). The internal temperature is monitored by a thermometer in this housing.

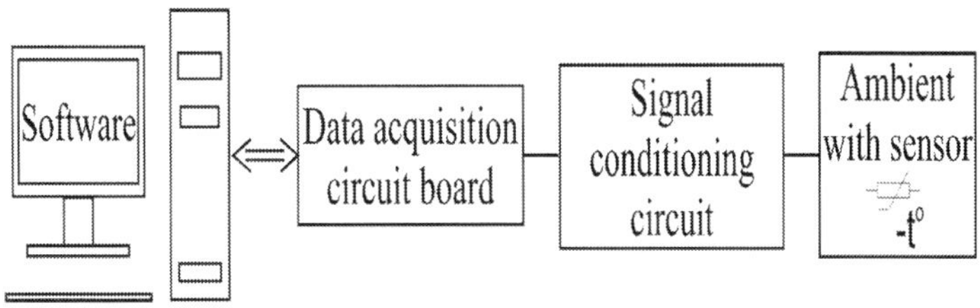

Fig. 1. Block diagram of experimental setup.

The *signal conditioning circuit* (Fig. 2) is intended to: provide the electrical decoupling between the data acquisition system and the sensor, adjust the *DAC* sensor excitation signal and adjust the sensor output signal to the *ADC* input requirements.

In this paper, the thermoresistive sensor circuit is highlighted to simplify the theoretical analysis (Fig. 3). The *data acquisition circuit board* is the *PCI6024E*, produced by National Instruments, and the man-machine interface is developed on *LABVIEW*. A virtual instrument was designed with this software, where a set of control functions allows quickly change the main program, insuring high flexibility on computer programming.

The manufacturer furnishes a table relating electrical resistance and temperature for the *NTC* used. This way, the values of the parameters ($A=R_0 \exp(-B/T_0)$) and B could be calculated from (Eq. 2): $A=1354.06e^{-5}\Omega$; $B=3342.21K$. These parameters characterize the static behaviour of the sensor. The feedback linearization developed in the present work makes possible to determine the dynamic parameter (time constant, τ) and another static parameter (DC gain, G_{th}) in a single test.

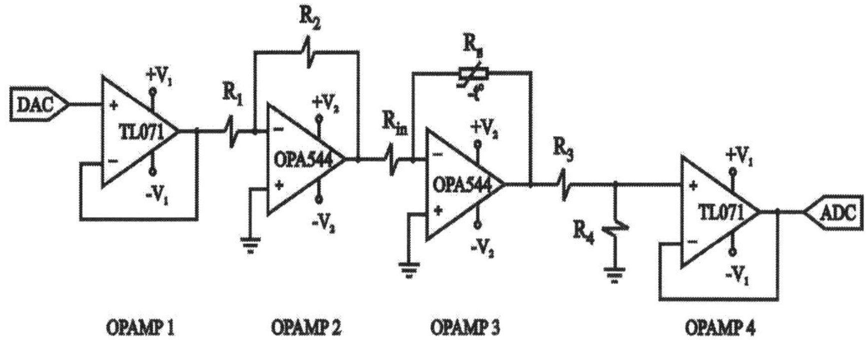

Fig. 2. Developed electronic circuit schematics

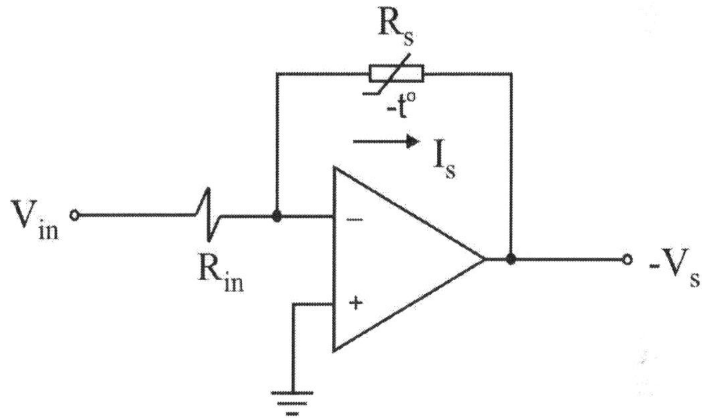

Fig. 3. Electronic sub-circuit.

4. Feedback Linearization

4.1 Theoretical Development

The proposed feedback linearization scheme is shown in Fig. 4. From this figure one can deduce:

$$I_s(t) = \frac{V_{in}(t)}{R_{in}} \tag{9}$$

$$V_{in}(t) = \frac{a(t)}{b(t)} = \frac{a(t)}{V_s(t)} \tag{10}$$

From (Eq. 9) and (Eq. 10),

$$P_s(t) = V_s(t)I_s(t) = \frac{a(t)}{R_{in}} \tag{11}$$

In this case, $a(t)$ is a virtual variable that is equivalent to the electric power dissipated by the sensor multiplied by the value of R_{in}. It is necessary to define a new input variable ($P_x(t)$) to distinguish this virtual product from the real electric power:

$$P_x(t) = \frac{a(t)}{R_{in}} \tag{12}$$

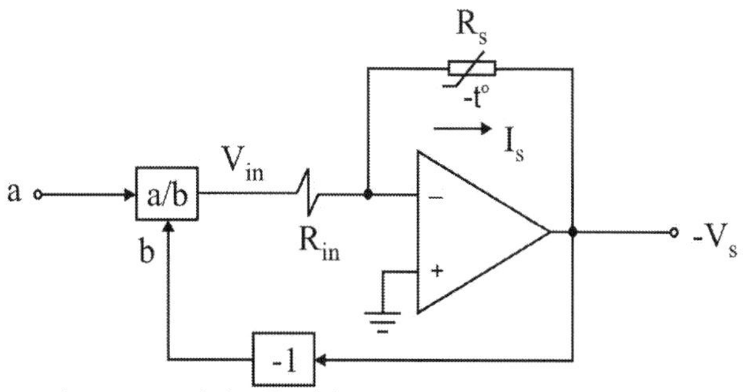

Fig. 4. Electronic sub-circuit with feedback linearization.

Considering null the incident radiation ($H=0$), assuming constant the ambient temperature, and adopting the new input variable defined in (Eq.12), (Eq.1) can be rewritten as:

$$P_x(t) = G_{th}T_\Delta(t) + C_{th}\frac{dT_\Delta(t)}{dt} \tag{13}$$

From this equation one can note that the feedback system is linear with respect to the new input variable $P_x(t)$. The transfer function in the linearized system (applying the Laplace Transform) in (Eq. 13) is a first order function:

$$\frac{T_\Delta(s)}{P_x(s)} = \frac{1/G_{th}}{\frac{C_{th}}{G_{th}}s+1} \tag{14}$$

4.2. Experimental results

The sensor response (Eq. 14) to a constant power P_{cte} is given by:

$$T_{\Delta}(t) = \frac{P_{cte}}{G_{th}}(1 - e^{-t/\tau}) \tag{15}$$

where $\tau = C_{th}/G_{th}$.

An increasing stair signal with *100s* per step (to guarantee that the sensor would be operating in steady state) was applied to both systems: without feedback linearization (non-linearized system), and with feedback linearization (linearized system). The responses are shown in Fig. 5 and Fig. 6.

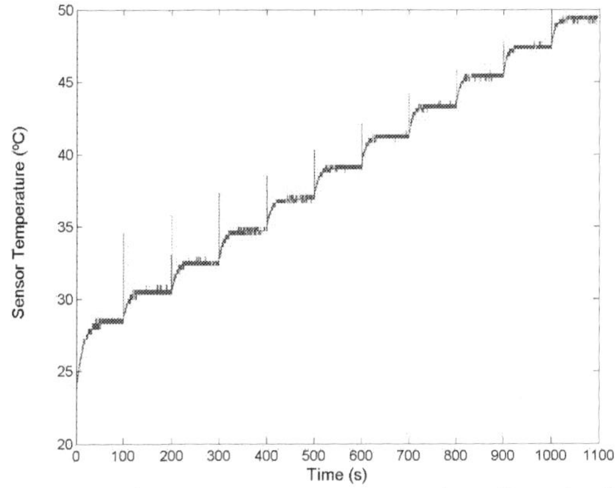

Fig. 5. Output temperature with input in increasing steps (non-linearized system).

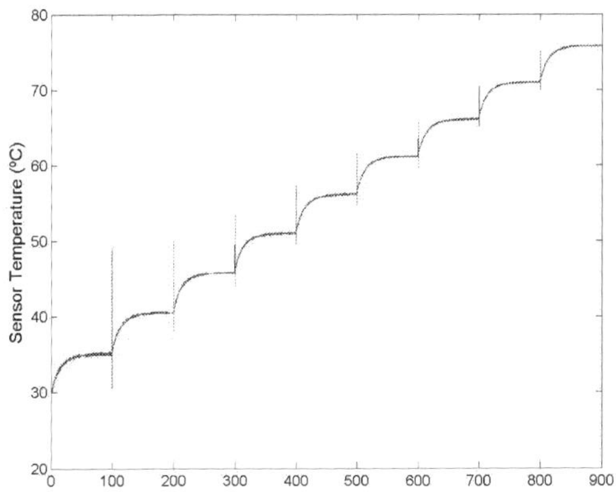

Fig. 6. Output temperature with input in increasing steps (linearized system).

The relation T_Δ/P_{cte} changes when the systems works at different operation points in the non-linearized systems. This time constant is called apparent (τ_a), and it does not correspond to the intrinsic time constant of the sensor (τ). The *DC* gain (G_{th}) is given by:

$$G_{th} = \frac{\overline{P_x}}{\overline{T_\Delta}} \tag{16}$$

where $\overline{P_x}$ and $\overline{T_\Delta}$ are the electrical power and the temperature difference in steady state, respectively.

The results for the static parameter (G_{th}) and dynamic parameter (τ) for both systems are presented in Table 1. In conclusion, for linearized systems it is possible to determine the static and dynamic behaviours with a single set of experimental data.

Step	Non-linearized		Linearized	
	G_{th} (mW/°C)	τ_a (s)	G_{th} (mW/°C)	τ (s)
1	0.80	14.3	1.56	12.0
2	0.80	12.4	1.40	11.3
3	0.82	11.1	1.30	11.3
4	0.80	10.6	1.29	11.3
5	0.81	10.1	1.26	11.6
6	0.82	9.5	1.24	11.4
7	0.83	9.2	1.22	11.5
8	0.81	9.1	1.21	11.4
9	0.81	8.1	1.19	11.7
10	0.82	8.5	-	-
11	0.82	7.8	-	-

Table 1. DC gain and time constant in non-linearized and linearized systems.

5. Control Design

5.1 Internal Model Control

The control objective is to make the temperature reach and stay in the setpoint as quick as possible. In this sense, the effects of external disturbances and variations at system parameters should be efficiently attenuated. The *IMC (Internal Model Control)* control design has the advantage of considering the internal model of the process. By assuming that the system operates in a closed-loop form, this control strategy makes it possible to mitigate the influence of variations and disturbances cited above.

The adopted *IMC* structure is shown in Fig. 7.

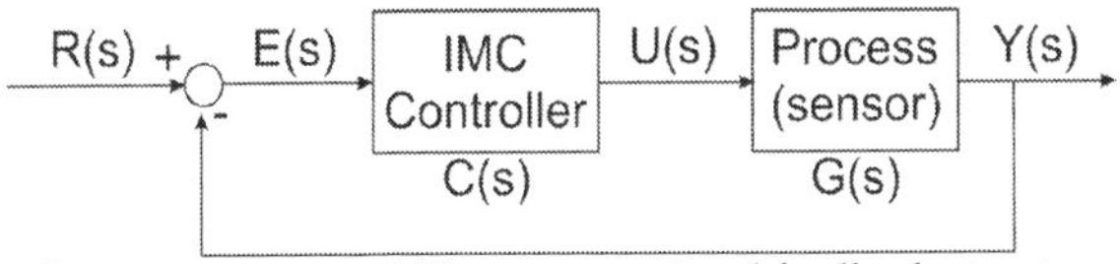

Fig. 7. *IMC* control structure arranged like a conventional feedback structure.

It turns out that for a simple first order linear system, the *IMC* controller results in a *PI* controller whose zero cancels the open-loop pole, i.e.

$$C(s) = K_P + \frac{K_I}{s} = K_P \frac{s + (K_I / K_P)}{s} \tag{17}$$

so that, by choosing $K_P = C_{th}/\tau_f$ and $K_I = G_{th}/\tau_f$ (where τ_f is the time constant of the *IMC* low pass filter), one has:

$$C(s)G(s) = \frac{C_{th}s + G_{th}}{\tau_f s} \frac{1}{C_{th}s + G_{th}} = \frac{1}{\tau_f s} \tag{18}$$

and the following closed loop transfer function:

$$\frac{Y(s)}{R(s)} = \frac{1}{\tau_f s + 1} \tag{19}$$

5.2. Experimental results

To compare the performance of linearized and non-linearized systems, two *PI* controllers have been implemented: one for each system. For the non-linearized system, a linear approximation around a specific operation point (*70°C*) has been made. The results for the following test conditions are presented:

(i) Setpoint increasing by five steps of *10°C*, varying from *50°C* up to *90°C*, step duration of *60s*, closed loop time constant=*5s* (Fig. 8);

(ii) Single setpoint by single step (*70°C*), duration of *60s*, closed loop time constant=*5s* (Fig. 9).

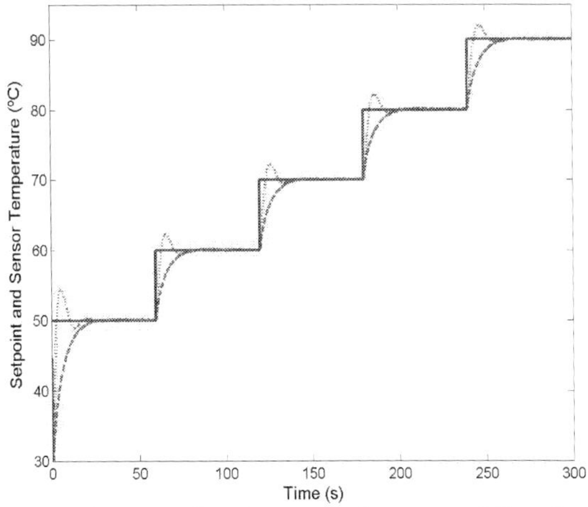

Fig. 8. Setpoint in increasing steps (continuous line), output of linearized system (dashed line) and output of non-linearized system (dotted line) with closed loop time constant=5*s*

It can be seen that both controllers could drive the temperature to the setpoint. However, in the non-linearized system overshoot occurs for all temperature steps, as well as for the designed operation point. Similar behaviour is observed in (Palma et al., 2003).

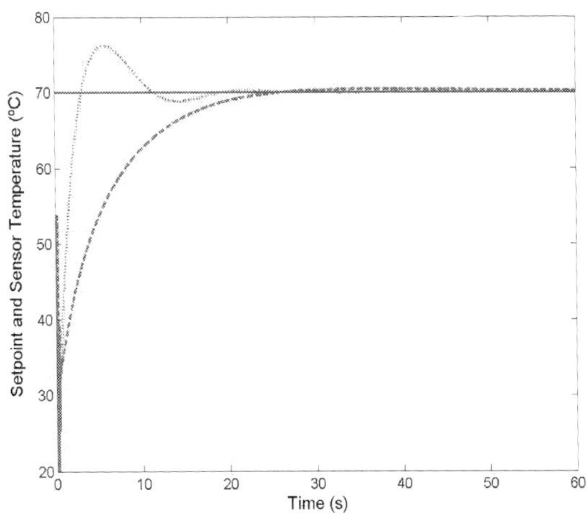

Fig. 9. Setpoint (continuous line), output of linearized system (dashed line) and output of non-linearized system (dotted line) with closed loop time constant=5*s*

Considering the linearized system, by adjusting the system response curves using the least squares method, for each temperature step (Fig. 8), it was possible to calculate the time constants for each step. These values are shown on Table 2.

Setpoint (°C)	Time constant (s)	Variation of time constant projected (%)
50	4.48	-10.4
60	5.41	+8.2
70	5.78	+15.6
80	5.92	+18.4
90	6.00	+20.0

Table 2. Calculated time constant for linearized system with setpoint in increasing steps and designed closed loop time constant for 5s.

Other different tests of the controller using several closed-loop time constants have been done. The system behaves similarly as for the *5s* time constant. The results for the closed time loop constant designed for *0.5s* are shown in Fig. 10, Fig. 11 and Table 3. The tests conditions are:

(i) Setpoint increasing by five steps of *10°C*, varying from *50°C* up to *90°C*, step duration of *10s* (Fig. 10);
(ii) Single setpoint by single step (*50°C*), duration of *10s*, (Fig. 11).

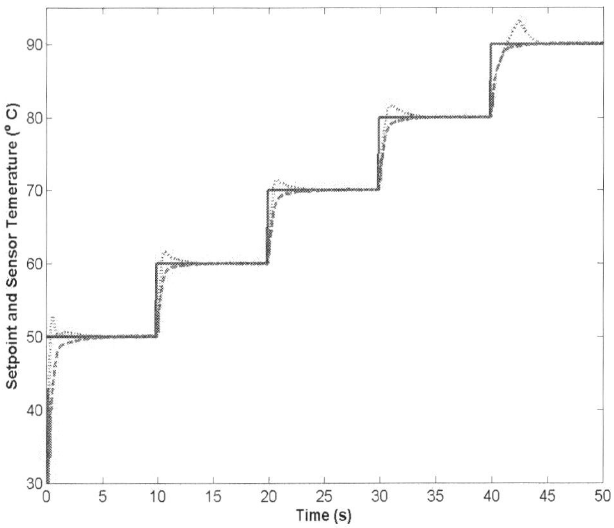

Fig. 10. Setpoint in increasing steps (continuous line), output of linearized system (dashed line) and output of non-linearized system (dotted line) with closed loop time constant=*0.5s*

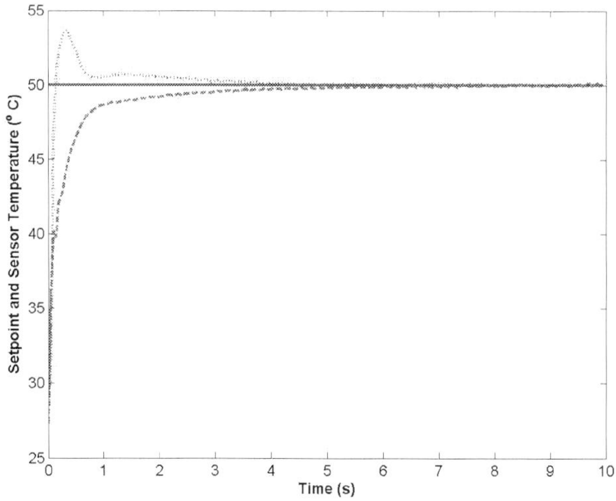

Fig. 11. Setpoint (continuous line), output of linearized system (dashed line) and output of non-linearized system (dotted line) with closed loop time constant equal to *0.5s*

Setpoint (ºC)	Time constant (s)	Variation of time constant projected (%)
50	0.44	-12
60	0.45	-10
70	0.47	-6
80	0.48	-4
90	0.55	+10

Table 3. Calculated time constant for linearized system with setpoint in increasing steps and designed closed loop time constant for *0.5 s*.

Variations between the designed and the calculated time constants have been observed. Some of the reasons for this difference could be:

(i) The expected pole-zero cancellation from the project of *IMC* controller is not precise;
(ii) The determination of the time constant depends on a curve fitting.

6. Conclusions

In this work, the direct use of electrical power as excitation parameter is proposed. This is due to the feedback linearization applied to the system, which also contributes for the characterization of sensor parameters in a single set of experimental test.

The *PI* controller designed by *IMC* technique was able to meet the imposed performance requirements. The results for the linearized system showed to be better when compared to the nonlinearized system, as the model used in the design is valid for all the operation range.

As perspective for future works it could be mentioned:

(i) The use of feedback linearization to characterize other kinds of thermoresistive sensors;
(ii) The application of another control strategy in combination with feedback linearization;
(iii) The use of the measurement system developed to measure physical variables.

7. References

Asch, G. Les capteurs en instrumentation industrielle, 5th ed., Dunod, 1999, chap. 6 - Capteurs de température.

Deep, G.S., R. C. S. Freire, P. C. Lobo, J. S. da Rocha Neto, A. M. N. Lima, "Dynamic response of thermoresistive sensors", *IEEE Transactions on Instrumentation and Measurement*, n° 6, pp. 815-819, 1992.

Doebelin, E.O. *Measurements systems: application and design*, McGraw Hill, 2004.

Fransis S. Tse & Ivan E. Morse, Measurement and Instrumentation in Engineering, Marcel Dekker, Inc.1989.

Freire, R.C.S., J. S. da Rocha Neto, G. S. Deep, A. M. N. Lima, P. C. Lobo, "Experimental determination of the time constant of a thermoresistive sensor", *IEEE Instrumentation and Measurement Technology Conference (IMTC'94)*, pp. 450-453, 1994.

Klaas B. Klaassen, Electronic Measurement and Instrumentation, Cambridge University Press, 1996.

Lima, A.M.N., G. S. Deep, J. S. da Rocha Neto, R. C. S. Freire, P. C. Lobo, "Identification of thermoresistive solar radiation sensors", *IEEE Transactions on Instrumentation and Measurement*, n° 2, pp. 133-138, 1994.

Lomas, C.G. *Fundamentals of hot wire anemometry*, Cambridge University Press, 1986.

Meijerand A. W. Herwaarden, Thermal Sensors, IOP Publishing Ltda, 1994, cap. 4 - Temperature-sensing Elements.

Middleton, R.H. & G. C. Goodwin, *Digital control and estimations - an unified approach*, Prentice Hall, 1990.

Pallas-Areny, R & J. G. Webster, *Sensors and signal conditioning*, John Wiley and Sons Ltd, 2001.

Palma, L.S., A. Oliveira, A. S. da Costa, A. Q. Andrade Jr., C. V. R. de Almeida, M. Zurita and R. C. S. Freire, "Implementation of a feedback I2-controlled constant temperature environment temperature meter", *Sensors*, n° 10, pp. 498-503, 2003.

Sarma, G.R. "Analysis of a constant voltage anemometer circuit", *IEEE Instrumentation and Measurement Technology Conference (IMTC'93)*, pp. 731-736, 1993.

11

Algal Biosensor-Based Measurement System for Rapid Toxicity Detection

Thi Phuong Thuy Pham, Chul-Woong Cho and Yeoung-Sang Yun
Chonbuk National University
Republic of Korea

1. Introduction

Microalgae are now widely used as relevant biological indicators in the field of environmental impact studies. Owing to their ubiquity, short life cycles, easiness of culture and high sensitivity to a number of pollutants, these organisms are frequently utilized in ecotoxicological screening of contaminated freshwater (Lewis, 1995). As primary producers, either directly or indirectly, of organic matter required by small consumers in aquatic food webs, microalgae serve an important role in nutrient recycling and equilibrium of aquatic ecosystems (Raja et al., 2008). The most important common biochemical attribute that unites algae is their ability to split water, producing molecular oxygen during photosynthesis and concomitantly assimilating carbon dioxide. Furthermore, the rest of biotic communities are strictly dependent upon the photosynthetic activity of these organisms. Perturbations of microalgal photosynthesis might lead to alterations of their primary production, which in turn causes severe repercussions on the aquatic biota (Morris, 1981).
Nowadays, the development of convenient methods or parameters for assessment of the presence of pollutants and their toxicity has become a major goal in environmental monitoring research. Growth rate, fluorescence induction and photosynthetic activity (through oxygen evolution or incorporation of ^{14}C) are the most popular endpoints studied (Jensen, 1984; Puiseux-Dao, 1989). Particularly, photosynthesis inhibition is a reliable indicator that rapidly demonstrates the toxic effect of hazardous contaminants (Overnell, 1976). Table 1 displays a brief review on the literature concerning research on application of microalgal photosynthesis to detect the effects of pollutants. The advantage of photosynthesis inhibition assay is the short duration of the test, usually 2-4 h compared to 48-96 h of chronic exposure (Hall et al., 1996). Bioassays involving photosynthesis process; however, are contingent upon light intensity and initial algal cell concentration. In this regard, light is the most critical element influencing phytoplanktonic photosynthetic activity (Aiba, 1982). In most cases, microalgae-based biosystems are restricted by light, which is easily absorbed and scattered by the microalgal cells (Yun & Park, 2001). It is therefore crucial to figure out and monitor the light dependence of microalgal activity for the sake of designing an efficient algal biosensor-based measurement system for toxicity assessment. Other than irradiance, algal cell concentration and initial dissolved oxygen level applied in the test are also important and should be evaluated to give a better performance of the proposed algal biosensor.

Organism	Test substances	Modulation	Ref.
Chlorella vulgaris	Potassium cyanide, methyl parathion, diuron (DCMU), paraquat	The detection of toxic agents in sunlight-exposed primary-source drinking waters based on fluorescence induction	Rodrigue z et al. (2002)
Chlorella vulgaris	DCMU, simazine, atrazine, alachlor, glyphosate	An algal biosensor using a fluorescence-based optical fiber for determination of herbicides	Naessens et al. (2000)
Chlorella vulgaris	Cadmium, lead	The detection of heavy metals using an optical algal biosensor based on alkaline phosphatase activity	Durrieu & Tran-Minh (2002)
Chlorella vulgaris	DCMU, atrazine, toluene, benzene	Development of a compact and disposable device for rapid toxicity testing on the basis of amperometric monitoring of O_2 generated photosynthetically by microalga *C. vulgaris* entrapped in an alginate gel or a polyion complex and immobilized directly on the surface of a transparent indium tin oxide electrode	Shitanda et al. (2005)
Selenastrum capricornutum	Zinc, lead, cadmium, benzene, phenol, chlorophenol, toluene, chlorobenzen,	Detection of the toxic effects of both organic and metallic toxicants using a closed-system algal toxicity test based on measuring dissolved oxygen production, algal growth rate and cell density	Lin et al. (2005)
Selenastrum capricornutum	Phenol. chlorophenol	Evaluation of the toxicity of chlorophenols using a closed-system algal toxicity test based on measuring dissolved oxygen production and growth rate	Chen & Lin (2006)
Selenastrum capricornutum	Zinc, cadmium	Development of a short-term (4 h) test based on uptake and assimilation of radio-labelled carbon in *S. capricornutum*	Pardos et al. (1998)
Selenastrum capricornutum	Copper	Determination of the response of phytoplankton to Cu in natural, soft lake waters to different dissolved organic carbon content, and to test the modifying effect of ultraviolet radiation on response to Cu using the kinetics of in vivo chlorophyll *a* fluorescence	West et al. (2003)

Chlamydomonas reinhardtii	Copper, nickel, zinc, lead	Determination if photosynthesis based on estimation of oxygen evolution and motility can be used as sensitive physiological parameters in toxicological studies of green unicellular alga	Danilov & Ekelund (2001)
Chlamydomonas reinhardtii	DCMU	The report on a biosensor system in which pH changes caused by the uptake or production of CO_2 by *Chlamydomonas* cells	Schnubnell et al. (1999)
Scenedesmus obliquus	Atrazine, metribuzin, diuron, paraquat, terbuthylazine	Measurement of the F684/F735 fluorescence ratio with a conventional fluorometer as an easy, rapid and sensitive assessment of the presence and toxicity of herbicides in a freshwater alga	Eullaffroy & Vernet (2003)
Plesiastrea versipora	Copper	Examination of the effects of copper on physiological interactions between symbiotic algae and their host, particularly with regard to two host signalling compounds that control algal carbon metabolism	Grant et al. (2003)
Spirogyra distenta	DCMU	Measurement of chlorophyll fluorescence from young and mature chloroplasts with a microscopic imaging system for monitoring the environmental degradation of aquatic ecosystems	Endo & Omasa (2004)
Spirulina subsalsa	Copper, mercury, atrazine, carbaryl	The monitoring of the evolution of photosynthetic O_2 and the detection of alterations due to toxic effects caused by environmental pollutants	Campanella et al. (2000)
Phytoplankton, bacteria	Cadmium	Evaluation of the effect of dredged sediment disposal using photosynthesis measured by the ^{14}C method and phosphate uptake by $^{32}P\text{-}PO_4$ and heterotrophy by uptake of ^{14}C labelled glucose	Nalewajko (1995)
Phytoplankton, periphyton, epipsammon	Paraquat, simazine	Comparison of sensitivities of three marine microalgal communities to two herbicides using photosynthesis (incorporation of ^{14}C) as a test parameter	Bonilla et al. (1998)

Table 1. Previous experimental approaches using algal photosynthetic activity to detect the effects of toxicants

2. Objective of the study

The inhibition of microalgal photosynthetic activity induced by different contaminants has been extensively investigated in the literature. Chen & Lin (2006) reported the investigation on hazardous impact of volatile organic compounds (VOCs) using an air-tight algal toxicity assay. Nonetheless, there has been no discussion on the effect of irradiance, although the tested BOD bottles could be influenced by shading effect at high algal densities. Also, the duration of the bioassay was not very short (i.e., 48 h) in their study. With regard to algal biosensor, a device for the monitoring of water toxicity in estuarine environments was reported in the work of Campanella et al. (2000). The developed biosensor provides a new approach to the research on harmful effects of heavy metals, herbicides and insecticides; however, no information on toxicity of volatile organic solvents was addressed using this system. Additionally, a biosensor with an oxygen electrode containing *Chlorella* cells immobilized on the membrane was well established to detect VOCs in the form of aerosols (Naessens & Tran-Minh, 1999). Nevertheless, one major drawback is that a controlled atmosphere chamber is required for the operation of this biosensor. Podola et al. (2004) described a non-selective sensor chips for the detection and identification of VOCs using different algal strains. Perhaps the disadvantage of the proposed multiple-strain biochip system is related to the complicate and expensive equipment, which might be regarded as limitations for practical utilization. Consequently, there is a need to design a simple and cost-effective indicator system that supports rapid toxicity detection of volatile and/or hazardous substances.

The aim of the present study is to design, construct and validate a new algal biosensor-based measurement system that provides a rapid toxicity determination of pollutants. The apparatus allows the monitoring of photosynthetic efficiency of the green alga *Selenastrum capricornutum* cells in the absence and presence of toxic agents by recording the oxygen produced. The new point of the work is that the biosensor was air-tight, with no headspace, thus prevents volatile organic toxicants from escaping into the environment as well as partitioning from the aqueous phase into the headspace until equilibrium was reached. In this aspect, the designed measurement system supports toxicity screening of volatile organic substances.

In this chapter, six common organic solvents including methanol, ethanol, isopropanol, acetone, acetonitrile, dimethylformamide and one ionic liquid (i.e., 1-butyl-3-methylimidazolium tetrafluoroborate, [BMIM] [BF_4]), a representative of non-volatile pollutants, were selected to check the system performance. The response of the proposed algal biosystem was studied in terms of light intensity, cell density and initial dissolved oxygen level. It was concluded that only 2 h was required to predict EC_{50} values (concentrations which result in a 50% reduction of the exposed organisms relative to controls) as compared to 96 h in a conventional algal assay based on algal growth rate.

3. Experimental Methods and Procedures

3.1 Microalgal strain and cultivation

The freshwater green alga *Selenastrum capricornutum* ATCC-22662 was used as the test organism and was obtained from the National Institute Environment Research (Incheon, Korea). Cells of *S. capricornutum* routinely have been propagated in a 250 mL Erlenmeyer flask containing 200 mL of Bold's Basal medium (Bold, 1950), which was nitrate-enriched by adding 58.8 mM $NaNO_3$ to avoid nitrogen limitation in a high-density culture (Yun & Park,

1997). Culture flask was shaken continuously at 170 rpm on a rotary shaking apparatus with air bubbling (1 vvm) without a sparger. Continuous illumination was provided at an average of 30 ± 5 $\mu Em^{-2}s^{-1}$ by warm-white fluorescent tubes (Korea General Electric, Yongin, Korea). The alga was subcultured every week with fresh medium (200 mL) and 10 mL of the cultured alga in order to keep algal cells in linear growth with doubling time of approximately 1 day at a controlled temperature of 25 ± 2°C.

3.2 Test reagents

The chemicals employed in the present study included an ionic liquid 1-butyl-3-methylimidazolium tetrafluoroborate ([BMIM] [BF_4]) and six common organic solvents (e.g. methanol, ethanol, isopropanol, acetone, acetonitrile and dimethylformamide). The ionic liquid was obtained at 98% of purity from C-Tri Company (Korea) whereas organic solvents (with purity > 99.5% for all compounds) were purchased from Samchun Pure Chemical Company (Korea).

3.3 Design of the algal biosensor-based measurement system and operating procedure

The system was constructed with a reaction cell, which was a double-jacket cylinder made of Pyrex® glass, an illuminator (A3200, Donan-Jenner, Boxborough, MA, USA), a quantum sensor (LI-190A, Licor, Lincoln, NE, USA), a light meter (LI-250, Licor), a dissolved oxygen meter (Hach, Loveland, CO, USA) and a computer for data acquisition using Hach software (Fig. 1). During experiments, microalgal suspension along with toxicants was injected into the reaction vessel with working volume of 3.58 mL and light path length of 1.8 cm. This mixture was made homogeneous by magnetic stirring with a small bar (0.5 cm in length). Opposite to the reaction vessel, a light beam was provided from a duck neck-like optical fiber connection to facilitate the algal photosynthesis. A convex lens was located on the head of optical fiber connection and was oriented to make the light beam parallel to the axial direction without dispersion. It was confirmed that the oxygen probe, inlet/outlet gates and stirring bar had minor effects on light penetration. A 150-W quartz halogen lamp (EKE, Tokyo, Japan) as a light source was equipped inside the optical fiber illuminator. The irradiance was controlled *via* the scale of illuminator aperture. The light absorption by Pyrex® glass, thermostating water and distilled water was negligible compared to absorption by microalgal cells. The quantum sensor connected to a light meter was positioned opposite to the illumination side in order to measure the transmitted light. Since algal photosynthesis is known to be temperature sensitive, cooling water of 25 ± 2°C from a water bath was circulated continuously through the double-jacket of the reaction cell. The oxygen probe was placed in the circular top of the reaction cell and used for measuring the concentration of dissolved oxygen generated by the algal photosynthesis.

Prior to the test, cell suspension was prepared by centrifuging algal cells in the late exponential phase at 3,000 × *g* for 5 min at room temperature and resuspending them in the fresh medium to yield different cell densities (0.048, 0.095 and 0.182 g cell/L). The estimation of cell densities based on algal dry cell weight was done by passing 5 mL of each suspension through a pre-dried and pre-weighted 0.45 μm cellulose nitrate membrane filter (Whatman, Ann Arbor, MI, USA), then drying in an oven at 70°C for 24 h. A correlation between algal dry cell weight versus optical density, DCW (g cell/ mL) = 0.139 × OD_{438}, was established to facilitate the measurement of cell density. Mixture of the earlier prepared algal broth and toxicant then was loaded to the reaction vessel after being exposed for 10 min and passed through a gas mixture at a rate of 75 ± 10 mL/min for 10 min to control the initial dissolved oxygen level. The gas combination used in this experiment included 99% N_2 and 1% CO_2 as an extra carbon source for algal growth. The photosynthetic oxygen released by algal cells was recorded every minute throughout a 10-min illumination period by the personal computer directly linked to the system. It took almost 2 h to conduct the entire experiment in order to obtain complete dose-response curves. A similar procedure was applied for controls in which deionized water rather than toxicants was used. In each experiment, the volumetric oxygen evolution rate was obtained from the slope of linearity between dissolved oxygen and time. The specific oxygen evolution rate was achieved by dividing the volumetric oxygen evolution rate by the cell concentration (Jeon et al., 2005).

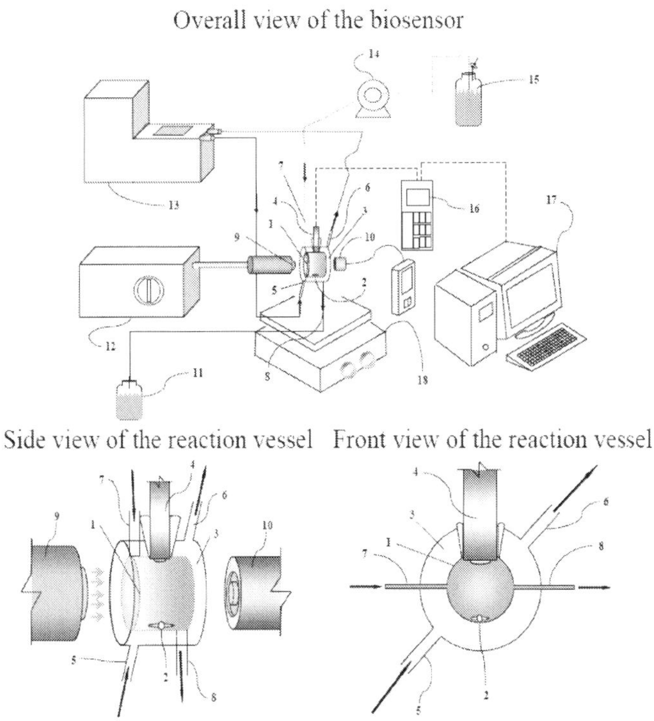

Fig. 1. Schematic diagram of the photosynthetic activity measurement system. *1* reaction cell, *2* magnetic bar, *3* cooling water jacket, *4* dissolved oxygen electrode, *5* inlet of cooling water, *6* outlet of cooling water, *7* inlet of sample, *8* outlet of sample, *9* convex lens, *10* quantum sensor, *11* wastewater, *12* quartz halogen illuminator, *13* water bath, *14* peristaltic pump, *15* sample reservoir, *16* dissolved oxygen meter, *17* computer and *18* magnetic stirrer.

3.4 Photosynthesis-irradiance model and parameter estimation

For estimation of algal photosynthetic activity, a general photosynthesis-irradiance model (Yun & Park, 2003) can be applied

$$A_X = \frac{A_m I_O}{K + I_O} - R_X \tag{1}$$

where A_X stands for the specific photosynthetic activity (g O_2/g cell min), A_m, K and R_X denote maximum specific activity (g O_2/g cell min), half constant (μ$Em^{-2}s^{-1}$) and specific respiration rate (g O_2/g cell min), respectively and I_O corresponds to the incident light intensity (μ$Em^{-2}s^{-1}$). The respiration rate (R_X) was obtained by measuring the specific oxygen consumption rate of algal broth in the dark. The maximum photosynthetic activity (A_m) and the half constant (K) were calculated based on the nonlinear regression with Marquardt-Levenberg algorithm (Marquardt, 1963).

3.5 Cell growth effect test

The conventional algal chronic toxicity assay was done according to the procedures set out in the U.S. Environmental Protection Agency (1996) and Organization for Economic Cooperation and Development (2002) guidelines. In this experiment, the algal cells were exposed to different concentrations of toxicants for 96 h and growth of cultures relative to optical density of algal suspension was determined at wavelength of 438 nm *via* a spectrophotometer (UV mini-1240, Shimadzu, Kyoto, Japan). The growth rate inhibition (I) was calculated from the below equation.

$$I(\%) = \frac{A_c - A_t}{A_c} \times 100 \tag{2}$$

where A_c and A_t indicate the mean value of area under the curve of the control and treatment groups, respectively.

3.6 Effect data modeling

The dose-response curves, where feasible, were fitted to the multinomial data with the nonlinear least-squares method adopting for the logistic model to determine the relationship of cell viability and inhibition to the decadic logarithm of the examined dosages, which can be written as:

$$P = \frac{1}{1 + (x/x_0)^b} \tag{3}$$

where x is the substrate concentration to which the cells are exposed, P represents the physiological response, normalized with negative controls to the interval from 1 (x = 0) to 0 (negative control), x_0 indicates the EC_{50} value whereas b stands for the slope of the function on a logit-log-scale. All calculations were performed using Sigmaplot® 10.0 (SPSS, Chicago, IL, USA). In particular cases, algal growth rate increased at low concentrations of toxicants instead of the expected decrease in response that was observed at higher doses. Therefore, the concentration-response curves were fitted with the linear-logistic model proposed by

Brain & Cousens (1989) and modified by van Ewijk & Hoekstra (1993) for the case of a subtoxic stimulus.

$$P = \frac{1 + fx}{1 + (2 \times fx_0 + 1)(x / x_0)^{b'}} \tag{4}$$

where b' is a parameter without intuitive interpretation and f is the parameter showing hormesis. If $f > 0$, then the curve exhibits an increase for low doses.

4. Results and Discussion

4.1 Effect of light intensity

It is well-known that physiological response to changes in light intensity is an important factor determining alteration in photosynthetic activity of microalgae in nature. In general, the photosynthetic performance of phytoplankton is enhanced as the light increase up to the point where photosynthetic apparatus comes to be saturated at higher photon flux densities. In the present study, various light incidents were adjusted to estimate the influence of light intensity on microalgal photosynthetic process in the presence and absence of a representative pollutant ([BMIM] [BF_4]).

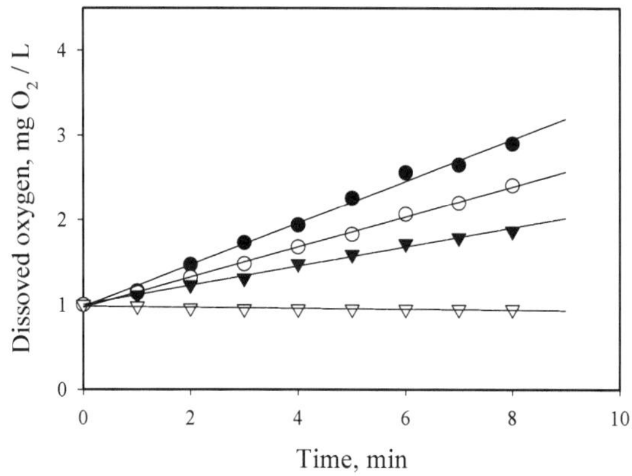

Fig. 2. Oxygen production by alga in different light intensities in the presence of toxicant. Volumetric oxygen evolution rates were evaluated using data in a linear range. The alga concentration was 0.095 g cell/L and the concentration of stimulated toxicant ([BMIM] [BF_4]) was 22.94 mg/L in all cases. The intensities of stimulated daylight were (▽) 0 μE m^{-2} s^{-1}, (▼) 100 μE m^{-2} s^{-1}, (○) 500 μE $m^{-2}s^{-1}$ and (•) 1,200 μE $m^{-2}s^{-1}$. Estimated volumetric activities were (▽) -0.0043 ± 0.0014, (▼) 0.1152 ± 0.0027, (○) 0.1778 ± 0.0027 and (•) 0.2498 ± 0.0040 mg O_2/L min.

As can be obviously observed in Figs. 2 and 3, the stronger the light intensity, the more oxygen will be produced. However, when no illumination was provided, dissolved oxygen concentration comparatively decreased as a result of algal respiratory process. In addition, the volumetric oxygen evolution rates were found to be lower in the presence of pollutant compared to the results of test medium without pollutant. This can be explained by the toxic effects of pollutant towards microalgal respiratory function. Also, the algal photosynthetic response was significantly different when exposed to toxicant at different light intensities with 55, 6.5, 31.7 and 25% of oxygen was generated at illumination power of 0, 100, 500 and 1,200 $\mu Em^{-2}s^{-1}$, respectively. This variability in the toxicity of the tested compound implies that the results of an algal photosynthesis inhibition assay can differ considerably under different light conditions. This of course makes the comparability more complicated and should be avoided by controlling stringent rationales for a light regime during the test. For this purpose, the optimum light intensity for phytoplankton photosynthetic efficiency was obtained by plotting the specific oxygen evolution rate against the light intensities. Figure 4 demonstrates that the generated oxygen initially increased with light intensity and attained a plateau at higher photon flux densities. As light intensities ranged between 1,000 and 1,200 $\mu Em^{-2}s^{-1}$, specific oxygen evolution rates were noticed to be almost constant. Therefore, the light intensity between 1,000 and 1,200 $\mu Em^{-2}s^{-1}$ was considered to be appropriate for examining the photosynthetic performance of *S. capricornutum* in the present apparatus.

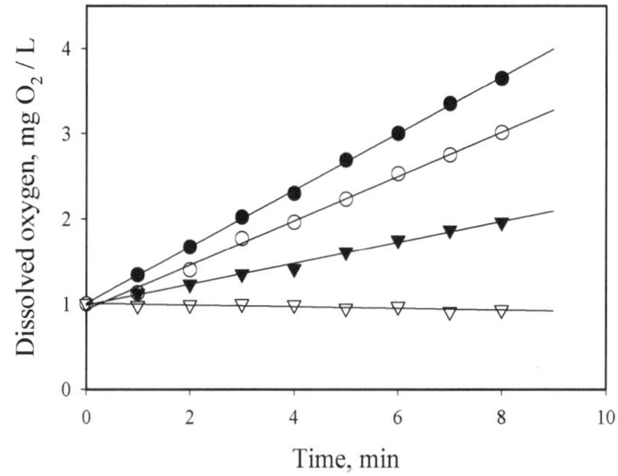

Fig. 3. Oxygen production by alga in different light intensities in the absence of toxicant. Volumetric oxygen evolution rates were evaluated using data in a linear range. The algal concentration was 0.0095 g cell/L in all cases. The intensities of stimulated daylight were (▽) 0 $\mu E\ m^{-2}\ s^{-1}$, (▼) 100 $\mu E\ m^{-2}\ s^{-1}$, (○) 500 $\mu E\ m^{-2}\ s^{-1}$ and (•) 1,200 $\mu E\ m^{-2}\ s^{-1}$. Estimated volumetric activities were (▽) -0.0097 ± 0.0030, (▼) 0.1232 ± 0.0039, (○) 0.2603 ± 0.0075 and (•) 0.3327 ± 0.0025 mg O_2/L min.

4.2 Effect of cell concentration

Regarding the effect of cell concentration, the experiment was performed with various concentrations of algal broth under light intensity set at 1,000 $\mu Em^{-2}s^{-1}$. The rates of photosynthetic oxygen evolution were observed to be very well correlated with the algal cell densities. It is apparent in Fig. 5 that the oxygen production was inhibited in the presence of [BMIM] [BF_4]. Also, the levels of inhibitory effects were different at different algal cell densities with 46, 34 and 34% of photosynthetic activity were hampered by this compound at cell concentrations of 0.048, 0.095 and 0.182 g cell/L, respectively. These data inferred that at low algal cell densities, the inhibitory percentages were rather higher. A possible explanation for this might be the relationship between toxicity and photoinhibition (Göksan et al., 2003), which is more likely to occur at low concentration due to the influence of mutual shading at high algal concentration (Contreras-Flores et al., 2003; Evers, 1991; Richmond, 2000). Though concentrations exceeding 0.182 g cell/L were not evaluated in the present study, it is supposed that mutual shading might be involved in high-density algal culture (Grobbelaar & Soeder, 1985). Taken together, 0.095 g cell/L was selected and employed for the subsequent experiments.

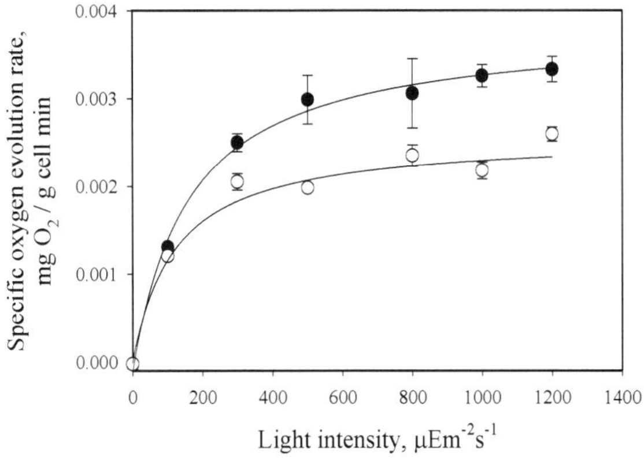

Fig. 4. Specific oxygen evolution rate as a function of incident photon flux density in the presence (○) and absence (●) of toxicant. Data points and error bars were average values and standard deviation of two or three replicated experimental results. Solid lines represent the calculated results from the photosynthesis-irradiance model (Eq. 1). The alga concentration was 0.095 g cell/L in all cases and the concentration of toxicant ([BMIM] [BF_4]) applied was 22.94 mg/L.

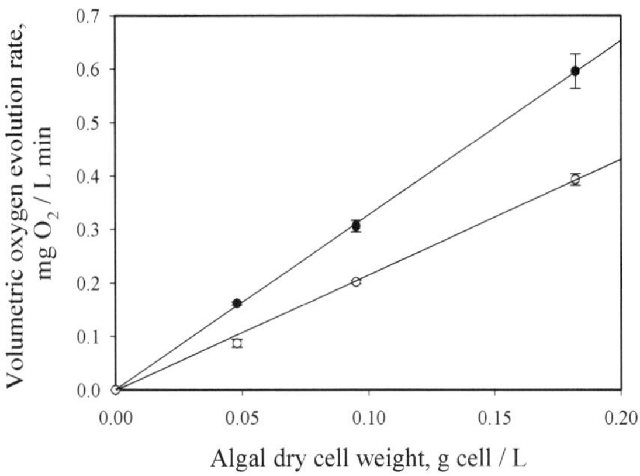

Fig. 5. Oxygen production by alga in different cell concentrations in the presence (○) and absence (●) of toxicant ([BMIM] [BF_4]).

4.3 Effect of initial dissolved oxygen concentration

Figure 6 depicts the influence of initial dissolved oxygen concentration on algal photosynthesis process with initial DO levels varied between 0.78 and 6.68 mg O_2/L. These concentrations (excluding the highest concentration of 6.68 mg O_2/L) were selected randomly by stripping with controlled amount of gas mixture containing N_2 and CO_2. Through preliminary studies, other conditions including light intensity, algal cell density and concentration of pollutant were fixed at 1,000 $\mu Em^{-2}s^{-1}$, 0.095 g cell/L and 22.94 mg/L, respectively. The data revealed that volumetric oxygen evolution rates were 0.2045 ± 0.0063, 0.1987 ± 0.0058, 0.2027 ± 0.0177 and 0.1315 ± 0.0169 mg O_2/L min corresponding to initial DO levels of 0.78, 3.37, 5.26 and 6.68 mg O_2/L. It should be pointed out that there was no effect of initial dissolve oxygen concentration towards algal photosynthetic response apart from the case of the highest DO value, in which CO_2 gas was not utilized. It can therefore be assumed that CO_2 plays an important role for microalgal photosynthesis process in the studied system.

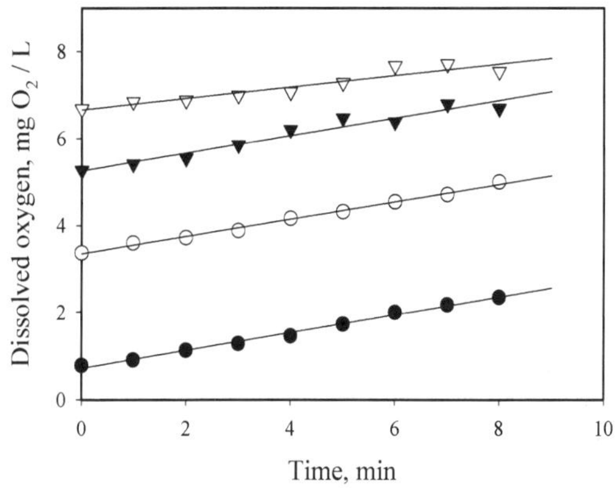

Fig. 6. Oxygen production by alga in different initial dissolved oxygen concentrations in the presence of toxicant. Volumetric oxygen evolution rates were evaluated using data in a linear range. The initial dissolved oxygen concentrations were (•) 0.78 mg O_2/L, (○) 3.37 mg O_2/L, (▼) 5.26 mg O_2/L, (▽) 6.68 mg O_2/L. The intensity of stimulated daylight was 1,000 $\mu Em^{-2}s^{-1}$ and the concentration of toxicant ([BMIM] [BF_4]) applied was 22.94 mg/L.

4.4 Toxicity testing

As a development of the research work carried out by our group on this topic, here we present the results of a short-term algal photosynthesis inhibition tests performed on a representative of imidazolium-based ionic liquids and commonly used organic solvents. For checking the validity of the present system, a conventional algal growth assay was conducted in cases of [BMIM] [BF_4] and methanol. According to the data obtained, the effective concentrations of [BMIM] [BF_4] were identical in both cases of short-term and traditional assays. From Fig. 7, the EC_{50} values of [BMIM] [BF_4] were determined to be 0.115 mM and 0.126 mM for inhibition of algal photosynthesis process and growth rate, respectively. For methanol, the corresponding results were 2,089 and 759 mM suggesting the hazardous impact of this compound was 2.75 times higher towards algal growth than photosynthetic activity. It seems possible that these results are owing to the longer exposure in growth rate assay compared to short-term test (96 h and 20 min, respectively), thus led to more critical injury to algal cells. Concerning the test data of the other commonly used organic solvents, it was found that all of these pollutants effectively inhibited algal photosynthesis with EC_{50} values varied between 589 and 2,089 mM (Table 2). Consequently, the toxicities of the tested organic compounds decreased in the order of isopropanol > acetone > acetonitrile > ethanol > dimethylformamide ≈ methanol.

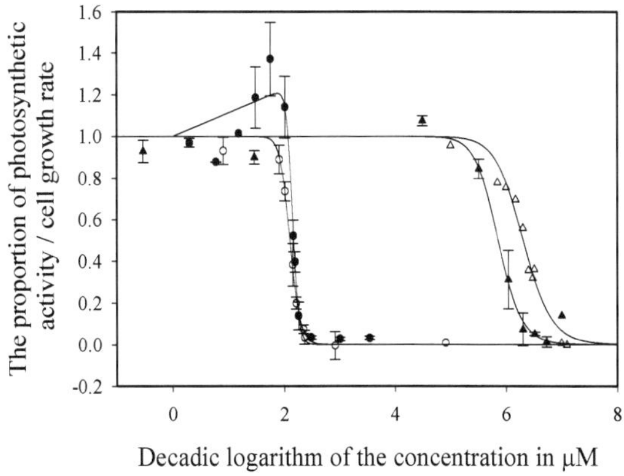

Fig. 7. Dose-response curves of algal toxicity test with respect to [BMIM] [BF_4] (○) and methanol (△) based on photosynthetic activity measurement whereas [BMIM] [BF_4] (●) and methanol (▲) based on growth rate.

Chemicals	$Log_{10}EC_{50}/\mu M^a$	EC_{50}/mM	95% confidence interval/mM
Methanol	6.32 ± 0.11	2089	1623 - 2692
Ethanol	5.94 ± 0.24	871	501 - 1514
Dimethylformamide	6.32 ± 0.25	2089	1175 - 3715
Acetone	5.85 ± 0.03	708	661 - 759
Isopropanol	5.77 ± 0.15	589	417 - 832
Acetonitrile	5.92 ± 0.11	832	646 - 1072

Table 2. Inhibition of photosynthetic activity induced by various pollutants

5. Conclusions

In this chapter, an algal biosensor-based measurement system for rapid toxicity detection was designed, constructed and validated for the use in detecting several toxic chemicals. The system provides a possibility to toxicological investigation by monitoring the photosynthetic activity of algal cells through the oxygen produced. In contrast to most biosensors previously reported, the present sensor offers some advantages such as fast response time, simple and low-cost instrument, and totally air-tight to prevent underestimation of VOCs due to their volatility. The validity of the system was verified in terms of light intensity, algal cell concentration and initial dissolved oxygen dosage. It was observed that illumination condition and algal cell density significantly affected the photosynthesis process, whereas initial oxygen level only caused effect when no CO_2 was supplied to the test suspension. At fixed light intensity of 1,000 $\mu Em^{-2}s^{-1}$ and algal broth concentration of 0.095 g cell/L, the device performance testing was conducted with an ionic liquid and six common organic solvents. Furthermore, for comparison, a standard assay based on algal growth rate was carried out for two representative toxicants (ionic liquid and methanol). Although there was a good correlation between the data obtained from the system and those of the conventional standard growth test only in case of ionic liquid, the proposed system can be considered as potential approach for rapid assessment of toxicants. For practical use, further improvement in sensitivity can be obtained by increasing the exposure time of algal cells to toxicants or using more sensitive photosynthetic strains to specific pollutants. Algal photosynthesis is one of the essential physiological phenomena that contribute to algal viability, which might affect the structure and functioning of the whole aquatic ecosystems. Therefore, the present system, which deals with the photosynthetic activity of phytoplankton, can serve as a beneficial tool in preliminary screening toxicity methods. It should be noticed that simple acute ecotoxicity measurements do not completely identify the full impact of pollutants released into the environment but are only part of the environmental impact assessment. In general, this system is applicable to the toxicity assessment of not only organic solvents, ionic liquids but also heavy metals, pesticides as well as other hazardous substances. Nonetheless, it is premature to make this claim based on the data obtained in this chapter. Whether the system can be expanded as mentioned or not remains to be further tested. If the system is valid for other polluted compounds, this rapid test can be used as an alternative for real time bio-monitoring, where immediate toxicity evaluation is required.

6. Acknowedgements

This work was supported by National Research Foundation of Korea Grant funded by the Korean Government (KRF-2007-521-D00106, NRL 2009-0083194).

7. References

Aiba, S. (1982). Growth kinetics of photosynthetic organisms. *Advances in Biochemical Engineering*, 23, 85-156.

Bold, H. C. (1950). Problems in the cultivation of algae, In: *The Culturing of Algae*, Brunel, J., Prescott, G. W. & Tiffany, L. H. (Eds.), pp. 11-17, Antioch, Yellow Springs, OH, USA.

Bonilla, S., Conde, D. & Blanck, H. (1998). The photosynthetic responses of marine phytoplankton, periphyton and epipsammon to the herbicides paraquat and simazine. *Ecotoxicology*, 7, 99-105.

Brain, P. & Cousens, R. (1989). An equation to describe dose responses where there is stimulation of growth at low doses. *Weed Research*, 29, 93-96.

Campanella, L., Cubadda, F., Sammartino, M. P. & Saoncella, A. (2000). An algal biosensor for the monitoring of water toxicity in estuarine environments. *Water Research*, 35, 69-76.

Chen, C.-Y. & Lin, J.-H. (2006). Toxicity of chlorophenols to *Pseudokirchneriella subcapitata* under air-tight test environment. *Chemosphere*, 62, 503-509.

Contreras-Flores, C., Peña-Castro, J. M., Flores-Cotera, L. B. & Cañizares-Villanueva, R. O. (2003). Advances in conceptual design of photobioreactor for microalgal culture. *Interciencia*, 28, 450-456.

Danilov, R. A. & Ekelund, N. G. A. (2001). Effects of Cu^{2+}, Ni^{2+}, Pb^{2+}, Zn^{2+} and pentachlorophenol on photosynthesis and motility in *Chlamydomonas reinhardtii* in short-term exposure experiments. *BMC Ecology*, 1, 1.

Durrieu, C. & Tran-Minh, C. (2002). Optical algal biosensor using alkaline phosphatase for determination of heavy metals. *Ecotoxicology and Environmental Safety*, 51, 206-209.

Endo, R. & Omasa, K. (2004). Chlorophyll fluorescence imaging of individual algal cells: effects of herbicide on *Spirogyra distenta* at different growth stages. *Environmental Science & Technology*, 38, 4165-4168.

Eullaffroy, P. & Vernet, G. (2003). The F684/F735 chlorophyll fluorescence ration: a potential tool for rapid detection and determination of herbicide phytotoxicity in algae. *Water Research*, 37, 1983-1990.

Evers, E. G. (1991). A model for light-limited continuous cultures : growth, shading, and maintenance. *Biotechnology and Bioengineering*, 38, 254-259.

Göksan, T., Durmaz, Y. & Gökpinar, S. (2003). Effects of light path lengths and initial culture density on the cultivation of *Chaetoceros muelleri* (Lemmermann, 1898). *Aquaculture*, 217, 431-436.

Grant, A. J., Graham, K., Frankland, S. & Hinde, R. (2003). Effect of copper on algal-host interactions in the symbiotic coral *Plesiastrea versipora*. *Plant Physiology and Biochemistry*, 41, 383-390.

Grobbelaar, J. U. & Soeder, C. J. (1985). Respiration losses in planktonic green algae cultivated in raceway ponds. *Journal of Plankton Research*, 7, 497-506.

Hall, N. E., Fairchild, J. F., La Point, T. W., Heine, P. R., Ruessler, D. S. & Ingersoll, C. G. (1996). Problems and recommendations in using algal toxicity testing to evaluate contaminated sediments. *Journal of Great Lakes Research*, 22, 545-556.

Jensen, A. (1984). Marine ecotoxicological tests with phytoplankton. In: *Ecotoxicological testing for the marine environment. Proceedings of the International Symposium on*

Ecotoxicological testing for the marine environment. Persoone, G., Jaspers, E. & Claus, C. (Eds.), pp. 195–213, Ghent, Belgium.

Jeon, Y.-C., Cho, C.-W. & Yun, Y.-S. (2005). Measurement of microalgal photosynthetic acitivity depending on light intensity and quality. *Biochemical Engineering Journal*, 27, 127–131.

Lewis, M. A. (1995). Algae and vascular plant tests, In: *Fundamentals of Aquatic Toxicology: Effects, Environment Fate, and Risk Assessment*, Rand, G. M. (2nd Ed.), pp 135–170, Taylors & Francis, Washington, DC, USA.

Lin, J.-H., Kao, W.-C., Tsai, K.-P., Chen, C.-Y. (2005). A novel algal toxicity of both metallic and organic toxicants. *Water Research*, 39, 1869–1877.

Marquardt, D. W. (1963). An algorithm for least square estimation of parameters. *Journal of the Society for Industrial and Applied Mathematics*, 11, 431–441.

Morris, I. (1981). Photosynthetic products, physiological state, and phytoplankton growth, In: *Physiological Bases of Phytoplankton Ecology*, Platt, T. (Ed.), *Canadian Bulletin of Fisheries and Aquatic Sciences*, 210, 83–102.

Naessens, M., Leclerc, J. C. & Tran-Minh, C. (2000). Fiber optic biosensor using *Chlorella vulgaris* for determination of toxic compounds. *Ecotoxicology and Environmental Safety*, 46, 181–185.

Naessens, M. & Tran-Minh, C. (1999). Biosensor using immobilized *Chlorella* microalgae for determination of volatile organic compounds. *Sensors and Actuators B*, 59, 100–102.

Nalewajko, C. (1995). Effects of cadmium and metal-contaminated sediments on photosynthesis, heterotrophy, and phosphate uptake in Mackenzie River Delta phytoplankton. *Chemosphere*, 30, 1401–1414.

Organization for Economic Cooperation and Development. (2002). Freshwater alga and cyanobacteria growth inhibition test, 2002 update. OECD guideline 201. Paris, France.

Overnell, J. (1976). Inhibition of marine algal photosynthesis by heavy metals. *Marine Biology*, 38, 335–342.

Pardos, M., Benninghoff, C. & Thomas, R. L. (1998). Photosynthetic and population growth response of the test alga *Selenastrum capricornutum* Printz to zinc, cadmium and suspended sediment elutriates. *Journal of Applied Phycology*, 10, 145–151.

Podola, B., Nowack, E. C. M. & Melkonian, M. (2004). The use of multiple-strain algal sensor chips for the detection and identification of volatile organic compounds. *Biosensors and Bioelectronics*, 19, 1253–1260.

Puiseux-Dao, S. (1989). « Phytoplankton model » in ecotoxicology. In: *Aquatic Ecotoxicology : fundamental, concepts and methodologies*, Boudou, A. & Ribeyre, F. (Eds.), pp. 163–185, CRC, Boca Raton.

Raja, R., Hemaiswarya, S., Kumar, N. A., Sridhar, S. & Rengasamy, R. (2008). A perspective on the biotechnological potential of microalgae. *Critical Reviews in Microbiology*, 34, 77–88.

Richmond, A. (2000). Microalgal biotechnology at the turn of the millennium: a personal view. *Journal of Applied Phycology*, 12, 441–451.

Rodriguez, M., Sanders, C. A. & Greenbaum, E. (2002). Biosensors for rapid monitoring of primary-source drinking water using naturally occurring photosynthesis. *Biosensors and Bioelectronics*, 17, 843–849.

Schubnell, D., Lehmann, M., Baumann, W., Rott, F. G., Wolf, B. & Beck, C. F. (1999). An ISFET-algal (*Chlamydomonas*) hybrid provides a system for eco-toxicological tests. *Biosensors and Bioelectronics*, 14, 465–472.

Shitanda, I., Takada, K., Sakai, Y. & Tatsuma, T. (2005). Compact amperometric algal biosensors for the evaluation of water toxicity. *Analytica Chimica Acta*, 530, 191–197.

U.S. Environmental Protection Agency. (1996). Ecological effect test guidelines. Algal Toxicity, Tiers I and II. OPPTs 850.5400. Washington, DC, USA.

van Ewijk, P. H. & Hoekstra, J. A. (1993). Calculation of the EC_{50} and its confidence interval when subtoxic stimulus is present. *Ecotoxicology and Environmental Safety*, 25, 25–32.

West, L. J. A., Li, K., Greenberg, B. M., Mierle, G. & Smith, R. E. H. (2003). Combined effects of copper and ultraviolet radiation on a microscopic green alga in natural soft lake waters of varying dissolved organic carbon content. *Aquatic Toxicology*, 64, 39–52.

Yun, Y.-S. & Park, J. M. (1997). Development of gas recycling photobioreactor system for microalgal carbon dioxide fixation. *Korean Journal of Chemical Engineering*, 14, 297–300.

Yun, Y.-S. & Park, J. M. (2001). Attention of monochromatic and polychromatic lights in *Chlorella vulgaris* suspensions. *Applied Microbiology and Biotechnology*, 55, 765–770.

Yun, Y.-S. & Park, J. M. (2003). Kinetic modeling of the light-independent photosynthetic activity of the green microalgal *Chlorella vulgaris*. *Biotechnology and Bioengineering*, 83, 303–311.

12

Error analysis and simulator in cylindrical near-field antenna measurement systems

Burgos Sara, Sierra-Castañer Manuel, Martín Fernando,
Cano Francisco and Besada José Luis
Technical University of Madrid (Universidad Politécnica de Madrid - UPM)
Spain

1. Abstract

Large antennas need special measurement systems because of their considerable dimensions. Typically, cylindrical near-field systems are appropriate measurement systems for omnidirectional antennas due to the characteristics of their radiation patterns. Furthermore, these systems are also appropriate for sizeable RADAR antennas, since they can be measured on their azimuthal positioner and the probe can be easily translated through a vertical linear slide. Thus, mechanical aspects of measurement systems are rather important since errors in the mechanical set-up can directly affect far-field radiation patterns.

This chapter presents an error estimation tool to analyze the most important errors for large L-band RADAR antennas in an outdoor cylindrical acquisition system and the effect of these errors in the calculated far-field radiation pattern. This analysis can be very convenient to evaluate the error budget of the Antenna Under Test (AUT).

The simulator computes the far-field with an array of vertical dipoles over a ground plane and compares an ideal infinite far-field with the electric field obtained using the cylindrical near-to-far-field (NF-FF) transformation algorithm. The influence of the inaccuracies on the final results is evaluated by introducing random and systematic sources of errors and then, analyzing the variations produced in the principal far-field patterns, antenna parameters and in the side lobe levels (SLL). Finally, this simulator can be employed to analyze the errors for L-band RADAR antennas.

One of the objectives of this investigation is thus to analyse how mechanical and electrical inaccuracies could affect the results of a cylindrical antenna measurement system, in order to minimize them as much as possible. This is highly important not only to meet the specifications, but also to reach high accurate results. There are several error sources studies for near-field patterns: the most complete are the ones developed by Joy and Newell in [Joy, 1988], [Newell, 1988], [Newell & Stubenrauch, 1988] and Hansen in [Hansen, 1988]. Later on, other investigations have been carried out analyzing precise error studies.

Another goal is the a-priori uncertainty analysis of these errors in the measurement of L-band RADAR antennas, detecting which are the main error sources for each antenna

parameter and calculating the uncertainty budget. Finally, ground reflections were analyzed by means of simulations.
These errors appear due to the grating lobes of the RADAR antennas when the peak value of the radiation pattern is orientated towards extreme angular positions (out of system specifications). Through simulations, it is possible to evaluate the effect of the grating lobes: distortion, ripple, influence of the reflections in the side lobes closer to the zenith or in the main lobe... To finish, a method based on a diagnostic technique for cancelling the effect of the reflections is presented. This work has been applied to an existing outdoor antenna measurement facility. This facility was designed by the authors and it is being used for the characterization of L-band RADAR antennas.

2. Introduction

Large antennas need special measurement system due to their considerable dimensions. This study presents the error analysis for large L-band RADAR antennas performed for a cylindrical outdoor measurement system. In the literature, there are some works previously published about cylindrical near-field systems - i.e. [Romeu et al., 1990], [Romeu et al., 1992], [Broquetas et al., 1994] –, although most of them deal only with the near-to-far-field transformation algorithm. This transformation can be based on different approaches.
In 1978, Borgiotti presented in [Borgiotti, 1978] an integral formulation using a superposition of plane waves to obtain the far-field from the measured near-field. Later, in the works published by Hansen in [Hansen J.A., 1980], by Yaghjian in [Yaghjian, 1986] and in [Rudge et al., 1982], a second methodology was detailed. In this case, the scattering matrix formulation is employed to derive the coupling equation. A third approach, introduced by Brown and Jull in [Brown & Jull, 1961] and Leach and Paris in [Leach & Paris, 1973], is based on the three-dimensional vector cylindrical wave expansion of an electromagnetic field which applies the Lorentz reciprocity theorem formulation to attain the complete vector far-field pattern of an arbitrary antenna. In addition to the procedure proposed by Leach, Bucci in [Bucci, 1988] and Hussein and Rahmat-Samii in [Hussein & Rahmat-Samii, 1993] studied some improvements in efficiency of this method. In the last years, some approaches to solve the problem of near-to-far-field transformation using equivalent currents have been presented, i.e. [Petre & Sarkar, 1992], [Sarkar & Taaghol, 1999], [Blanch et al., 1995], [Las Heras et al., 2002], [Las Heras et al., 2005].
The first objective of this investigation is to analyse how the mechanical and electrical inaccuracies could affect the results of a cylindrical antenna measurement system, in order to minimize them as much as possible. This is highly important not only to fulfil the specifications, but also to reach a high accuracy in the results. There are several studies of the error sources for near-field patterns: the most complete are the ones developed by Joy in [Joy, 1988], Newell in [Newell, 1988] and in [Newell & Stubenrauch, 1988] and by Hansen in [Hansen J.E, 1988].
Later on, other investigations have been carried out analyzing particular error studies. The one presented in the First AMTA Europe Symposium - [Pivnenko et al., 2006] – could be an example. The second objective is the analysis of these errors in the measurement of L-band RADAR antennas, detecting which are the main source of errors for each parameter. This study has been applied to the facility described in [Martín, 2006] and in [Burgos, 2006]. The maximum length of the array antenna (up to 12 meters) requires a particularly large antenna measurement system.

As mentioned before, since cylindrical near-field ranges are appropriate to measure large antennas, the system considered is a cylindrical near-field range. In such facility, the RADAR antennas rotate on its own positioner and the probe (double-polarized probe) moves along a 15.5 meters linear slide, stopping on each defined position to acquire the near-field. In addition, the AUT can work in reception and transmission and can operate with a sum or a difference monopulse pattern.

The last aim of this research is to evaluate the effect of ground reflections on the final results. This is particularly important for this kind of antennas and antenna measurement system: when peak values of the radiation patterns of the RADAR antenna are oriented towards extreme angular positions, grating lobes of the RADAR antennas can appear and produce errors due to reflections in the facility ground. To analyze the grating lobes influence, simulations were carried out. Finally, a method based on a diagnostic technique for eliminating the effect of the reflections is applied, the reflections are located and the radiation pattern is improved.

This chapter is divided in the following parts. First, *section 2* details the introduction. Then, *section 3* summarizes three different points. The first one explains the near-to-far-field transformation algorithm and its validation. Next, the error simulator for a cylindrical near-field system and the evaluation of the results are presented. The last point of *section 3* describes an error analysis of the L-Band RADAR antenna measurement system. After that, *section 4* analyses the effect of the grating lobes and studies an algorithm for cancelling the effect of reflections. Finally, in *section 5* the conclusions and future researches are drawn.

3. Error Simulator for a Cylindrical Near-Field System and Evaluation of the Results

In order to evaluate the performances of this error simulator, first the near-to-far-field transformation algorithm is studied and validated. Then, some of the results achieved from the simulations are presented. Finally, a detailed error analysis of a real L-Band RADAR antenna measurement system based on Montecarlo simulations is commented.

3.1 Near-to-Far-Field Transformation Algorithm and its Validation

In [Burgos, 2008], the description of the cylindrical near-to-far-field transformation, considering probe correction, its validation and its application is explained. By using Reciprocity Theorem, the electric field can be expressed as a combination of the four weighting functions (for AUT and probe), obtaining the expression proposed by Leach and Paris in [Leach & Paris, 1973]:

$$v(\rho_o,\varphi_o,z_o)=\frac{\kappa^2}{4\pi^2k^2}\sum_{n=\infty}^{\infty} e^{jn\varphi_0}\cdots$$
$$\int_{\infty}^{\infty}\left(a_n(h)\sum_{m=\infty}^{\infty} c_m(-h)H_{n+m}{}^{(2)}(k\rho_o)+b_n(h)\sum_{m=\infty}^{\infty} d_m(-h)H_{n+m}{}^{(2)}(k\rho)\right)\cdots \quad (1)$$
$$e^{jhz_0}dh$$

where:

- "n" indicates the modal index (integer number),

- "h" symbolizes the propagation constant in the z-direction (real number) h=k·cosθ,
- "k" characterizes the wave number in free space,
- $\kappa=\sqrt{(k^2-h^2)}=k\cdot\sin\theta$ is the projection of the propagation constant in the xy-plane,
- $H_{n+m}^{(2)}(k\rho_0)$ denotes the cylindrical Hankel function of 2nd kind and order n+m,
- $a_n(h)$ and $b_n(h)$ correspond to the AUT weighting functions of the TE- and TM-modes,
- $c_m(h)$ and $d_m(h)$ represent the probe modal coefficients.

The model developed for this application introduces the probe compensation method proposed by Hansen in [Hansen J.A., 1980], and usually applied to the spherical near-to-far-field transformation algorithms. Assuming the rotationally symmetry of the antenna probe (μ=±1), the probe pattern is calculated from the main planes of the probe diagram. For this particular case, the antenna probe is a conical corrugated horn that fulfils the previous condition. The first step is the calculation of the co-polar far-field of the antenna probe in the principal planes $E_e(\theta)$ and $E_h(\theta)$. The θ- and ϕ-components of the electric field at each angular position are obtained through:

$$
\begin{aligned}
E_{1,\theta}(\theta,\phi) &= E_e(\theta)\cos\phi + P_1\cdot E_e(\theta)\sin\phi \\
E_{1,\phi}(\theta,\phi) &= -P_1\cdot E_h(\theta)\cos\phi + E_h(\theta)\sin\phi \\
E_{2,\theta}(\theta,\phi) &= \frac{1}{A_{xy}}\left[P_2\cdot E_e(\theta)\cos\phi + E_e(\theta)\sin\phi\right] \\
E_{2,\phi}(\theta,\phi) &= \frac{1}{A_{xy}}\left[-E_h(\theta)\cos\phi + P_2\cdot E_h(\theta)\sin\phi\right]
\end{aligned}
\tag{2}
$$

where E_1 and E_2 represent the probe patterns for acquiring both polarizations. P_1 and P_2 symbolize the linear polarization ratio of the antenna probes, and A_{xy} the amplitude and phase factors of the signals coming from both probes. If single probe is used, P_1=-P_2 and A_{xy}=1. If an ideally polarized antenna is employed, P_1=P_2=0. If the coordinate systems of the measurement of the antenna probe are different to the coordinate systems of the cylindrical acquisition (i.e. a classical spherical far-field system is used), several angular rotations of the probe radiation pattern are required - using the method proposed in [Rahmat-Samii, 1979] - in order to obtain the probe compensated far-field radiated pattern. In addition, this method employs the Discrete Fourier Transform (DFT) instead of the Fast Fourier Transform (FFT). With this change, the far-field in the final coordinate system could be achieved without needing an interpolation of the radiation pattern results, although the processing time is a bit larger. *Fig. 1* summarizes the transformation algorithm applied for this application.

The previous tool has been validated using both simulations and measurements. The implemented procedure starts with the modelling of the transmitting and receiving antenna. The simulated AUT is an array of 28x16 λ/2 dipoles vertically displaced, over a ground plane at a distance equal to λ/4, and assumed to be infinite, so Image Theory can be applied. The excitation is separable in vertical and horizontal planes, and with decreasing amplitude to reduce the side lobe levels. Besides, in the horizontal plane, a phase error is added, in order to simulate a more realistic AUT. The total size of the simulated AUT is 5.3 m x 2.1 m. On the other hand, the antenna probe is a vertically polarized conical corrugated horn, and its main planes radiated fields obtained through the integration of the electric field on the aperture are:

$$\vec{E}_{ap} = \hat{y} J_0 \left(\frac{2.405 r'}{a} \right) e^{-j\pi r'^2 / \lambda L} \quad (3)$$

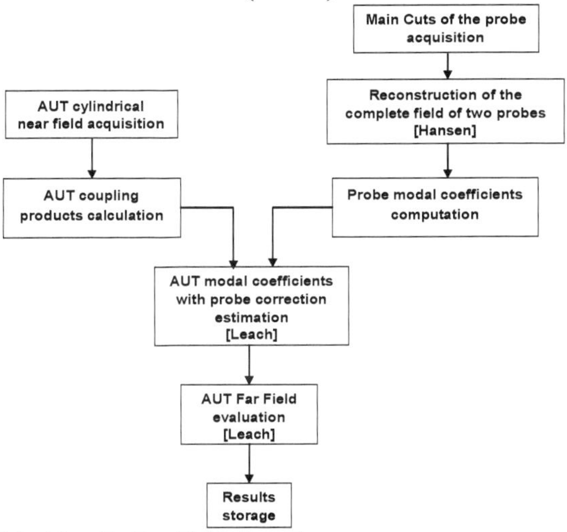

Fig. 1. Diagram of the Near-To-Far-Field algorithm

The acquisition is simulated on a cylinder where the distance from the AUT to the probe tower is 5 meters, the vertical path of the probe is 15 meters. The number of samples in the azimuth is 256 (to satisfy the Nyquist sampling theorem), while the distance between samples at the vertical axis is 10 cm. The frequency selected for the simulations 1215 MHz. The field in each point of the grid was calculated taking into account the field radiated by all the dipoles modified by the probe pattern. The field from each dipole in each point of the grid is given by the sum of 3 spherical waves, in the way explained in [Elliot, 1981]:

$$E_z = -j30 I_{mn} \left(\frac{e^{-jkR_1}}{R_1} + \frac{e^{-jkR_2}}{R_2} - 2\cos(kL_1) \frac{e^{-jkr}}{r} \right) \quad (4)$$

The induced voltage at the antenna probe can be calculated applying the expression (5), since each element of the array could be considered in far-field respect the antenna probe:

$$v \approx k.I_{mn} \left(\frac{e^{-jkR_1}}{R_1} f_s(\theta_1) + \frac{e^{-jkR_2}}{R_2} f_s(\theta_2) - 2\cos(kL_1) \frac{e^{-jkr}}{r} f_s(\theta) \right) \quad (5)$$

where k is a complex constant, I_{mn} is the dipole excitation, the angles and distances are shown in Fig. 2, and $f_s(\theta)$ is the radiation pattern of the probe. The far-field radiation patterns achieved for the main planes are compared with the infinite far-field, obtained by multiplying the array factor by the element radiation pattern, as Fig. 3 and Fig. 4 show. Fig. 3 and Fig. 4 show a significant concordance between both calculated patterns, so the NF-FF transformation could be considered validated. The small discrepancies in the extremes of the horizontal angular range are due to the approximation of the acquisition model, shown in expression (4). This simulation tool is also useful for testing the elevation validity range of the NF-FF transformation, which is often approximated by the method proposed by Yaghjian [Yaghjian, 1975] and Newell [Newell & Crawford, 1974]:

$$\theta_0 = \tan^{-1} \frac{L_z - D}{2x_o} \tag{6}$$

where L_z indicates the path performed by the probe, D corresponds to the height of the AUT and x_0 is the separation between the probe and the AUT, as shown in Fig. 5.

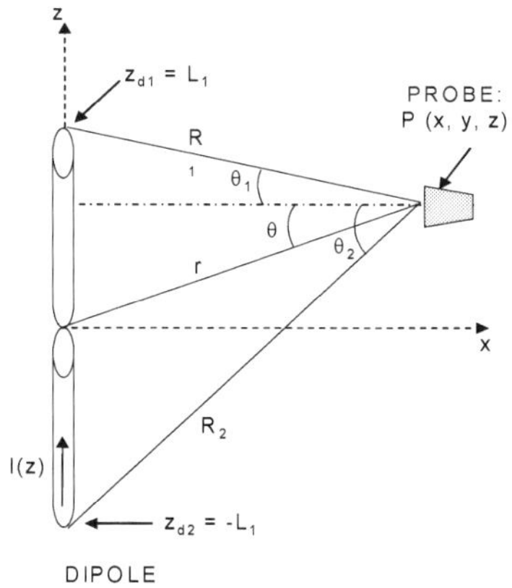

Fig. 2. Geometry of the dipole and probe

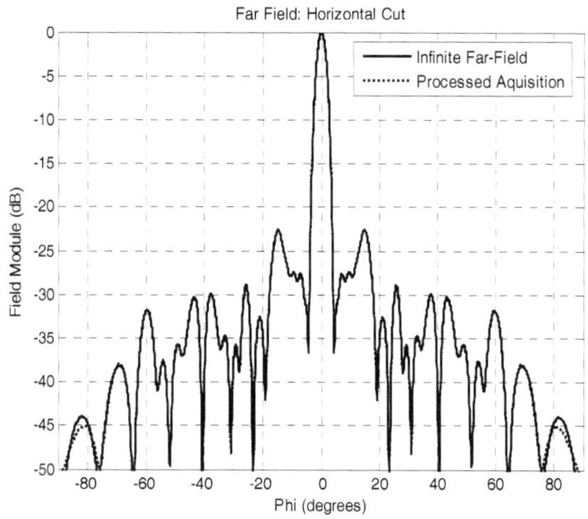

Fig. 3. Horizontal cut: Theoretical Far-field versus ideal NF-FF transformed pattern.

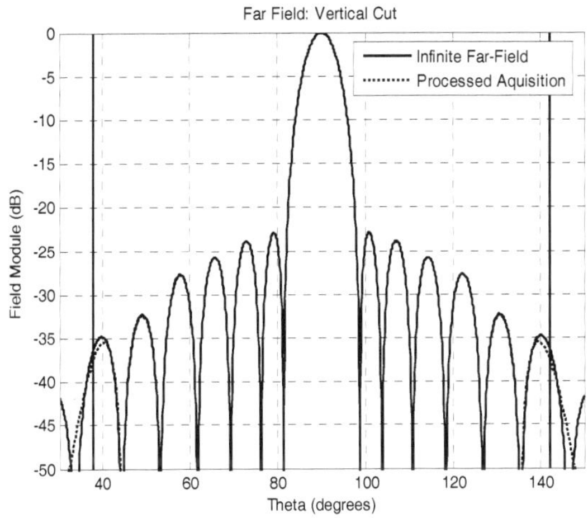

Fig. 4. Vertical cut: Theoretical Far-field versus ideal NF-FF transformed pattern.

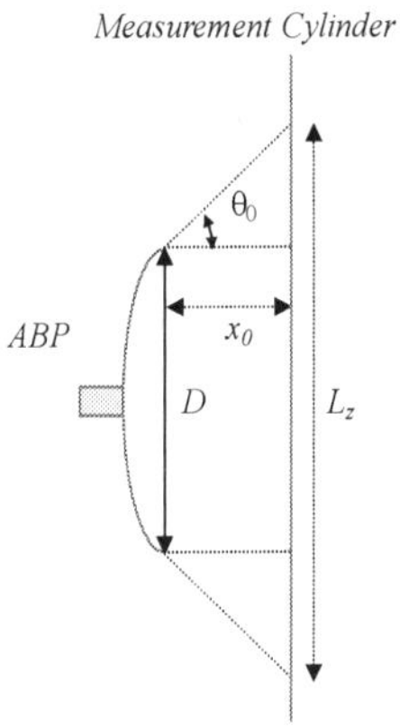

Fig. 5. Angular validity margin

With this configuration of the cylindrical system and the modelling of this specific AUT, the angular range of validity goes from 37.7 to 142.3 degrees. Fig. 4 confirms that the angular margin of the measurement is valid.

A measurement in the cylindrical near-field system of the Technical University of Madrid was performed to validate this algorithm. The AUT utilized was the Ku-band reflector shown in Fig. 6. This antenna was measured in both the spherical near-field system (using TICRA SNIFTD Software) and the cylindrical near-field system. The results from the comparison of the horizontal and the vertical cuts are illustrated in Fig. 7 and Fig. 8.

Fig. 6. Ku-band reflector employed for the validation measurements

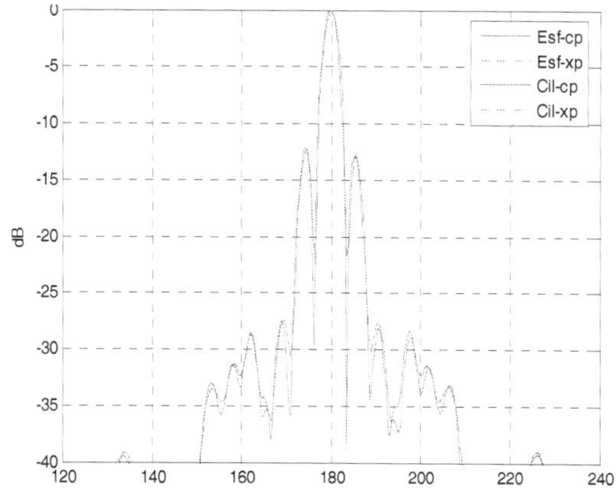

Fig. 7. Horizontal cut for the measurement with the Ku-reflector

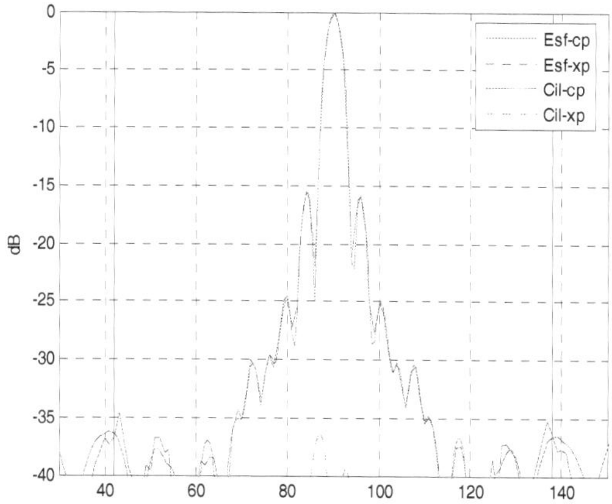

Fig. 8. Vertical cut for the measurement with the Ku-reflector

3.2 Error simulations

To evaluate how errors could affect the final results, a model of the antennas and a simulation of the acquisition process including errors has been performed. The simulator compares the outcomes achieved from the reference data (i.e. far-field obtained from an acquisition free from additional errors) with the ones including the deviations, as Fig. 9.

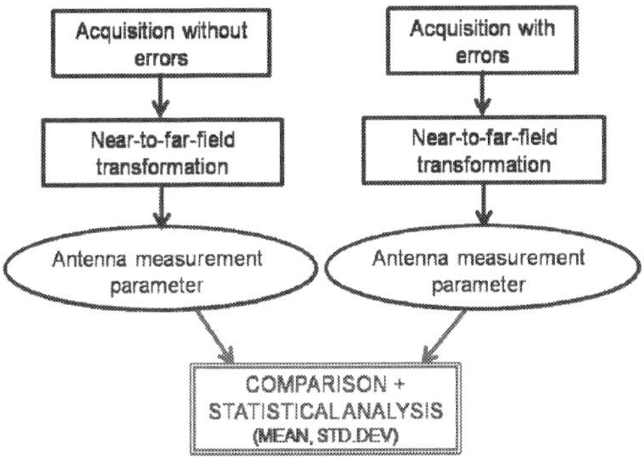

Fig. 9. Simulator diagram

For the uniform random error in probe x-, y- and z-position, an exhaustive study was completed and 10 iterations for each peak to peak amplitude were achieved for a RADAR L-Band antenna of 5.32 metres by 2.08 metres. From these simulations, it could be calculated the mean of the difference in peak directivity, which gives an estimation of the error, and the standard deviation of the difference in peak directivity, which gives an estimation of the uncertainty, as shown in Fig. 10 and Fig. 11. From these results, it can be observed that in both cases the deviations in the x-position of the probe have clearly larger influence on the mean and on the standard deviation of the differences in peak directivity. This is due to the fact that the AUT and the probe are aligned along the x-axis and thus the deviations in this direction produce larger errors.

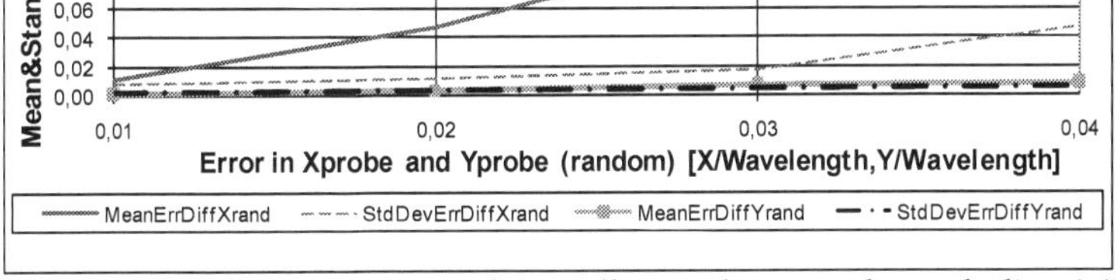

Fig. 10. Mean and standard deviation of the difference between the peak directivity with and without error, for errors in the x- and y-position of the probe

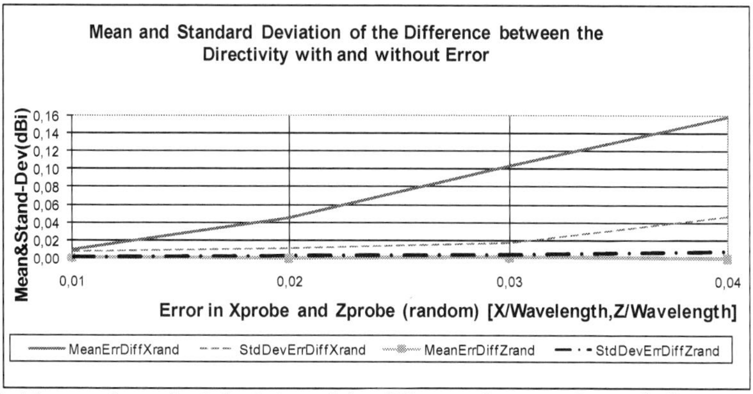

Fig. 11. Mean and standard deviation of the difference between the peak directivity with and without error, for errors in the x- and z-position of the probe

In order to verify that the errors in the x-position of the probe and the phase errors are comparable, the equivalent phase errors for the deviations in the x-position of the probe was calculated, as shown in Table 1. Fig. 12 and Fig. 13 represents the mean and the standard deviations of the directivity differences according to all the considered phases (the ones calculated from the deviations in the x-position of the probe and the ones taken into account in the phase deviations simulations).

Error in x-probe	0.01λ		0.02λ		0.03λ	0.04λ
Corresponding Phase	3.6		7.2		10.8	14.4
Phase Error	2	4	6	8		

Table 1. Evaluated phase errors and comparison with longitudinal errors along x-axis

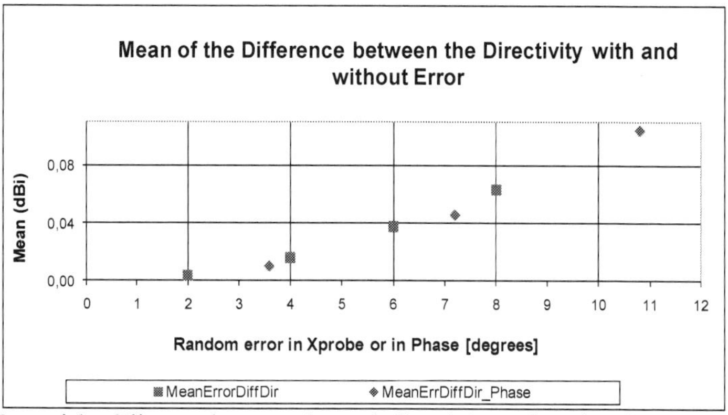

Fig. 12. Mean of the difference between the peak directivity with and without an error in the x-position of the probe and with a random phase error

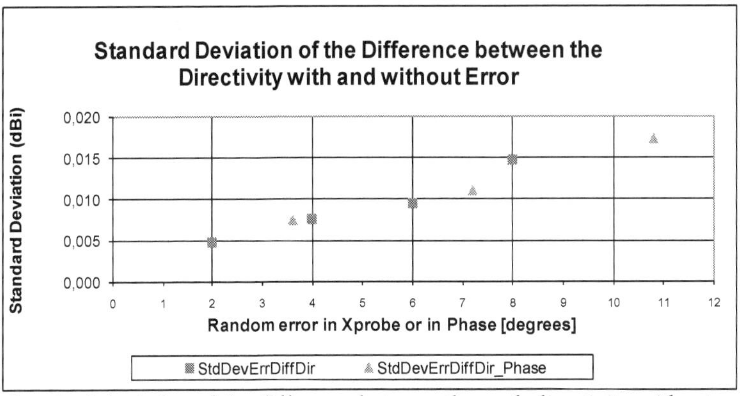

Fig. 13. Standard deviation of the difference between the peak directivity without and with an error in the x-position of the probe and with a random phase error

Since the mean and the standard deviation linearly vary as the errors increase, it can be deduced that the errors in the x-position of the probe and in the phase are comparable. As the second ones are easier to compute, the uncertainties due to mechanical deviations can be calculated in a faster way using the equivalent phase errors.
In order to analyze the influence of the antenna size in the statistical parameters (i.e. mean and standard deviation), achieved from the difference between the antenna parameter with and without error, simulations with five different antennas - shown in Table 2 - have been carried out.

	L (m)	H (m)	Excitation
Ant 1	1,52	0,52	Uniform
Ant 2	1,90	0,78	Uniform
Ant 3	2,66	1,04	Uniform
Ant 4	5,32	2,08	Uniform
Ant 5	7,22	2,86	Uniform

Table 2. Sizes of the antennas considered for the simulations

In this study, 10 iterations per case have been completed. This number is not enough to get error probability distributions, although it is good enough to analyze the trends. Fig. 14. and Fig. 15. correspond to representative results of the standard deviations of the difference between the directivity with and without error, when considering a random error in the x-position of the probe, and a random phase error or a white Gaussian noise for 5 antennas with uniform excitation, after 10 iterations. From the obtained results, it can be concluded that the uncertainty is lower as the antenna size is larger.

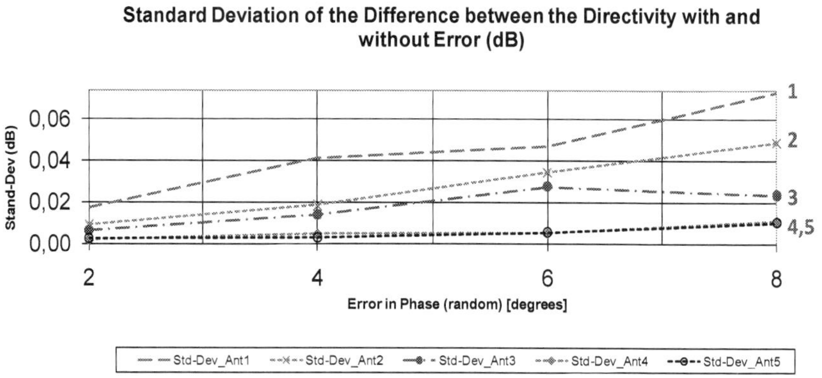

Fig. 14. Standard deviation of the difference between the differences in directivity without and with a random phase error

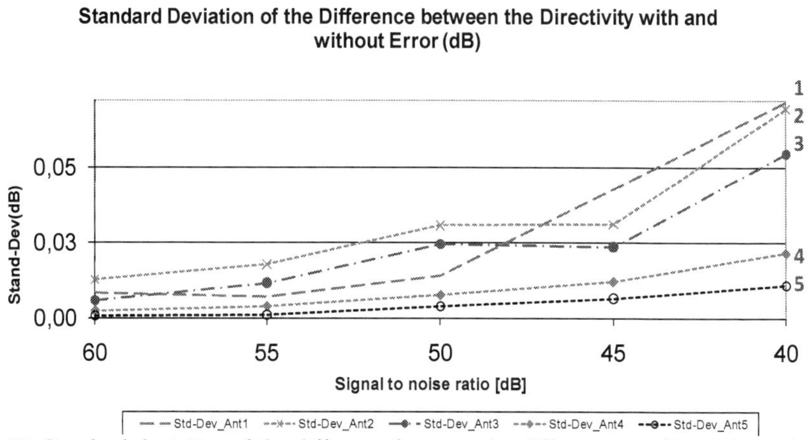

Fig. 15. Standard deviation of the difference between the differences in directivity without and with a white Gaussian noise (WGN)

Fig. 16 and Fig. 17 represent some examples of the statistical analysis in the SLL when considering a random phase error or a WGN for 5 antennas with different sizes and with uniform excitation after 10 iterations.

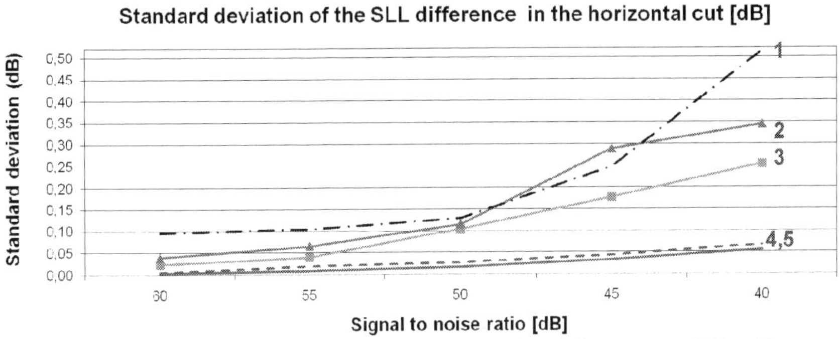

Fig. 16. Standard deviation of the difference between the differences in SLL without and with WGN

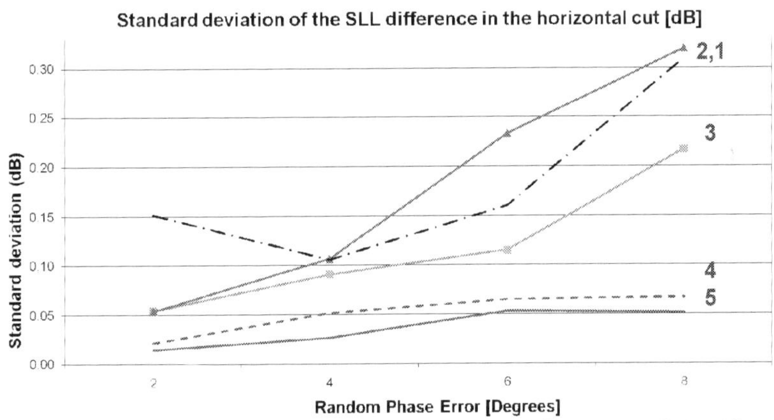

Fig. 17. Standard deviation of the difference between the differences in SLL without and with a random phase error

From the results obtained, it can be observed that the means and the standard deviations are larger for the smallest antennas (antenna 1 and antenna 2) and smaller for the biggest antennas (antennas 4 and 5).

3.3 Error Analysis of the L-Band RADAR Antenna Measurement System: Montecarlo Simulations

Typical calculations of uncertainties and systematic errors in antenna measurements are based on the calculation or estimation of the standard deviation and mean of each error term respectively. In this particular case, a Montecarlo study with one hundred iterations per antenna excitation was implemented, including some error sources. The study calculates the inaccuracies in the next main antennas parameters: directivity, pointing directions, SLL and beamwidth. In order to carry out a more detailed analysis, two different antennas were simulated. The antennas are not centered in the measurement range.

Their main dimensions (length and height) and some of the parameters of the measurement range (position of the tower respect to the antenna, length of the tower and valid angular range of the measurement results) of these two scenarios are detailed in Table 3:

RADAR	Long. [m]	Height [m]	x_o [m]	L_z [m]	Mech. Tilt [deg]	$\theta_{v,min}$ [deg]	$\theta_{v,max}$ [deg]
Antenna 1	5.32	2.08	5.0	15.5	16.1	20.9	123.2
Antenna 2	11.78	6.82	7.0	15.5	10	42.1	110.1

Table 3. Antenna dimensions and parameters of the measurement system

Besides, it is worth noting that for this case, different amplitude distribution - sum and difference patterns, electric tilt… - were examined:

- Sum amplitude distribution in the horizontal and in the vertical planes,
- Sum amplitude distribution in the horizontal plane and difference amplitude distribution in the vertical plane,
- Difference amplitude distribution in the horizontal plane and sum amplitude distribution in the vertical plane.

These cases were studied with the antenna 1 mechanically tilted 16.1 degrees and with electrical tilt of 0 or 40 degrees in the vertical plane. In the case of antenna 2, this is mechanically tilted 10 degrees, and there is not any electrical tilt.

As commented before, the results obtained with the simulator show the comparison between the ideal far-field radiation pattern (no errors were added) with the far-field radiation pattern including errors. In this analysis, the probe correction was omitted to reduce the computational time of the simulations. Besides, the AUT was mechanically tilted from the vertical axis, as it happens in the measurement. The simulator allows representing the histograms of the different parameters. For instance, Fig. 18 to Fig. 20 show the histograms of the sum pattern distribution for antenna 1.

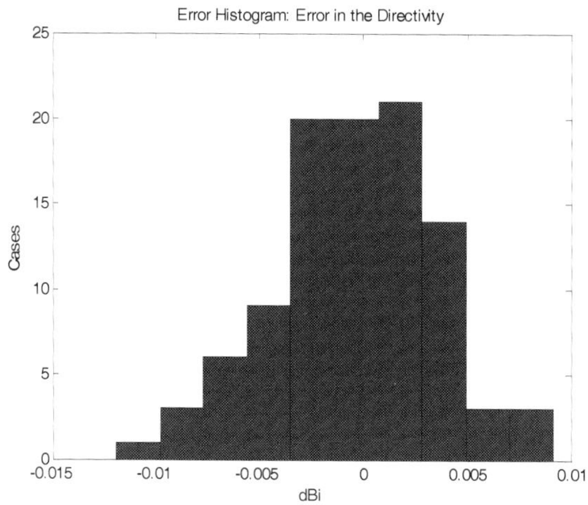

Fig. 18. Histogram of the error in the directivity

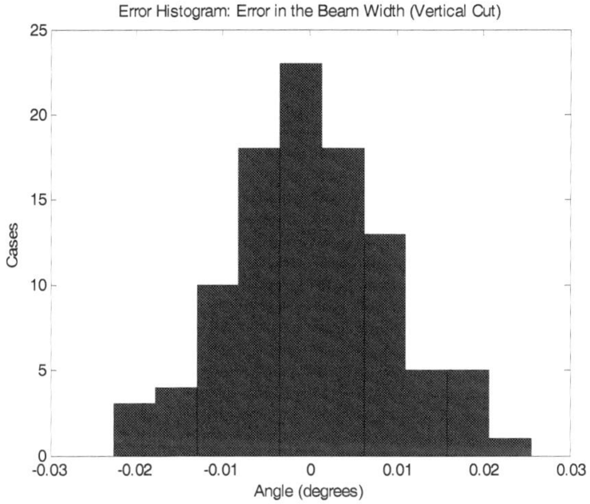

Fig. 19. Histogram of the error in the beam width in the vertical cut

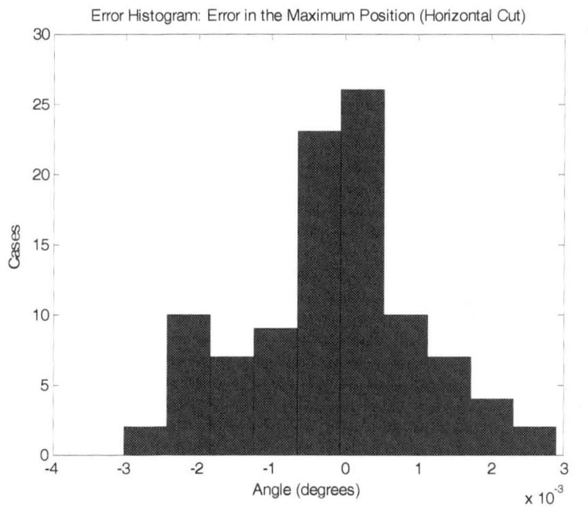

Fig. 20. Histogram of the error in the position of the maximum in the horizontal cut

In addition, the Montecarlo study was completed with the analysis of the statistics achieved from the 100 iterations. The next table summarizes the results obtained for Antenna 1 and Antenna 2 when no electrical tilt is designed:

PARAMETER	QUANTITY	RMS ERROR	MAXIMUM ERROR
Directivity (dBi)	39.86	0.0036	0.2762
Beam Width in the Horizontal Cut (degrees)	1.51	$3.85\ 10^{-5}$	0.0016
Beam Width in the Vertical Cut (degrees)	2.46	$6.30\ 10^{-5}$	0.0026
SLL (-20 dB to -30 dB)	-	-	1.1 dB
SLL (-30 dB to -40 dB)	-	-	2.0 dB
Position of the maximum in the Horizontal Cut (degrees)	0	$3.65\ 10^{-5}$	$7.85\ 10^{-4}$
Position of the maximum in the Vertical Cut (degrees) for the sum pattern	80.0	$5.52\ 10^{-5}$	0.0014
Position of the null in the Vertical Cut (degrees) for the difference pattern	80.0	$1.00\ 10^{-5}$	$3.47\ 10^{-4}$

Table 4. Results obtained in the Montecarlo study for the antenna 2

In conclusion, when the antenna is bigger, the errors found with the Montecarlo simulations were smaller for all the parameters, except for the directivity and side lobe level, where they were larger.

4. Reflections study

Since the cylindrical near-field system under study is an outdoor range, an important error factor is the reflections on the environment, particularly the effect of the ground. In this section, an analysis of the effect of the reflections in the ground plane and a method to reduce this effect are presented. Measurements on an L-Band RADAR antenna are shown.

4.1 Effect of Grating Lobes

During a measurement campaign some negative effects due to the ground reflection were detected. These errors appear as a ripple in the far-field obtained. These errors are due to the grating lobes of the RADAR antennas when the peak value of the radiation pattern is orientated towards extreme angular positions and they are operating at highest frequencies (out of the operation frequency range).

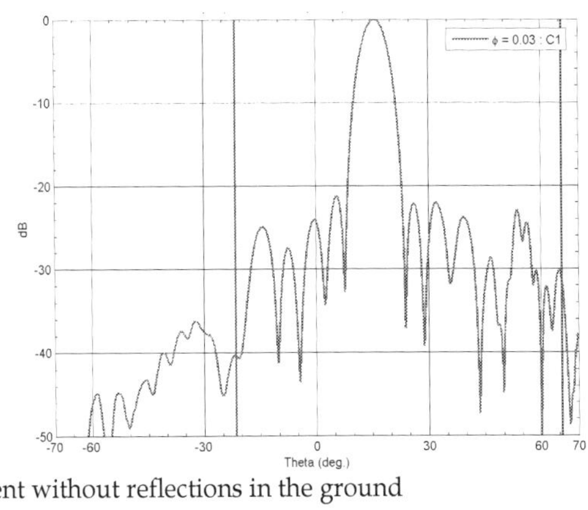

Fig. 21. Measurement without reflections in the ground

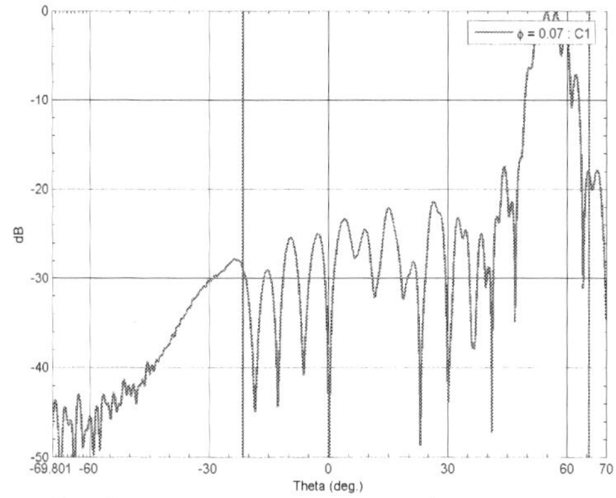

Fig. 22. Measurement with reflection in the ground (Higher tilt and higher frequency)

Several simulations have been carried out to confirm that the reflections are the cause of ripple problem. The first one consisted in increasing the antenna beam pointing at elevation, from 0 to 40 degrees at the centre of the frequency band, using different reflection coefficients (from 0 to -1). Results depicted in Fig. 23 show the higher the pointing is, the more affected the acquired near-field appears. This perturbation creates a ripple in the far-field, even noticeable in the main lobe, when the electrical tilt is 40 degrees. In fact, at 40 degrees the near-field is so affected that the main lobe of the far-field is completely degraded.

For the electrical tilt of 40 degrees, the frequency was varied. Fig. 24 concludes that the higher the frequency is, the more important the ripple is. Therefore, the grating lobes and their reflections on the ground plane are the responsible of the far-field ripple observed in the measurements. According with equation (7), these grating lobes appear at higher tilt and frequency for this antenna structure.

$$\theta = a\cos\left(\frac{N2\pi - \alpha}{kd}\right) \tag{7}$$

- k is the wave number,
- d represents the distance between array elements in vertical,
- α corresponds to the progressive phase,
- N symbolizes an integer number.

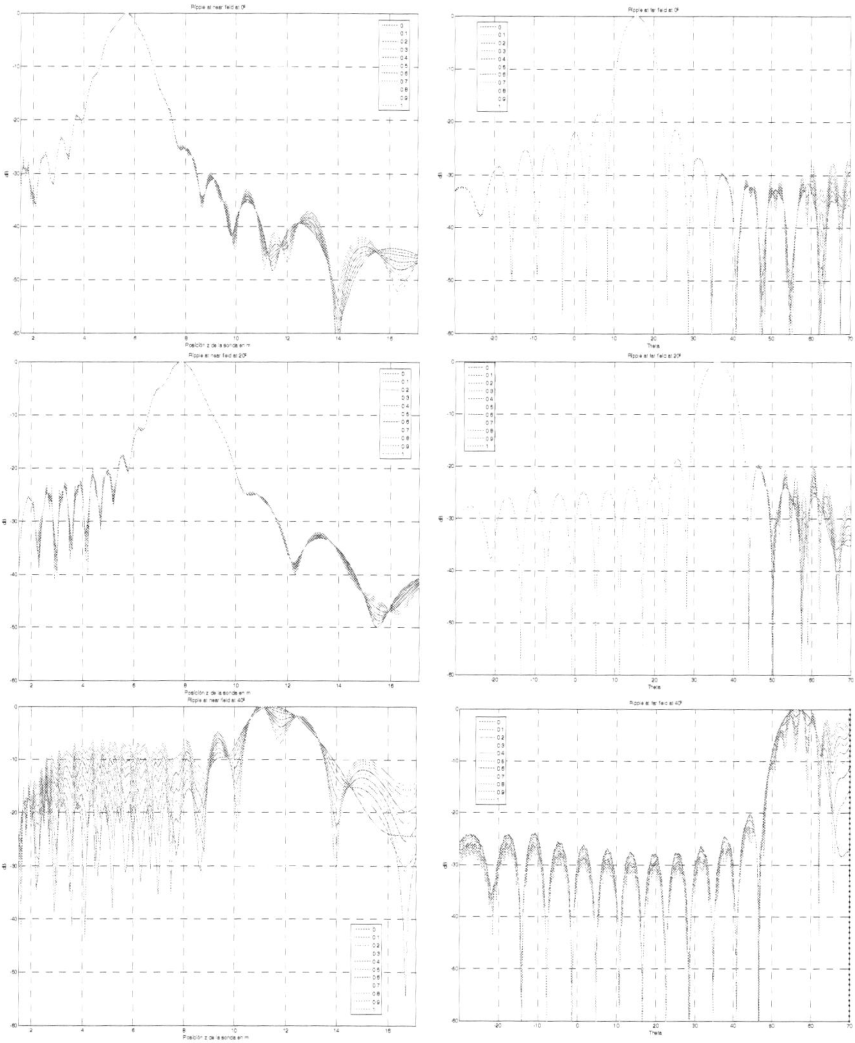

Fig. 23. Simulations of vertical plane near-field acquisitions and their transformed far-field at the centre of the frequency band and varying electrical pointing from 0 to 40 degrees (mechanical tilt 16 degrees)

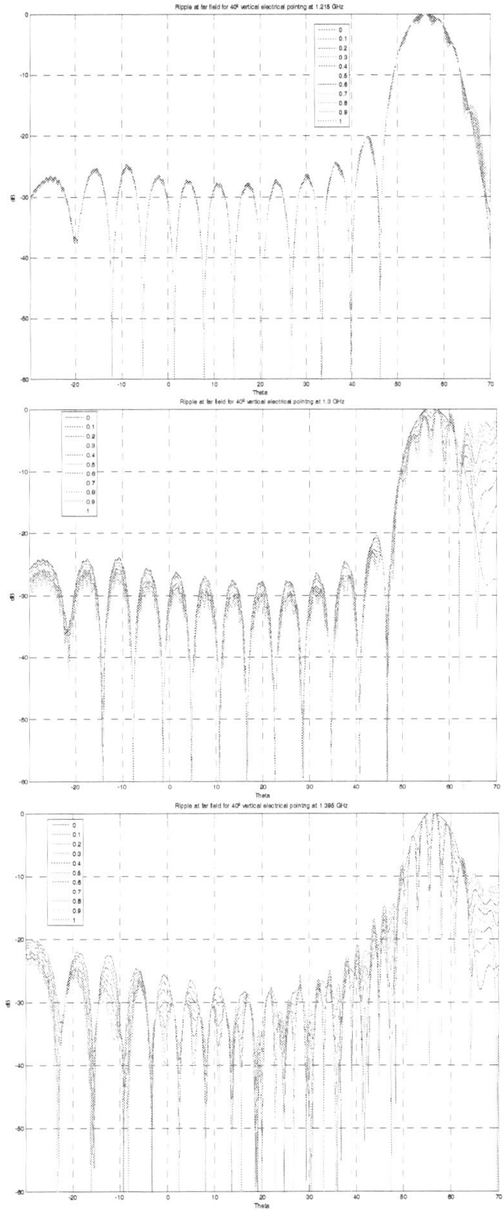

Fig. 24. Simulations of vertical plane far-field at 40 degrees of vertical pointing varying the frequency (1215, 1300 and 1395 MHz) along the band (mechanical tilt 16 degrees)

4.2 Diagnostic techniques for Cancelling the Effect of Reflections

As it has been seen in the previous section, the effect of reflections disturbs the results of the radiation pattern, especially when the frequency and the tilt angle are high. There are several ways to reduce such effect, but the most popular and always used in anechoic chamber, is the use of radiation absorbent material (RAM). Since the measurement range is an outdoor cylindrical near-field system, it is not the most suitable solution, and it is necessary to use a measurement processing technique to reduce the effect of the reflections. Different kinds of these processing techniques have been proposed and tested, being two of them the most effective. The first one is based on the "Matrix-Pencil Method", used to approximate signals into a sum of complex exponentials. Then, the exponential functions are processed to identify the terms corresponding to the different propagating components. Once identified, the terms corresponding to the reflected components are removed. The theory of this technique was presented in [Sarkar & Pereira, 1995] and applied to the reflection cancelling problem in [Fourestié et al., 1999]. The second technique is called "FFT-Based Method" and estimates the impulse response of the measured environment from its frequency response by using the inverse Fourier transform. In the time domain, the direct contribution is detected, eliminating the undesired echoes. This method was studied by [Loredo et al., 2003] and compared with the first one in [Loredo et al., 2004]. The main problem of both techniques is the necessity to measure in a frequency band that produces an important increase of the measured time.

To reduce the acquisition time, an algorithm which only needs the information in a frequency is applied. Such algorithm is based on a diagnostic technique which is a method to obtain the extremely near-field or the equivalent currents distribution of an antenna from the knowledge of its radiated field (near or far-field). With this information, it is possible to detect errors, and also to identify which are the causes of such errors, for example, electrical errors in arrays or mechanical errors in reflectors. The presence of these errors is always detected in the far-field pattern, when it is seen that a difference between the real case and the design case exists. However, to identify errors in the AUT could be a very hard and long task for the antenna designer. So, in this case it is interesting to use one of these techniques to reduce the difficulty and the time in the design stage. This is the typical and classical application of a diagnostic process. However, it is also possible to use a diagnostic technique in other applications, like reflection cancelling, as shown in [Cano et al., 2009]. The steps of the proposed method to reduce the effect of the reflections are indicated below.

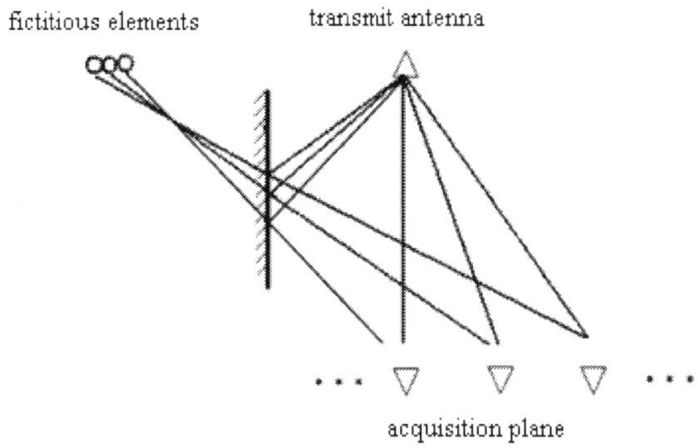

Fig. 25. Fictitious elements which appear when there are reflections.

- First of all, a diagnostic process is applied and the extremely near-field in a zone larger than the antenna dimensions is obtained. Out of the antenna, a "fictitious field" appears due to the reflection on the environment, as it can be seen in Fig. 26.
- After that, the field value out of the antenna dimension is set to zero, removing the effect of the reflections.
- Finally, from the filtered field, a new plane wave spectrum is recalculated where the ripple due to the reflections has disappeared. With this information the θ and ϕ components of the far-field is obtained by means of the expressions (8) and (9).

$$E\theta = j\lambda \frac{e^{-jkr}}{r}\left(P_x \cos\phi + P_y sen\phi\right) \tag{8}$$

$$E\theta = j\lambda \frac{e^{-jkr}}{r}\left(P_x sen\phi - P_y \cos\phi\right) \tag{9}$$

where "P_x" and "P_y" are the components of the plane wave spectrum, "λ" is the wavelength and "k" is the wavenumber.

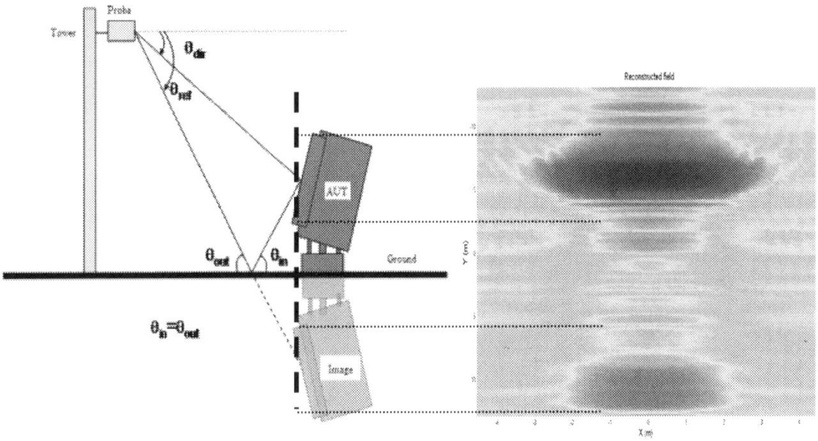

Fig. 26. Reconstructed field over the antenna plane (f = 1.22 GHz, tilt = 30º).

Once the reflection cancelling technique has been explained, the following stage is to apply it to the RADAR measurements and to verify that the effect of reflections is less important. As it has been indicated in the previous section, the reflection level depends on the frequency and the tilt angle, so the proposed technique will be applied to two different cases, varying both mentioned parameters. Firstly, the measurement with a frequency equal to 1.22 GHz and a tilt angle equal to 30 degrees is taken, and after applying the technique, the results depicted in the following figures are obtained.

The Fig. 27 represents the reconstructed field over the antenna plane, where it is possible to observe the presence of reflections. The next figure, Fig. 28, shows the effect of the reflections in the radiation pattern and also the result after applying the reflection cancelling technique, where it can be seen that the ripple due to reflections disappears.

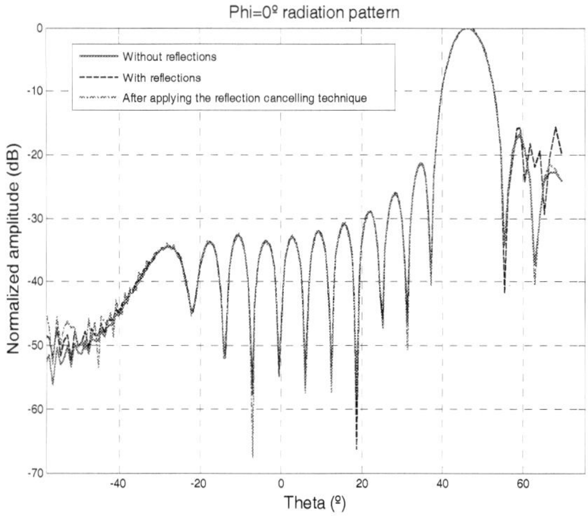

Fig. 27. Radiation pattern comparison with and without reflections and applying the reflection cancelling technique (f = 1.22 GHz, tilt = 30°).

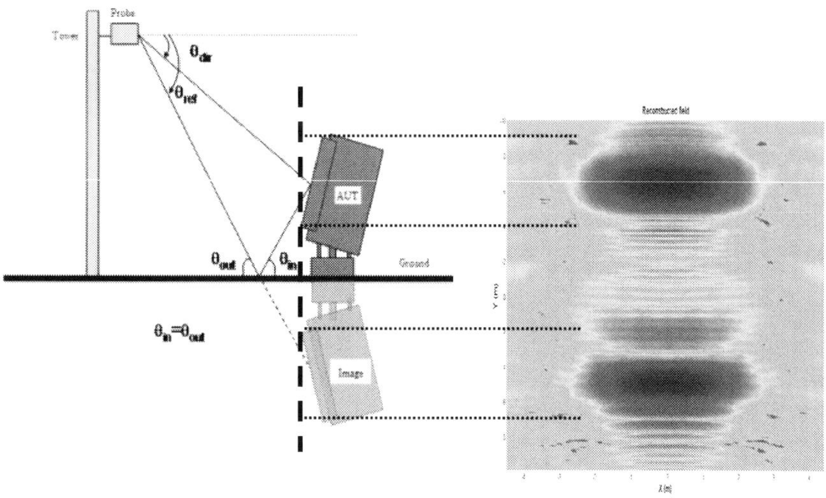

Fig. 28. Reconstructed field over the antenna plane (f = 1.4 GHz, tilt = 40°).

Next, the same results are presented for the case of a frequency equal to 1.4 GHz and a tilt angle equal to 40 degrees. As it can see in Fig.29, the fictitious field amplitude is higher because the measurement has been performed with a higher frequency and a larger tilt, so the grating lobe is directly steered to the ground plane.

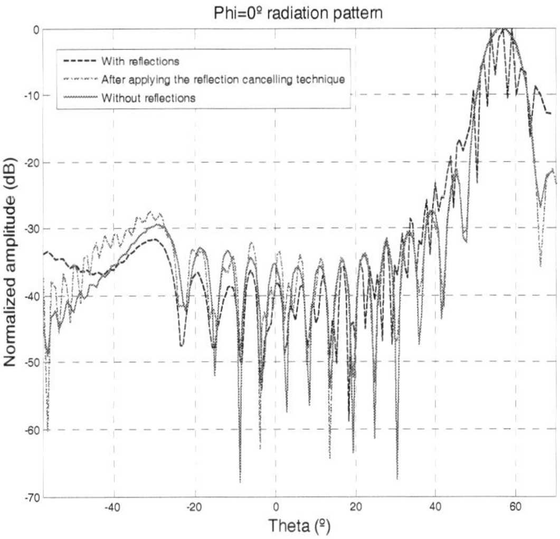

Fig. 29. Radiation pattern comparison with and without reflections and applying the reflection cancelling technique (f = 1.4 GHz, tilt = 40°).

5. Conclusions and Future Researches

The implemented simulator for this investigation allows characterizing uncertainties in the antenna parameters according to the kind of antenna and the mechanical and electrical system performances. With the error simulation in acquisition achieved using this tool, the analysis of their effects on some of the main antenna parameters (i.e. far-field radiation pattern, directivity, SLL) has been carried out.

While considering random errors, if several iterations are carried out, a statistical analysis allows obtaining the mean and the standard deviation of the errors that respectively gives an estimation of the error and the uncertainty produced. Since the AUT and the probe are aligned along the x-axis of the probe, the random errors in the x-direction are the ones that have a larger influence in the antenna parameters. This effect can be simulated as an error phase equivalent to $\Delta\phi \cong -k_0 \Delta x$.

The uncertainty in the peak directivity decreases if the antenna size increases when considering a random deviation in the x-probe position a random phase or a WGN. In addition, the mean and the uncertainty of the error in SLL in both vertical and horizontal cuts are larger for the smallest antennas and less significant for the largest antennas when including a random change in the x-position of the probe, a random phase or a WGN.

Furthermore, Monte-Carlo simulations with one hundred iterations were performed to check the effect of the sum of different error sources in the antenna parameters. The results showed that whereas the errors in the directivity or SLL were larger with a bigger antenna, almost all the errors in the other type of parameters were smaller with this antenna.

Apart from mechanical and electrical deviations, in this work the effect of ground reflections has been analyzed. These errors appear due to the grating lobes of the RADAR antennas when the peak value of the radiation pattern is orientated towards extreme angular positions and in the upper extreme of the frequency band. It was observed that the grating lobes and their reflections on the ground plane were the responsible of ripples in the far-field radiation patterns.

In order to cancel the effect of the ground reflections, a method based on a diagnostic technique has been applied. Such algorithm allows obtaining the extremely near-field or the equivalent currents distribution of an antenna from the knowledge of its radiated field. With such information, it is possible to detect errors, and also to identify the causes of these deviations. As seen from the results, the implemented technique reduces the ripple due to the reflections.

Some interesting future researches could be evaluate other sources of errors, such as for example AUT support scattering, the amplitude non-linearity or the truncation errors. In addition, other cancelling techniques could be analyze with the intention of minimizing errors, such as mechanical errors, electrical errors, acquisition errors, processing errors or stray signals.

6. References

Blanch, S.; Jofre, L.; Romeu, J. (1995). Comparation Between Classical and Equivalent Current Apprach Near-Field to Far-Field Transformation, Antenna and Propagation Society International Symposium, AP-S. Digest, Vol.1, Iss., 18-23, Jun. 1995, pp. 260-263.

Borgiotti, G. V. (1978). Integral Equation Formulation for Probe Corrected Far-Field Reconstruction from Measurements on a Cylinder, IEEE Transactions on Antennas and Propagation, Vol. AP-26, No. 4, July 1978, pp. 572-578.

Broquetas, A.; Romeu, J.; Rius, J. M.; Elias-Fuste, A. R.; Cardama, A.; Jofre, L. (1994). An approximate expression to estimate signal-to-noise improvement in cylindrical near field measurements, IEEE Transactions on Antennas and Propagations, Vol 42, N°7, June 1994, pp. 1007-1010.

Brown, J. & Jull, E. V. (1961). The Prediction of Aerial Radiation Patterns From Near-Field Measurements, Inst. Elec. Eng., Paper No. 3649 E, Nov. 1961.

Bucci, O. M. (1988). Use of Sampling Expansions in Near-Field-Far-Field Transformations: The Cylindrical Case, IEEE Transactions on Antennas and Propagation, Vol. 36, No. 6, June 1988, pp. 830-835.

Burgos, S.; Martín, F.; Sierra-Castañer, M.; Besada, J. L. (2006). Error Estimations In Cylindrical Near Field System For Large RADAR Antennas, European Conference on Antennas and Propagation (EuCAP) symposium proceedings, November 2006, Nice, France.

Burgos, S.; Martín, F.; Sierra-Castañer, M.; Besada, J. L. (2008). Cylindrical near-to-far-field transformation system for RADAR antennas: Design, validation and application, *Microwave and Optical Technology Letters*, Vol. 50, No. 10, October 2008, pp. 2527 - 2531.

Cano, F.; Sierra-Castañer, M.; Besada, J.L. (2009). Application of diagnostic technique for reflection cancelling in antenna measurement, European Conference on Antennas and Propagation (EuCAP) symposium proceedings, March 2009, Berlin, Germany.

Elliot, R. S. (1981). *Antenna Theory and Design*, Ed. Prentice-Hall, Inc., Englewood Cliffs, New Jersey.

Fourestié, B.; Altman, Z.; Wiart, J.; Azoulay, A. (1999). On the Use of the Matrix-Pencil Method to Correlate Measurements at Different Test Sites, IEEE Transaction on Atennas and Propagation, Vol. 47, No. 10, October 1999.

Joy, E. B. (1988). Near-Field Range Qualification Methodology, *IEEE Transactions on Antennas and Propagation*, Vol. 36, No. 6, June 1988, pp. 836-844.

Hansen, J. E. (1988). *Spherical Near-Field Antenna Measurements*, Peter Peregrinus Ltd. on behalf of IEE, ISBN: 0 86341 110 X, London, UK.

Hansen, J. A. (1980). On Cylindrical Near-Field Scanning Techniques, IEEE Transactions on Antennas and Propagation, Vol. AP-28, No. 2, March 1980, pp. 231-234.

Hussein, Z. A. & Rahmat-Samii, Y. (1993). Probe Compensation Characterization in Cylindrical Near-Field Scanning, IEEE, 1993, pp. 1808-1811.

Las Heras, F.; Galocha, B.; Besada, J. L. (2002). Far-Field Performance of Linear Antennas Determined From Near-Field Data, IEEE Transactions on Antennas and Propagation, Vol. 50, No. 3, March 2002, pp. 408-410.

Las Heras, F.; Galocha, B.; Besada, J. L. (2005). Circular Scanning and Equivalent Magnetic Currents for Main Plane Near-Field Transformation, International Journal of Numerical Modelling: Electronic Networks, Devices and Fields, Vol. 15, No. 4, July/August 2005, pp. 329-338.

Leach, W. M. Jr. & Paris, D. T. (1973). Probe Compensated Near Field Measurements on a Cylinder, IEEE Transactions on Antennas and Propagation, Vol. AP-21, No. 4, July 1973, pp. 435-445.

Loredo, S.; Rodríguez, M.; Las-Heras, F.; Sarkar, T.P. (2003). Cancelación de ecos en cámaras de medida no anecoicas, XXIV Simposium Nacional de la Unión Científica Internacional de Radio, URSI 2009.

Loredo, S.; Rodríguez, M.; Las-Heras, F.; Sarkar, T.P. (2004). Echo identification and Cancellation Techniques for Antenna Measurement in Non-Anechoic Test Sites, IEEE Antennas and Propagation Magazine, Vol. 46, No. 1, February 2004.

Martín, F.; Burgos, S.; Sierra-Castañer, M.; Besada, J. L. (2006). Design of a cylindrical near field system for RADAR antennas, European Conference on Antennas and Propagation (EuCAP) symposium proceedings, November 2006, Nice, France.

Newell, A. C. & Crawford, M. L. (1974). Planar near-field measurements on high performance array antennas, *Nat. Bur. Stand.*, NBSIR 74-380, July 1974.

Newell, A. C. (1988). Error Analysis Techniques for Planar Near-field Measurements, *IEEE Transactions on Antennas and Propagation*, Vol. 36, No. 6, June 1988, pp. 754-768.

Newell, A. C.; Stubenrauch, C. F. (1988). Effect of Random Errors in Planar Near-Field Measurements, *IEEE Transactions on Antennas and Propagation*, Vol. 36, No. 6, June 1988, pp. 769-773.

Petre, P. & Sarkar, T.K. (1992). Planar Near-Field to Far-Field Transformation Using an Equivalent Magnetic Current Approach, IEEE Transactions on Antennas and Propagation, Vol. 40, No. 11, Nov. 1992, pp. 1348-1356.

Pivnenko, S.; Nielsen, J. M.; Breinbjerg, O. (2006). Electrical Uncertainties In Spherical Near-Field Antenna Measurements, Proceedings of the First Antenna Measurements Techniques Association Europe (AMTA Europe) Symposium, pp.183-186, May 2006, Munich, Germany.

Rahmat-Samii, Y. (1979). Useful Coordinate Transformations for Antenna Applications, *IEEE Transactions on Antennas and Propagation*, Vol. AP-27, No. 4, July 1979, pp. 571-574.

Romeu, J.; Baquero, M.; Ferrando, M.; Jofre, L.; Alemany, J.; González, V. (1990). A Cylindrical near field test facility, Microwave Engineering Europe, September/October 1990, pp. 25-31.

Romeu, J.; Jofre, L.; Cardama, A. (1992). Far-field errors due to random noise in cylindrical near-field measurements, IEEE Transactions on Antennas and Propagations, Vol. 40, N°1, June 1992, pp. 79-84.

Rudge, W.; Milne, K.; Olver, A.D.; Knight, P. (1982). The Handbook of Antenna Design, Vol 1, pp. 609-614, Peter Peregrinus Ltd. on behalf of IEE, ISBN: 0-906048-82-6, London, UK.

Säily, J.; Eskelinen, P.; Räisänen, A. V. (2003). Pilot Signal-Based Real-Time Measurement and Correction of Phase Errors Caused by Microwave Cable Flexing in Planar Near-Field Tests, *IEEE Transactions on Antennas and Propagation*, Vol. 51, No. 2, Feb. 2003.

Sarkar, T.K. & Pereira, O. (1995). Using the Matrix Pencil Method to Estimate the Parameters of a Sum of Complex Exponentials, IEEE Antennas and Propagation Magazine, Vol. 37, No. 1, February 1995.

Sarkar, T.K. & Taaghol, A. (1999). Near-field to Near/Far-field Transformation for Aribitrary Near-Field Geometry Utilizing an Equivalent Electric Current and MoM Antennas and Propagation, IEEE Transactions on Antennas and Propagation, Vol. 47, Iss. 3, Mar. 1999, pp. 566-573.

Yaghjian, A. D. (1975). Upper-bound errors in far-field antenna parameters determined from planar near-field measurements, Part 1: Analysis, *Nat. Bur. Stand., and Tech. Note 667*, 1975.

Yaghjian, A. D. (1986). An Overview of Near-Field Antenna Measurements, IEEE Transactions on Antennas and Propagation, Vol. AP-34, No. 1, January 1986, pp. 30-45.

13

Nano-metrology based on the Laser Interferometers

Saeed Olyaee and Samaneh Hamedi
Nano-Photonics and Optoelectronics Research Laboratory
Shahid Rajaee Teacher Training University
Iran

1. Introduction

In recent years, demands for non-contact displacement measurement systems with sub-nanometer uncertainty have significantly increased. One technique to reach the non-contact displacement measurements is based on the use of interferometers. Since 1887 when Michelson and Morley did the famous experiment to detect possible changes in the speed of light in different directions, optical interferometry has been widely used in length-related measurements (Wu et al., 2002). Low-coherence interferometry (LCI) is an optical technique that may be used for industrial surface metrology with accuracy in the micron range. Displacement measurement systems based on the coherent methods provide a high accuracy measurement in different axes. The laser interferometers can measure the absolute positions as well as displacements in a wide dynamic range by measuring the changes of laser interference fringes.

A high precision displacement measurement is necessary in many applications such as photolithography, transducer calibration, geodesy, semiconductor fabrication, precision cutting, shape measurement, and robotic systems. The fabrication of semiconductor chips requires lithographic stepper machines in order to measure high accuracy displacement (Brink et al., 1996; Demarest, 1998). The planar lithographic fabrication process is used to fabricate many nano-systems like electronic, photonic, magnetic, mechanical, chemical, and biological devices. Semiconductor manufacturing tools depend on accurate metrology frames such as lithography scanners and steppers, lithography mask (reticle) writers, circuit and mask repair tools, CD metrology tools, pattern placement, and overlay metrology tools. Pattern placement metrology measures and/or controls pattern location and overlay. A metrology frame consists of three components stable structure, length scale (e.g., wavelength of light), and means to compare work piece with length scale (e.g., microscope). The minimum feature of the integrated circuits can be decreased by improvement in accuracy in the displacement measurements. The most important requirements for future photolithography and masking process such as accuracy of nano-displacement are annually reported by ITRS (ITRS, 2008). Industrial development of semiconductor devices and circuits is asking for improved resolution and reached the high accuracy of nano-metrology systems.

The laser interferometers are mainly divided into two categories; homodyne and heterodyne. The laser heterodyne interferometers have been widely used in displacement measuring systems with sub-nanometer resolution. During the last few years nanotechnology has been changed from a technology only applied in semiconductor industry to the invention of new production with micro and nanometer size until in future picometer size such as, nano electro mechanical systems (NEMS), semiconductor nano-systems, nano-sensors, nano-electronics, nano-photonics and nano-magnetics (Schattenburg & Smith, 2001).

In this chapter, we investigate some laser interferometers used in the nano-metrology systems, including homodyne interferometer, two-longitudinal-mode laser heterodyne interferometer, and three-longitudinal-mode laser heterodyne interferometer (TLMI). Throughout the chapter, we use the notations described in Table 1.

2. Principles of the Laser Interferometers as Nano-metrology System

2.1 Interference Phenomenon

Everyone has seen interference phenomena in a wet road, soap bubble and like this. Boyle and Hooke first described interference in the 17th century. It was the start point of optical interferometry, although the development of optical interferometry was stop because the theory of wave optics was not accepted.

A beam of light is an electromagnetic wave. If we have coherence lights, interference phenomenon can be described by linearly polarized waves. The electrical field E in z direction is represented by exponential function as (Hariharan, 2003):

$$E = \mathrm{Re}\{a \exp[i2\pi\nu(t - z/c)]\} \tag{1}$$

where a is the amplitude, t is the time, ν is the frequency of the light source and c is the speed of propagation of the wave. If all equations on E are linearly assumed, it can be renewed as:

$$\begin{aligned} E &= \mathrm{Re}\{a \exp(-i2\pi\nu z/c)\exp(i2\pi\nu t)\} \\ &= \mathrm{Re}\{a \exp(-i\varphi)\exp(i2\pi\nu t)\} \end{aligned} \tag{2}$$

The real part of this equation is:

$$\begin{aligned} E &= A \exp(i2\pi\nu t) \\ A &= a \exp(-i\varphi) \end{aligned} \tag{3}$$

where

$$\varphi = \frac{2\pi\nu z}{c} = \frac{2\pi nz}{\lambda} \tag{4}$$

In this formula λ is the wavelength of light and n is the refractive index of medium. According to Fig. 1, if two monochromic waves with the same polarization propagate in the same direction, the total electric field at the point P is given by:

$$E = E_1 + E_2 \tag{5}$$

where E_1 and E_2 are the electric fields of two waves. If they have the same frequency, the total intensity is then calculated as:

$$I = \left| A_1 + A_2 \right|^2 \tag{6}$$

Abbreviations		**Constants & Symbols**	
APD	avalanche photodiode	$\tilde{E}$	amplitude of leakage electrical field
BPF	Band pass filter	$\hat{E}$	amplitude of main electrical field
BS	non-polarizing beam splitter	f_s	secondary beat frequency
CCP	corner cube prism	$f_{bH} = \nu_3 - \nu_2$	higher intermode beat frequency
DBM	double-balanced mixer	$f_{bL} = \nu_2 - \nu_1$	the lower intermode beat frequency
FP	frequency-path	I_b	base photocurrent
IVC	I to V converter	I_m	measurement photocurrent
LCI	low-coherence interferometry	n	refractive index of medium
LP	linear polarizer	$\overline{V}$	target velocity
OPD	optical path difference	α	rotation angle of the PBS with respect to the laser polarization axis
PBS	polarizing-beam splitter	α and β	non-orthogonality of the polarized beams
TLMI	three-longitudinal-mode interferometer	δ_{ε_l} and δ_{ε_r}	ellipticity of the central and side modes
Vectors & Jones Matrices		Δf	Doppler shift
LP	matrix of LP	Δz	the displacement measurement
RCCP	matrix of reference CCP	$\Delta\varphi$	the phase change
RPBS	matrix of reference PBS	$\Delta\Phi$	the phase change resulting from optical path difference
TCCP	matrix of target CCP	φ_0	the initial phase corresponding to the electrical field of E_i
TPBS	matrix of target PBS	λ	the wavelength of input source
$\vec{\mathbf{E}}_{\mathbf{LP}_X}$	X component of the total electric field	Λ_{II}	the synthetic wavelength in two-mode laser heterodyne interferometer
$\vec{\mathbf{E}}_{\mathbf{LP}_y}$	Y component of the total electric field	Λ_{III}	the synthetic wavelength in three-mode laser heterodyne interferometer
Constants & Symbols		ν	the optical frequency
c	the speed of light in vacuum	θ	the deviation angle of polarizer referred to 45°
$d\varepsilon$	ellipticity of the polarized beams	ρ	the reflection coefficients of the PBS
$\vec{E}$	electrical field vector	ω	optical angular frequency
ℓ	the number of distinct interference terms	τ	the transmission coefficients of the PBS
ψ	nonlinearity phase	$\aleph$	the number of active FP elements

Table 1. Nomenclatures

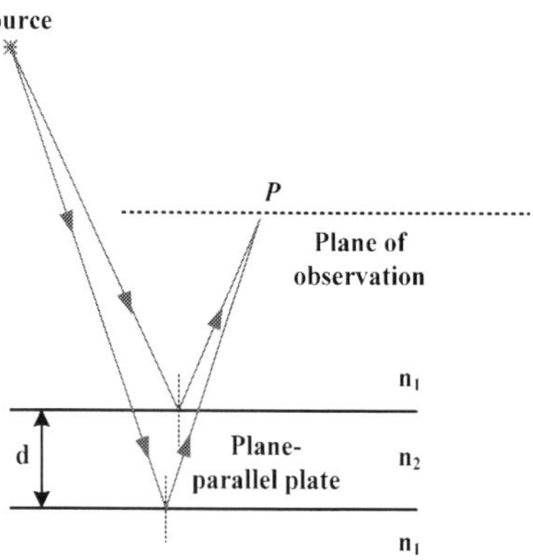

Fig. 1. Formation of interference in a parallel plate waves

$$\begin{aligned} I &= A_1^2 + A_2^2 + A_1 A_2^* + A_1^* A_2 \\ &= I_1 + I_2 + 2(I_1 I_2)^{1/2} \cos \Delta\varphi \end{aligned} \tag{7}$$

where I_1 and I_2 are the intensities at point *P*, resulting from two waves reflected by surface and

$$A_1 = a_1 \exp(-i\varphi_1) \tag{8}$$

$$A_2 = a_2 \exp(-i\varphi_2) \tag{9}$$

The phase difference between two waves at point *P* is given by:

$$\Delta\varphi = \varphi_1 - \varphi_2 = \frac{2\pi v \Delta z}{c} = \frac{2\pi n \Delta z}{\lambda} \tag{10}$$

According to Eq. (10), the displacement can be calculated by detecting the phase from interference signal. An instrument which is used to measure the displacement based on the interferometry phenomenon is interferometer. Michelson has presented the basic principals of optical displacement measurement based on interferometer in 1881. According to using a stabilized He-Ne as input source (Yokoyama et al., 1994; Eom et al., 2002; Kim & Kim, 2002; Huang et al., 2000; Yeom & Yoon, 2005), they are named laser interferometers. Two kinds of laser interferometers depending on their detection principles, homodyne or heterodyne methods, have been developed and improved for various applications.

Homodyne interferometers work due to counting the number of fringes. A fringe is a full cycle of light intensity variation, going from light to dark to light. But the heterodyne interferometers work based on frequency detecting method that the displacement is arrived from the phase of the beat signal of the interfering two reflected beams. On the other hand, heterodyne method such as Doppler-interferometry in comparison with homodyne method provides more signal-to-noise ratio and easier alignment in the industrial field applications (Brink et al., 1996). Furthermore, the heterodyne interferometers are known to be immune to environmental effects. Two-frequency laser interferometers are being widely used as useful instruments for nano-metrology systems.

2.2 Homodyne Interferometer

Commercial homodyne laser interferometers mainly includes a stabilized single frequency laser source, two corner cube prisms (CCPs), a non-polarizing beam splitter (BS), two avalanche photodiodes (APDs), and measurement electronic circuits. The laser frequency stabilization is many important to measure the displacement accurately. A laser source used in the interferometers is typically a He-Ne laser.

An improved configuration of the single frequency Michelson interferometer with phase quadrature fringe detection is outlined in Fig. 2. A 45° linearly polarized laser beam is split by the beam splitter. One of the two beams, with linear polarization is reflected by a CCP_r which is fixed on a moving stage. The other beam passes through a retarder twice, and consequently, its polarization state is changed from linear to circular. The electronics following photodetectors at the end of interferometer count the fringes of the interference signal (see section 3.2). With interference of beams, two photocurrent signals I_x and I_y are concluded as:

$$I_y = a\sin\left(\frac{4n\pi}{\lambda}\Delta z\right) \tag{11}$$

$$I_x = b\cos\left(\frac{4n\pi}{\lambda}\Delta z\right) \tag{12}$$

where Δz is the displacement of CCP_t which is given by:

$$\Delta z = \frac{\lambda}{4n\pi}\tan^{-1}\left(\frac{I_y}{I_x}\right) \tag{13}$$

This is called a DC interferometer, because there is no dependency to the time in the measurement signal (Cosijns, 2004).

2.3 Heterodyne Interferometer

A heterodyne laser interferometer contains a light source of two- or three-longitudinal-mode with orthogonal polarizations, typically a stabilized multi-longitudinal-mode He-Ne laser. The basic setup of a two-mode heterodyne interferometer is shown in Fig. 3. The electric field vectors of laser source are represented by:

$$\vec{E}_1 = \hat{E}_{01}\exp(2\pi\nu_1 t + \varphi_{01})\vec{e}_1 \tag{14}$$

$$\vec{E}_2 = \hat{E}_{02}\exp(2\pi\nu_2 t + \varphi_{02})\vec{e}_2 \tag{15}$$

where $\hat{E}_{01}$ and $\hat{E}_{02}$ are the electric field amplitudes, ν_1 and ν_2 are the optical frequencies stabilized in the gain profile and φ_{01} and φ_{02} represent the initial phases. As it can be seen from Fig. 3, the optical head consists of the base and measurement arms. The laser output beam is separated by a non-polarizing beam splitter from which the base and measurement beams are produced. The base beam passing through a linear polarizer is detected by a photodetector. Consequently, in accordance with Eq. (6), the base photocurrent I_b with $\nu_2 - \nu_1$ intermode beat frequency is obtained as:

$$I_b = 2\hat{E}_{01}\hat{E}_{02}\cos\left(2\pi(\nu_2 - \nu_1)t + (\phi_{02} - \phi_{01})\right) \tag{16}$$

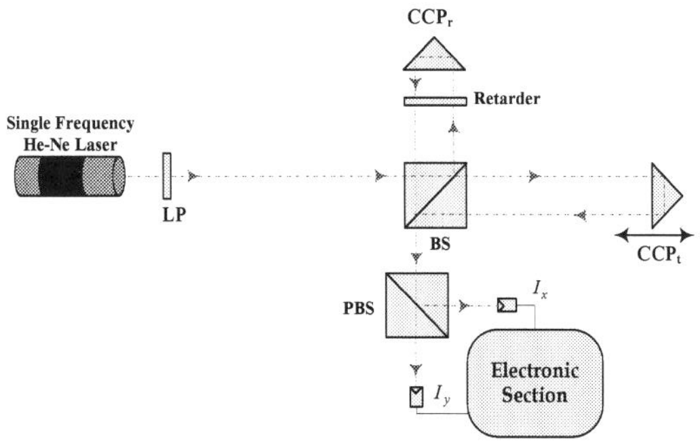

Fig. 2. The schematic representation of homodyne laser interferometer

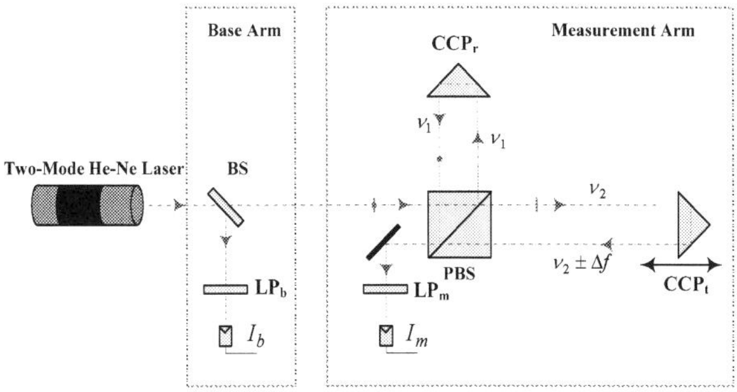

Fig. 3. The schematic representation of heterodyne laser interferometer

As it is concluded from Eq. (16), the heterodyne interferometer works with the frequency ($\nu_2 - \nu_1$), therefore it is called an AC interferometer. The measurement beam is split into two beams namely target and reference beams by the polarizing-beam splitter (PBS) and are directed to the corner cube prisms. The phases of modes are shifted in accordance with the optical path difference (OPD). To enable interference, the beams are transmitted through a linear polarizer (LP) under 45° with their polarization axes. After the polarizer, a photodetector makes measurement signal I_m :

$$I_m = 2\hat{E}_{01}\hat{E}_{02}\cos\left(2\pi(\nu_2 - \nu_1)t + (\varphi_{02} - \varphi_{01}) + (\varphi_t - \varphi_r)\right) \tag{17}$$

The phase difference between base and measurement arms represents the optical path difference which is dependent to the displacement measurement. As the CCP$_t$ in the measurement arm moves with velocity $\overline{V}$, a Doppler shift is generated for ν_2 :

$$\Delta f = 2\pi\left(\frac{2n\overline{V}v_2}{c}\right) \tag{18}$$

The phase change in the interference pattern is dependent on the Doppler frequency shift:

$$\Delta\Phi = \int_{t_1}^{t_2} 2\pi\Delta v t = \frac{4n\pi v_2}{c}\Delta z \tag{19}$$

Finally, the displacement measurement of the target with vacuum wavelength λ_2 is given as:

$$\Delta z = \frac{\Delta\Phi\lambda_2}{4n\pi} \tag{20}$$

3. Comparison Study between Two- and Three-Longitudinal-Mode Laser Heterodyne Interferometers

3.1 The Optical Head

To reach higher resolution and accuracy in the nanometric displacement measurements, a stabilized three-longitudinal-mode laser can replace two-longitudinal-mode laser. In the two-mode interferometer, one intermode beat frequency is produced, whereas in three-mode interferometer three primary beat frequencies and a secondary beat frequency appear. Although the three-longitudinal-mode interferometers (TLMI) have a higher resolution compared to two-longitudinal-mode type, the maximum measurable velocity is dramatically reduced due to the beat frequency reduction. Yokoyama et al. designed a three-longitudinal-mode interferometer with 0.044 nm resolution, assuming the phase detection resolution of 0.1° (Yoloyama et al., 2001). However, limitation of the velocity in the displacement measurement can be eliminated by a proper design (Yokoyama et al., 2005).

The source of the multiple-wavelength interferometer should produce an appropriate emission spectrum including of several discrete and stabilized wavelengths. The optical frequency differences determine the range of non-ambiguity of distance and the maximum measureable velocity. The coherence length of the source limits the maximal absolute distance, which can be measured by multiple-wavelength. If we consider a two-wavelength interferometry using the optical wavelengths λ_1 and λ_2 with orthogonal polarization, the phase shift of each wavelength will be:

$$\Phi_i = \frac{4\pi\Delta z}{\lambda_i} \tag{21}$$

where Δz is the optical path difference and Φ_i is the phase shift corresponding to the wavelength λ_i. Therefore, the phase difference between λ_1 and λ_2 is given by:

$$\Delta\Phi = 4\pi\Delta z\left(\frac{1}{\lambda_1} - \frac{1}{\lambda_2}\right) \tag{22}$$

And the synthetic wavelength, Λ_{II}, can be expressed as:

$$\Lambda_{\mathrm{II}} = \frac{\lambda_1 \lambda_2}{|\lambda_1 - \lambda_2|} = \frac{c}{|\nu_1 - \nu_2|} \tag{23}$$

where ν_1 and ν_2 are the optical frequencies corresponding to λ_1 and λ_2, and c is the speed of light in vacuum. If the number of stabilized wavelengths in the gain curve increase to three-longitudinal-mode, the synthetic wavelength is obtained as:

$$\Lambda_{\mathrm{II}} = \frac{\lambda_1 \lambda_2 \lambda_3}{|\lambda_1 \lambda_2 - 2\lambda_1 \lambda_3 + \lambda_2 \lambda_3|} = \frac{c}{f_s} \tag{24}$$

where f_s is the secondary beat frequency in the three-mode laser heterodyne interferometers. Therefore, the synthetic wavelength in the three-longitudinal-mode interferometer comparing to two-mode system is considerably increased (Olyaee & Nejad, 2007c). The stabilized modes in the gain profile of the laser source and optical head of the nano-metrology system on the basis of two- and three-longitudinal-mode lasers are shown in Fig. 4. As it is represented three wavelengths for which the polarization of the side modes λ_1 and λ_3 is orthogonal to the polarization of the central mode λ_2. The electric field of three modes of laser source is obtained as:

$$E_i = \hat{E}_i \sin(2\pi \nu_i t + \varphi_i) \quad , \quad i = 1,2,3 \tag{25}$$

where φ_i is the initial phases corresponding to the electric field E_i. In both cases, the optical head consists of the base and measurement arms. First, the laser output is separated by BS, so that base and measurement beams are produced. Then, the beam is split into two subsequent beams by PBS and directed to each path of the interferometers. Two reflected beams are interfered to each other on the linear polarizer. Because of orthogonally polarized modes, the linear polarizer should be used to interfere two beams as shown in Fig. 5. The stabilized multimode He-Ne lasers are chosen in which the side modes can be separated from the center mode due to the orthogonal polarization states.

But in reality, non-orthogonal and elliptical polarizations of beams cause each path to contain a fraction of the laser beam belonging to the other path. Hence, the cross-polarization error is produced. In the reference path (path.1) of TLMI, ν_1 and ν_3 are the main frequencies and ν_2 is the leakage one, whereas in the target path (path.2), ν_2 is the main signal and the others are as the leakages.

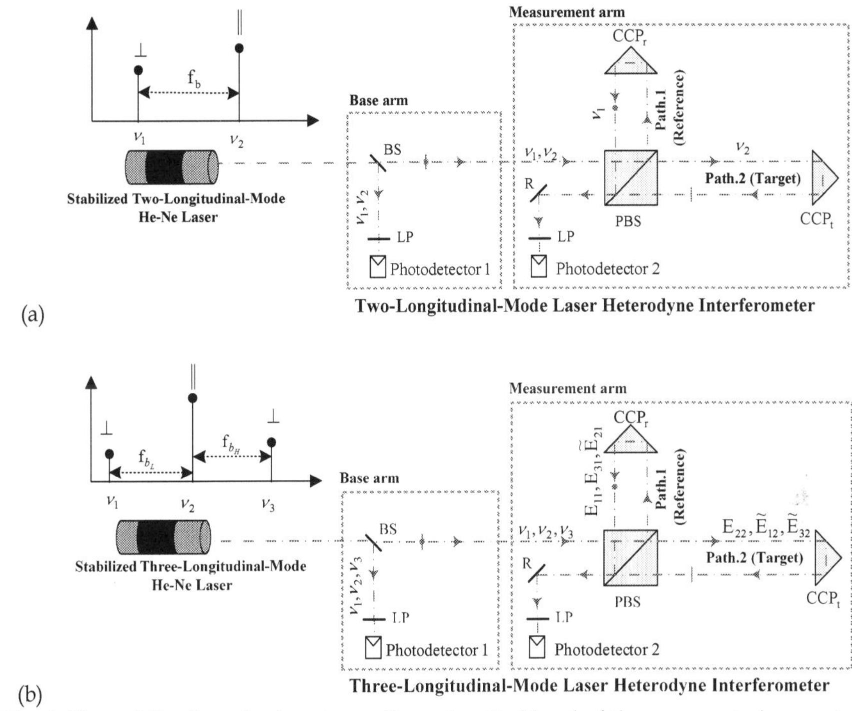

Fig. 4. The stabilized modes in gain profile and optical head of the nano-metrology system based on (a) the two- and (b) three-longitudinal-mode He-Ne laser interferometers

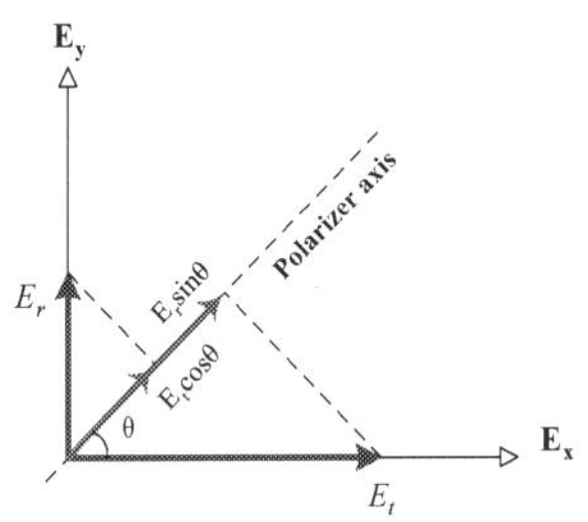

Fig. 5. Combination of orthogonally polarized beams on the linear polarizer

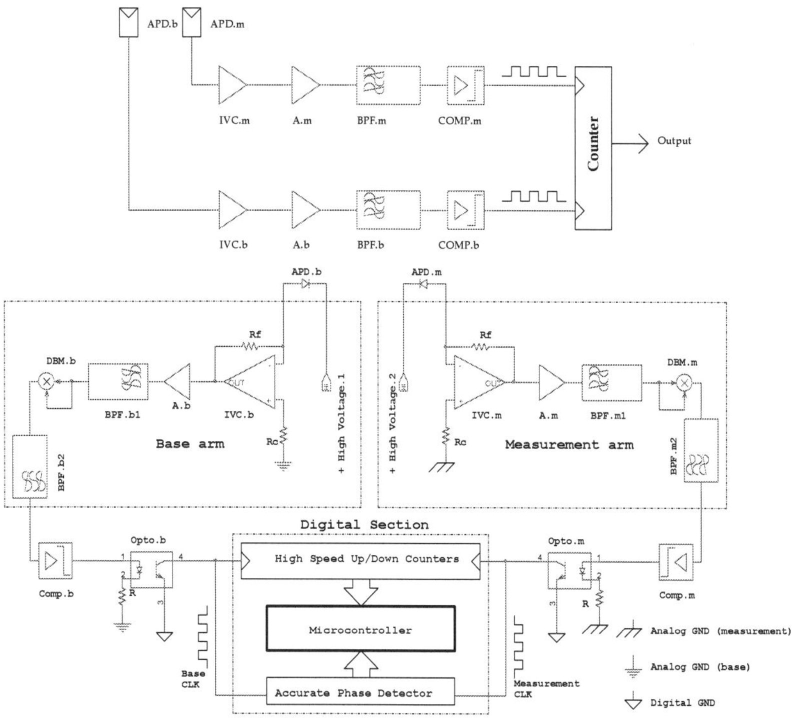

(a)

(b)

Fig. 6. The schematic of the electronic circuits of the nano-metrology system based on the (a) two- and (b) three-longitudinal-mode laser interferometers

Owing to the square-law behavior of the photodiodes, the reference signal is expressed as:

$$
\begin{aligned}
I_{APD_b} &= A\cos(2\pi(\nu_3 - \nu_1)t) + B\cos(2\pi(\nu_3 - \nu_2)t) + C\cos(2\pi(\nu_2 - \nu_1)t) + D \\
&= A\cos(2\pi(f_{bH} + f_{bL})t) + B\cos(2\pi f_{bH}t) + C\cos(2\pi f_{bL}t) + D
\end{aligned}
\tag{26}
$$

Similarly, the output current of the measurement avalanche photodiode, APD_m, is:

$$
I_{APD_m} = A\cos(2\pi(f_{bH} + f_{bL})t) + B\cos(2\pi(f_{bH} \mp \Delta f)t) + C\cos(2\pi(f_{bL} \pm \Delta f)t) + D \tag{27}
$$

where A, B, C, and D are constant values and Δf is the frequency shift due to the Doppler effect and its sign is dependent on the moving direction of the target. To extract the phase shift from Eqs. (26) and (27), two signals are fed to the proper electronic section as described in the following.

3.2 The Electronic Sections

The schematic diagram of the electronic circuits of the two- and three-longitudinal-mode laser interferometers are shown in Fig. 6. In both systems, the photocurrents of the avalanche photodiodes are amplified and converted to voltage signals. In two-mode system, the amplified signals pass through the band-pass filters (BPFs) involving the intermode beat frequency (typically several hundred MHz which can be reduced by heterodyne technique).

Parameter		Two-Longitudinal-Mode Laser Interferometer		Three-Longitudinal-Mode Laser Interferometer		Unit
Wavelength		632.8		632.8		nm
Cavity length		25		35		cm
Synthetic wavelength		0.5		1000		m
Maximum absolute distance		0.25		500		m
Intermode beat frequency		600		435.00, 435.30, 870.30		MHz
Secondary beat frequency		---		300		kHz
Maximum measurable velocity		21		0.047		m/s
Phase detection accuracy (similar circuit)		11.8		5.9		pm
Cross-talk and intermodulation distortion error		100		18		pm
The number of active frequency-path elements		4		6		
The number of distinct interference terms	Optical power	Total: 10	4	Total: 21	6	
	AC interference		2		6	
	DC interference		2		3	
	AC reference		2		6	

Table 2. A comparison between two- and three-longitudinal-mode He-Ne laser interferometers with typical values (Olyaee & Nejad, 2007c)

Two signals from base and measurement arms are then fed to a counter to measure the target displacement resulting from optical path difference.

But in TLMI, the amplified signals are self-multiplied by two double-balanced mixers, DBM_b and DBM_m. As a result, the secondary beat frequency generates (typically several hundred kHz). The high frequency and DC components are eliminated by two band-pass filters, BPF_{b2} and BPF_{m2}. The input signals of the comparators for base and measurement arms are respectively described as:

$$V_{ob} = k'\cos(2\pi f_s t) \tag{28}$$

$$V_{om} = k'\cos(2\pi f_s t - 2\Delta\Phi) \tag{29}$$

The phase shift resulting from optical path difference is measured by a high-speed up/down counter. The base and measurement signals can be exerted to a half exclusive-or gate and the pulse width is measured by a high speed counter. The phase difference between the base and measurement signals is proportional to the output pulse width. The resolution of the phase detector is proportional to the clock pulse of the counter. The phase shift due to optical path difference is given by:

$$\Delta\Phi = \frac{4n\pi}{\lambda_2}\Delta z \tag{30}$$

where λ_2 is the central wavelength. It should be noted that in Eq. (29), the phase shift is multiplied by 2 which indicates the resolution in TLMI is doubled compared to two-mode type (see Eq. (17)).

On the other hand, the maximum measurable velocity corresponding to Eq. (18) is dependent on the intermode beat frequencies. In the TLMI, because we use super-heterodyne method to extract the secondary beat frequency (that is much smaller than primary beat frequencies produced in the TLMI or than intermode beat frequency in two-mode type), the maximum measurable velocity to be considerably reduced.

A comparison between two- and three-mode laser interferometers with typical values is summarized in Table 2. The maximum measurable velocity for two-mode type is about 21m/s, whereas in TLMI it is limited to 47.46mm/s. But according to Table 2, the resolution of the displacement measurement and synthetic wavelength in the three-longitudinal-mode is considerably increased. The output signals of the measurement double-balanced mixer and band-pass filter for fixed target, -47 mm/s, -20 mm/s, and +47 mm/s target velocities are shown in Fig. 7.

3.3 The Frequency-Path Modeling

A multi-path, multi-mode laser heterodyne interferometer can be described by a frequency-path (FP) model. The frequency-path models of two- and three-longitudinal-mode interferometers are shown in Fig. 8. In the measurement arm of TLMI, there are three frequency components and two paths namely the reference and the target (the bold lines are the main signal paths and the dashed lines are the leakage paths), whereas in two-longitudinal-mode interferometer, there are two frequency components and two paths. The number of active frequency-path elements, $\aleph$, is obtained by multiplying the number of frequency components by paths (Schmitz & Beckwith, 2003). Consequently, in two-path, two- and three-mode interferometers, the number of active FP elements is 4 and 6, respectively.

Figure 9 shows the identification of the physical origin of each frequency-path element for the measurement arms of two- and three-mode interferometers. Because the wave intensity being received by an APD is proportional to the square of the total electrical field, the number of distinct interference terms is equal to:

$$\mathrm{M} = \frac{\aleph(\aleph+1)}{2} = 21 \tag{31}$$

In TLMI, the reference path field can be described by:

$$E_i = E_{i1}\cos(\omega_i t - k_i z_1 + \varphi_{i1}) \quad , \quad i = 1,2,3 \tag{32}$$

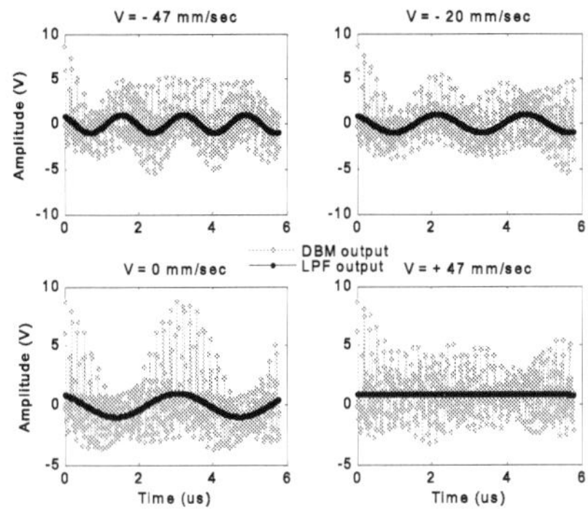

Fig. 7. The output signals of the measurement BPF and DBM (Olyaee & Nejad, 2007b)

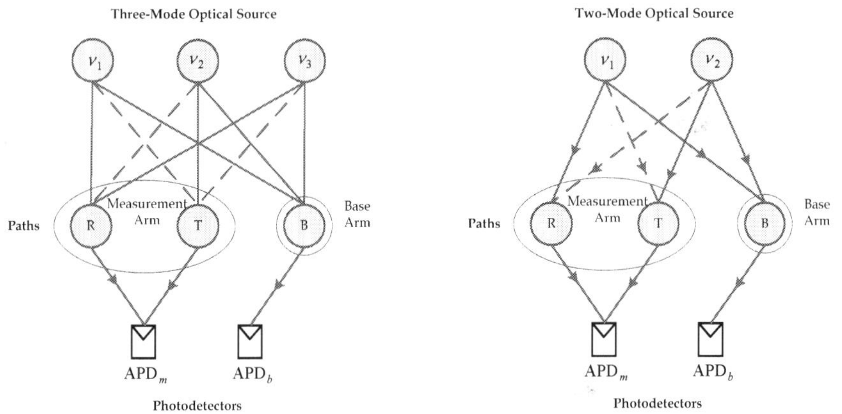

Fig. 8. The frequency-path model in two- and three-longitudinal-mode laser interferometers

where φ_{ij} is the initial phases corresponding to the electrical field E_{ij}, k_i is the propagation constant or wave number $(=2\pi/\lambda_i)$, $\omega_i = 2\pi\nu_i$ is the optical angular frequency, and z_1 is the motion of the corner cube prism in the reference path (CCP$_r$). Similar to Eq. (32), the target path field is described by:

$$E_i = E_{i2}\cos(\omega_i t - k_i z_2 + \varphi_{i2}) \quad , \quad i = 1,2,3 \tag{33}$$

where z_2 is the motion of the corner cube prism in the target path (CCP$_t$). In this system, the CCP$_r$ is fixed and hence, $z_1 = 0$. Furthermore, the wavelengths are so close that propagation constants become almost equal to each other ($k_1 = k_2 = k_3 = k$). The high frequency components such as ω_i, $2\omega_i$ and $\omega_i + \omega_j$ are eliminated by the avalanche photodiodes ($i, j = 1,2,3$). Therefore, ignoring the high frequencies in the fully unwanted leaking interferometers, there are 21 distinct interference terms for three-longitudinal-mode interferometer and 10 distinct interference terms for two-mode type (see Table 2). The distinct interference terms can be divided into four groups namely dc interference (DI), ac interference (AI), ac reference (AR), and optical power (OP). These components in the three-longitudinal-mode interferometer are respectively given by (Olyaee & Nejad, 2007a):

$$I_{DI}/K = E_{11}\widetilde{E}_{12}\cos(kz_2) + \widetilde{E}_{21}E_{22}\cos(kz_2) + E_{31}\widetilde{E}_{32}\cos(kz_2) \tag{34}$$

$$\begin{aligned} I_{AI}/K = {} & E_{11}E_{22}\cos(\omega_{bL}t - kz_2) + E_{11}\widetilde{E}_{32}\cos((\omega_{bH} + \omega_{bL})t - kz_2) + \widetilde{E}_{21}\widetilde{E}_{12}\cos(\omega_{bL}t + kz_2) + \\ & \widetilde{E}_{21}\widetilde{E}_{32}\cos(\omega_{bH} - kz_2) + E_{31}\widetilde{E}_{12}\cos((\omega_{bH} + \omega_{bL})t + kz_2) + E_{31}E_{22}\cos(\omega_{bH}t + kz_2) \end{aligned} \tag{35}$$

$$\begin{aligned} I_{AR}/K = {} & \left(E_{11}\widetilde{E}_{21} + \widetilde{E}_{12}E_{22}\right)\cos(\omega_{bL}t) + \left(\widetilde{E}_{21}E_{31} + E_{22}\widetilde{E}_{32}\right)\cos(\omega_{bH}t) \\ & + \left(\widetilde{E}_{12}\widetilde{E}_{32} + E_{11}E_{31}\right)\cos((\omega_{bH} + \omega_{bL})t) \end{aligned} \tag{36}$$

$$I_{OP}/K = \frac{1}{2}\left(E_{11}^2 + \widetilde{E}_{21}^2 + E_{31}^2 + \widetilde{E}_{12}^2 + E_{22}^2 + \widetilde{E}_{32}^2\right) \tag{37}$$

Figure 10 shows the combination of the frequency-path elements in the measurement arm of two systems. The main signals and leakages are depicted by large and small solid circles, respectively. On the other hand, the diameter of the circles presents the amplitude of the signals. The small solid circles are exaggerated for clarification. All of the distinct interference terms are shown in Fig. 10 by different lines.

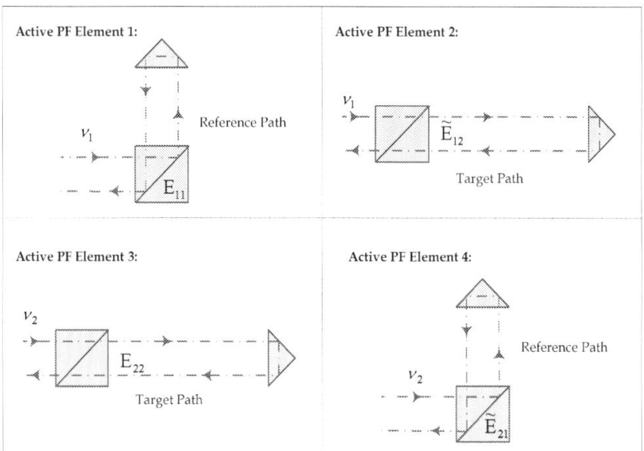

(a)

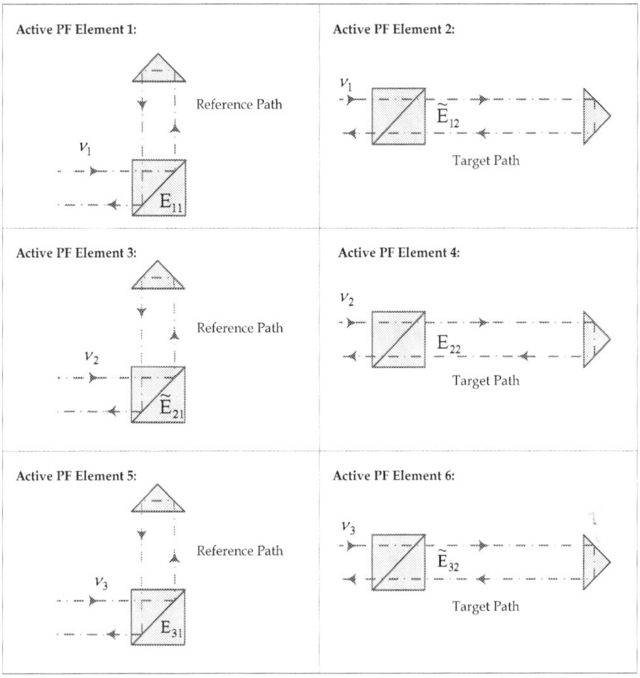

(b)

Fig. 9. Identification of the physical origin of each frequency-path element in the reference and target paths (measurement arms). (a) Two- and (b) Three-longitudinal-mode laser interferometers

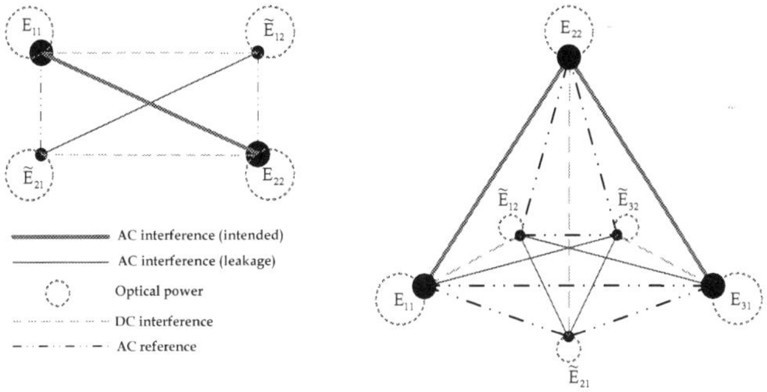

(a) (b)

Fig. 10. The combination graph of the frequency-path elements in (a) two- and (b) three-longitudinal-mode interferometers

4. Nonlinearity Analysis

In section 2, we have shown that for a heterodyne interferometer the displacement could be determined by measuring the phase change between measurement and base signals. From Eq. (20), it can be seen that the accuracy of the displacement depends on the accuracy of the determination of the phase change, the wavelength of light and the refractive index of the medium. But, the displacement accuracy is also influenced far more by setup configuration, instrumentation section and environmental effects (Cosijns, 2004).

The first group is related to misalignment and deviations in the optical setup and components such as polarizer, polarizing-beam splitter and laser head. These can be minimized or even eliminated by using a correct setup and alignment procedures. In the second group related to electronic and instrumentation section includes laser frequency instability, phase detection error and data age uncertainty (Demarest, 1998). Instability in the mechanical instruments, cosine error and Abbe error are directly related to the setup configuration. The accuracy of refractive index determination, turbulences and thermal instability are the environmental parameters affecting the accuracy of the displacement (Bonsch & Potulski 1998; Wu, 2003; Edlen, 1966). In the mentioned errors, several of them are considered as linear errors that can be simply reduced or compensated.

When the measured displacement with a non-ideal interferometer is plotted against the real displacement of the moving target an oscillation around the ideal straight line is observed. This effect is known as a periodic deviation of the laser interferometer (Cosijns, 2004). The stability of the laser source, alignment error, vibration, temperature variation and air turbulence are the main sources of error for the optical interferometer. If all of the above conditions can be kept good enough, then the practical limitations will be given by the photonic noise and the periodic nonlinearity inherent in the interferometer (Wu & Su, 1996). The nonlinearity of one-frequency interferometry is a two-cycle phase error, whereas in heterodyne interferometry is mainly a one-cycle phase error as the optical path difference changes from 0 to 2π. Although the heterodyne interferometers have a larger nonlinearity than do the one-frequency interferometers, with first-order versus second-order error, the first-order nonlinearity of heterodyne interferometers can be compensated on-line (Wu et al, 1996).

In the ideal heterodyne interferometers, two beams are completely separated from each other and traverse with pure form in the two arms of the interferometer. Although the heterodyne method compared to the homodyne method provides more signal to noise ratio and easy alignment, in contrast, because of using two separated beams in the heterodyne method, the nonlinearity errors especially cross-talk and cross-polarization dominate. The polarization-mixing happens within an imperfect polarizing-beam splitter. This is nonlinearity error which is often in the one frequency interferometer. Meanwhile in case of heterodyne laser interferometer, frequency mixing error which arises from non-orthogonality of the polarizing radiations, elliptical polarization and imperfect alignment of the laser head and other components produce periodic nonlinearity error (Cosijns et al., 2002; Freitas, 1997; Eom et al. 2001; Hou & Wilkening 1992; Meyers et al., 2001; Sutton, 1998). The two waves, which are regarded as orthogonal to each other in a heterodyne interferometer, are not perfectly separated by the polarizing-beam splitter, with the result that the two frequencies are mixed. The mixing leads to a nonlinear relationship between the measured phase and the actual phase, and limits the accuracy of the heterodyne interferometer with a two-frequency laser to a few nanometers. The periodic nonlinearity can be analytically modeled by both Jones calculus and plane wave which will be described in the next two sections.

4.1 Analytical Modeling of the Periodic Nonlinearity based Jones Calculus

According to the setup of TLMI in Fig. 4b, in both paths of the reference (reflected) and target paths (transmitted), there are small fractions of oppositely polarized beams as a leakage caused by the ellipticity of the laser mode polarization and misalignment of the polarization axes between the laser beam and PBS. The leakage beams result in the frequency mixing and produce the periodic nonlinearity in the detected heterodyne signal. To have a model of nonlinearity in the TLMI, we first assume that the beam emerged from the laser is to be elliptically polarized. The ellipticity of the central and side modes are denoted by δ_{ε_t} and δ_{ε_r}, respectively, as usual. Then, the electric fields of three longitudinal modes are respectively given as:

$$\vec{\mathbf{E}}_1 = \begin{pmatrix} \cos\delta_{\varepsilon_r} \\ i\sin\delta_{\varepsilon_r} \end{pmatrix} \exp i\left(\omega_1 t + \varphi_{0_1}\right) \tag{38}$$

$$\vec{\mathbf{E}}_2 = -\begin{pmatrix} i\sin\delta_{\varepsilon_t} \\ \cos\delta_{\varepsilon_t} \end{pmatrix} \exp i\left(\omega_2 t + \varphi_{0_2}\right) \tag{39}$$

$$\vec{\mathbf{E}}_3 = \begin{pmatrix} \cos\delta_{\varepsilon_r} \\ i\sin\delta_{\varepsilon_r} \end{pmatrix} \exp i\left(\omega_3 t + \varphi_{0_3}\right) \tag{40}$$

If the rotation angle of the PBS with respect to the laser polarization axis is denoted by α, the matrix representing the PBS for reference and target beam directions respectively can be calculated as :

$$\mathbf{RPBS} = \begin{pmatrix} \cos\alpha & -\sin\alpha \\ \sin\alpha & \cos\alpha \end{pmatrix} \begin{pmatrix} 1 & 0 \\ 0 & 0 \end{pmatrix} \begin{pmatrix} \cos\alpha & \sin\alpha \\ -\sin\alpha & \cos\alpha \end{pmatrix} \tag{41}$$

$$\mathbf{TPBS} = \begin{pmatrix} \cos\alpha & -\sin\alpha \\ \sin\alpha & \cos\alpha \end{pmatrix} \begin{pmatrix} 0 & 0 \\ 0 & 1 \end{pmatrix} \begin{pmatrix} \cos\alpha & \sin\alpha \\ -\sin\alpha & \cos\alpha \end{pmatrix} \tag{42}$$

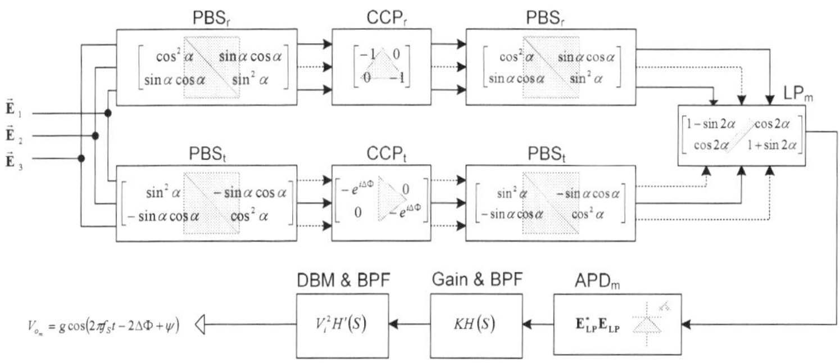

Fig. 11. Jones matrix components of the measurement arm of the TLMI with leakage fields. The dashed lines indicate optical leakage

The Jones matrix for linear polarizer oriented at 45° relative to the polarization directions is then described by:

$$\mathbf{LP} = \begin{pmatrix} \cos^2(45+\alpha) & \cos(45+\alpha)\sin(45+\alpha) \\ \cos(45+\alpha)\sin(45+\alpha) & \sin^2(45+\alpha) \end{pmatrix} \tag{43}$$

Figure 11 shows the simplified matrices for optical components of the target and reference paths including the optical leakages in the TLMI. The optical leakages are shown by dashed lines (Olyaee et al., 2009).

According to Fig. 11, the Jones vector of the reference electrical field incident upon the linear polarizer is obtained as:

$$\vec{\mathbf{E}}_{\mathbf{r}} = \mathbf{RPBS.RCCP.RPBS} \sum_{l=1}^{3} \vec{\mathbf{E}}_l \tag{44}$$

where **RCCP** is the Jones matrix for reference corner cube prism (CCP_r) in the reference path which is given by:

$$\mathbf{RCCP} = \begin{pmatrix} -1 & 0 \\ 0 & -1 \end{pmatrix} \tag{45}$$

If the reflection coefficient of the PBS in the reference path direction is denoted as ρ, by substitution of the related Jones matrices, Eq. (44) can be rewritten as:

$$\vec{\mathbf{E}}_{\mathbf{r}} = \rho \begin{pmatrix} -\cos^2\alpha\cos\delta_{\varepsilon_r} - i\sin\alpha\cos\alpha\sin\delta_{\varepsilon_r} & i\cos^2\alpha\sin\delta_{\varepsilon_t} + \sin\alpha\cos\alpha\cos\delta_{\varepsilon_t} \\ -\sin\alpha\cos\alpha\cos\delta_{\varepsilon_r} - i\sin^2\alpha\sin\delta_{\varepsilon_r} & i\sin\alpha\cos\alpha\sin\delta_{\varepsilon_t} + \sin^2\alpha\cos\delta_{\varepsilon_t} \end{pmatrix} \times \begin{pmatrix} \exp i(\omega_1 t + \varphi_{0_1}) + \exp i(\omega_3 t + \varphi_{0_3}) \\ \exp i(\omega_2 t + \varphi_{0_2}) \end{pmatrix} \tag{46}$$

Similar to Eq. (44), the target electrical field is given by:

$$\begin{aligned} \vec{\mathbf{E}}_{\mathbf{t}} &= \mathbf{TPBS.TCCP.TPBS} \sum_{l=1}^{3} \vec{\mathbf{E}}_l \\ &= \tau \begin{pmatrix} -\sin^2\alpha\cos\delta_{\varepsilon_r} + i\sin\alpha\cos\alpha\sin\delta_{\varepsilon_r} & i\sin^2\alpha\sin\delta_{\varepsilon_t} - \sin\alpha\cos\alpha\cos\delta_{\varepsilon_t} \\ \sin\alpha\cos\alpha\cos\delta_{\varepsilon_r} - i\cos^2\alpha\sin\delta_{\varepsilon_r} & -i\sin\alpha\cos\alpha\sin\delta_{\varepsilon_t} + \cos^2\alpha\cos\delta_{\varepsilon_t} \end{pmatrix} \times \\ &\quad \begin{pmatrix} \exp i(\omega_1 t + \varphi_{0_1} + \Delta\Phi) + \exp i(\omega_3 t + \varphi_{0_3} + \Delta\Phi) \\ \exp i(\omega_2 t + \varphi_{0_2} + \Delta\Phi) \end{pmatrix} \end{aligned} \tag{47}$$

where τ is the transmission coefficient of the PBS in the target path direction and **TCCP** is the Jones matrix for corner cube prism in the target path which is given by:

$$\mathbf{TCCP} = \begin{pmatrix} -\exp(i\Delta\Phi) & 0 \\ 0 & -\exp(i\Delta\Phi) \end{pmatrix} \tag{48}$$

The reflected beam of the reference path and the transmitted beam from the target path at the output port of the PBS interfere with each other through the linear polarizer. Therefore, the final electric field vector after passing through the linear polarizer (LP) is obtained by:

$$\vec{\mathbf{E}}_{\mathbf{LP}} = \mathbf{LP}\left(\vec{\mathbf{E}}_{\mathbf{t}} + \vec{\mathbf{E}}_{\mathbf{r}}\right) \tag{49}$$

The intensity of the laser beam which is proportional of the photocurrent at the detector can be obtained by pre-multiplying the Jones vector with its complex conjugate of the matrix transpose. Consequently, the photocurrent detected by an avalanche photodiode can be given as:

$$I_{APD_m} \propto \vec{\mathbf{E}}^*_{\mathbf{LP}_X}\vec{\mathbf{E}}_{\mathbf{LP}_X} + \vec{\mathbf{E}}^*_{\mathbf{LP}_Y}\vec{\mathbf{E}}_{\mathbf{LP}_Y} \tag{50}$$

where $\vec{\mathbf{E}}_{\mathbf{LP}_X}$ and $\vec{\mathbf{E}}_{\mathbf{LP}_Y}$ are the X and Y components of the total electric field vector, respectively. By expanding Eq. (50) and eliminating the optical frequencies and dc component, the Fourier spectrum components and phase terms of the time-dependent photocurrent is obtained as:

$$\begin{aligned} I_{APD_m} = {} & (e+ia)\left(e^{-i\omega_{bL}t} + e^{i\omega_{bH}t}\right) + (e-ia)\left(e^{i\omega_{bL}t} + e^{-i\omega_{bH}t}\right) + \\ & + (b+ic)\left(e^{-i(\omega_{bL}t+\Delta\Phi)} + e^{i(\omega_{bH}t-\Delta\Phi)}\right) + (d-ic)\left(e^{i(\omega_{bL}t-\Delta\Phi)} + e^{-i(\omega_{bH}t+\Delta\Phi)}\right) + \\ & + (d+ic)\left(e^{-i(\omega_{bL}t-\Delta\Phi)} + e^{i(\omega_{bH}t+\Delta\Phi)}\right) + (b-ic)\left(e^{i(\omega_{bL}t+\Delta\Phi)} + e^{-i(\omega_{bH}t-\Delta\Phi)}\right) \end{aligned} \tag{51}$$

where

$$\begin{aligned} e+ia &= A_X B_X^* + C_X D_X^* + A_Y B_Y^* + C_Y D_Y^* \\ e-ia &= B_X A_X^* + D_X C_X^* + B_Y A_Y^* + D_Y C_Y^* \\ b+ic &= A_X D_X^* + A_Y D_Y^* \\ d-ic &= B_X C_X^* + B_Y C_Y^* \\ d+ic &= C_X B_X^* + C_Y B_Y^* \\ b-ic &= D_X A_X^* + D_Y A_Y^* \end{aligned} \tag{52}$$

and

$$\begin{aligned} A_X/\rho = {} & (\sin 2\alpha - 1)\left(\cos^2\alpha\cos\delta_{\varepsilon_r} + i\sin\alpha\cos\alpha\sin\delta_{\varepsilon_r}\right) \\ & - \cos 2\alpha\left(\sin\alpha\cos\alpha\cos\delta_{\varepsilon_r} + i\sin^2\alpha\sin\delta_{\varepsilon_r}\right) \end{aligned} \tag{53}$$

$$\begin{aligned} B_X/\rho = {} & (1-\sin 2\alpha)\left(i\cos^2\alpha\sin\delta_{\varepsilon_t} + \sin\alpha\cos\alpha\cos\delta_{\varepsilon_t}\right) \\ & + \cos 2\alpha\left(i\sin\alpha\cos\alpha\sin\delta_{\varepsilon_t} + \sin^2\alpha\cos\delta_{\varepsilon_t}\right) \end{aligned} \tag{54}$$

$$\begin{aligned} C_X/\tau = {} & (1-\sin 2\alpha)\left(-\sin^2\alpha\cos\delta_{\varepsilon_r} + i\sin\alpha\cos\alpha\sin\delta_{\varepsilon_r}\right) \\ & + \cos 2\alpha\left(\sin\alpha\cos\alpha\cos\delta_{\varepsilon_r} - i\cos^2\alpha\sin\delta_{\varepsilon_r}\right) \end{aligned} \tag{55}$$

$$\begin{aligned} D_X/\tau = {} & (1-\sin 2\alpha)\left(i\sin^2\alpha\sin\delta_{\varepsilon_t} - \sin\alpha\cos\alpha\cos\delta_{\varepsilon_t}\right) \\ & + \cos 2\alpha\left(-i\sin\alpha\cos\alpha\sin\delta_{\varepsilon_t} + \cos^2\alpha\cos\delta_{\varepsilon_t}\right) \end{aligned} \tag{56}$$

$$\begin{aligned} A_Y/\rho = {} & -\cos 2\alpha\left(\cos^2\alpha\cos\delta_{\varepsilon_r} + i\sin\alpha\cos\alpha\sin\delta_{\varepsilon_r}\right) \\ & -(1+\sin 2\alpha)\left(\sin\alpha\cos\alpha\cos\delta_{\varepsilon_r} + i\sin^2\alpha\sin\delta_{\varepsilon_r}\right) \end{aligned} \tag{57}$$

$$B_Y / \rho = \cos 2\alpha\left(i\cos^2\alpha\sin\delta_{\varepsilon_t} + \sin\alpha\cos\alpha\cos\delta_{\varepsilon_t}\right) \\ +(1+\sin 2\alpha)\left(i\sin\alpha\cos\alpha\sin\delta_{\varepsilon_t} + \sin^2\alpha\cos\delta_{\varepsilon_t}\right) \tag{58}$$

$$C_Y / \tau = \cos 2\alpha\left(-\sin^2\alpha\cos\delta_{\varepsilon_r} + i\sin\alpha\cos\alpha\sin\delta_{\varepsilon_r}\right) \\ +(1+\sin 2\alpha)\left(\sin\alpha\cos\alpha\cos\delta_{\varepsilon_r} - i\cos^2\alpha\sin\delta_{\varepsilon_r}\right) \tag{59}$$

$$D_Y / \tau = \cos 2\alpha\left(i\sin^2\alpha\sin\delta_{\varepsilon_t} - \sin\alpha\cos\alpha\cos\delta_{\varepsilon_t}\right) \\ +(1+\sin 2\alpha)\left(-i\sin\alpha\cos\alpha\sin\delta_{\varepsilon_t} + \cos^2\alpha\cos\delta_{\varepsilon_t}\right) \tag{60}$$

Finally, Eq. (51) can be simplified as:

$$I_{APD_m} = a(\sin(\omega_{bL}t) - \sin(\omega_{bH}t)) + b(\cos(\omega_{bH}t - \Delta\Phi) + \cos(\omega_{bL}t + \Delta\Phi)) \\ +c(\sin(\omega_{bL}t + \Delta\Phi) - \sin(\omega_{bH}t - \Delta\Phi) + \sin(\omega_{bL}t - \Delta\Phi) - \sin(\omega_{bH}t + \Delta\Phi)) \\ +d(\cos(\omega_{bL}t - \Delta\Phi) + \cos(\omega_{bH}t + \Delta\Phi)) + e(\cos(\omega_{bL}t) + \cos(\omega_{bH}t)) \tag{61}$$

As shown in Fig. 6b, the photocurrent of the APD is further processed for the super-heterodyning by using a double-balanced mixer (DBM) and band-pass filter (BPF). Because of using the BPF, the higher frequency component and the dc component of the photocurrent can be effectively filtered out, resulting in the detection of a heterodyne signal oscillating only at the secondary beat frequency with a high signal-to-noise ratio. Thus, the output voltage is proportional to:

$$V'_{o_m} = \left(-a^2 + 2bd - 2c^2 + e^2\right)\cos(\omega_S t) - 2(ab+ce)\sin(\omega_S t - \Delta\Phi) \\ -2(ac-be)\cos(\omega_S t - \Delta\Phi) - 2(cb+cd+ae)\sin(\omega_S t) - 2(ad+ce)\sin(\omega_S t + \Delta\Phi) \\ -2(ac-de)\cos(\omega_S t + \Delta\Phi) - 2bc\sin(\omega_S t - 2\Delta\Phi) - 2cd\sin(\omega_S t + 2\Delta\Phi) \\ +\left(b^2 - c^2\right)\cos(\omega_S t - 2\Delta\Phi) + \left(d^2 - c^2\right)\cos(\omega_S t + 2\Delta\Phi) \tag{62}$$

To extract the phase nonlinearity from the above analytical formula, Eq. (62) could be simplified as:

$$V_m = D\cos(\omega_S t - 2\Delta\Phi) + N\sin(\omega_S t + 2\Delta\Phi) = \sqrt{D^2 + N^2}\cos\left(\omega_S t - 2\Delta\Phi - \tan^{-1}\frac{N}{D}\right) \tag{63}$$

where

$$D = \left(-a^2 + 2bd - 2c^2 + e^2\right)\cos 2\Delta\Phi - 2(ab+ce)\sin\Delta\Phi \\ -2(ac-be)\cos\Delta\Phi - 2(cb+cd+ae)\sin 2\Delta\Phi - 2(ad+ce)\sin 3\Delta\Phi \\ -2(ac-de)\cos 3\Delta\Phi - 2cd\sin 4\Delta\Phi + \left(b^2 - c^2\right) + \left(d^2 - c^2\right)\cos 4\Delta\Phi, \tag{64}$$

$$N = \left(a^2 - 2bd + 2c^2 - e^2\right)\sin 2\Delta\Phi - 2(ab+ce)\cos\Delta\Phi \\ +2(ac-be)\sin\Delta\Phi - 2(cb+cd+ae)\cos 2\Delta\Phi - 2(ad+ce)\cos 3\Delta\Phi \\ +2(ac-de)\sin 3\Delta\Phi - 2bc - 2cd\cos 4\Delta\Phi - \left(d^2 - c^2\right)\sin 4\Delta\Phi, \tag{65}$$

And the phase nonlinearity can be finally calculated as:

$$\psi = -\tan^{-1}\frac{N}{D} \tag{66}$$

This nonlinearity causes to appear the periodic nonlinearity in the displacement measurement.

4.2 Analytical Modeling of the Periodic Nonlinearity based Plane Waves

Another method for nonlinearity modeling is based on the plane wave method. By using these approaches, a similar model for periodic nonlinearity is obtained. Here, the electrical field vectors of side modes, $\vec{\mathbf{E}}_{1,3}$, and central mode, $\vec{\mathbf{E}}_2$, emerging from the three-longitudinal-mode laser source with ellipticity of the reference and target beams are respectively given as:

$$\vec{\mathbf{E}}_{1,3} = \cos\delta_{\varepsilon_r} E_{01}\left(\sin(2\pi v_1 t + \varphi_{01}) + \sin(2\pi v_3 t + \varphi_{03})\right)) - \sin\delta_{\varepsilon_r} E_{02}\cos(2\pi v_2 t + \varphi_{02}) \tag{67}$$

$$\vec{\mathbf{E}}_2 = \sin\delta_{\varepsilon_t} E_{01}(\cos(2\pi v_1 t + \varphi_{01}) + \cos(2\pi v_3 t + \varphi_{03})) - \cos\delta_{\varepsilon_t} E_{02}\sin(2\pi v_2 t + \varphi_{02}) \tag{68}$$

Considering the non-orthogonality of the polarized beams in combination with the deviation angle of PBS with respect to the laser head, α and β, a leakage of modes appears. The electrical field magnitudes of the transmitted and reflected beams by the PBS are given by:

$$\vec{\mathbf{E}}_t = \tau\begin{pmatrix}\sin\beta\\ \cos\alpha\end{pmatrix}\times\left[E_{13}(x_i + \phi_t)E_2(x_i + \phi_t)\right] \tag{69}$$

$$\vec{\mathbf{E}}_r = \rho\begin{pmatrix}\sin\beta\\ \cos\alpha\end{pmatrix}\times\left[E_{13}(x_i + \phi_r)E_2(x_i + \phi_r)\right] \tag{70}$$

Here the terms $2\pi v_i t + \varphi_{0i}$ denoted by x_i, $i = 1,2,3$ and τ and ρ are the transmission and reflection coefficients of the PBS and φ_t and φ_r are the optical phase shift in the target and reference paths, respectively. If the deviation angle of polarizer referred to 45° is represented as θ, the output of measurement photocurrent is corrected as:

$$I_{APD_m} \propto \left|E_r\cos(45-\theta) + E_t\sin(45-\theta)\right|^2 \tag{71}$$

With detection of secondary beat frequency by the super-heterodyne detection, the nonlinearity phase error is concluded similar to Eq. 66:

$$\begin{aligned}V_m = C_1\sin x_s + C_2\cos x_s + C_3\sin(x_s - \Delta\Phi) + C_4\cos(x_s - \Delta\Phi) + C_5\sin(x_s + \Delta\Phi) + \\ C_6\cos(x_s + \Delta\Phi) + C_7\sin(x_s - 2\Delta\Phi) + C_8\cos(x_s - 2\Delta\Phi) + C_9\sin(x_s + 2\Delta\Phi) + C_{10}\cos(x_s + 2\Delta\Phi)\end{aligned} \tag{72}$$

where

$$\begin{aligned}C_1 &= 2(A_1A_2 + B_1B_3 + B_1B_2)\\ C_2 &= A_1^2 - A_2^2 + 2B_1^2 - 2B_2B_3\\ C_3 &= 2(A_1B_2 + A_2B_1)\\ C_4 &= 2(A_1B_1 - A_2B_2)\\ C_5 &= 2(A_1B_3 + A_2B_1)\\ C_6 &= 2(A_1B_1 - A_2B_3)\end{aligned} \tag{73}$$

$$
\begin{aligned}
C_7 &= 2B_1B_2 \\
C_8 &= B_1^2 - B_2^2 \\
C_9 &= 2B_1B_3 \\
C_{10} &= B_1^2 - B_3^2
\end{aligned}
$$

and

$$
\begin{aligned}
A_1 &= \sin\delta_{\varepsilon_r}\cos\delta_{\varepsilon_t}(K_2\cos^2\alpha + K_1\sin^2\alpha) - \cos\delta_{\varepsilon_r}\sin\delta_{\varepsilon_t}(K_2\sin^2\beta + K_1\cos^2\beta) \\
A_2 &= (K_1\sin\alpha\cos\beta + K_2\cos\alpha\sin\beta)\cos(\delta_{\varepsilon_t} - \delta_{\varepsilon_r}) \\
B_1 &= K_3(-\cos\beta\sin\beta\cos\delta_{\varepsilon_r}\sin\delta_{\varepsilon_t} + \sin\alpha\cos\alpha\sin\delta_{\varepsilon_r}\cos\delta_{\varepsilon_t}) \\
B_2 &= K_3(\sin\alpha\sin\beta\sin\delta_{\varepsilon_r}\sin\delta_{\varepsilon_t} + \cos\alpha\cos\beta\cos\delta_{\varepsilon_r}\cos\delta_{\varepsilon_t}) \\
B_3 &= K_3(\cos\alpha\cos\beta\sin\delta_{\varepsilon_r}\sin\delta_{\varepsilon_t} + \sin\alpha\sin\beta\cos\delta_{\varepsilon_r}\cos\delta_{\varepsilon_t}) \\
K_1 &= \rho^2 E_{0_{1,3}} E_{0_2}\cos^2(45-\theta) \\
K_2 &= \tau^2 E_{0_{1,3}} E_{0_2}\sin^2(45-\theta) \\
K_3 &= \tau\rho E_{0_{1,3}} E_{0_2}\sin(45-\theta)\cos(45-\theta)
\end{aligned} \tag{74}
$$

Here the nonlinearity is given by Eq. (66) and following equations:

$$
\begin{aligned}
N = {} & C_1\cos(2\Delta\Phi) - C_2\sin(2\Delta\Phi) + C_3\cos(\Delta\Phi) \\
& - C_4\sin(\Delta\Phi) + C_5\cos(3\Delta\Phi) - C_6\sin(3\Delta\Phi) + C_7 + C_9\cos(4\Delta\Phi) - C_{10}\sin(4\Delta\Phi)
\end{aligned} \tag{75}
$$

$$
\begin{aligned}
D = {} & C_1\sin(2\Delta\Phi) + C_2\cos(2\Delta\Phi) + C_3\sin(\Delta\Phi) + C_4\cos(\Delta\Phi) \\
& + C_5\sin(3\Delta\Phi) + C_6\cos(3\Delta\Phi) + C_8 + C_9\sin(4\Delta\Phi) + C_{10}\cos(4\Delta\Phi)
\end{aligned} \tag{76}
$$

4.3 Error Analysis in the TLMI

Deviation angle and unequal transmission- reflection coefficients of the PBS

Considering the orthogonal linearly polarized modes, the effect of the PBS deviation angle is described by $\alpha = -\beta$. The periodic nonlinearities in terms of the nanometric displacement in the TLMI and two-mode interferometer are respectively illustrated in Fig. 12a and 12b. The range of displacement is equal to the one wavelength of the He-Ne laser at 632.8 nm. The result appears as the second-order nonlinearity. If the transmission and reflection coefficients of the PBS are equal, then the periodic nonlinearity does not change. The unequal coefficients and the mismatch between the factors causes the first-order nonlinearity dominates, as shown in Fig. 12c (curve 3).

Non-orthogonality associated with deviation angle of the PBS

If there is non-orthogonality even in the absence of deviation angle, there will be a leakage of modes. The effect of non-orthogonality of the polarized modes in combination with the deviation angle of PBS is represented by $\alpha \neq -\beta$ and is shown in Fig. 12d. Unlike Fig. 12a, the periodic nonlinearity due to the non-orthogonality of polarized modes is represented as the first order nonlinearity (see curve 1 in Fig. 12d for orthogonality).

Non-orthogonality in combination with ellipticity of adjacent polarized modes

Considering the TLMI free of nonlinearity except for one-ellipticity of the polarizations $\delta_{\varepsilon_t} = 0$, and two-elliptical polarized modes, the periodic nonlinearity is the first-order type of nonlinearity. But, if ellipticity of the polarized modes occurs simultaneously with orthogonality, it may reproduce the second-order nonlinearity (see curves 2 and 3 in Fig. 12e). The nonlinearity generated in the displacement measurement in one fourth of the laser wavelength range (158.2 nm) resulting from the ellipticity of laser polarization states for ideal alignment ($\alpha = 0$ deg) and non-ideal alignment ($\alpha = 3$ deg) are compared in Fig. 13. The nonlinearity error increases with increasing misalignment.

Rotation angle of the polarizer

A small rotation angle of the polarizer can provide nonlinearity, it can be appropriately utilized for nonlinearity compensation in a proper optical setup. The nonlinearity resulting from the two-ellipticity of the adjacent orthogonal and non-orthogonal polarized modes associated with the rotation angle of the polarizer is shown in Fig. 12f.

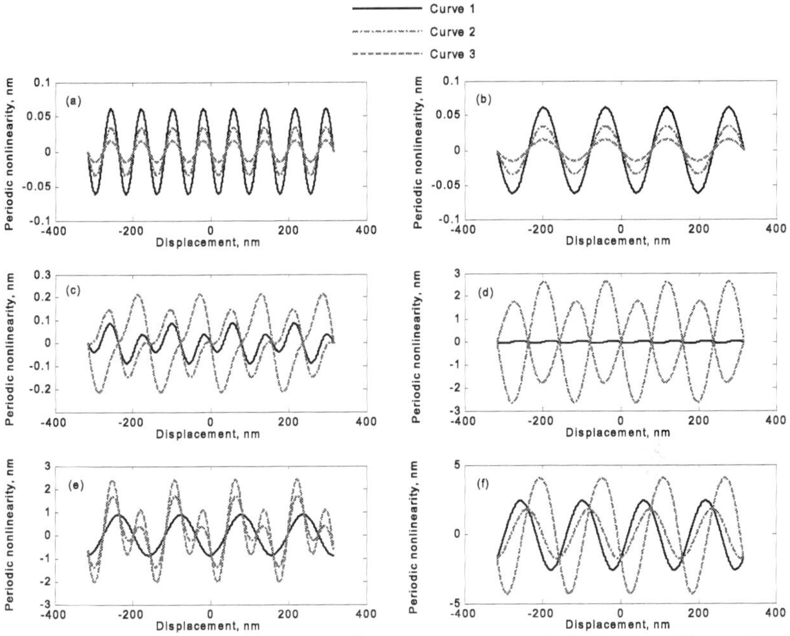

Fig. 12. Periodic nonlinearity resulting from various optical deviations. The orthogonal linearly polarized modes in the (a) three-mode interferometer and (b) two-mode interferometer. (c) Orthogonal linearly polarized modes in combination with unequal transmission and reflection coefficients. (d) The orthogonal (curve 1) and non-orthogonal (curves 2 and 3) linearly polarized modes in combination with deviation angle of PBS. (e) One-ellipticity of the adjacent orthogonal polarized modes. (f) Two-ellipticity of the adjacent orthogonal (curve 2) and non-orthogonal (curves 1 and 3) polarized modes in combination with the rotation angle of polarizer

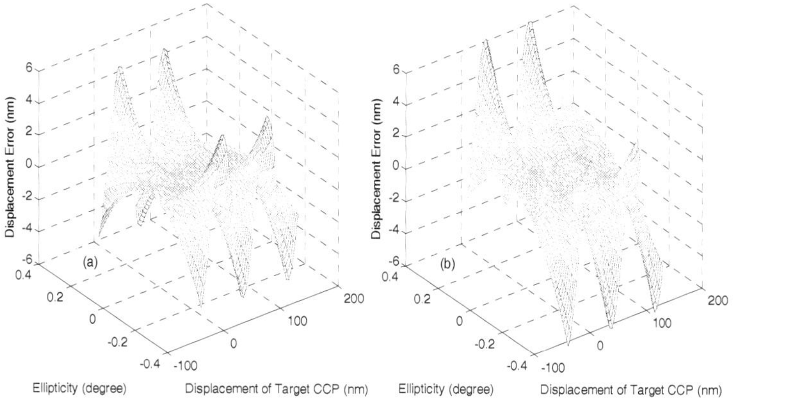

Fig. 13. The error of displacement measurement resulting from ellipticity of laser polarization states for (a) ideal alignment ($\alpha = 0^{\circ}$) and (b) non-ideal alignment ($\alpha = 3^{\circ}$)

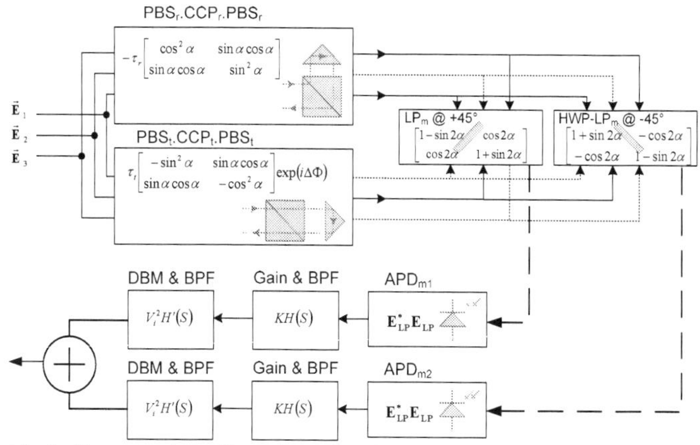

Fig. 14. The block diagram of nonlinearity reduction system

4.4 Nonlinearity Reduction

To reduce the nonlinearity in two-mode heterodyne interferometers, various kinds of heterodyning systems and methods are still being developed (Wuy & Su, 1996; Freitas, 1997; Wu, 2003; Badami & Patterson, 2000; Lin et al., 2000; Hou, 2006; Hou, & Zhaox, 1994). A basic block diagram of the optical and electrical nonlinearity reduction system designed for TLMI is schematically represented in Fig. 14. Two linear polarizers oriented at +45° and -45°, two avalanche photodiodes and a half-wave plate is used in the measurement arm. The current-to-voltage converter, pre-amplifier and band-pass filter is denoted by transfer function KH(S).

The output voltage is led to the double-balanced mixer and other band-pass filter to extract the secondary beat frequency. The electrical cross-talk error occurs due to the unwanted induction between the electrical section of the base and measurement arms. This error can be reduced by: (i) coupling reduction between the base and measurement arms using a proper shielding, grounding and utilizing two isolated power supplies, (ii) noise reduction by using electromagnetic interference shield, (iii) electrical power reduction and (iv) utilizing the band pass filters with high quality factors.

A schematic diagram of the nano-displacement measurement based on the TLMI with noise and electrical error reduction is shown in Fig. 15. The unwanted electrical induction between the base and measurement paths is reduced by using two isolated power supplies for APD biasing and by separating the grounds (analogue-base, analogue-measurement and digital grounds). Due to switching noise of digital section including frequency and phase measurement circuits and microcontroller, two high-speed opto-couplers isolate the analogue circuits from digital section.

In order to reduce the periodic nonlinearity, two output signals can be averaged as shown in Fig. 14. Therefore noise and electrical cross-talk are considerably reduced.

The corresponding compensated nonlinearity signals are depicted in the right panels parallel to the left panels of Fig. 16. The four-cycle peak-to-peak periodic nonlinearity can be effectively reduced from 38 pm to 1.2 pm, as shown in Fig. 16d. But 124 pm eight-cycle nonlinearity cannot be reduced (see Fig. 16c).

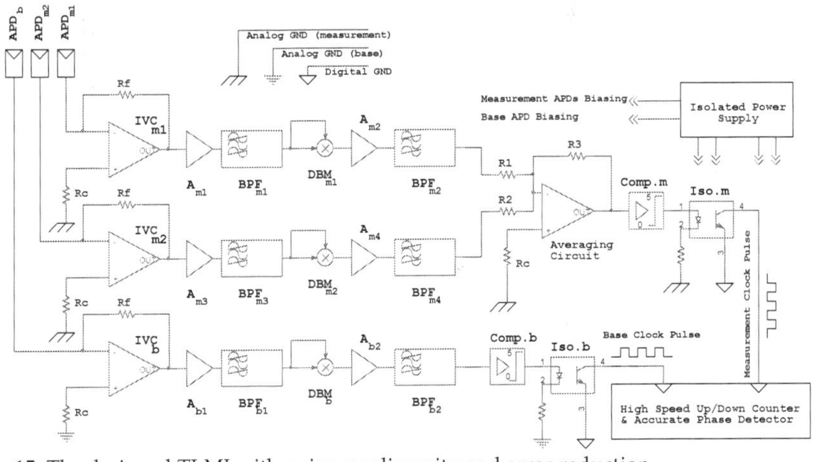

Fig. 15. The designed TLMI with noise, nonlinearity and error reduction

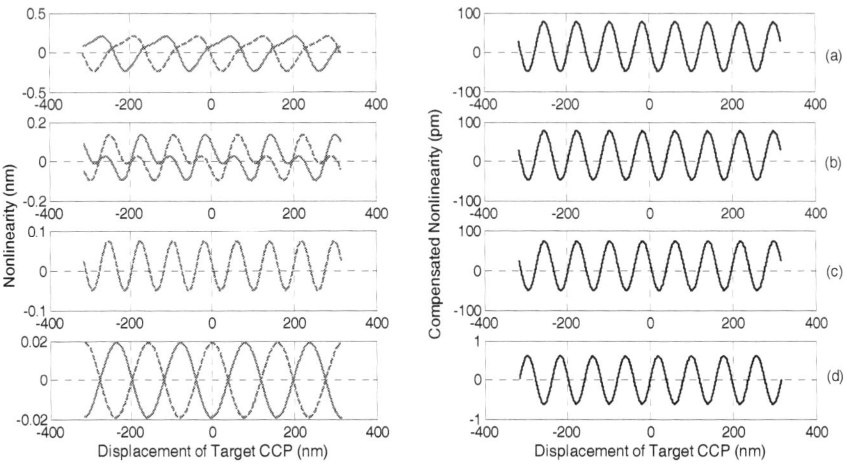

Fig. 16. The periodic nonlinearity (left) and compensated nonlinearity (right) in terms of the displacement of target corner cube prism. (a) $\alpha = 2°$, $\delta_{\varepsilon_r} = 0.28°$, $\delta_{\varepsilon_t} = 0.20°$, $\tau_r = 0.95$, $\tau_t = 0.90$, (b) $\alpha = 2°$, $\delta_{\varepsilon_r} = 0.28°$, $\delta_{\varepsilon_t} = 0.20°$, $\rho = 1$, $\tau = 1$, (c) $\alpha = 2°$, $\delta_{\varepsilon_r} = 0.20°$, $\delta_{\varepsilon_t} = 0.20°$, $\rho = 1$, $\tau = 1$ and (d) $\alpha = 0°$, $\delta_{\varepsilon_r} = 0.20°$, $\delta_{\varepsilon_t} = 0.20°$, $\rho = 0.95$, $\tau = 0.90$

5. Conclusion

The optical interferometry is the most frequent technique which has been widely used in metrology systems with sub-nanometer resolution. In this chapter, first we have discussed an overview of principals of interferometry, interfere phenomenon in the laser interferometers and their applications. A comparison of homodyne interferometers, advantages and disadvantages with heterodyne interferometers was also investigated. The optical setup and electronic sections of two- and three-longitudinal-mode heterodyne interferometers was then described. A frequency-path model of the nano-displacement measurement system based on the laser heterodyne interferometers was presented. This model is described by the AC reference, AC interference, DC interference and optical power components of the multi-mode, multi-path interferometer.

The main parameters affecting the displacement measurement accuracy were discussed in the fourth section. The most important errors are related to imperfect alignment of the optical setup and non-ideal polarized modes. The periodic nonlinearity is the principal limitation of the laser interferometers involving polarized light. Therefore, we have analyzed and modelled the periodic nonlinearity resulting from ellipticity and non-orthogonality of the polarised modes, deviation angle and unequal transmission–reflection coefficients of the PBS and a rotation angle of polarizer with two main approaches; plane wave and matrix method based on the Jones calculus. Then by using two polarizers oriented under +45° and -45°, a half-wave plate, and another avalanche photodiode, the periodic nonlinearity has been considerably decreased. In addition it was shown that the first-order nonlinearity can be effectively reduced compared to the second-order nonlinearity.

6. References

Badami, V. G. & Patterson, S. R. (2000). A frequency domain method for the measurement of nonlinearity in heterodyne interferometry. *Precision Eng.*, Vol. 24, 41–49.

Bonsch, G. & Potulski, E. (1998). Measurement of the refractive index of air and comparison with modified Edlen's formula, *Metrologia*, Vol. 35, 133–139.

Brink, M.; Jasper, H.; Slonaker, S.; Wijnhoven, P. & Klaassen, F. (1996). Step-and-scan and step-and-repeat; a technology comparison, Proceeding of SPIE, Vol. 2726, 1996, pp. 734–753.

Cosijns, S. J. A. G. (2004). Displacement laser interferometry with sub-nanometer uncertainty, PhD diss., *Eindhoven Technology Univ.*

Cosijns, S. J. A. G.; Haitjema, H. & Schellekens, P.H. (2002). Modeling and verifying non-linearities in heterodyne displacement interferometry, *Precis. Eng.*, Vol. 26, 448–455.

Demarest, F. C. (1998). High-resolution, high-speed, low data age uncertainty, heterodyne displacement measuring interferometer electronics, *Meas. Sci. Technol.*, Vol. 9, 1024–1030.

Edlen, B. (1966). The refractive index of air, *Metrologia*, Vol. 2, 71–79.

Eom, T. B.; Choi, H.S.; & Lee, S.K. (2002). Frequency stabilization of an internal mirror He-Ne laser by digital control, *Rev. Sci. Instrum.*, Vol. 73, 221–224.

Eom, T. B.; Kim, J.Y, & Jeong, K. (2001). The dynamic compensation of nonlinearity in a homodyne laser interferometer', *Meas. Sci. Technol.*, Vol. 12, 1734–1738.

Freitas, J. M. (1997). Analysis of laser source birefringence and dichroism on nonlinearity in heterodyne interferometry', *Meas. Sci. Technol.*, Vol. 8, 1356–1359.

Hariharan, P. (2003). *Optical interferometry.* USA: Elsevier Science, 9-18, United States of America.

Hou, W. & Zhaox X. (1994). Drift of nonlinearity in the heterodyne interferometer. *Prec. Eng.*, Vol. 16, 25–35.

Hou, W. (2006). Optical parts and the nonlinearity in heterodyne interferometers. *Prec. Eng.*, Vol. 30, 337–346

Hou, W., & Wilkening, G. (1992). Investigation and compensation of the non-linearity of heterodyne interferometers. *Precis. Eng.*, 14, 91–98

Huang, T. L.; Chen, Y. S.; Shy, J. T. & Liu, H. P. (2000) Two-mode frequency stabilization of an internal-mirror 612 nm He-Ne laser, *Proceeding of Proc. Natl. Sci. Counc.*, , Vol. 24, No. 4, 274–278.

ITRS, http://public.itrs.net, (2008) edition metrology, The international roadmap for semiconductors, accessed November 2009.

Kim, M. S. & Kim, S. W. (2002). Two-longitudinal-mode He-Ne laser for heterodyne interferometry of displacement measurement, *Appl. Opt*, Vol. 41, 5938–5942.

Lin, D.; Jiang, H. & Yin, C. (2000). Analysis of nonlinearity in a high resolution grating interferometer. *Opt. Laser Technol.*, Vol. 32, 95–99.

Meyers, J. F.; Lee, J. W., & Schwartz, R. J. (2001). Characterization of measurement error sources in Doppler global velocimetry, *Meas. Sci. Technol.*, Vol. 12, 357–368.

Olyaee, S. & Nejad, S. M. (2007a). Nonlinearity and frequency-path modelling of three-longitudinal-mode nanometric displacement measurement system, *IET Optoelectron.*, Vol. 1, No. 5, 211-220.

Olyaee, S. & Nejad, S. M. (2007b). Design and simulation of velocity and displacement measurement system with sub nanometer uncertainty based on a new stabilized laser Doppler- interferometer. *Arabian J. Sci. Eng*, Vol. 32, 90-99.

Olyaee, S. & Nejad, S. M. (2007c). Error analysis, design and modeling of an improved heterodyne nanodisplacement interferometer, *Iranian Journal of Electrical & Electronic Engineering*, Vol. 3, 53-63

Olyaee, S.; Yoon, T. H. & Hamedi, S. (2009). Jones matrix analysis of frequency mixing error in three-longitudinal-mode laser heterodyne interferometer. *IET Optoelectron.*, Vol. 3, 215-224.

Schattenburg, M. L. & Smith, H. I. (2001). The critical role of metrology in nanotechnology, *Proceeding of SPIE*, Vol. 4608, 116-124.

Schmitz, T. & Beckwith, J. F. (2003). An investigation of two unexplored periodic error sources in differential-path interferometry, *Precis. Eng.*, Vol. 2, 311-322.

Sutton, C. M. (1998). Nonlinearity in the length measurement using heterodyne laser Michelson interferometry. *J. Phys. E: Sci. Instrum.*, 1987, 20, pp. 1290-1292.

Wu, C. (2003). Periodic nonlinearity resulting from ghost reflections in heterodyne interferometry, *Opt. Commun.*, Vol. 215, 17-23.

Wu, C.; Lin, M. S. T.; & Fu, J. (2002). Heterodyne interferometer with two spatial separated polarization beams for nanometrology. *Optical and Quantum Electron.*, Vol. 34, 1267-1276.

Wu, Ch. M. & Su, Ch. Sh. (1996). Nonlinearity in measurements of length by optical interferometry. *Meas. Sci. Technol.*, Vol. 7, 62-68.

Wu, Ch. M.; Su, Ch. Sh. & Peng, G. Sh. (1996). Correction of nonlinearity in one-frequency optical interferometry, *Meas. Sci. Technol.* Vol. 7, 520-524.

Wuy, C.; & Su, C. (1996). Nonlinearity in measurements of length by optical interferometry, *Meas. Sci. Technol.*, Vol. 7, 62-68.

Yeom, J. Y. & Yoon T. H. (2005). Three-longitudinal-mode He-Ne laser frequency stabilized at 633 nm by thermal phase locking of the secondary beat frequency. *Appl. Opt.*,Vol. 44, 266-270.

Yokoyama, S.; Araki, T. & Suzuki, N. (1994) Intermode beat stabilized laser with frequency pulling, *Appl. Opt*, , Vol. 33, No. 3, 358-363.

Yokoyama, S.; Yokoyama, T. & Araki, T. (2005). High-speed subnanometre interferometry using an improved three-mode heterodyne interferometer, *Meas. Sci. Technol*, Vol. 12, 157-162.

Yokoyama, T.; Araki, T.; Yokoyama, S. & Suzuki, N. (2001). A subnanometre heterodyne interferometric system with improved phase sensitivity using a three-longitudinal-mode He-Ne laser, *Meas. Sci. Technol*,Vol. 12, 157-162.

14

Inductive Telemetric Measurement Systems for Remote Sensing

Daniele Marioli, Emilio Sardini and Mauro Serpelloni
Department of Information Engineering:
Electronics, Informatics, Telecommunications and Control,
University of Brescia
Italy

1. Introduction

Inductive telemetric measurement systems can be used for contactless measurement using passive sensors that may be interrogated by an external unit without a physical contact.
The main appealing features with respect to traditional sensors come from the absence of any cable linking to the acquisition unit, the reduced ecological impact, since they are battery-less, and from the extension of their application fields from traditional to harsh environments. Inductive telemetric systems can be adopted in situations where the measurement information should be acquired or in presence of incompatibility with the electronics requirements or in inaccessible environments.
Their use widens to applications where wires, connecting a data acquisition unit and the sensor element, cannot be used such as, for examples, in implantable devices inside the human body to avoid risk of infections or skin damage or in hermetic environments.
Inductive telemetric systems are applied in industrial fields, for example, when the measuring environment is not accessible, since inside, for example, a hermetic box. Other typical applications are measurement of physical quantities in environments at high temperature where it is not possible to use silicon microelectronic circuits, with consequent unavailability of active blocks within the sensing head for the electronic processing and transmission of the signals. In this case the sensor can be only of passive type and requires to be interrogated contactless.
A solution can be offered sometimes by standard wireless techniques that anyway does not present, in many cases, a definitive solution, because the wireless electronic circuits need powering energy supplied by a battery that can periodically be substituted. Furthermore, exhausted batteries are important environmental problems and they require appropriate disposal. In the previous examples, inductive telemetric measurement systems can represent a valid solution, since they do require neither connections through cables or electronic circuits to process and transmit the signal and avoids the use of batteries.
Inductive telemetric measurement systems are usually constituted by two inductors: one (sensing inductor) connected to the sensitive element, commonly a capacitive transducer, and the other to the measurement circuit (readout inductor).

Different examples of application using the telemetric measurement system are reported in the literature. In (Fonseca et al., 2002), the telemetric system is used to monitor the pressure inside high temperature environments. In (Hamici et al., 1996), an example in the field of biomedical applications for the monitoring of internal package humidity for either in vitro or in vivo testing is described. The system consists of a high-sensitivity capacitive humidity sensor that forms an LC tank circuit together with a hybrid coil wound around a ferrite substrate. The resonant frequency of the circuit depends on the humidity sensor capacitance. This sensor uses the hybrid coil for coupling through the magnetic field with an external inductor, and then the information of humidity is transduced to outside the package.

Another example of humidity monitoring using a passive telemetric system is quoted in (Harpster et al., 2002). The proposed telemetric technique advantageously allows the insertion of a cheap sensor inside every package and executes the measurement without the necessity of opening the hermetic box.

Inductive telemetric measurement systems are also adopted in many other fields: in the literature, several applications in harsh environments are described, such as under high temperatures, cold, humidity or corrosive conditions (Akar et al., 2001; Todoroki et al., 2003; Birdsell et al., 2006; Jia et al., 2006, Tan et al., 2007). Furthermore, examples of inductive telemetric measurement systems for biomedical applications are reported in (Fonseca et al., 2006; Schnakenberg et al., 2000).

In (Takahata et al., 2008) the coil core of a wire wound inductor is a micromachined capacitive pressure device; the sensor operates in harsh or protected environments and can be remotely interrogated by a wireless setup. The inductive telemetric system has been tested in a plastic chamber full of water; the resonant frequency of the tank is monitored outside by an antenna connected to an impedance analyzer. In (Ong et al., 2001) a system for environmental wireless monitoring consists of a LC sensor and two loop antennas (transmitter and receiver). A change of the L and/or C parameters is reflected as mutual impedance on the receiver antenna.

This chapter is basically aim at describing telemetric systems, their electrical models and the measurement techniques used. Then a general architecture of an inductive telemetric system is presented.

The model and its simplified version are shown and two measurement techniques that offer different performance characteristics are reported. Furthermore, an electronic circuit that implements the technique fullest extent is described and tested. Finally, the chapter reports two applications for harsh or hermetic environments: two examples of passive sensors that are related by an inductive telemetric system are described, the first one for humidity measurements and the second for high temperatures. The first application is the measurements of relative humidity in hermetic environments, for example in logistic and biomedical fields. The second application is the temperature monitoring of an isolated system inside a harsh environment, like ovens.

The telemetric temperature systems can be a solution for measurements into harsh or hermetic environments, thereby eliminating the need for physical contacts, active elements, or power supplies, which cannot withstand harsh or hermetic environments.

2. Architectures and Measurement Techniques

2.1 Telemetric System

This section describes an inductive telemetric measurement system and its components. Inductive telemetric measurement systems have the capability, through an inductive coupling, to link wirelessly two devices, one of them being just a passive element. A general architecture of a telemetric measurement system is schematically shown in Figure 1.

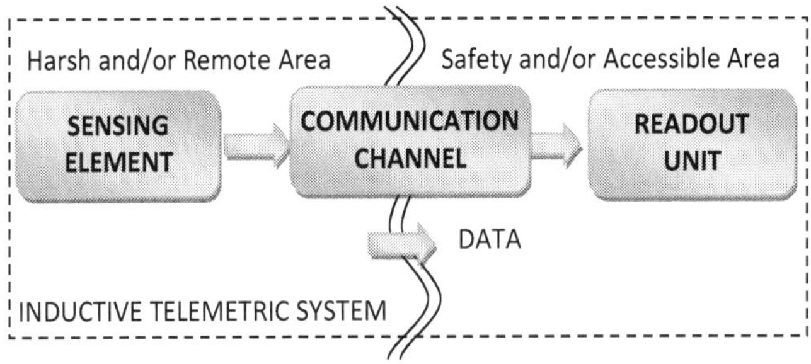

Fig. 1. General architecture of a telemetric system.

Basically it is a measurement system with two different areas: the harsh and/or remote area, where the sensing element transduces the quantity under measurement and the safety or accessible area, where the read out unit is hosted. Between the sensing element and the readout unit there is usually a barrier separating the two areas. The sensing element and the readout unit are connected by a wireless communication channel that uses a magnetic link. The measurement information is acquired by the sensing element, sent through the communication channel using the magnetic coupling, read as reflected impedance by the conditioning electronics and elaborated by the readout unit. The characteristics of the barrier material (mainly material type and dimensions) influence the system's performance. A strong influence is due to the conductive property of the material; it constitutes a serious obstacle since the induced eddy currents reduce the magnetic field amplitude. The frequency of the magnetic field depends on the specific application: different examples of telemetric systems use frequencies ranging from hundreds up to tens of Mega Hertz.

As shown in Figure 2, an inductive telemetric system consists of two inductors: one, labelled sensing inductor, is connected to the sensing element, and the other, readout inductor, to the readout unit. The two inductors are coupled by a magnetic field, and they are placed at a distance that can be fixed or can change. The sensing element is a passive device that, normally, does not require any additional power supply. The front-end electronic circuit is linked to the terminals of the read out inductor: the reflected impedance at these terminals is correlated by the quantity under measurement.

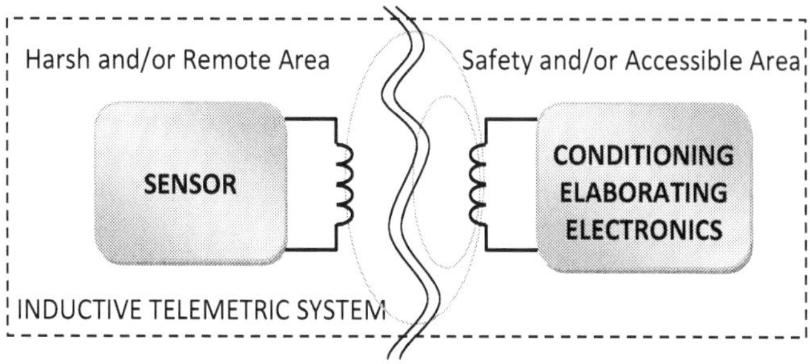

Fig. 2. General architecture of an inductive telemetric measurement system.

2.2 Model of the system

2.2.1 Physical model

Models of different complexity are reported in literature. For sake of completeness, a physical model should also include the effects of parasitic capacitances of each inductor, the coupling capacitance between the two inductors and the leakage magnetic fluxes. The reason to keep into consideration these elements are that the values of the parasitic elements can be equal to the value of the transducer capacitance, and a changing in the geometry of the system produces a change in the coupling and leakage fluxes.

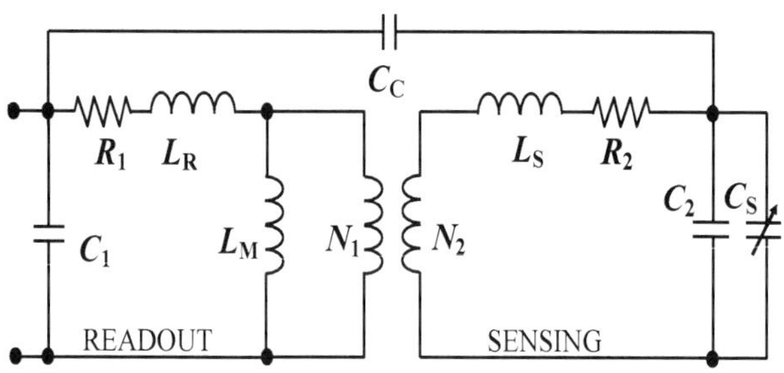

Fig. 3. The proposed physical model to analyze the inductive telemetric systems.

A physical model of an inductive telemetric system is shown in Figure 3. The parameters have the following meaning:

1) R_1, R_2 are the equivalent resistances of readout and sensor;
2) C_1, C_2 are the parasitic capacitances of the readout and sensor inductor respectively;
3) C_S is the sensor capacitance;
4) L_R, L_S are the readout and sensor leakage inductances;
5) L_M is referred to coupled flux;
6) N_1 and N_2 are the equivalent number of the inductor windings; they kept into considerations the possible different number of turns of each inductor.
7) C_C is the coupling capacitance.

The coupling capacitance C_C keeps into consideration the effects of the electric field coupled between the conducting tracks of the two inductors; its value is generally low even for close distances and can be further minimized increasing the distance between the two inductors. The coupling capacitance value depends on the inductor construction type and on the geometry. For coil inductors, the area faced between the two inductors is small and the coupling capacitance value is very low and can be neglected. Flat type inductors are different since the faced area can be very large. But also in this case, making the external diameter of the sensible inductor equal to the internal diameter of the readout inductor, the faced area can be reduced and the coupling capacitance can be neglected.

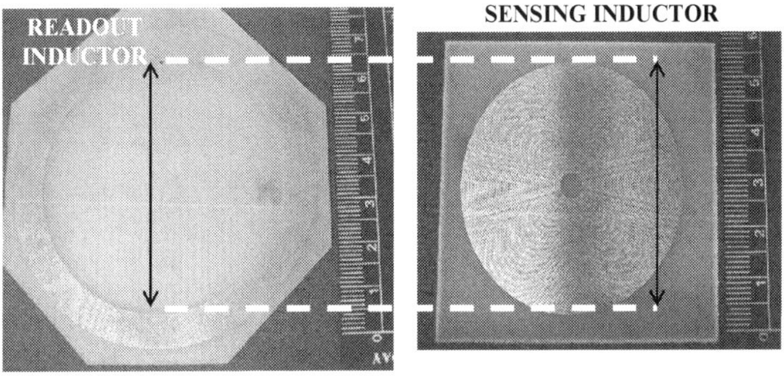

Fig. 4. The realized planar inductors.

An example of flat type inductors with an internal diameter of the readout inductor equal to the external diameter of the sensing inductor is reported in Figure 4. For these reasons, the coupling capacitance Cc can be neglected without loss of generality. Since the working frequency range is high, the resistances R_1, R_2 can be neglected since they have impedances of much less value than those of the series inductor L_R, L_S. The model of Figure 3 can be simplified into that shown in Figure 5, where the elements of the sensing circuit are brought to the readout circuit. L and C are L_S and C'_S as seen from the primary of the ideal transformer and parameter "n" is the ratio between N_1 and N_2. C'_S is the parallel of C_2 and C_S.

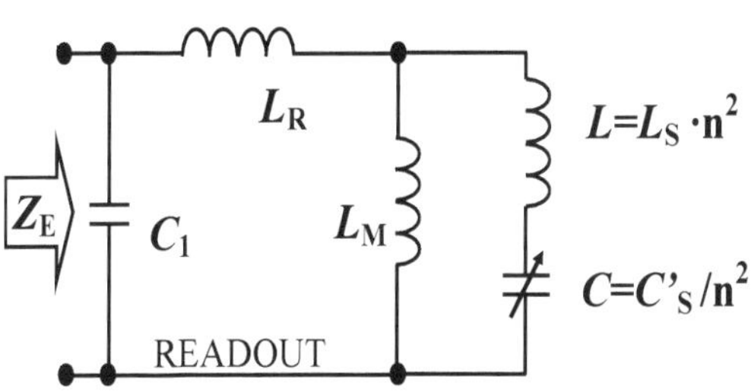

Fig. 5. A simplified version of Figure 3 without the coupling capacitance and the resistances.

If the simplified model is considered, the impedance at the terminal is:

$$Z_E(s) = \frac{s^3(L_M LC + L_R C(L_M + L)) + s(L_M + L_R)}{s^4 C_1(L_M LC + L_R C(L_M + L)) + s^2(C_1(L_M + L_R) + C(L_M + L)) + 1} \quad (1)$$

The impedance function of equation (1) has a diagram that agrees with the data obtained by experimental measurements. Figure 6 shows typical impedance measured on a real telemetric system and obtained when the measuring probes are applied to the readout circuit terminals. There are three noticeable frequencies f_{ra}, f_{rb} and f_a: f_{rb} is mainly due to the self resonance of the readout inductor. f_{min} is the frequency at which the phase of the impedance reaches its minimum as indicated in Figure 6.

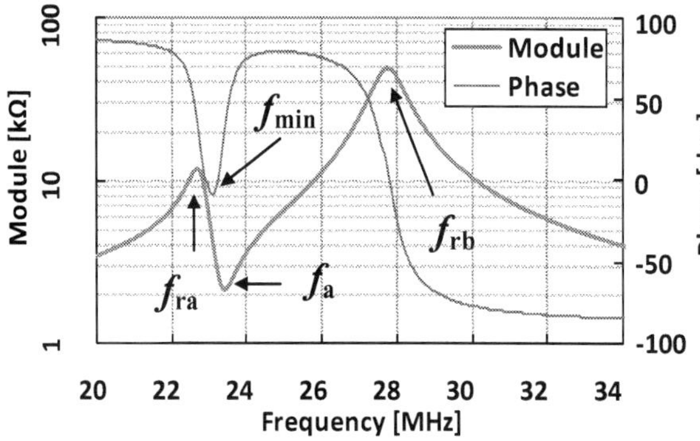

Fig. 6. Impedance as seen from the terminal of the readout inductance.

First (f_{ra}) and second resonant frequency (f_{rb}) have complicated expressions that are reported below (2) and (3). They are both influenced by C_1 and C. The resonant frequency (f_a) is influenced only by C and it is more sensitive to C than the other two frequencies. The expression for f_a, f_{ra} and f_{rb} are:

$$f_a = \frac{1}{2\pi\sqrt{C\cdot\left(L + \frac{L_M L_R}{L_R + L_M}\right)}} \quad (2)$$

$$(2\pi f_{ra,b})^2 = \frac{1}{2C\left(L + \frac{L_R L_M}{L_R + L_M}\right)} + \frac{1}{2C_1\left(L_R + \frac{L L_M}{L + L_M}\right)} \pm \sqrt{\frac{1}{4C^2\left(L + \frac{L_R L_M}{L_R + L_M}\right)^2} - \frac{1}{4C_1^2\left(L_R + \frac{L L_M}{L + L_M}\right)^2} - \frac{L_M^2 - L_R L_M - L L_R - L L_M}{2CC_1(L L_R + L L_M + L_R L_M)^2}} \quad (3)$$

f_a depends on C, and also on distance, since the distance changes the inductance.

The other two frequencies (f_{ra} and f_{rb}) have dependence more complex and dependent also on the parasitic capacitance of the readout inductor C_1.
Because f_a depends only on the transducer capacitance, while f_{ra} and f_{rb} depend also on C_1, it seems more useful to measure f_a. A sensitivity analysis has been conducted on the equivalent system of the telemetric circuit with the simulation software PSpice using the circuit of Figure 5. The simulation results show that the changes of C influence both the three resonances, as expected, but with different sensitivity: the f_a resonance is more sensitive. A confirmation of the previous considerations can also be obtained by analyzing the experimental results reported in paragraph 4.
Anyway, even if the transducer capacitance is kept constant, but the distance between the readout and sensing inductances changes, the three resonant frequencies change. In fact, when the distance changes, the coupled and leakage flux change too modifying the value of L_M, L_R and L: for f_a this effect is clearly visible by analyzing the terms into parenthesis of equation 2. The results obtained from theoretical analysis are confirmed from experimental measurements.

2.2.2 Simplified model

A simplified model of a telemetric system is reported in Figure 7. This model neglects the parasitic capacitances and, also, the leakage fluxes and their changing due to a distance variation, but M coefficient keeps into consideration the coupled magnetic flux. Even if the two previously effects are neglected, the model can be accurately used in many applications when the transducer capacitance is much greater than the parasitic capacitance of the inductor and the distance between the two inductors is kept constant.

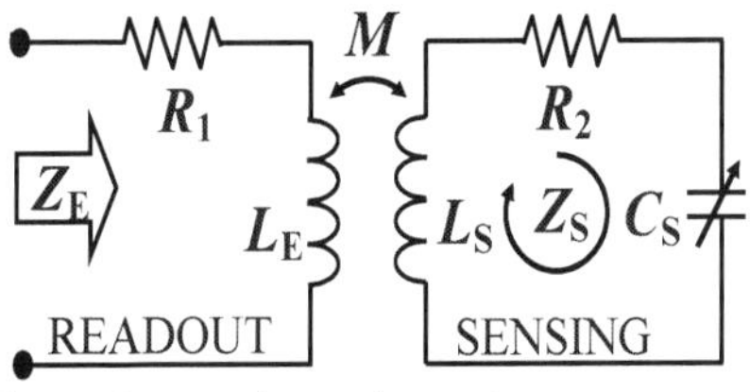

Fig. 7. A model, commonly used, to analyze telemetric systems.

Figure 7 reports the equivalent circuit: the planar inductance is modelled with an inductor (L_S), a series parasitic resistance (R_2) and a variable capacitor (C_S) representing the capacitance sensor. The readout circuit is modelled with an inductor (L_E) and a series parasitic resistance (R_1). When the two circuits are close, there is a mutual inductance coupling (M) between the inductor L_S and the inductor L_E.

2.3 Measurement techniques.

The measurement techniques reported in literature are based on impedance measurements, to identify a particular resonant frequency, or a frequency point that has a particular property such as a minimum of the phase. The measurement techniques are:

a) minimum phase measurement technique (Min-Phase),
b) three resonances (3-Resonances).

As reported in the following in this paragraph, the analysis of the two techniques leads to the conclusion that: the minimum phase measurement technique is simpler since it is requested to measure the minimum of a phase and the related conditioning circuit is simpler, but the measurement system operates at fixed distance. The technique of the three resonances is more complicated since it requires the measurement of three resonant frequencies and added electronic circuits to execute a formula but is able to compensate for variations in distance.

2.3.1 Min phase technique

One method, reported in (Harpster et al, 2002; Wang et al., 2008), measures the frequency at which the phase of the impedance reaches its minimum (Min-phase method). This frequency called f_{min} is shown in Figure 6. In the papers, the inductive telemetric systems consist of a planar inductance, in the sensing circuit, and a coil inductance, in the readout circuit. Considering the model reported in Figure 7, the readout circuit measures the impedance and the frequency at which the phase, in a short frequency interval, is at its minimum value (f_{min}); this frequency is related to the resonant frequency (f_0) of the sensing circuit that corresponds to:

$$f_0 = \frac{1}{2\pi\sqrt{L_S C_S}} \tag{4}$$

In fact the total impedance, as seen from the terminals of the readout circuit, is given by:

$$Z_E(\omega) = R_1 + j\omega L_E + \frac{\omega^2 M^2}{Z_S(\omega)} \tag{5}$$

In which the sensing circuit impedance Z_S is seen as reflected impedance to the readout impedance, with:

$$Z_S(\omega) = R_2 + j\omega L_S - j\frac{1}{\omega C_S} \tag{6}$$

The total impedance reported in (5) can be rewritten considering the following expressions for the k (coupling coefficient) and Q (quality factor of the sensing circuit):

$$k = \frac{M}{\sqrt{L_E L_S}} \tag{7}$$

$$Q = \frac{1}{R_2}\sqrt{\frac{L_S}{C_S}} \tag{8}$$

Then, considering expression (4) the two following expressions can be derived:

$$L_S C_S = 1/\omega_0^2 \tag{9}$$

$$R_2 = \omega_0 L_S / Q \tag{10}$$

Substituting (7), (9), (10) and (6) in (5), the total impedance is given by:

$$Z_E(\omega) = R_1 + j\omega L_E\left[1 + k^2 \frac{\omega^2/\omega_0^2}{1-\omega^2/\omega_0^2 + j\omega/\omega_0 Q}\right] \tag{11}$$

The phase of the impedance can be evaluated as:

$$\angle Z_E(\omega) = arctg\frac{\mathrm{Im}(Z_E(\omega))}{\mathrm{Re}(Z_E(\omega))} \tag{12}$$

To find the frequency at the phase minimum ($f_{\min}$), equation (12) is first differentiated and then equated to zero, thus:

$$\frac{d\left\{ arctg\left[\mathrm{Im}(Z_E(\omega))/\mathrm{Re}(Z_E(\omega))\right] \right\}}{d\omega} = 0 \tag{13}$$

The frequency at the phase minimum ($f_{\min}$) is related by f_0 from the expression (14), which corresponds to the Taylor expansion of $f_{\min}$ in k (coupling coefficient) and Q^{-1} (quality factor of the sensing circuit) (Fonseca et al., 2002).

$$f_{\min} = f_0\left(1 + \frac{k^2}{4} + \frac{1}{8Q^2}\right) \tag{14}$$

The expression denotes that a distance change, considered equation (7), affects the coupling factor and consequently the $f_{\min}$. This method is valid only in the interval near $f_{\min}$ and the self-resonance of the readout inductor is much higher.

2.3.2 The three resonances method

In the recent years a more accurate method has been reported in the literature. It measures all the three resonances and compensates the distance variation between the two inductors (3-Resonances Method) (Marioli et al., 2005).

This method is based on a parameter, called "F", whose value depends only on distance. According to the symbols reported in Figure 5, "F" parameter is defined equal to:

$$F = \frac{1}{C_1\left(L_R + \frac{L_M L}{L_M + L}\right)} \tag{15}$$

"F" parameter derives from the sum of the squares of three resonant frequencies according to the following expression:

$$\left(2\pi f_{ra}\right)^2+\left(2\pi f_{rb}\right)^2-\left(2\pi f_{a}\right)^2 \tag{16}$$

Since from equations (3):

$$\left(2\pi f_{ra}\right)^2+\left(2\pi f_{rb}\right)^2=\frac{1}{C\left(L+\dfrac{L_R L_M}{L_R+L_M}\right)}+\frac{1}{C_1\left(L_R+\dfrac{L L_M}{L+L_M}\right)} \tag{17}$$

And from equation (2)

$$\left(2\pi f_{a}\right)^2=\frac{1}{C\cdot\left(L+\dfrac{L_M L_R}{L_R+L_M}\right)} \tag{18}$$

It is immediate to demonstrate that:

$$\left(2\pi f_{ra}\right)^2+\left(2\pi f_{rb}\right)^2-\left(2\pi f_{a}\right)^2=\frac{1}{C_1\left(L_R+\dfrac{L_M L}{L_M+L}\right)} \tag{19}$$

Or in other word:

$$F=\left(2\pi f_{ra}\right)^2+\left(2\pi f_{rb}\right)^2-\left(2\pi f_{a}\right)^2=\frac{1}{C_1\left(L_R+\dfrac{L_M L}{L_M+L}\right)} \tag{20}$$

If C_1 is fixed, "F" depends only on coupled and leakage fluxes: these values are related only to the distance and not to the transducer capacitance. Moreover, the parameter "F" is obtained by a direct measurement since it can be calculated by elaborating the measurement of the three f_{ra}, f_{rb} and f_a resonant frequencies.
Introducing the following expressions:

$$L=L_S\cdot n^2 \tag{21}$$

$$C=\frac{C'_S}{n^2} \tag{22}$$

$$L_1=L_R+L_M \tag{23}$$

and

$$L_2=L_S+\frac{L_M}{n^2} \tag{24}$$

Substituting equations (21), (22), (23) and (24) into (18) and in (20) and re-arranging the expressions, the following expressions are obtained:

$$F = \frac{1}{C_1\left(L_R + \frac{L_M L_S \cdot n^2}{L_M + L_S \cdot n^2}\right)} = \frac{1}{\frac{C_1}{n^2}\left(\frac{L_R L_S \cdot n^2 + L_M L_S \cdot n^2 + L_M L_R}{L_2}\right)} \tag{25}$$

$$(2\pi f_a)^2 = \frac{1}{\frac{C'_S}{n^2}\cdot\left(L_S \cdot n^2 + \frac{L_M L_R}{L_R + L_M}\right)} = \frac{1}{\frac{C'_S}{n^2}\cdot\left(\frac{L_R L_S \cdot n^2 + L_M L_S \cdot n^2 + L_M L_R}{L_1}\right)} \tag{26}$$

The ratio of equation (25) with equation (26) is equal to:

$$\frac{F}{(2\pi f_a)^2} = \frac{L_2 C'_S}{L_1 C_1} \tag{27}$$

Re-arranging (27), a straightforward expression of the sensor capacitance (C'_S) is:

$$C'_S = \frac{L_1 C_1}{L_2}\frac{F}{(2\pi f_a)^2} \tag{28}$$

C'_S is obtained as a product between a constant term and a second one calculated from the three measured f_{ra}, f_{rb} and f_a frequencies. The constant term can be automatically obtained from a calibration or can be calculated measuring the equivalent circuit parameters of the each single planar inductor: L_1 and L_2 are the self-inductances of the read-out and sensing inductances, while C_1 is the parasitic capacitance (or any other added capacitance) of the readout circuit.

Equation (28) has been derived with no restricting hypothesis and maintains its significance also when the self resonance of the readout circuit is lower than f_0.

3. Measurement Circuit

In this paragraph, an electronics system that measures the three resonant frequencies and implements the 3-Resonances technique is presented.

The three resonant frequencies (f_{ra}, f_{rb} and f_a) are evaluated measuring the squared impedance module, in a defined frequency range, as seen from the terminals of the readout inductor, while the capacitance value of the sensor is calculated using equation (28) by a digital elaboration unit.

The module of the impedance is obtained applying a sinusoidal voltage to the inductor terminals and measuring the current flowing into. The sinusoidal current is subsequently squared and low-pass filtered such as to extract only the square of the amplitude value. Also the imposed voltage is squared and low-pass filtered such as to extract only the square of the amplitude value. The ratio between the squared modules of the imposed voltage and that of the measured current gives the squared impedance module. From this signal the value of the three resonant frequencies are recognized as minimum and maximum values: f_{ra} and f_a are the first maximum and minimum respectively, while f_{rb} is the second maximum.

An experimental circuit has been designed, fabricated and tested. A block diagram of this circuit is shown in Figure 8. In this scheme the telemetric system is reported as "Z-block". The circuit consists of a block for the generation of the reference signal (synthesizer), of a block for the analysis of the impedance module (impedance analyzer) and of another block for the command and elaboration of the data (elaboration system). The synthesizer generates a sinusoidal variable-frequency signal (V_{DDS}) that drives the impedance analyzer module. The value of the frequency is controlled by the digital unit. The impedance module analyzer is driven by the signal coming from the synthesizer and gives a d.c. signal (V_{out}) proportional to the squared impedance module of the telemetric system.
The elaboration system acquires V_{out}, identifies f_{ra}, f_{rb} and f_a, and calculates C's according to equation (28).

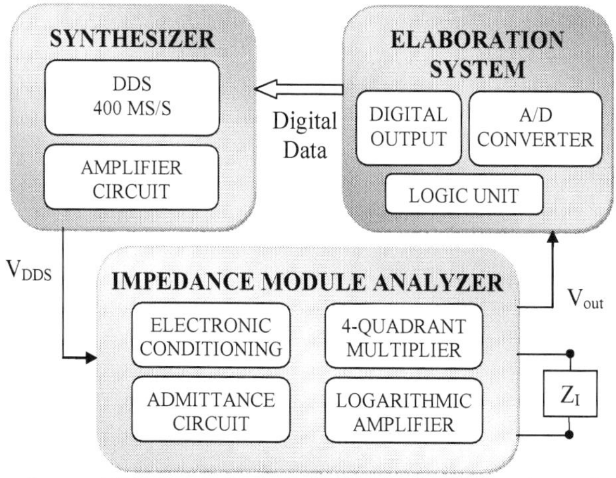

Fig. 8. A block scheme of the electronic circuit for resonant frequencies analyzer.

The synthesizer contains a DDS (Direct Digital Synthesizer), produced by Analog Device (AD9954). The elaboration system consists of a logic unit that drives the DDS, acquires the impedance module through the analog to digital converter, calculates the resonant frequencies and the C'_S value. For practical reason the elaboration system of the fabricated circuit consists of a PC equipped with an acquisition card of the National Instruments (PCI-6024E) and a dedicated software (LabVIEW), but a microcontroller could be used as well. The software LabVIEW controls the frequency and amplitude of the sinusoidal voltage generated by the synthesizer and subsequently samples the output signal V_{out}.

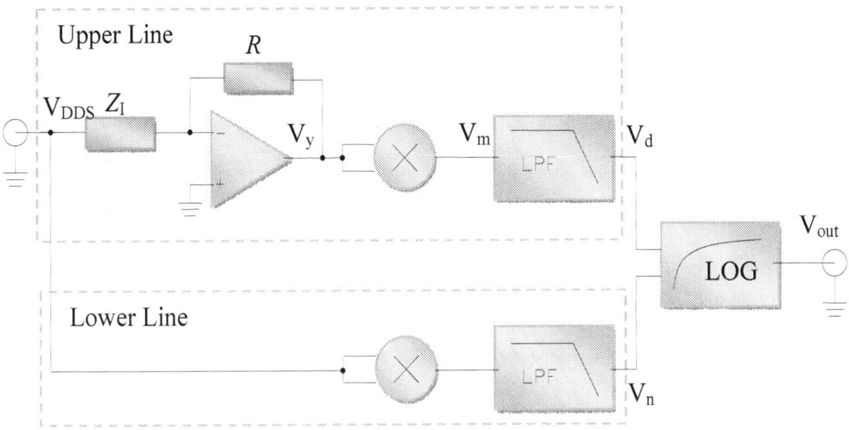

Fig. 9. Block diagram of the impedance module analyzer.

In Figure 9 the block diagram of the impedance analyzer is reported. The input signal V_{DDS} is the sinusoidal voltage generated by the synthesizer and represents the input of two different paths: the upper line and the lower line. In the upper line the readout inductor has one terminal connected to the input of the operational amplifier and the other is driven by the V_{DDS} signal coming from the DDS. The output V_y is proportional to the current, since Z_I is the impedance of telemetric sensor: V_y is proportional to the current flowing into the telemetric sensor, thus:

$$V_y = -R\frac{V_{DDS}}{Z_I} = -RI \tag{29}$$

Subsequently the signal is squared by a four-quadrant multiplier (AD835): V_m has a mean value proportional to the squared current module and a double frequency component. Then the third order low-pass filter extracts the d.c. value of the signal (V_d), thus:

$$V_d = R^2|I|^2 = R^2\frac{|V_{DDS}|^2}{|Z_I|^2} \tag{30}$$

Also V_{DDS} is elaborated and a signal V_n equal to the square module of V_{DDS} is obtained.

$$V_n = |V_{DDS}|^2 \tag{31}$$

The following block, using a logarithmic amplifier (LOG104) produced by Texas Instruments, calculates the difference of the logarithmic of V_n and V_d. The logarithmic amplifier permits also to implement the ratio between the voltage of the lower line and the current of the upper line, then the output is an impedance diagram. Since the difference among two logarithms corresponds to the logarithm of the ratio, the signal V_{out} is proportional to the logarithm of the impedance Z_I.

$$V_{out} = \frac{1}{2} LogV_n - LogV_d = \frac{1}{2}\left(Log\frac{V_n}{V_d} \right) = \frac{1}{2} Log\frac{|V_{DDS}|^2 |Z_I|^2}{|V_{DDS}|^2 R^2} = \\ = \frac{1}{2} Log\frac{|Z_I|^2}{R^2} = Log|Z_I| - b \quad (32)$$

The impedance module of the telemetry system has wide variations and in order to keep the signals into the linear range of each block the V_{DDS} voltage can vary. Moreover V_{DDS} voltage can also slightly change due to problems of nonlinearity or temperature shift of the DDS circuit's output. The logarithmic block, according to equation (32), compensates for V_{DDS} change. Furthermore, the constant term b of equation (32) can be neglected because the resonant frequencies are evaluated as relative maximum and minimum quantities. The whole system has been tested in the laboratory applied to an inductive telemetric system for humidity measurement; several results are reported in the following paragraph (5).

4. An Inductive Telemetric System for Temperature Measurements

In this paragraph an inductive telemetric system measures high-temperature in harsh industrial environments. The sensing inductor is a hybrid device constituted by a MEMS temperature sensor developed using the Metal MUMPs process (Andò et al., 2008) and a planar inductor fabricated in thick film technology by screen printing over an alumina substrate a conductive ink in a spiral shape. The MEMS working principle is based on a capacitance variation due to changing of the area faced between the two armatures. The area changing appears as a consequence of a structural deformation due to temperature variation. The readout inductor is a planar inductor too.

An impedance analyzer measures the impedance at the terminals of the readout inductor, and the MEMS capacitance value is calculated by applying the methods of the three resonances and minimum phase. Moreover, the capacitance value of similar MEMS is also evaluated by another impedance analyzer through a direct measurement at the sensing inductor terminals. The values obtained from the three methods have been compared between them.

The inductive telemetric system for high temperature measurement is shown schematically in Figure 10. On the left side of the figure a diagram of the inductive telemetric system is reported: the sensing element that consists of a planar inductor and a MEMS sensor is placed in an oven, while outside, separated by a window of tempered glass with a thickness of 8 mm, there is the readout inductor. The readout inductor was positioned axially to the hybrid sensor at about one centimetre to the hybrid sensor inside the chamber, while outside the readout was connected to the impedance analyzer. The two inductors represent an inductive telemetric system.

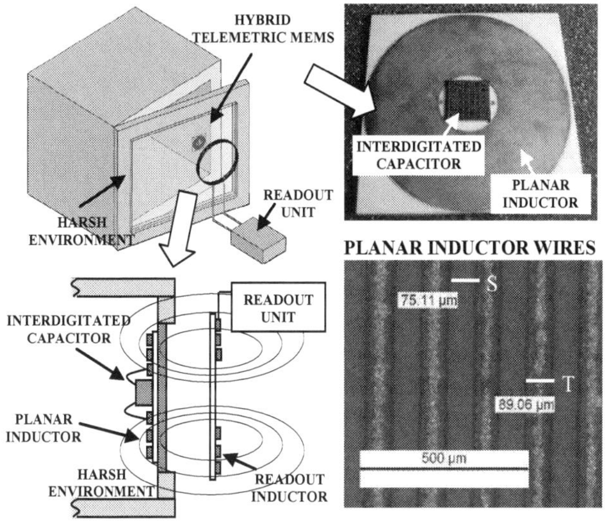

Fig. 10. The inductive telemetric system for high temperature measurement.

The planar inductor, reported on the right, has been obtained by a laser micro-cutting of a layer of conductive thick films (Du Pont QM14) screen printed over an alumina substrate (50 mm x 50 mm x 0.63 mm). The micro-cutting process consists of a material ablation by a laser. The inductor has the external diameter of 50 mm, 120 windings each of about 89 μm width and spaced 75 μm from the others: an enlargement is reported below on the left of Figure 10. The readout inductor is a planar spiral; it has been realized by a photolithographic technology on a high-temperature substrate (85N commercialized by Arlon). The readout inductor has 25 windings, each of 250 μm width and spaced 250 μm from the others. The internal diameter is 50 mm wide.

The experimental apparatus is schematically reported in Figure 11 and consists of an oven, three Fluke multimeters, three Pt100 references, two impedance analyzers, a PC and a power interface. In the measurement chamber (in the centre of the figure) an IR heater of 500 W rises the temperature up to 350 °C. Three Pt100 thermo-resistances (only one is shown in the Figure) measure the internal temperature in three different points, and each one is connected to a multimeter (Fluke 8840A). The three values are used to assure that the temperature is uniformly distributed.

A Personal Computer, over which runs a developed LabVIEW™ virtual-instrument, monitors the temperature inside the oven and controls the IR heater by turning alternatively on and off the power circuit. Two MEMS sensors are placed in the oven. The first one is directly connected to the impedance analyzer (HP4194A) to measure its capacitance; the second one is connected to the external readout inductor for the telemetric measurement. The experimental measurement has been conducted to a temperature up to 330 °C in a temperature-controlled measurement oven.

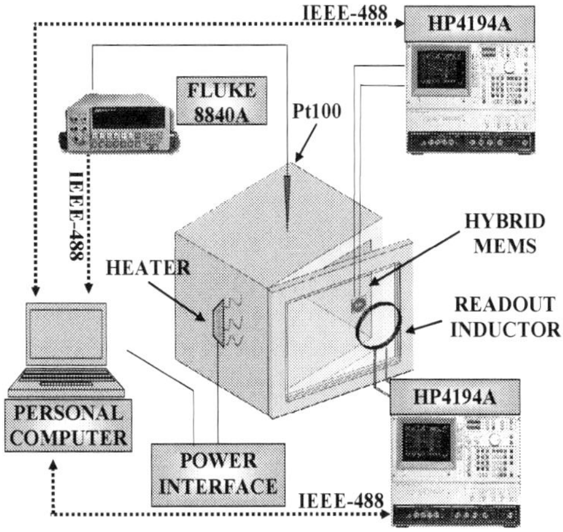

Fig. 11. A diagram of the experimental setup.

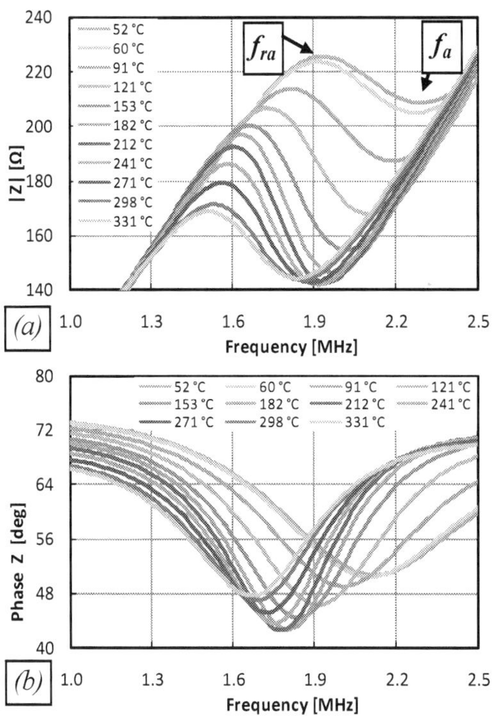

Fig. 12. Modulus (a) and phase (b) of the hybrid MEMS measured with the impedance analyzer at different temperatures.

In Figure 12 modulus (a) and phase (b) diagrams of the impedance, as measured by the impedance analyzer at the readout terminal, for different temperatures are reported. The frequency interval of the abscissa has been chosen to make visible he resonant frequencies f_{ra}, f_a. As expected an increasing in temperature generates a decreasing of the values of the resonant frequencies, since the sensor capacitance value increases.

TEMP. [°C]	f_{ra} [MHz]	f_a [MHz]	f_{rb} [MHz]
52	1.9323	2.2904	4.9825
60	1.9120	2.2760	4.9883
91	1.8125	2.1866	4.9853
121	1.7348	2.1005	4.9783
153	1.6625	2.0146	4.9903
182	1.6308	1.9710	4.9928
212	1.6040	1.9414	4.9943
241	1.5828	1.9156	5.0053
271	1.5580	1.8918	5.0255
298	1.5328	1.8557	5.0285
301	1.5243	1.8798	5.0485
331	1.5130	1.8386	5.0463

Table. 1. Frequencies values of f_{ra}, f_{rb} and f_a measured for different temperatures.

In Table 1, f_{ra}, f_{rb} and f_a values are reported. The two frequencies f_{ra}, f_a, shown also in Figure 12, move down in frequency with increasing temperature as expected. The third frequency f_{rb} is sensitive to temperature, but less than the previous two.

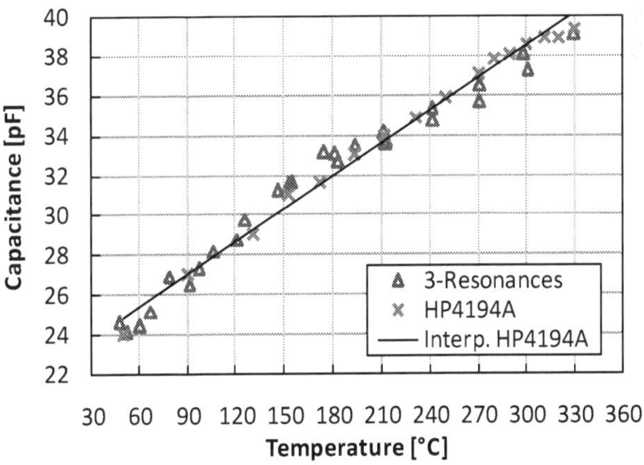

Fig. 13. Sensor's capacitance is reported as a function of the temperature.

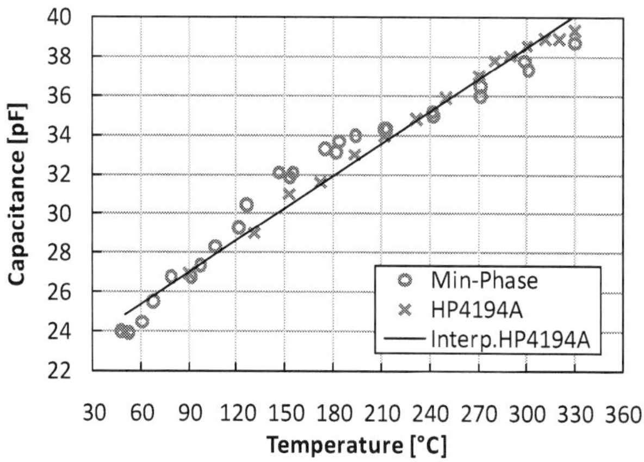

Fig. 14. Sensor's capacitance is reported as a function of the temperature.

In Figure 13 the sensor's capacitance is reported as a function of the temperature: cross points are the values directly measured on the sensor terminals, while the triangle are values calculated using the 3-Resonances method and measuring the impedance from the external inductor terminals. The straight line represents the linear interpolation of the data obtained by the impedance analyzer and it is reached as reference line. The calculated values using the 3-Resonances method (Figure 13) shows a quasi linear behaviour of the sensor: the maximum deviation is about 1.61 pF. Same consideration can be done for the data obtained using the Min-phase method: the maximum deviation is about 2.15 pF; a comparison is shown in Figure 14. Then, both the values calculated with the two methods are closely to the reference one measured with the impedance analyzer (HP4194A).

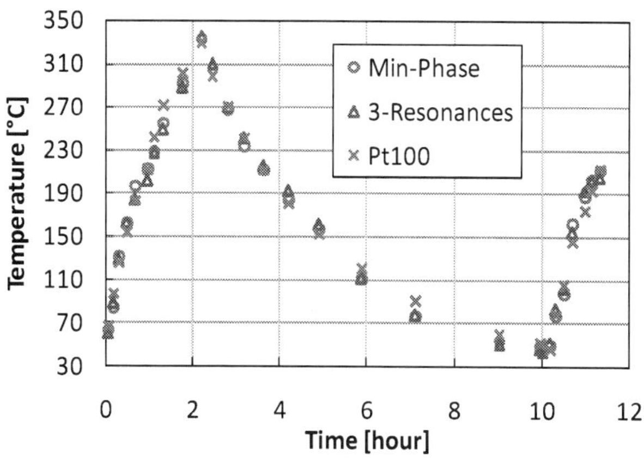

Fig. 15. Temperature values measured with the Pt100 and compared with the Min-Phase and 3-Resonances calculated values.

In Figure 15 the temperatures measured with the reference sensor (Pt100) are compared with the values calculated by the Min-Phase and 3-Resonances methods. The temperature values are obtained using the sensitivity of about 54.6 fF/°C, calculated using the linear interpolation previously reported. Figure 15 shows a good agreement of the temperature values during both the heating and the cooling process. The hybrid MEMS follows the trend of the temperature signal that it has estimated of about 1.9 °C/min and 0.6 °C/min during the heating and cooling process, respectively.

5. An Inductive Telemetric System for Relative Humidity Measurements

This paragraph describes a telemetric system to measure the relative humidity (RH). A telemetric system can be useful in hermetic environments since the measurement can be executed without violating the integrity of the protected environment.
The telemetric system presented here has an interesting characteristic: the sensing inductor does not have any transducer, since the parasitic capacitance of the sensing inductor is the sensing element. In this paragraph, the measurement technique of the three resonances has been used to analyse the effectiveness of compensation in the distance.
In this system the sensing inductor consists only of the planar inductor over which a polymer, humidity sensitive, is deposited. This polymer is sensitive to the humidity and changes its dielectric permittivity causing a variation of the inductor parasitic capacitance. The terminals of the readout inductor are the input of the conditioning electronics reported in paragraph 3. The electronics measures the frequency resonances, extracts the corresponding capacitance values and compensates the distance variation as well.

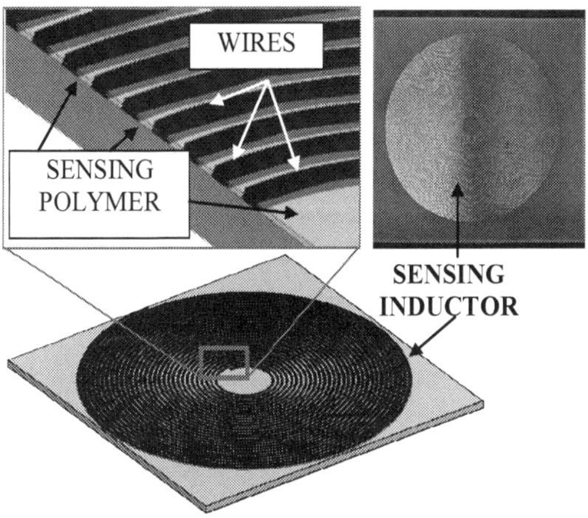

Fig. 16. The inductive sensor, on which a polymer, humidity sensitive, is deposited.

In Figure 16 the passive inductive sensor is reported, which is a standalone planar inductor, fabricated in PCB technology of 25 windings with an external diameter of 50 mm covered by polyethylene glycol (PEG). Polyethylene glycol (PEG) was chosen for the highest sensitivity, but other polymer sensitive to the RH can be used as well. Differently from the others tested in laboratory, this polymer is soluble in water: this characteristic influences the sensitivity positively, but increases the hysteresis as well. Its dielectric constant changes from 2.2 to 4 and depends on temperature and humidity. The characteristics of the telemetric system have been verified with a humidity-controlled hermetical measurement chamber changing also the distance between the sensing and readout inductors.

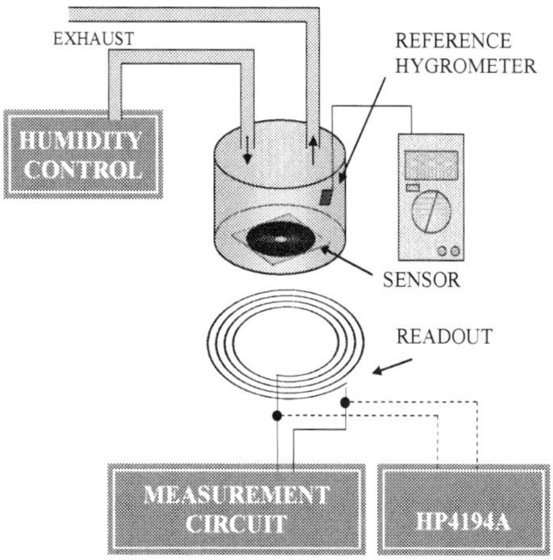

Fig. 17. Block scheme of the experimental system.

In Figure 17 the experimental apparatus to test the telemetric system is schematically represented. The sensor is positioned inside a Plexiglas chamber, which is used as a hermetic container for the damp air. Two pipes are linked to the measurement chamber, one of which introduces controlled damp air. The damp air is produced by a system that compounds dry air and wet air using two flux-meters. The time required to reach the new RH value is about one hour and half. In the chamber there is a hygrometric sensor (HIH-3610 Honeywell) for reference measurements. The inductances are positioned parallel and their axes are coincident. The distance of the readout from the sensor is controlled by a micrometric screw with resolution 10 μm and runs up to 25 mm. The terminals of the readout inductor are connected to the input of the conditioning electronics or, alternatively, to the input of the impedance analyzer. The use of the impedance analyzer is used only for test purposes. The proposed electronics measures the frequency resonances and calculates the corresponding capacitance values according to formula (28). The formula compensates the distance variation as well.
The capacitance values measured at a distance of 20 mm between the readout and sensing inductors the calculated capacitance values are reported in Figure 18: the square point are the value obtained by the electronics while the values obtained using the impedance analyzer (HP4194A) are reported as cross points. All the measurement points are a function of the RH values as measured by the reference sensor. Interpolating the two sets of measurement data the maximum difference between the two curves is less than 15 fF, corresponding to less than 8% of the capacitance measurement range.
In Figure 19 the capacitance values as a function of distance are reported over a distance variation from 15 to 30 mm. The maximum variation of the capacitance is, in the worst case, limited to 20 fF corresponding to about of 1% of FS for each millimetre of distance variation.

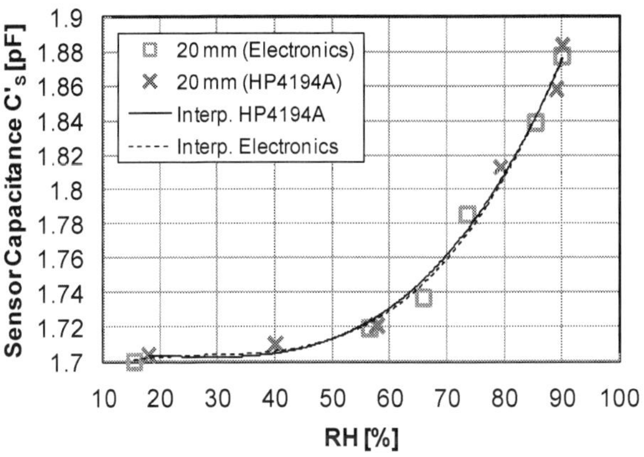

Fig. 18. The calculated capacitance values as a function of RH and for different distance values.

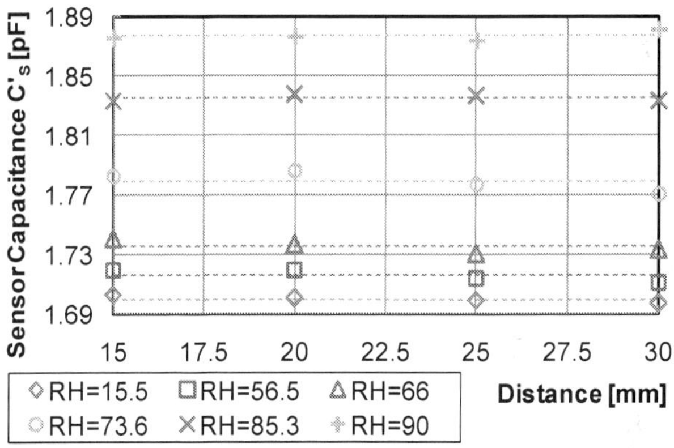

Fig. 19. The capacitance values as a function of distance for different RH values.

6. Conclusion

Inductive telemetric systems offer solutions to specific applications where the measurement data should be acquired in environments that are incompatible with the active electronics or are inaccessible. They also work without batteries, consequently reducing the problem of environmental impact. The general architecture of an inductive telemetric system, the measurement techniques, commonly used, were presented, along with the description of developed telemetric systems applied in harsh or hermetic environments. Two examples of passive inductive telemetric systems were reported, the first one for humidity measurements which presents a distance interval of about 30 mm and the possibility to compensate the distance variation. The second one can measure high temperatures with a maximum limit of about 350 °C, guaranteeing the inviolability of the harsh environment.

7. References

Akar, O.; Akin, T. & Najafi, K. (2001). A wireless batch sealed absolute capacitive pressure sensor, *Sensors and Actuators A*, Vol. 95 pp. 29-38.

Andò, B.; Baglio, S.; Pitrone, N.; Savalli, N. & Trigona, C. (2008). Bent beam MEMS temperature sensors for contactless measurements in harsh environments, *Proceedings of IEEE I2MTC08*, Victoria BC, Canada, pp. 1930-1934.

Birdsell, E. & Allen, M.G.; (2006). Wireless Chemical Sensors for High Temperature environments, *Tech. Dig. Solid-State Sensor, Actuator, and Microsystems Workshop*, Hilton Head Island, SC, USA, pp. 212-215.

Fonseca, M.A.; Allen, M.G.; Kroh, J. & White, J. (2006). Flexible wireless passive pressure sensors for biomedical applications, *Tech. Dig. Solid State Sensor, Actuator, and Microsystems Workshop*, Hilton Head Island, South Carolina, June 4-8, pp. 37-42.

Fonseca, M.A.; English, J.M.; Von Arx, M. & Allen, M.G. (2002). Wireless micromachined ceramic pressure sensor for high temperature applications, *Journal of Microel. Systems*, Vol. 11, pp. 337-343.

Hamici, Z.; Itti, R. & Champier, J. (1996). A high-efficiency power and data transmission system for biomedical implanted electronic device, *Measurement Science and Technology*, Vol. 7, pp. 192-201.

Harpster, T.; Stark, B. & Najafi, K. (2002). A passive wireless integrated humidity sensor, *Sensors and Actuators A*, Vol. 95, pp. 100-107.

Jia, Y.; Sun, K.; Agosto, F.J. & Quinones, M.T. (2006). Design and characterization of a passive wireless strain sensor, *Measurement Science and Technology*, Vol. 17, pp. 2869-2876.

Marioli, D.; Sardini, E.; Serpelloni, M. & Taroni, A. (2005). A new measurement method for capacitance transducers in a distance compensated telemetric sensor system, *Measurement Science and Technology*, Vol. 16, pp. 1593-1599.

Ong, K.G.; Grimes, C.A.; Robbins, C.L. & Singh, R.S. (2001). Design and application of a wireless, passive, resonant-circuit environmental monitoring sensor, *Sensors and Actuators A*, Vol. 93, pp. 33-43.

Schnakenberg, U.; Walter, P.; Vom Bogel G.; Kruger C.; Ludtke-Handjery H.C.; Richter H.A.; Specht W.; Ruokonen P. & Mokwa W. (2000). Initial investigations on systems for measuring intraocular pressure, *Sensors and Actuators A*, Vol. 85, pp. 287-291.

Takahata, K. & Gianchandani, Y.B. (2008). A micromachined capacitive pressure sensor using a cavity-less structure with bulk-metal/elastomer layers and its wireless telemetry application, *Sensors*, Vol. 8, pp. 2317-2330.

Tan, E.L.; Ng, W.N.; Shao, R.; Pereles, B.D. & Ong, K.G. (2007). A wireless, passive sensor for quantifying packaged food quality, *Sensors*, Vol. 7, pp. 1747-1756.

Todoroki, A.; Miyatani, S. & Shimamura, Y. (2003). Wireless strain monitoring using electrical capacitance change of tire: part II-passive, *Smart Materials and Structures*, Vol. 12, pp. 410-416.

Wang, Y.; Jia, Y.; Chen, Q. & Wang, Y. (2008). A Passive Wireless Temperature Sensor for Harsh Environment Applications, *Sensors*, Vol. 8, pp. 7982-7995.

15

Measurement of Voltage Flicker: Application to Grid-connected Wind Turbines

J.J. Gutierrez and J. Ruiz and A. Lazkano and L.A. Leturiondo
University of the Basque Country
Spain

1. Introduction

Electric power is an essential commodity for most industrial, commercial and domestic processes. As a product, electric power must be of an acceptable quality, to guarantee the correct behavior of the equipment connected to the power distribution system. Low-frequency conducted disturbances are the main factors that can compromise power quality. The IEC 61000-2-1 standard classifies low-frequency conducted disturbances in the following five groups: harmonics and interharmonics, voltage dips and short supply interruptions, voltage unbalance, power frequency variations and voltage fluctuations or flicker.

Voltage fluctuations are defined as cyclic variations in voltage with amplitude below 10% of the nominal value. Most of the connected equipment is not affected by voltage fluctuations, but these fluctuations may cause changes in the illumination intensity of light sources, known as *flicker*. Flicker may produce a very unpleasant visual sensation, leading to complaints from utility customers. The annoyance level depends on the type of lamp and amplitude, frequency and duration of the voltage fluctuations. Its precise quantification is a complex task that must be statistically approached to characterize adequately the perception of a large number of people. A flickermeter must characterize the behavior of the lamp-eye-brain set that represents most people and must provide an indication of the discomfort, or flicker severity. In 1986, The International Electrotechnical Commission (IEC) published the first standard describing the functional and design specifications for the measurement of flicker.

The main sources of flicker are large industrial loads, such as arc furnaces, or smaller loads with regular duty cycles, such as welding machines or electric boilers. However, from the point of view of power generation, flicker as a result of wind turbines has gained attention in recent years. Rapid variations in wind speed produce fluctuating power, which can lead to voltage fluctuations at the point of common coupling (PCC), which in turn generate flicker. The IEC 61400-21 standard establishes the procedures for measuring and assessing the power quality characteristics of grid-connected wind turbines. The section dedicated to flicker proposes a complex model for calculating the flicker coefficient that characterizes a wind turbine. This coefficient must be estimated from the current and voltage time series obtained for different wind conditions. The wind turbine being tested is usually connected to a medium-voltage network, having other fluctuating loads that may cause significant voltage fluctuations. In addition, the voltage fluctuations imposed by the wind turbine depend on the characteristics of the grid conditions. The most relevant block of the model is responsible for simulating the voltage fluctuations on a fictitious grid with no source of flicker other than the wind turbine. This chapter is organized in two related sections. The first section deals with the IEC flickermeter. First, the main research enabling modeling of the lamp-eye-brain set is summarized. A description of the IEC 61000-4-15 standard follows, as well as a detailed account of a high-precision digital implementation of the flickermeter, after which the ability of the IEC flickermeter to assess the actual annoyance produced by flicker in people is critically analyzed. This analysis is based on field measurements obtained from analytically generated test signals and subjective experimental data obtained from a small group of people. In the second section, the IEC flickermeter is used to characterize flicker caused by wind turbines. The section contains a detailed description of the part of the IEC-61400-21 standard dedicated to flicker, together with a critical analysis of the different methods used to solve the fictitious grid. The chapter concludes by analyzing how the errors in the estimation of the fictitious grid affect the calculation of flicker severity.

2. Measurement of flicker

2.1 Historical perspective

Flicker is defined as the variation in the luminosity produced in a light source because of fluctuations in the supply voltage. Fig. 1 shows an example of rectangular fluctuation at a frequency of 8.8 Hz and an amplitude ΔV = 0.4 V (i.e., $\frac{\Delta V}{V}$ = 40 %), which modulates a mains signal of 50 Hz and amplitude V = 1 V.

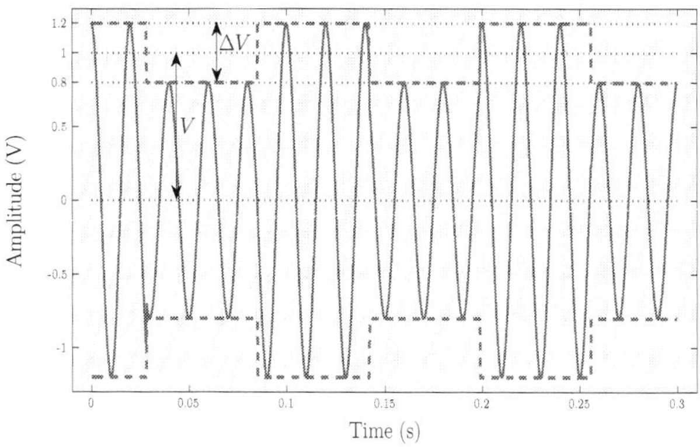

Fig. 1. Example of rectangular fluctuation in voltage supply.

Variations in luminosity can annoy humans. A flicker measuring device or flickermeter must assess the annoyance, or the flicker severity, caused to people exposed to variations in luminosity. The measurement of the annoyance caused should be done starting from the supply voltage of the light source.

It is obvious that the annoyance caused is a subjective phenomenon, related to the sensitivity of each individual to light fluctuations. In this sense, the measurement of annoyance can only be performed on a statistical basis; that is, by carrying out experiments involving a large number of people. A flickermeter has to provide an acceptable model of the behavior of the lamp-eye-brain set responsible for converting the voltage fluctuations into annoyance.

The voltage fluctuations are converted in the lamp into light fluctuations. The response depends, to a great extent, on its construction, power and nominal voltage. Consequently, in order to define the specifications of a flickermeter, it is necessary to select a suitable reference lamp. The analysis of the lamp-eye system requires carrying out statistical studies to enable characterization of the behavior of the human eye when exposed to light fluctuations. Lastly, the eye-brain set constitutes a complex, nonlinear system, and its neurophysiological study also requires a statistical basis. Complex characteristics of the brain, such as its memory capacity and its inertia when faced with consecutive variations in luminosity, must be modeled. The first research into the behavior of the lamp-eye set was carried out by K. Simons (Simons, 1917). More detailed studies on the behavior of the lamp-eye set were carried out by P. Ailleret, at the end of the 1950s (Ailleret, 1957). These experiments were based on various subjective tests on representative groups of people, and they analyzed the behavior of the lamp-eye set with various lamp types. They demonstrated that the lamp-eye system has a band-pass-type response with maximum sensitivity around 10 Hz for incandescent lamps. This work also defined the response of the incandescent lamp under small variations in voltage:

$$\frac{\Delta L}{L_n} = \gamma \frac{\Delta V}{V_n}, \tag{1}$$

where V_n represents the root mean square (rms) value of the nominal voltage, L_n is its corresponding luminosity and γ is a proportionality constant. This expression leads to the conclusion that the level of annoyance calculated in flicker measurement must be proportional to the relative level of voltage fluctuation. That is, double the amplitude of voltage fluctuation corresponds to double the amplitude of luminosity fluctuation and, therefore, double the annoyance.

In a second experiment, P. Ailleret related the annoyance to the amplitude of the fluctuation and its duration. The results demonstrated that the annoyance depends on the product of two factors, the duration and square of the amplitude, according to the following expression:

$$Annoyance = f(L^2 \cdot t), \tag{2}$$

where L represents the fluctuation amplitude and t the duration.

That is, a continuous variation in luminosity with a specific voltage amplitude and frequency, during a particular interval, provokes the same annoyance as three-quarters of the interval without fluctuation and a quarter of the interval with double the amplitude.

Finally, P. Ailleret studied the combination of annoyance provoked by light fluctuations with different frequencies. He demonstrated that the combination of the amplitudes follows a quadratic law. If the annoyance at frequency f_1 has equivalent amplitude, L_1, at 20 Hz, and at another frequency f_2 it has equivalent amplitude L_2, the overall effect of the combined presence of the two frequencies is given by:

$$\Delta L = \sqrt{\Delta L_1^2 + \Delta L_2^2} \tag{3}$$

In parallel with the previous works, H. de Lange considered that the ambient luminosity is an important factor in the evaluation of the annoyance and characterized the response of the human eye by taking into account the influence of the illumination level of the retina. Fig. 2 shows the relation between the amplitude of the luminous fluctuation and the average ambient luminosity against frequency, at the perceptibility threshold (de Lange, 1961) for an incandescent lamp. The variation of this relationship with frequency is provided on a logarithmic scale for different illuminations of the retina. From the figure, it can be deduced that for high levels of illumination, the frequency response of the optical system behaves as a band-pass filter, with a maximum sensitivity at a frequency of 8.8 Hz, making it the reference of sensitivity for human visual perception of flicker.

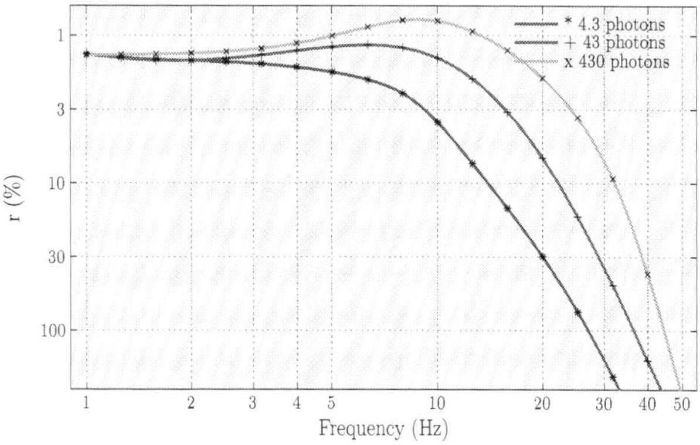

Fig. 2. Frequency characteristics of the human optical system at the threshold of perception for different illumination levels. Source: (de Lange, 1952).

Once the lamp-eye set had been studied, to complete the model of perception, it was essential to analyze the behavior of the eye-brain system. During the 1970s, a series of experiments were conducted, aimed at mathematical modeling of the neurophysiological processes caused by light fluctuations.

The first such research, undertaken by C. Rashbass, obtained the lowest intensity at which the rectangular changes of luminance of a specific duration are perceptible (Rashbass, 1970). The results demonstrated that the relative intensity decreases with increasing flash duration, with a minimum at 64 ms, supporting the band-pass characteristic postulated by H. de Lange and Ailleret.

In the second study, C. Rashbass combined two flashes of the same duration but with intensities that were not necessarily the same. The results demonstrated that the response to any combination of two intensities obeyed a quadratic law, which could be modeled using three elements:

a. a band-pass filter coinciding with the one previously used by H. de Lange to model eye behavior;

b. a second element reproducing the quadratic response of the system, which is modeled using a squaring circuit; and

c. a third element to model the effect of the brain's memory using a first-order band-pass filter and a time constant between 150 and 250 ms [1].

Fig. 3 shows the analog model of the eye-brain set produced from Rashbass' experiments. This model constitutes the nucleus of the current specification of the IEC flickermeter (IEC-61000-4-15, 2003; IEC-868, 1986).

[1] This constant was definitively fixed at 300 ms starting from the studies of Koenderink and Van Doorn (Koenderink & van Doorn, 1974).

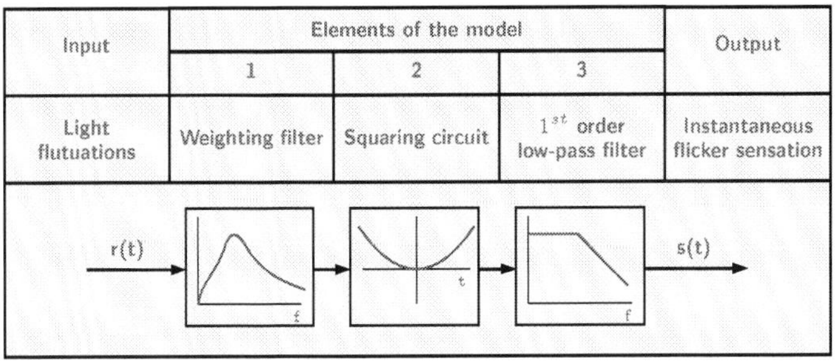

Fig. 3. Model of visual perception (eye-brain set) based on studies by H. de Lange and C. Rashbass. Source: (UIE, 1992).

2.2 Description of the IEC flickermeter

At the end of the 1970s, the UIE [2] perturbations working group started to prepare a specification for the measurement of flicker that was universally accepted. The first results of this work were presented to the international community at the 1984 UIE congresses (Nevries, 1984). The definitive version was standardized in 1986 through the IEC 868 standard (IEC-868, 1986), which provided the functional and design specifications of a flicker measuring device. Currently, the standard containing the specifications of the flickermeter is IEC 61000-4-15 (IEC-61000-4-15, 2003).

Fig. 4 shows the block diagram defined by IEC 61000-4-15. The simulation of the response of the lamp-eye-brain system is carried out in the first four blocks, based on the physiological experiments described previously. In addition, the standard requires integration of the sensation experienced by the observer during a specific period in a single value. Block 5 is responsible for this, through a statistical evaluation of the output from block 4.

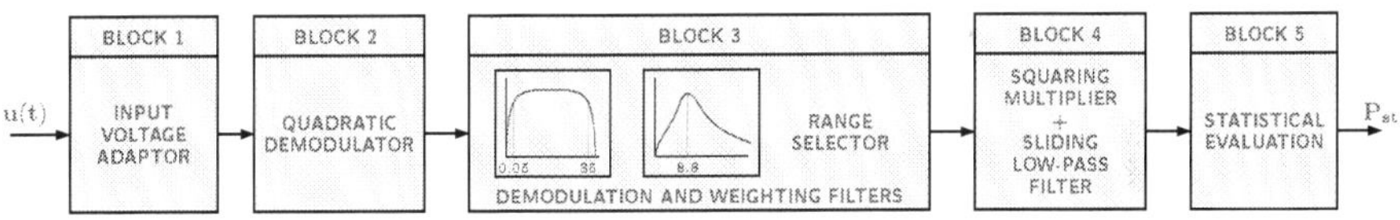

Fig. 4. Block diagram of the flickermeter specified in the IEC 61000-4-15 standard.

Next, a brief description is given of each block shown in Fig. 4 for 50 Hz systems. The main characteristics of a high-precision digital implementation developed as a reference for the results found in the rest of this chapter are described in the following sections.

2.2.1 Block 1: Input voltage adaptor

Given that the flicker measurement must be made from the relative fluctuations in voltage, expressed in percentages, it is necessary to guarantee the independence of the input voltage measurement. In this block, the input is scaled with respect to its average value. This operation can be done through automatic adjustment of the gain at the *rms* value of the input voltage, with a constant time of 1 min.

[2] International Union for Electrical Applications.

In our reference implementation, the input signal is scaled to an internal reference value proportional to the 1 min *rms* value, using a half-cycle sliding window.

2.2.2 Block 2: Quadratic demodulator

Voltage fluctuations normally appear as a modulation in amplitude of the fundamental component. Thus, the input to block 2 can be understood as a modulated signal with a sinusoidal carrier of 50 Hz. Block 2 is responsible for carrying out the quadratic demodulation of the input.
The light source chosen by IEC as reference for the construction of the flickermeter is an incandescent lamp filled with inert gas with a spiral tungsten filament and a nominal power of 60 W at 230 V for 50 Hz systems, and 120 V for 60 Hz systems.
The processing required by this block is simply to square the samples from the signal obtained in block 1. This operation generates a signal containing frequency components corresponding to the fluctuation in luminosity and other frequencies that have to be suitably eliminated.

2.2.3 Block 3: Demodulation and weighting filters

To select the frequency components that generate flicker from the output of block 2, it is necessary to suppress the continuous and 100 Hz components generated in the demodulation process. This is done through the demodulation filters, which consist of the cascade connection of:

a. a first-order high-pass filter with a cutoff frequency of 0.05 Hz; and
b. a sixth-order low-pass Butterworth filter with a cutoff frequency of 35 Hz, which introduces an attenuation of 55 dB at 100 Hz.

The human eye has a selective, frequency-dependent behavior toward variations in luminosity. For this reason, the second stage of filtering consists of a weighted band-pass filter that follows the frequency response of the lamp-eye set. This filter is based on the threshold curve of perceptibility obtained experimentally by H. de Lange (de Lange, 1952; 1961). The standard provides the transfer function in the continuous domain of this filter. With respect to attenuation, at 100 Hz, this filter adds 37 dB to what has already been achieved using the band-pass demodulation.
Finally, block 3 contains a measurement scale selector that determines the sensitivity of the instrument. It modifies the gain depending on the amplitude of the voltage fluctuation to be measured. The scales, expressed as the relative changes in voltage, $\frac{\Delta V}{V}(\%)$, for a sinusoidal modulation of 8.8 Hz, are 0.5%, 1%, 2%, 5% and 10%, the 20% scale being optional.
For the discrete implementation of the filters, Infinite Impulse Response (IIR) systems were selected. The demodulation filters were designed through the impulsive invariance method, and the weighting band-pass filter using bilinear transformation. All the filters were implemented using the direct-form II transpose.
The reference flickermeter works with sampling frequencies (f_s) of 1600, 3200, 6400, 12800 and 25600 $\frac{samples}{s}$, and demodulation filters were designed for these frequencies.
Given that the bandwidth of the output signal of the low-pass demodulation was practically reduced to 35 Hz, maintaining such high sampling rates is not necessary. For this reason, a decimation process is implemented at the output of the low-pass demodulation filter, which reduces f_s to $f_p = 800\ \frac{samples}{s}$. This decimation process does not require low-pass filtering, as the signal to be decimated is band limited. In this way, the weighting filter designed for f_p is the same for all the input sampling frequencies.

2.2.4 Block 4: Nonlinear variance estimator

To complete the model of visual perception defined by C. Rashbass, it is necessary to add two new functions: the modeling of the nonlinear perception of the eye-brain set and the effect of the brain's memory. These two functions are introduced using a quadratic multiplier and a first-order sliding low-pass filter with a time constant of 300 ms.

As for the low-pass filter in block 3, this filter has also been designed using the impulsive invariance method for 800 $\frac{samples}{s}$, and it was also implemented in the transposed form of the direct II form.

The output of this block represents the instantaneous sensation of flicker. It should be stressed that this signal must not be evaluated as an absolute indicator. On the contrary, it must be referred to the unit, taking this as the maximum value of the output of this block if the supply voltage is modulated by a sinusoidal frequency fluctuation of 8.8 Hz and an amplitude 0.25%, corresponding to the threshold of perceptibility.

2.2.5 Block 5: Statistical evaluation

Block 5 has the aim of assessing the level of annoyance starting with the values of the instantaneous sensation of flicker that are exceeded during a certain percentage of the observation time. It is important to choose a suitable assessment period that is characteristic of the reaction of an observer confronted with different types of light fluctuations. Because of the disparity in the characteristics of flicker-generating loads, the standard defines two observation periods:

a. short term, normally fixed at 10 min, during which short-term flicker severity, P_{st}, is assessed; and

b. long term, usually 2 h, during which long-term flicker severity, P_{lt}, is assessed.

It should be noted that the annoyance threshold corresponds to P_{st} = 1. When $P_{st} > 1$, the observer is understood to suffer annoyance; when $P_{st} < 1$, the light fluctuations may be perceivable but not annoying.

2.2.5.1 Evaluation of short-term flicker severity, P_{st}

Because of the random nature of flicker, it must be assumed that the instantaneous sensation of flicker may be subject to strong and unpredictable variations. For this reason, not only the maximum value reached but also the levels exceeded during specific parts of the observation period must be taken into account. Therefore, it seems best to design a method based on a statistical evaluation of the instantaneous sensation. The standard specifies a multipoint adjustment method according to the following expression:

$$P_{st} = \sqrt{k_1 P_1 + k_2 P_2 + ... + k_n P_n}\,, \tag{4}$$

where k_n are weighting coefficients and P_n are levels corresponding to the percentiles [3] $1, 2, \ldots, n$ of the output of block 4. The values k_n and P_n were adjusted starting with the annoyance threshold curve or P_{st} = 1 curve, obtained experimentally from a large group of people undergoing rectangular light fluctuations at more than one change per minute (cpm). The results providing values lower than 5% for all cases were as follows:

[3] Level of instantaneous sensation of flicker that is surpassed during a specific part of a time period.

$$\begin{aligned} k_1 &= 0.0314 \quad P_1 = P_{0.1} \\ k_2 &= 0.0525 \quad P_2 = P_{1s} = \frac{P_{0.7} + P_1 + P_{1.5}}{3} \\ k_3 &= 0.0657 \quad P_3 = P_{3s} = \frac{P_{2.2} + P_3 + P_4}{3} \\ k_4 &= 0.2800 \quad P_4 = P_{10s} = \frac{P_6 + P_8 + P_{10} + P_{13} + P_{17}}{5} \\ k_5 &= 0.0800 \quad P_5 = P_{50s} = \frac{P_{30} + P_{50} + P_{80}}{3} \end{aligned} \tag{5}$$

The index s refers to values or averages, and $P_{0.1}$ is the value of the instantaneous sensation of flicker exceeded during 0.1% of the observation time.

For implementation of this block, the standard specifies sampling the output of block 4 at a constant frequency of 50 Hz or above. The statistical analysis starts by subdividing the amplitude of the output of block 4 into an appropriate number of classes. For each sample, the counter of the corresponding class increases by one. Using the classified samples, at the end of the observation period, the curve of accumulated probability of the instantaneous sensation of flicker, which provides the appropriate percentiles, is obtained. Nevertheless, it should be taken into account that the classification introduces errors, basically because of the number of classes utilized and the resolution of the accumulated probability function within the range of values corresponding to each class.

In the reference flickermeter, the classification is not carried out, but the accumulated probabilities are calculated starting with all the stored samples of the output of block 4 during the 10 min evaluation of P_{st}. This procedure provides total precision in the measurement of P_{st}, given that the errors derived from the classification of the samples are avoided.

2.2.5.2 Evaluation of long-term flicker severity, P_{lt}

The method for calculating P_{lt} is based on the cubic geometric average of the 12 values of P_{st} in a period of 2 h, according to the expression:

$$P_{lt} = \sqrt[3]{\frac{1}{12}\sum_{i=1}^{12} P_{st,i}^3} \tag{6}$$

2.3 A deep review of the annoyance assessment by the IEC flickermeter

Over the last 25 years a very few studies have reported doubts about the goodness of the IEC flickermeter's annoyance assessment. The main problems described are related to the accuracy requirements specified by the standard. In this sense, it has been reported that different flickermeters, all compliant with IEC 61000-4-15, report different P_{st} values for the same input signal (Key et al., 1999; Szlosek et al., 2003). These deviations are a result of the limited number of accuracy requirements specified by the standard (WG2CIGRÉ, 2004). They should be solved with the new edition of the standard, planned for 2010, which includes a higher number of accuracy requirements. Other studies have analyzed the nonlinear behavior of the IEC flickermeter when subject to rectangular voltage fluctuations (Ruiz et al., 2007).

We have analyzed the annoyance assessment performed by the IEC flickermeter by means of P_{st}. We have considered three aspects of the question. Firstly, we analyzed several reports providing field measurements on domestic lines to study the relation between flicker severity levels and the existence of complaints from the users. Secondly, we studied the behavior of block 5 when the IEC flickermeter was subjected to nonuniform rectangular voltage fluctuations. Finally, because it is not easy to find a consistent relationship between the true annoyance and flicker severity, we performed some laboratory tests to correlate the values provided by the IEC flickermeter and the sensation that was experienced by several people, previously trained and qualified.

2.3.1 Field measurements vs complaints

Power quality objectives must be based on statistical limits defined by the regulators according to long-term measurements. In the context of the European electricity market, the standard EN 50160 specifies the index to be used as the weekly P_{lt} for 95% of the time, $P_{lt,95}$. The objective for this index is a value of $P_{lt,95} < 1$. Standard IEC 61000-4-30 also establishes that the minimum measurement period should be one week, defining limits for $P_{st,99} = 1$ and $P_{lt,95} = 0.8$.
There have been very few studies contrasting field measurements of flicker and the level of annoyance perceived by people. The work (Arlt et al., 2007) compares the international planning levels for flicker in high-voltage networks and the flicker requirements that must be fulfilled by customers running flicker-generating equipment. They show that in many cases, the real flicker values in high-voltage networks, which are supplying towns with industrial areas, are much higher than the planning levels without causing complaints by residential customers who are supplied via medium voltage and low voltage from these systems. However, other industrial loads that produce P_{st} levels quite similar to previous examples, but clearly over the planning levels, generate complaints by customers and require corrective actions.
Another study of this issue was elaborated by the joint working group CIGRE C4.07/ CIRED and presented in their report (WGC4CIGRé, 2004). This group was formed in 2000 to research available power quality measurement data with the intention of recommending a set of internationally relevant power quality indices and objectives. One of the sections is, obviously, dedicated to the analysis of the flicker indices. In many sites, characterized by strong and meshed networks, the actual flicker disturbance is sometimes more than double the planning levels without known problems. Some studies suggest as causes of this divergence the conservative character of the objectives defined by the regulatory standards, or the decreasing use of incandescent lamps. However, we wanted to study the next hypothesis: that the short-term flicker severity assessment made by block 5 of the IEC flickermeter may not be the most appropriate way to characterize the annoyance.

2.3.2 Behavior of block 5 when subject to nonuniform rectangular voltage fluctuations

The multipoint algorithm for P_{st} assessment (see Equation 6) was adjusted by the standard to provide the flicker severity caused by rectangular voltage fluctuations that remained completely uniform throughout the 10 min period. In this section we will analyze the P_{st} assessment when the rectangular voltage fluctuation is not homogeneous; that is, when there are several voltage fluctuations with different frequencies and amplitudes during the observation period.
Next, we will describe the experiments that we carried out to analyze the behavior of the IEC flickermeter when subject to nonuniform rectangular voltage fluctuations.

2.3.2.1 Experiment 1

Fig. 5 shows the type of fluctuation used for this case. During a certain period t_1, in seconds, we applied to the reference flickermeter a rectangular fluctuation of frequency f_1 cpm and amplitude $A_1 = \frac{\Delta V}{V}\Big|_{P_{st}=2}$; that is, the amplitude that would produce $P_{st} = 2$ for the frequency f_1 if it were applied during the complete 10 min period. For the rest of the time up to 10 min, the input signal is a 50 Hz sinusoidal without fluctuations.

Fig. 5. Outline of the fluctuation used in Experiment 1.

Because A_1 is the fluctuation amplitude that produces $P_{st} = 2$ for the frequency f_1, the assessment of the annoyance should not depend on f_1 for different values of t_1. Considering Equation 2, the diagram of Fig. 5 is equivalent to a fluctuation applied during the whole 10 min period, of frequency f_1 and amplitude:

$$A = A_1 \cdot \sqrt{\frac{t_1}{600}} \,. \tag{7}$$

Because the amplitude A_1 applied during 10 min produces $P_{st} = 2$, the flicker severity value corresponding to the situation showed in Fig. 5 should follow:

$$P_{st} = 2 \cdot \sqrt{\frac{t_1}{600}} \,. \tag{8}$$

According to Equation 8, Table 1 provides the theoretical P_{st} values for the values of t_1 used in this experiment.

t_1	P_{st}	t_1	P_{st}	t_1	P_{st}
5	0.183	50	0.577	120	0.894
10	0.258	55	0.606	180	1.095
15	0.316	60	0.632	240	1.265
20	0.365	65	0.658	300	1.414
30	0.447	70	0.683	360	1.549
35	0.483	75	0.707	420	1.673
40	0.516	80	0.730	480	1.789
45	0.548	90	0.775	540	1.897

Table 1. Theoretical P_{st} values for Experiment 1.

Fig. 6 shows the results that were obtained with the IEC reference flickermeter for the values of t_1 compiled in Table 1 and values of f_1 from $f_{1,min}$ to 2400 cpm. The value of $f_{1,min}$ depends on t_1 and was adjusted to generate a rectangular fluctuation with at least five voltage changes during the period t_1.

The analysis of Fig. 6 reveals three main conclusions, all of them contrary to the expected results.

a. The P_{st} values depend on the fluctuation frequency, f_1, basically up to 800 cpm.

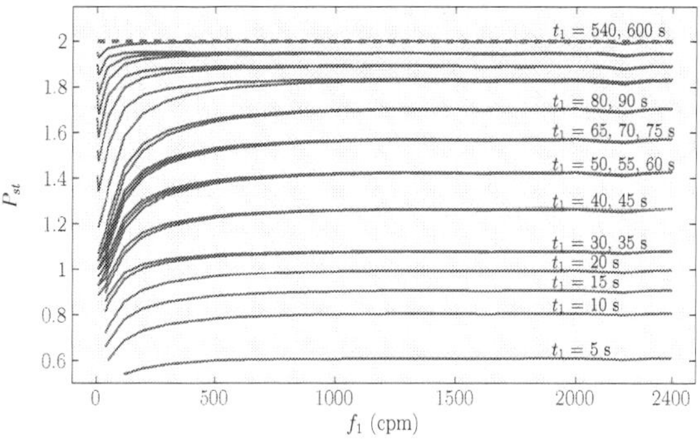

Fig. 6. P_{st} values obtained with the reference flickermeter for Experiment 1.

b. The P_{st} values for each t_1 show an important deviation from the theoretical values compiled in Table 1.

c. Additionally, regarding the relation between the annoyance and the duration of the fluctuation, it is possible to distinguish abrupt variations in P_{st} as a function of t_1.

The origin of the discrepancies between the experimental and the expected results for the above experiment is located in block 5 of the IEC flickermeter. The multipoint algorithm, assessing the P_{st} by the calculation of the percentiles of the instantaneous flicker sensation, provides accurate results when the rectangular fluctuation is applied uniformly during the whole period of 10 min. When the fluctuation is not applied in that way, the evolution of the percentiles becomes unpredictable, and P_{st} presents abrupt changes in terms of the duration of the fluctuation, t_1.

2.3.2.2 Experiment 2

Fig. 7 shows the fluctuations used for this case. During the first 5 min, the amplitude of the rectangular fluctuation is A_1 and the frequency is f_1, whereas for the last 5 min, the amplitude is A_2 and the frequency is f_2. Both A_1 and A_2 correspond to the amplitudes that would produce P_{st} = 2 for the frequencies f_1 and f_2 respectively if they were applied independently during the complete 10 min period.

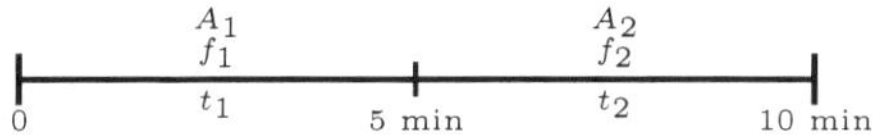

Fig. 7. Outline of the fluctuation used in Experiment 2.

It is obvious that the expected flicker severity value when computing the whole period should be P_{st} = 2, independently of f_1 and f_2. Fig. 8 shows the percentage of the P_{st} deviation from the theoretical value of 2 for f_1 = 1, 2, 3, 5, 7, 10 and 20 cpm and a range of f_2 from 30 to 2400 cpm.

The deviations become quite important for small values of f_1 and large values of f_2. For f_1 = 1 cpm and f_2 = 1000 cpm, the deviation is 22%. This means that the IEC flickermeter does not compute the annoyance properly when the rectangular fluctuations consist of two

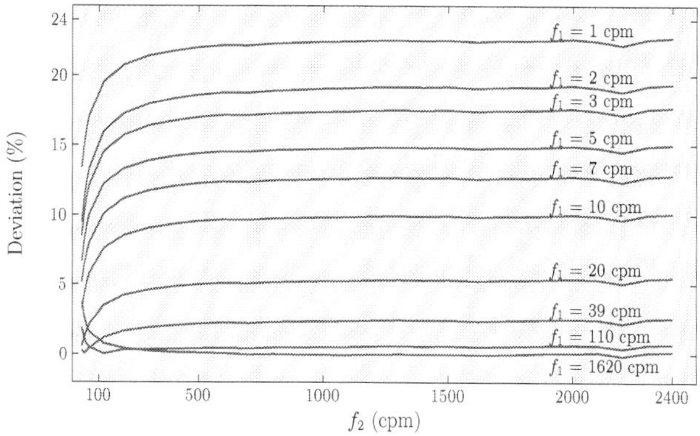

Fig. 8. P_{st} deviations calculated with the reference flickermeter for Experiment 2.

frequencies during the 10 min period. The origin of this error is located in the multipoint algorithm that is implemented in block 5.

2.3.2.3 Experiment 3

In this experiment, we worked with rectangular voltage fluctuations following the diagram outlined in Fig. 9. The fluctuation frequency, f, is the same for the 10 min period. During the first 5 min, t_1, the fluctuation amplitude is A_1, and A_2 for the rest of the time, t_2. Both A_1 and A_2 correspond to the amplitude of the fluctuations that would produce P_{st_1} and P_{st_2} for the complete 10 min period, respectively.

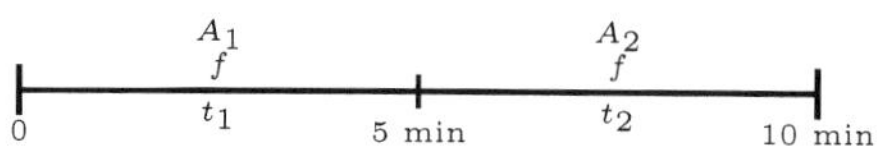

Fig. 9. Outline of the fluctuation used in Experiment 3.

The annoyance produced by a 5 min fluctuation of frequency f and amplitude A_1 is equivalent to the annoyance produced by a 10 min fluctuation of the same frequency and amplitude $A_{1e} = A_1 \cdot \sqrt{\frac{1}{2}}$. In the same way, the annoyance corresponding to the last 5 min is equivalent to the annoyance produced by a 10 min fluctuation of frequency f and amplitude $A_{2e} = A_2 \cdot \sqrt{\frac{1}{2}}$. By applying the amplitude quadratic composition of Equation 3, the fluctuation equivalent amplitude would be:

$$A_{eq} = \sqrt{A_{1e}^2 + A_{2e}^2} = \sqrt{\frac{A_1^2 + A_2^2}{2}}, \tag{9}$$

and therefore the equivalent flicker severity for the complete 10 min period would be:

$$P_{st_{eq}} = \sqrt{\frac{P_{st_1}^2 + P_{st_2}^2}{2}}, \tag{10}$$

independently of the fluctuation frequency, f.

We carried out simulations for values of A_1 and A_2 corresponding to the values of P_{st_1} and P_{st_2} compiled in Table 2 and for a range of the fluctuation frequency, f, from 1 to 2400 cpm.

P_{st_1}	P_{st_2}	$P_{st_{eq}}$
1	2	$\sqrt{2.5} = 1.581$
1	3	$\sqrt{5} = 2.236$
1	4	$\sqrt{8.5} = 2.915$
1	5	$\sqrt{13} = 3.606$
1	6	$\sqrt{18.5} = 4.301$

Table 2. Values of the total P_{st} for different combinations of P_{st_1} and P_{st_2}.

Fig. 10 shows the results of this experiment. Each curve of the figure reveals that the total P_{st} value depends on the fluctuation frequency in the interval from 1 to approximately 200 cpm. Additionally, the P_{st} values are always far from the expected ones, detailed in Table 2. In general, all the results are closer to the P_{st_2} value, although for 50% of the time, the fluctuation amplitude corresponds to P_{st_1}. Once again, the reason for this behavior is located in Block 5 of the IEC flickermeter.

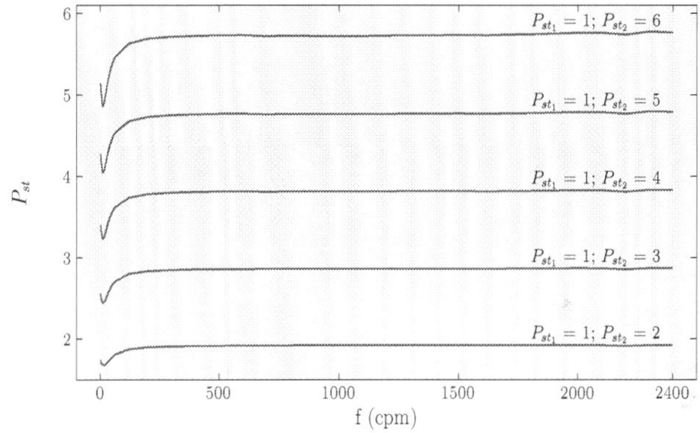

Fig. 10. P_{st} values obtained with the reference flickermeter for Experiment 3.

2.3.3 Laboratory subjective experiments using field registers

To quantify the relationship between P_{st} and the true annoyance perceived by people, we performed some laboratory tests using actual registered signals that were applied to a group of 11 people. These experiments do not try to be statistically significant but are intended to point out an agreement or discrepancy with the measurements carried out by the IEC flickermeter. The experiment consisted of two main tasks, training and assessment, based on a digital-to-analog conversion system that reproduced very accurately the voltage fluctuations corresponding to analytical signals or to field-registered signals.

Fig. 11 shows the physical layout of the laboratory. The luminance conditions were quite similar to those used for the characterization and specification of the IEC flickermeter (Cornfield,

1988). Those experiments were performed by using incandescent lamps of 60 W and a luminance of 125 lx. We used the same type of lamp, covered by a light diffuser that generates an average luminance of 125 lx over the table where the subjects spent most of the time during the experiments reading newspapers that were laid flat on the desk in front of them.

Fig. 11. The subjective experiments laboratory during the performance of the tests.

The main purpose of these experiments was to create a group of people with the ability to quantify the annoyance when subjected to light fluctuations produced by actual voltage signals. To achieve that objective, it was first necessary to perform a training task so that the 11 subjects could identify a numeric level of annoyance in particular light fluctuations.
The training task was carried out using rectangular voltage fluctuations, generated analytically by the reproduction system. The frequencies of the fluctuations were 1, 10, 100 and 1000 cpm with amplitudes corresponding to P_{st} = 0.5, 1, 1.5, 2, 2.5 and 3.
Originally, the assessment task was designed to evaluate a single value of annoyance for a 10 min period. This objective was not fulfilled because all the subjects reported the impossibility of averaging the annoyance over such a long period. Because of that, the assessment period was redefined to 1 min, maintaining the 10 min period for the complete duration of each test. We selected eight windows of 10 min from the field registers. The selection was carried out in terms of the P_{st} values and the distribution of the annoyance during the 10 min.
The results obtained during the laboratory subjective experiments are presented in Fig. 12. Each figure corresponds to a 10 min interval and shows, by using circles as the plot symbols, the average of the annoyance assessed by the 11 subjects for each minute and the flicker severity calculated by the IEC flickermeter with a short-term period of 1 min, $P_{st,1min}$. The solid lines in the figures show the flicker severity assessed by the IEC flickermeter for the usual short-term period of 10 min, $P_{10,min}$, and the subjective annoyance for the 10 min period, calculated by applying Ailleret's quadratic laws (see Equations 2 and 3) to the 1 min average values evaluated by the 11 subjects. The graphics have been organized for increasing P_{st} values, from P_{st} = 0.9 in case of Test 1 to P_{st} = 3.8 in case of Test 8. The distribution along the interval of the $P_{st,1min}$ values is quite regular for the tests 1, 2, 3 and 5. However, the rest of the tests show sudden changes for the $P_{st,1min}$ values.
The figures demonstrate that the $P_{st,1min}$ values were clearly larger than the subjective annoyance for the eight tests. Moreover, $P_{st,10min}$ is quite dependent on the largest $P_{st,1min}$ values.

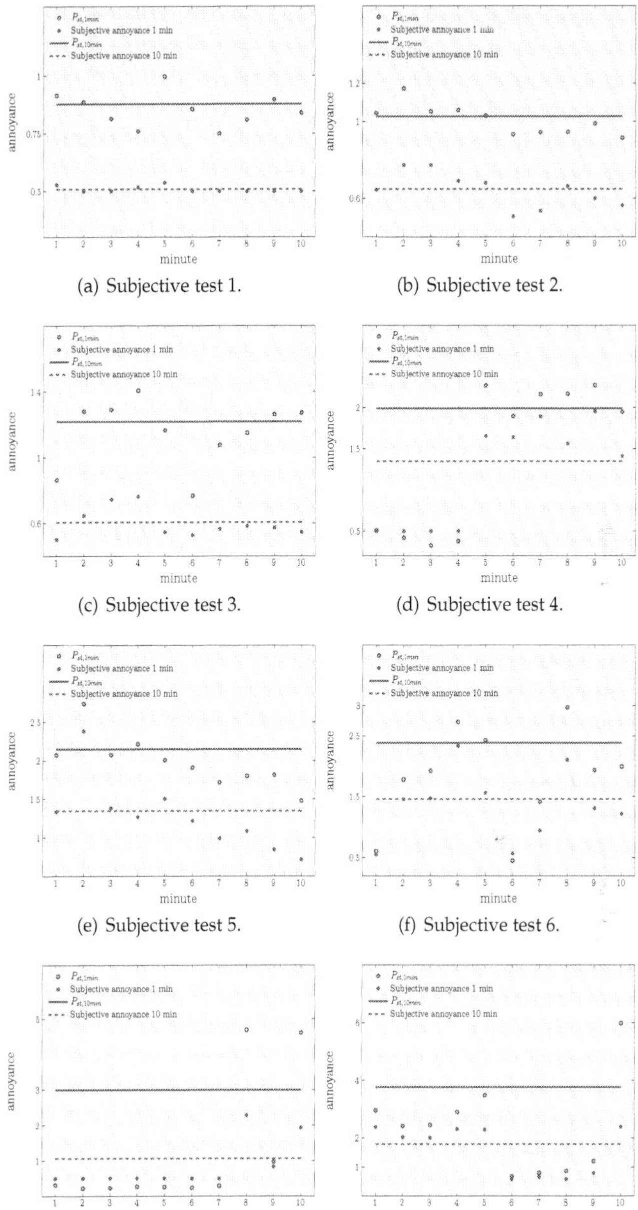

(a) Subjective test 1.
(b) Subjective test 2.
(c) Subjective test 3.
(d) Subjective test 4.
(e) Subjective test 5.
(f) Subjective test 6.
(g) Subjective test 7.
(h) Subjective test 8.

Fig. 12. Results for the laboratory subjective experiments.

This fact produces, for all the flicker levels, the important difference between the $P_{st,10min}$ and the average of the 10 min subjective annoyance. This difference becomes more important when the annoyance during the 10 min period is not uniform.

As a summary, in realistic conditions, when the voltage fluctuations are not uniform but have varying frequency and amplitude, the multipoint algorithm does not assess the true flicker annoyance precisely.

We think it is not possible to obtain a correct assessment by using a method based on the percentiles of the output of block 4. Another relevant conclusion refers to the observation period used for the calculation of the P_{st}. The standard established this time considering the duty cycles of the loads generating flicker. Our experiments proved that the human optical system does not average properly the annoyance caused by nonuniform fluctuations during the 10 min period. An alternative short-term period could be obtained from studies related to the assessment of the flicker annoyance by the people. From this perspective, our work suggests a short-term period of 1 min approximately.

3. Measurement of flicker in grid-connected wind turbines

3.1 Introduction. IEC 61400-21 Standard

Wind energy can be considered the source of a series of perturbations that alter the ideal form of the voltage signal at the common connection point. This is because of the relation between the wind speed and the power generated by the wind system, as well as the technical characteristics of the turbine itself.

Among the perturbations generated by the wind turbines, the fluctuations in voltage (Ackerman, 2005) are the most notable. These fluctuations in voltage and the associated flicker are linked to the connection and disconnection operations; the tower shadow effect; errors in the angle of incidence of the sail; yaw errors; wind shears and slight fluctuations in the wind.

The definitions, rules and procedures for the characterization of waveform quality in a wind turbine connected to the grid are defined by IEC 61400-21 (IEC-61400-21, 2001). This standard aims to define and specify the magnitudes that need to be determined to characterize wave quality; the measurement procedures to quantify the characteristics; and the procedures to assess the fulfillment of supply quality requirements.

The application framework of the standard is focused on individual wind turbines connected to the grid at medium or high voltage, that can regulate to a sufficient extent the active and reactive power.

With respect to the measurement of flicker, the standard requires that the wind turbine be characterized for the following situations:

a. Continuous operation: the specifications establish a processing and statistical evaluation scheme to obtain the flicker coefficients defining the wind turbine being studied. The phase-to-neutral voltage and the line current, $u_m(t)$ and $i_m(t)$, need to be processed for at least 10 registers of 10 min duration for each wind velocity between the cut-in of the wind turbine and 15 $\frac{m}{s}$.

b. Switching operations: the specifications establish an alternative processing and statistical evaluation scheme to assess the consequences of the start-up and shut-down maneuvers of the wind turbine. In this case, signals $u_m(t)$ and $i_m(t)$ need to be processed during these connection operations.

For both the continuous operation and the switching operations, the specifications establish a model to make the results independent of the characteristics and conditions of the grid to which the wind turbine is connected. This model is based on a fictitious grid that enables evaluation of voltage fluctuations caused exclusively by the wind turbine.

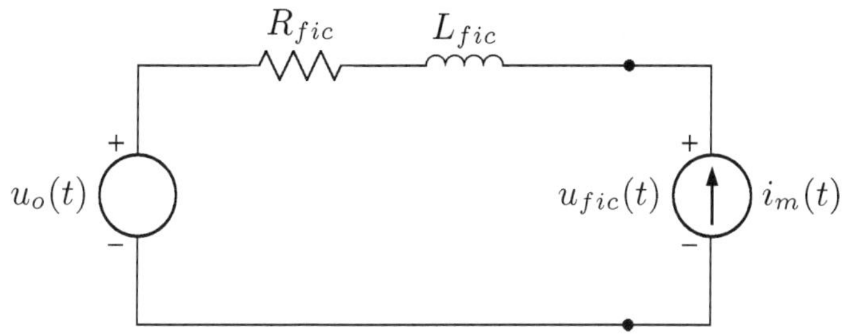

Fig. 13. Fictitious grid used to define $u_{fic}(t)$ for flicker measurements in wind turbines.

The fictitious grid is shown in Fig. 13. The turbine is represented by a current generator with a value of $i_m(t)$ (phase current of the turbine), and the grid to which it is connected is represented by the equivalent Thevenin, which is made up of the following components.

a. An ideal voltage supply $u_0(t)$, neutral-phase voltage with no fluctuation:

$$u_0(t) = \sqrt{\tfrac{2}{3}} U_n \sin(\alpha_m(t)), \tag{11}$$

where U_n is the *rms* nominal voltage, $\alpha_m(t)$ is the instantaneous phase of the fundamental component of the voltage and $u_m(t)$ is the phase-to-neutral voltage, measured in the turbine's terminals.

b. A fictitious impedance composed of the values R_{fic} and L_{fic}. The standard specifies angles of $30°, 50°, 70°$ and $85°$ for this grid impedance.

With this model, the fluctuations in voltage resulting from the turbine will be represented by the fictitious voltage fluctuations obtained using the following expression.

$$u_{fic}(t) = u_0(t) + R_{fic} \cdot i_m(t) + L_{fic} \cdot \frac{di_m(t)}{dt}, \tag{12}$$

The flicker measurement procedure established in the IEC 61000-4-15 standard is applied to this fictitious voltage, the first step being the calculation of its P_{st}. The value of $u_{fic}(t)$ determines the value of P_{st}. The main error source in the calculation of $u_{fic}(t)$ appears because $u_0(t)$ is obtained from $u_m(t)$. For this reason, the different techniques for obtaining $u_0(t)$ are described next.

3.2 Techniques for estimation of $u_0(t)$

The voltage $u_0(t)$ must be composed only of the fundamental component of the voltage in the terminals of the turbine, $u_m(t)$, suitably scaled. The estimation of $u_0(t)$ requires two operations:

a. Calculation of the instantaneous phase of the fundamental component of $u_m(t)$.

b. Adjustment of the amplitude of $u_0(t)$ in agreement with the nominal voltage of the turbine, U_n.

A small error in the estimation of the phase of the fundamental component of $u_m(t)$ can generate important changes in $u_{fic}(t)$ that significantly affect the P_{st} value calculated (Gutierrez et al., 2008).

To calculate the instantaneous phase of the fundamental component of $u_m(t)$, it is important to understand that this signal is limited in band and that most of its power is concentrated around its fundamental component of frequency, which is equal or very close to 50 Hz. Under this hypothesis, four techniques for obtaining the instantaneous phase of $u_m(t)$ are described next.

3.2.1 Techniques for calculating the instantaneous phase of $u_m(t)$

3.2.1.1 Zero-Crossing Method

The estimation of the frequency of the power system using the zero-crossing technique has been well known for a long time (Lee & Devaney, 1994). Starting with the frequency or period of each cycle of $u_m(t)$, constructing the instantaneous phase of the signal $u_m(t)$ is straightforward.

Working in the discrete domain, the algorithm searches for the positions of the contiguous samples of $u_m(t)$ that mark a transition of values from positive to negative. To achieve a more precise approximation to the zero-crossing point, a linear interpolation between the points of the transition is used, as is shown in Fig. 14.

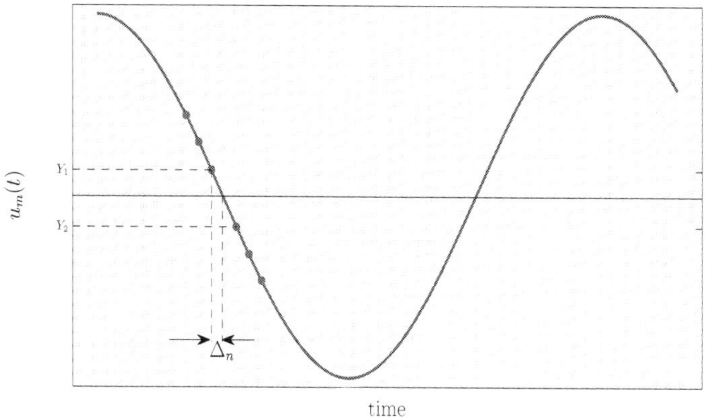

Fig. 14. The zero-crossing scheme.

The fraction of the sampling period that places the zero-crossing can be obtained from the following expression:

$$\Delta_n = \frac{Y_1}{Y_1 - Y_2} \tag{13}$$

Knowing the number and the fraction of the samples that make up a period, reconstruction of the instantaneous phase of the fundamental component is done, sharing the 2π radians uniformly for each sampling instant.

3.2.1.2 Hilbert Transform

If $u_m(t)$ is considered to be a band-pass signal, the estimation of the instantaneous phase of its fundamental component can be approached using the Hilbert transform (Gabor, 1946). In the context of signal processing and communications, the analytical representation of real signals facilitates some mathematical manipulations of the signal (Oppenheim & Shafer, 1999; Proakis & Manolakis, 2006). Given a band-pass signal, $u_m(t)$, it is possible to obtain the positive analytical signal corresponding to $u_m(t)$ in terms of its Hilbert transform with the composition:

$$u_{m+}(t) = u_m(t) + j\hat{u}_m(t)\,, \tag{14}$$

where $\hat{u}_m(t)$ represents the Hilbert transform of $u_m(t)$.
Starting from the analytical representation, the instantaneous phase of the signal can be obtained as:

$$\alpha_m(t) = \arg\{u_{m+}(t)\} \tag{15}$$

3.2.1.3 Short Time Fourier Transform (STFT)

To obtain the instantaneous phase of $u_m(t)$, a spectral estimation technique such as the short-time Fourier transform that is very widely applied in digital signal processing can be used. Basically, it consists in windowing the signal $u_m(t)$ and calculating the fundamental component at 50 Hz using the DFT (discrete Fourier transform) of the windowed signal. The window duration establishes the trade-off between time and frequency resolution. The minimum window necessary to characterize the fundamental component correctly is of 1 cycle and the maximum corresponds to the 10 min of $u_m(t)$.
When using 1 cycle windows, a very high time resolution can be achieved, and if advancing the window sample by sample, the apparition of sudden phase changes can be minimized. This method implies a large computational load and has a large associated spectral leakage, so it is possible that interharmonic components present in $u_m(t)$ close to the fundamental component distort the phase estimation. To improve the precision in these cases, spectral leakage can be reduced by increasing window length. In the extreme case of using a 10 min window, the frequency resolution between two points of the DFT would be 1.6 mHz, but a single average value of the phase of the fundamental component would be obtained. This means that the instantaneous phase of $u_m(t)$ is not followed; instead, its average value over 10 min is obtained.

3.2.1.4 Digital Phase-Locked Loop (DPLL)

A DPLL is based on the fundamentals of analogue systems that enable the synchronization of the reference signal generated by an oscillator, and with a specific input signal in both frequency and phase.
The working principle of a PLL is based on a feedback loop that can lock two waveforms in frequency and phase. The block diagram of the PLL is composed of three basic functional blocks: Phase Detector, Loop Filter+Amplifier and Voltage Controlled Oscillator (VCO). A block diagram of the PLL system is shown in the figure.
The PLL system generates a reference signal in phase with the fundamental component of $u_m(t)$, but it does not have a constant amplitude as required by $u_0(t)$. Therefore, it is necessary to complement the PLL system with an automatic gain control (AGC) at the VCO output, as shown in Fig. 15.
The loop filter output generates small phase errors in stationary working conditions all the time; therefore, the model's closed-loop transfer function can be linearized and its discrete implementation simplified through a DPLL system.

If the measured voltage, $u_m(t)$, was a single component at 50 Hz with variable amplitude, any one of the four aforementioned methods would enable the calculation of the voltage $u_0(t)$

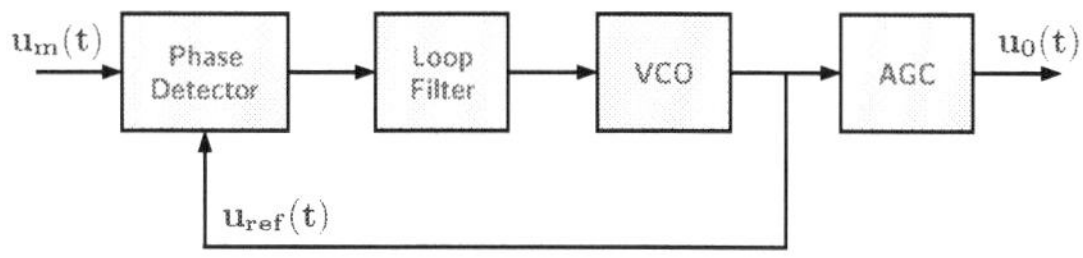

Fig. 15. The phase-locked loop scheme.

practically without error. However, the signal $u_m(t)$ can contain harmonic and interharmonic components, and its fundamental component may not be exactly 50 Hz. In this case, errors will be produced in the estimation of $u_0(t)$ that depend on the technique used. To minimize these errors, the $u_m(t)$ signal can be filtered using a very narrow band-pass filter that only selects its fundamental component. Given that this filtering step is critical from the precision viewpoint, it will be analyzed in more detail in the next subsection.

3.2.2 Filtering of the fundamental component of $u_m(t)$

When a signal has very narrow band interference, the traditional method of eliminating it consists in filtering the signal using a notch filter. Our case is the inverse, given that the objective is the fundamental component of the signal $u_m(t)$. Working in a discrete domain, a very narrow band-pass filter needs to be designed around the discrete pulsation corresponding to the fundamental frequency $\Omega_0 = 2\pi\frac{f_0}{f_s}$ with f_0 = 50 Hz.

In this section, the narrow band-pass filter design is analyzed first through an adaptive scheme based on the LMS[4] algorithm. This design enables the fundamental component at 50 Hz to be obtained without distortion and without any delay at the output with respect to the input. Nevertheless, when the fundamental component is at a frequency close to, but not exactly 50 Hz, the response of the filter is delayed a few samples with respect to the fundamental component of $u_m(t)$. This delay causes an error when calculating the P_{st} of $u_{fic}(t)$. To solve this problem, a zero-phase IIR filter implementation is used.

3.2.2.1 Adaptive Design of a Narrow Band Pass Filter

The adaptive interference canceller is a classical application of the adaptive filter, LMS (Widrow & Stearns, 1985). It is designed to eliminate an undesired sinusoidal interference. In reality, it implements a notch filter with the advantages of offering easy control of bandwidth and capacity to adaptively follow the frequency and phase of the interference. The Fig. 16 shows a discrete implementation of the adaptive interference canceller.

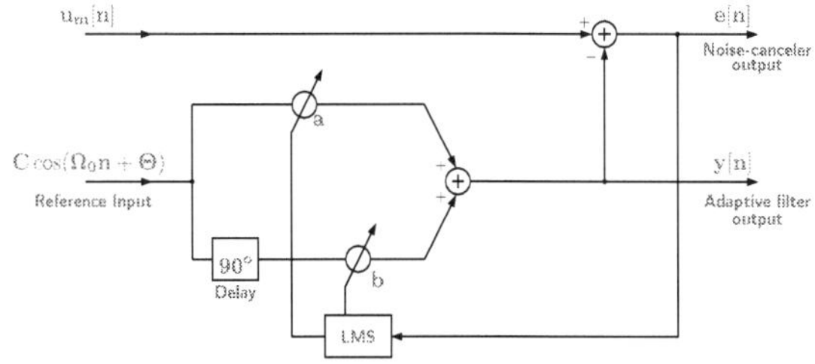

Fig. 16. The Adaptive Interference Canceller scheme.

[4] Least Mean Square

The transfer function of this output scheme $e[n]$ with respect to $u_m[n]$ is given by:

$$H_1(z) = \frac{E(z)}{U_m(z)} = \frac{z^2 - 2z\cos(\Omega_0) + 1}{z^2 - 2(1-\mu C^2)z\cos(\Omega_0) + 1 - \mu C^2} \tag{16}$$

It can be observed that this function does not depend on the phase of the reference signal, and that the amplitude of the reference signal, C, and the step-size of the algorithm LMS, μ, are interchangeable values given that $H_1(z)$ only depends on the product μC^2. The frequency response of $H_1(z)$ corresponds to a notch filter with a 3 dB bandwidth $BW = 2\mu C^2$ rad $= \frac{f_s \mu C^2}{\pi}$ Hz.

In the same scheme, if the transfer function of the adaptive filter output, $y[n]$, is calculated with respect to the input, $u_m[n]$, the following expression is obtained:

$$\begin{aligned} H(z) &= \frac{Y(z)}{U_m(z)} = \frac{U_m(z) - E(z)}{U_m(z)} = 1 - H_1(z) \\ &= 2\mu C^2 \cdot \frac{z\cos(\Omega_0) - 1}{z^2 - 2(1-\mu C^2)z\cos(\Omega_0) + 1 - \mu C^2} \end{aligned} \tag{17}$$

The frequency response $H(\Omega)$ corresponds to a narrow band-pass filter that enables obtaining the fundamental component of $u_m[n]$. When working with $C = 1$, $f_s = 3200$ and $\mu = 0.0003$, a bandwidth of approximately 0.3 Hz is found around the 50 Hz component. The Fig. 17 shows the module and the phase delay $\tau_f(\Omega) = \frac{-\phi(\Omega)}{\Omega}$ of $(1-\mu C^2)H_1(\Omega)$ and $H(\Omega)$ scaling the axis of frequency in Hz.

The main problem of $H(z)$ in obtaining the fundamental component of $u_m[n]$ centers on the abrupt behaviour of the phase delay around 50 Hz. If the fundamental component of $u_m[n]$ were of 50.05 Hz, the output of $H(z)$ would be displaced with respect to the input by three samples. This causes an appreciable error in the P_{st} of $u_{fic}(t)$. To solve this problem, the Anticausal Zero-Phase Filter Implementation can be used.

3.2.2.2 Anticausal Zero-Phase Filter Implementation

For IIR filters, such as $H(z)$, the phase distortion is usually highly nonlinear. To eliminate phase distortion, anticausal zero-phase filter implementation can be used. Consider the processing scheme in Fig. 18.

After filtering in the forward direction, the filtered sequence is reversed and run it back through the filter. The result has exactly zero-phase distortion. In fact, in the frequency domain, $Y(\Omega) = U_m(\Omega) \cdot |H(\Omega)|^2$. The magnitude is the square of the filter's magnitude response, and the filter order is double the order of $H(z)$.

This implementation can only be used in cases such as ours, in which $u_m[n]$ is a finite duration signal known before being filtered. From the signal obtained, $y[n]$, it is necessary to eliminate the transitory at both its ends.

3.3 Results

3.3.1 Results Using Test Signals

A comparison was made of different methods of estimation of $u_0(t)$ under the same test signals. To do this, analytical signals were defined corresponding to $i_m(t)$ and $u_m(t)$. Moreover, the nominal working conditions of the wind turbine defining the parameters making up the fictitious grid were established:

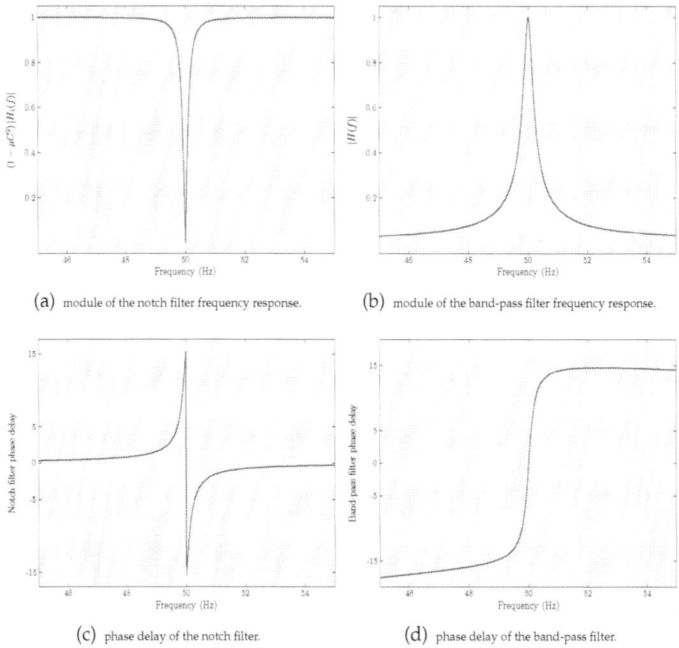

(a) module of the notch filter frequency response.

(b) module of the band-pass filter frequency response.

(c) phase delay of the notch filter.

(d) phase delay of the band-pass filter.

Fig. 17. Frequency responses of the notch and band-pass filters.

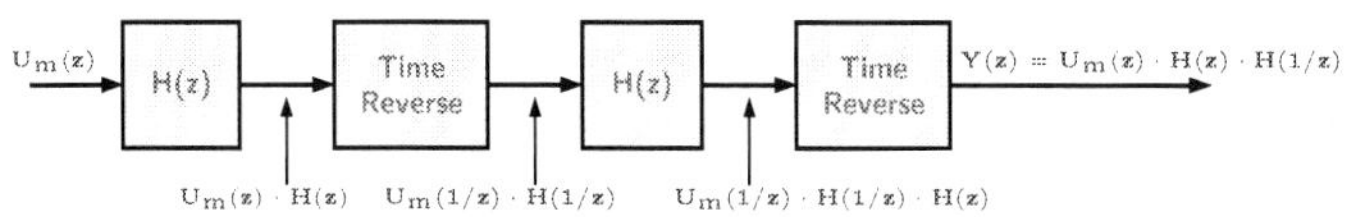

Fig. 18. The anticausal zero-phase filter scheme.

Parameters of the wind turbine.

The characteristic parameters of the wind turbine affect the fictitious grid and the waveforms of the test signals. In this sense, a turbine with a rated apparent power of S_n = 600 kVA and nominal voltage of U_n = 690 V were considered. Furthermore, a ratio between the short-circuit apparent power of the fictitious grid, S_{kfic} and S_n of 20 and a grid angle ψ_k = 85° was considered.

Test voltage, $\mathbf{u_m(t)}$

To model an analytical voltage at the terminals of the wind turbine, $u_m(t)$, near real conditions, different studies into the measurement of perturbations produced by wind turbines (Sørensen, 2001; Vilar et al., 2003) were analyzed and a test voltage was configured according to the following analytical expression.

$$u_m(t) = \sqrt{\tfrac{2}{3}}U_n \cos\left(2\pi f_0 t + 2\pi \tfrac{\Delta f}{2} \sin(2\pi f_r t)\right) + \sqrt{\tfrac{2}{3}}U_n \sum_{i=1}^{N} a_i \cos(2\pi f_i t) \tag{18}$$

In this way, the test voltage is made up of a fundamental component of frequency f_0 that undergoes a sinusoidal variation with a rate $f_r = \frac{1}{60}\,\frac{Hz}{s}$ and a range Δf =0.2 Hz. Furthermore, N frequency components of amplitude a_i and frequency f_i were added including, $3^{rd}, 5^{th}, 7^{th}$, 9th and 11th harmonic components of f_0 with a_i = 0.5% of the fundamental's amplitude and, the interharmonic components from 40 to 60 Hz, with 1 Hz separation and with amplitudes a_i = 0.1% of the fundamental's amplitude.

Test current, $\mathbf{i_m}(t)$
From the same bibliographic analysis used for $u_m(t)$, a test current was configured and derived from the following expression:

$$i_m(t) = \sqrt{2}I_n \cos\left(2\pi f_0 t + 2\pi \tfrac{\Delta f}{2}\sin(2\pi f_r t) + \alpha_{iu}\right) + \sqrt{2}I_n \sum_{k=1}^{M} b_k \cos(2\pi f_k t) \qquad (19)$$

The configuration of the fundamental component is identical to the $u_m(t)$ case, except for the inclusion of phase α_{iu}. This is calculated after considering a power factor of 0.95. In this case, M interharmonic components different from those of $u_m(t)$ were added including the 3rd, 5th, 7th, 9th and 11th harmonic components of f_0 with b_k = 1.5% of the fundamental's amplitude; the $20^{th}, 21^{th}, 22^{th}, 23^{th}, 24^{th}$ and 25th harmonic components of f_0 with b_k = 1.75% of the fundamental's amplitude and, the interharmonic components from 1025 to 1225 Hz, with 50 Hz separation and with amplitude b_k = 1.75% of the fundamental's amplitude.

Results
Starting from the test signals and the wind turbine parameters, the theoretical $u_{fic}(t)$ and the P_{st} produced were calculated, giving a value of $P_{st,fic_{ref}}$ = 0.536 for a fundamental frequency f_0 = 50 Hz and $P_{st,fic_{ref}}$ = 0.535 for a frequency f_0 = 50.05 Hz. This value is taken as the reference for the comparison of all of the methods of estimation of $u_0(t)$ described previously. On the other hand, starting from the same test signals and wind turbine data, a different $u_0(t)$ was estimated for each method described, and the corresponding $u_{fic}(t)$ and their P_{st} were calculated. Tables 3 and 4 show the results for each method in relation to the reference values. In all cases, a sampling frequency of $f_s = 3200\ \frac{m}{s}$ was used. Both tables give details about the value of $P_{st,fic_{est}}$ obtained for the $u_{fic}(t)$ calculated from the estimation of $u_0(t)$, as well as the error with respect to the theoretical reference value.
Table 3 corresponds to the phase estimation techniques without filtering $u_m(t)$, whereas table 4 shows the results for the methods that use a band-pass filter before estimating the phase.

Method	50 Hz		50.05 Hz	
	$P_{st,fic_{est}}$	Error (%) [1]	$P_{st,fic_{est}}$	Error (%) [2]
Zero-Crossing	0.548	2.24	0.547	2.24
Hilbert Transform	0.553	3.12	0.552	3.12
STFT 1 cycle	0.556	3.78	0.555	3.66
DFT 10 min	0.546	1.78	0.651	21.70
PLL	0.592	10.42	0.585	9.21

[1] Percentage deviation from $P_{st,fic_{ref}} = 0.536$
[2] Percentage deviation from $P_{st,fic_{ref}} = 0.535$

Table 3. Simulated results for analytical signals by non-filtering methods.

Method	50 Hz		50.05 Hz	
	$P_{st,fic_{est}}$	Error (%) [1]	$P_{st,fic_{est}}$	Error (%) [2]
LMS+ Zero-Crossing	0.537	0.22	0.564	5.36
Zero-Phase filter+ Zero-Crossing	0.536	0.00	0.535	0.00

[1] Percentage deviation from $P_{st,fic_{ref}} = 0.536$
[2] Percentage deviation from $P_{st,fic_{ref}} = 0.535$

Table 4. Simulated results for analytical signals by band-pass filtering methods.

It can be observed that the phase estimation of $u_m(t)$ is affected by the presence of interharmonic content, which causes errors in the calculation of $P_{st,fic_{est}}$. These errors are appreciable in the techniques that do not eliminate those components through filtering (less than 3.5% in the zero-crossing case and Hilbert and reaching 4% and 10% in the STFT and PLL cases, respectively). However, most of these techniques are not substantially affected by the deviation of the fundamental frequency f_0. Most of them provide values quite similar for both frequencies, expect in the case of DFT with 10 min window that provides an error over the 21% when the fundamental frequency corresponds to 50.05 Hz. Moreover, it can be seen that the phase estimation using zero-crossing or Hilbert provides practically identical results in all cases.
Table 4 shows how the errors in the estimation of $u_0(t)$ are drastically reduced by applying a narrow band-pass filter. Nevertheless, the deviation in f_0 = 50.05 Hz produces important errors (more than 5%) as a result of the nonlinear phase of the narrow band filter. Once again, it can be observed that there are no differences between the zero-crossing and Hilbert methods, although the signal was previously filtered. The problem of the nonlinear phase can be solved by applying a band-pass filtering using a zero-phase IIR filter, thus obtaining a very precise estimation of $u_0(t)$ and leaving the errors in P_{st} practically nil.

3.3.2 Results using real signals

To finalize the study, a comparison of the most relevant techniques brought to light in the previous analysis was carried out using the signals from the real registers. To achieve this, several measurements were performed on two different wind turbines[5]. The first corresponds to a machine with a double speed asynchronous generation system (4 and 6 poles); fixed sail passage and a fixed generator velocity; nominal power of 600 kW and nominal voltage of 690 V. The second machine has a four-pole, synchronous generation system and electronic power control; variable sail passage and variable generator velocity, and provides a nominal power of 800 kW and a nominal voltage of 1000 V.
A total of three-phase 45 registers were taken of voltage, current and wind velocity with a duration of 10 min in the first machine and 25 registers in the second. These registers were used for the comparison of the four most relevant methods for estimation of $u_0(t)$ through the calculation of the P_{st} of the $u_{fic}(t)$. The techniques employed for the comparison of the real registers were zero-crossing, STFT with a 1 cycle window, zero-crossing after narrow band filtering using LMS and zero-crossing after filtering using a zero-phase filter. For comparison

[5] The authors would like to thank SOTAVENTO GALICIA S.A. (Spain) for making the signals available free of charge for the purpose of this work.

purposes, the values of P_{st} obtained with the zero-crossing method after filtering using a zero-phase filter were used as reference.

Fig. 19 shows the results for the first machine. This figure shows the percentage deviation in the P_{st} for each register corresponding to the different techniques selected with respect to the reference value, as a function of the corresponding mean power of each 10 min register.

It can be observed that the average percentage of deviation is not excessive at less than 2%, when techniques without filtering are used. However, on filtering the fundamental component of $u_m(t)$ using a narrow band-pass filter, LMS, the average percentage of variation is greater than 5%, and even in numerous cases over 10%. This is the result of the variability of the fundamental component f_0 of $u_m(t)$ in the generation of the first machine. In these cases, the application of a nonlinear filtering of phase causes errors in the estimation of the phase of $u_m(t)$, which affect the formation of $u_{fic}(t)$ and thus the calculation of the P_{st}.

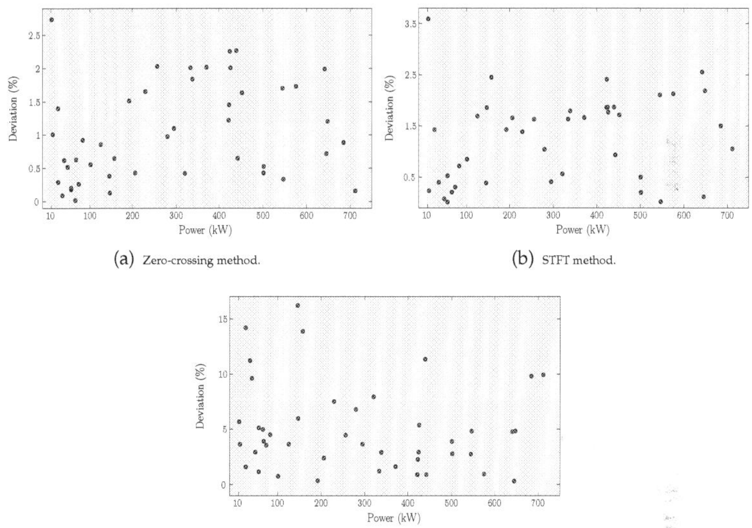

Fig. 19. Results for actual registers in first turbine of Sotavento wind farm.

Fig. 20 shows the results for the second machine. This figure shows the values of P_{st} obtained for each register corresponding to the different techniques selected, as a function of the corresponding mean power of each 10 min register.

It can be observed in this case that the zero-crossing without filtering provides very large variations with respect to the other methods as a result of the important interharmonic components present in $u_m(t)$. As the fundamental component of $u_m(t)$ in this machine varies only slightly with respect to its nominal value, the LMS, narrow-band filtering does not generate as many variations as in the first machine, fitting with the reference method very precisely.

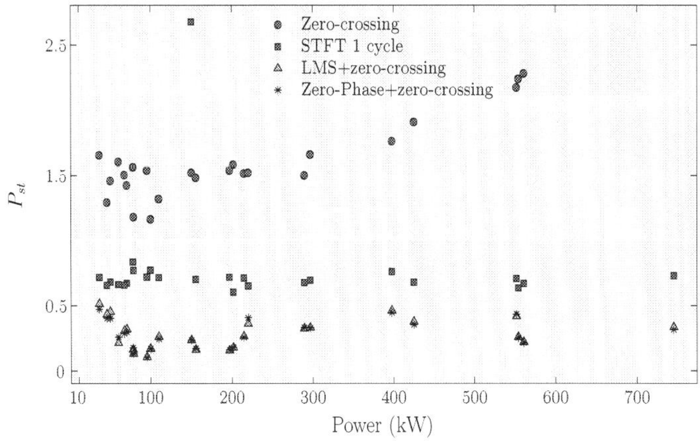

Fig. 20. Results for actual registers in the second turbine of the Sotavento wind farm.

4. Conclusion

This chapter dealt with the voltage flicker and the current investigations around the main instrument used to characterize the annoyance produced by the light flicker, the IEC flickermeter. After a brief description of the IEC 61000-4-15 standard we have critically analyzed the ability of the IEC flickermeter to assess the actual annoyance produced by flicker in people. All the studies we presented lead to the same conclusion. The assessment of flicker severity performed by the IEC flickermeter does not match the true flicker annoyance. Block 5 of the IEC flickermeter calculates the annoyance accurately only for uniform fluctuations during the 10 min period. In realistic conditions, when the voltage fluctuations are not uniform but have varying frequency and amplitude, the multipoint algorithm does not assess the true flicker annoyance precisely. We think it is not possible to obtain a correct assessment by using a method based on the percentiles of the instantaneous flicker sensation. Moreover, the introduction of a higher accuracy for the adjustment of the multipoint algorithm does not seem to be a proper solution.

In the last section, the IEC flickermeter is used to characterize flicker caused by wind turbines. The section contains a description of the part of the IEC-61400-21 standard dedicated to flicker. We demonstrated the sensitivity of the P_{st} of the fictitious voltage when $u_0(t)$ is estimated by different discrete techniques. Our results show, using both analytical signals as real signals, how the filtering of the measured voltage, $u_m(t)$, is essential for an accurate estimation of the phase of the fundamental component of $u_m(t)$. Moreover, a typical band pass filter could not be efficient enough to obtain good results in some real conditions. These filtering techniques produce small phase delays in the estimation of $u_0(t)$ that could cause appreciable errors in the P_{st} of $u_{fic}(t)$. To solve this problem, the Anticausal Zero-Phase Filter Implementation can be used, eliminating the phase distortion.

5. References

Ackerman, T. (2005). Wind Power in Power Systems., John Wiley & Sons, Ltd.

Ailleret, P. (1957). Détermination des lois expérimentales du papillotement (flicker) en vue de leur application aux réseaux basse tension sur lesquels les charges varient périodiquement ou aléatoirement (soudeuses démarrages de moteurs), Bulletin de la Société Française des Electriciens **7**(77): 257–262.

Arlt, D., Stark, M. & Eberlein, C. (2007). Examples of international flicker requirements in high voltage networks and real world measurements, Proceedings of the 9th International Conference on Electrical Power Quality and Utilisation, pp. 1–4.

Cornfield, G. (1988). Definition and measurement of voltage flicker, IEE Colloquium on Electronics in Power Systems Measurement .

de Lange, H. (1952). Experiments on flicker and some calculations on an electrical analogue of the foveal systems, Physica **18**(11): 935–950.

de Lange, H. (1961). Eye's response at flicker fusion to square-wave modulation of a test field surrounded by a large steady field of equal mean luminance, Journal of the Optical Society of America **51**(4): 415–421.

Gabor, D. (1946). Theory of communication, Journal of IEE **93**: 429–457.

Gutierrez, J., Ruiz, J., Leturiondo, L. & Lazkano, A. (2008). Filcker Measurement System for Wind Turbine Certification, IEEE Transactions on Instrumentation and Measurement **57**(12): 375–382.

IEC-61000-4-15 (2003). Electromagnetic compatibility (emc) Ű part 4: Testing and measurement techniques - section 15: Flickermeter functional and design specifications.

IEC-61400-21 (2001). Wind turbine generator systems. Part 21: Power quality requirements for grid connected wind turbines.

IEC-868 (1986). Flickermeter. functional and design specifications.

Key, T., Nastasi, D., Sakulin, H., Harding, J. & Cooke, T. (1999). System Compatibility Research Project Final Report, Task 21: Power Line Monitors, Part II: Flickermeters, EPRI PEAC Corporation .

Koenderink, J. & van Doorn, A. (1974). Visibility of unpredictably flickering lights., Journal of the Optical Society of America **64**(11): 1517–1522.

Lee, J. & Devaney, M. (1994). Accurate measurement of line frequency in the presence of noiseusing time domain data, Proceedings of the 10th IEEE Instrumentation and Measurement Technology Conference., pp. 1016–1019.

Nevries, K. (1984). Netzrückwirkungen verursacht durch den Betrieb von Elektrwarmeanlagen - Tätigkeitbericht des UIE Studienkomitees Netzrückwirkungen, Proceedings of the UIE Congress, Stockholm.

Oppenheim, A. & Shafer, R. (1999). Discrete-Time Signal Processing, Prentice Hall.

Proakis, J. & Manolakis, D. (2006). Digital Signal Processing, Prentice Hall.

Rashbass, C. (1970). The visibility of transient changes of luminance, The Journal of Physiology **210**(1): 165.

Ruiz, J., Gutierrez, J. & Irusta, U. (2007). Singular Frequencies in Rectangular Fluctuations in the IEC Flickermeter, IEEE Transactions on Power Delivery **22**(2): 1255–1256.

Simons, K. (1917). Das Flackern des Lichtes in elektrischen Beleuchtungsanlages, ETZ (37): 453–455.

Sørensen, P. (2001). European Wind Turbine Testing Procedure Developments. Task 2: Power Quality, Risø National Laboratory .

Szlosek, M., Piekarz, M., Hanzelka, Z., Bien, A., Piatek, A., Loziak, W., Pietrucha, R., Hashad, M., Wolski, M., Olczykowski, Z. & Smajek, L. (2003). Comparative tests of flickermeters, Proceedings of the 17th International Conference on Electricity Distribution, Barcelona.

UIE (1992). Flicker measurement and evaluation.

Vilar, C., Usaola, J. & Amaris, H. (2003). A frequency domain approach to wind turbines for flicker analysis, IEEE Transactions on Energy Conversion **18**(2): 335–341.

WG2CIGRÉ (2004). Test Protocol for IEC Flickermeter used in Power System Voltage Monitoring.

WGC4CIGRé (2004). Report on Power Quality Indices and Objectives.

Widrow, B. & Stearns, S. (1985). Adaptive Signal Processing, Prentice Hall.

16

Wideband MIMO Measurement Systems for Antenna and Channel Evaluation

Carlos Gómez-Calero, Jonathan Mora, Luis Cuéllar
Leandro de Haro and Ramón Martínez
Technical University of Madrid (UPM)
Spain

1. Introduction

The introduction of new generations of mobile communication systems, such as the current 3G and the future 4G, along with the provision of multimedia services, has exponentially increased the number of users demanding higher data rate. This fact has aroused the interest in studying possible solutions to improve system performance without making use of new frequency band allocations in the already crowded frequency spectrum. As an interesting option, wireless communication systems with Multiple-Inputs and Multiple-Outputs (MIMO) in which many antennas are used for transmission and reception have received increasing attention from the research community over the past few years (Paulraj et al., 2004, Goldsmith et al., 2003).

On the other hand, the OFDM (Orthogonal Frequency Division Multiplexing) technique has been adopted for the transmission of signals over wireless channels in several standards of systems such as DAB, DVB-T and IEEE 802.11a/g for LAN and IEEE 802.16a, for MAN. Moreover, this technique is presented as a potential candidate for the 4G mobile communications. Regarding this, different studies and schemes have been realized to use MIMO systems with OFDM as in (Sampath et al, 2002). In the last two years, the standard IEEE 802.11n is been developing to introduce MIMO with OFDM in WLAN systems.

The performance achieved with a MIMO system depends on factors such as the propagation channel, the space-time processing algorithm, or the number and characteristics of antennas. In order to see MIMO systems deployed over real communication networks, a thorough knowledge of the MIMO channel is a must. Therefore, a significant amount of measurements under real conditions and their detailed analysis are needed. Furthemore, in order to evaluate the performances of MIMO systems regarding the type of antenna array and channel propagation characteristics, two MIMO-OFDM measurement systems have been created in Universidad Politécnica de Madrid (UPM) and described in this chapter.

2. MIMO testbeds

Since MIMO systems begins to be a promising technology for increasing spectral efficiency in the new coming wireless communications systems, many research institutions and universities have been working towards developing some type of MIMO testbeds or prototypes that validated some of the benefits of using multiple antennas at both ends of the radio link.

Development of those platforms is focus on several research topics like MIMO channel for different scenarios, algorithm testing, hardware and software platform employed, coding and modulation techniques, and scope of applications like cellular or wireless LAN.

One of the most dynamic topics has been prototypes showing performance of different MIMO schemes for the mobile Third Generation, 3G. In (Adjoudani et al., 2003), a prototype of 4x4 antennas is integrated in a base station for UMTS system allowing up to 1 Mbps downlink capacity.

Many researchers have been focus in topics regarding wireless LAN standards and applications. Currently, several solutions have been presented as MIMO channel sounders, some of them even at a commercial level. A straightforward implementation is an offline mode operation, which allows implementing the MIMO algorithms and most of the baseband signal processing in a PC-based environment. This operation mode simplifies the algorithms testing, since no real-time constrains have to be fulfilled. In (Batariere et al., 2001) a 2x2 MIMO prototype based in this idea is reported. The signal is pre-calculated, loaded in the transmitter memory and transmitted through the MIMO channel. At the receiver, the raw data is post-processed after being sent to a PC connected to the receiver module. However, this testbed does not include real-time performance evaluation or feedback analysis. Several MIMO testbeds working in offline basis have been presented in (Seeger et al., 2003), (Aschbacher et al., 2004), (Pietsch et al., 2005), (van Zelst et al., 2004).

One of the most world-accepted MIMO channel sounders and with an excellent measurement accuracy and broadband operation is the RUSK channel sounder with MIMO extension, by MEDAV (GmbH, 2001). This channel sounder is based on fast switches to change from one antenna to another. An important drawback of this channel sounder is its very high cost and its low flexibility. Another wideband MIMO testbed is the owned by the Virginia Polytechnic Institute (MCMS testbed) (Ellingson, 2005). It allows up to 4x16 antenna configuration with capabilities like tuning frequency parameter from 250 MHz to 6 GHz with a bandwidth of up to 40 MHz. The signal processing module works over a real 802.11b wireless LAN. A further step in MIMO prototypes systems is carried out by the Georgia Tech (Kaiser et al., Sept. 2004) which is designed for 802.11a and 802.16 standards. It is designed for working in real time basis performing some signal processing tasks over an FPGAs. A similar testbed called "STARS" prototype is presented in (Kaiser et al., 2004) which improves flexibility by configuring multiusers scenarios. Also a few MIMO prototypes with feedback capabilities have been reported. An example of them is presented in (Samuelsson et al., 2005) where a real-implementation of a space-time scheme with wired feedback is presented. Its main advantage is the adaptive modulation and MIMO algorithm testing capabilities. However the system bandwidth is very small.

Some commercial prototypes can be considered as general purpose platforms for MIMO testing, measuring and developing of algorithms. An interesting example is the HaLo prototyping platform, from (Signalion). It consist of a flexible and scalable hardware platform that includes some of the basic modules to be used in a real-time implementation and easy-to-use interface with MATLAB/Simulink flow, and several reference modules for the simulation of the physical layer of 802.11a/n standards, among others. An implementation of a 4x4 MIMO-OFDM testbed using the HaLo prototype was presented in (Stege et al., 2004). For more extensive resume of MIMO testbeds the reader can review the first part of (Kaiser et al., 2004).

3. Measurement system for 2.45 GHz MIMO channels

In this section, a description of design and implementation of a measurement system (testbed) at 2.45 GHz is detailed. The testbed UMATRIX (UPM Multi-Antenna Testbed for Research in Indoor eXperiments) is based on Software-Radio and it works at 2.45 GHz. One of the most important advantages of this testbed is that it can realize measurements in MIMO-OFDM channels with different types of antenna and configurations, such as reconfigurable antennas.

The general scheme of MIMO testbed UMATRIX is shown in Fig. 1. It allows the use up to 4 antennas at the transmitter and the receiver side, so a 4x4 MIMO scheme can be evaluated. The way of working is the following: in the first place, the OFDM signals for all the antennas are generated in an offline mode in the PC by Matlab. Once the signals are generated, they are introduced to online processing module, which consists of SDR (Software-Defined Radio) platforms and it sends the signals to RF subsystem. This is the IF/RF module and its objective is to amplifier, filter and upconvert the signals to RF (2.45 GHz). The last module in the transmitter is the antenna one, which receives the signals from RF stage and send them to the receiver through MIMO channel.

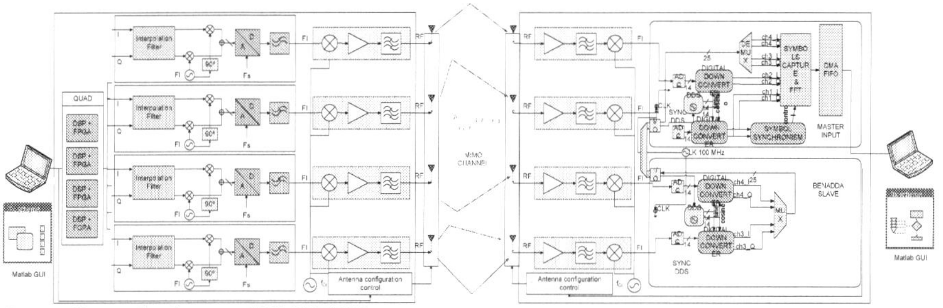

Fig. 1. General scheme of UMATRIX

At the receiver side, the signals are received by the antennas and passed to RF/IF stage which downconverts the signals to IF (40.625 MHz). From here, the signals go to SDR which are based on a FPGA design with 2 boards: one for receivers 1 and 2, and another for 3 and 4. This module process the OFDM signals in time and send them in frequency domain to the PC in order to do a pos-process (offline) to obtain the channel response. Moreover, a new module has been added to control the different configurations of reconfigurable antenna under test.

3.1 Signal processing

The signal processing subsystem deals with all the different modules which are related to the OFDM signal, the software-radio and the channel estimation.

3.1.1 OFDM structure

Due to have a demonstrator of wideband, the OFDM technique will be used, since is very efficient to transmit data over selective frequency fading channels. The main idea is to divide in frequency a wideband channel in narrowband subchannels. Likewise, each subchannel is a channel with flat fading despite of frequency-selective feature of a wideband radio channel. To generate these subchannels in OFDM, an inverse of Fourier Fast Transformation (IFF) is applied to one block of N data symbols:

$$x(n) = \frac{1}{N}\sum_{k=0}^{N-1} X(K)e^{j\frac{2\pi f_c kn}{N}} \tag{1}$$

In order to avoid inter-symbol interference due to the spreading of channel delay, a cyclic-prefix block is inserted. This prefix is known as guard interval (GI), where the number of samples of th prefix should be higher than the length of channel impulse response. The effects of cyclic-prfix delete the ISI and convert the convolution between transmitted symbols and channel in a circular convolution. Thus, the FFT is used at the receiver to recover the block of data symbols. The synchronism module in the FPGA of the receiver is based on (van de Beek et al., 1997). In Table 1 the most important parameters of the system are detailed.

Parameter	Symbol	Value
Sampling frequency	Fs	6.25 MHz
Useful symbol time	Tu	1024/Fs=163.84 μs
Guard time	Tg	Ts/8=40.96 μs
Symbol time	Ts	184.32 μs
Spacing between carriers	Δf	1/Tu≈6.1 KHz
Number of carriers	N	768
Bandwith	BW	4687500 Hz

Table 1. Main testbed OFDM parameters

In the transmitter, a frame with 8 OFDM symbols is continuously generated, as shown in Fig. 2. The first symbol is used for receiver synchronism and is a null symbol. After that, a reference symbol which will be used to estimate the channel is introduced. And finally, 6 data symbols are included. These symbols are randomly generated since they are not going to be evaluated, only the reference symbol to obtain the channel response.

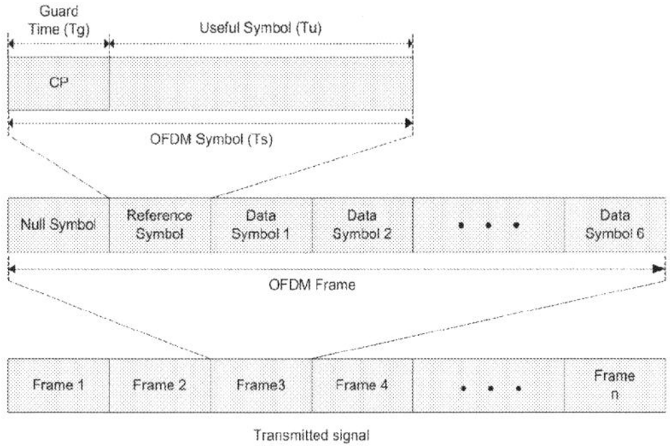

Fig. 2. OFDM frame structure

3.1.2 Channel estimation

The channel estimation in MIMO systems is a very important stage, since in MIMO systems the performances of algorithms depend on the accuracy of this estimation. The received signal in each carrier is given by

$$\mathbf{R}_k = \mathbf{H}_k \mathbf{X}_k + \mathbf{N}_k \tag{2}$$

where **X** is the vector of transmitted signals by each antenna, **H** indicates the MIMO channel matrix and **N** represents the noise in the channel, all for each k-th subcarrier. The MIMO channel matrix can be computed by

$$\mathbf{H}_k = \begin{bmatrix} h_{1,1,k} & \cdots & h_{1,M_T,k} \\ \vdots & \ddots & \vdots \\ h_{M_R,1,k} & \cdots & h_{M_R,M_T,k} \end{bmatrix} \tag{3}$$

where each element of the matrix represents the channel response between each pair of transmitter and receiver antennas.

On the other hand, different ways of obtaining the channel response have been studied. In UMATRIX, orthogonal codes are used as pilots to let the receiver split the different contributions from each antenna. Due to the maximum number of antennas is 4, a 4x4 matrix is needed. In our case, we use the following pilot matrix:

$$\mathbf{P} = \begin{bmatrix} 1 & -1 & 1 & 1 \\ 1 & 1 & -1 & 1 \\ 1 & 1 & 1 & -1 \\ -1 & 1 & 1 & 1 \end{bmatrix} \tag{4}$$

where the number of rows represents the space and the columns can represent either the time or the frequency. In a firs option, the frequency axis was chosen, so in this way, the channel is assumed invariant in 4 subcarriers. However, and with the aim of measuring frequency selective channels, the time was as chosen axis in columns.

In order to get a better synchronization at the receiver, the pilot matrix **P** is multiplied by a pseudorandom sygnal (S). Thus, at the receiver, for each *k*-subcarrier, we will have (2) with

$$\mathbf{X}_k = S_k \mathbf{P} \tag{5}$$

And if it is chosen

$$\mathbf{Y}_k = \mathbf{X}_k^H \left(\mathbf{X}_k \mathbf{X}_k^H\right)^{-1} \tag{6}$$

to estimate the channel, then the channel is multiplied by received signal **Y**, obtaining:

$$\begin{aligned} \hat{\mathbf{H}}_k &= \mathbf{R}_k \mathbf{Y}_k \\ &= \mathbf{H}_k \left(S_k \mathbf{P}\right) \mathbf{Y}_k + \mathbf{N}_k \mathbf{Y}_k \\ &= \mathbf{H}_k + \mathbf{N}_k \mathbf{Y}_k \end{aligned} \tag{7}$$

In Fig. 3. the estimated MIMO channel is plotted using the previous scheme of testbed. To do it, each transmitter antenna was connected to each correspondent receiver antenna (h_{11}=h_{22}=h_{33}=h_{44}=1), with the aim of testing the orthogonality of pilots.

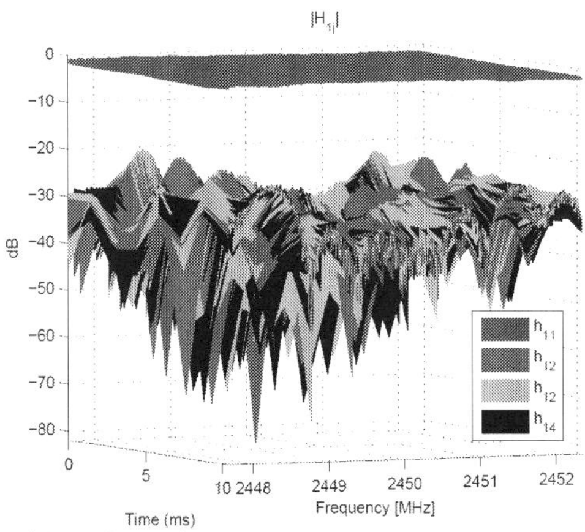

Fig. 3. Orthogonality of the pilot code to estimate the channel

3.2 Antennas

For each combination of transmitter and receiver locations, three types of antennas have been used: firstly the monopole array were utilized at both the receiver and the transmitter, in order to evaluate the system performance when only vertical polarization is used.

Afterwards, the dual-polarized antennas (crossed dipoles) were used, so polarization diversity is included in the system, to the cost of reducing spatial diversity (since the dual-polarized dipoles are co-located). And finally, a planar inverted-F antenna (PIFA) array with 2 elements was placed to be evaluated (Gómez et al, 2008).

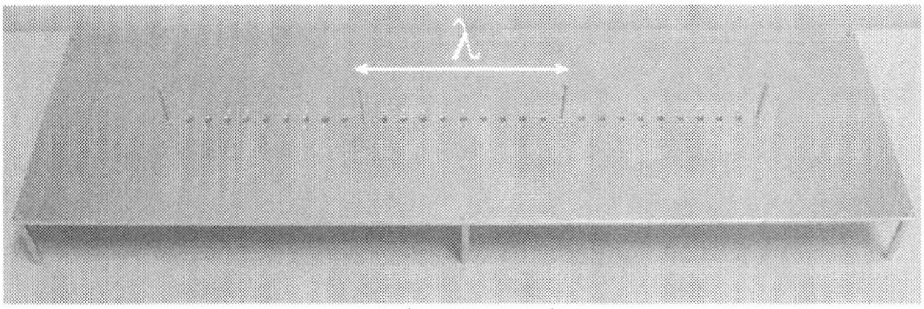

a) Monopoles

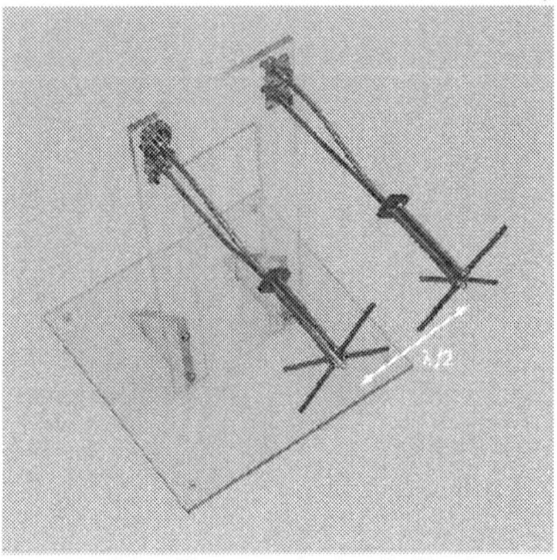

b) Cross-polarized dipoles

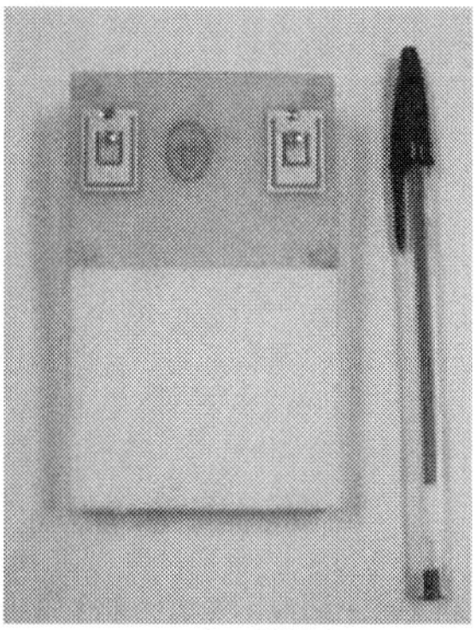

c) PIFAs

Fig. 4. Antennas under test

3.3 Measurements

3.3.1 UMATRIX application

One of the main objectives of the UMATRIX is that it has to allow measurements of reconfigurable antennas in different environments. Thus, a tool with a friendly-user interface and easy to use has been development in Matlab for the integration of processing and measurement parts. In Fig. 5. the main window of the application is shown, where the user goes checking the measured points, received signals and MIMO channel capacity obtained.

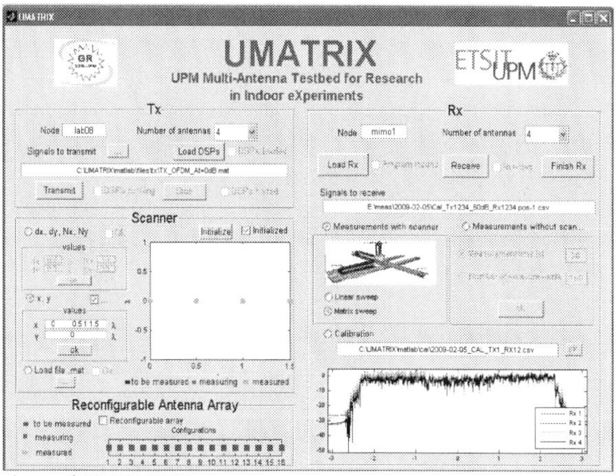

Fig. 5. Main window of UMATRIX

Fig. 6.a) shows the transmitter of UMATRIX where the antenna array is located at the top. All the transmitter is placed in a mobile platform to put it in several locations. On the other hand, the receiver has a scanner which can sweep any point within an area of 6λx6λ (Mora et al., 2008). Fig. 6.b) depicts the receiver with the scanner and the antenna array in the scanner.

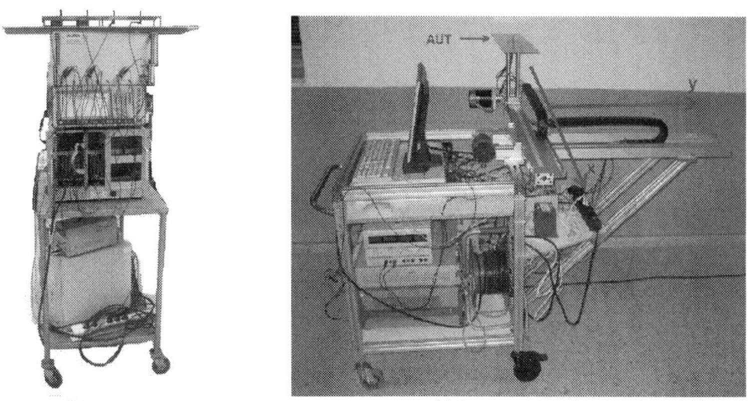

a)Transmitter b)Receiver
Fig. 6. Implementation of the testbed

3.3.2 Locations

All the measurements were taken in the ETSI de Telecomunicación (Madrid), in the fourth floor of building C. In Fig. 7. different types of measurements can be distinguished regarding the scenario: office and corridor. For the corridor environment, the transmitter was put at the end of the corridor and the receiver was located in position 1 for LoS and positions 2 and 3 for NLoS situations. In the case of office scenario, the transmitter was placed in a laboratory (Tx B in Fig. 7.) and the receiver in another office.

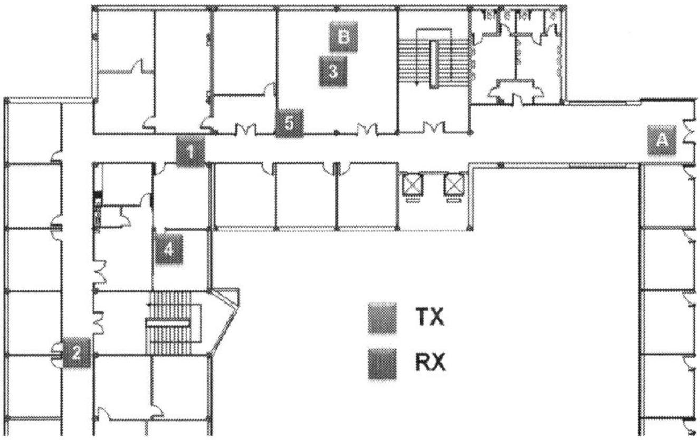

Fig. 7. Map of locations in the measurements campaigns

3.3.2 Results

Once the channel is obtained in the receiver, the MIMO channel capacity is calculted. Previously, the **H** matrix is normalized with the Frobenius norm. In order to remove the path loss effect and study the diversity characteristics of the MIMO propagation channel, the channel matrix **H** is usually normalized to obtain a fixed local signal to noise ratio for each measured point. The use of this normalization is equivalent to considering a perfect power control in the system. This is interesting to characterize the multipath richness and diversity offered by the propagation environment, but it does not take into account the path loss, shadow fading and penetration losses. Then the normalized channel will be

$$\mathbf{H}_{norm} = \sqrt{M_T \cdot M_R}\frac{\mathbf{H}}{\|\mathbf{H}\|_F} = \sqrt{M_T \cdot M_R}\frac{\mathbf{H}}{\sqrt{\sum_{i=1}^{M_T}\sum_{j=1}^{M_R} h^{ref}{}_{ij} \cdot h_{ij}^{ref*}}} \tag{8}$$

where $\mathbf{I}_{MR}$ is the identity matrix of size $M_R x M_R$, M_T is the number of transmitter antennas. To compare the capacities for different types of antenna, a normalization with one antenna array in each type of scenario has been done. On the other hand, as the channel state information is not known at the transmitter, the capacity (in bps/Hz) in each k subcararier is calculated from

$$C_k = \max_{\mathbf{Q}} \log_2\left[\det\left(\mathbf{I}_{M_R} + \frac{\rho}{M_T}\mathbf{H}_k\mathbf{Q}\mathbf{H}_k{}^H\right)\right] \tag{9}$$

where **Q** is the covariance matrix of transmitted signals, such that $\mathrm{Tr}\{\mathbf{Q}\} \leq M_T$ to account for power constraint, ρ is the signal to noise ratio at the receiver, $(\cdot)^H$ denotes Hermitian and

$|\mathbf{A}|$ is the determinant of matrix **A**. Two cases were considered in this analysis: no channel state information (CSI) at transmitter and total CSI at transmitter. In the first case, the power allocation strategy is assumed to be uniform, so that the channel capacity expression may be simplified to

$$C_k = \log_2\left[\det\left(\mathbf{I}_{M_R} + \frac{\rho}{M_T}\mathbf{H}_k\mathbf{H}_k^{\ H}\right)\right] \tag{10}$$

When total CSI at transmitter is considered, the optimum waterfilling scheme is assumed to allocate power, so the singular value decomposition (SVD) of **H** is realized, and the capacity is computed as

$$C_{WF} = \sum_{i=1}^{K} \ln(\mu\lambda_i)^{+} \tag{11}$$

where $(\cdot)^{+}$ denotes taking only those terms which are positive, and λ_i is the i (out of k) non-zero eigenvalue of the correlation channel matrix $\mathbf{R}=\mathbf{H}\mathbf{H}^{H}$. The parameter μ is chosen to satisfy the power constraint

$$\rho = \sum_{i=1}^{K}\left(\mu - \frac{1}{\lambda_i}\right)^{+} \tag{12}$$

For 4x4 MIMO channel measurements, a comparison of single with dual-polarization performances was realized for each scenario. Fig. 8. shows the capacity obtained for the corridor scenario with LoS (position 1 of the receiver in Fig. 7.). As it is shown in Fig. 8.a), the capacity increases with the spacing between elements, except for the case of 0.3λ. The knowledge in the transmitter can give an extra capacity, as Fig. 8.b) depicts. The use of Waterfilling scheme improve the performances in all the SNR range.

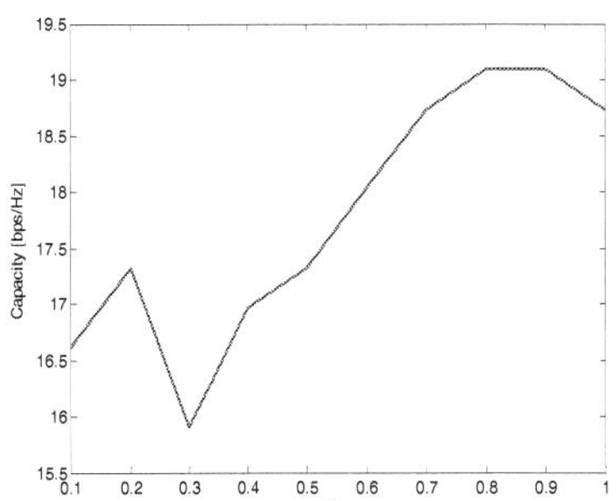

a) Capacity of monoples array as a function of spacing with a SNR=20dB

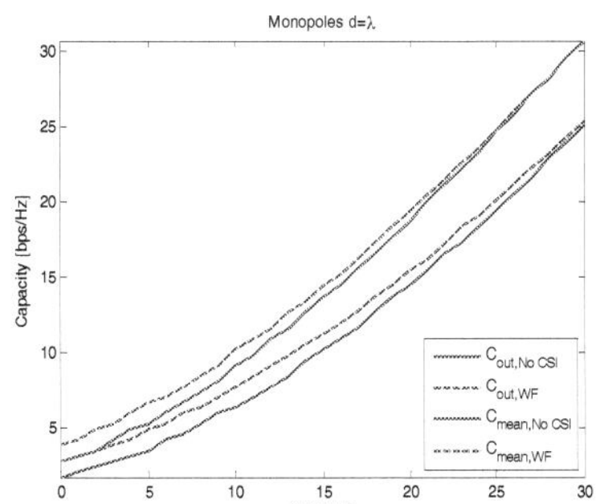

b) Comparison of monopoles capacity with a spacing of λ, as a function of SNR and CSI

Fig. 8. Capacity of monopole array.

On the other hand, the importance of using single or dual polarization has been also studied. Fig. 9. represents the CDF of the capacity for all the monopole array spacings and the cross-polarized dipole array. It is shown that for LoS the employ of dual polarization enhances the performances with respect to the MIMO channel capacity. However, for NLoS cases, the use of dual polarization does not have a great impact on the performances.

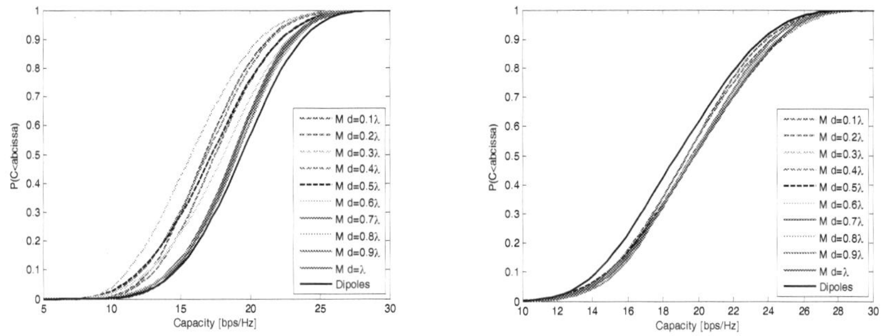

a) CDF capacity of corridor LoS (position 1 of Fig. 7.)

b) CDF capacity in NLoS scenario (position 3 of Fig. 7.)

Fig. 9. Comparison of the CDF capacity for single and dual polarization antennas.

Moreover, 4x2 MIMO channel measurements were carried out to compare the MIMO channel capacity by using different radiating elements. Fig.10. compares the CDF of the capacity obtained for monopoles, dipoles and PIFAs in different scenarios, with LoS and NLoS. It can be concluded that for two radiating elements at the receiver side, MIMO channel capacity strongly depends on the antenna characteristics, such as radiation pattern, mutual coupling and spacing between elements.

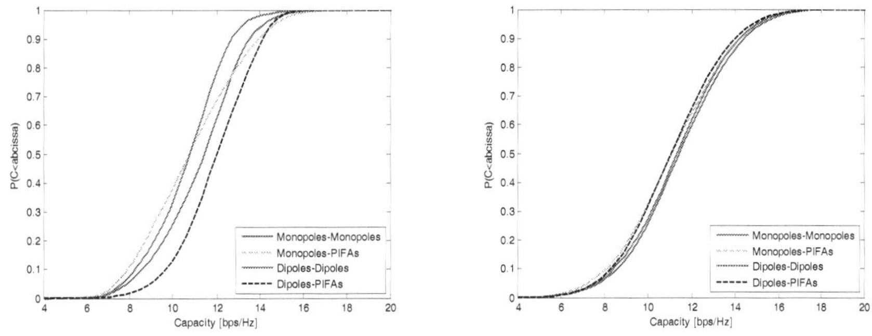

a) CDF capacity of corridor LoS (position 1 of Fig. 7.)

b) CDF capacity in NLoS scenario(position 4 of Fig. 7.)

Fig. 10. Comparison of the CDF capacity for 4x2 MIMO channels with different antennas

4. MIMO prototype for DVB-T2 system

DVB-T2, the second generation of the DVB proposal for digital terrestrial TV, has been recently proposed by DVB project (dvb) as an evolution of DVB-T when the shutdown of analog television process will be finished. In order to give a newer technical response to the necessity the digital dividend, process by which some free frequencies at UHF used by analog TV will be assigned to different services (3G/4G), DVB-T2 will improve frequency efficiency to provide multicast in HD with the same 8 MHz channel.

As DVB-T, DVB-T2 expects to be received in plugged TV terminals in mobile environment or with unplugged terminals in indoor or in low speed (pedestrian) environments, so a MISO scheme has been included, transmitting with a distributed Alamouti block code. However, in order to go further a full MIMO scheme is proposed in this paper, which may be similar to the one that will be included in NGH (second generation of DVB-H) in the next future, obtaining a very efficient performance in highly Doppler environments, that is to terminals (unplugged or not) operating in high speed vehicles.

On the other hand, DVB-T2 will provide higher efficiencies in frequency than the nowadays DVB standard DVB-T. DVB-T2 proposal considers the inclusion of MISO technology but not MIMO. MIMO will be considered in future revisions and it will provide a further increment of frequency efficiency mainly in harsh scenarios as strong multipath environments or highly Doppler radio channels.

In order to evaluate the performances of a DVB-T2 system in realistic scenarios, the use of a real platform is of great interest, since it enables to include several aspects that are not usually addressed in theoretical studies or simulations, such as the effect of different antennas or scenarios (Gómez-Calero et al., 2006). In this section, a novel 2x2 MIMO testbed for DVB-T2 has been designed and implemented in order to test the enhancement obtaining by the using of multiple antennas at each side of the radio link for UHF band, particularly at frequency of 594 MHz.

The general architecture of the testbed is depicted in Fig. 11., where 2 antennas can be placed at the transmitter and the receiver side. The DVB-T2 signals are generated off-line in a PC (e.g. using Matlab) and then they are sent to the Software-Defined-Radio (SDR) platform. This platform receives the signals and transmits them in real-time and in Intermediate Frequency (IF) to the RF module. Finally, signals are upconverted to RF frequency, amplified and filtered, and then transmitted to the radio channel by the antenna array. In the receiver, the signals are captured by the antenna array and downconverted, amplified and filtered by the RF module. Finally, the SDR realizes the synchronization and FFT previous to send the signals to the PC.

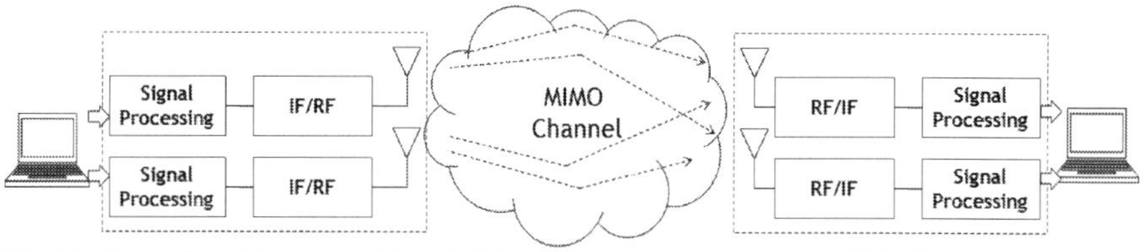

Fig. 11. General architecture of the MIMO measurement system for DVB-T2

4.1 Signal processing

The most complex part of the testbed is the signal processing module, since it supports all the digital to analog and analog to digital conversions (DAC and ADC) and the processing of the OFDM signal with the synchronization and the use of the Fast Fourier Transmorm (FFT). In the following subsections the transmitted DVB-T2 signals and SDR platform are explained.

4.1.2 DVB-T2 structure

The DVB-T2 frame structure is divided in three different parts (ETSI, 2008). The first one is the P1 symbol which is used to do a faster detection and frequency synchronization. Then, the P2 symbol is transmitted to indicate the type of encoders and data configuration of the data symbols. However, for the sake of simplicity in this testbed the P2 symbol is removed. Finally the data symbols are sent with the user data and pilots for the channel estimation.

The block diagram of the transmitter is shown in Fig. 12. Data are generated and passed to the MIMO encoder which applies the distributed Alamouti to transmitt the signals by each antenna and then to recover the data symbols in the receiver. The Alamouti scheme (Alamouti, 1998) is modified with the scheme detailed in Table 2.

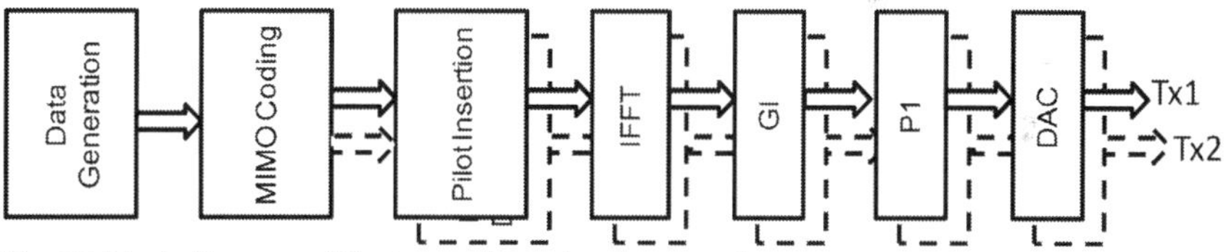

Fig. 12. Block diagram of the transmitter signal processing

Subcarrier	Tx antenna 1	Tx antenna 2
k_i	s_1	$-s_2^*$
k_{i+1}	s_2	s_1^*

Table 2. Alamouti modified scheme

After that, the continual and scattered pilots are inserted to estimate the channel at the receiver. Then, the Inverse-FFT (IFFT) is done and the Guard Interval (GI) is inserted in order to avoid inter-symbol interference (ISI) due to the channel delay spread. The cyclic-prefix removes the ISI and converts the convolution between transmitted symbols and channel into a circular convolution. Finally, the P1 symbol is inserted and the two signals are converted from digital to analog and sent to RF module.

4.1.3 Sowtfare-Radio

For the real-time process, two XtremeDSP boards based on the BenADDA module of Nallatech have been used: one for the transmitter and one for the receiver. Each board has a FPGA VirteX II Pro V2P30 and 4 MBytes of SRAM memory with 2 DACs of 160 MHz and 2 ADCs of 105 MHz (Nallatech).

Fig. 13. shows the architecture of the transmitter. The data symbols are sent to the board via DMA through the PCI bus. The DMA SRAM IFACE module stores them in the board external memory ZBT SRAM. The maximum capacity is 512 samples per channel. Once the data write cycle is finished, the same module reads them from the memory and extracts them in a continuous and cyclic way. The I/Q data streams of each channel go to *Digital Up Converter* module, where the signals are interpolated by a factor of 10 and are upconverted to an IF of 36 MHz. The data are sent to DACs, which operate at a frequency of 91428571 Hz. This frequency is selected for being 10 times the inverse of the sample period of a 8 MHz channel, which according to the DVB-T2 standard (ETSI, 2008) is T=7/64 μs.

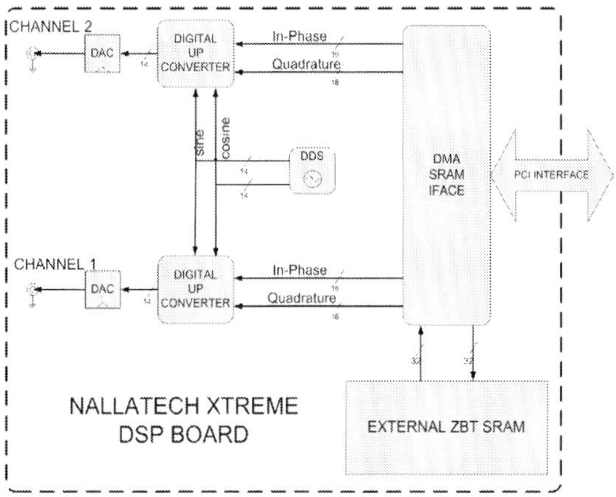

Fig. 13. Architecture of SDR transmitter

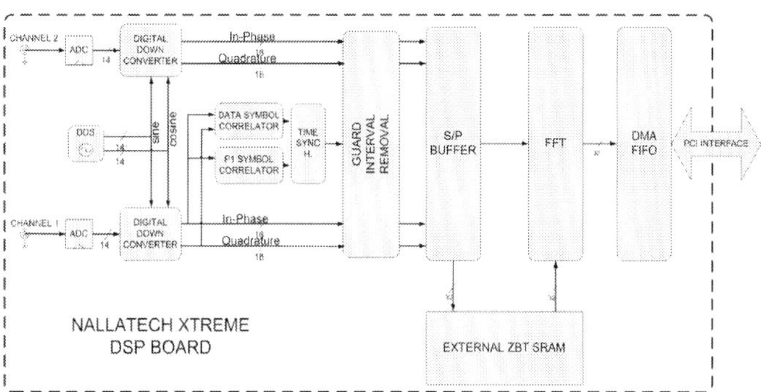

Fig. 14. Architecture of SDR receiver

On the other hand, Fig. 14. represents the main blocks of receiver subsystem. The signal from channel 1 is used to obtain the synchronism in time domain of received symbols. The synchronism algorithm is based on the correlation of the cyclic prefix of OFDM symbols (van de Beek et al., 1997). Then, the signals go to *Time Synch* module which generates the synchronism signal which allows to obtain the OFDM symbols in the next module, named *Guard Interval Removal*. Finally, the FFT is done to recover the block of data symbols.

In order to estimate the beginning of the frame, the signal passes through a correlator with the P1 symbol. Fig. 15. represents the correlation of a frame made up a P1 symbol and 9 data symbols.

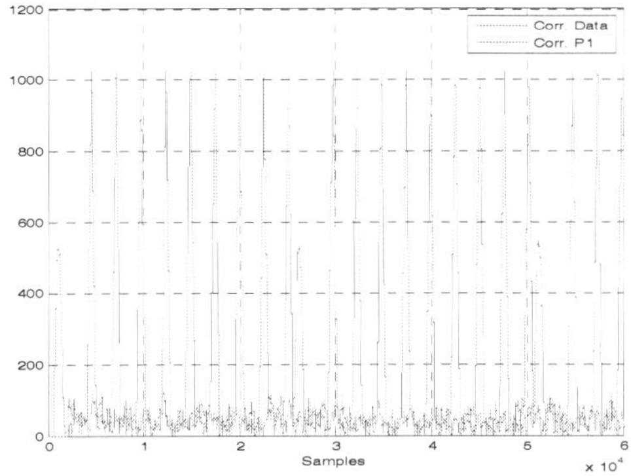

Fig. 15. Correlation of received signals. Red line represents the correlation with the P1 symbol and blue line shows the autocorrelation with the data symbol

4.1.5 Channel estimation

One of the key aspects in MIMO is the channel estimation, since MIMO system performances depend on the accuracy of the estimated channel matrix. Due to the selection of 2K mode of FFT and a GI of 1/8, the DVB-T2 standard (ETSI, 2008) proposes the PP1 pattern for scattering pilots for MISO.

The scattered pilots, for each OFDM symbol are placed each 12 subcarriers along the frequency axis. Attending to the temporal axis, the pilots start in the first subcarrier and the initial point is shifted in 2 subcarriers for the following 3 OFDM symbols, as Fig. 16.a) and Fig. 16.b) show for antenna 1 and 2, respectively. Thus, the pilots structure is done in 4-symbol blocks in time domain and it depends on the used antenna. In antenna 2 case (Fig. 16.b), the corresponding pilots to symbols 1 and 3 are inverted to distinguish the transmitter antenna. Moreover, it is worth mention here that the scattering pilots are generated according to a pseudorandom sequence and PRBS (ETSI, 2008).

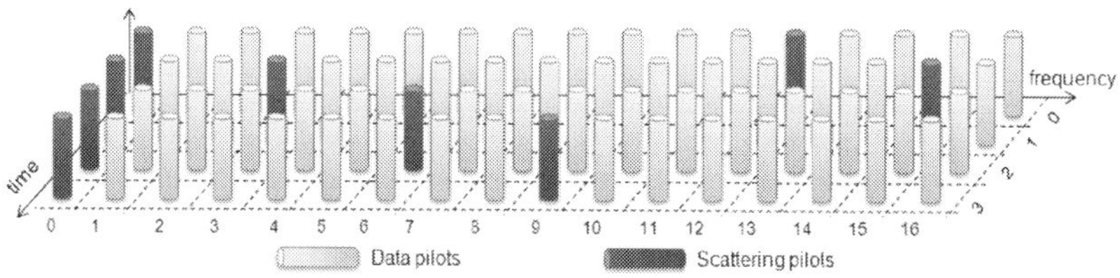

a) Antenna 1

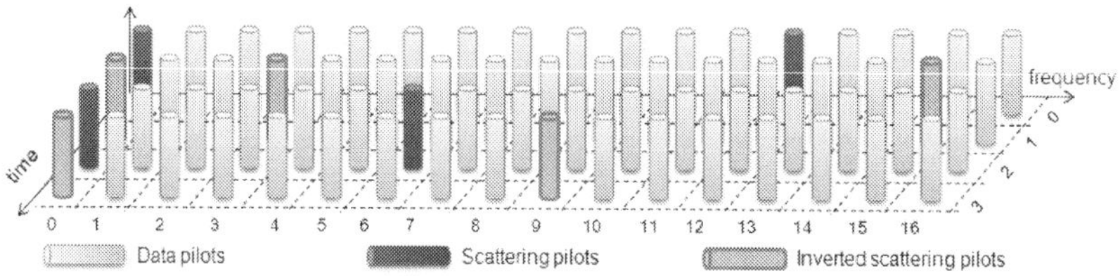

b) Antenna 2

Fig. 16. Scattered pilots distribution

Thus, if $\mathbf{X}_k$ represents the transmitted symbols vector for both antennas for the *k*-subcarrier, the received vector $\mathbf{R}_k$ is given by (2). Due to the testbed has two antennas in the transmitter and other two in the receiver, the MIMO channel matrix is represented by

$$\mathbf{H}_k = \begin{bmatrix} h_{1,1,k} & h_{1,2,k} \\ h_{2,1,k} & h_{2,2,k} \end{bmatrix} \tag{13}$$

where each matrix element indicates the subchannel between each pair of transmitter-receiver antenna.

In the literature a MIMO channel estimator for DVB-T2 has not been proposed, so in this section an original scheme is presented to estimate the channel for MIMO case. First of all, the channel is asummed not to vary in time domain for 4 data symbols, similar to other channel estimators for DVB-T (Lee et al., 2007, Palin & Rinne 1999, Chen et al., 2003, Fang & Ma, 2007). Therefore, the pilots are gathered in groups of 4 OFDM data symbols in only one symbol, obtaining one pilot subcarrier out of 3 subcarriers. Fig. 17. depicts the association for both antennas.

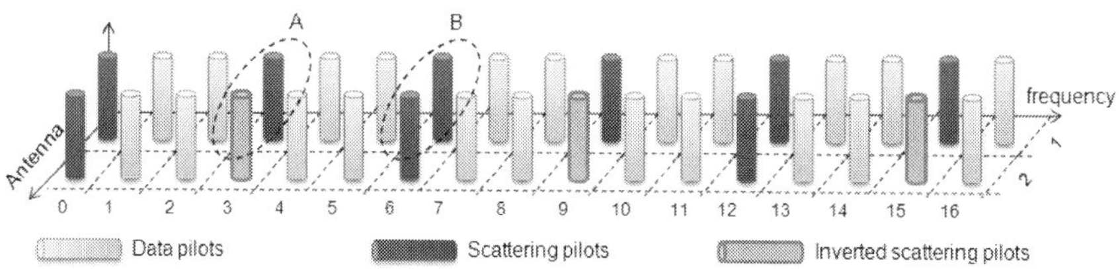

Fig. 17. Scattered pilots association for antennas 1 and 2

The received signals in antennas 1 and 2, $r_{1,k}$ and $r_{2,k}$, respectively, for the k-subcarrier are given by

$$r_{1,k}=x_{1,k}\cdot h_{11,k}+x_{2,k}\cdot h_{12,k}+n_{1,k}$$
$$r_{2,k}=x_{1,k}\cdot h_{21,k}+x_{2,k}\cdot h_{22,k}+n_{2,k} \tag{14}$$

where $x_{i,k}$ represents the pilot signal transmitted by antenna i, and $n_{i,k}$ is the noise contribution in antenna i. From this point Here, two cases are considered to estimate the channel. The first one is the case in which the pilots of both antennas are scattered pilots for a given subcarrier. This case is marked with ***A*** in Fig. 17. In the other case, the pilot for one antenna is a scattered pilot and for the other antenna is an inverted scattered pilot, for a given subcarrier. This is case ***B*** in Fig. 17.

Thus, for case A the received signals are

$$\begin{aligned} r^a_{1,k} &= x^a_{1,k}\cdot h_{11,k} - x^a_{1,k}\cdot h_{12,k} + n^a_{1,k} \\ &= x^a_{1,k}\cdot (h_{11,k} - h_{12,k}) + n^a_{1,k} \end{aligned} \tag{15}$$

$$\begin{aligned} r^a_{2,k} &= x^a_{1,k}\cdot h_{21,k} - x^a_{1,k}\cdot h_{22,k} + n^a_{2,k} \\ &= x^a_{1,k}\cdot (h_{21,k} - h_{22,k}) + n^a_{2,k} \end{aligned} \tag{16}$$

And for case B

$$\begin{aligned} r^b_{1,k} &= x^b_{1,k}\cdot h_{11,k} + x^b_{1,k}\cdot h_{12,k} + n^b_{1,k} \\ &= x^b_{1,k}\cdot (h_{11,k} + h_{12,k}) + n^b_{1,k} \end{aligned} \tag{17}$$

$$\begin{aligned} r^b_{2,k} &= x^b_{1,k}\cdot h_{21,k} + x^b_{1,k}\cdot h_{22,k} + n^b_{2,k} \\ &= x^b_{1,k}\cdot (h_{21,k} + h_{22,k}) + n^b_{2,k} \end{aligned} \tag{18}$$

Operating with the obtained signals from (15)-(18), channel coefficients can be estimated as

$$\tilde{h}_{1,1,k} = \frac{r^a_{1,k} + r^b_{1,k}\left(\dfrac{x^a_{1,k}}{x^b_{1,k}}\right)}{2x^a_{1,k}} \tag{19}$$

$$\tilde{h}_{1,2,k} = \frac{r^a_{1,k} - r^b_{1,k}\left(\dfrac{x^a_{1,k}}{x^b_{1,k}}\right)}{-2x^a_{1,k}} \tag{20}$$

$$\tilde{h}_{2,1,k} = \frac{r^a_{2,k} + r^b_{2,k}\left(\dfrac{x^a_{1,k}}{x^b_{1,k}}\right)}{2x^a_{1,k}} \tag{21}$$

$$\tilde{h}_{2,2,k} = \frac{r^a_{2,k} - r^b_{2,k}\left(\dfrac{x^a_{1,k}}{x^b_{1,k}}\right)}{-2x^a_{1,k}} \tag{22}$$

And the estimated MIMO channel matrix is

$$\widetilde{\mathbf{H}}_k = \begin{bmatrix} \widetilde{h}_{1,1,k} & \widetilde{h}_{1,2,k} \\ \widetilde{h}_{2,1,k} & \widetilde{h}_{2,2,k} \end{bmatrix} \tag{23}$$

4.2 RF-FI

In the transmitter, the RF stage receives the signal from signal proccesing module and upconverts from IF (36 MHz) to RF frequency (594 MHz). Then, the signal is filtered and amplified, with a transmitted power of +2 W rms. In order to upconvert the signals of both branches 1 and 2, a direct digital synthesizer (DDS) is used.

In the receiver, the signals are received from antenna ports and then are amplified, filtered and downconverted to IF. In this case, a voltage controlled attenuator is placed to adapt the recevied signal power to the best range of levels. The variation of the attenuator is from 3 dB to 38 dB in steps of 5 dB.

4.3 Antenna array

The MIMO testbed has two antennas at the transmitter and two at the receiver, respectively. The antennas have been designed for the testbed to work at 594 MHz. The radiating element is a dipole with a reflector element. The reflection coefficient of the transmitter and receiver antennas is depicted in Fig. 18.a). It shows that all the implemented antennas have a good matching at desired frequency band, obtaining a reflection coefficient lower than -17 dB. On the other hand, the radiation pattern is presented in Fig. 18.b), where the XPD is higher than 25 dB in the maximum direction of θ=0°.

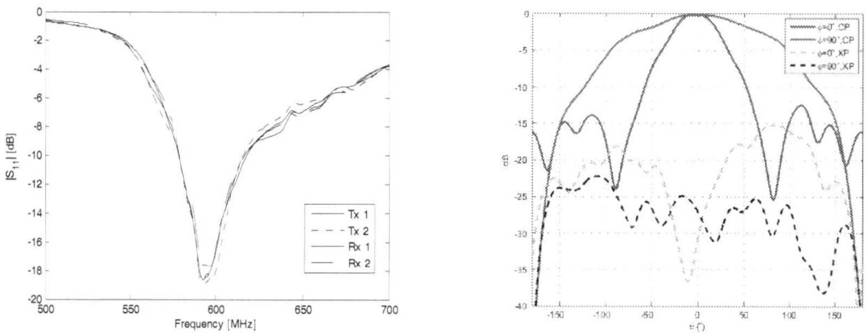

a) Reflection coefficient of the antennas

b) Measured radiation pattern of an antenna

Fig. 18. Antenna performances

4.5 Measurements in indoor/outdoor scenarios for different polarizations

Once the different modules of transmitter and receiver are implemented and work properly, the integration of all of them is done. In Fig. 19. the transmitter and receiver are illustrated. The transmitter is placed on a 19'' rack and the received is mounted on a mobile platform in order to measure different environments.

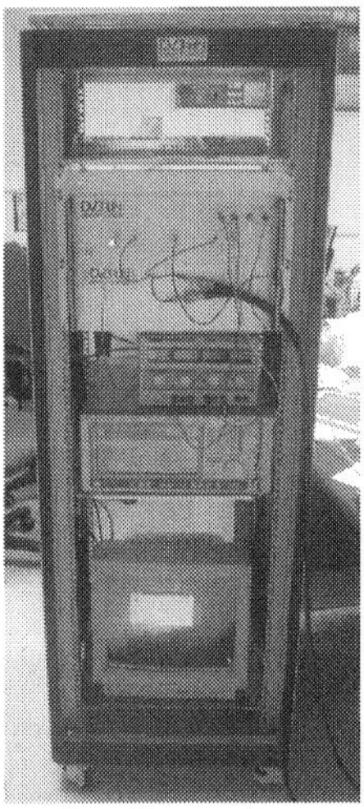
a) Transmitter

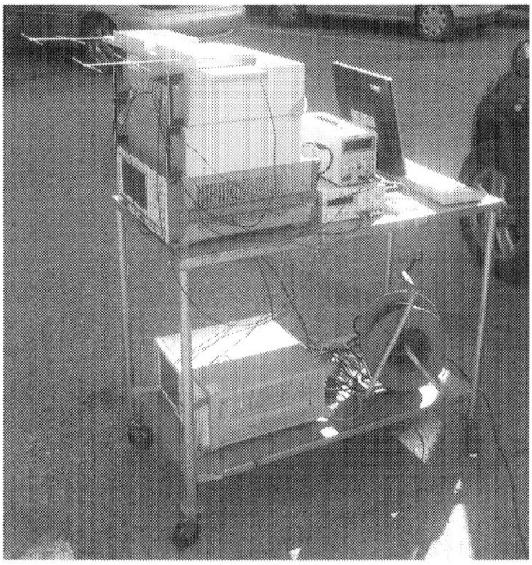
b) Receiver

Fig. 19. Integration of the testbed

As in any real system, implementation issues such as frequency errors by using different local oscillators internal clocks at each side of the radio link appear. In general, all the errors can be represented as

$$\mathbf{R}_k = \mathbf{H}_k \mathbf{X}_k \cdot \left(\exp\left(j \left[2\pi k \frac{Td}{Tu} + 2\pi k \frac{\Delta t}{Tu} + \phi_0 + 2\pi k \Delta f Ts \right] \right) \right) + \mathbf{I}_k + \mathbf{N}_k \quad (24)$$

where T_d represents the symbol temporal offset, Δ_t the sampling temporal offset, φ_0 the phase offset, Δ_f the frequency offset and $\mathbf{I}_k$ is the ICI (Inter-Carrier Interference) due to frequency offset for the *k*-carrier. These errors must be taken into account and have been mitigated in the signal processing module.

Once the MIMO testbed has been realized, a measurement campaign has been carried out in order to evaluate the enhancement obtained by using the MIMO scheme. The measurements were carried out in the E.T.S.I of Telecommunications School, at Universidad Politécnica de Madrid, Spain. The transmitter was situated on the rooftop of building C with a spacing between elements of λ. Fig. 20. depicts the topview of the measurement locations. In order to measure several scenarios, the receiver was located in three different positions. In position 1, the receiver was placed in the parking area of the building in LoS (Line of Sight). The receiver for NLoS (Non Line of Sight) was situated in position 2, in the parking of the next building. Finally, for the indoor scenario, the receiver was located in the third floor of the building, as Fig. 20. shows. For each scenario, three different polarization schemes were evaluated: HH, HV and VV, where H and V means horizontal and vertical polarization for antennas 1 and 2 both in the transmitter and the receiver, respectively.

Fig. 20. Top view of the Tx and Rx positions

4.5.1 MIMO channel

The first step for evaluating the MIMO channel is to normalize channel power. After chanel estimation, the channel power is calculated for all the measured cases. The channel power is averaged over all frequencies f_v and temporal snapshots t_u and is calculated by

$$P_{avr} = \frac{\sum_{i=1}^{N_t} \sum_{j=1}^{N_f} P(t_u, f_v)}{M_T M_R} \tag{25}$$

where N_t represents the number of OFDM symbols (100 in this case), N_f indicates the number of subcarriers (2048) and P represents the mean power in the measured point and is given by

$$P(t_u, f_v) = \frac{\| P(t_u, f_v) \|_F^2}{M_T M_R} \tag{26}$$

Table 3 shows the comparison of mean power for each scenarios as a function of antenna polarization. It is shown in LoS scenarios, the received power is increased. However, in indoor case the power only is reduced in 6 dB because the distance between the transmitter and receiver is about 23 m smaller than in the case 3.

Measured scenario	Mean channel power (dB)		
	HH	HV	VV
1 - Outdoor LoS	-63.5	-62.8	-63.2
2 - Outdoor NLoS	-81.1	-81.8	-73.7
3 - Indoor NLoS	-67.9	-68.8	-69.5

Table 3. Comparison of measured mean power

4.5.2 MIMO capacity

Besides the diversity gain, the other important of MIMO channels is the increase of the channel capacity. In the case of DVB-T2, the transmitter does not know the Channel State Information (CSI), and the MIMO capacity can be calculated from

$$C_k = \sum_{k=1}^{N} \log_2 \left[\det \left(\mathbf{I}_{M_R} + \frac{\rho}{M_T} \mathbf{H}_k \mathbf{H}_k^{\ H} \right) \right] \text{ bps/Hz} \tag{27}$$

where M_R and M_T represent the number of receiver and transmitter antennas, respectively, k is the given subcarrier, N represents the total number of subcarriers and ρ is the Signal to Noise Ratio (SNR). The measurements have been realized with M_T=M_R=2.

With the aim of comparing the capacity of SISO and MIMO, Fig. 21. represents the cummulative distribution function of the capacity calculated for the LoS case with polarization diversity. The capacity increases in the 2x2 case.

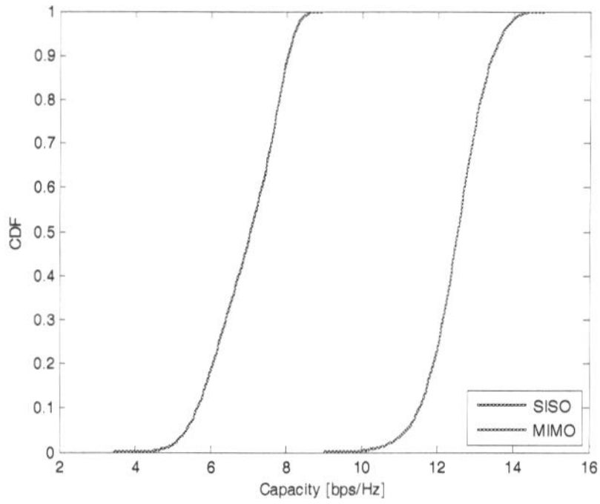

Fig. 21. Comparison of CDF of capacity with 1x1 and 2x2 in outdoor LoS case with polarization diversity

On the other hand, Fig. 22. shows the comparison of all the cases comparing the use of different polarization schemes. The highest values of capacity are obtained for indoor NLoS, followed by outdoor NLoS and the, outdoor LoS, except when polarization diversity (HV) is used in LoS case, since the obtained outage capacity at 10% is the highest.

Attending to outage capacity, the highest values are obtained for the outdoor LoS HV case and the indoor situations for all the polarization schemes.

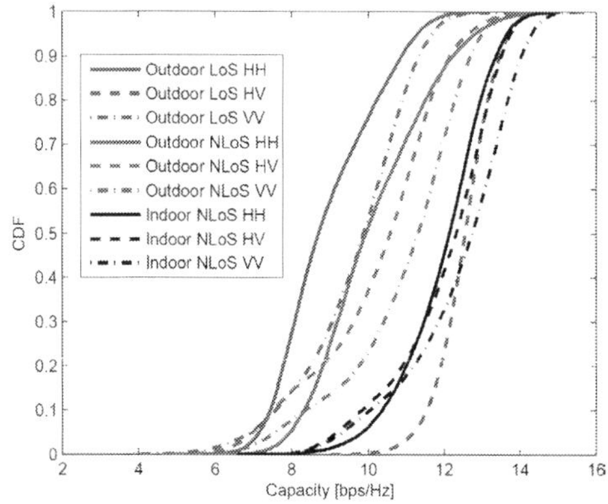

Fig. 22. CDF of capacity for all the measured scenarios

5. Conclusions

In the last decade, Multiple-Input Multiple-Output (MIMO) systems have created a great interest in research. Many works shows an increase in terms of data bit rate by using several antennas at each side of the radio link. In this chapter, two novel measurement systems for MIMO channels in indoor environment are presented. In order to study the propagation characteristics of these systems, using both polarization and spatial diversity in multi-antenna systems can be evaluated, thanks to the use of different types of antennas.

On one hand, a novel MIMO-OFDM testbed (UMATRIX), designed and implemented in the UPM, has been used to accomplish the measurements at 2.45 GHz. The UMATRIX has several characteristics which makes it good for antenna reconfigurable measurements. OFDM technique is introduced to measure the wideband MIMO channel response, so a FPGA-based receiver has been developed. A measurement campaign to compare the system performance with either single-polarized or dual-polarized antennas was conducted in an indoor scenario, including multiple locations for the transmitter module. From the capacity analysis, it may be concluded that in indoor environment for corridor scenario, dipoles present better performances than monopoles. However, in NLoS office scenario, monopoles outperforms the dual-polarized antennas. Finally, higher capacity is obtained with higher spacing between radiating elements, due to less correlation among the MIMO subchannels.

On the other hand, a new 2x2 MIMO testbed has been designed and implemented for the future digital television system DVB-T2. The testbed is based on software radio platforms where the signal processing is implemented according to the standard. The testbed has been designed to carry out measurements at the frequency of 594 MHz. A measurements campaign has been done in outdoor and indoor scenarios. Results show the importance of using multiple antennas at each side of the radio link for increasing the capacity of the MIMO system. Moreover, it has also been shown that polarization diversity provides an additional capacity gain especially for outdoor LoS cases (up to 4 bps/Hz in outage capacity).

6. References

Adjoudani, A.; Beck, E.; Burg, A.; Djuknic, G. M.; Gvoth, T.; Haessig, D.; Manji, S.; Milbrodt, M.; Rupp, M.; Samardzija, D.; Siegel, A.; Sizer II, T.; Tran, C.; Walker, S.; Wilkus, S.A.; Wolniansky, P., "Prototype Experience for MIMO BLAST over Thrid Generation Wireless System," Special Ussue JSAC on MIMO Systems, vol 21, pp. 440-451, April 2003.

Alamouti, S., "A simple transmit diversity technique for wireless communications," IEEE Journal on Selected Areas in Communications, vol. 16, pp. 1451-1458, October 1998.

Aschbacher, E.; Caban, S.; Mehlfuhrer, C.; Maier, G.; Rupp, M., "Design of a flexible and scalable 4x4 MIMO testbed," IEEE 11th Digital Signal Processing Workshop, 2004 and the 3rd IEEE Signal Processing Education Workshop. 1-4 Aug. 2004, pp. 178-181.

Batariere, M. D.; Kepler, J. F.; Krauss, T. P.; Mukthavaram, S.; Porter, L. W.; Vook, F. W., "An experimental OFDM system for broadband mobile communications," IEEE Vehicular Technology Conference, v 4, n 54ND, 2001, p 1947-1951.

Chen, S.-H.; He, W.-H. ; Chen, H.-S. & Lee, Y., "Mode detection, synchronization, and channel estimation for DVB-T OFDM receiver," in IEEE Global Telecommunications Conference, 2003. GLOBECOM '03, vol. 5, Dec. 2003, pp. 2416–2420 vol.5.

Dvb http://www.dvb.org.

Ellingson S. W., "A flexible 4x16 MIMO testbed with 250 MHz - 6 GHz tuning range," IEEE Antennas and Propagation Symposium, Washington, DC, July 2005.

ETSI, "Digital Video Broadcasting (DVB);Frame structure channel coding and modulation for a second generation digital terrestrial television broadcasting system (DVB-T2)," Draft ETSI EN 302 755 V1.1.1 (2008-04), 2008.

Fang, R.-D. & Ma, H.-P., "A DVB-T/H Baseband Receiver for Mobile Environments," in 2007 WSEAS International Conference on Circuits, Systems, Signal and Telecommunications, Queensland, Australia, January 2007.

Foschini, G. & Gans, M., "On Limits of Wireless Communications in a Fading Enviroment when Using Multiple Antennas," Wireless Personal Communications, vol. 6, pp. 311–335, March 1998.

GmbH, "RUSK MIMO: Broadband Vector Channel Sounder for MIMO Channels," MEDAV 2001. Available at http://channelsounder.de/

Goldsmith, A.; Jafar, S.; Jindal, N & Vishwanath, S., "Capacity limits of MIMO channels," IEEE Journal on selected areas in communications, vol. 21, no. 5, June 2003.

Gómez-Calero, C.; García-García, L.; Martínez, R. & de Haro, L., "Comparison of antenna configurations in different scenarios using a wideband MIMO testbed," IEEE Antennas and Propagation Society International Symposium 2006, pp. 301–304, July 2006.

Gómez-Calero, C.; González, L. & Martínez, "Tri-Band Compact Antenna Array for MIMO User Mobile Terminals at GSM 1800 and WLAN bands," Microwave and Optical Technology Letters, vol. 50, no. 7, pp. 1914–1918, July 2008.

Kaiser, T; Wilzeck, A.; Tempel, R., "A modular multi-user MIMO test-bed," IEEE Radio & Wireless Conference 2004, Atlanta Georgia, USA, September, 2004.

Kaiser, T; Wilzeck, A.; Berentsen, M.; Rupp, M., "Prototyping for MIMO Systems - an Overview," Proc. Of the XII European Signal Processing Conference, Vienna (Austria), Oct 2004, pp. 681-688.

Lee, Y.-S.; Kim, H.-N. & Son, K. S., "Noise-Robust Channel Estimation for DVB-T Fixed Receptions," IEEE Transactions on Consumer Electronics, vol. 53, no. 1, pp. 27-32, February 2007.

Mora-Cuevas, J.; Gómez-Calero, C.; Cuéllar, L. & de Haro, L., "A Wideband OFDM MIMO Measurement System for Antenna Evaluation," Antennas and Propagation International Symposium, 2008 IEEE, 5-12 July 2008.

Nallatech http://www.nallatech.com.

Palin, A. & Rinne, J., "Symbol synchronization in OFDM system for time selective channel conditions," in Proceedings of The 6th IEEE International Conference on Electronics, Circuits and Systems, vol. 3, 1999, pp. 1581-1584 vol.3.

Paulraj, A.; Gore, D.; Nabar, R & Bölcskei, H., "An overview of MIMO communications - A key to gigabit wireless," Proceedings of the IEEE, vol. 92, no. 2, pp. 198 - 217, 2004.

Pietsch, C.; Teich W. G.; Lindner, J.; Waldschmidt, C.; Wiesbeck, W., "A highly flexible MIMO demonstrator," International ITG/IEEE Workshop on Smart Antennas WSA 2005. Duisburg, Germany. April 2005.

Sampath, H.; Talwar, S.; Tellado, J.; Erceg, V. & Paulraj A., "A fourthgeneration MIMO-OFDM broadband wireless system: design, performance, and field trial results," IEEE Communications Magazine, vol. 40, no. 9, pp. 143-149, September 2002.

Samuelsson, D.; Jaldén, J.; Zetterberg, P.; Ottersen B., "Realization of a Spatially Multiplexed MIMO System," EURASIP Journal on Applied Signal Processing, March 2005.

Seeger R.; Brotje L.; Kammeyer K. D., "A MIMO hardware demonstrator: application of space-time block codes," Proc. of 3rd IEEE International Symposium on Signal Processing and Information Technology, 2003. ISSPIT 2003. Proceedings, 14-17 Dec. 2003, pp. 98-101.

Signalion www.signalion.com

Stege, M.; Schafer, F.; Henker, M.; Fettweis, G., "Hardware in a loop-a system prototyping platform for MIMO-approaches," ITG Workshop on Smart Antennas, 2004, pp. 216-222.

Telatar, E., "Capacity of multi-antenna Gaussian channels," Eur. Trans. Telecomm. ETT, vol. 10, no. 6, pp. 585-596, November 1999.

van de Beek, J; Sandell, M. & Borjesson, P., "ML estimation of time and frequency offset in OFDM systems," IEEE Transactions on Signal Processing, vol. 45, no. 7, pp. 1800-1805, Jul 1997.

van Zelst, A.; Schenk, T. C. W., "Implementation of a MIMO OFDM-based wireless LAN system," IEEE Transactions on Signal Processing, vol. 52, Issue 2, Feb. 2004, pp. 483-494.

Passive All-Fiber Wavelength Measurement Systems: Performance Determination Factors

Ginu Rajan, Yuliya Semenova, Agus Hatta and Gerald Farrell
Photonics Research Centre, Dublin Institute of Technology
Dublin, Ireland

1. Introduction

Passive all-fiber edge filter based devices are very often used for wavelength demodulation system for many sensors such as fiber Bragg gratings. The advantage of passive linear edge filter systems are their low cost, ease of fabrication and high measurement speed compared to active wavelength measurement systems. The accuracy and resolution of a fiber edge filter based wavelength measurement system (WMS) is determined by three main factors: the noise in the system, polarization dependence and temperature dependence of the system.

Passive edge filters used for wavelength measurements are commonly employed in a ratiometric scheme which makes the system independent of input signal power variations. A ratiometric optical wavelength measurement system's operation is perturbed by both the inherent optical noise of the input signal as well as the electrical noise due to optical-to-electrical conversion at the receivers. At the receivers, even though the measurement is performed by taking the power ratio of the signal levels, because of the uncorrelated random nature of noise, the effect of noise sources will not be eliminated and will adversely affect the system's performance. An optimization of the slope of the system considering the total noise of the system is required to achieve the best possible resolution for the widest possible wavelength range.

Since a ratiometric wavelength system contains concatenated polarization dependent loss (PDL) elements, the net effect of the PDL is different from the effect caused by individual PDL components and because of this an estimation of the range of the wavelength error is necessary to determine the accuracy of the system. PDL is an important factor determining the accuracy of fiber edge filter based WMS and needs to be minimized to improve the performance of the system. Another significant contributor that can degrade the performance of the system is temperature drift. Commonly a ratiometric WMS is calibrated and the ratio response is obtained at a fixed temperature and hence a change in temperature can alter the ratio response. It is important to know the nature of ratio variation with temperature at different wavelengths in order to evaluate and mitigate its impact on the measurement system.

This chapter focuses on these issues and their impact on the performance of a passive fiber edge filter based wavelength measurement system. It is also intended to introduce new types of passive fiber edge filters to the engineering community, which have applications in the optical sensing area where there is an increasing demand for fast wavelength measurements at lower cost. The two new edge filters introduced in this chapter are the macro-bend fiber edge filter and the Singlemode-multimode-singlemode fiber edge filter .

2. Passive all-fiber edge filters and wavelength measurement systems

Optical wavelength detection in sensing can be generally categorized into two types: passive detection schemes and active detection schemes. In passive schemes there are no power driven components involved. A passive detection scheme refers to those that do not use any electrical, mechanical or optical active devices in the optical part of the system. Most of the passive devices are linearly wavelength dependent devices such as bulk edge filters (Mille et al., 1992), biconical fiber filters (Ribeiro et al., 1996), wavelength division couplers (Davis & Kersey, 1994), gratings (Fallon et al., 1999), multimode interference couplers (Wang & Farrell, 2006) etc. In active detection schemes the measurement depends on externally powered devices and examples of these schemes include those based on tunable filters (Kersey et al., 1993) and interferometric scanning methods (Kersey et al., 1992).

2.1 Linear edge filters

The simplest way to measure the wavelength of light is to use a wavelength dependent optical filter with a linear response. This method is based on the usage of an edge filter, which has a narrow linear response range with a steep slope or a broad band filter, which has a wide range with less steep slope. In both cases, the wavelength interrogator is based on intensity measurement, i.e., the information relative to wavelength is obtained by monitoring the intensity of the light at the detector. For intensity based demodulators, the use of intensity referencing is necessary because the light intensity may fluctuate with time. This could occur not only due to a wavelength change but also due to a power fluctuation of the light source, a disturbance in the light-guiding path or the dependency of light source intensity on the wavelength. Generally, because of these factors, most of the edge filter based systems use a ratiometric scheme which renders the measurement system independent of input power fluctuations. Fig. 1 shows a schematic of a ratiometric wavelength measurement system based on an edge filter. The input light splits into two paths with one passing through the wavelength dependent filter and the other used as the reference arm. The wavelength of the input signal can be determined using the ratio of the electrical outputs of the two photo detectors, assuming a suitable calibration has taken place.

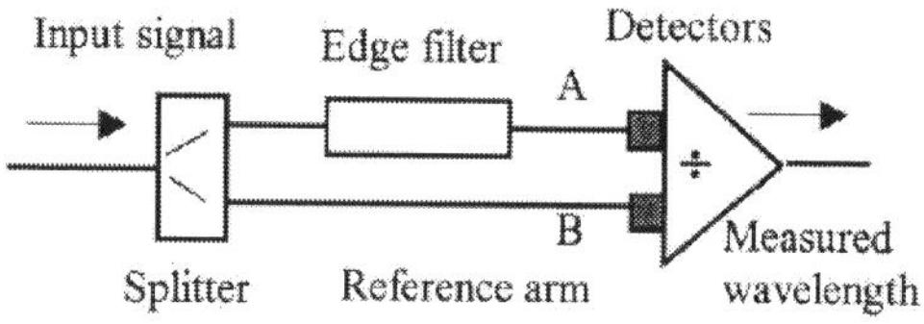

Fig. 1. Schematic of an edge filter based ratiometric wavelength measurement system

For the edge filter used in a ratiometric system, the two important parameters are its discrimination range (wavelength attenuation range) and baseline loss (transmission loss at the starting wavelength). An ideal edge filter will have a very low baseline loss and a high discrimination range. In Fig. 2 three spectral responses (A, B and C) with different discrimination ranges and baseline losses are shown and a selection of a proper response requires the knowledge of the impact of noise in the system, polarization and temperature dependences of the edge filter and their influence on the system.

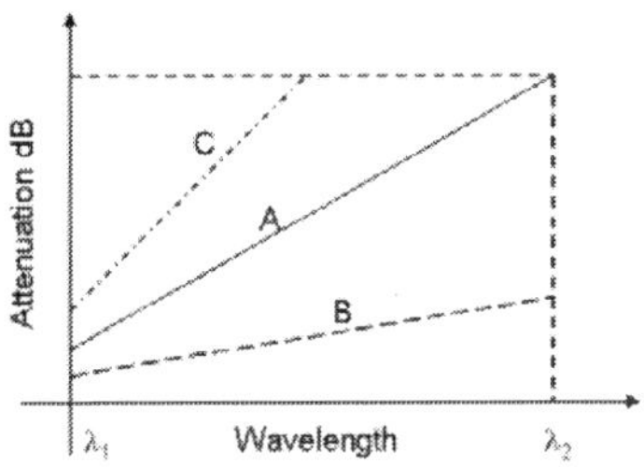

Fig. 2. Spectral response of edge filters with different discrimination range and baseline loss

The first experiment based on a ratiometric scheme was reported in 1992 and used a bulk edge filter, a commercial infrared high-pass filter (RG830), which had a linearly wavelength dependent edge in the range of 815 nm - 838 nm (Mille et al., 1992). Later the use of a biconical fiber filter was proposed as an edge filter (Ribeiro et al., 1996). This filter is made from a section of single mode depressed-cladding fiber, which consists of a contracting tapered region of decreasing fiber diameter followed by an expanding taper of increasing fiber diameter. The wavelength response of the filter is oscillatory with a large modulation depth propagating only a certain wavelengths through the fiber while heavily attenuating others. The reported filter was designed with an oscillation period of 45 nm and an extinction ratio of 8 dB. Over the range 1520 nm - 1530 nm the filter showed a near linear response with a slope of 0.5 dB/nm. Another type of passive wavelength filter is the one based on a wavelength division multiplexing coupler which was first proposed by Mille et. al. and demonstrated by Davis and Kersey. In this scheme the WDM coupler has a linear and opposite change in coupling ratios between the input and two output ports. Another reported edge filter is the one based on long period gratings (LPG) (Fallon et al., 1998). An LPG utilizes the spectral rejection profile to convert wavelength into intensity encoded information. The latest addition to linear fiber edge filters are fiber bend loss filters and single-mode-multimode fiber filters which are explained in the section below.

2.2 A fiber bend loss edge filter

One of the recent additions to the range of available fiber edge filters are bend fiber filters (Wang et al., 2006). This filter comprises of multiple macro bends of standard singlemode fiber (eg. SMF28) coated with an absorption layer. The filter can be made further compact by using a single turn of a buffer stripped bend sensitive fiber (eg. 1060XP) with an applied absorption coating (Wang et al., 2007). The cross sections of both the fiber filters are shown in Fig 3(a) and Fig 3(b) respectively. Prototypes of the filters are shown in Fig 3(c) and Fig 3(d) respectively. To use a macro-bend fiber as an edge filter for wavelength measurement, the optimal design of the bend radius and surface processing methods are required to achieve a linear transmission response with wavelength.

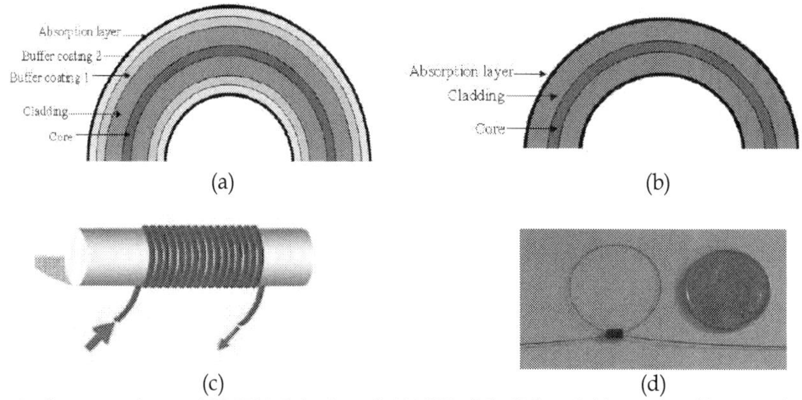

Fig 3. Cross sections of SMF28 (a& c) and 1060XP (b& d) bend fiber edge filters and their prototypes

It is known that for an optical waveguide with an infinite cladding the bend loss will increase as the bend radius decreases. However, a practical fiber contains one or two coating layers outside the cladding to offer mechanical protection. Because of the reflection of the radiated field at the interface between the cladding layer and the coating layer, so-called whispering gallery modes (WGM) are created and the fiber shows significantly different bend loss characteristics as compared to the simple case of an infinite cladding. The theoretical and experimental investigations (Wang et al., 2005), which considered the bend loss in SMF28 standard single-mode fiber, taking into account two coating layers, show that the bend loss is in the measurable range (40 dB) of photo detectors for bend radii in the range of 9.5 mm - 11 mm. To use the macro-bend loss fiber as an edge filter it is also

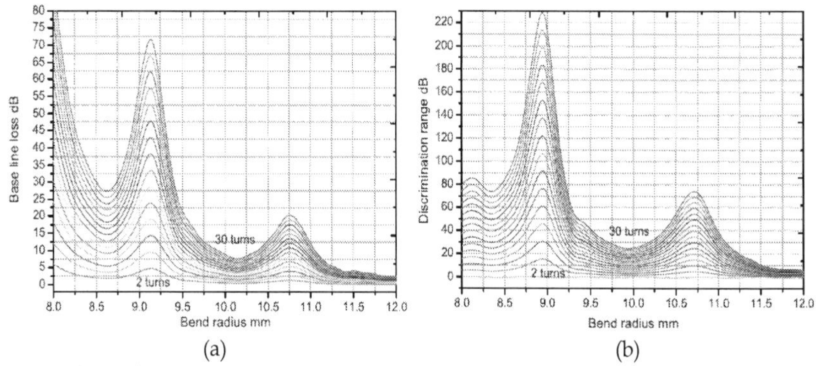

Fig. 4. (a) Baseline loss and (b) discrimination range for filters with differnent numbers of bend turns and bend radii

important to achieve a linear wavelength response and an acceptable discrimination range. Fig. 4(a) and Fig. 4(b) show the predicted baseline loss (at 1500 nm) and discrimination range (from 1500-1600 nm) for different fiber filters. From this response one can determine the baseline attenuation and the discrimination range of the filter. By changing the bending length or by increasing or decreasing the number of bend turns, at a fixed bend radius baseline attenuation and discrimination range can be varied.

As mentioned earlier in the case of a bend fiber, when the radiated field escapes from the cladding layer, some of the radiated field is reflected back and forms whispering- gallery modes, while the rest penetrates into the coating layers. Most of the radiated field reaching the coating is absorbed or scattered in the coating layers, but a small amount of the radiated field reaches the fiber surface. Because of the strong reflection at the interface between the outer coating layer and the air, the field reflected back toward the core affects propagation in the bending fiber, resulting in a transmission spectrum which displays non-linear variations. To remove such variations and to make the bending fiber suitable as an edge filter, one has to eliminate the back reflections at the coating-air interface. One simple method to remove the reflections at the air boundary is to apply an absorption layer which absorbs light in the wavelength range of 1500 nm - 1600 nm. The simplest example of such a coating is a black pigment ink (Indian ink) or a carbon paste. Indian ink is generally considered as a good optical absorber (Mourant et al., 1997). The transmission response of the filter before and after applying the absorption layer is shown in Fig. 5(a) and Fig. 5(b) respectively for filters with different bend radii. It can be seen that the transmission responses in Fig. 5(b) are much smoother after applying the absorption layer and quasi linear in comparison to those in Fig. 5(a), thus allowing a macro-bend fiber to be reliably used as an edge filter.

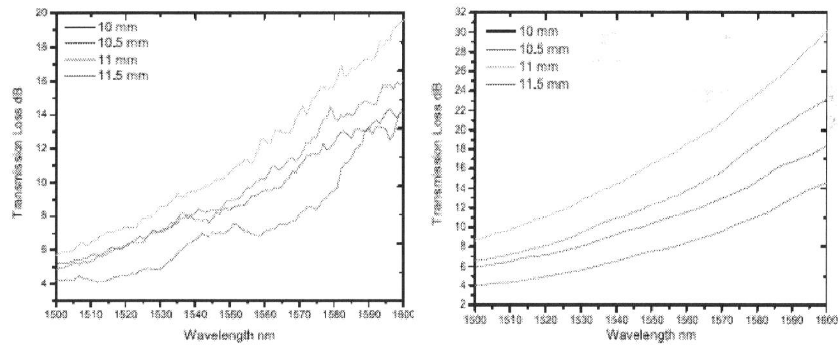

(a) (b)

Fig. 5. Transmission response of the fiber filter (a) before and (b) after absoprtion coating

2.3 A singlemode-multimode-singlemode (SMS) fiber edge filter

Another recently inducted fiber edge filter is the singlemode-multimode-singlemode fiber filter (Wang et al., 2008). It is formed by splicing a step-index multimode fiber (MMF) between two standard singlemode fibers (SMF). An SMS edge filter structure is shown in Fig. 6. The operating mechanism for this edge filter can be described as follows: the light field propagating along the input SMF enters the MMF section and excites a number of guided modes in the MMF. Interference between the different modes occurs while the light field propagates through the MMF section. By choosing a suitable length for the MMF section, the light is coupled into the output SMF in a wavelength dependent manner due to interference. The input-to-output transmission loss is expected to increase/decrease monotonically, as the wavelength of the propagating light increases in a certain wavelength range.

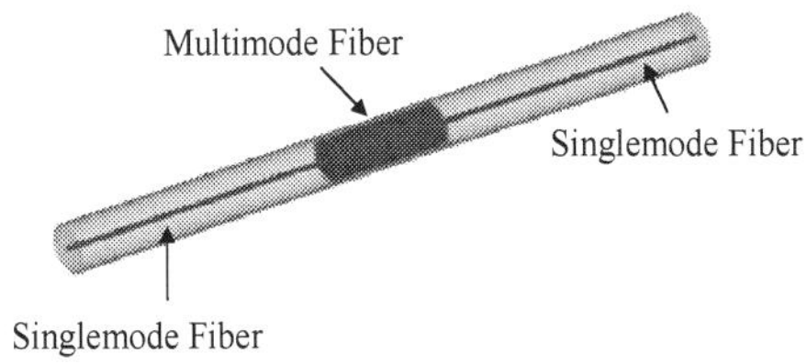

Fig. 6. Schematic of an SMS edge filter

A modal propagation analysis for linearly polarized (LP) modes can be used to investigate light propagation in the SMS fiber structure. To design the SMS based edge filter, the MMF length needs to be accurately determined. At a re-imaging distance (the transmission loss will reach a peak at a self image of the input) the filter is highly wavelength dependent. If re-coupling of light into the SMF takes place at the re-imaging distance, then the SMS structure operates as a band-pass filter as shown in Fig. 7. However for the purpose of designing an edge filter, the band-pass response can be considered as two edge-type responses, on the either side of a center wavelength. Consequently the device can behave as an edge filter with either a positive or a negative slope for a selected wavelength range.

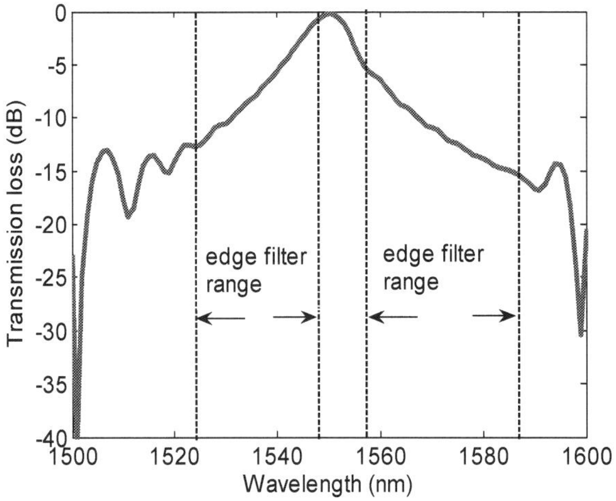

Fig. 7. Typical spectral response of an SMS fiber filter

3. Factors determining the performance of a linear edge filter based WMS

Precision, accuracy and resolution are extremely important for wavelength measurements involved in multi-channel dense wavelength division multiplexing (DWDM) optical communication systems and fiber Bragg grating based optical sensing systems (Hill & Meltz, 1997). The core component of a passive wavelength measurement system for such applications is a linear edge filter (Zhao & Liao, 2004). Therefore it is important to study the factors that influence the performance accuracy of such a system.

The three main factors that determine the performance of a linear edge filter based wavelength measurement system are:

1. Noise in the receiver system and the signal-to-noise ratio (SNR) of the source.
2. Polarization dependency of the components involved in the system.
3. Temperature dependence of the components involved in the system.

4. Modelling and analysis of impact of noise on the resolution and accuracy of an edge filter based WMS

Many of the interrogation schemes employed in optical sensing are based on a ratiometric wavelength approach which makes the system independent of input signal power variations. However a ratiometric optical wavelength measurement system's operation is perturbed by both the inherent optical noise in the input signal as well as the electrical noise due to optical-to-electrical conversion at the receivers. At the receivers, even though the measurement is performed by taking the power ratio of the signal levels, because of the uncorrelated random nature of receiver noise, the effect of noise sources will not be eliminated, but will affect the system's performance adversely.

4.1 Influence of SNR of the source on the ratio response of a WMS

In practice a narrow-band input to the wavelength measurement system with a center wavelength λ_0 could originate from a tunable laser source or a reflection from a fiber Bragg grating. Such an input signal can be approximated as a Gaussian function with a spectral

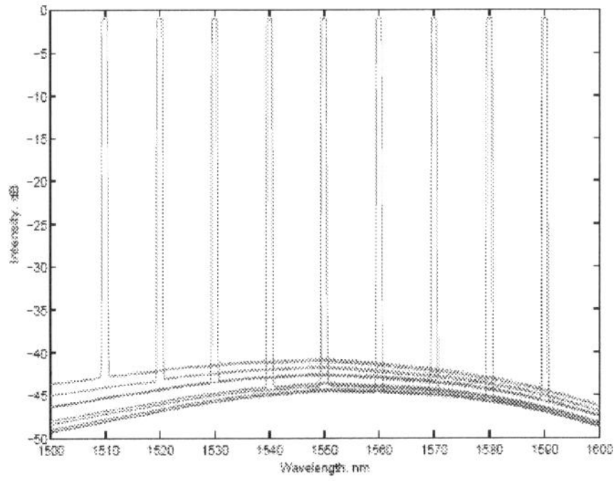

Fig. 8. Intensity distribution of a typical tunable laser source in the wavelength region 1500 nm to 1600 nm.

width $\Delta\lambda$ and center wavelength λ_0. An input signal from any source generally has a limited SNR, which means that there is measurable power even far from the center wavelength of the spectrum. For example in Fig. 8, the typical spectral distributions of the output intensity from a tunable laser at different central wavelengths, measured by an optical spectral analyzer in a wavelength range from 1500 nm - 1600 nm are shown. From these measured spectral distributions it can be seen that for this tunable laser the source spontaneous emission ratio is >40 dB and has different values for different output center wavelengths. To take account of the SNR, the output spectral response of the source (Xu et al, 1996) can be described as (the power at the peak wavelength is assumed as 0 dBm),

$$10\log_{10}[I_{\lambda 0}(\lambda)] = \begin{cases} 10\log_{10}\left[\exp(-4\ln 2\dfrac{(\lambda-\lambda_0)^2}{\Delta\lambda_0^{\ 2}}\right], & |\lambda-\lambda_0| \le \Omega \\ -S(\lambda) + Rand.Rs, & |\lambda-\lambda_0| > \Omega \end{cases} \tag{1}$$

where S is the SNR of the source. To describe the random fluctuations in the noise floor of the optical source, the term *Rand.Rs* is used, where *Rand* is a random number (between +0.5 and -0.5) and *Rs* is a parameter in dB which dictates the peak fluctuation in the SNR and is dependent on the nature of the source. Ω is a parameter which is determined by the noise level and can be found for a source with a given SNR from the relation:

$$10\log_{10}\left[\exp\left(-4\ln 2\frac{\Omega^2}{\Delta\lambda_0^{\ 2}}\right)\right] = -S \tag{2}$$

For real-world wavelength measurements there is always a possibility that the input signal's SNR may change from that used in the initial calibration. This can happen, for example, when the system is switched to a different source and in such cases the measured wavelength will have an error as the calibration is done for a different SNR. Fig. 9(a) shows

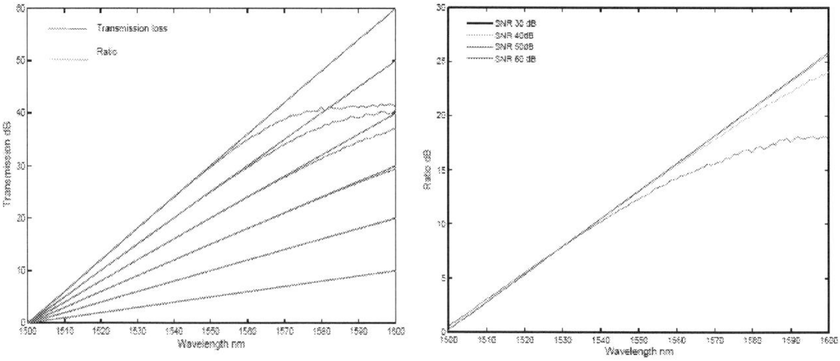

(a) (b)

Fig. 9. (a) Trasmission responses of edge filters of different slope at a fixed SNR (50 dB) (b) Transmission response of an edge filter with a fixed slope (0.25 dB/nm) at different SNRs the ratio response of a system for edge filters of different slope at a fixed SNR while in Fig. 9(b) the ratio response of an edge filter at a fixed slope at different SNRs are shown. From the figure it can be seen that a decrease in SNR of the source produces a great level of non-linearity in the discrimination characteristic, in particular for low values of SNR, the slope of the edge filter is reduced significantly at longer wavelengths, impairing accuracy and resolution. For any edge filter based ratiometric wavelength measurement system, source SNR is an important factor that needs to be considered as the limited SNR of the optical source affects the system by limiting the slope of the filter and also the measurable wavelength range (Rajan et al., 2007).

4.2 Noise mechanisms in an optical receiver

An understanding of the origin of receiver noise is required for a receiver performance to be accurately characterized. The amount of noise present in a receiver will be the primary factor that determines the receiver sensitivity. Fig. 10 shows a simple but adequate noise model of a trans-impedance optical receiver (Motechenbacher & Connelly, 1993), where I_{th}, I_{sh}, I_{amp} and e_{amp} are the thermal noise, shot noise, amplifier current noise and amplifier voltage noise respectively. C_f and C_d are the feedback and photodiode capacitance respectively. In addition, in optical sensing applications, it is very likely that every receiver has an analog-to-digital converter (ADC) associated with it. So the receiver system noise will also contain the quantization noise from the ADC. For simplicity the main noise mechanisms considered in this analysis are (a) shot noise of the photodiode (b) thermal noise and (c) quantization noise. If a low noise pre-amplifier is used which has a very low I_{amp} (in the order of fA) and e_{amp} (15 nv/√Hz, f=1 kHz, shunt resistance=100 Ω) then the effect of amplifier noise can be neglected.

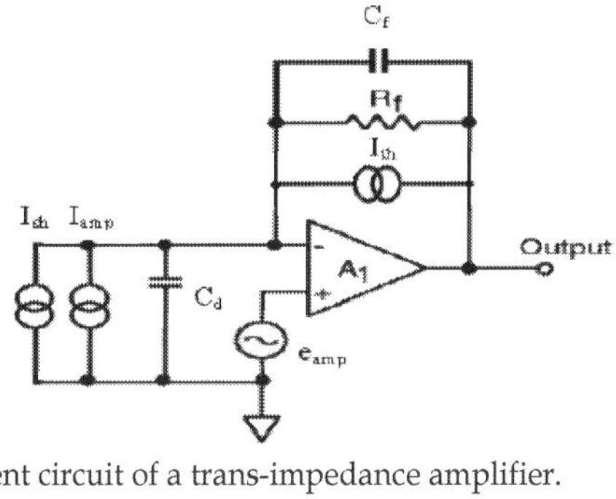

Fig. 10. Noise equivalent circuit of a trans-impedance amplifier.

Shot noise: In a photodiode the conversion of photons to charge carriers is a random process and the noise generated in this process is called shot noise. Because shot noise is associated with current flow, it is naturally modeled as a current-noise source in parallel with the photodiode, whose mean square value is given as

$$I^2_{sh} = 2eI_{ph}B \tag{3}$$

where I_{ph} represents the average photon generated current flowing through the photodiode, B is the post-conversion electrical bandwidth of the system and e is the elementary charge. When a photodiode is used together with a trans-impedance pre-amplifier, the shot noise produced in the photodiode is amplified together with the original signal. If A_{sig} is the amplification factor of the pre-amplifier, the rms shot noise voltage will be

$$e_{sh} = \sqrt{2eI_{ph}B}.A_{sig} \tag{4}$$

Thermal noise: Thermal noise is a result of thermally induced random fluctuations in the number of charge carriers in an electrical resistance element. Carriers are in random motion in all resistances at a temperature higher than absolute zero. The amount of motion is a direct function of the absolute temperature of the resistance. Nyquist showed that the open circuit rms voltage produced by a resistance R is given by

$$e_{th} = \sqrt{4KTR_f B} \tag{5}$$

where K is the Boltzmann's constant, T is the absolute temperature in Kelvin and B is the electrical system bandwidth. In the case of a trans-impedance amplifier front end the main source of thermal noise is the feed back resistor R_f.

Quantization noise: Quantization noise is a noise error introduced by quantization in the analog-to-digital converter used to convert an analog signal to a digital signal. A digital signal representation of an analog signal involves discrete levels and thus has an accuracy that depends on the quantization resolution or number of bits used to represent samples in the analog to digital process. Therefore, when converting a continuous analog level into a discrete digital level, the actual analog value must be approximated as a digital value. The error arising during this conversion is called quantization error. Quantization error can be treated as random noise added to the converted signal and is referred to as quantization noise.

4.3 Noise model of a ratiometric wavelength measurement system

The ultimate precision and resolution of wavelength measurement is determined by the electrical noise in the receiver and the signal-to-noise ratio of the input optical signal. As a starting point to determine the effects noise on a ratiometric system a basic noise source representation of the system is considered (Rajan et al., 2008) and is presented in Fig. 11.

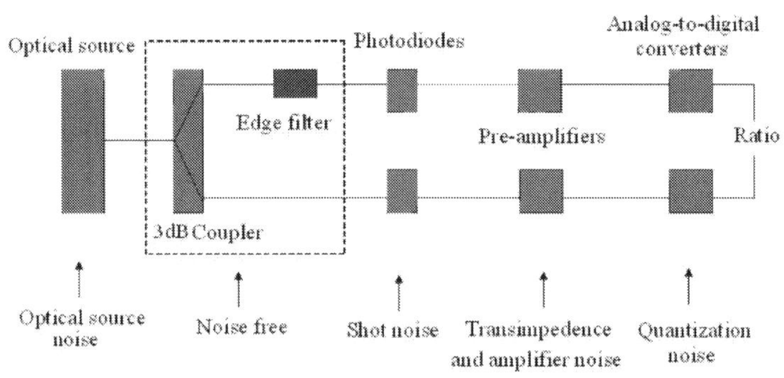

Fig. 11. Noise model of an edge filter based ratiometric system.

Since the different noise mechanisms are statistically independent the total mean square electrical noise that contributes to each arm of the system can be expressed as the sum of the individual mean square noises. Assuming a low noise pre-amplifier, the amplifier noise can be neglected. Thus the total rms noise voltage can be expressed as:

$$e_o = (e_{sh}^{\ 2} + e_{th}^{\ 2} + e_{adc}^{\ 2})^{1/2} \tag{6}$$

The calculated total rms noise voltage from all sources for a wavelength range from 1500 nm to 1600 nm for the receivers connected to both the reference arm and edge filter arm for different filters are shown in Fig. 12. At higher wavelengths the main noise contribution is the quantization noise of the analog-to-digital converter.

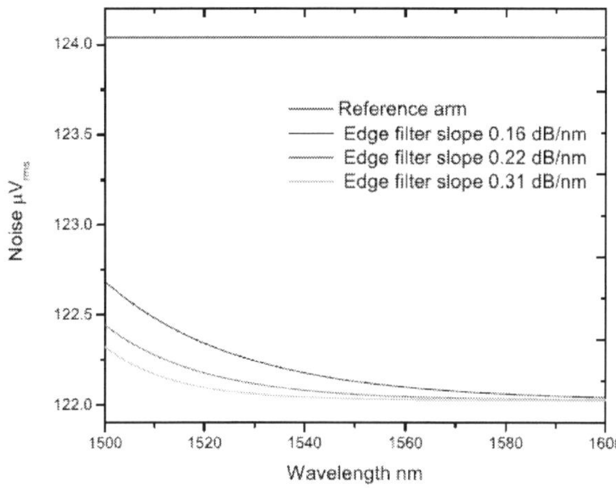

Fig. 12. Total rms noise of the receivers connected to reference arm and edge filter arms with filters of different slope.

4.4 Ratio Response of the system considering noise effects

In the previous section the ratiometric system was analyzed in the context of the limited signal-to-noise ratio of the input signal and it was shown that for a given measurable wavelength range, the slope of the edge filter is effectively limited due to the noise level of the input signal. To take account of the noise of the receivers connected to the reference and the edge filter arms, together with the signal-to-noise ratio of the input signal, the ratio of the system at a wavelength λ_0 can be expressed as

$$R(\lambda_0) = -10\log_{10}\left[\frac{\int I_{\lambda 0}(\lambda)T_f(\lambda)d\lambda + Grand_{e\,\lambda 0}}{\int I_{\lambda 0}(\lambda)d\lambda + Grand_r}\right] \tag{7}$$

T_f is the transmission response of the edge filter which in the simple case is a linear function within a wavelength range (λ_1, λ_2). Gaussian statistics is used to model the electrical noise.

$Grand_e$ and $Grand_r$ are Gaussian random numbers used to represent the receiver's noise with a mean value 0 and standard deviation equal to the rms noise of the receivers connected to the edge filter and reference arms respectively and are uncorrelated to each other. As the shot noise of the receiver connected to edge filter is a function of wavelength it makes $Grand_e$ also wavelength dependent.

To understand the separate effects of optical noise and receiver noise on the ratio fluctuation, the ratio fluctuation is calculated for the system with different edge filters in the presence of receiver noise by assuming the source SNR is infinite (100 dB). Then to calculate the ratio fluctuation that arises due to the noise of the input signal, the receivers are assumed to be noise free. The ratio fluctuations in both cases are shown in Fig. 13(a) and Fig. 13(b) respectively. For a system with an infinite source SNR the ratio fluctuation is caused due to the receiver noise and is very low at lower wavelengths as shown in Fig 13(a). For an ideal receiver with zero noise the ratio fluctuation is due to the source's noise and changes as the SNR of the source changes as shown in Fig. 13(b).

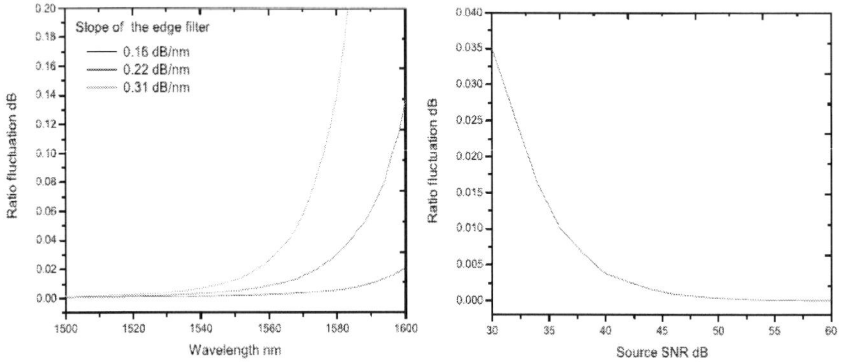

(a) (b)

Fig. 13. (a) Ratio fluctuation of the system arises from the receiver noise (b) ratio fluctuation arises from the limited SNR of the source.

In practice all optical sources have a limited signal-to-noise ratio and the net ratio fluctuation of the system is determined by the combined effect of optical noise and the receiver noise. Taking in account both a limited source SNR and the presence of receiver noise, the net ratio fluctuation can be calculated using equation (7). Fig. 14(a) shows the total ratio fluctuation of the system with different edge filters and with input signals with different SNRs. Fig. 14(b) shows a comparison of the modelled ratio fluctuation and the experimentally measured ratio fluctuation for edge filters of different slope at a fixed SNR. From the figures it can be seen that the ratio fluctuation is high at longer wavelengths which is the effect of the receiver noise and also the level of noise fluctuation increases irrespective of wavelength when the SNR of the input signal changes. Thus in the design of a ratiometric system it is important to consider the inaccuracy in measurement due to any changes in the SNR of the source and the noise in the receiver system. Ultimately to maintain accuracy it may be necessary to ensure that the SNR of the input signal is the same as that used during calibration.

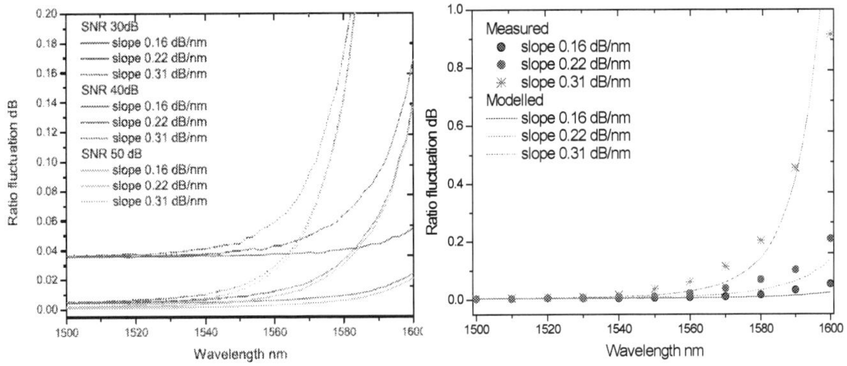

(a) (b)

Fig. 14. (a) Ratio fluctuation of the system for different optical SNR and (b) for different edge filters of different slopes at a fixed SNR

4.5 Influence of Noise on the Resolution of the System

From the Fig. 14 it is seen that the ratio fluctuation increases as the wavelength increases, which limits the wavelength resolution of the system in addition to the wavelength range. In any measurement system perturbed by noise, the resolvable step change should be higher than the peak-to-peak fluctuation of the parameter being measured. In the case of ratiometric wavelength measurement system the minimum reliable detectable ratio variation corresponding to a wavelength shift is assumed equal to the peak-to-peak ratio fluctuation and can be expressed as

$$\Delta R(\lambda_0)_{\min} = [R(\lambda_0)]_{p-p} \tag{8}$$

For a system of known slope m_r the wavelength resolution can be determined from the ratio fluctuation, which is given as

$$resolution = {[R(\lambda_0)]_{p-p}}\Big/{m_r} \tag{9}$$

Assuming a linear slope the estimated wavelength resolution for systems containing different edge filters (slopes of 0.16dB/nm, 0.22 dB/nm and 0.31 dB/nm) are shown in Fig. 15(a), Fig. 15(b), Fig. 15(c) respectively at different wavelengths (different receiver noise levels) and for different optical source SNRs. It is clear that a 10 pm resolution is possible for all systems below 1520 nm if the SNR of the source is above 50 dB. When the SNR drops, for a system with edge filter having a higher slope, a high resolution is achieved only at lower wavelengths. For example, for the systems with slopes of 0.16 dB/nm, 0.22 dB/nm and 0.31 dB/nm a 10 pm resolution can be achieved in the range of 36 nm, 22 nm, 16 nm respectively starting from 1500 nm in the presence of receiver noise and an input signal SNR of 50 dB. Thus the slope of the filter used within a system in the presence of receiver noise and optical noise depends on the required resolution and wavelength range. High resolution in a very narrow wavelength range can be obtained with an edge filter with a higher slope, while for a wide wavelength range a reasonable resolution can be obtained with an edge filter of lower slope.

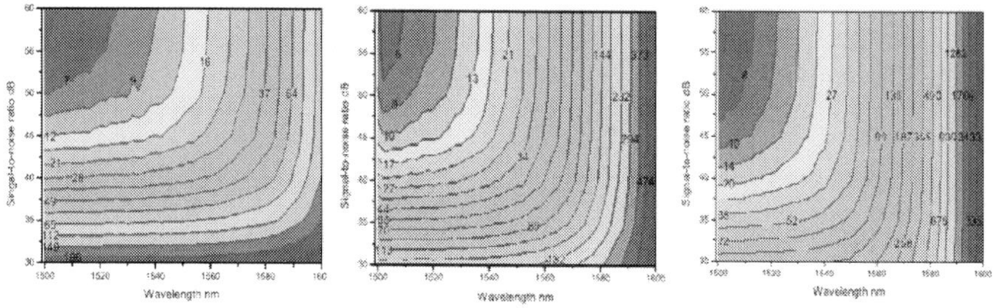

(a) (b) (c)

Fig. 15. Contour plots of achievable resolution for system contains edge filters of slope (a) 0.16 dB/nm (b) 0.22 dB/nm (c) 0.31 dB/nm

For further illustration, experimentally measured ratio variations for a wavelength step of 20 pm from 1550 nm are shown in Fig. 16 (a), Fig. 16 (b), Fig. 16 (c) for the systems with three different edge filters. From the figure it is seen that a wavelength change can be resolved more readily for the system with an edge filter of 10 turns than by systems that use edge filters with a larger numbers of turns (and thus higher slopes). This confirms the assertion arising from the model that increasing the slope of filter itself will not improve the resolution. Also it is not possible to achieve the same resolution throughout the wavelength range and for a system with a given edge filter a higher resolution will always be achieved at lower wavelengths.

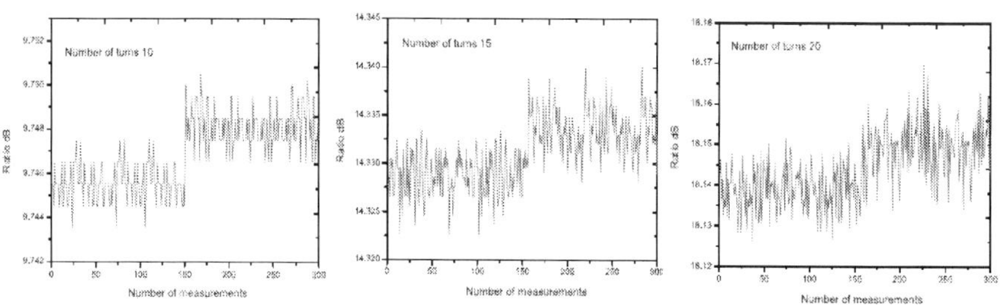

(a) (b) (c)

Fig. 16. Ratio variation for 20 pm wavelength step at 1550 nm for the systems with edge filters (a) 10 turns (b) 15 turns (c) 20 turns.

5. Polarization analysis of an edge filter based WMS

A typical ratiometric detection system consists of a 3 dB coupler and an edge filter, both of which are polarization sensitive devices. If the system contains more than one polarization sensitive component connected by standard optical fibers, its global PDL may fluctuate and hence the measured ratio also fluctuates. The fluctuation in the ratio will have an impact on the performance accuracy of the system. The fluctuation arises because the state of polarization of the input light to the system may vary randomly, for example because of stress variations in the connecting fibres or a change of source etc. A system calibrated at a particular input polarization state gives errors as the power attenuation will be different because of the polarization dependency of the elements.

Prior to the introduction of systems which demand very high wavelength accuracy (such as DWDM), the role of PDL in a ratiometric system was not considered important. DWDM systems require an accurate control of each channel's wavelength, e.g., the allowable maximal frequency deviation for 25 GHz channel spacing (2.5Gbit/s) is 4-5 GHz, i.e., 0.032-0.04 nm according to the ITU-T Recommendation G.652. In a system where high accuracy is needed errors from all sources must be calculated in order to estimate the accuracy and resolution and in such a case the error from the PDL is important (Rajan et al., 2008).

The PDL of an optical component is commonly defined as the difference between the maximum and the minimum insertion losses for all the possible states of polarization. If T_{min} and T_{max} are the minimum and maximum transmission coefficients of an optical element, then PDL can be expressed as

$$PDL_{dB} = 10\log(T_{\max} / T_{\min}) \tag{10}$$

5.1 Polarization sensitivity of a 3 dB coupler

It is very well known that a 3 dB coupler is a polarization sensitive device. The common fabrication technique for a coupler is the fused-fiber method which involves twisting, melting, and pulling two single-mode fibers so they are fused together over a uniform length section. During fabrication a deviation of the coupler geometry from a circular symmetry to the elliptical shape in the fused tapered region can occur and this changes the coupling characteristics of polarization states and leads to birefringence, which makes the coupler polarization sensitive and therefore a device exhibiting PDL (Wu & Chang, 1995). The polarization sensitivity of a 3 dB coupler mainly depends on the geometrical structure parameters and the degree of fusion (Wu, 1999). The PDL of any device is commonly characterized at a specific wavelength. But because of the wavelength dependency of PDL and especially for wavelength measurements, the PDL of the components has to be considered over the whole range of wavelengths from 1500 - 1600 nm. Typical commercially available 2x2 fused couplers have a PDL of 0.1 - 0.15 dB at the central wavelength, but the value does vary with wavelength as shown in Fig. 17.

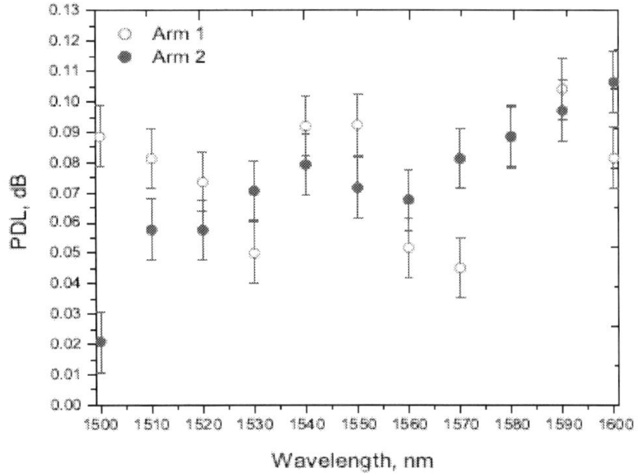

Fig. 17. Measured PDL of the arms of a 3dB coupler for a wavelength range 1500-1600 nm

5.2 Polarization sensitivity of edge filters

In this section the polarization sensitivity of macro-bend fiber filter and SMS fiber filter are explained. There are a couple of reasons which underpin the polarization sensitivity of a bent fiber. The state of polarization of light propagating along a single-mode fiber can be influenced by perturbations, such as bending and twisting. In a macro-bend fiber filter it can originate from the anisotropic nature of the refractive index caused by the bending stress. Single-mode fibers with nominal circular symmetry about the fiber axis are in fact bimodal and they can propagate two nearly degenerate modes with orthogonal polarizations, the TE and TM modes. The bend induced mechanical stress can induce birefringence by changing the modes of polarization. In the case of a standard single-mode fiber such as SMF28, the stress induced refractive index variation is very small. As a result, the bend induced birefringence will have only a small impact on the polarization modes propagating along the fiber. A practical single-mode fiber consists of a core, a cladding and a polymer coating layer which offers mechanical protection. The difference in the refractive index between the polymer coating layer and the cladding layer is much higher as compared to that between the cladding and core. The reflectance of the radiated field occurring at the interface between the coating layer and cladding layer is believed to be different for different polarization states. This can lead to polarization dependence of the bend loss, which originates because of the coupling between the reflected radiated field and the guided fundamental mode (Wang et al., 2007). Thus, the main contributor to the polarization sensitivity of a bend single-mode fiber is the polymer coating.

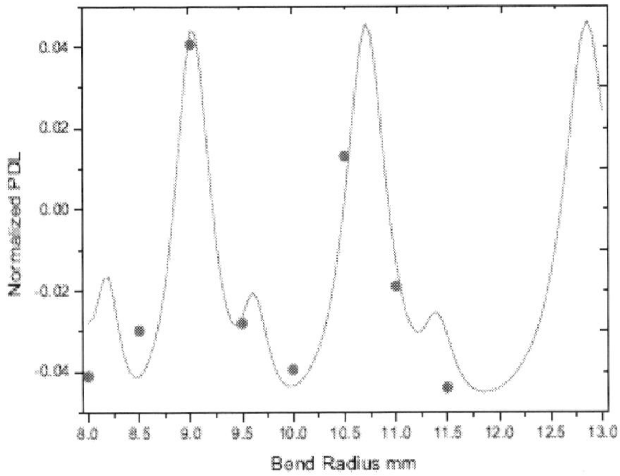

Fig. 18. Normalized PDL of 10 turns of SMF28, simulated and measured at 1550 nm

The theoretical model of polarization sensitivity of bend fiber shows that the TE and TM modes propagating along the fiber experience different values of bend losses. Similar to the bend loss behaviour of a single-mode fiber (Fig. 4) the absolute PDL does not increase monotonically with the bend radius. The normalized polarization dependence has a quasi-periodical characteristic with bending radius and this quasi-periodical behaviour is very close to that of the bend loss vs. bending radius of a standard single mode fiber. Since the bend loss varies with wavelength, the PDL also changes with wavelength. The normalized PDL for 10 turns of SMF28 for different bend radii is shown in Fig. 18.

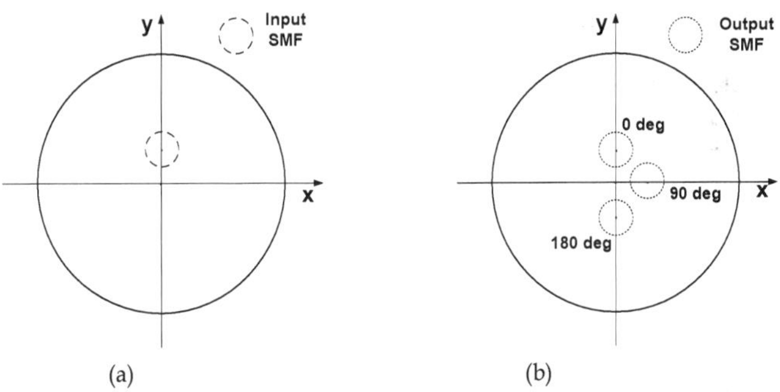

Fig. 19. Interfaces of input/output SMF core to the MMF core (a) position of input SMF core, and (b) position of output SMF core.

The main source of PDL in a SMS edge filter originates from the lateral core offset that occurs during the splicing between the singlemode fiber and the multimode fiber, i.e. the input/output SMFs may have lateral core offsets relative to the centre of the MMF core.
In addition the two lateral core offsets at each end of the MMF may also have a different orientation relative to each other which can be referred to as a rotational offset. In order to

analyze the PDL of a SMS fibre structure, a means to precisely describe lateral and rotational offsets is needed. Fig. 19(a) and Fig. 19(b) show the interfaces between the input SMF and the MMF section cores and the MMF section and the output SMF cores, respectively.
Assuming some lateral core offset of the input SMF is as in Fig 19(a), the field profile at the output end of the MMF section depends on the input field polarization state of the quasi TE mode (x-directed) or quasi TM mode (y-directed). In turn, the transmission loss for each mode depends on the overlap between the field profile at the output end of the MMF section and the eigen-mode profile of the output SMF. Thus the PDL can be calculated from the difference in the transmission loss between the quasi TE and quasi TM modes in dB using the equation;

$$PDL = \left| Ls_{TE} - Ls_{TM} \right| \tag{11}$$

The PDL of the SMS fiber edge filter for different lateral and rotational offsets is shown in Fig. 20. Generally, a larger lateral core offset induces a higher PDL.

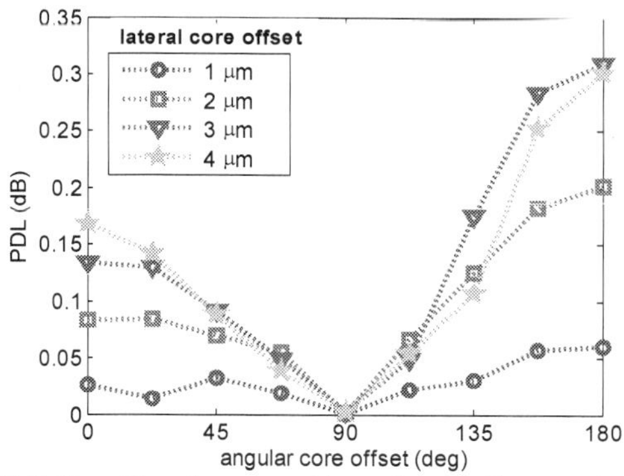

Fig. 20. PDL of the SMS fiber filter for several lateral core offsets at the rotational core offset from 0 to 180^0.

5.3 Theoretical model to estimate the range of PDL and ratio fluctuation

As mentioned above, in a fiber edge filter based ratiometric wavelength measurement system, the PDL components are the edge filter itself and the 3 dB coupler. The total PDL of the system is not simply the sum of the contribution of each PDL element. If the polarization sensitive axis of the 3 dB coupler and the fiber filter are not aligned with each other, the resulting PDL depends on the relative orientation of the PDL axes at each connection. In an all-fiber ratiometric configuration the PDL will fluctuate in the filter arm as it effectively contains two PDL elements. The total attenuation ratio fluctuation of the system depends on the PDL of both the filter arm and the reference arm. The edge filter considered in this analysis is a bend fiber edge filter.

The general expression for PDL of any optical component is given by Equation (10). But a more convenient approach is to express the PDL as a 3-dimensional vector $\vec{\Gamma}$ (Gisin, 1995) of length Γ as

$$\Gamma = \frac{T_{\max} - T_{\min}}{T_{\max} + T_{\min}} \tag{12}$$

If Γ_{3dB} and Γ_{bend} are the PDL of the 3 dB coupler and the fiber filter, the net PDL of the filter arm can be expressed as (Almari et al., 1998)

$$\vec{\Gamma}_{3dBbend} = \frac{\sqrt{1-\Gamma_{bend}^{\ 2}}}{1+\vec{\Gamma}_{3dB}\vec{\Gamma}_{bend}}\vec{\Gamma}_{3dB} + \frac{1+\vec{\Gamma}_{3dB}\vec{\Gamma}_{bend}\left(1-\sqrt{1-\Gamma_{bend}^{\ 2}}\right)/\Gamma_{bend}^{\ 2}}{1+\vec{\Gamma}_{3dB}\vec{\Gamma}_{bend}}\vec{\Gamma}_{bend} \tag{13}$$

The maximum value of the PDL will occur when Γ_{3dB} and Γ_{bend} are parallel and minimum occurs when they are anti parallel. The maximum and minimum global PDL of the fiber filter arm (Γ_{Gmax} and Γ_{Gmin}) can thus be expressed as

$$\Gamma_{G\max} = \frac{\Gamma_{3dB} + \Gamma_{bend}}{1+\Gamma_{3dB}\Gamma_{bend}} \quad \text{and} \quad \Gamma_{G\min} = \frac{\left|\Gamma_{3dB} - \Gamma_{bend}\right|}{1-\Gamma_{3dB}\Gamma_{bend}} \tag{14}$$

Using these equations the range of PDL of the filter arm can be predicted and together with the PDL of the reference arm of the system the ratio error of the system can be estimated. The maximum variation in the ratio from the calibrated value occurs when one arm gives the maximum attenuation and the other gives the minimum attenuation for a given state of polarization. Since the photodiodes measure the integral power over the wavelength range, by knowing the maximum and minimum power at the arms of the system when the polarization state changes, the maximum possible polarization dependent change in ratio of the system for any wavelength can be obtained and is expressed as

$$\Delta R_{\max} = 10\log_{10}\left[\frac{\int P_{1\max}(\lambda)I_{\lambda0}(\lambda)d\lambda}{\int P_{2\min}(\lambda)I_{\lambda0}(\lambda)d\lambda}\right] - 10\log_{10}\left[\frac{\int P_{1\min}(\lambda)I_{\lambda0}(\lambda)d\lambda}{\int P_{2\max}(\lambda)I_{\lambda0}(\lambda)d\lambda}\right] \tag{15}$$

where P_{1max} and P_{2min} are the maximum and minimum output powers of the filter arm when the polarization state changes and P_{2max} and P_{2min} are that of the reference arm. $I_{\lambda0}$ is the narrow band input signal with a central wavelength λ_0 which could be from tunable laser source. By knowing the variation in the ratio the corresponding wavelength variation can be calculated.

5.4 Ratio and wavelength error due to PDL

The ratio responses of an edge filter based ratiometric system at different polarization states are shown in Fig. 21. The ratio response for polarization state A is the calibrated response, while the response curves for the two other states B and C show the variation in ratio response from the calibrated response as the polarization state changes. The experimental arrangement to measure the polarization induced errors is shown in Fig. 22. A manual fiber polarization controller is used to change the polarization state of the input signal from a tunable laser source. The polarization controller allows for the complete control of the output polarization state and the output power varies with the change in polarization states for a particular wavelength. It is assumed that the fiber polarization controller covers all the possible polarization states at a fixed wavelength.

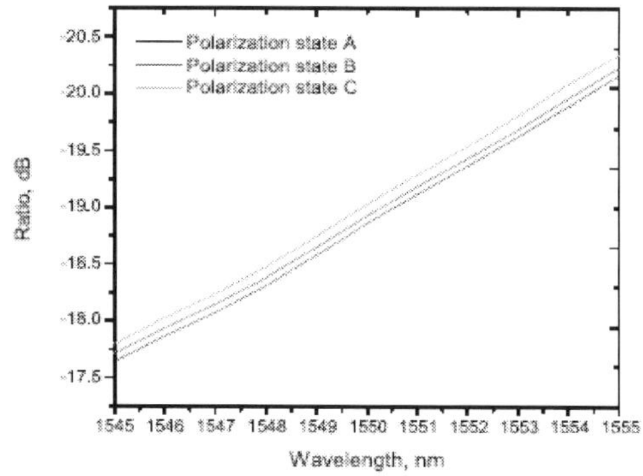

Fig. 21. Ratio response of the system at different polarization states

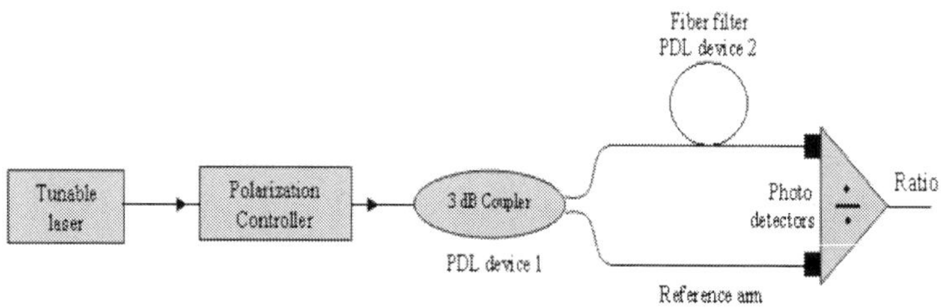

Fig. 22. Experimental arragement to study the impact of PDL on wavelength measurements

Estimation of the maximum variation in the measured wavelength is important as we can determine the system's worst case performance. The maximum and minimum values of the fluctuation of PDL of the filter arm due to the 3 dB coupler and the fiber filter can be calculated using Equation (14). A comparison of the estimated maximum and minimum of the PDL of the filter arm with the measured PDL of the filter arm is shown in Fig. 23.

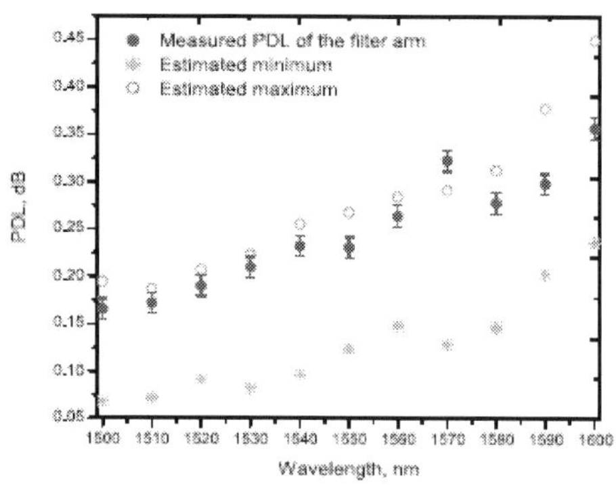

Fig. 23. Maximum and minimum PDL of the fiber filter arm and its comparison with the measured PDL.

For a ratiometric system both of the arms contribute to the total ratio variation. The PDL of the reference arm and filter arm obtained from the experiment provides the maximum and minimum power levels of each arm. Based on that, a numerical simulation can be carried out to find the maximum ratio variation and can be estimated using Equation (15). The estimated variation in ratio and wavelength of the system and its comparison with the measured versions are shown in Fig. 24(a) and Fig. 24(b) respectively. To estimate the wavelength error, the local slope of the ratio spectrum is used. The wavelength error, which is a consequence of ratio variation, in practice depends on the slope of the system which is low at shorter wavelengths and high at longer wavelengths which results in a larger error at shorter wavelengths than at longer wavelengths. In the example shown in Fig. 24 for a fiber filter of 10.5 mm radius and 15 turns it is estimated that the maximum wavelength error at 1500 nm is 1.9 nm from the original value. Any measured error in wavelength should be within this estimated wavelength error range.

From the figure it is clear that the measured ratio and wavelength variation of the system are well within the estimated limits. The effect of fluctuation in the attenuation due to PDL, which leads to the variation in ratio and measured wavelength, is confirmed by the results.

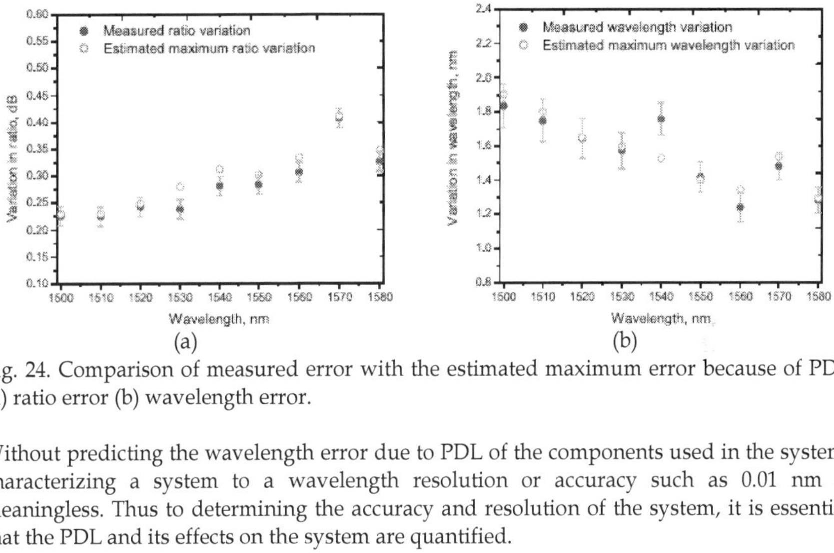

(a) (b)

Fig. 24. Comparison of measured error with the estimated maximum error because of PDL (a) ratio error (b) wavelength error.

Without predicting the wavelength error due to PDL of the components used in the system, characterizing a system to a wavelength resolution or accuracy such as 0.01 nm is meaningless. Thus to determining the accuracy and resolution of the system, it is essential that the PDL and its effects on the system are quantified.

6. Polarization dependent loss minimization techniques

In the case of macro-bend fiber filter since the PDL of the filter originates from the difference in bend loss for TE and TM modes one method to compensate the bend loss of the modes is to split the fiber filter into two bending sections with equal length and introduce a 90^0 twist in the middle of the filter between the two sections (Rajan et al., 2008). This changes the polarization state for the second bending section, i.e., the TE (TM) mode is turned to be the TM (TE) mode for the second bending section. The net effect is that the individual losses for the input TE and TM modes are equalized over the total length of the fiber so that the PDL can be minimized for the whole bending section. The schematic of the configuration is shown in Fig. 25.

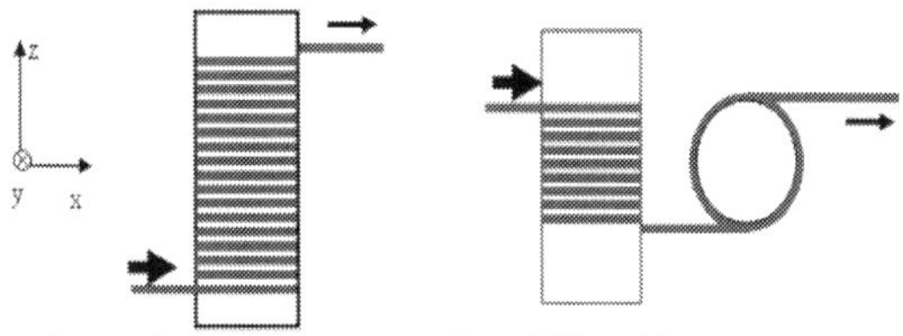

Fig. 25. Bending configurations of the macro-bend fiber filter: conventional bending and a 90^0 twist between the bending sections.

To demonstrate how the 90^0 twist reduces the PDL at higher bend lengths, the PDL of the filter is measured for different bend lengths and is shown in Fig. 26(a). For comparison the PDL of fiber filters without a twist is also presented for the same number of turns. From the figure it is clear that PDL is not eliminated completely in the fiber filter due to physical inaccuracies such as small variations in the bend length of the two sections of the filter and variations in the twist angle from 90^0 leading to residual PDL. It should be noted that a twist in the fiber induces circular birefringence and can make the fiber polarization dependent. However, such stress induced birefringence is very low in SMF28 fiber which means that the twist induced birefringence is negligible and its contribution to the PDL of the fiber filter is very small. Overall from the figures it is clear that the PDL of the fiber filter decreases considerably with a 90^0 twist at higher bend lengths which in turn allows the filter to utilize a larger number of turns to obtain the required steepness and thus increase the measurement resolution of the system without reaching an unacceptable level of PDL.

The PDL of an SMS structure can be reduced/eliminated by using accurate splicing methods which reduce the lateral offset between the SMF and the MMF at both ends. However conventional fusion splicers cannot guarantee a perfect splicing without lateral offset. In such cases by introducing a rotational offset of 90^0 will minimize the PDL as shown in Fig. 20. This is because at a rotational core offset of 90^0, the orientation between the input/output SMF and the input field direction of TE/TM are parallelized. Thus the overlap between the field profile at the output end of the MMF section and the eigen-mode profile of the output SMF for both TE and TM modes are similar and thus the PDL will be minimised.

Minimizing the polarization dependency of the fiber filter alone will not minimize the polarization dependency of the whole system. As the system contains another PDL component, the 3 dB coupler, it is important to minimize the PDL of the coupler also. One way to minimize the total polarization dependency of the system is using a polarization insensitive (PI) 3 dB couplers (couplers with very low PDL, in the range of 0.01 - 0.02 dB). The wavelength inaccuracy of a macro-bend fiber filter together with low PI 3 dB coupler and its comparison with conventional system are shown in Fig. 26(b). Thus, for wavelength measurements based on macro-bend fiber filters the polarization dependency can be significantly reduced by the 90^0 twisted fiber filter together with low PI 3 dB coupler configuration and can deliver measurements with high wavelength accuracy irrespective of the input state of polarization.

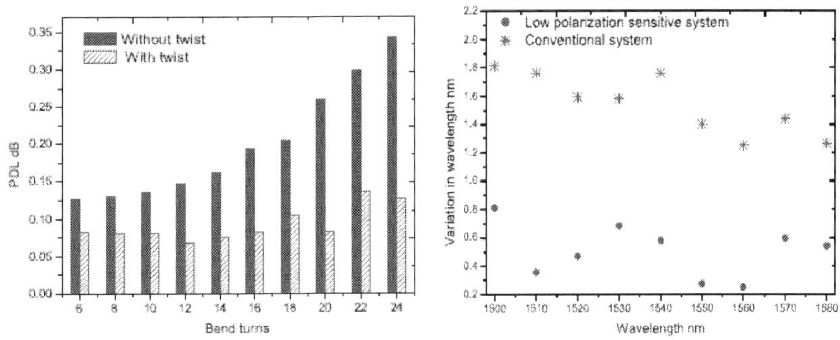

(a) (b)

Fig. 26. (a) PDL of the fiber filters with 90^0 twist and its comparison with the PDL of the filters without twist (b) Comparison of wavelength errors in a low polarization system vs. conventional system

7. Temperature induced inaccuracies in a macro-bend fiber filter based WMS

When a single-mode fiber forms a macro-bend, WGMs may be created, which propagate in the cladding or buffer. These WGMs can interfere with the guided core mode to produce interference induced oscillations in the bend loss spectral response (Morgan et al., 1990). The dominant source of WGMs is the buffer-air interface and also the cladding-buffer interface. The formation of such whispering gallery modes effectively creates an interferometer within the fiber, with the core and buffer/cladding as the two arms. To utilize a macro-bend fiber as an edge filter, an absorption layer is applied to the buffer coating to eliminate these WG modes, which makes the bend loss spectral response smoother and ideally achieves a linear response versus wavelength as explained earlier.

The temperature sensitivity of such a fiber filter arises mainly from the temperature sensitive properties of the buffer coating, characterized by the thermo-optic coefficient (TOC) and thermal expansion coefficient (TEC). The TOC and TEC of the buffer coatings, such as acrylates, are much higher than those of fused silica which forms the core and the cladding of the fiber. Macro-bend fiber edge filters can be based on low bend loss fiber such as SMF28 fiber or high bend loss fiber such as 1060XP as explained in section 2.

The most common single-mode fiber, SMF28 fiber, has two buffer coating layers. Due to the coating layers, even with the absorption layer a low level of reflection from the cladding-primary coating boundary will still exist and interfere with the core mode. As a result of this when there is a change in temperature which changes the refractive index and thickness of the buffer coating, the path length variation of the WG modes and phase difference between the WG mode and the core mode leads to constructive and destructive interference between the WG mode and the core mode. This results in oscillatory variations in the spectral response of the bend loss. In a macro-bend fiber filter without a buffer coating but with an applied absorption layer the temperature induced periodic variations in the bend loss can be eliminated.

A fiber filter based on SMF28 fiber requires multiple bend turns with small bend radii to achieve a better slope and high wavelength resolution. The removal of the buffer coating over a meter or more of fiber is beyond practical limits as the fiber breaks if it is wrapped for more than one turn at small bend radii without a buffer. However, a fiber such as 1060XP is highly sensitive to bend effects due its low normalized frequency (V). The V parameter for 1060XP fiber is 1.5035 while for SMF28 fiber it is 2.1611. Since the normalized frequency of the 1060XP is smaller, power will be less confined in the core and will be more susceptible to bending loss and the bend loss will be higher when compared to SMF28. As a result an edge filter based on a bend sensitive 1060XP fiber requires only one bend turn and hence the buffer can be stripped easily and an absorption layer can be applied directly to the cladding.

After removing the buffer coating from the sensor head, the only negative TOC material is eliminated and the sensor head consists of only positive TOC materials; the cladding and core, which are made of silica. For the silica core and cladding the thermally induced effective change in refractive index is linear in nature, resulting in a linear variation of bend loss with temperature. Since the temperature dependent loss is proportional to the bend loss in the fiber filter, 1060XP fiber shows higher temperature induced loss, when compared to its SMF28 counterpart, for the case of a single bend turn. For a system with this configuration, a temperature corrected calibration is feasible. A temperature corrected calibration means that temperature of the fiber filter is continually measured and therefore, the measurement system can apply correction factors to the calibration in use. This allows the system to be used over a wide range of ambient temperatures (Rajan et al., 2009).

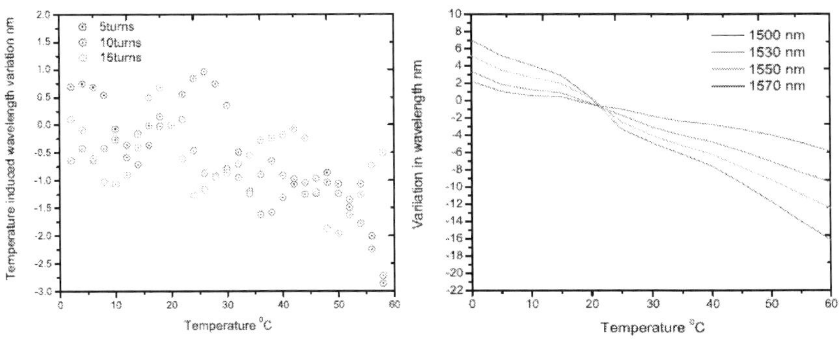

(a) (b)

Fig. 27. Temperature induced wavelength error (a) SMF28 fiber filter (b) 1060XP fiber filter

A comparison of wavelength errors due to ambient temperature variation in the case of edge filters fabricated from standard singlemode fiber (SMF28) and bend sensitive fiber (1060XP) are shown in Fig. 27(a) and Fig. 27(b) respectively. While it is apparent that the SMF28 fiber filter based system is less temperature sensitive, nevertheless the oscillatory nature of the bend loss and ratio of the system makes correction of the calibrated response unfeasible. For the SMF28 based filter the only option is to use active temperature stabilization of the filter temperature. Whereas for the bend sensitive fiber based filter temperature compensation requires a sensor and compact electronics only, temperature stabilization will additionally demand a Peltier cooler, heat sinks, a complex feedback control system and, depending on the ambient temperature variation to be dealt with, will involve significantly higher power consumption by the system.

The temperature stabilization approach will thus require more physical space, as well as higher complexity and cost than the temperature compensation approach. Using high bend loss fibers such as 1060XP will mean that the fiber filter will have higher temperature dependence than the SMF28 fiber filter, but due to the linear nature of the ratio variation with temperature, the temperature induced error can be compensated by adding correction factors to the calibration ratio response. The wavelength accuracy can be improved by obtaining the correction in the ratio response with smaller temperature intervals or by extrapolating the correction response between the required temperature intervals. Thus, irrespective of the temperature dependence of the 1060XP fiber filter, such a filter can be operated over a wide temperature range, if the correction in ratio response is added to the original ratio response and thus precise wavelength measurements can be obtained.

8. Summary

A brief review of all-fiber passive edge filters for wavelength measurements is presented in this chapter. Along with the review two recently developed fiber edge filters: a macro-bend fiber filter and a singlemode-multimode-singlemode fiber edge filter are also presented. For the macro-bend fiber filter an optimization of the bend radius and the number of bend turns together with the application of an absorption coating is required in order to achieve a desired edge filter spectral response. For the SMS fiber filter, the length of the MMF section sandwiched between the singlemode fibers is important. The length of the MMF section determines the operating wavelength range of the filter.

The main factors that affect the performance accuracy of edge filter based ratiometric wavelength measurement are also discussed in this chapter. Due to the limited SNR of the optical source and the noise in the receiver system, the measurable wavelength range is limited and also it is not possible to achieve a uniform resolution throughout the wavelength range. The resolution of the system depends on the filter slope and the noise in the system.

The origin of the polarization sensitivity of the components of a ratiometric system is also analysed in this chapter. The polarization sensitivity of a 3 dB coupler, a macro-bend fiber filter and a SMS fiber filter are explained. Since a ratiometric wavelength measurement system consists of more than one PDL component, the net PDL depends on the relative orientation of the PDL axes of each component. A theoretical model to predict the ratio and wavelength fluctuation due to the polarization dependence of the components involved in the system is presented. It is concluded that for determining the accuracy and resolution of the system the PDL of the system and its effects on the system performance have to be quantified. To minimize the effect of PDL on a macro-bend and a SMS fiber filters, methods to minimize the polarization dependence are also presented. In the case of a macro-bend fiber filter, PDL can be minimized by dividing the filter into two sections and by introducing a 90^0 twist between the two bending sections. For SMS fiber filters PDL can be minimized by reducing the lateral core offset and also by introducing a 90^0 rotational offset.

The influence of temperature on a macro-bend fiber based wavelength measurement system is also presented in this chapter. The temperature dependencies of two types of macro-bend fiber filters based on SMF28 and 1060XP fibers are presented. In the case of SMF28 fiber based filter, the temperature dependence is lower, but the response is oscillatory in nature, which makes correction to the temperature calibration too complex to be feasible. In the case of 1060XP fiber based system, the temperature dependence is higher but since it is linear in nature a temperature correction to the calibration response is feasible.

9. References

Davis, M. A. & Kersey, A. D. (1994). All-fiber Bragg grating strain sensor demodulation technique using a wavelength division coupler, *Electron. Lett.*, 30, 75-77

El Amari, A.; Gisin, N.; Perny, B.; Zbinden, H. & Zimmer, W. (1998). Statistical prediction and experimental verification of concatenations of fiber optic components with polarization dependent loss, *IEEE J. Lightwave Technol.*, 16, 332-339

Fallon, R. W.; Zhang, L.; Everall, L. A. & Williams, J. A. R. (1998). All fiber optical sensing system: Bragg grating sensor interrogated by a long period grating, *Meas. Sci. Technol.*, 9, 1969-1973

Fallon, R. W.; Zhang, L.; Gloang, A. & Bennion, I. (1999). Fabricating fiber edge filters with arbitrary spectral response based on tilted chirped grating structures, *Meas. Sci. Technol.*, 10, L1-L3

Gisin, N. (1995). The statistics of polarization dependent losses, *Optics Communications*, 114, 399-405

Hill, K. O. & Meltz, G. (1997). Fiber Bragg grating technology fundamentals and overview, IEEE *J. Lightwave Technol.*, 15, 1263-1276

Kersey, A. D.; Berkoff, T. A. & Morey, W. W. (1992). High resolution fiber grating sensor with interferometric wavelength shift detection, *Electron. Lett.*, 28, 236-138

Kersey, A. D.; Berkoff, T. A. & Morey, W. W. (1993). Multiplexed fibre Bragg grating strain-sensor system with a fibre Fabry Perot wavelength filter, *Opt. Lett.*, 18, 1370-1372

Mille, S. M.; Liu, K. & Measures, R. M. (1992). A passive wavelength demodulation system for guided wave Bragg grating sensors, *IEEE Photon. Tech Lett.*, 4, 516-518

Morgan, R.; Barton, J. S.; Harper, P. G. & Jones, J. D. C. (1990) Temperature dependence of bending loss in monomode optical fibers, *Electron. Lett.*, 26, 937-939

Motechenbacher, C. D. & Connelly, J. A. (1993). *Low-Noise Electronic System Design*, John Wiley and Sons, Inc

Mourant, J. R.; Bigio, I. J.; Jack, D. A.; Johnson, T. M. & Miller, H. D. (1997). Measuring absorption coefficients in small volumes of highly scattering media: source-detector separations for which path lengths do not depend on scattering properties, *Appl. Opt.*, 36, 5655-5661

Rajan, G.; Wang, Q.; Farrell, G.; Semenova, Y. & Wang, P. (2007). Effect of SNR of input signal on the accuracy of a ratiometric wavelength measurement system, *Microwave and Optical Technology Letters*, 49, 1022-1024

Rajan, G.; Semenova, Y.; Freir, T.; Wang, P. & Farrell, G. (2008). Modeling and analysis of the effect of noise on an edge filter based ratiometric wavelength measurement system, *IEEE J. Lightwave Technol.*, 26, 3434-3442

Rajan, G.; Wang, Q.; Semenova, Y.; Farrell, G. & Wang, P. (2008). Effect of polarization dependent loss on the performance accuracy of a ratiometric wavelength measurement system, *IET Optoelectron.*, 2, 63–68

Rajan, G.; Semenova, Y.; Farrell, G.; Wang, Q. & Wang, P. (2008). A low polarization sensitivity all-fiber wavelength measurement system, *IEEE Photon. Technol. Lett.*, 20, 1464–1466

Rajan, G.; Semenova, Y.; Wang, P. & Farrell, G. (2009). Temperature induced instabilities in macro-bend fiber based wavelength measurement systems, *IEEE J. Lightwave Technol.*, 27, 1355-1361

Ribeiro, A. B. L.; Ferreira, L. A.; Tsvetkov, M. & Santos, J. L. (1996). All fiber interrogation technique for fiber Bragg sensors using a biconical fiber filter, *Electron. Lett.*, 32, 382–383

Wang, Q.; Farrell, G. & Freir, T. (2005). Theoretical and experimental investigations of macro bend losses for standard single mode fibers, *Optics Express*, 13, 4476–4484

Wang, Q. & Farrell, G. (2006). Multimode fiber based edge filter for optical measurement and its design, Microwave and Optical Technology Letters, 48, 900–902

Wang, Q.; Farrell, G.; Freir, T.; Rajan, G. & Wang, P. (2006). Low cost wavelength measurement based on macrobending singlemode fiber, *Opt. Lett.*, 31, 1785–1787

Wang, Q.; Rajan, G.; Wang, P. & Farrell, G. (2007). Polarization dependence of bend loss in a standard singlemode fiber, *Optics Express*, 1, 4909–4920

Wang, P.; Farrell, G.; Wang, Q. & Rajan, G. (2007). An optimized macrobending fiber-based edge filter, *IEEE Photon. Technol. Lett.*, 19, 1136–1138

Wang, Q.; Farrell, G. & Yan, W. (2008). Investigation on singlemode- multimode-singlemode fiber structure, *IEEE J. Lightwave Technol.*, 26, 512–519

Zhao, Y & Liao, Y. (2004). Discrimination methods and demodulation techniques for fiber Bragg grating sensors, *Optics and Lasers in Engg.*, 41, 1–18

Wu, T. L. & Chang, H. C. (1995). Rigorous analysis of form birefringence of weakly fused fiber-optic couplers, *IEEE J. Lightwave Technol.*, 13, 687- 691

Wu, T. L. (1999). Vectorial analysis for polarization effect of wavelength flattened fiber-optic couplers, *Microwave and Optical Technology Letters*, 23, 12–16

Xu, M. G.; Geiger, H. & Dakin, J. P. (1996). Modeling and performance analysis of a fiber Bragg grating interrogation system using an acousto-optic tunable filter, *IEEE J. Lightwave Technol.*, 14, 391-396

The idea of the measurement system for quick test of thermal parameters of heat-insulating materials

Stanislaw Chudzik
Czestochowa University of Technology
Poland

1. Introduction

For the sake of climate and atmosphere conservation the emission of gasses must be bounded. A significant reduction of emission can be obtained by rational heat energy consumption, which is a substantial percent of the world's consumed energy. One of the ways is using in the building engineering and industry suitable insulating materials: foamed polystyrene, mineral wool, glass fiber, polyurethane foam, synthetic clothes, foam glass or cellular concrete. The existing methods of determination of material's thermal parameters are based mainly on stationary heat transfer conditions (Bayazitoğlu & Özişik, 1988; Bejan, 1993; Janna, 2000; Minkina & Chudzik 2004; Platunov, 1986). These methods allow determining in the experiment only a single thermophysical parameter of the tested material. They require the use of big and heavy measuring systems and a long period of time to conduct the measurement. Author do not know a commercial solution of portable measuring system which in relatively short time could assess fulfilling the requirements of insulating materials delivered to building site or leaving the factory from the point of view of thermal conductivity. Therefore, it seems to be crucial to work on design of such a measuring system. The research in this field concentrates, among other things, on possibility of application of artificial neural networks to solve the coefficient inverse problem of diffusion process (Alifanov et al., 1995; Beck, 1985). To determine the usability of network an analysis of its response for known values of thermal parameters is needed. It is necessary to generate input data for network training process using mathematical model of the tested sample of heat insulation material. The discrete model of a nonstationary heat flow process in a sample of material with hot a probe and an auxiliary thermometer based on a two-dimensional heat-conduction model was presented. The minimal acceptable dimensions of the material sample, the probe and the auxiliary thermometer were determined. Furthermore, the presence of the probe handle was considered in the heat transfer model. The next stage of the research is solving the inverse problem in which the thermal parameters will be estimated on the basis of recorded temperatures. Methods employing the classical algorithm of the mean square error minimization in the inverse problem of the heat conduction equation have an advantage of making it possible to take into consideration the arbitrary, varying boundary conditions that occur during the measurement (Aquino & Brigham, 2006; Chudzik & Minkina, 2001; Chudzik & Minkina, 2001a). Temperature changes of the input can be unbounded and they are taken into consideration in the calculations. The main basic disadvantage of it is the requirement of a portable computer or specially made measuring equipment based on powerful microprocessor (e.g. ARM core). It is conditioned by a great deal of iterative computations conducted in the inverse problem solution algorithm. To reduce the amount of computations and solution time of the inverse problem the application of artificial neural networks was proposed, which would determine a thermophysical parameter on the basis of the time characteristic recorded in the sample of tested material (Daponde & Grimaldi, 1998; Hasiloglu & Yilmaz, 2004; Mahmoud & Ben-Nakhi, 2003; Turias et al., 2005; Chudzik, 1999; Chudzik et al., 2001; Minkina & Chudzik 2004;).

2. Model of Heat Diffusion in the Sample of Insulating Material for Different Probe Designs

In the classic transient line heat source method (LHS), called also hot wire method or the probe method (Boer et al., 1980; Bouguerra et al., 2001; Gobbé et al., 2004; Kubicar & Bohac, 2000; Cintra & Santos, 2000; Tavman, 1999; Ventkaesan et al., 2001), a heated wire is initially inserted into a sample of insulating material at uniform and constant temperature, T_0. Constant power is then supplied to the line heater element starting at time t=0 and temperature adjacent to the line heat source is recorded with respect to time during a short heating interval. The principle of the method is based on the solution of the heat conduction equation in the cylindrical co-ordinate system:

$$\frac{\partial^2 T}{\partial r^2} + \frac{1}{r}\frac{\partial T}{\partial r} = \frac{1}{a}\frac{\partial T}{\partial t} \tag{1}$$

with the following initial and boundary conditions:

$$\begin{array}{lll} t = 0 & r > 0 & T = T_0 = 0 \\ t > 0 & r \to \infty & T = 0 \\ t > 0 & r \to 0 & -2\pi r k \dfrac{\partial T}{\partial r} = q' = constant \end{array} \tag{2}$$

where: a - thermal diffusivity (m^2/s), k - thermal conductivity (W/(m·K)), q'- linear power density (W/m). Several variations of the hot wire method are known. The theoretical model is the same as described by (1) and the basic difference among them lies in the temperature measurement procedure. This technique was standardized in 1978 by DIN 51046 Standard-Part 2. The approximate solution of (1) is given by the temperature rise T(t). The thermal conductivity is calculated according to the following equation (Boer et al., 1980):

$$k = \frac{-q'}{4\pi T(t)} E_i\left(\frac{-\rho c_p r^2}{4kt}\right) \tag{3}$$

where: ρ - material bulk density (kg/m^3), c_p - specific heat of the material at constant pressure (J/(kg·K)), r - distance between the hot wire and the thermocouple (m), t - time elapsed after start of heat release (s), $T(t)$ - temperature rise registered by the thermocouple related to the initial reference temperature (K), $E_i(-x)$ - exponential integral function given by:

$$E_i(-x) = -\int_x^{\infty} \frac{e^{-t}}{t} dt. \tag{4}$$

In the mathematical formulations given by (3), the following assumptions were made: the hot wire (that is the heat source) has negligible mass and heat capacity, it is infinitely thin and long, and the material whose thermal conductivity is determined is half-infinite (Boer et al., 1980; Cintra & Santos, 2000). In the case of measurement of thermoinsulation material properties using the hot probe (Al-Homoud, 2005; Ventkaesan et al., 2001), the approximate solutions of (1) are inaccurate. The conditions mentioned above can not be satisfied in general. Moreover, the heat capacity of the hot probe and the testing sample of material are comparable.

Our proposition of measurement system with hot probe consists in evaluating three thermal parameters simultaneously. It is sufficient to determine two of them, because they are related by equation:

$$a = \frac{k}{\rho \cdot c_p}. \tag{5}$$

The measurement system should record the temperature changes at the heat probe T_H and auxiliary thermometer T_X. The proposed distance between the hot probe and the auxiliary thermometer is 8 mm, the hot probe diameter is 2 mm and diameter of the auxiliary thermometer is 1 mm. A predesign of such thermal probe is presented in Fig. 1. The model of the heat diffusion in the sample of material with hot probe and auxiliary thermometer is given by (Quinn, 1983):

$$\rho c_p \frac{\partial T}{\partial t} - \nabla \cdot (\lambda \nabla T) = Q \tag{6}$$

where Q is the volume power density (W/m^3). The equation was implemented in the Matlab environment using the Partial Differential Equation Toolbox. Several alternative designs were considered and the results are obtained in the next subsections.

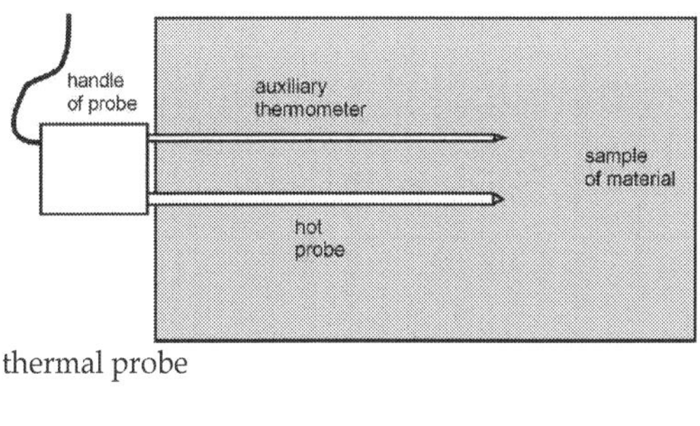

Fig. 1. Predesign of thermal probe

2.1 Heat diffusion in the sample of material for uniform hot probe and auxiliary thermometer references

To obtain the temperature field in the sample the finite element method (FEM) was applied (Alifanov et al., 1995; Aquino & Brigham, 2006; Augustin & Bernhard 1996; Beck, 1985; Jurkowski et al., 1997). In a two-dimensional XY co-ordinate model of the material sample, treated as a square plate, the simplified boundary condition $\partial T/\partial x=0$ was assumed. The values of thermal parameters were set to: a = 2.3·10^{-6} m^2/s, k = 0.04 W/(m·K) of sample of material ensure negligible influence of the boundary condition. Therefore, the modeled sample can be treated as infinitely extensive. The additional assumptions are as follows: the probe is made of copper with diameter Ø = 2 mm and thermal parameters a = 116·10^{-6} m^2/s, k = 401 W/(m·K), heating power is generated in the whole volume of the hot probe, the line power density of the heat source is $P_G(T_G=0)$ = P_0 = 9 W/m and depends on the instantaneous value of temperature increment of heater T_G built-in the probe. The heat power of the probe can be expressed as:

$$P_G(T_G) = \frac{U^2}{R_0 + \alpha T_G} \cdot \frac{1}{l} \tag{7}$$

where: a - average increment of heater resistance, U - supply voltage, R_0 - heater resistance in initial conditions, l - length of probe. Zero values of initial conditions were assumed. It means that the initial temperature of the sample, probe and thermometer equal to ambient temperature. The auxiliary thermometer placed in the tested probe can disturb the thermal field, therefore the temperature measured will not properly indicate real temperature in the sample. For that reason the real thermometer placed at a distance of 8 mm from probe was assumed. The modeled thermometer could be made of stainless steel with the following parameters: diameter Ø = 1 mm, a = 3.8·10^{-6} m^2/s, k = 15 W/(m·K). The thermal parameters of the sample, probe and thermometer were taken from (Grigoryev, 1991). A half section of the sample with the thermal probe and discrete mesh is presented in Fig. 2 for two cases: the probe without and with the auxiliary thermometer.

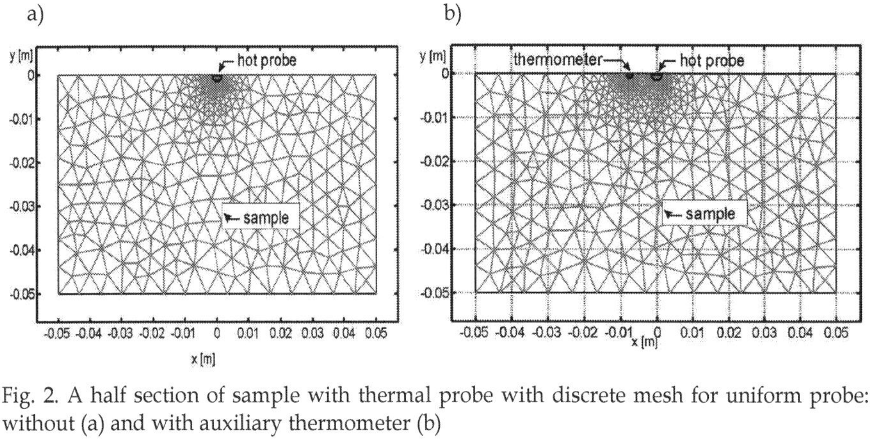

Fig. 2. A half section of sample with thermal probe with discrete mesh for uniform probe: without (a) and with auxiliary thermometer (b)

Fig. 3 presents the temperature profile of the sample after 100 s, where values are related to ambient temperature (difference). The comparison of these figures shows the influence of presence of the auxiliary thermometer on the temperature field in sample, particularly visible in the place of thermometer's location - enlarged part in Fig. 3b.

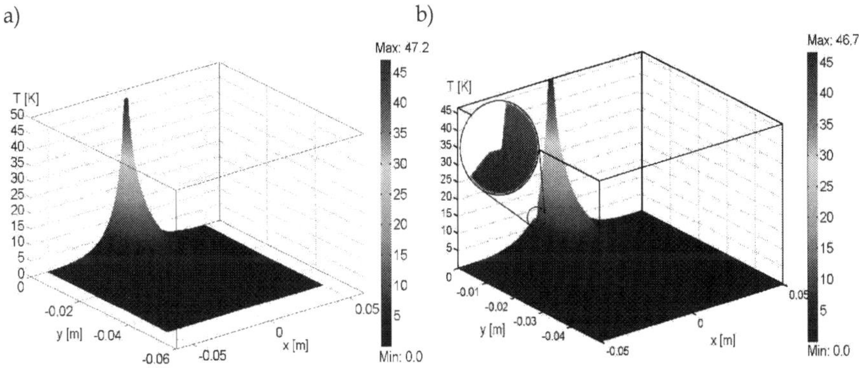

Fig. 3. The temperature profile of sample after 100 s for uniform probe: without (a) and with auxiliary thermometer (b)

Fig. 4 presents changes in temperature of the probe (curve 1), un-disturbing heat diffusion auxiliary thermometer (curve 2) and real auxiliary thermometer (curve 3) in the time period of 0-100 s after the start of sample heating. Curves 2 and 3 in Fig. 4 differ significantly similarly to the previous figure, hence the presence of real auxiliary thermometer must be taken into consideration in the mathematical model of heat diffusion in the sample of the tested material.

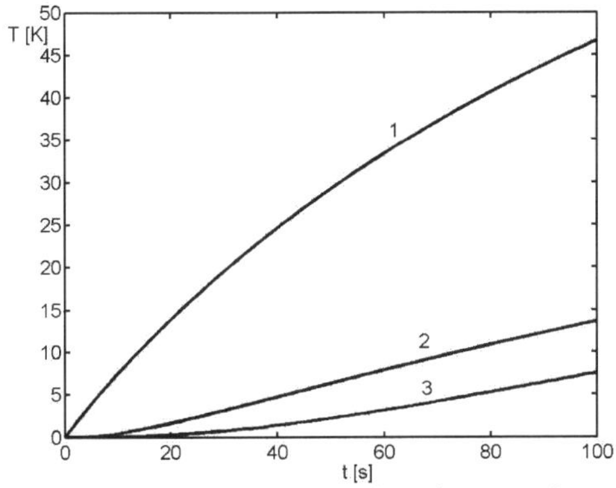

Fig. 4. Changes in temperature of probe (1), un-disturbing auxiliary thermometer (2) and real auxiliary thermometer (3) placed in a distance of 8 mm from the probe

2.2 Heat diffusion in the sample of material for nonuniform (multi-layer) hot probe and auxiliary thermometer

A real probe consists of three-layers: heater, filling material and shield. It must be checked how three-layer construction will have effect on temperature field. Assumed parameters of the modeled probe are: heater (copper) Ø = 1 mm, $a = 116 \cdot 10^{-6}$ m²/s, $k = 401$ W/(m·K), filling material (epoxide gum) $d = 0.5$ mm, $a = 7.8 \cdot 10^{-7}$ m²/s, $k = 1.3$ W/(m·K), shield (brass) $d = 0.5$ mm, $a = 34.2 \cdot 10^{-6}$ m²/s, $k = 111$ W/(m·K) where d is thickness of the layer. Other simulation conditions are similar to those mentioned in subsection 2.1. Again, a half section of the sample with the thermal probe and discrete mesh are presented in Fig. 5.

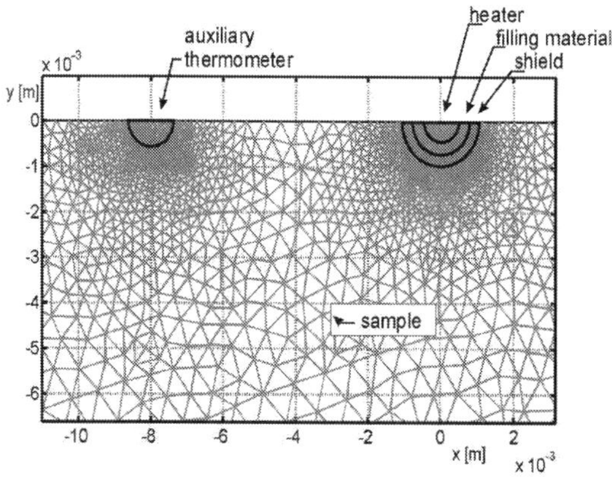

Fig. 5. A half section of sample with thermal probe and discrete mesh for multi-layer probe

Fig. 6 shows the temperature profile of the sample after 100 s (a) and top view of it with marked points 1-4 used to analyze the temperature difference (b).

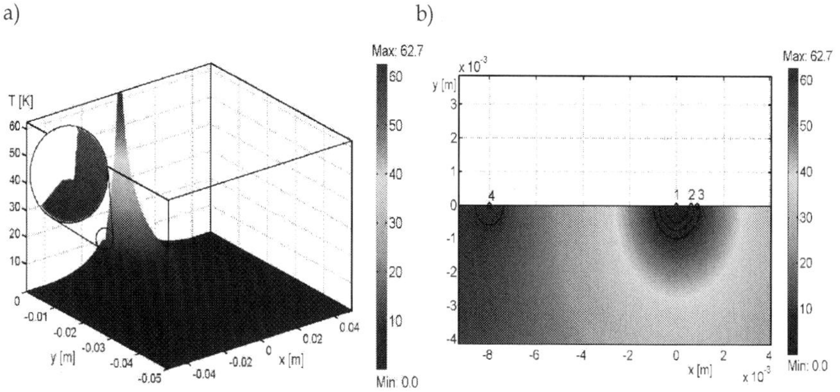

Fig. 6. The temperature profile of sample after 100 s (a) and top view with marked points 1-4 (b)

Fig. 7 presents changes in temperature in arbitrary chosen points 1-4. These numbers correspond to the following curves: heater (curve 1), filling material (curve 2), shield (curve 3) and real auxiliary thermometer (curve 4) in time period 0-100 s after the start of sample heating. In this case the curves 1, 2 and 3 dedicated to the three layers of the probe, overlap each other. It means that the assumption about nonuniform probe is not necessary. The higher temperature value (curve 1) for time instant 100 s in comparison to the value presented in Fig. 4 (also curve 1) is the result of less total heat capacity of nonuniform probe.

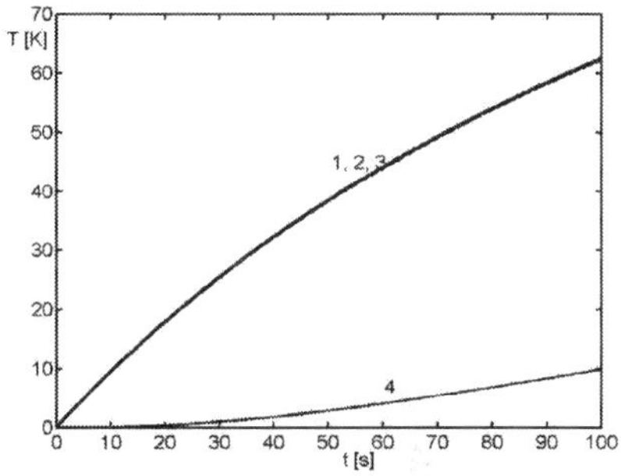

Fig. 7. Changes in temperature of probe parts: heater (1), filling material (2), shield (3) and real auxiliary thermometer (4) placed in a distance of 8 mm from the probe

3. Model of heat diffusion in the sample of insulating material for probe with handle

A typical method of temperature measurement of solid is the contact method, where the sensor is placed into material or has good thermal contact with material surface. Usually, simple sensors are used. They consist of long metal pipe working as a shield and active part assembled inside the pipe. One of the pipes is ended by a header or a handle with wires. Placement of the sensor into checked material causes some disturbance in the temperature field. The case of stationary temperature field measurement needs sufficiently long waiting for transient state to fade. In general, the dynamical error caused by the sensor presence must be taken into consideration. Usually contact temperature sensors have length much bigger than diameter and therefore the heat transfer along the sensor is neglected. This simplification can be erroneous in the case of small heat transfer coefficient of active sensor surface, because the temperature of the probe handle can have relevant impact on sensor measured temperature.

3.1. Model of heat diffusion in the sample of insulating material for probe with significant heat capacity handle

In Fig. 8 a model of uniform probe (copper) of a diameter of 2 mm and length of 10 cm long with handle (plastics) of a diameter of 5 mm and length of 2 cm is presented.

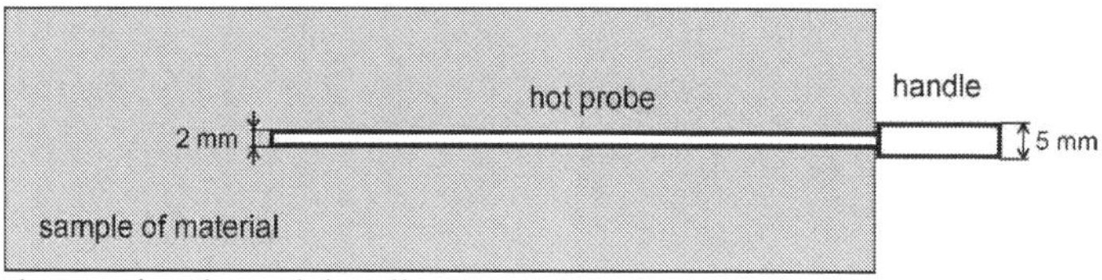

Fig. 8. Predesign of probe with handle

Fig. 9. presents a quarter of symmetrical model of the probe with handle in XYZ co-ordinates: discrete mesh (a), temperature field after 100 s (b). The considered sample of material is treated as a cubicoid which base dimensions are 10 x 10 cm and the height is 15 cm. The third kind of boundary condition (Fourier-Robin) on lateral surfaces of the sample and the probe handle was assumed. The typical value of heat transfer coefficient α=5 W/(m^2·K) for laminar, natural heat flow close to surface was taken from (Grigoryev, 1991).

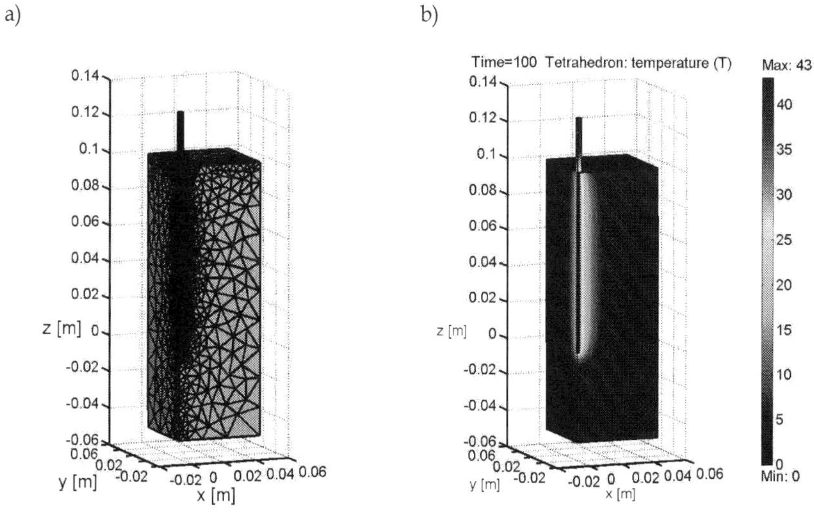

Fig. 9. Quarter of symmetrical model of probe with handle in XYZ co-ordinates: discrete mesh (a), temperature field after 100 s (b)

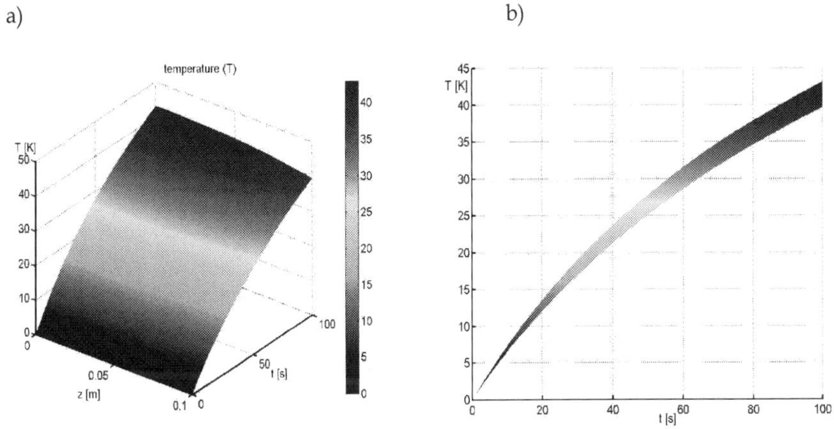

Fig. 10. Temperature changes along probe length (Z axis) after 100 s (a) and Z axis view presenting the changes in temperature along probe (b)

Zero values of initial conditions were assumed. The values of thermal parameters of the probe and the sample of material are the same as those considered in chapter 2. Fig. 10a presents the temperature profile along probe length (Z axis) after 100 s (a). For better visibility the Z axis view presenting the changes in temperature gradient along the probe was additionally showed in Fig. 10b.

It follows from Fig. 10 that change in temperature along probe after 100 s is about 3.5 K. For the probe made of copper this value is relatively big. The probe handle is made of plastic whose thermal conductivity is several times less than for metals. The amount of heat absorbed by handle is considerable in comparison to the heat absorbed by the sample of material. Taking into consideration the presence of the probe handle in model is difficult. The boundary conditions on handle surfaces depend on ambient conditions and generally are not predictable in real measurements. To eliminate this undesirable effect being an additional source of measurement error, the thermal probe handle compensation can be used.

3.2. Model of heat diffusion in the sample of insulating material for probe with temperature compensated handle

If the handle is temperature compensated, its presence in mathematical model can be neglected. Other simulation conditions are the same as in subsection 3.1. Fig. 11 presents a quarter of symmetrical model of probe without handle (equivalently to temperature compensated handle) in XYZ co-ordinates: discrete mesh (a), temperature field after 100 s (b).

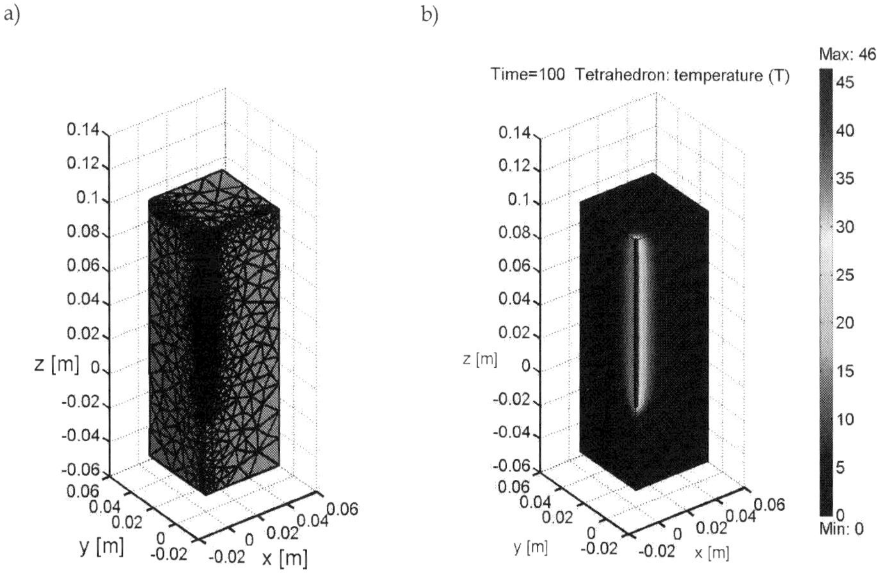

Fig. 11. Quarter of symmetrical model of probe with temperature compensated handle in XYZ co-ordinate: discrete mesh (a), temperature field after 100 s (b)

Fig. 12 presents the changes in temperature gradient along Z-axis after 100 s: probe with handle (curve 1) and probe without handle (curve 2) or temperature compensated handle. It is evident that temperature gradients after 100 s differ from each other significantly. The increase of average temperature in the middle of the probe length for probe with handle is 42.0 K and for the probe without handle is 46.4 K. It shows how presence of the handle influences the temperature field in the sample of material. The curve 2 is almost flat. It let us state that finite length of probe (z=0) and boundary condition on sample top surface (z=10) have small impact on the temperature field. Similar simulations for the probe with handle for another boundary condition on lateral and bottom surface of the sample were conducted. In this case, the sample is completely thermally insulated from surroundings with the exception of the top surface. No visible difference between corresponding curves 2 was observed hence they are not presented in the paper.

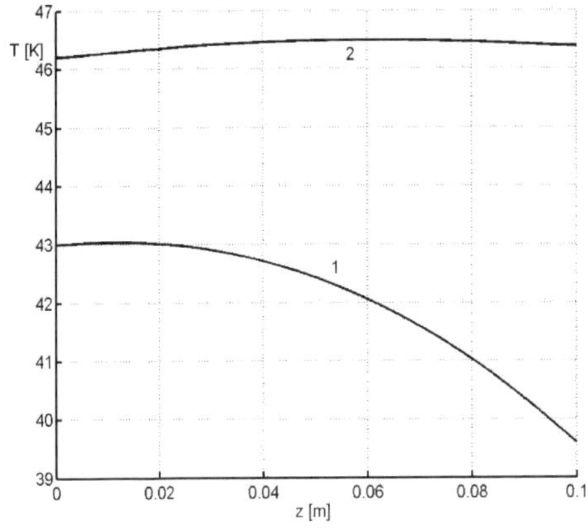

Fig. 12. Changes in temperature along Z-axis after 100 s: probe with handle (curve 1), probe without handle (curve 2)

4. Neural network in inverse problem solution

A neural network can learn a phenomenon model when analytic description is complicated or unknown. It is sufficient to present at the training stage the values of input quantities of the modeled phenomenon or system at the network inputs. There are assumed values of output quantities (responses) of the modeled phenomenon or system at the network outputs. The key problem is optimal selection of the network architecture. Different types of neural network have specific limitations in terms of functions they can represent. A lot of possible applications of neural networks have not been investigated yet or are still under research. In this work, we attempt to investigate the usability of a neural network in solving the coefficient inverse problem. In computer simulations simple network architecture was initially defined and next, its performance for model of heat conduction was tested. The idea of the coefficient inverse problem solution is presented in Fig. 1. The network determines the value of heat diffusivity a and heat conductivity k on the basis of the temperature responses recorded at the hot probe $T_H(t)$ in the symmetry axis and the auxiliary thermometer $T_X(t)$ in the sample and assuming repeatable boundary conditions. The network can be trained with temperature values (or its increments with respect to the initial condition) calculated for given values of a and k which have been taken from predicted ranges of its possible variations.

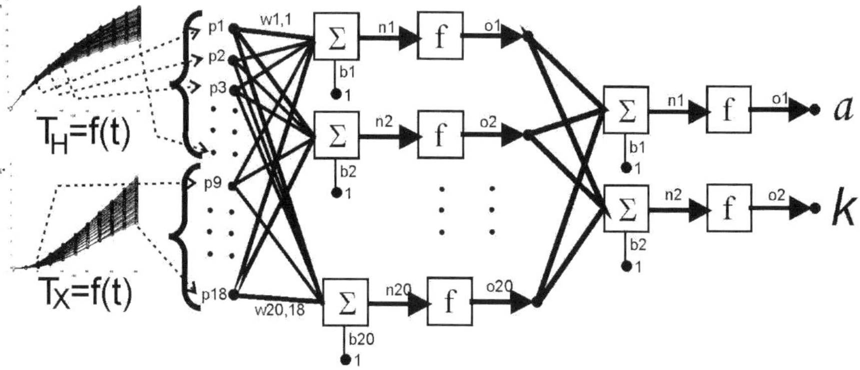

Fig. 1. Hypothetical architecture of the neural network with input and output signals

4.1. Discussion on the neural network architecture

Using the model and the FEM presented in part 1 of the paper there were generated training vectors of nine selected instantaneous values of the temperature responses of the hot probe $T_H(t)$ and the auxiliary thermometer $T_X(t)$ in the sample for 10x10 combinations of values of $a \in \langle 1.0 \div 3.0 \rangle \cdot 10^{-6}$ m²/s and $k \in \langle 3.0 \div 5.0 \rangle \cdot 10^{-2}$ W/(m·K) for time interval 100 s. The training input vectors of the instantaneous values of the temperature $T_H(t)$ and $T_X(t)$ are shown in Fig. 2 and Fig. 3.

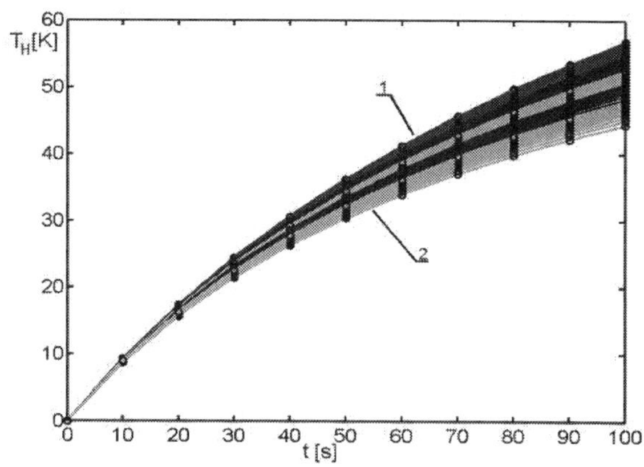

Fig. 2. Temperature changes in the symmetry axis of the hot probe $T_H(t)$: 1) a=1.0·10^{-6} m²/s, k=3.0·10^{-2} W/(m·K) and 2) a=3.0·10^{-6} m²/s, k= 5.0·10^{-2} W/(m·K)

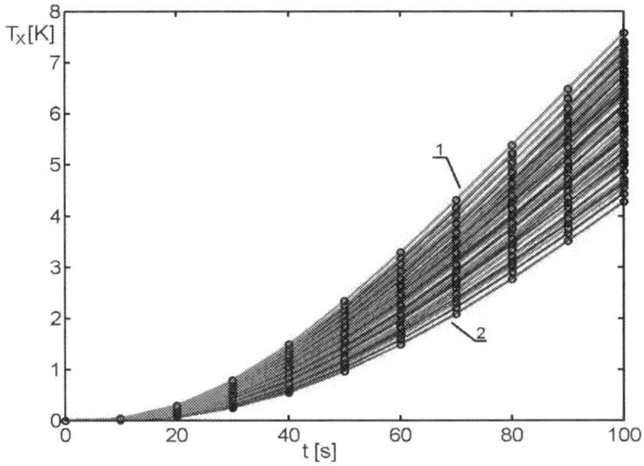

Fig. 3. Temperature changes of the auxiliary thermometer $T_X(t)$: 1) a=3.0·10^{-6} m^2/s, k=5.0·10^{-2} W/(m·K) and 2) a=1.0·10^{-6} m^2/s, k= 3.0·10^{-2} W/(m·K)

Several architectures of the neural network were tested (all with 2 linear neurons in output layer) (Hagan et al., 1996; Caudill & Butler 1992) :

- two-layer classical nonlinear network with 10 neurons in input layer for 20, 50 and 1000 training epochs,
- two-layer classical nonlinear network with 20 neurons in input layer for 20, 50 and 1000 training epochs,
- three-layer classical nonlinear network with 20 neurons in input layer and 10 neurons in hidden layer for 25 and 1000 training epochs,
- radial basis functions RBF,
- radial basis function RBF with given error goal,
- generalized regression GRNN.

The neural networks were trained by presenting successive training vectors including the values of T_H and T_X at their inputs (input vectors) and the corresponding values of a and k coefficients at their outputs (target vectors).

In the case of classical radial basis function neural network, the results confirm its possibility to appreciatively solve the inverse problem. For some learning parameters the output error is negligible for training and testing data. However, network structure consisting of 100 RBF neurons is relatively big. In the case of GRNN the "overfitting effect" was occurring. The network answered with small error for training vector, but for intermediate values, that are included in testing vector, the output error was very big. Reduction of the number of the neurons or decreasing the size of training vector can remove this disadvantage. The Matlab environment has in this case limited possibilities of parameter selection. Furthermore, the input training vector preprocessing or output vector postprocessing can also be helpful. This extra processing of this data increases the algorithm complexity. A better solution is application of the RBF network with given error goal which automatically choose the number of neurons to draw output error with error goal. Such a solution facilitates looking for the optimal network structure because the amount of neurons is automatically selected. Better results were obtained for classical nonlinear network with hyperbolic tangent transfer function in input and hidden layers. In the case of hidden layer, it is sufficient to use the linear activity function. Three-layer network reached good performance after 25 epochs. Thanks to its flexibility the overfitting did not occur. Taking 1000 epochs, the output error was very small, both for learning and testing vectors. Performance of two-layer network was also investigated. The output error in this case was somewhat larger than for three-layer network but it can be compensated by longer learning period (more epochs).

4.2. Results of simulation for "optimal" neural network search

The best of considered architectures of the neural network was the classical nonlinear feedforward two-layer neural network: 20 neurons with hyperbolic-tangent transfer function in input layer and two neurons with linear transfer function in output layer. The network has 18 inputs and 2 outputs. The vector of nine instantaneous values of the temperature of the heating probe T_H and the vector of nine instantaneous values of the temperature of the auxiliary thermometer T_X are loaded into 18 inputs of the neural network. The network achieved satisfactory results already after about 50 training epochs. An example of the network training process for the traditional error back propagation algorithm is presented in Fig. 4.

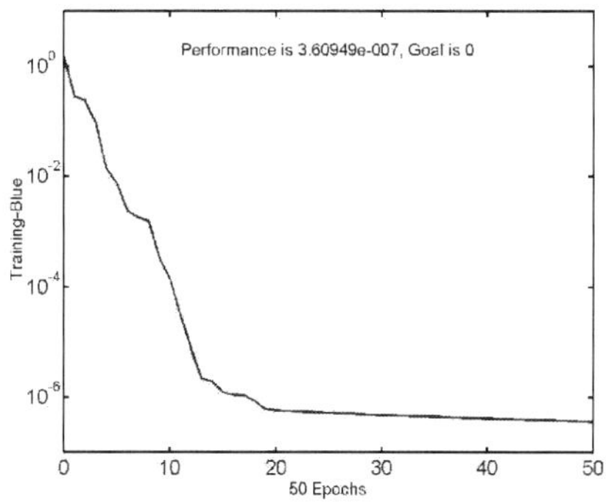

Fig. 4. Error of network learning during 50 epochs

The network outputs were compared to the values of heat diffusivity and heat conductivity coefficients given in the training stage. The relative errors of the network response are presented in Fig. 5.

To verify whether the network response is correct for intermediate values of a and k from the ranges defined above, the responses were simulated for 100 values of heat diffusivity coefficient a from the range $a \in \langle 1.0;3.0 \rangle \cdot 10^{-6}$ m^2/s and 100 values of heat conductivity coefficient k from the range and $k \in \langle 3.0;5.0 \rangle \cdot 10^{-2}$ m^2/s. Consequently, there 10000 different testing vectors were generated. Fig. 7 and Fig. 8 show the relative error of the network response versus target value of the coefficients a and k.

The performance of a trained network was additionally measured using regression analysis between the network response and the corresponding targets. In the posttraining analysis the "Linear regression method" implemented in Matlab was used for network validation.

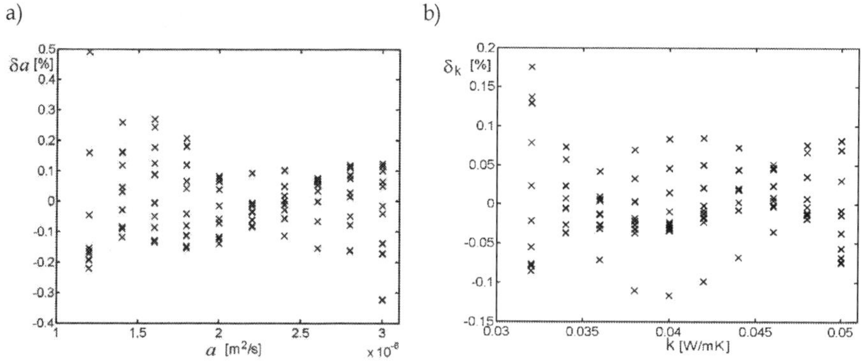

Fig. 5. Relative error of the network response for thermal diffusivity *a* (a) and for thermal conductivity *k* (b) for the training stage

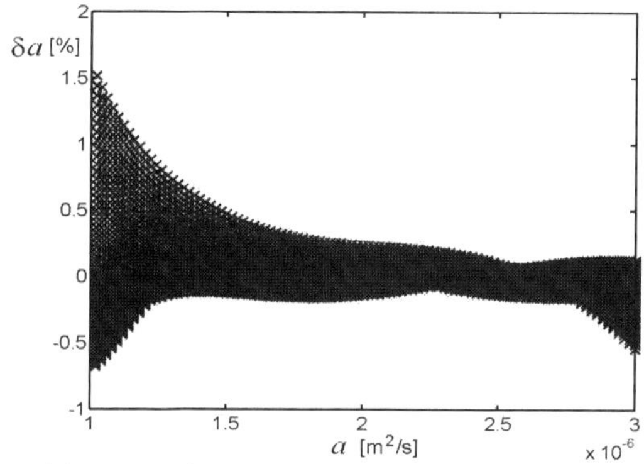

Fig. 7. Relative error of the network response for thermal diffusivity *a* for the testing stage

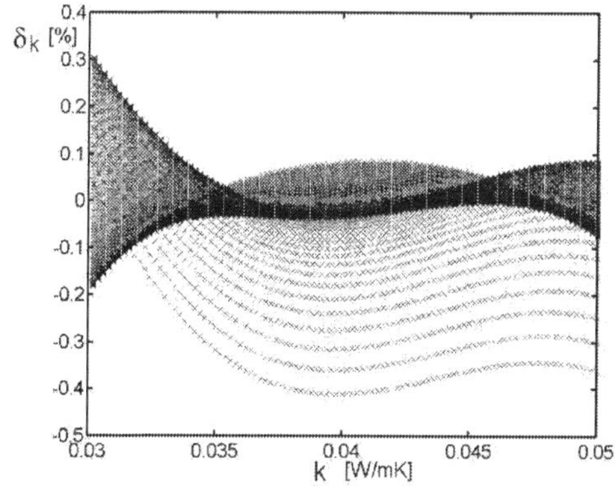

Fig. 8. Relative error of the network response for thermal conductivity *k* for the testing stage

In Fig. 9 and Fig. 10 the perfect fit (outputs exactly equal to targets) can be seen, the slope is almost 1, and the y-intercept is 0.

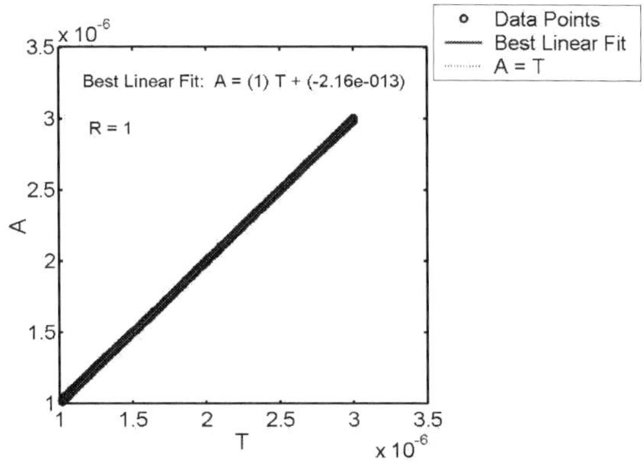

Fig. 9. Linear regression method matching for heat diffusivity: T - given training output value of a, A - network answer

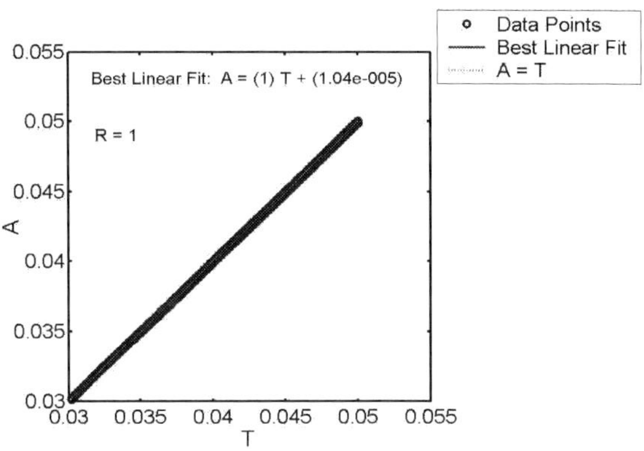

Fig. 10. Linear regression method matching for heat conductivity: T - given training output value of k, A - network answer

Furthermore, the correlation coefficient R that indicates how well the variation in the output is explained by the targets, is equal to 1. Thus, there is perfect correlation between the targets and the outputs. It means that approximation quality of the neural network is very good.

As mentioned earlier, the chosen network architecture was classical nonlinear two-layer neural network. On this stage of the research this architecture cannot be objectively treated as optimal. Generalization ability is one of many conditions a network must satisfy only. A very important issue is determination of sensitivity of the neural network to input quantities disturbance. Good ability of generalization of the chosen network going hand in hand with its small sensitivity must be proved. If it fails, then another network candidate for "optimal network architecture" must be selected and checked.

5. Sensitivity analysis of neural network on input quantities disturbance

Sensitivity analysis of the neural network on existing in real measurement uncertainties is a very essential stage of method validation. The designed instrumentation is dedicated to immediate measurements, hence uncertainty on a level of a few percents is sufficient. In this study, there was assumed, that the following input quantities had influence on the output quantities a and k:

- temperature of the hot probe $T_H(t)$,
- temperature of the auxiliary thermometer $T_X(t)$,
- heat power P_G supplying the hot probe,
- distance r between the hot probe and the auxiliary thermometer.

5.1. Application of Monte Carlo technique in sensitivity analysis

In Ref. (Joint Committee for Guides in Metrology, 2004), treated as supplement to Guide to the Expression of Uncertainty in Measurement (BIPM et al., 1995) an interesting procedure is proposed. It gives recommendation how the uncertainty could be evaluated if the conditions for „law of propagation of uncertainty" are not fulfilled, because of the complexity of the model, for example (Joint Committee for Guides in Metrology, 2004).
The procedure applies to evaluation of 95% coverage interval for the output quantity value.
The described procedure consists of the following stages (BIPM et al., 1995):
a) define the output quantity, the quantity required to be measured,
b) decide the input quantities upon which the output quantity depends,
c) develop a model relating the output quantity to these input quantities,
d) on the basis of available knowledge assign probability density functions to the values of the input quantities,
e) propagate the probability density functions for the values of the input quantities through the model to obtain the probability density function for the output quantity value,
f) obtain from the probability density function for the output quantity value:
 1) its expectation, taken as the estimate of the output quantity value;
 2) its standard deviation, taken as the standard uncertainty associated with the estimate of the output quantity value,
 3) the coverage interval containing the unknown output quantity value with a specified probability.

Generation of input quantities for 10^6 trials (this number is strongly recommended by (Joint Committee for Guides in Metrology, 2004)) and model developing for one pair given thermal parameters were taken over 6 days using modern PC computer. Hence, the analysis was constrained to the following five combinations of values treated as "true" values:

1. a=2.5·10^{-6} m^2/s and k= 4.0·10^{-2} W/(m·K),
2. a=1.5·10^{-6} m^2/s and k= 3.3·10^{-2} W/(m·K),
3. a=3.5·10^{-6} m^2/s and k= 4.7·10^{-2} W/(m·K),
4. a=3.5·10^{-6} m^2/s and k= 3.3·10^{-2} W/(m·K),
5. a=1.5·10^{-6} m^2/s and k= 4.7·10^{-2} W/(m·K).

Uniform symmetric distribution of probability of input quantities was assumed with the following half widths: $\Delta T_H(t)$=0.1 K, $\Delta T_X(t)$=0.05 K, ΔP_G=1 mW/m, Δr=0.1 mm. To obtain distribution function of network output quantities a and k the Monte Carlo simulation technique was applied. For each pair of a and k the output value of the network was computed and estimates of distribution function $G(a)$ and $G(k)$ were carried out. According to f) point of the above mentioned procedure (underlined text) the recommended parameters: expectation, standard deviation and 95% coverage interval were obtained and presented in Table I.

No.	Heat diffusivity a m²/s			Heat conductivity k W/(m·K)		
	Expectation	Standard deviation	95 % coverage interval	Expectation	Standard deviation	95 % coverage interval
1	$2.53 \cdot 10^{-6}$	$2.74 \cdot 10^{-8}$	$[2.48 \cdot 10^{-6}; 2.58 \cdot 10^{-6}]$	$4.05 \cdot 10^{-2}$	$4.29 \cdot 10^{-4}$	$[3.97 \cdot 10^{-2}; 4.13 \cdot 10^{-2}]$
2	$1.49 \cdot 10^{-6}$	$2.30 \cdot 10^{-8}$	$[1.45 \cdot 10^{-6}; 1.54 \cdot 10^{-6}]$	$3.29 \cdot 10^{-2}$	$4.16 \cdot 10^{-4}$	$[3.21 \cdot 10^{-2}; 3.37 \cdot 10^{-2}]$
3	$3.41 \cdot 10^{-6}$	$2.95 \cdot 10^{-8}$	$[3.35 \cdot 10^{-6}; 3.45 \cdot 10^{-6}]$	$4.56 \cdot 10^{-2}$	$4.05 \cdot 10^{-4}$	$[4.49 \cdot 10^{-2}; 4.64 \cdot 10^{-2}]$
4	$3.33 \cdot 10^{-6}$	$3.02 \cdot 10^{-8}$	$[3.27 \cdot 10^{-6}; 3.39 \cdot 10^{-6}]$	$3.27 \cdot 10^{-2}$	$4.01 \cdot 10^{-4}$	$[3.19 \cdot 10^{-2}; 3.34 \cdot 10^{-2}]$
5	$1.49 \cdot 10^{-6}$	$2.24 \cdot 10^{-8}$	$[1.44 \cdot 10^{-6}; 1.53 \cdot 10^{-6}]$	$4.70 \cdot 10^{-2}$	$4.65 \cdot 10^{-4}$	$[4.61 \cdot 10^{-2}; 4.79 \cdot 10^{-2}]$

Table 1. Estimated parameters of probability for five combination of parameters a and k and for two layer classical neural network

This table is valid for the previously tested classical nonlinear two-layer neural network: 20 neurons with hyperbolic-tangent transfer function in input layer and two neurons with linear transfer function in output layer. Analysis of data from Table I let us state that the relative extended uncertainty (95% coverage interval) of estimated thermal parameters a and k for all five cases did not exceed 3% and 2.5%, respectively.

Similar analysis was conducted for other network architectures and configurations mentioned and tested in the chapter II. The best results were obtained for RBF network with error goal. They are presented in Table II.

No.	Heat diffusivity a m²/s			Heat conductivity k W /(m·K)		
	Expectation	Standard deviation	95 % coverage interval	Expectation	Standard deviation	95 % coverage interval
1	$2.50 \cdot 10^{-6}$	$2.73 \cdot 10^{-8}$	$[2.45 \cdot 10^{-6}; 2.55 \cdot 10^{-6}]$	$4.00 \cdot 10^{-2}$	$2.32 \cdot 10^{-4}$	$[3.96 \cdot 10^{-2}; 4.04 \cdot 10^{-2}]$
2	$1.50 \cdot 10^{-6}$	$1.56 \cdot 10^{-8}$	$[1.45 \cdot 10^{-6}; 1.53 \cdot 10^{-6}]$	$3.30 \cdot 10^{-2}$	$2.07 \cdot 10^{-4}$	$[3.26 \cdot 10^{-2}; 3.33 \cdot 10^{-2}]$
3	$3.49 \cdot 10^{-6}$	$3.75 \cdot 10^{-8}$	$[3.43 \cdot 10^{-6}; 3.56 \cdot 10^{-6}]$	$4.70 \cdot 10^{-2}$	$2.56 \cdot 10^{-4}$	$[4.66 \cdot 10^{-2}; 4.75 \cdot 10^{-2}]$
4	$3.48 \cdot 10^{-6}$	$4.98 \cdot 10^{-8}$	$[3.40 \cdot 10^{-6}; 3.57 \cdot 10^{-6}]$	$3.30 \cdot 10^{-2}$	$2.07 \cdot 10^{-4}$	$[3.26 \cdot 10^{-2}; 3.33 \cdot 10^{-2}]$
5	$1.50 \cdot 10^{-6}$	$1.22 \cdot 10^{-8}$	$[1.48 \cdot 10^{-6}; 1.52 \cdot 10^{-6}]$	$4.70 \cdot 10^{-2}$	$2.46 \cdot 10^{-4}$	$[4.66 \cdot 10^{-2}; 4.74 \cdot 10^{-2}]$

Table 2. Estimated Parameters of Probability for Five Combination of Parameters a and k and for RBF neural Network

Analysis of data from Table II let us state that the relative extended uncertainty (95% coverage interval) of estimated thermal parameters a and k for all five cases did not exceed 3% and 1.5%, respectively.

6. Conclusion

The aim of the presented analysis was checking how several details of the considered model of heat diffusion affects the temperature field. The recommendations for the next research stage are:

- assumed dimensions of measuring stand are accurate, hot probe can be treated as uniform (fewer elements of discrete mesh),
- presence of auxiliary thermometer and probe handle must be considered in mathematical model,
- the sample dimensions checked in simulations let treat a real sample as infinitely widespread.

It allows for conducting measurement with no specially prepared sample assuming that the sample dimensions are greater or equal than considered in the paper, i.e. 10 x 10 cm (base) x 15 cm (height). This is the great advantage of proposed method.

The usability of an artificial neural network (ANN) to estimate the coefficients of inverse heat conduction problem for solid will be presented. The network should determine the value of effective thermal conductivity and effective thermal diffusivity on the basis of temperature response of hot probe and auxiliary thermometer. In developing of the ANN model, several configurations of ANN architecture will be checked.

The results obtained in simulations show that is possible to determine the parameters of insulating material using trained neural networks for repeatable input function of the probe and measurement conditions. Such a solution could considerably reduce the cost of the measuring system.

The recommended network architecture for further study is the classical nonlinear two-layer neural network: 20 neurons with hyperbolic-tangent transfer function in input layer and two neurons with linear transfer function in output layer. 50 epochs should be sufficient to achieve a good result of the network training. The relative extended uncertainty (95% coverage interval) of estimated thermal parameters a and k will not exceed 3% and 2.5%, respectively. This network was chosen as a compromise between generalization ability, architecture simplicity, small output error and small sensitivity on input signal disturbances.

The main aim of the research is to determine the three basic thermal parameters on a single portable measuring system in a relatively short time and not only in laboratory (fully controlled) conditions.

The future work will be experimental verification of the proposed method on real samples of insulating material taking into consideration sample size, network architecture, training recommendation and the use of a real probe consisting of a hot probe and an auxiliary thermometer with temperature compensated handle.

Acknowledgment

This work was supported by Ministry of Science and Higher Education of Poland under research grant No. N N505 3129 33 realized in years 2007-2009.

7. References

Boer J., Butter J., Grosskopf B. & Jeschke P. (1980), Hot wire technique for determining high thermal conductivities, *Refractories Journal*, vol. 55, 1980.

Quinn T.J. (1983) *Temperature.* Academic Press Ltd., London, 1983.

Beck J.V. (1985). *Inverse Heat Conduction.* A Wiley-Intersc. Publ., 1985.

Platunov C.E. (1986). *Thermophysical Measurements and Instrumentation* (title in Russian is Теплофизические измерения и приборы), Изд. Машиностроение, San Petersburg, 1986.

Bayazitoğlu Y. & Özişik M.N. (1988) *Elements of Heat Transfer.* McGraw-Hill Book Company, New York, 1988.

Grigoryev I.C. (1991). *Physical quantities.* Handbook (title in Russian is Физические величины, Справочник). Энергоатомиздат, Moscow, 1991.

Caudill M. & Butler C. (1992). *Understanding Neural Networks: Computer Explorations*, Vols. 1 and 2, Cambridge, MA The MIT Press, 1992.

Bejan A., (1993). *Heat Transfer.* John Wiley & Sons, New York, 1993.

BIPM, IEC, IFCC, ISO, IUPAC, IUPAP & OIML, (1995). *Guide to the Expression of Uncertainty in Measurement.* 1995.

Alifanov O., Artyukhin E. & Rumyantsev S. (1995). *Extreme Methods for solving Ill-Posed Problems with Applications to Inverse Heat Transfer Problems*, New York, 1995.

Augustin S. & Bernhard F. (1996). Numerical calculation of static and dynamic parameters of industrial temperature sensors, *Measurement*, vol. 17, pp. 217-228, 1996.

Hagan M.T., Demuth H.B. & Beale M.H. (1996) *Neural Network Design*, Boston, MA PWS Publishing, 1996.

Jurkowski J., Jarny Y. & Delanuay Y. (1997). Estimation of thermal conductivity of thermoplastics under moulding conditions: an apparatus and an inverse algorithm, *Int. J. Heat Mass Transfer*, vol. 17, pp. 4169-4181, 1997.

Daponde P. & Grimaldi D. (1998). Artifical neural networks in measurements, *Measurement*, vol. 23, pp. 93-115, 1998.

Chudzik S. (1999). Determination of thermal diffusivity of heat-insulating material using neaural networks (title in Polish is Okreslenie wspolczynnika dyfuzyjnosci cieplnej materialow termoizolacyjnych z wykorzystaniem sieci neuronowych) *Proc. of 3th Scientific-Technical Conference on Methods and Computer Systems in Automatics and Electrical Engineering*, Poraj (Poland), pp. 126-128, September 1999.

Tavman I.H. & Tavman S. (1999). Measurement of thermal conductivity of dairy products, *Journal of Food Engineering*, vol. 41, pp. 109-114, 1999.

Janna W.S. (2000) *Engineering Heat Transfer*. CRS Press, Washington DC, 2000.

Kubicar L. & Bohac V. (2000) A Step-wise method for measuring thermophysical parameters of materials, *Meas. Sci Technol.*, vol. 11, pp. 252-258, 2000.

Sylos Cintra J. & Santos W. (2000) Numerical analysis of sample dimensions in hot wire thermal conductivity measurements, *Journal of the European Ceramic Society*, vol. 20, pp. 1871-1875, 2000.

Bouguerra A., Ait-Mokhtar O., Amiri M. & Diop B. (2001) Measurement of thermal conductivity, thermal diffusivity and heat capacity of highly porous building materials using transient plane source technique, *Int. Comm Heat Mass Transfer*, vol. 28, pp. 1065-1078, 2001.

Chudzik S. & Minkina W. (2001). Quick quality inspection of thermal parameters of heat-insulating materials, *International Conference Material Testing and Research*, Nuremberg ,Germany, pp. 341-347, May 2001.

Chudzik S. & Minkina W. (2001a). Dynamic method to determine thermal parameters of heat-insulating materials, *TEMPMEKO'2001, 8th International Symposium on Temperature and Thermal Measurements in Industry and Science IMEKO TC 12*, vol. II, pp. 963-968, Berlin, Germany, June 2001.

Chudzik S., Grys S. & Bąbka R. (2001). Identification of thermal parameters of heat-insulating materials using neural networks" (title in Polish is Identyfikacja parametrow cieplnych materialow termoizolacyjnych z wykorzystaniem sieci neuronowych), *Proc. of 4th Scientific-Technical Conference on Methods and Computer Systems in Automatics and Electrical Engineering*, Poraj (Poland), pp.46-48, September 2001.

Ventkaesan G., Guang-Pu Jin, (2001). Measurement of thermophysical properties of polyurethane foam insulation during transient method, *Int. J. Therm. Sci.*, vol. 40, pp. 133-144, 2001.

Mahmoud M. & Ben-Nakhi A. (2003). Architecture and performance of neural networks for efficient A/C control in buildings," *Energy Conversion and Management*, vol. 44, pp. 3207–3226, 2003.

Hasiloglu A., Yilmaz M., Comakli O. & Ekmekci I. (2004). Adaptive neuro-fuzzy modeling of transient heat transfers in circular duct air flow, *International Journal of Thermal Sciences*, vol. 43, pp. 1075–1090, 2004.

Joint Committee for Guides in Metrology (2004). *Guide to the Expression of Uncertainty in Measurement. Supplement 1. Numerical Methods for the Propagation of Distributions.*, 2004.

Gobbé C., Iserna S. & Ladevie B. (2004). Hot strip method: application to thermal characterisation of orthotropic media, *International Journal of Thermal Sciences*, vol. 43, pp. 951–958, 2004.

Minkina W. & Chudzik S. (2004). *Measurement of Thermal Parameters of Thermoinsulating Materials - Instrumentation and Methods* (title in Polish is *Pomiary parametrow cieplnych materialow termoizolacyjnych - przyrzady i metody*). Publishing House of Czestochowa University of Technology, ISBN 83-7193-216-2, Czestochowa (Poland), 2004.

Al-Homoud M. (2005). Performance characteristics and practical applications of common building thermal insulation materials, *Building and Environment*, vol. 40, pp. 353–366, 2005.

Turias I., Gutie´rrez J. & Galindo P. (2005). Modelling the efective thermal conductivity of an unidirectional composite by the use of artificial neural networks, *Composites Science and Technology*, vol. 65, pp. 609–619, 2005.

Aquino W. & Brigham J. (2006), Self-learning finite elements for inverse estimation of thermal constitutive models, *International Journal of Heat and Mass Transfer*, vol. 49, pp. 2466–2478, 2006.

New Technologies For Measurement Systems Distributed On A Wide Area

Giovanni Bucci, Fabrizio Ciancetta and Edoardo Fiorucci
Università dell'Aquila (Dip. Ing. Elettrica e dell'Informazione)
ITALY

1. Introduction to Distributed Measurement Systems

In different technological and information fields there is the need for multipoint measurement systems, to monitor a process evolution and to take proper controlling actions; several applications require for a *Distributed Measurement System* (DMS) able to measure the same or different parameters at different points.

Traditional test and measurement applications have been based on the use of centralized control and data management; usual implementations involve a central controller that handles instruments and sensors, polls for any results and processes the acquired data.

In these systems, the essential determination of the system's behaviour resides just in the central controller. The primary advantage of centralized systems is their simplicity: because all data is concentrated in one place, these systems are easily managed and have no problems of data consistency or coherence.

However, in some cases the physical area over which carry out the measurements is so wide that the implementation of a single measurement system will be unacceptable; main reasons are cost, reliability, and the distance between the measurement points. The measurement system must be split in a number of measurement devices, able to carry out each single measurement, to process the acquired raw data and to transmit the information to the final user.

Today, with the advent of dedicated microcontrollers and digital signal processors, complex data processing and transmission can be performed in a fast and easy way. These new technologies permitted the implementation of DMSs that addresses these issues: instead of having one single powerful system, distributed systems employ multiple systems communicating to each other via a common network.

DMSs usually refer to systems comprising one or more controllers each with one-to-one connections to sensors, instruments and possible actuators. This approach is always based on a number of smart measuring systems, where the nodes determine the system behaviour; the control protocol must therefore support each node internally, managing the application details occurring at that node. In addition, nodes can communicate directly with each other or with groups of other nodes, without any restriction to one-to-one communication links. The control protocol must support the transmission of synchronization messages between nodes to produce the correct overall system behaviour. Synchronization includes not only the timing of measurements, but the overall progress of the application from one sequence of events to another.

In a DMS the management of data must be more structured than in traditional ones: in centralized systems many data management tasks, such as identifying the source and time of a measurement, are based on the properties of point-to-point communication links; in distributed systems using multicast, other techniques must be used for linking the various pieces of information in the system.

This chapter discusses about the new technologies that have been proposed, in recent years, for applications involving measurement systems distributed on a wide area.

1.1 Basic of DMS

A DMS can be seen as a collection of autonomous measurement systems linked by a network and equipped with distributed system software (Coulouris et al., 1994). The system employs intelligent controllers to perform message editing, data collection, dialogue with remote computer clients, some security functions, and message packaging.

The DMS software enables the measurement systems to coordinate their activities and to share system resources. A well-developed distributed system software provides the illusion of a single and integrated environment, although it is actually implemented by multiple measurement systems positioned in different places. In other words, is the software that gives a distribution transparency to the systems. Given the independent and distributed nature of these systems, it is important to underline the importance of having distributed software to provide a common and transparent view of the systems (Tari & Bukhre, 2001).

1.2 Classification of DMSs

Most of the architectures of DMSs include local or remote measurement systems, linked by communication connections with a central unit that provides for the system management and data acquisition. A centrally controlled distributed system is viewed as a collection of peripheral small measurement system, which might be capable to completely execute the measurement task and process the obtained results. Each of them is subordinate to a higher level central unit in the overall system structure. The main interconnection topologies are: the star, hierarchical or ring structure.

As shown in Fig. 1, each remote measurement system in the star structure is connected to the central unit via a network front end (e.g. a modem).

Fig. 2(a) illustrates a DMS interconnected in a ring structure, consisting of autonomous measurement systems linked in a peer-to-peer fashion.

Fig. 2(b) depicts a DMS with hierarchical structure. One or more locations have their own pre-processing unit, each of which performs the data processing and allows the communication of the measurement system with the central unit; the pre-processing unit periodically sends the required summary data to the central unit.

The idea of distributed systems represents a structure ranging from separate, geographically dispersed applications cooperating with each other, to a single application formed by relatively independent, stand-alone components. Another classification for the DMSs refers to four types of structural configurations.

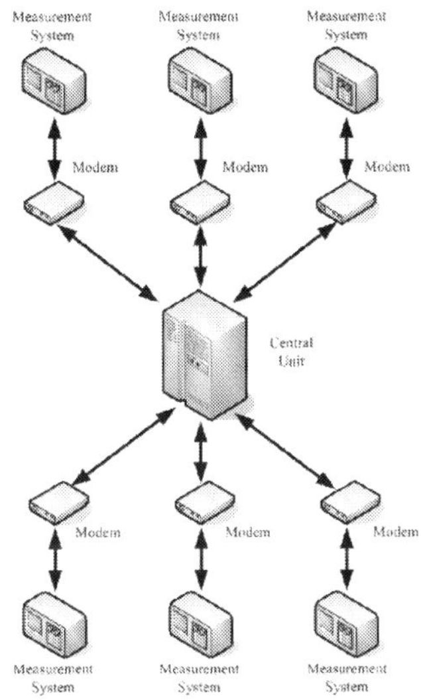

Fig. 1. Star structure

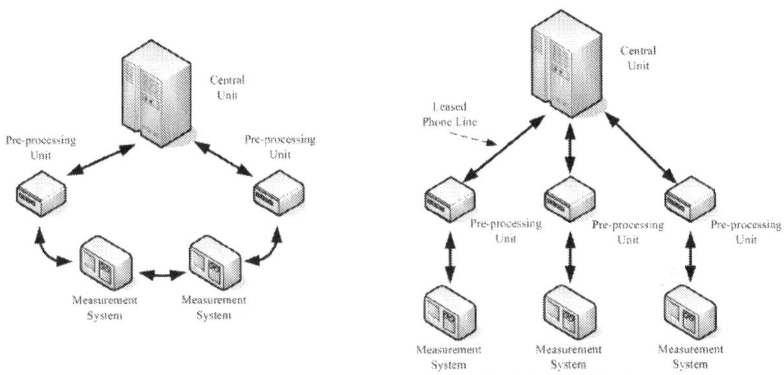

(a) Ring structure (b) Hierarchical structure

Fig. 2. Architectures of distributed measurement systems.

The first one, which is shown in Fig. 3(a), is a central unit connected to measurement systems with certain functions or applications.

Another type is a measurement system connected to mid-range units, which, in turn, are connected to a central unit; as depicted in Fig. 3(b), all of them are organized in a hierarchical structure. The mid-range units deliver data to the central unit which handles the central level processing.

Fig. 3(c) illustrates a distributed system arranged as a connection of peers which have different details of interconnection and dependency; however, it is no evident a central control unit in this system.

The last type, shown in Fig. 3(d), is a collection of peer hierarchies. The central unit, or each of the mid-range units, exhibits a hierarchical configuration and is connected to a collection of measurement systems.

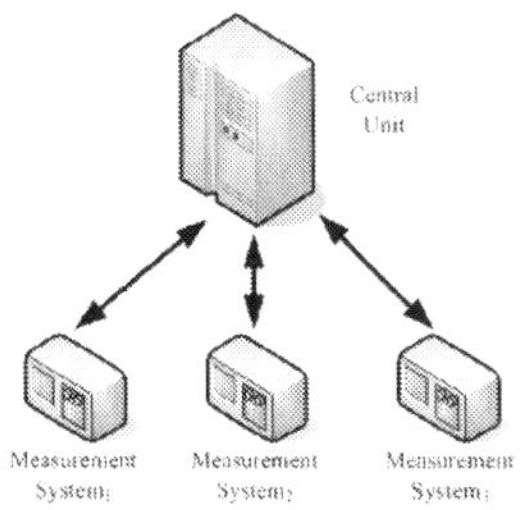

(a) Central unit and interconnected measurement systems

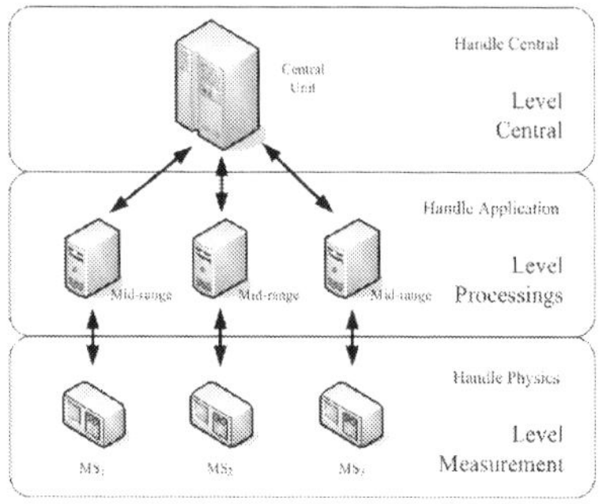

(b) Hierarchical structure and interconnection configuration

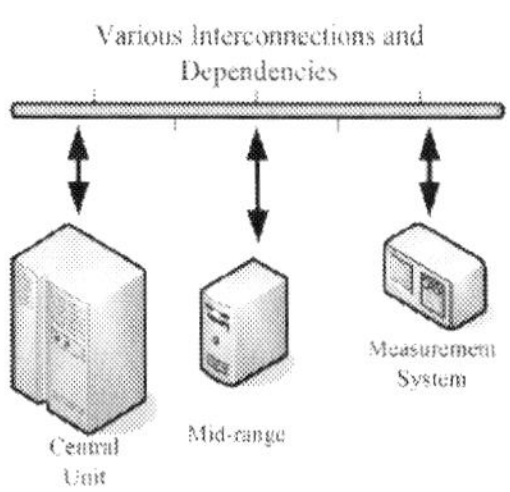

(c) Connection of peers structure and interconnection configuration

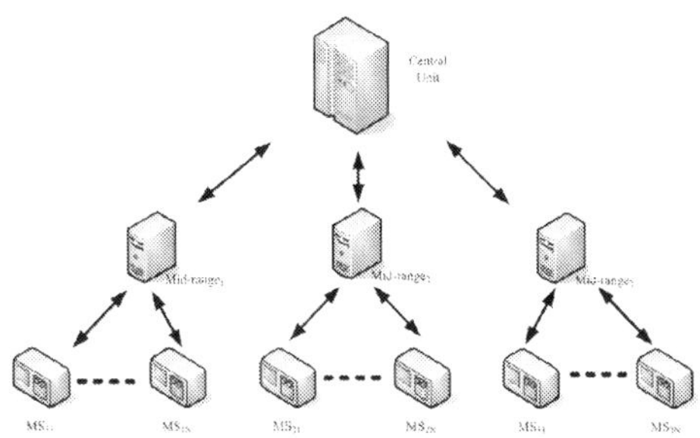

(d) Connection of hierarchical peers structure and inter-connection configuration

Fig. 3. Structural configurations

1.3 Main characteristics of a DMS

The trend of DMSs is motivated by the potential benefits that they could yield (Ozsu & Valduriez, 1991); (Tanenbaum, 1992). The first advantage of a DMS over a centralized system is cost effectiveness; other important characteristics are (Coulouris et al., 1994): (1) resource (provided by the processors) and data (measured parameters) sharing, (2) system expandability (hardware and software), (3) concurrency (simultaneous execution of multiple tasks), (4) scalability, (5) fault tolerance (carried out by hardware redundancy and software recovery), and (6) transparency (the users perceive the DMS as a whole system rather than as a collection of independent components). These characteristics are not automatic consequences of distribution; instead, they are acquired as a result of a careful design and implementation.

Beside these advantages, a DMS has some disadvantages: network reliance (problems on the network would disrupt activities in the system as a whole); complexities (a DMS

manipulates resources of computers with a wide range of heterogeneities); security (private resources would be exposed to a wider range of potential hackers, with unauthorized accesses).
Decentralization contributes to the extensibility, fault-tolerance, and lawsuit proofing of the system, while the partial centralization makes the system more coherent than a purely decentralized system (Makarenko et al., 2004); (Berkes, 2003).
In a DMS all concurrent access must be synchronized to avoid problems such as lost update (two concurrent accesses update the same data, but one of the updates is lost), dirty read (one access updates the data read by another access, but the former fails and affects the latter), incorrect summary (a set of data is updated by an access while the set is being processed by another access), and unrepeatable read (an access reads data twice, but the data are changed by another access between the two reads) (Elmasri & Navathe, 1994).

2. Implementation techniques

2.1 Smart Web sensors

In this field, a new revolutionary technology is that of the microprocessor driven (smart) sensors (Yong et al., 2004), (Hamrita et al., 2005). The increased availability of communications and networking systems (both wired and wireless) is likely to bring about a crossing over of price and functionality between sensor networking technology and communications technology. This makes possible the installation of smart sensors on remote places, transmitting the measured information to the final user (a client).
A simplified block diagram of hardware implementation of a *remote smart sensor* is shown in Fig. 4. In the first section the physical quantity under measurement is transduced and then conditioned for the A/D converter input. The samples, acquired by the A/D converter, are pre-processed (i.e. averaging the measurements, changing the scale) by a local processor (microcontroller). A smart sensor can also include high hierarchical level applications, such as electronic data sheets, self-identification, self-testing, self-adaptation, smart calibration and compensation.

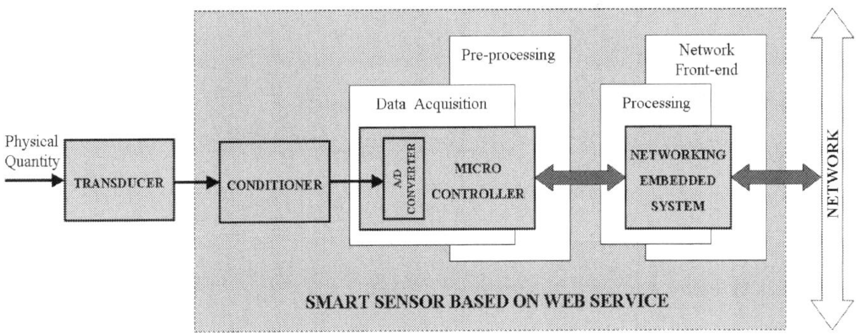

Fig. 4. Simplified block diagram of a Smart Sensor

The sensor needs to be able to communicate for remote monitoring and remote configuration; in Fig. 4 is the last block, the networking embedded system, that performs the final processing, storing and data transmission.

These new sensors are smaller, cheaper and more versatile than any other measurement device; their deployment in intelligent, interacting networks is opening up a new age in the DMS field.

The more revolutionary aspect is that related to the communication capability, which is addressed to link the sensors using a data transmission network. A lot of effort has been invested in order to overcome the obstacles associated with connecting and sharing these heterogeneous sensor resources. An interesting solution is to embody each sensor with tools capable to transfer the processed information through the network of sensors to a central computer.

Smart sensors can use different communication technologies. From network side, Internet is a widely adopted network where any user is uniquely identified with its IP (Internet Protocol) address. So, to implement a DMS, it is necessary to use a common and open communication protocol to exchange information and a methodology to auto-configure any smart sensor is linked to the network. This technology, known as *smart Web sensor*, makes various types of Web-resident sensors, instruments, image devices, and repositories of sensor data, discoverable, accessible, and controllable via the World Wide Web (Chu et al., 2006), (Lee et al., 2007).

DMSs based on smart Web sensors represent an interesting solution to many different measurement problems (Castaldo et al., 2004). These smart devices can transmit data to a remote processor for implementing remote monitoring of production processes, R&D experiments, environmental, security, or a wide-range of other, sensor-based monitoring tasks. In these applications, a computer client interacts with one or more smart sensors to download the measured parameters with a browser or an application capable to receive information from the Web server.

Interfacing transducers to all communication networks and supporting the wide variety of protocols is time-consuming and costly for manufacturers. To simplify this problem a standardized connection methods to interface smart transducers to the existing control networking technology has been proposed by the IEEE 1451 family of standards (IEEE Std 1451.1-4). The heart of the IEEE 1451.4 standard is the definition of the TEDS (Transducer Electronic Data Sheets), the information structure that contains the critical sensor information to enable plug-and-play operation.

2.2 Architecture of a DMS based on smart Web sensors

The architectures of a DMS based on smart Web sensors can be basically grouped in two categories.

The first, widely adopted, approach is based on a number of devices, linked with a centralized system, a central server keeping a list of users and shared resources (Fig. 5**Error! Reference source not found.**(a)). The primary advantage of centralized systems is their simplicity. Because all data is concentrated in one place, centralized systems are easily managed and have no problems of data consistency or coherence. During a search, every client sends a request to the central server that consults its lists providing results of IP user addresses. The file downloading happens between the two interested users from outside-centralized network. So, the server does not keep up any files. Each system is an independent server and must be selectively interrogated by the clients. The client needs to know the server position on the network (IP address) before starting the operations.

The second approach, the decentralized system shown in Fig. 5(b), is still based on a number of smart measuring devices, but presents the advantage to make easier the interrogation by the clients. This gives more extensibility to the network in which any node can join the network and instantly make new data available to the whole network. Another important feature of decentralized networks is that the failure or shutdown of any particular node does not influence the rest of the system. On the other hand, the intrinsic nature of this network gives two problems: the difficulties to manage the network because all the nodes have the same hierarchic level and the possibility to establish a packet loop that causes useless traffic.
To implement a network of smart Web sensors, two are the main problems to solve: i) how discover the smart Web sensors and ii) which interface is published to consume the services. From developer side, smart Web sensors present always a closed approach to interact with them, so Web services are adopted (Viegas et al., 2007), in order to give a standard approach in developing a Service Oriented Architecture. The user can search all the measurement related information available on the network using dedicated (special) services. Then he can ask to transfer the needed data form one or more of the measuring systems able to carry out the required measurement. There is no need to know any information related to the server (address list) before starting the search.
A hybrid schema is also possible combining centralized and decentralized systems (Fig. 5(c)).

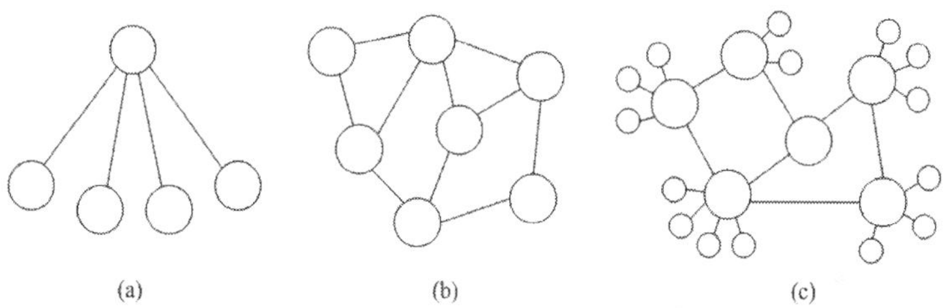

Fig. 5. Type of distributed measurement networks

2.3 Technologies for the DMSs

Ethernet is a cost-effective communication medium for DMSs; the growing dissemination and use of World Wide Web browsers and Java programming language also has a significant impact on sensor networking. DMS is frequently based on software packages that support Web-based data interfaces.
Despite the adoption of smart Web sensors is becoming a more and more suitable solution for different metrological applications (Grimaldi & Rapuano, 2005), the devices proposed in these years are still limited in the performance (Coulouris et al., 1994).
If the protocol used to exchange information between server and client is developed by the manufacturer, the smart sensor becomes a close system in which only the producer can modify the service for the client (Bucci et al., 2005). Moreover, proprietary solutions make impossible to correlate different measures supplied by two different systems using a single system.
Software tools play a fundamental role to define the performance of advanced DMSs; for this reason in the following paragraphs we will analyse the different possibilities.

2.3.1 HTML, Java and XML

One of the most popular solutions adopts the HTML (*HyperText Markup Language*), the encoding scheme used to create and format a web document. All strings which constitute the markup either begin with the character "<" and end with a ">", or begin with the character "&" and end with a ";".

```
<html>
    <body>
        <p>Here is a paragraph.
        <p>And here is another.
    </body>
</html>
```

Fig. 6. Example of HTML style

HTML is used to generate static Web pages in which both formatting tags and information requests are present (Morelli et al., 2004). This involves that, usually, only the manufacturer knows the exact protocol and data format used by server and client to exchange information (Bucci et al., 2005); this implies insurmountable problems to develop different user applications (Bertocco et al., 1998).

A smart Web sensor based on the HTML technique creates, at every request or page refresh, a new complete Web page, rather than update only a specific page field, with a consequent useless data retransmission. Moreover, because the information is transmitted only in text format, other data formats (e.g. an image) require ad hoc solutions (Hrushal et al., 2005).

For example, to obtain a graphical representation of a data streaming it is necessary to leave HTML and use a system that allows the creation of sockets to exchange data. The socket is a standard technique to send data over a communication system; its use is more efficacious because the server sends only the specific data requested by the client.

Another limitation of HTML technique is evident in multisensor metrological applications where it is necessary to synchronize all the measurements; in this case it is often necessary to adopt a further protocol to synchronize the data transfer between servers and clients. But the synchronization protocol is non standard, complicating the implementation of some client applications.

The reach of these browsers is complemented by their support of the *Java* programming language, which can help implement sharing and distribution of functionality across a network (Grimaldi et al., 1997); (Michal & Wieslaw, 2001); (Knyziak & Wieslaw, 2003). Although usually perceived as a technology to animate web pages or to develop portable client applications, Java was designed to be a portable, clean, object-oriented language for small embedded systems. Java is network-aware (TCP/IP, Transmission Control Protocol/Internet Protocol) and provides support for dynamically downloadable code, as well as for communication between applications. Java gives more advantages, compared with HTML, and offers a method to delegate the graphical processing to the client with the Java Virtual Machine (JVM).

The combination of browser- and Java-based systems has tremendous benefits for integration of networked sensor applications into the enterprise.

2.3.2 XML structure of the sensor and communication protocol

On the Web, the *XML technology* is growing in importance, spreading as a new system for the data exchange between different platforms and different software tools (Han et al., 2000). XML stands for the eXtensible Markup Language (Han et al., 2000); (W3C, 2006) and is an application profile or restricted form of SGML, the Standard Generalized Markup Language [ISO 8879]. A markup language is a text language that enables to describe the data content in a way independent of hardware, software, formats, or operating system. Markup encodes a description of the document's storage layout and logical structure (Rusty, 2004).

XML describes a class of data objects called XML documents and partially describes the behaviour of computer programs which process them. XML documents are made up of storage units called entities, which contain either parsed or unparsed data. Parsed data is made up of characters, some of which form character data, and others form markup. XML provides a mechanism to impose constraints on the storage layout and logical structure.

XML gained widespread acceptance because of its simplicity in the application development. Even if XML's rules are strict, they make the burden of parsing and processing the files more predictable and simple.

XML presents an important advantage: the tags are not defined and the user itself can define the set of markup necessary for its application. In fact, the XML does not use a closed, limited and fixed set of markup tags, as defined for HTML, interpreted and used by the browser to represent the information. In XML the tags are defined by the user, which is free to represent any kind of information and any tree structure of the data source for the application that requires the service.

Defining a scheme that represents the information provided by the smart sensor in a complete and exhaustive way and defining the policy to follow for the correct exchange of information between the server (smart Web sensor) and the client (high level application) all based on XML, we have a complete Web product offering a measurement service through the internet network, platform and application-oriented independent (Amiano et al., 2006) .

The smart sensor needs to be visible to the client through an XML structure that describes its functionality, its use and its characteristic. In fact, the smart Web sensor architecture uses a hierarchical approach: the smart sensor can manage many transducers becoming a sensors node in a sensors network. The client will receive not only the information about the smart Web sensor but also the information about the transducers connected to it.

The smart Web sensor packages the measurements using the XML structure and sends all to the client, via the internet network. This device is charged for the communication between the client and the transducers and executes the commands imparted via the Web.

As an example of application we suggest a communication protocol that can be implemented for the definition of the packet sequence during a request. This protocol defines the structure of every XML streaming file involved in the communication and has been developed to reduce the overhead of XML tags indispensable for the communication (Benz & Durant, 2003). The main packets of the communication protocol are:

- CONNECTION: When a host needs to download data from the smart Web sensor, it will establish a connection resolving the IP address of the server.
- SETUP: When the client performs a CONNECTION, it receives a XML SETUP data streaming reporting the number of transducers connected to the smart Web sensor, its nature and the data type that they use as response for a request. In this streaming file more general information are present such as the description of the application, the

location of the sensor, the date and the release of the last firmware upgrade, the author of the system and the organization that developed the device.

- ACK: If the SETUP packet has been received correctly by the client and it interprets the data in the right way, the client sends an acknowledgment packet to the server.
- REQUEST: When the server receives an acknowledgment packet, it waits for a request packet from the client for the accessing of the data provided by a transducer connected to the server.
- DATA-SEND: After receiving the request packet, the server begins to process the acquired data from the sensor involved in the calling and then transmits a new datagram with the result. The form of the datagram depends on the dimension of the data type that represents the measurement.

3. The Web service technology

The use of the XML as streaming support of measurement results is a good solution for all the remote measuring applications. However, XML presents a limitation: even if the streaming support is open, well organized and cross platform, the way used by client and server to exchange XML streaming data is proprietary. These problems present important limitations in the development of complex sensors network (Ferrari et al., 2003).

The basic requirement beyond smart Web sensor is the needing to have in some way the accessibility to some measured value (Bucci et al., 2003). The supplying of this value can be seen as a service done by an embedded server that is accessible on Internet. Every server allows the client to access the information acquired from a sensor.

A different approach to Web sensor development is based on the new concept of server that has been developed by the W3C (World Wide Web Consortium) (http://www.w3.org/): the idea is to consider a Web server not only as a stand alone server that a client can access to download files or HTML pages, but also a Web component that supply a service on the Internet network (Mielcarz & Winiecki, 2005). This solution, known as *Web service* approach, transforms a smart Web sensor into a server of measurement functions. In this way it is possible to offer great possibilities in terms of easy access for measurement data, integration of large complex Web sensors networks, realization of flexible custom applications and services reusability. Every client or developer can use this service to obtain information or to develop new complex services starting from the received information.

It is important to underline that Web services are similar to the local components used to build Windows applications (COM Object) with the method and attribute that the COM (Component Object Model) Object provides to the developer, but they aren't physically present in the local machines.

In the past, clients accessed these services using a tightly coupled, distributed computing protocol, such as DCOM (Distributed Component Object Model), CORBA (Common Object Request Broker Architecture), or RMI (Remote Method Invocation). While these protocols are very effective for building a specific application, they limit the flexibility of the system. Specifically, is the tight coupling used in these protocols (dependencies on vendor implementations, platforms, languages, or data encoding schemes) that limits the reusability of individual services.

The Web service architecture takes all the best features of the service-oriented approach and combines it with the Web, supporting universal communication using loosely coupled connections. Web protocols are completely vendor-, platform-, and language-independent. Web services support Web-based access, easy integration, and service reusability.

3.1 Smart Web sensors based on Web services

As previously discussed, the today's smart Web sensors present in literature adopt a micro-embedded Web server to transfer data and information to the clients that perform the request. As an application, starting form a low cost widely adopted smart Web sensor (Castaldo et al., 2003) ; (Castaldo et al., 2004) (Testa et al. , 2004) , a new kind of smart Web sensor with the Web service functionality is proposed; its simplified block diagram is shown in Fig. 7.

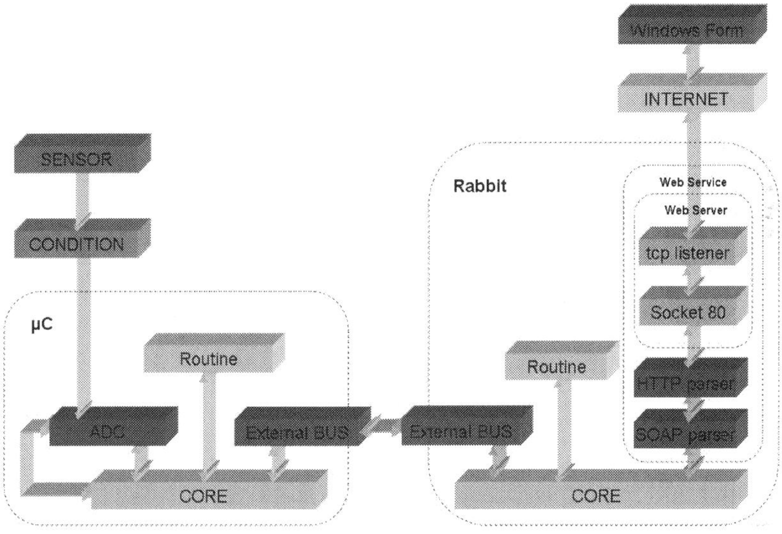

Fig. 7. Simplified block diagram of a smart Web sensor based on Web services

For including a Web service in a server environment, the main and widely adopted software architecture is ASP.NET, available in Microsoft Visual Studio .NET.

However, the use of a real Web service determines hard constraints on a general embedded architecture in term of cost, portability and power consumption. For these reasons, a possible solution for the developed of embedded Web service server is the use of a low cost embedded Web server.

In general a Web server does not have the same functionally of a Web service because of the use of HTTP (as protocol for sending data packets), HTML (to display information to a browser) and SOAP, Simple Object Access Protocol, (to exchange data with a client or with a Web service), while a Web server manages only HTTP and HTML.

As reported in Fig. 7, the communication system emulates a Web service opening a socket on port 80 for the listening of all the packets; then, a HTTP and SOAP parser controls and responses to the SOAP messages.

The most remarkable aspect of the entire flow is the waiting time of the Windows Form during a request; this time depends on the network load and on the number of samples acquired by the microcontroller. When the Windows Form sends a request on HTTP with a SOAP message to the Light Web service, it waits a SOAP response (an XML streaming file) in which the waveform is serialized. During this time, the Windows Form doesn't execute any other thread and it waits for the SOAP message.
To continue to use the Windows Form, it is necessary to control the thread of the Windows Form otherwise the process seizes up and any operation can be run (see Fig. 8).

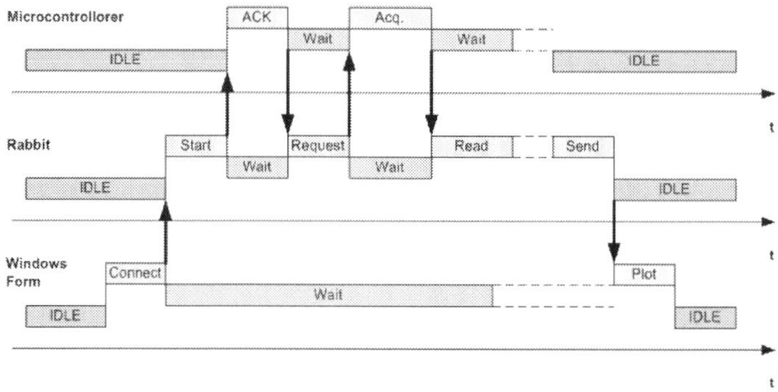

Fig. 8. Time analysis of the tasks present in the whole system

4. Plug-n-play smart Web sensors based on Web services

In a DMS, based on this technology, the services published by a Web service are reported in the WSDL (Web Services Definition Language) file.
Unfortunately, the Web service technology does not give any mechanism to refresh the services published and to manage dynamically the new services exported or deleted (Mielcarz & Winiecki, 2005). For instance, the access to a deleted service by a distributed application can generate an exception, collapsing the whole system and switching off the application. This is a stiff limitation, especially for a network of sensors that are often reconfigured to perform different measurements (Bucci et al., 2001), that require an appropriate run-time control for managing these service problems. Therefore, it is very important to develop a methodology to create a network in which smart Web sensors (network nodes) can be plugged without the need for an external configuration (Bucci et al., 2007).
A suitable solution is that every sensor sets an IP address and communicates its ability to the network master, who has two functions: master of the entire network and gateway (Ciancetta et al., 2007) Besides, the network master provides a Web service interface to every client that wants to use the sensors network: the entire network is equivalent to a single dynamic Web service (Ciancetta et al., 2006).
The core of this new approach is the adoption of two different tables in the smart sensors network: IP Routing Table and Services Table. The IP Routing Table is a table necessary to route a request from a client. This table stores the IP address and the services of every node; allowing the server to join the network node with its services. So, every request from a client

can be sent to the specific node. However, the client request has a different approach: the client sends a request to the server that, consulting its IP table routing, decides if it can execute the request. Next, the server sends a request to the network node present in the table to await the response and re-sends it to the client. This operation works well if there is a request to a specific service present in the network.

The main advantage of this solution is the possibility to merge more services to implement another new service. For example, we can imagine having a sensors network with two nodes: a voltage measurement sensor and a current measurement sensor. Besides the voltage or current services, the server can create other "virtual" services by fusion of the existing services. As an example, power or resistance can be "virtually" measured starting from these two services and the server can show four different services stored in the Service Table. This table, showing all the services available to the client and how they can be implemented, is upgraded every time a new sensor, executing new services, is plugged.

The service table describes whether the service is direct (not virtual) or virtual as shown in Fig. 9. A direct service is directly connected to a node, so, the Web service consults its IP table routing to resolve it. On the contrary, if a client sends a virtual request, the Web service consults an execution table, where the service is linked with a specific function related to actual devices.

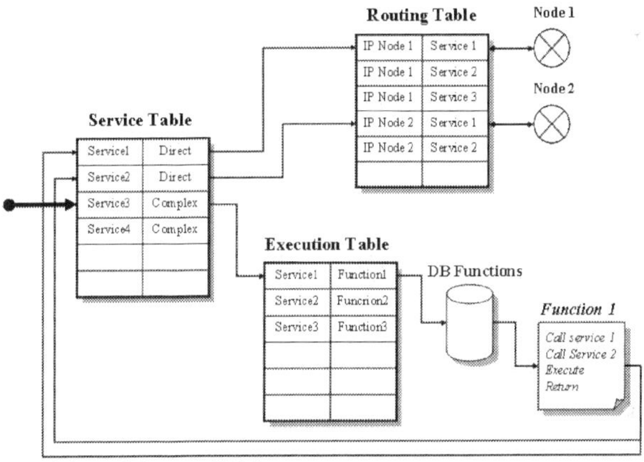

Fig. 9. IP Routing Table and Service Table

The Web service presents a DataBase (DB) storing all the executable functions. A typical function executes these tasks: i) reserves the required memory to every element; ii) receives all the values from the services involved in the function; iii) performs all the operations necessary to have the correct result; iv) gives the result to the Web service that resends it to the client, using SOAP.

The Fig. 9 illustrates how a Web service deals with a virtual service received from a client. The execution table is consulted to know whether the Web service can perform the function. Then the service table is consulted, to find the services it requires. Adopting this technique, it's possible to execute a virtual service by means of other virtual services. The service table has an important role in this approach. Every time a new network node is plugged in a sensors network, the Web service maps all the direct services available on the node, upgrading the IP routing table. Moreover, it scans all the DB functions that can be performed, to correctly execute virtual services.

5. A peer-to-peer distributed system for multipoint measurement techniques

To implement a DMS based on smart Web sensors, it is necessary to use a common and open communication protocol to exchange information and a methodology to auto-configure any smart sensor is linked to the network. Peer-to-peer networks allow individual computers to communicate directly with each other and to share information and resources without using specialized servers. A common characteristic of this new breed of applications is that they build, at the application level, a virtual network with its own routing mechanisms. The topology of this virtual network and the adopted routing mechanisms has a significant influence on the application properties such as performance and reliability (Ripenanu, 2001). Significant advantages can be gained using a freeware and widely adopted technology, such as the *Gnutella.*

The Gnutella protocol (The Gnutella protocol specification v4.0) .is an open, decentralized group membership and search protocol, mainly used for file sharing. The term Gnutella also designates the virtual network of Internet accessible hosts running Gnutella-speaking applications (this is the Gnutella network) and a number of smaller, and often private, disconnected networks. The graph in Fig. 10 depicts the topology of peers forming a connected segment of the Gnutella network.

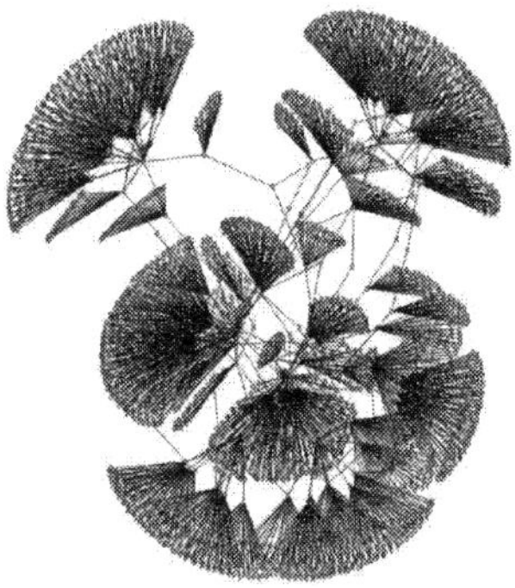

Fig. 10. A representation of the topology of Gnutella Network

Like most peer-to-peer file sharing applications, Gnutella was designed to meet the following goals:

- *Ability to operate in a dynamic environment.* Peer-to-peer applications operate in dynamic environments, where hosts may join or leave the network frequently. They must achieve flexibility in order to keep operating transparently despite a constantly changing set of resources.
- *Performance and Scalability.* The peer-to-peer paradigm shows its full potential only on large-scale deployments where the limits of the traditional client/server paradigm become obvious. Moreover, scalability is important as peer-to-peer

applications exhibit what economists call the "network effect" (Makarenko et al. 2004): the value of a network to an individual user scales with the total number of participants. Ideally, when increasing the number of nodes, aggregate storage space and file availability should grow linearly, response time should remain constant, while search throughput should remain high or grow.

- *Reliability*. External attacks should not cause significant data or performance loss.
- *Anonymity*. Anonymity is valued as a means of protecting the privacy of people seeking or providing unpopular information.

Gnutella nodes, called *servents* by developers, perform tasks normally associated with both SERVers and cliENTS. They provide client-side interfaces through which users can issue queries and view search results, accept queries from other servents, check for matches against their local data set, and respond with corresponding results. These nodes are also responsible for managing the background traffic that spreads the information used to maintain network integrity.

The *Ultrapeer* is an important concept that was not specified in the original Gnutella protocol, but which has now become a prominent feature of the Gnutella network. The Ultrapeer scheme improves network efficiency and scalability by categorizing nodes into regular clients and super nodes. A super node is a reliably connected host with plenty of network bandwidth that can act as a proxy for a large number of connecting clients. The super node removes the burden of extensive network message routing from the client, which may be a low bandwidth modem user. With this scheme, the Gnutella network mimics the Internet itself: low bandwidth nodes are connected to larger routers (the super nodes) that transmit the majority of the data over high bandwidth backbones.

As an example of using the Gnutella network, we describe a network that allows linked hosts to share arbitrary resources. This is a decentralized peer-to-peer system, consisting of hosts connected to one another using TCP/IP. In this network a client request for a measurement application is addressed to a computer which performs a particular Web service (Gnutella Web Service). This systems use the Gnutella network to search all the users able to perform the specific measurement, called Gnutella Embedded Clients (GECs) as reported in Fig. 11.

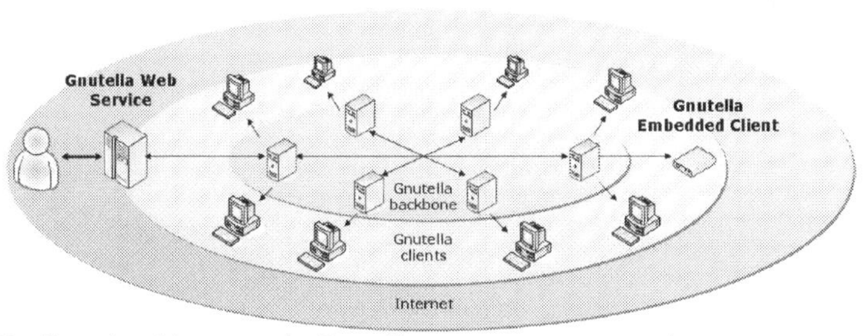

Fig. 11. Distributed architecture of a Gnutella measurement network

The name client for GEC is because it is a client of the Gnutella network. To execute the user search the request (query message) is repeated to all the Gnutella network computers (Fig. 12). When the suitable user is found, this network sends back the GEC address to the client.

At this point, the client can download the measures directly from the GEC, without overloading the Gnutella network (Bucci et al., 2005).
In this system, the measurement points are the GECs; each GEC can perform special measurements, depending on the kind of sensors embodied. This network creates an Internet over-structure from which all clients can perform a free access without external configuration and the GECs are visible without special operations. In order to implement this kind of system, a special Gnutella Web Service, a kind of interface between the client and the Gnutella network (Fig. 13) has been implemented, because the current implementations, referring exclusively on files sharing, cannot support a measurement process. When a measurement operation is asked, GEC sends the results to the *Gnutella Web Service* (GWS). One of the advantages of the proposed solution is the simplification of the activities to search and locate the measurement systems (GECs).

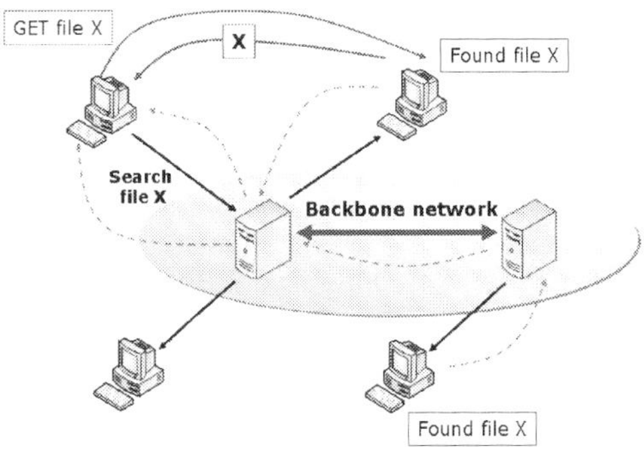

Fig. 12. The measurement server search, route and download

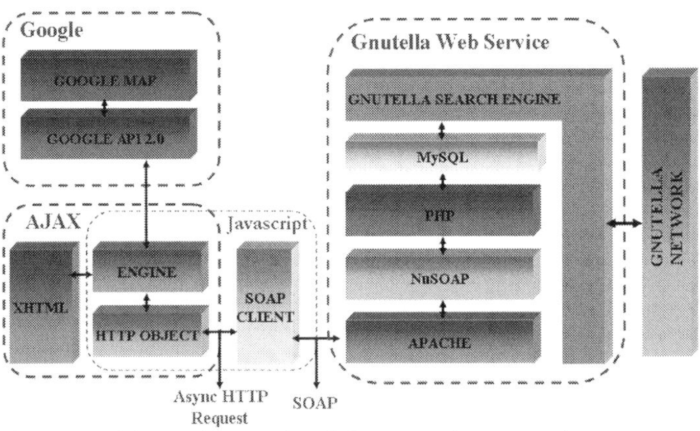

Fig. 13. Architecture of the implemented Web Service and User Interface

The GWS provides a particular implementation of typical Gnutella software, developing an ad-hoc Gnutella Search Engine. The methods are specifically developed for a measurement application; in particular the exported methods are:

1. GetStations: to obtain information about the stations present in a limited geographic area defined by GPS coordinates, in order to restrict the searching. The output of the method gives an array of stations in which every one reports.
2. GetCurrentData: the user calls the method passing the HASHID (hash identification)of the remote station and the service request to obtain the current data.
3. GetHistoryData: is similar to GetCurrentData, but accesses to stored DB data.

The Gnutella network is time consuming during the searching. In order to reduce this time, we adopted a caching system: at the end of a search, the authenticated stations are cached and their IP address stored in a DB for a limited period. Therefore, to obtain some information from a particular station, it is not necessary to start a new search, but it is possible to directly perform the download.

5.1 Environment monitoring application

In order to evaluate the feature of the proposed architecture, we implemented a monitoring application able to measure atmospheric values (Manuel et al., 2005), (Simic & Sastry, 2003) developing a remote measurement system (GEC), a GWS and a Web interface between the server and the operator (Ciancetta et al., 2007), (Ciancetta, Bucci et al. 2007).
The Web user interface has been implemented as a XHTML (eXtensible HyperText Markup Language) page that sends a request to Web Service and displays the results using Google Map (Fig. 14).

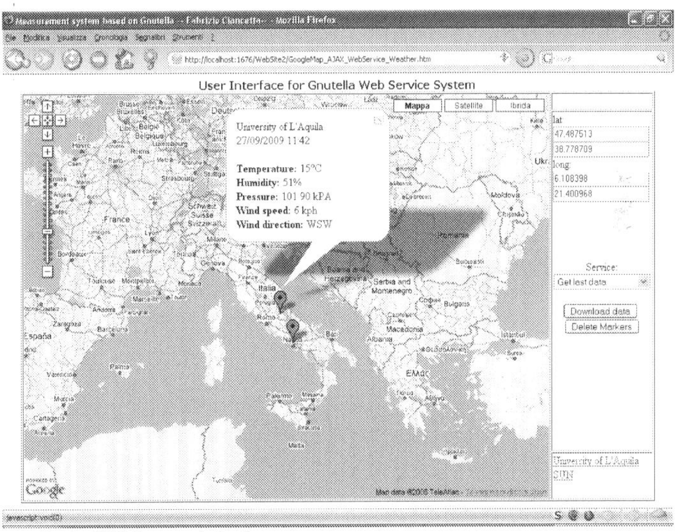

Fig. 14. Screenshot of Web user interface.

The Web user interface gives a more degree of freedom to the whole system, allowing the user to directly access measurement information with a common browser. We used Ajax (Asynchronous JavaScript and XML) technology to create interactive Web applications.

The XHTML page sends asynchronous requests to the Web Service and installs a callback function on the XMLHttpRequest. All the management of the function is done in JavaScript.
To interface the XHTML-JavaScript page with GWS, we adopted a SOAP client, a JavaScript class able to receive/create XML data form XHTML page and create/receive SOAP packet to GWS. In particular, on the remote station we implemented the services: temperature, humidity, pressure, wind direction and speed as shown in the Google Map Balloon accessible directly on the map.
To provide a more powerful mode to represent data from Gnutella Embedded Client we suggest a Windows Form user interface, based on Framework .NET 2.0. In the example, the user interface is divided in two parts: the first part, placed on the right side of the Windows Form, in which the user can: i) list the GECs present in the geographic area limited by the GPS coordinates; ii) select a station, looking at the available services and its GPS coordinates; iii) see a geographic view of all the station involved in the search. On the left Windows Form side there are two panels, reporting the downloaded data.
In the Current Data Panel (Fig. 15) there is a current view of the station with the last stored data acquired by the Gnutella Embedded Client and a graphical view of all the data of the current day from the 0:00 to the current hour retrieved form the GEC DB. The History Data Panel (Fig. 16) performs a direct access to the Gnutella Embedded Client DB, downloading the data.
In this example, all data are accessible directly to the GEC, without using the Gnutella network to reduce the traffic. In order to reduce space there are two DBs: one for the values accumulated during the day and another for an historical trend of the measurements.

6. Sensor synchronization

In a DMS time synchronization is a very important feature; many applications need local clocks of sensor nodes to be synchronized, requiring various degrees of precision. Unfortunately clock devices generate signals with some relative time uncertainties: local clock signals may drift from each other in time, hence sampling time or durations of time intervals may differ for each node in the network.
In general, a DMS can require different clock synchronization. The simplest case is the need to order the measures, that is to determine whether a measure m_1 carried out by a sensor has occurred before or after another measure m_2 carried out by another one. This problem presents simple solutions, because it is just required to compare the local clocks rather than to synchronize them.
Another more important occurrence is when each node embodies an independent clock and it is necessary to obtain information about the deviation from the other clocks in the network. In this way each node has its own local clock, but it is possible to convert a local time to the local times of other nodes. The majority of the synchronization procedures proposed for sensor networks use this technique (Elson et al., 2002); (Greunen &, Rabaey, 2003); (Sichitiu & Veerarittiphan, 2003)
The most complex situation is when all nodes must maintain a local clock synchronized to a remote reference clock. This is, for example, the case of two sensors sampling voltage and current that must be synchronized for calculating the electrical power. The synchronization scheme of (Ganeriwal et al., 2003) conforms to this model.

The synchronization methods are generally based on message exchange between nodes. In effect the problem is complicated by the nondeterminism in the network data access time, typical of Ethernet, characterized by a random access time, and in the variable packet transmission time. If a node transmits a measure with the local timestamp to another node or client, the packet can have a variable amount of delay before it is delivered, precluding the possibility of comparing and synchronizing the two clocks. Other access techniques, such as the TDMA (Time Division Multiple Access) can eliminate the uncertainty on the access time, but not on the transmission time.

Traditional synchronization techniques such as the use of a global positioning system (GPS) are not suitable for use in sensor networks; a GPS device may be too expensive to attach on a small sensor devices, and GPS service may not be available everywhere, such as inside a building. Moreover, this problem becomes important especially for a network of wireless smart sensors, because of their intrinsic properties such as limited resources of energy, storage, and computation.

To solve this problem, several solutions are under study in terms of synchronization algorithms, specifically designed for sensor networks.

The most diffused protocol is the Reference Broadcast Synchronization (RBS) (Elson & Estrin, 2001) where the sensors are divided in clusters, each with a cluster-head that transmit a synchronization packet (beacon). A reference beacon does not include a timestamp, but instead, its time of arrival is used by receiving nodes as a reference for comparing clocks. All receivers record the packet arrival time. The receiver nodes then exchange their recorded timestamps and estimate their relative phase offsets. RBS also estimates the clock skew by using a least-squares linear regression. The interesting feature of RBS is that it records the timestamp only at the receivers, thus, all timing uncertainties, including MAC (Media Access Control) medium access time, on the transmitter's side are eliminated. This characteristic makes it especially suitable for hardware that does not provide low-level access to the MAC layer. The main disadvantage of RBS is that it does not synchronize the sender with the receiver directly and that, when the programmers have low-level access at the MAC layer, simpler methods can achieve a similar precision to RBS.

Another protocol is the Flooding Time Synchronization Protocol (FTSP) or Tiny-Sync. FTSP, designed for applications requiring very high precision, utilizes a customized MAC layer time-stamping and calibration to eliminate unknown delays (Mar´oti et al., 2004). Linear regression from multiple timestamps is used to estimate the clock drift and offset. The main drawback of FTSP is that it requires calibration on the hardware actually used in the deployment (it is not a simply software algorithm). FTSP also requires intimate access to the MAC layer for multiple timestamps. However, if well calibrated, the FTSP's precision is less than 2 μs.

The Precision Time Protocol (PTP) is a high precision time synchronization protocol, defined in the IEEE 1588 standards "Standard for a Precision Clock Synchronization Protocol for Networked Measurement and Control Systems". There are two steps for synchronizing devices using PTP: (1) determine which device serves as the master clock, and (2) measure and correct time skew caused by clock offsets and network delays. When a system is initialized, the protocol uses an algorithm to determine which clock (Master Clock) in the network is the most precise. All other clocks become slaves and synchronize their clocks with the master. Because the time difference between the master clock and slave clock is a combination of the clock offset and message transmission delay, correcting the clock skew is done in two phases: offset correction and delay correction. Accuracy in the sub-microsecond range may be achieved with low-cost implementations.

7. Conclusions and future trends

Smart sensors are an enabling technology that will influence the future applications of measurement and data acquisition systems distributed on a wide area. The main revolutionary aspect of DMSs is the advanced integration of many state-of-the-art enabling technologies, mainly sensor, wireless communication, positioning, tracking and information technologies.

The first consequence of present trends is the supposition that in the future all sensors will be smart to some degree. Certainly a much higher percentage of them will be self-identifying and communicating. Communication is an important requirement for these devices and Internet, with either wired or wireless links, another widely shared solution.

It will be hard to solve all the problems in a "standard" way, also because there are several different applications with conflicting requirements. Proprietary solutions will be proposed again, especially for industrial applications. A plentiful supply of software tools for information and communication applications can help the DMS developers; even if the needs of a network of measurement systems are substantially different from a network of computers or communication devices.

In this chapter we tried to give an overview of the actual possibilities and trend in this field, even if the evolution run very fast and every day new standards and tools are available.

8. References

Amiano, M., Cruz, C., D., Ethier, K. and Thomas, M., D. (2006), *XML Problem - Design - Solution*, Wiley, 2006. ISBN-13: 978-0-471-79119-5, ISBN-10: 0-471-79119-9.

Benz, B. and Durant, J., R. (2003) *XML Programming Bible*, Wiley, 2003. ISBN-10: 0-7645-3829-2

Berkes, J., E. (2003), *Decentralized Peer-to-Peer Network Architecture: Gnutella and Freenet*, University of Manitoba, Winnipeg, Manitoba, Canada, April, 2003.

Bertocco, M., Ferraris, F., Offelli C. and Parvis, M. (1998), *A Client-Server Architecture for Distributed Measurement Systems* , Proceedings of IEEE Instrumentation and Measurement Technology Conference St. Paul, Minnesota, USA, May 18-21, 1998 pp 67–72.

Bucci, G., Ciancetta F., Fiorucci, E., Gallo, D. and Landi, C. (2005), *A low cost embedded Web Services for measurements on power system*, Proceeding of IEEE International Conference on Virtual Environments, Human-Computer Interfaces, and Measurement Systems, Giardini Naxos, Italy, 18-20 July, 2005.

Bucci, G., Ciancetta, F. and Fiorucci E. (2005), *A DSP-Based Wireless and Modular Data Acquisition Unit for Real-Time Measurement*, TechOnline Technical Papers, March 16, 2005 www.techonline.com.

Bucci, G., Ciancetta, F. and Fiorucci, E. (2003), *Unità d'acquisizione dati remota per sistemi di misura e controllo su rete TCP/IP*, Proceedings of Convegno Misure & Energia: l'importanza della metrologia nellindustria energetica italiana, Milano, 25 Novembre 2003.

Bucci, G., Ciancetta, F. and Rotondale, N. (2007), *Rete di sensori Plug-N-Play basata sui servizi Web: applicazioni al controllo di processi industriali*, Proceeding of LI Convegno Nazionale Motion Control, ANIPLA 2007, Milano, Italy, 10-11 Maggio 2007.

Bucci, G., Fiorucci, E. and Landi, C. (2001), *Digital Measurement Station for Power Quality Analysis in Distributed Enviroments*, Proceeding of IEEE International Conference on Instrumentation and Measurement Technology Conference, Budapest, Hungary, May 21-23,2001, pp 368–373.

Castaldo, D., Gallo, D. and Landi, C. (2004), *Collaborative Multisensor Network Architecture Based On Smart Web Sensor for Power Quality Applications*, Proceedings of IEEE International Conference on Instrumentation and Measurement Technology Conference, Como, Italy, 18-20 May, 2004, pp 1361– 1366.

Castaldo, D., Gallo, D., Landi, C., Langella, R. and Testa, A. (2003), *A Distributed Measurement System for Power Quality Analysis*, Proceedings of IEEE Power Tech 2003, Bologna, Italy, June 23-26, 2003.

Chu X., Kobialka T., Durnota B., and Buyya R. (2006). Open Sensor Web Architecture: Core Services, *Proceedings of the 4th International Conference on Intelligent Sensing and Information Processing (ICISIP 2006)*. ISBN 1-4244-0611-0, pp.:98-103. Bangalore, India.

Ciancetta, F., Bucci, G., Fiorucci, E., D'Apice, B. and Landi, C. (2007), *Proposta di un sistema di misura distribuito basato su una rete Peer-To-Peer*, Proceeding of XXIV Congresso nazionale GMEE (Gruppo Nazionale di Coordinamento Misure elettriche ed Elettroniche), Torino, Italy, 5-8 Settembre 2007.

Ciancetta, F., D'Apice, B., Gallo, D. and Landi, C. (2006), *Sistema di misura distribuito basato sui sensori smart e servizi Web*, Proceeding of XXIII Congresso nazionale GMEE (Gruppo Nazionale di Coordinamento Misure elettriche ed Elettroniche), L'Aquila, Italy, 11-13 Settembre 2006.

Ciancetta, F., DApice, B., Landi, C. and Pelvio, A. (2007), *Sistema di misura distribuito per il monitoraggio di rete di potenza*, Proceeding of XXIV Congresso nazionale GMEE (Gruppo Nazionale di Coordinamento Misure elettriche ed Elettroniche), Torino, Italy, 5-8 Settembre 2007.

Ciancetta, F., Fiorucci, E., D'Apice, B. and Landi, C. (2007), *A Peer-to-Peer Distributed System for Multipoint Measurement Techniques*, Proceedings of IEEE Instrumentation and Measurement Technology Conference, Warsaw, Poland, May 1-3, 2007, pp 1–6

Coulouris, G., Dollimore, J. and Kindberg, T. (1994) *Distributed Systems, Concepts and Design*, Addison-Wesley, Reading, MA, 1994.

Elmasri, R., and Navathe, S., B. (1994) *Fundamentals of Database Systems*, Addison-Wesley, Reading, MA, 1994.

Elson, J. and Estrin, D. (2001), *Time synchronization for wireless sensor networks*, In Proc. of the 2001 International Parallel and Distributed Processing Symposium (IPDPS), Workshop on Parallel and Distributed Computing Issues in Wireless Networks and Mobile Computing. San Francisco, CA.

Elson, J., Girod, L. and Estrin, D. (2002), *Fine-Grained Time Synchronization using Reference Broadcasts*, Proceedings of the Fifth Symposium on Operating Systems Design and Implementation (OSDI 2002), Boston, MA, December 2002.

Ferrari, P., Flammini, A., Marioli, D., Sisinni, E. and Taroni, A. (2003), *Sensor integration in Industrial Environment: From Field-bus to web-sensors*, Computer standards & Interfaces, 25, 2003.

Ganeriwal, S., Kumar, R. and Srivastava, M. (2003), *Timing Sync Protocol for Sensor Networks*, Proceedings of ACM SenSys, Los Angeles, November 2003.

Greunen, J. and V., Rabaey, J. (2003), *Lightweight Time Synchronization for Sensor Networks*, Proceedings of the 2nd ACM International Conference on Wireless Sensor Networks and Applications (WSNA), San Diego, CA, September 2003.

Grimaldi, D., Nigro, L. and Pupo, F. (1997), *Java based distributed measurement systems*, Proceedings of IEEE Instrumentation and Measurement Technology Conference, 19-21 May 1997, Ottawa, Canada, pp 686–689.

Grimaldi, D., Rapuano, S. and Laopoulos, T. (2005) *State of Art of the Distributed Measurement Systems for Industrial and Educational Purposes*, IEEE Workshop on Intelligent Data Acquisition and Advanced Computing Systems: Technology and Applications, 5-7 September 2005, Sofia, Bulgaria pp 289–294.

Hamrita, T.K. Kaluskar, N.P. Wolfe, K.L. (2005). Advances in smart sensor technology. *Proc. Of Industry Applications Conference, 2005*. ISBN: 0-7803-9208-6. Volume: 3, pp.: 2059 - 2062.

Han, R., Perret, V. and Naghshineh M. (2000), *WebSplitter: A Unified XML Framework for Multi-device Collaborative Web Browsing*, Computer Supported Cooperative Work, pp 21–23.

Hrushal, V., Osolinskiyl, O., Daponte P. and Grimaldi D.(2005), *Distributed Web-based Measurement System*, Proceedings of IEEE Workshop on Intelligent Data Acquisition and Advanced Computing Systems: Technology and Applications, 5-7 September 2005, Sofia, Bulgaria pp 355–358.

IEEE Standard for a Smart Transducer Interface for Sensors and Actuators, IEEE Std 1451.1-4, 1997, http://ieee1451.nist.gov/.

Knyziak, T. and Winiecki, W. (2003), *The New Prospects of Distributed Measurement Systems Using JavaTM 2 Micro Edition Mobile Phone*, Proceedings of IEEE International Workshop on Intelligent Data Acquisition and Advanced Computing System: Technology and Applications, 8-10 September 2003, Lviv, Ukraine, pp 291–295.

Makarenko, A., Brooks, A., Williams, S., Durrant-Whyte, H. and Grocholsky B. (2004), *A decentralized architecture for Active Sensor Networks*, Proceedings of IEEE International Conference on Robotics and Automation, New Orleans, LA, USA, April 26-May 1, 2004, pp 1097–1102.

Manuel, A., DelRio, J., Shariat, S., Piera, J. and Palomera, R. (2005), *Software Tools for a Distributed Temperature Measurement Systems*, Proceedings of Instrumentation and Measurement Technology Conference Ottawa, Ontario, Canada, May 17-19, 2005, pp 1566–1570.

Mar´oti, M., Kusy, B., Simon, G., and L´edeczi, A. (2004) *The flooding time synchronization protocol*, Proceedings of the 2nd international conference on Embedded networked sensor systems, SenSys '04, ACM Press, 39–49.

Michal, K. and Wieslaw, W. (2001), *A New Java-Based Software Environment for Distributed Measurement Systems Designing*, Proceedings of IEEE Instrumentation and Measurement Technology Conference, 21-23 May 2001, Budapest, Hungary, pp 397–402.

Mielcarz, T. and Winiecki, W. (2005), *The Use of Web-services for Development of Distributed Measurement Systems,* Proceedings of IEEEWorkshop on Intelligent Data Acquisition and Advanced Computing Systems: Technology and Applications, 5-7 September 2005 ,Sofia, Bulgaria, pp 320–324.

Morelli, S., Morelli, R., Ciancetta, F., Vasile, A., D'Intino, A., Di Donato, M., A., Di Gioacchino, M. and Boscolo P. (2004), *Monitoraggio dei campi elettromagnetici nelle aree urbane di Chieti e Pescara,* Proceedings of LXVII Congresso Nazionale S.I.M.L.I.I., Sorrento, Italy, 3-6 Novembre 2004, pp 301-302.

Ozsu, T. and Valduriez, P. (1991) *Principles of Distributed Database Systems,* Prentice-Hall, Englewood Cliffs, NJ, 1991.

Ripeanu, M. (2001), *Peer-to-Peer Architecture Case Study: Gnutella Network Analysis,* 1st International Conference in Peer-to-Peer Networks, Aug. 2001, Linkpings Universitet, Sweden.

Rusty, H., E. (2004) *XML 1.1 Bible,* Wiley, 2004. ISBN-10: 0-7645-4986-3.

Sichitiu, M.,L. and Veerarittiphan, C. (2003), *Simple, Accurate Time Synchronization for Wireless Sensor Networks,* Proceedings of IEEE Wireless Communications and Networking Conference, WCNC 2003.

Simic, S. and N., Sastry, S. (2003), *Distributed environmental monitoring using random sensor networks,* Proceedings of the 2nd International Workshop on Information Processing in Sensor Networks, Palo Alto, California, April 22-23, 2003, pp 582–592.

Tanenbaum, A., S. (1992) *Modern Operating Systems,* Prentice-Hall, Englewood Cliffs, NJ, 1992.

Tari, Z. and Bukhres, O. (2001) *Fundamentals of Distributed Object Systems: The CORBA Perspective,* Wiley, 2001.

Testa, A., Castaldo, D., Gallo, D. and Landi, C. (2004), *A Digital Instrument for non-Stationary Disturbance Analysis in Power Lines,* IEEE Transactions on Instrumentation and Measurement , 53, no. 5, August, 2004, pp 1353–1361.

The Gnutella protocol specification v4.0. http://dss.clip2.com/GnutellaProtocol04.pdf., 2004.

Viegas V., Dias Pereira J. M., Silva Girão P. (2007). Framework and Web Services: A Profit Combination to Implement and Enhance the IEEE 1451.1 Standard. *IEEE Transactions on Instrumentation and Measurement, Volume 56*.NET, Issue 6, pp. 2739-2747, December 2007.

W3C, *Extensible Markup Language (XML) 1.0 (Fourth Edition),* http://www.w3.org/TR/2006/REC-xml-20060816/ 2006.

Yong Z.; Yikang G.; Vlatkovic, V.; Xiaojuan W. (2004). Progress of smart sensor and smart sensor networks. *Proc. of Intelligent Control and Automation,* 2004. WCICA 2004. Digital Object Identifier 10.1109/WCICA. 2004.1343265. Volume 4, pp.: 3600 - 3606.

A methodology for measuring intellectual capital. A structural equations modelling approach

Mariolina Longo and Matteo Mura
Department of Management, University of Bologna
Italy

1. Introduction

The past decade has been characterized by a process of growing dematerialization of the strategic resources possessed by firms. The relational capabilities of the firm, technology connected with the planning and management of firm processes, know-how, as well as the decisional autonomy and technical competencies of the employees all represent intangible assets that are determining in the value creation process of a firm (Longo & Mura, 2007; Roos et al., 2005).

The relevance of this topic is supported by the attention that financial markets attribute to the accounting of these assets. In January 2007 the International Accounting Standard Board issued a technical document in support of the 'Intangible Assets' project, which is examining the possibility of adding to the balance sheet the intangible assets that are generated internally to the firm and that are not subject to any negotiation on active markets (IASB, 2007). This 'opening up' in the accounting system has important effects on the economic evaluation of a company and on its ability to gain access to credit, in that it provides the market, the institutional investors and the financial analysts very precious information regarding the development of fundamental resources for the value creation process of a firm.

Furthermore, performance management literature has highlighted the need for specific tools for the measurement of internally-generated intangible assets, defined in managerial literature as intellectual capital (IC) (Tayles et al., 2002). These tools have been shown to greatly support management activity (Roos et al., 2005; Carlucci et al., 2004). As a matter of fact, the integration of information related to company's intellectual capital together with quantitative information relative to the firm's strategic policies, offers management a display of important indicators for the definition and the control of corporate objectives. Numerous intellectual capital frameworks have been proposed in the literature (e.g. Edvinsson & Malone, 1997; Roos et al, 2005; Sveiby, 1997), however, further research is still needed to investigate the challenges and opportunities of designing intellectual capital measurement tools that are grounded in relevant measurement theory (Bollen, 1989; M'Pherson & Pike, 2001).

The chapter we propose describes the development and implementation of an IC measurement system within an Italian company that is leader in the agricultural food product sector. Since IC creation and development is primarily founded on the actions and capabilities of the employees (Roslender et al., 2006; Roslender & Fincham, 2001), the individual employee has been used as the unit of analysis of this study. This element constitutes an innovative factor with respect to other studies which instead use MBA students (Bontis, 1998; Bontis et al., 2000), or adopt the managers' perceptions as proxy of the company they work for (Staples, 1999; Youndt & Snell, 2004). The measurement system has been developed based on two surveys that were conducted in 2005 and 2006 on all the employees of the 13 business units of the company. About 3,400 questionnaires were distributed and, with an average redemption of 35%, the sample consists of 1,117 observations. Structural equations modelling (SEM) methodology was used to validate the IC measurement model and to identify and test the effect that two specific company's human resource practices have on IC.

The chapter is structured as follows: next section describes the concept of intellectual capital as emerges from academic and practitioners' literature, followed by the theoretical model we propose in this study. The third section illustrates the methodology adopted and the data-gathering process and the following section presents the analyses of the data and the results obtained. The managerial implications of our study, together with the limitations and the future developments of the tool are described in the closing section.

2. Intellectual capital: definition and dimensions

Numerous studies have extensively reviewed and discussed the IC literature (Allee, 2000; Andriessen, 2004; Hunter et al., 2005; Roos et al., 2005; Serenko & Bontis, 2004). Therefore, the focus of this section will efficiently turn to defining the constructs we intend to measure. The following definitions summarize some of the highlights of this field.

IC scholars have generally identified three main dimensions of IC that include human capital, structural capital, and relational capital. Human capital represents the individual knowledge stock of an organization as represented by its employees (Bontis, 2002). Employees generate IC through their competence, in terms of skills and knowledge, and their attitude, and in terms of the behavioural components of employees' work (Roos et al., 2005). Structural capital consists of mechanisms and organizational procedures which support the employees in completing their tasks, and includes all non-human storehouses of knowledge in organizations like databases, process manuals, routines, strategies, and anything whose value to the company is higher than its material value (Bontis, 2000). Relational capital is associated with the network of relations that the organization and its members are able to establish both inside and outside the working environment. The resources that emerge, that are transferred and are made connatural with these multifarious relations constitute the relational capital of the organization (Adler & Kwon, 2002).

In developing a theoretical foundation for the three dimensions of IC, we have draw primarily from human capital theory (e.g., Becker, 1964; Flamholtz & Lacey, 1981; Schultz, 1961), knowledge-based theory (eg., Grant, 1996; Polanyi, 1962; Spender, 1996), and social capital theory (e.g., Jacobs, 1965; Loury, 1977; Baker, 1990). We have chosen these three theories for their explicit theoretical relevance concerning IC. As a matter of fact, each

perspective offers a different lens for understanding how firms can measure and manage their IC, giving insights of each different dimension of the IC construct.
Below, we briefly discuss the contribution that each of the three theories gives to its respective IC dimension. Specifically, we adopt the human capital theory to discuss the human dimension of the IC construct, the knowledge-based theory to examine the structural dimension, and the social capital theory to analyze the relational dimension.

2.1 Human capital

Human capital theory focuses on the concept that people possess skills, experience, and knowledge that have economic value for firms. For the purpose of this study we propose two arguments, previously discussed by Snell and Dean (1992), that expands on this proposition.
The *productivity* argument emphasizes that employee skills and knowledge represent capital because they enhance productivity, adding value to a firm. Even if part of this value is tangible, in that it is created through the transformation of firm's product, much of it is intangible, and consist in problem solving skill, in the ability to identify the key aspects of the work from those of less importance, and in the capability to be innovative and creative in performing the job (Hitt et al., 2001; Nahapiet & Goshal, 1998). A firm can choose to invest in human capital both internally developing employee skills or acquiring them on the market (Hatch & Dyer, 2004; Lepak & Snell, 1999). Internalizing employment is more desirable when firm can do so without investing in employee development, on the contrary, if employee productivity is not expected to exceed investment costs, a firm will acquire these skills on the labour market. Therefore, the decision to internalize or outsource human capital is based on a comparison of the expected returns of employee productivity (Becker, 1964).
The *transferability* argument suggests that human capital has a price on the labour market because it is valuable from other firms, and, more important, it is transferable. This argument is based on the fact that firms don't own human capital, because it is embodied in employees, who are free to move from one firm to another (Becker, 1964; Hatch & Dyer, 2004). Low employee turnover therefore, represents an important element in the firm's value creating process in that secures the firm from loosing key skills, knowledge, and expertise (Arthur, 1994; Hudson, 1993). Notwithstanding, even if employees stay with a firm, their contribution depends on their willingness to perform. For this reason employee satisfaction, motivation, and commitment are decisive components in the development of human capital (Arthur, 1994; Lepak & Snell, 1999).

2.2 Structural capital

In their analysis of the intellectual capital concept, Nahapiet & Goshal (1998) clearly distinguish between two types of knowledge that form the base of IC: the people knowledge and the social knowledge. While the former represents a part of the human capital dimension, and was discussed in the previous section, we delve into the latter in defining the structural dimension of IC.
Knowledge management scholars (Polanyi, 1962; Weick & Roberts, 1993; Spender, 1996) define social knowledge as the knowledge that is shared and embedded in the organization, and suggest it comprises two elements: social explicit knowledge, and social tacit

knowledge. Social explicit knowledge, also called "objectified knowledge" (Spender, 1996), corresponds to the shared corpus of knowledge of the organization, and it depends on effective use of the institutional mechanisms, such as databases, patents, registered designs, process manuals, and information systems, that contribute to distribute knowledge and intellect (Youndt & Snell, 2004). As an evidence of the relevance of this element, Quinn, Anderson and Finkelstein (1996) show that an increasing number of organizations make major investments in the development of procedures and systems to pool and to leverage such objectified knowledge.

Social tacit knowledge, also called "collective knowledge" (Spender, 1996), corresponds to the knowledge that is embedded in the form of social practice and resides in the tacit experience of the collective (Brown & Duguid, 1991). Nelson and Winter (1982) define this form of knowledge as the organization's genetic material that may reside in bureaucratic and formal rules, or in organization's norms and culture, and call it "routines". This collective knowledge is produced internally (Penrose, 1959) and may represents the outcome of firm's evolving methods and policies that: foster and support employees in the development of new ideas and innovative approaches that give rise to extrarational learning processes (Nelson, Winter, 1982); give emphasis to quality procedures; or contribute to align employees and organizational goals (Schiemann, 2006).

2.3 Relational capital

Relational, or social[1], capital is defined as the sum of actual and potential resources embedded within, available through, and derived from the networking relationships developed by an individual or an organization (Lin, 2001; Nahapiet & Goshal, 1998). Therefore, social capital encompass both the network of relations and the assets that may be mobilized through that network (Bourdieu, 1986; Burt, 1992). The networking relationships provide value for actors (e.g. individuals or organizations) by allowing them to tap into the resources embedded in such relationships for their benefit (Acquaah, 2007). Researchers at organizational level have suggested that the greater the uncertainty in the firm's business environment, the more likely the firm will rely on networking relationships when entering into economic exchange relations (Pfeffer & Salancik, 1978; Peng & Heath, 1996).

Dyer & Nobeoka (2000) propose that networking relationships between the firm and its external stakeholders, such as customers, suppliers, and business partners, stimulate the creation, acquisition, and exploitation of knowledge and IC. In particular, networking with customers may develop both customer and brand loyalties (Park & Luo, 2001), those with suppliers may give access to quality raw materials, better service, and fast and reliable deliveries (Peng & Luo, 2000), while those with business partners reduce the possibility of opportunistic behaviour (Pisano, 1989), increase inter-firm trust (Kale et al., 2000), and enhance the evolution of inter-partner relationships in terms of freer and greater exchange of information, skills and know-how, and of development of new competences (Kale et al., 2000; Walker et al., 1997).

Also intra-firm relations, like teamwork and department integration, represent a source of knowledge development and acquisition and, consequently, contribute to the development of IC (Nahapiet & Goshal, 1998). Collaborations and teamwork have been shown to be

[1] Accordingly to Kale, Singh & Perlmutter (2000), in this article we use relational capital and social capital as synonyms.

important elements for the development of collective and shared knowledge in biotechnology firms (Zucker et al., 1996), and, in general, in knowledge-based firms (McGill & Slocum, 1994). As a matter of fact, cooperation among employees, rather than competition, contributes on open disclosure of information, and on building loyalty to the firm (Starbuck, 1992).

2.4 The proposed model

In this chapter we first aim to develop an IC measurement model that is coherent with the theoretical framework proposed. Secondly, we explore the effect that two specific human resource (HR) policies pursued by the company we analyzed have on the IC dimensions.
The HR policies were identified by means of numerous focus groups conducted with a panel of company managers. By means of an in-depth analysis of the firm's vision and mission we selected two strategic policies of the company that could have an effect on IC.
The first policy identified, which we define as *stakeholder management policy*, refers to the numerous corporate social responsibility and stakeholder management activities and actions that have been carried out by the firm in recent years with regard to its human resources.
The second policy identified, which we define as *perception of customers/consumers policy*, represents the perception, held by the employees, of the role played by the company within its own economic environment.
In relation to the internal strategic policy we propose the following hypothesis:

Hp1: Stakeholder management policy positively affects IC.

Which can be divided into the following:

H1a: Stakeholder management policy positively affects human capital;
H2b: Stakeholder management policy positively affects structural capital;
H3c: Stakeholder management policy positively affects relational capital.

In relation to the *perception of customers/consumers policy* we propose the following:

H2: Perception of customers/consumers policy positively affects IC.

Which can be divided into the following:

H2a: Perception of customers/consumers policy positively affects human capital;
H2b: perception of customers/consumers policy positively affects structural capital;
H2c: perception of customers/consumers policy positively affects relational capital.

Figure 1 shows the theoretical model proposed.

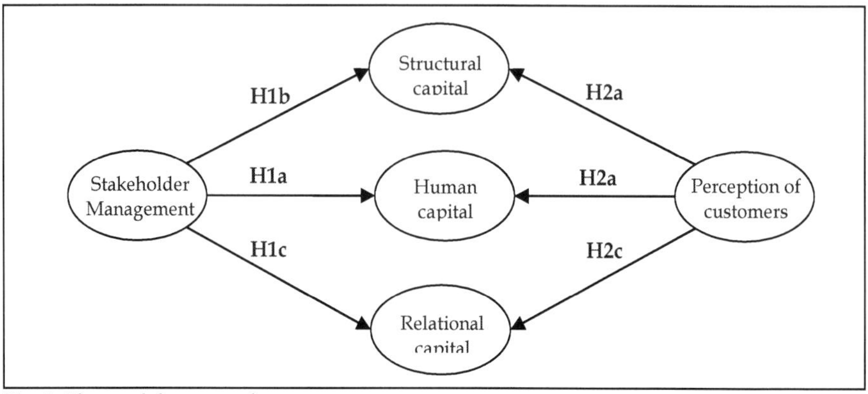

Fig. 1. The model proposed

3. Method

3.1 Measures

In order to measure intellectual capital at the individual level of analysis we employed measurement scales that were already developed in the literature. Moreover, we specifically developed two scales that measure the human resource practices adopted by the company and that were suggested to affect intellectual capital dimensions. All the scales have been fully discussed together with company management. Thirteen scales have thereby been identified and the questionnaire that has been prepared comprises 58 items on a 5 point Likert scale. Those interviewed were asked to express how they felt about the stated items, which varied from "strongly disagree" (level 1 on the scale) to "strongly agree (level 5 on the scale). The following scales were used in order to measure intellectual capital (Table 1):

Intellectual capital dimensions	Scales	Authors
Human capital	Low turnover propensity	Chalykoff and Kochan (1989)
	Job satisfaction	Zeitz et al. (1997)
	Intrinsic work reflection	Holman et al. (2001)
	Practical application	Holman et al. (2001)
Structural capital	Trust	Zeitz et al. (1997)
	Sense of belonging	Kidwell and Robie (2003)
	Responsibility	George (1992)
Relational capital	Ability to work in a group	Kidwell and Robie (2003)
	Networking and communication	Zeitz et al. (1997)
HR practices	Customers	Developed by authors
	Stakeholder management	Developed by authors

Table 1. Measures

The following control variables were used in this study: professional category, seniority within the company, gender, area of work, business unit.

3.2 Sample

Two surveys were conducted in 2005 and in 2006 taking the employees of the company as the unit of analysis. The questionnaire was sent to employees by putting it in with the employees' pay slips. Along with board managers was considered non opportune to count in 2005 survey three business units acquired few months before the administration of the survey; on the contrary these plants were included in 2006 survey. In 2005 a total of 1.310 questionnaires were sent out and 460 were returned which meant that a percentage of 35% replied. In 2006, 1928 questionnaires were sent out and 657 were returned (34 percent). The data collected are significant in that they are uniformly split up into production plants and job levels in relation to the company's population (Table 2 and Table 3).

Business Unit (BU)	BU Population (2005)	Questionnaires Returned (2005)	Redemption (%) (2005)	BU Population (2006)	Questionnaires Returned (2006)	Redemption (%) (2006)
No. 1	427	176	41.2	479	145	30.3
No. 2	144	38	26.4	117	26	22.2
No. 3	47	14	29.8	48	21	43.8
No. 4	145	29	20.0	165	30	18.2
No. 5	61	20	32.8	60	8	13.3
No. 6	40	17	42.5	41	25	61.0
No. 7	208	57	27.4	194	50	25.8
No. 8	85	37	43.5	91	44	48.4
No. 9	28	6	21.4	22	3	13.6
No. 10	125	66	52.8	108	69	63.9
No. 11	-	-	-	362	100	27.6
No. 12	-	-	-	172	84	48.8
No. 13	-	-	-	69	52	75.4
Total	**1310**	**460**	**35. 1**	**1928**	**657**	**34.1**

Table 2. Survey redemption for each business unit

Job Level	Population (2005)	Questionnaires Returned (2005)	Redemption (%) (2005)	Population (2006)	Questionnaires Returned (2006)	Redemption (%) (2006)
Executives	40	10	25.0	47	15	31.9
Middle-Managers	75	34	45.3	94	30	31.9
Office workers	458	198	43.2	718	214	29.8
Manual workers	737	201	27.3	1017	359	35.3
Total	**1310**	**443**	**33.8**	**1928**	**617**	**32.0**

Table 3. Survey redemption for each job level

4. Analyses and results

4.1 Validity and reliability of the scales

The development of the measurement system comprised four consecutive stages of analysis. Using the data provided by the 2006 survey, an exploratory factor analysis was carried out on the items relative to each established category of resources, in order to define the measurement of all the constructs. We then adopted a confirmatory factor analysis, using the same sample, to evaluate how the data supplied by the employees related to the second-order measurement model composed of 11 first-order latent factors, representing the measurement scales as proposed in Table 1, and 4 second-order factors that consists of human, structural and relational capital constructs. The measurement properties of convergent validity, discriminant validity and internal consistency were all supported (analyses are available under request).

In the final step of the analyses, the adequacy of the model is formally cross-validated by means of a confirmatory factor analysis, using the 2005 survey data. The confirmatory data analysis was performed using LISREL, structural equation modelling program (Jöreskog & Sörbom, 2004). The exploratory analysis reduced the number of items adopted from 58 to 50. Tables 4 shows means, standard deviations and reliabilities for first-order latent variables both for the 2005 and the 2006 sample.

	2005 sample			2006 sample		
Variables	Means	s.d.	Cronbach's alpha	Means	s.d.	Cronbach's alpha
Sense of Belonging	3.70	1.04	0.86	3.76	1.09	0.88
Trust	3.09	1.31	0.81	3.25	1.31	0.80
Responsibility	4.03	0.99	0.62	4.01	0.90	0.79
Low turnover propensity	2.93	0.94	0.80	3.90	0.98	0.78
Job satisfaction	2.26	1.09	0.87	2.39	1.22	0.90
Intrinsic work reflection	3.94	0.80	0.76	3.91	0.88	0.79
Practical application	3.38	0.90	0.48	3.33	0.99	0.60
Ability to work in a group	3.26	1.00	0.93	3.40	1.01	0.89
Networking / Communication	2.59	0.94	0.89	2.57	0.99	0.88
Stakeholder management	2.66	0.93	0.90	2.81	0.97	0.91
Customers	3.30	0.94	0.86	3.62	0.92	0.84

Table 4. Means, standard deviations, and reliabilities

4.2 The effect of human resource policies on IC

In order to understand the effect that the human resource practices and the control variables have on intellectual capital dimensions, a hierarchical regression analysis was conducted in two consecutives steps (George, 1992). This technique allowed us to calculate, separately, the influence that the two groups of independent variables have on intellectual capital. Tables 5 and 6 synthetically show the results of the regression analysis. The values given in the column ΔR^2 indicate the variance explained by the groups of independent variables.

	Step 1 Context factors (ΔR^2)	Step 2 HR practices (ΔR^2)
Sense of belonging	13%	24%
Trust	7%	48%
Responsibility	12%	14%
Low turnover propensity	7%	19%
Job satisfaction	7%	51%
Intrinsic work reflection	10%	16%
Pratical application	6%	5%
Ability to work in a group	7%	36%
Network and Communication	11%	63%

Table 5. Results of regression analysis using 2005 data

	Step 1 Context factors (ΔR^2)	Step 2 HR practices (ΔR^2)
Sense of belonging	8%	30%
Trust	6%	46%
Responsibility	5%	19%
Low turnover propensity	6%	20%
Job satisfaction	6%	54%
Intrinsic work reflection	4%	16%
Pratical application	3%	10%
Ability to work in a group	5%	29%
Network and Communication	5%	64%

Table 6. Results of regression analysis using 2006 data

From the results it emerges that the human resource practices identified have a significant effect, even though its intensity varies, on most of the intellectual capital resources, and in particular on *Trust, Job Satisfaction, Networking and Communication,* and *Ability to work in a group*. Moreover, no significant differences have been found between the 2005 and the 2006 values and this further strengthens the results obtained. Figure 1 illustrates the placing of the intellectual capital resources and the impact of the company's human resource practices on the same in terms of variance explained.

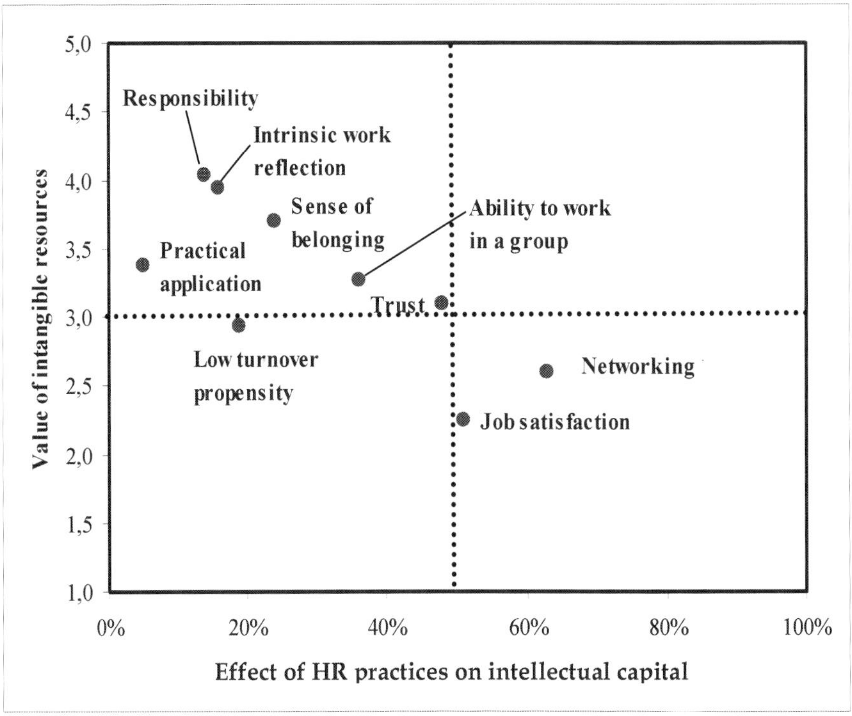

Fig. 2. Evaluation grid of intellectual capital resources - 2005 survey

The subdivision into quadrants shown in Figure 2 has been carried out exclusively to indicate the mean value of the scales and does not intend to be a reference to any objectives the company might have regarding the results of its human resource practices. The majority of the resources are positioned in the top left quadrant and they are therefore characterized by high values and low impact from human resource practices. *Networking and Communication* together with *Job Satisfaction* are to be found, on the other hand, in the bottom right quadrant which shows resources with low values and high impact from human resource practices.
The subdivision into quadrants constitutes the information that will guide all decision making within the company in relation to the human resource practices it intends to pursue. The horizontal line enables the company to establish the intervention threshold below which it can consider feedback from its human resource practices unsatisfactory. The vertical line informs the company on the impact threshold of its practices on its employees below which any input from the company does not produce significant results in terms of intellectual capital resources.
In Figure 3 a different subdivision into quadrants is adopted in order to clarify how the graph has to be used. According to this hypothesis for defining the threshold, the company might consider it interesting to focus its human resource practices on the resources shown in the bottom right quadrant, considering them to be highly influenced by the policies and at present unsatisfactory in terms of value. Its target would therefore be to move the resources

in question towards the top right quadrant. On the other hand, the company might judge it inopportune to pay attention to the resources that fall within the bottom left quadrant, due to the fact that they are only slightly influenced by human resource practices and that they would not deliver any significant increase in value.

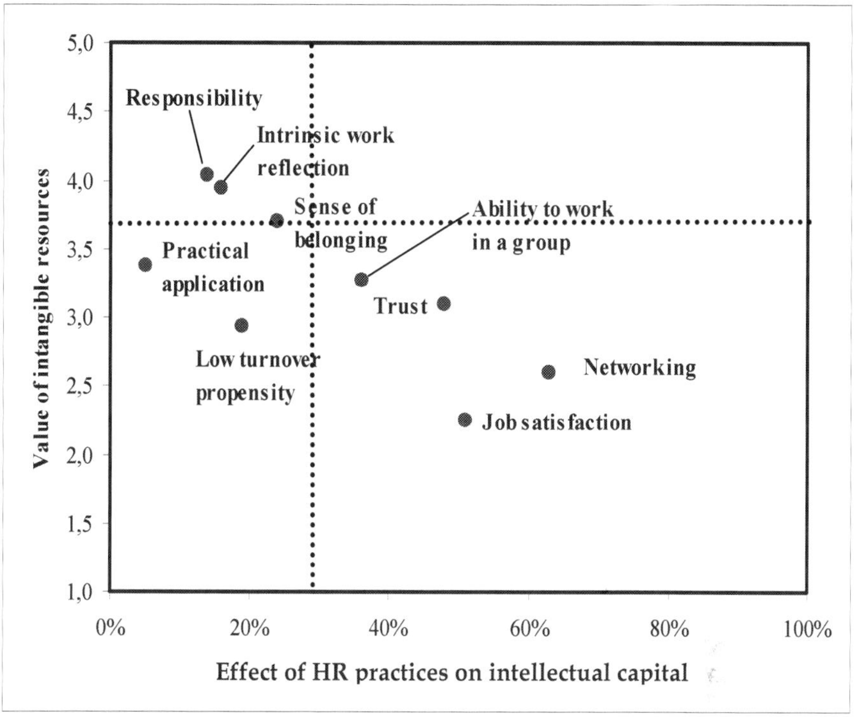

Fig. 3. Evaluation grid of intellectual capital resources - 2005 survey

In order to have a broad understanding of the effect the two human resource policies analyzed have on IC dimensions, a causal model has been developed using a structural equation modelling methodology (Figure 3). To determine the fit of the model we used the following indices: the normed fit index (NFI), the non-normed fit index (NNFI), the comparative fit index (CFI) and the chi-square statistics (Bollen, 1989) (Table 7).

Index	Sample 2005	Sample 2006
NFI	0.95	0.97
NNFI	0.97	0.97
CFI	0.97	0.98
Chi-square (d.f.)	2475.4 (718)	2689.91 (718)

Table 7. Goodness of fit indices

The results suggest a significant coherence between the model proposed and the data collected. In fact, fit indices higher than 0.9 are usually considered indicative of a good fit. The fact that the fit indices calculated using data collected in two different surveys are similar means that there is a reduced risk of there having been any non-random bias in the delivery, filling in and collection of the questionnaires over the two years. It therefore gives increased validity and strength to the results obtained.

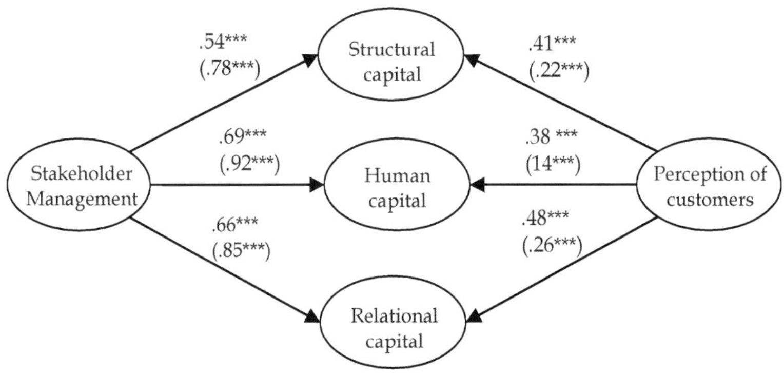

Numbers can be interpreted as standardized regression coefficients (β).
Not in brackets 2005 data, in brackets 2006 data; *** significant at $p < .001$.

Fig. 4. The effect of human resource practices on intellectual capital dimensions

The results suggest that the perception employees have of the company's stakeholder management practices has a much greater impact than the perception of customers on all three intellectual capital dimensions and, in particular, with respect to human capital and to relational capital (Figure 4). These results underline the importance that communication has within the company. The adoption and implementation of stakeholder management practices have limited worth if they are not adequately communicated to the company's employees. The possibility of capitalizing on the value created by investments human resource practices does in fact derive from the ability of company's managers to effectively communicate to employees the actions undertaken and the results attained.

5. Managerial implications

The intellectual capital measurement system we propose can be adopted by managers as a control tool for the company's intangible assets and may also support investments on company's human resource practices. In this section we suggest some implementations of the tool within the company analyzed. Specifically, we propose a cross-section analysis of company's intellectual capital, together with a longitudinal analysis using split-samples of diverse professional categories. ANOVA analyses, t-tests and non-parametric tests have been used in order to statistically validate the results.

Figure 5 shows the intellectual capital of the company in each year of analysis. In both the surveys we found a higher level of development for the human capital and the structural capital dimensions. The company appears to be capable of increasing the motivation of its employees, and these show a desire to continue to work within the organization, and of favouring their professional advancement in terms of skill development (*Low turnover propensity* and *Intrinsic work reflection* give a very high value of approximately 4 over 5). The critical factor seems to be linked to the relational capital dimension: *Communication* is little developed and this could also help to explain the low values of the resources *Trust* and *Job Satisfaction* which are closely related to *Communication* (correlation coefficients are 0.67 and 0.77 respectively and are statistically significant at $p < 0.01$).

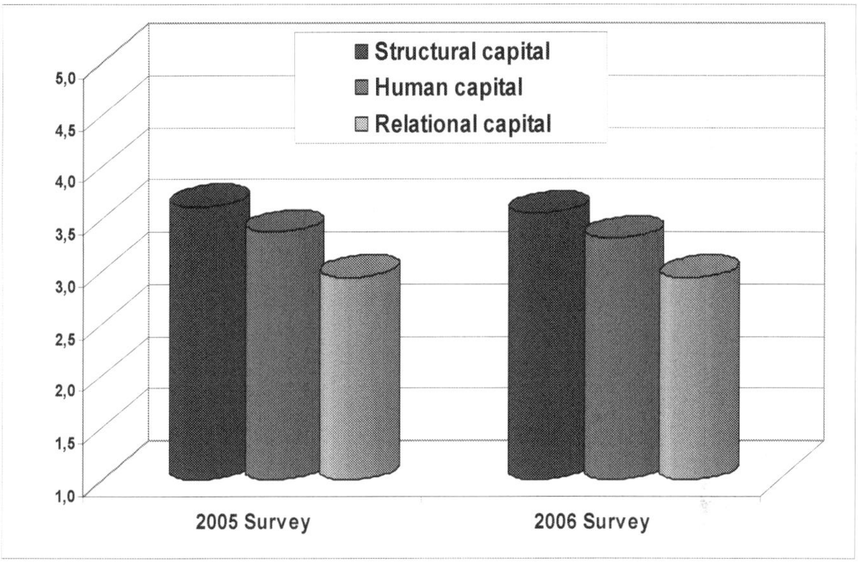

Fig. 5. Cross section analysis of the intellectual capital of the company

Comparison of the results obtained in the two consecutive years (longitudinal analysis) show decrements, although not significant in statistical terms, in the categories structural capital (-1.3%) and human capital (-1.2%). On examining the specific intellectual capital resources the t-tests show significant decrements in *Low turnover propensity* (-4%) and *Sense of Belonging* (-3%); strong decrements are also to be found in *Responsibility* (-2%) and *Communication* (-2%) (Figure 6).

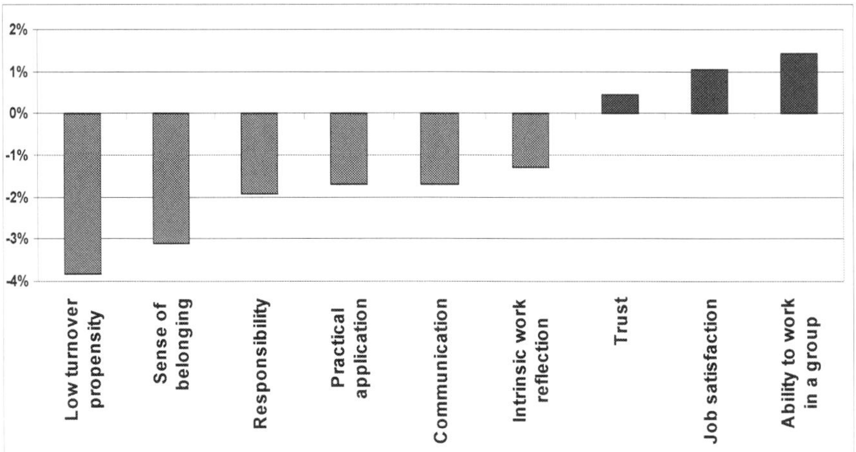

Fig. 6. Longitudinal analysis of the intellectual capital resources of the company

Besides being used as a company-wide measurement tool, the model can be adopted in order to monitor specific employees' categories. Statistical tests show a significant difference between managers and non-managers in the development of all the IC resources with the exception of *Practical application*. For each year of the survey the Managers have developed higher IC resources, particularly with regard to the resources *Trust, Job satisfaction, Responsibility and Communication*. However, the longitudinal analysis shows much more significant decrements in the IC resources within the Managerial category and, in particular, in *Communication, Job Satisfaction, Sense of* belonging and *Ability to work in a group* (Figure 7).

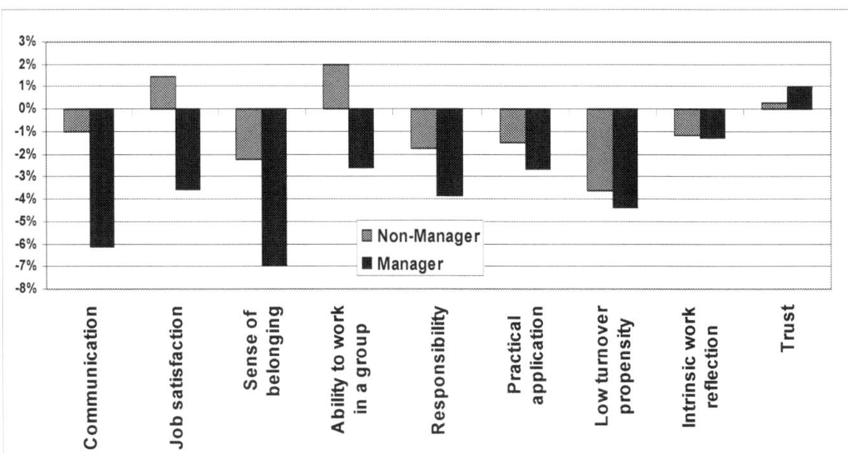

Fig. 7. Professional category - longitudinal analysis

The signal coming from the managerial category reveals an evident state of unease and widespread professional discontent. The split-samples analysis between middle-managers and executives (Figure 8) shows that middle-managers manifest relevant decrements in relational capital (-5%) and structural capital (-4%) which underlines the non-alignment of executives and middle-managers with regard to the development of intellectual capital.

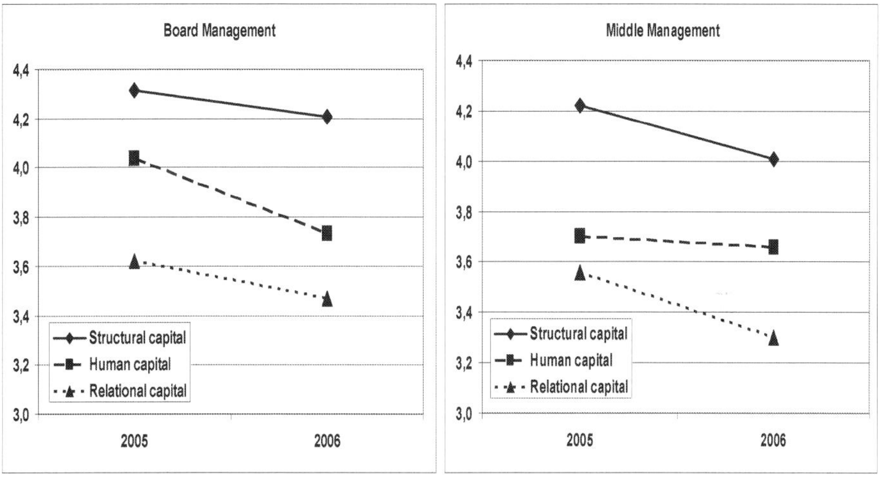

Fig. 8. Managers - longitudinal analysis

Other analyses can be proposed in relation to the plant, the seniority of the employee and other control variables, but go beyond the purpose of this chapter.

5. Conclusions

The measurement system proposed in this chapter represents a useful control tool for any organization that considers IC as a key driver for its value creation process. Measuring and monitoring over time the IC possessed by the company and assessing the effect that specific HR policies have on IC provide an important support for managers in setting targets and planning company's objectives in terms of its human resources.

By means of a quantitative case study we highlighted some implementations of the proposed model within an Italian company with a network of numerous production plants and distribution centres. The analyses conducted have underlined the existence of varied orientations in the organization's development of the IC and the presence of some critical situations that merit further examination.

The questionnaire developed in this study can be used for any kind of company that intends to measure IC.

Further research should attempt to overcome some of the limits of this study, like the exclusive adoption of perceptual indicators and the focus on a single organization, that certainly inhibits the generalizability of the findings.

6. References

Allee, V. (2000) The value evolution - Addressing larger implications of an intellectual capital and intangibles perspective. *Journal of Intellectual Capital,* Vol. 1, No. 1 pp. 17-32.

Acquaah, M. (2007) Managerial social capital, strategic orientation, and organizational performance in an emerging economy. *Strategic Management Journal,* Vol. 28, No. 12 pp. 1235-1255.

Adler, P. S. & Kwon, S. (2002) Social capital: prospects for a new concept. *Academy of Management Review,* Vol. 27, No. 1 pp. 17-40.

Andriessen, D. (2004) IC valuation and measurement: classifying the state of the art, *Journal of Intellectual Capital,* Vol. 5, No. 2 pp. 230-242.

Arthur, J. B. (1994) Effects of human resource systems on manufacturing performance and turnover, *Academy of Management Journal,* Vol. 37, No. 1 pp. 670-687.

Baker, W. (1990) Market networks and corporate behaviour *American Journal of Sociology,* Vol. 96 pp. 589-625.

Becker, G. S. (1964) *Human Capital.* Columbia University Press, New York.

Bollen, K. A. (1989) *Structural equations with latent variables.* New York: Wiley.

Bontis, N. & Fitz-enz, J. (2002) Intellectual capital ROI: a causal map of human capital antecedents and consequents, *Journal of Intellectual Capital,* Vol. 3, No. 3 pp. 223-247.

Bontis, N. (1998) Intellectual capital: An exploratory study that develops measures and models, *Management Decisions,* Vol. 36, pp. 63-76.

Bontis, N., Keow, W. C. C. & Richardson, S. (2000) Intellectual capital and business performance in Malaysian Industry, *Journal of Intellectual Capital,* Vol. 1, No. 1 pp. 85-100.

Bourdieu, P. (1986) The forms of capital. In *Handbook of theory and research for the sociology of education* (Richardson JG, Ed.), pp. 241-258, Greenwood, New York.

Brown, J. S. & Duguid, P. (1991) Organizational learning and communities-of-practice: Toward a unified view of working, learning and innovation, *Organization Science* , Vol. 2, pp. 40-57.

Burt, R. S. (1992) *Structural holes: The social structure of competition.* Harvard University Press, Cambridge, MA.

Carlucci, D., Marr, B. & Schiuma, G. (2004) The knowledge value chain - How intellectual capital impacts business performance, *International Journal of Technology Management,* Vol. 27, No. 6-7 pp. 575-590.

Chalykoff, J. & Kochan, T. A. (1989) Computer aided monitoring: its influence on employee job satisfaction and turnover, *Personnel Psychology,* Vol. 42 No. 4, pp. 807-834.

Dyer, J. H. & Nobeoka, H. (2000) The relational view: cooperative strategy and sources of interorganizational competitive advantage, *Academy of Management Review,* Vol. 23, No. 4 pp. 660-679.

Edvinsson, L. & Malone, M. S. (1997) *Intellectual capital: Realizing your company's true value by finding its hidden brainpower,* 1st edition, Harper Collins, New York.

Flamholtz, E. & Lacey, J. (1981) *Personnel management: Human capital theory and human resource accounting,* Institute of Industrial Relations, UCLA, Los Angeles.

George, J. M. (1992) Extrinsic and intrinsic origins of perceived social loafing in organizations, *Academy of Management Journal,* Vol. 35 No. 1, pp. 191-202.

Grant, R. M. (1996) Prospering in dynamically-competitive environments: organizational capability as knowledge integration, *Organization Science, Vol.* 7, No. 4 pp. 375-387.

Hatch, N. W. & Dyer, L. H. (2004) Human capital and learning as a source of sustainable competitive advantage, *Strategic Management Journal,* Vol. 25, pp. 1155-1178.

Hitt, M. A., Bierman L., Shimizu, K. & Kochhar ,R. (2001) Direct and moderating effect of human capital on strategy and performance in professional firms: a resource-based perspective, *Academy of Management Journal,* Vol. 44, No. 1 pp. 13-28.

Holman, D., Epitropaki, O. & Fernie, S. (2001) Understanding learning strategies in the workplace: a factor analytic investigation, *Journal of Occupational and Organizational Psychology,* Vol. 74, pp. 675-681.

Hudson, W. (1993) *Intellectual Capital. How to build it, enhance it, use it.* John Wiley, New York, NY.

Hunter, L., Webster, E. & Wyatt, A. (2005) Measuring intangible capital: A review of current practice, *Australian Accounting Review,* Vol. 15, No. 2 pp. 4-21.

IASB (2007) *Projet Intangibles. Tecnical Paper 23.1.2007.* available at: www.iasb.org

Jacobs, J. (1965) *The death and life of great American cities.* Penguin Books, London.

Jöreskog, K. & Sörbom, D. (2006) *Lisrel 8.* Lincolnwood, IL: Scientific Software International.

Kale, P., Singh, H. & Perlmutter, H. (2000) Learning and protection of proprietary assets in strategic alliances: Building relational capital, *Strategic Management Journal,* Vol. 21, pp. 217-237.

Kidwell, R. E. & Robie, C. (2003) Withholding efforts in organizations: toward development and validation of a measure, *Journal of Business and Psychology,* Vol. 17, No. 4 pp. 537-561.

Lepak, D. P. & Snell, S. A. (1999) The human resource architecture: towards a theory of human capital allocation and development, *Academy of Management Review,* Vol. 24, No. 1 pp. 31-48.

Lin, N. (2001). *Social capital: A theory of structure and action.* Cambridge University Press: New York.

Longo, M. & Mura, M. (2007) A multidimensional measure of employees' intangibles. A managerial implementation of the tool, *Management Research News,* Vol. 30, No. 8 pp. 548-569.

Loury, G. C. (1977) A dynamic theory of racial income differences. In P. A. Wallace & A. M. LaMonde (Eds.), *Women, minorities and employment discrimination,* pp. 153-186. Lexington, MA: Lexington Books.

M'Pherson, P. K. & Pike S. (2001) Accounting, empirical measurement and intellectual capital, *Journal of Intellectual Capital,* Vol. 2, No. 3, pp. 246-260.

McGill, M. E. & Slocum, J. W. (1994) *The smarter organization: How to build a business that learns and adapt to marketplace needs,* Wiley, New York, NY.

Nahapiet, J. & Goshal, S. (1998) Social capital, intellectual capital, and the organizational advantage, *Academy of Management Review,* Vol. 23, No. 2 pp. 242-266.

Nelson, R. R. & Winter, S. G. (1982) *An evolutionary theory of economic change,* Belknap Press, Cambridge, MA.

Park, S. H. & Luo, Y. (2001) Guanxi and organizational dynamics: organizational networking in Chinese firms, *Strategic Management Journal,* Vol. 22, No. 5 pp. 455-477.

Peng, M. W. & Heath, P. S. (1996) The growth of the firm in planned economies in transition: institutions, organizations, and strategic choice, *Academy of Management Journal*, Vol. 21, No. 2 pp. 492-528.

Peng, M. W. & Luo, Y. (2000) Managerial ties and firm performance in a transition economy: the nature of a micro-macro link, *Academy of Management Journal*, Vol. 45, No. 3 pp. 486-501.

Penrose, E. T. (1959) *The Theory of the Growth of the Firm*. Blackwell, Oxford.

Pfeffer, J. & Salancik, G. R. (1978) *The external control of organizations: A resource dependence perspective*. Harper & Row, New York.

Pisano, G. P. (1989) Using equity participation to support exchange: Evidence from the biotechnology industry, *Journal of Low, Economics, and Organization*, Vol. 5, pp. 109-126.

Polanyi, M. (1962) *Personal knowledge: Towards a post-critical philosophy* (corrected ed.). University of Chicago Press, Chicago, IL.

Roos, G., Pike, S. & Fernström, L. (2005), *Managing intellectual capital in practice*. Elsevier Publications, Burlington (MA).

Roslender, R. & Fincham, R. (2001) Thinking critically about intellectual capital accounting, *Accounting, Auditing & Accountability Journal*, Vol. 14, No. 4, pp. 383-398.

Roslender, R., Stevenson, J. & Kahn, H. (2006) Employee wellness as intellectual capital: an accounting perspective, *Journal of Human Resource Costing and Accounting*, Vol. 10, No. 1 pp. 48-64.

Schiemann, W. A. (2006) People equity: A new paradigm for measuring and managing human capital HR, *Human Resource Planning*, Vol. 29, No. 1 pp. 34-44.

Schultz, T. W. (1961) Investment in human capital, *American Economic Review, Vol.* 51(March), pp. 1-17.

Serenko, A. & Bontis, N. (2004) Meta-Review of Knowledge Management and Intellectual Capital Literature: Citation Impact and Research Productivity Rankings, *Knowledge and Process Management*, Vol. 11, No. 3 pp. 185-198.

Snell, S. A. & Dean, J. W. (1992) Integrated manufacturing and human resource management: A human capital perspective, *Academy of Management Journal*, Vol. 35, No. 3 pp. 467-504.

Spender, J. C. (1996) Making knowledge the basis of a dynamic theory of the firm, *Strategic Management Journal*, Vol. 17(winter special issue), pp. 45-62.

Staples, D. S., Hulland, J. S. & Higgins, C. A. (1999) A self-efficacy theory exploitation for the management of remote workers in virtual organizations, *Organization Science*, Vol. 10, No. 6 pp. 758-776.

Starbuck WH (1992) Learning by knowledge intensive firms Journal of Management Studies 29, 713-740.

Sveiby, K. (1997) *The invisible balance sheet: Key indicators for accounting, control, and evaluation of know-how companies*, The Konrad Group, Stockholm.

Tayles, M., Bramley, A. & Farr, A. J. (2002) Dealing with the management of intellectual capital. The potential role of strategic management accounting, *Accounting, Auditing & Accountability Journal*, Vol. 15, No. 2 pp 251-267.

Walker, G., Kogut, B. & Shan, W. (1997) Social capital, structural holes, and the formation of an industry network, *Organization Science*, Vol. 8, pp. 109-125.

Weick, K. E. & Roberts, K. H. (1993) Collective mind in organizations: heedful interrelating on flight decks, *Administrative Science Quarterly*, Vol. 38, pp. 357-381.

Youndt, M. A. & Snell, S. A. (2004) Human resource configurations, intellectual capital, and organizational performance, *Journal of Managerial Issues*, Vol. 16, No. 3 pp. 337-360.

Zeitz, G., Johannesson, R. & Ritchie, J. E. (1997) An employee survey measuring total quality management practices and culture. Development and validation, *Group & Organization Studies*, Vol. 22, No. 4 pp. 414-444.

Zucker, L. G., Darby, M. R., Brewer, M. B. & Peng, Y. (1996) Collaboration structures and information dilemmas in biotechnology: Organization boundaries as trust production. In *Trust in organizations: Frontiers of theory and research* (Kramer RM and Tyler TR, Eds), pp. 90-113, Sage, Thousand Oaks, CA.

21

SIMEFAS: Wide Area Measurement, Protection and Control System in Mexico

Enrique Martínez Martínez
Comisión Federal de Electricidad
México

1. Introduction

This chapter reports on the development of a Wide-Area Measurement System in Mexico entitled SIMEFAS (*Sistema de Medición Fasorial*). It has been designed to increase the security and integrity of Mexico's electric power system. Current development of SIMEFAS includes five regional PDCs to which phasor data is sent by PMUs spread over the National Electrical System. Developed applications include model-validation, computation of operation and performance curves, automatic generation and load shedding, adaptive relaying, and CCVT monitoring. SIMEFAS's architecture as well as future development plans are also presented. The remainder of this chapter is organized as follows. Section 2 gives a description of SIMEFAS and the initial deployment stage of the system. Section 3 discusses the structure of the Mexican National Electrical System and of SIMEFAS. In Section 4, the main applications that have been deployed and considerations for the deployment of future applications are presented. In Section 5, the architecture and hierarchy levels of SIMEFAS is presented. Section 6 discusses the Local Control and Protection Actions Using Synchronized Phasor, and in the section 7 the Wide Area Protection Scheme using PMUs in SIMEFAS is presented.

2. Description of SIMEFAS

During the 1990's Mexico's Federal Electricity Commission (*Comisión Federal de Electricidad,* CFE) envisioned a project that consisted of the deployment of a synchronized phasor measurement system for contingency analysis, and visualization of the operational state of the National Electrical System (*Sistema Eléctrico Nacional,* SEN). The initial stage of this project consisted in the installation of fifteen Phasor Measurement Units (PMUs) in two of the four subsystems that comprise the National Electrical System. The installation criteria for the PMUs considered included larger generation complexes, major load points, and critical power transfer interfaces. In the initial stage post-fault analysis and model validation of power system simulator models have been of primary interest. In the second stage, a synchronized phasor measurement system for the security and integrity of Mexico's electric power system, namely SIMEFAS, has been proposed. SIMEFAS includes visualization and real-time decision applications whose main aim is to guarantee reliability and security for the National Electrical System. With the current fiber optic infrastructure installed by CFE,

SIMEFAS acquires major importance, enabling the application of special protection schemes and discrete controls. SIMEFAS has also promoted international participation within NASPI (North American Synchropahsor Initiative), which proposes the integration of phasor measurement information between Canada, USA, and Mexico. The infrastructure developed by CFE allows the installation of a phasor data concentrator (PDC) at each transmission region giving rise to regional wide-area measurement systems at very low cost. This infrastructure also allows the integration of PMUs from different manufacturers and models and to take advantage of the PMU functions implemented in digital relays within discrete control schemes and substation control modules. SIMEFAS has provided data for fault recording analysis, database validation, contingency simulation, and prototype adaptive protection (E. Martínez 2002) and discrete control applications through the calculation of angular differences (E. Martinez 2008).

3. Structure of the National Electric System of Mexico and SIMEMAS

The functionality of a wide-area measurement system is based on the operational philosophy and structure of electric power systems. This is the main reason why many of the WAMS that operate successfully worldwide have been built by transmission system operators and utilities. Based upon the state of the transmission network and the requirements from economic dispatch, these entities are able to adapt phasor measurement data to the visualization and information needs of their operators and users of specific applications (S. H. Horowitz 1987 and D. Karlsson 2001).

At CFE the key to the development of SIMEFAS has been the integration of PMUs, digital relays, and disturbance recorders with PMU functionalities to regional PDCs.

As of November 2007, the National Electrical System is formed by three main electric power systems that operate independently (see Fig. 1):

- National Interconnected Power System (*Sistema Interconectado Nacional*)
- North Baja California Electric Power System (*Sistema Eléctrico de Baja California Norte*)
- South Baja California Electric Power System (*Sistema Eléctrico de Baja California Sur)*

Based on the topology of the National Electrical System, three phasor data concentrators were installed one for each independent power system. The number of PMUs in each concentrator will depend on the information requirements from each substation and each system, and also on the utilization of the phasor measurement data which may be used by protection specialists, analysts, system operators, etc.

For off-line analysis, the installation of GPS-synchronized PMUs represents 90% of the project. Meanwhile, for real-time visualization and real-time decisions in the power system, the communications infrastructure surpasses 95% of the project. The main reason is that to obtain reliable angle measurements in the PDC information has to be transmitted in a pre-defined bandwidth that allows synchronization of PMUs with minimum delay and continuous flow (V. Centeno 1993, E. Martinez 2006).

4. Main Applications within SIMEFAS

CFE has established short- and mid-term goals to enable the following four applications within SIMEFAS:

1.1. SIMEFAS Off-Line, Post-fault analysis
1.2. SIMEFAS RT, Wide Area Measurement System
1.3. Wide-Area Protection and Control Schemes

4.1. SIMEFAS Off-Line Analysis

Phasor Measurement Units have been in use in Mexico since 1990s. Initial interest was in observing the power system through the eyes of the high precision synchronized measurement system provided by the PMUs. The data obtained from the PMUs over the years has been invaluable in observing under-damped low frequency oscillations on the network, and the performance of protection systems during power system disturbances.

On March 10, 2005 two of the four electric systems in Mexico (Northwest and the National Grid), which had operated asynchronously until that date were synchronized. See Figure 1. As soon as the tie-lines were connected and the systems synchronized, sustained inter-area oscillations were observed throughout the system which eventually led to separating the systems again. These low frequency oscillations spread throughout the whole power system of the country, especially affecting the most remote points in the network, subjecting the generators and turbines to extraordinary control actions and putting at risk the integrity of the whole Electric System.

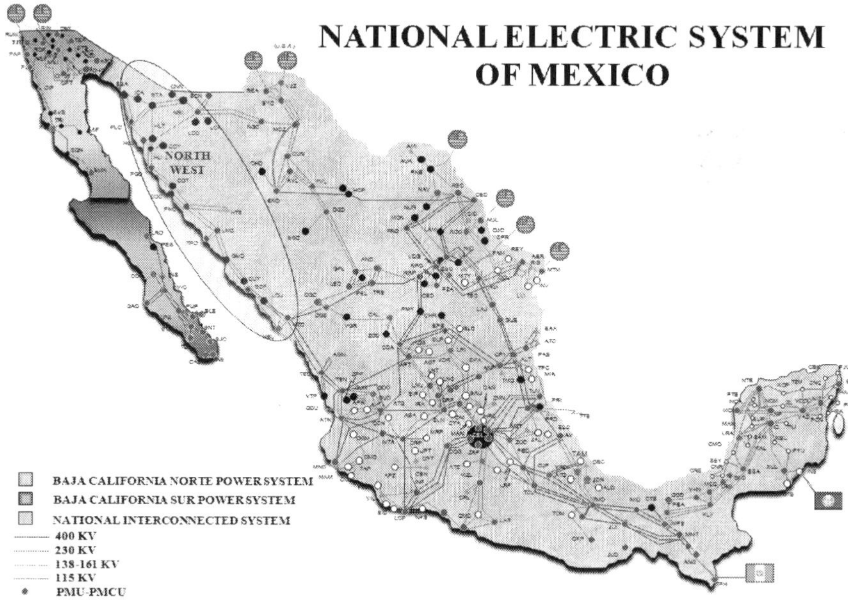

Fig. 1. Three independent systems in the National Electric Network of Mexico 2009

The challenge for the engineers of CFE was to determine the origin of these oscillations, to determine the oscillation modes and the participation factors of main generators in the two systems. This process required many simulations for different conditions of generation, load and network configurations. The PMU recordings were the most important tool in investigation and analysis performed to determine safe procedures for the interconnection.
Thanks to Wide Area Measurement System implemented in the CFE, it was possible to record and simulate the oscillations that were present during the synchronous operation. The strategic location of the PMU´s was a decisive factor in the analysis and the visualization of the oscillations in real time. PMUs also make possible direct visualization of power system frequency, voltage, phase angle, active and reactive powers etc. Figures 2 & 3.

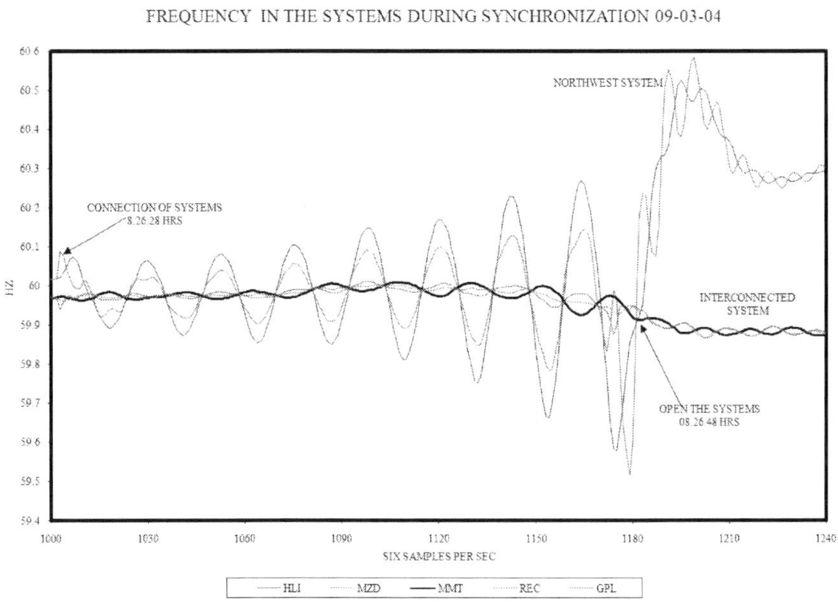

Fig. 2. Undamped inter-area frequency oscillations in both systems during synchronization in March 09 2004.

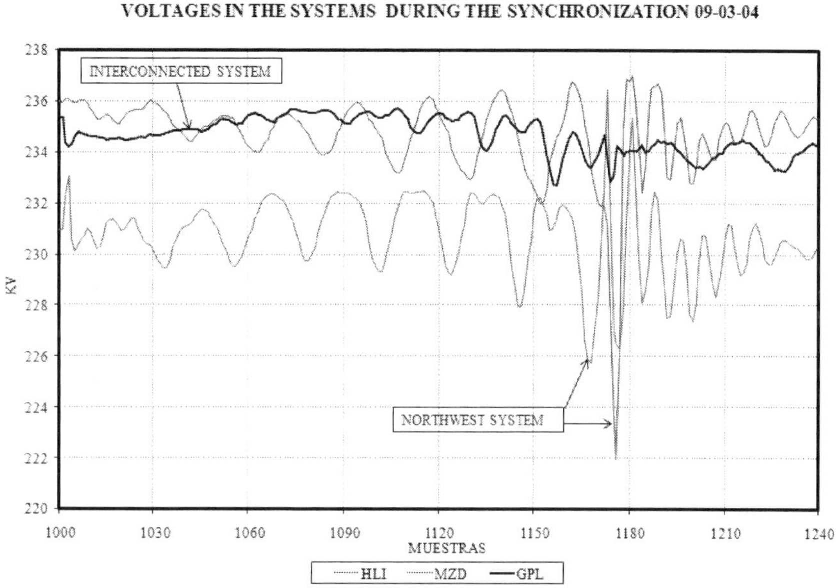

Fig. 3. Undamped inter-area voltage oscillations in both systems during synchronization in March 09 2004.

However, to find the origin of the oscillations and to the correct settings for excitation controls, power system stabilizers of generators it was necessary to calculate the curves of Real Power vs Frequency (P-F) and Reactive Power vs Voltage (Q-V) of machines and connection circuit among systems. Figures 4 & 5 show the behavior of speed and voltage regulators of generators with strong participation in an un-damped oscillations present during 20 seconds of systems interconnection.

At the moment, the systems operate interconnected through different points and voltages levels, however we continue the studies that allow us to detect power oscillations that put in risk the stability and integrity of the electric system in Mexico.

Using the records obtained with the PMUs during diverse interconnection tests, and contingencies records that have been presented in the interconnected systems, now we are studying the third dimension curves P-F and Q-V.

In this three dimensional analysis, figs. 6 & 7 shows the same graphs as figs. 4 & 5 but, in the time. Let us observe that the circuit that interconnected the systems only has frequency when it is opened (without real power in the circuit), this is due to the fact that the voltage is connected to bus PT.

In the future, the off line analysis will make it possible to implement in real time the detection of risks of instability, using Phasors Measurement Units, and establishing limits with alarms for oscillations of P-F, Q-V, P-V, etc curves, for generators, tie lines or critical points of systems during various levels of load, generation or for different system configurations.

In this process of interconnection of systems, the PMUs also has been very important in analyzing the effect of the oscillations in the transmition line protection schemes. See fig. 8

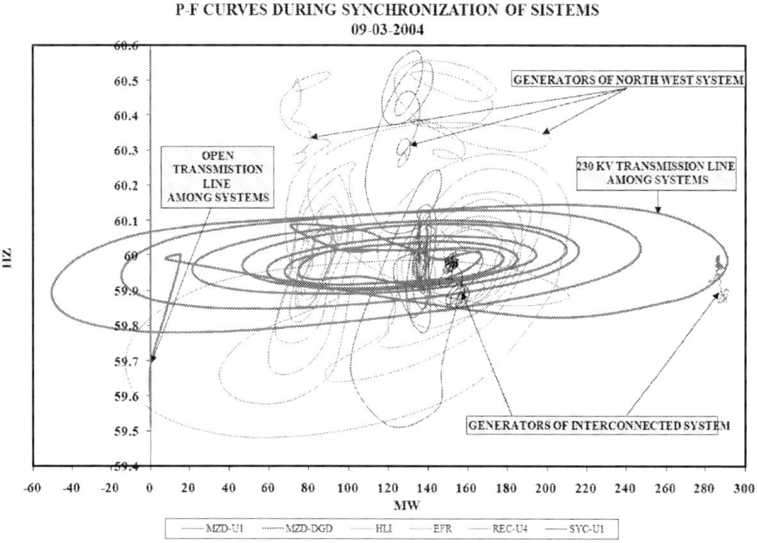

Fig. 4. P-F curves of the generators with larger participation in the oscillations during the interconnection of systems. March 09 2004.

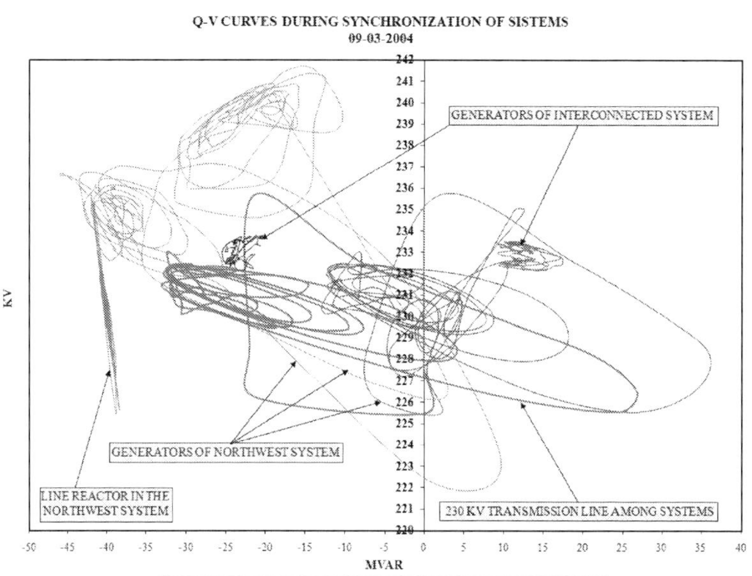

Fig. 5. Q-V curves of the generators with larger participation in the oscillations during the interconnection of systems. March 09 2004.

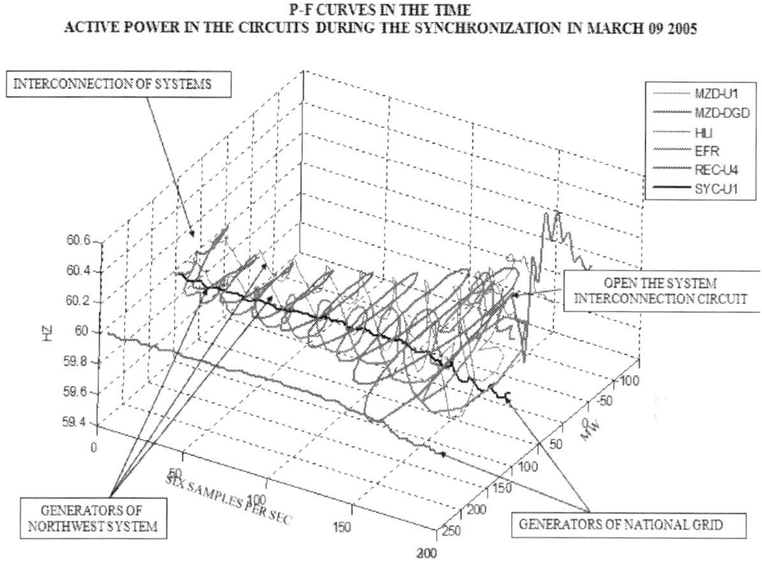

Fig. 6. P-F curves during 20 seconds of systems interconnection.

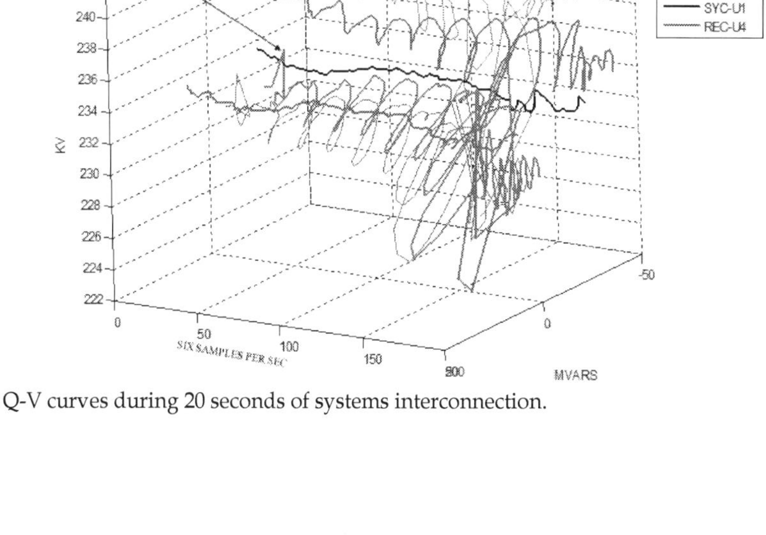

Fig. 7. Q-V curves during 20 seconds of systems interconnection.

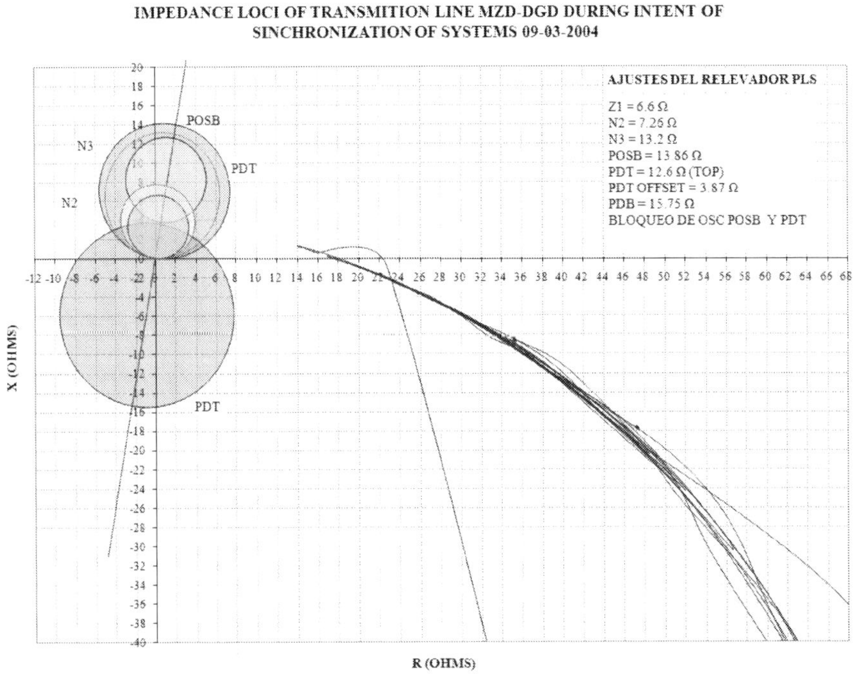

Fig. 8. R-X curves during 20 seconds of systems interconnection.

Currently data recording no longer requires triggering, and is done continuously, with all the information sent to regional PDCs. A sampling frequency of 20 samples per second for all the PMUs installed within the National Interconnected Power System, and of 30 samples per second for the PMUs at the North Baja California Electric Power System. The sampling rate at North Baja California was selected to share the same sampling rate as the one proposed by NASPI that will coordinate all the phasor measurement from NERC. The information is stored in batches of 5000 samples. However, these information packets can be made smaller if the number of installed PMUs is larger or if the archival procedure requires large amounts of memory or time.

For off-line data analysis and applications it is important to have a sufficiently high sampling frequency, reliable data capture, and precise signal processing. These requirements intrinsically depend on the technology of the PMUs. For these applications the communications infrastructure does not affect the reliability of the analyses.

4.2. SIMEFAS-RT Wide Area Measurement System

For this application it is important to have high-quality PMUs and a reliable and secure communications system, preferably based on fiber optic, interfaces and routers with sufficient bandwidth.

To provide operators with adequate signals for corrective or preventive actions, state visualization requires high speed data transmission because the computation of angle differences is done in real-time as the samples of each PMU reach the PDCs. See fig. 9

Sampling frequency is determined by the requirements of each specific real-time application, dynamic or transient. In other countries there are applications that use two samples per cycle requiring high bandwidth from the communication channels and large memory capacity of PDCs for data storage.

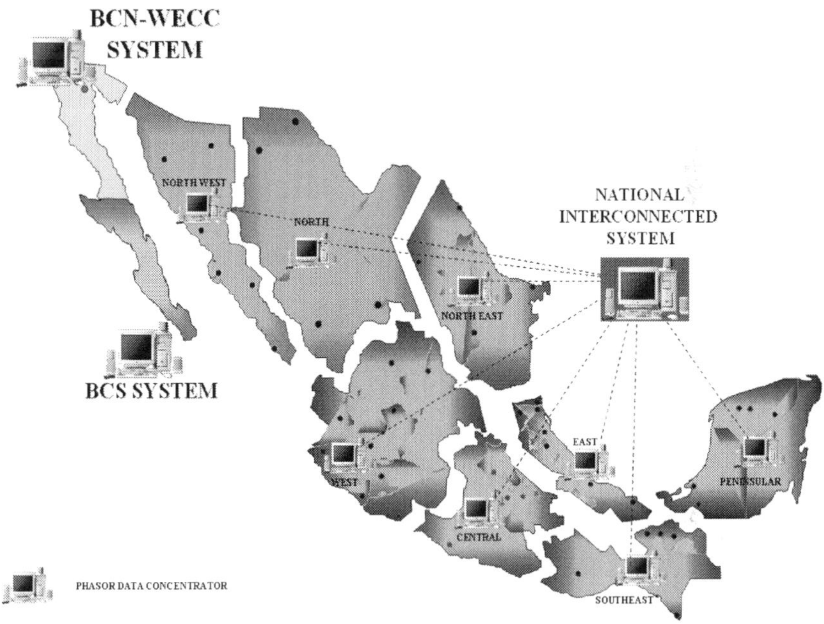

Fig. 9. Regional PDCs and integration of few strategic PMUs into the single PDC of the National Electrical System

4.3. Wide Area Protection and Control Schemes

Currently CFE is developing an adaptive protection scheme based on the angular difference between subsystems. CFE is also performing field test data analysis on the behavior of a prototype generation shedding scheme which assesses the transmission capability among two hydro generation stations. The scheme has a decision logic that uses signals from the active power flow, voltage, frequency, breaker state, and angular difference between the two stations. Both the adaptive protection and generation shedding schemes are fully independent from the WAMS with regards of decision making for maintaining stability in the system. Nevertheless, phasor information is continuously sent to the WAMS providing visibility of isolated areas in case of islanding. See section 6.

5. Architecture and hierarchy level of SIMEFAS

WAMS of CFE have been designed to satisfy the requirements from each phasor measurement data user or potential client. This has been done to cater to the speed, quantity, and type of information needs of each client depending on their responsibility area and security.

Fig. 10 shows pyramidal architecture. The base is formed by the network entities that manage local data, *i.e.* substations and generation stations. At this level the local data transfer speed requirements to PDCs are lower than local data management requirements. At the highest level less local PMU information is required for obtaining a broader visibility of the overall system, this requires a higher information transfer speed than at the lower levels.

Fig. 10. Data management structure. SIN: National Interconnected Power System (*Sistema Interconectado Nacional,* SIN), Regional Control Task Manager (*Gerencia Regional de Transmisión*), Interarea Ties (*Enlaces interareas*), Substations (*Subestaciones*), Generation Stations (*Centrales Eléctricas*).

5.1. First level: Substations and Generation Stations

At this level data transfer speed is not of primary concern and applications are unlimited. Applications are local and independent; however, it is possible to send a lower amount of information with higher speed to PDCs to enable the functioning of a Real-Time WAMS. At CFE we have seen that at this level the use of PMUs or disturbance recorders with PMU functions have shown more advantages over digital relays with PMU functionalities. The main current and voltage phasors being monitored in a substation will share the same time tags, which is important when analyzing the response of machine clusters, automatic voltage regulators, stabilizers, and the AGC. Common monitored variables from PMUs at this level are frequency, voltage, phase angle, active and reactive power. These variables have been used to compute *R-X*, *P-f*, *Q-V*, and *PV* curves. See fig. 11

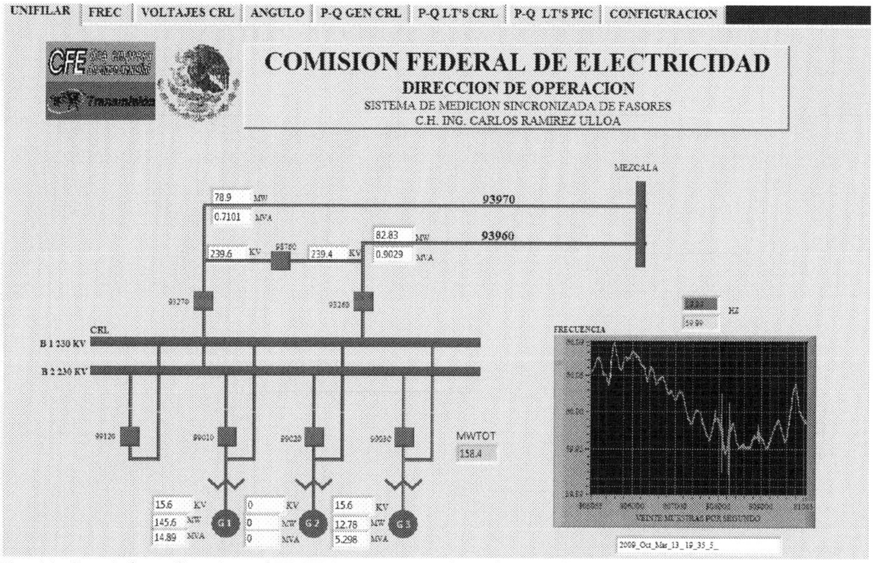

Fig. 11. Special application of PMU in a power plant for oscillation and PSS actions analysis.

5.2. Second level: Inter Area Links

This level is important for the analysis of the behavior of critical inter-area ties which have shown small signal oscillations or power flow inversions that have impact on transmission line protection schemes.

At CFE, one of the dilemmas that protection engineers have faced is on how protection schemes are enabled or disabled when power oscillations appear. Triggering of protection schemes requires monitoring of the magnitude, speed, and frequency of oscillation of the modes involved. This permits maintaining stability and load-generation balance when large disturbances occur in the system.

Also, CFE has developed other special applications for the monitoring of connections critics with restrictions in the power transmission or low frequency oscillations circuits by radial topology of systems, is the case of the interconnection Mexico- Central America through Guatemala in 400 kV. Fig. 12

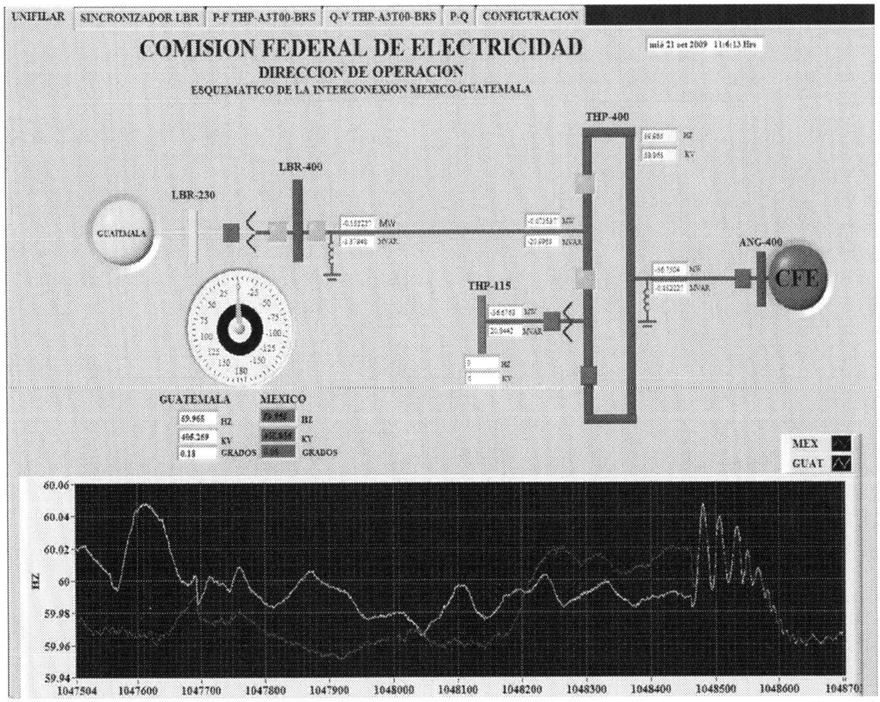

Fig. 12. Special application to study inter systems oscillations, during interconnection of Mexico-Guatemala systems.

5.3. Third Level: Regional PDCs

At this level the main goal is to guarantee efficient information management and to provide appropriate maintenance to the PMUs and PDCs installed by each Regional Transmission Manager (*Gerencia Regional de Transmisión*). These are shown in Fig. 9 Location of PDCs.

CFE has considered using the wealth of available information for real-time visualization of each regional area independently and to apply it in transformer bank and transmission line loading studies, energy interchange, power quality, and the behavior of protection and relief schemes. Fig. 13

Fig. 13. Using PMUs in real time operation of power system.

5.4. Fourth level: Central PDC SIMEFAS

CFE has considered using the wealth of available information for real-time visualization of each regional area independently and to apply it in transformer bank and transmission line loading studies, energy interchange, power quality, and the behavior of protection and relief schemes. CFE has designed to integrate all the information from the regional PDCs and from strategically placed PMUs into a single PCD for each island system. This PDC will hold the most selective information from a large number of PMUs allowing it to synchronize frequency and voltage phasor measurements with ease and to calculate phase angles accurately, giving the operator a broader view of the system state from the measurements from each selected location. To enable a state estimator it is also necessary to have active and reactive power measurements from the networks, and PMUs can provide this information. At the moment each PMU can simultaneously send the information to four different concentrators and the applications are specific of each user. The information that Integra in the central PDC does not go through the local concentrator, since this one is sent directly from each PMU through independent channels. See Fig. 14 Architecture SIMEFASNet.

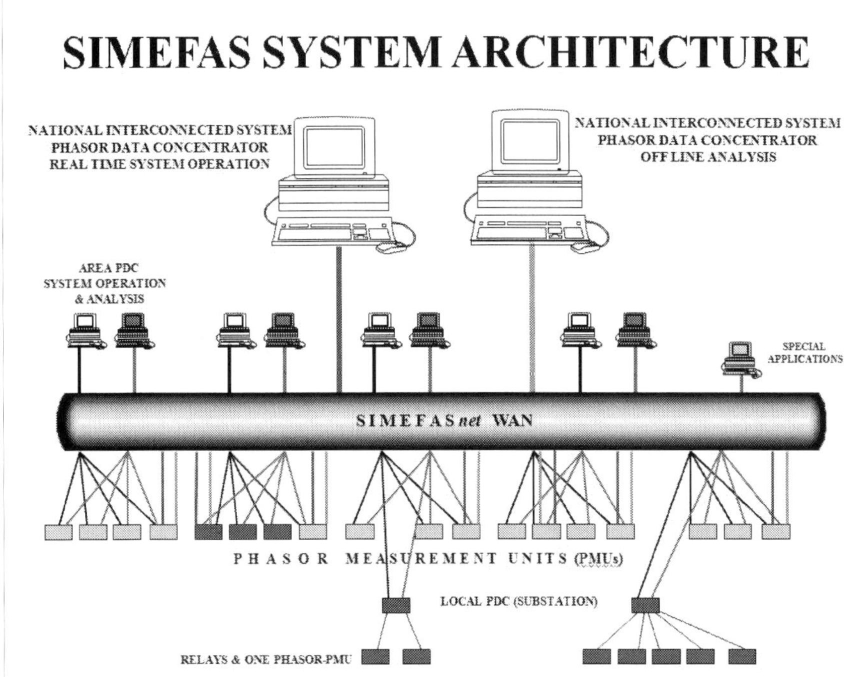

Fig. 14. Schematic of SIMEFASnet.

6. Local Control and Protection Actions Using Synchronized Phasor

Another application in use at CFE is the monitoring of CCVTs. Some regions in Mexico experience extreme heat and humidity conditions. CFE has observed that under these conditions CCVTs may explode. Voltage differences in the CCVT are monitored in real-time. When abnormal conditions are detected an alarm will enable and the equipment is taken out of service to protect the device, the installation, and the personnel.

Based on registers obtained by PMUs, we have identified the CCVT behavior model minutes before its explosion and we determined the alarm times or transmition lines opening for the change of the transformers that put in risk the people and adjacent equipment during an explosion, which in addition, affects the service by generation or load trip, since during this phenomenon all the substation can be lost. See fig. 15

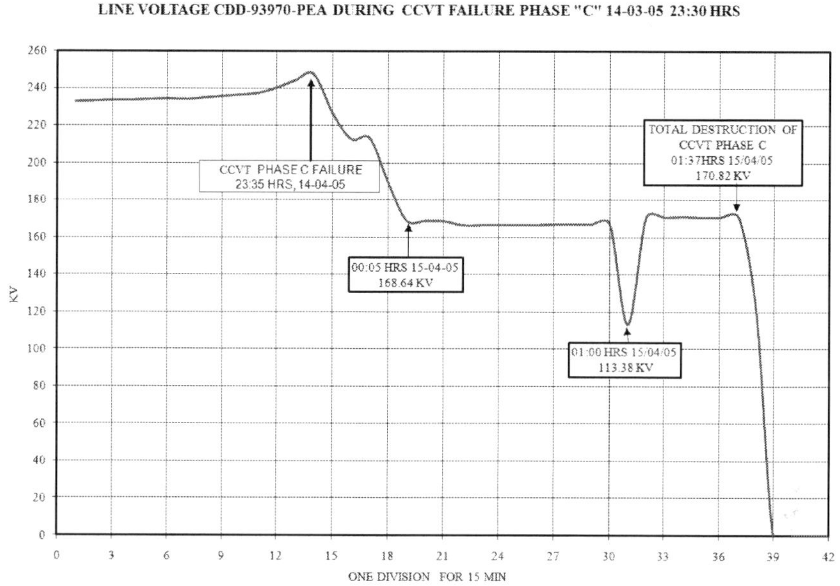

Fig. 15. Application of PMUs for substations local control actions, based on analysis of voltage behavior CCVT.

7. Wide Area Protection & Control Schemes

Recently some relay manufacturers have implemented PMU functionalities in distance and overcurrent protective schemes, these prototype schemes or Phasor Measurement and Control Units (PMCUs) provide the assignment of logical variables in devices. In angular instability triggered-based automatic generation or load shedding applications the Phasor Measurement and Control Units PMCUs must take control actions independently from the GPS signal, similarly as is done in differential protection schemes. This new special protection scheme, called "Angular Difference Protection Scheme", should be able to operate as a discrete control scheme and at the same time transmit measurements at the same sampling frequency and under the standard IEEE C37.118 protocol (IEEE Sinchrophasors for Power Systems 2006). In this section, we will analyze the application of the PMCUs in a remedial action called Automatic Generation Shedding Schemes (AGSSs), however, this same application principle can be used in dynamic breaker control or a load shedding action when there is a loss of a circuit in the multiple connections among areas, systems, countries or electric companies. Currently in industry, programmable logic controllers (PLCs) are used to make control actions through dedicated communication channels that allow decision making based on the pre-programmed logic. These control actions enable automatic generation shedding, load shedding, or transmission line switching. However, with this method when the system is separated through opening of tie

lines the system operator losses control and visibility of isolated areas making the event analysis and resynchronization process complex and slow (E. Martinez 2006).
As shown in several studies, wide area monitoring, protection, and control systems (WAMPAC) are required to measure, evaluate the measurement, and return the control action commands. This process requires high reliability and speed in the communications system to provide control actions when necessary, or to detect false triggers. Thus measurement, protection and control concepts need to be integrated within these systems. CFE's WAMS integrates different PMU brands and models. CFE is installing these relays to perform control actions based on voltage angle differences calculated at different locations in the power system. The success of these applications depends on the ability of the relays and communication networks to perform these tasks.

7.1. Automatic Generator Shedding Using Synchronized Measurement

The real power transfer, P, between two network buses connected by a reactance, X_L, is determined by the phase angle difference δ, the voltage magnitudes at the buses, E_A and E_B, and the reactance, X_L (see Fig. 16). Notice that the angle at Bus B (reference) is 0. The two buses exchange real power according to Equation 1.

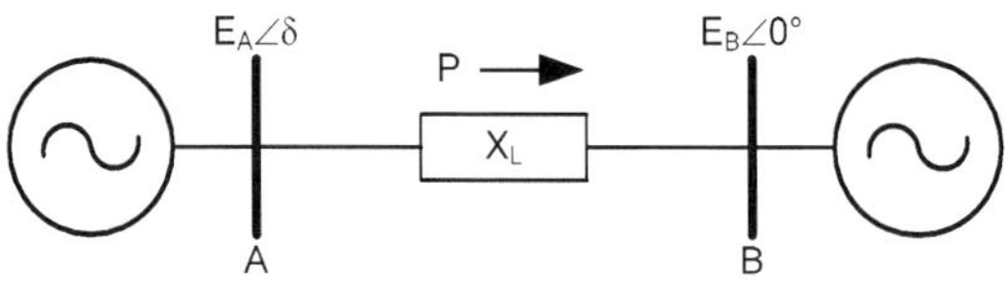

Fig. 16. δ, E_A, E_B, and X_L Determine the Real Power Transfer, P, Between Bus A and Bus B

$$P = \frac{E_A \bullet E_B}{X_L} \bullet \sin\delta \qquad (1)$$

During steady state operating conditions, the voltage magnitudes of the network buses are close to one per unit. That is, the real power transfer capability mainly depends on the phase angle difference, δ, and the transmission link reactance, X_L. X_L depends of the number of lines and transformers in service between the two buses. When transmission lines are lost during a system disturbance, X_L increases and the angle difference also increases to maintain the same amount of real power exchange between the two buses. Fig. 17 illustrates the real power transfer capability and the real power transfer operating point as a function of the angle difference during normal operating conditions and after transmission links are lost because of a system disturbance. Notice that the increase in impedance between the system buses reduces the system maximum power transfer capability (E. Martinez 2006 and J. Rasmussen 2006).

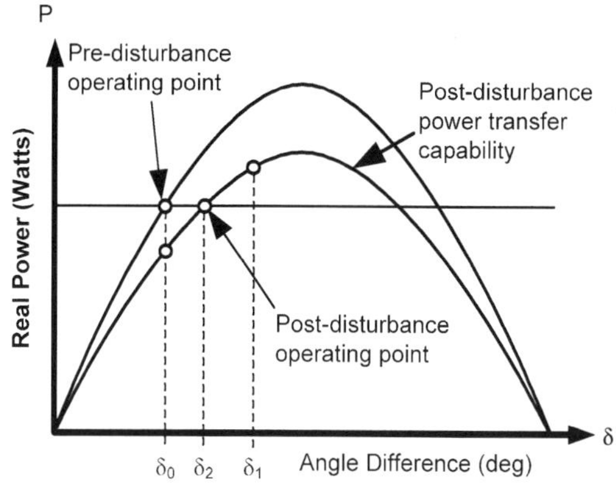

Fig. 17. Real Power Transfer Capability Mainly Depends on the Angle Difference δ, or the Transmission Link Reactance X_L, When E_A and E_B Are Close to Nominal Values

For transmission links with several lines and intermediate substations, existing AGSSs monitor network topology and power transfer capability using open line detectors for arming themselves, selecting generators to trip, or activating tripping commands. Open line detectors are based on breaker auxiliary contact signals (52A or 52B,) undercurrent, and/or under-active-power elements. Usually, these AGSSs use information from both ends of each transmission line to determine if the line is open. With this angle difference information, the SPS can take action instantaneously.

The angle difference information between two buses can perform the following tasks:

- Arm an AGSS
- Trip generation
- Supervise present AGSSs to increase security

For these reasons, we propose an AGSS based on the positive-sequence voltage angle difference between two buses at different locations of the power system.

7.2. Location AGSS in the Power System

There are several SPSs in service in the Southeast region of Mexico because the largest load on the national system is located at the center of the country and 4820 MW of hydroelectric generation is located at the Southeast part of the country. The distance between the heavy load region and the large generation region is 2,000 km. The Grijalva River Hydroelectric Complex is depicted in Fig. 18. One of the SPS in service at Angostura Hydroelectric Power Plant monitors the loss of the transmission link in 400 kV between Chicoasen and Angostura. During normal conditions, Angostura can generate up to 180 x 5 = 900 MW while the total load of Tapachula and South Chiapas region does not exceed 100 MW. The excess power in the region flows from Angostura to Chicoasen and from there to the rest of the system. If two 400 kV parallel lines are lost between Angostura and Chicoasen, both areas remain connected through the 115 kV network with the following consequences:

- The transfer impedance between Angostura and Chicoasen power plants increases, causing the Angostura machines to accelerate. This machine acceleration may lead to angular instability.
- The 115 kV network is overloaded until line or transformer overload protection operates. When this happens, Angostura and Tapachula area form a network isolated from the rest of the system.

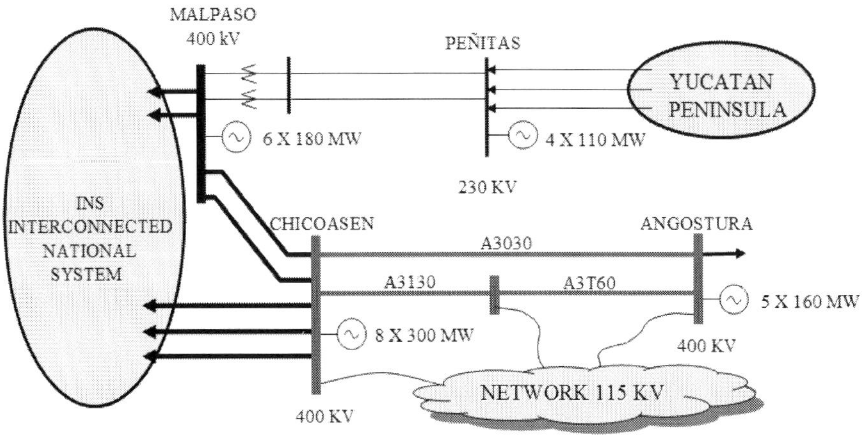

Fig. 18. Grijalva River Hidroelectric complex, Chicoasen-Angostura Transmission link with parallel 115 kV network

For some operating and fault conditions, this double contingency could lead to a blackout at Tapachula City and south of the State of Chiapas. The following simulation results show angle differences between Angostura and Chicoasen for single (loss of one tie line) and double (loss of two tie lines) contingencies on this link with maximum generation at Angostura and Chicoasen if there are no protection or AGSS control actions taken.

Table I show PSS/E™ simulation results for steady-state and transient conditions for single and double contingencies.

Based on the following results, an angle difference threshold of 10 degrees can detect double contingencies and does not operate for single contingencies. This threshold could be used in the AGSS to trip part of the generation in Angostura.

Case	Prefault Angle Diff. δ	Contingency	δ at Line Trip	Additional Comments
1	3.38°	Single Chicoasen-Angostura	6.1°	Max. δ during oscillation 8.7°
2	3.38°	Single Angostura-Sabino	5.25°	Max. δ during oscillation 6.56°
3	3.38°	Single Chicoasen-Sabino	4.11°	Max. δ during oscillation 4.56°
4	3.38°	Chicoasen-Angostura and Sabino-Angostura	14.69°	No AGSS trip, system lost stability
5	3.38°	Chicoasen-Angostura and Sabino-Angostura	14.69°	AGSS trip generation after 100 ms, δ at AGSS trip 27.28°
6	3.38°	Chicoasen-Angostura and Chicoasen-Sabino	10.72°	AGSS trip generation after 200 ms, δ at AGSS trip 25.55°

Table. 1. Simulation Results for Different Steady-State, Single and Double Contingencies

From the results shown in Table 1, the loss of one 400 kV line on this link does not cause stability problems Fig. 19a. However, if two parallel lines are lost, simultaneously or sequentially, the system stability is lost because of power transfer limitations on the 115 kV network. Fig. 19b shows the angle difference between these buses without control actions and when the AGSS trips one or two generators 300 ms after the double contingency occurs.

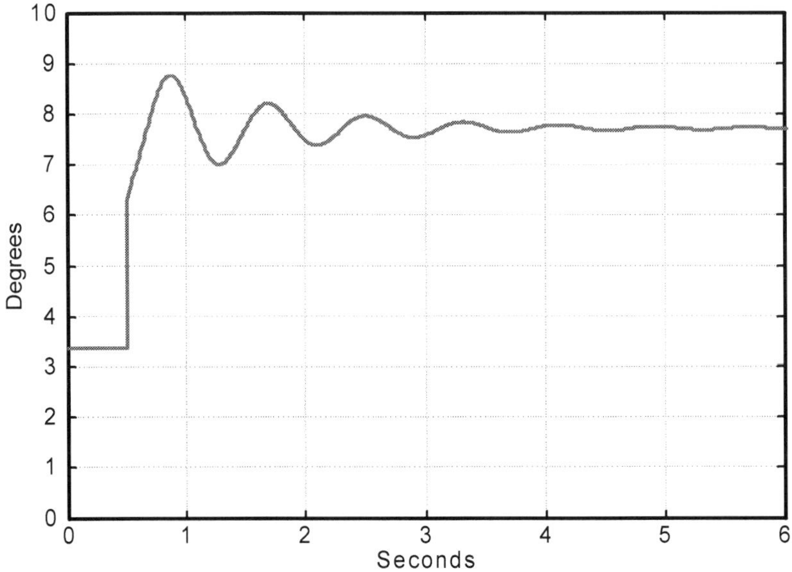

Fig. 19a Angle difference between Angostura and Chicoasen for a single contingency without AGSS Protective Action (case 1).

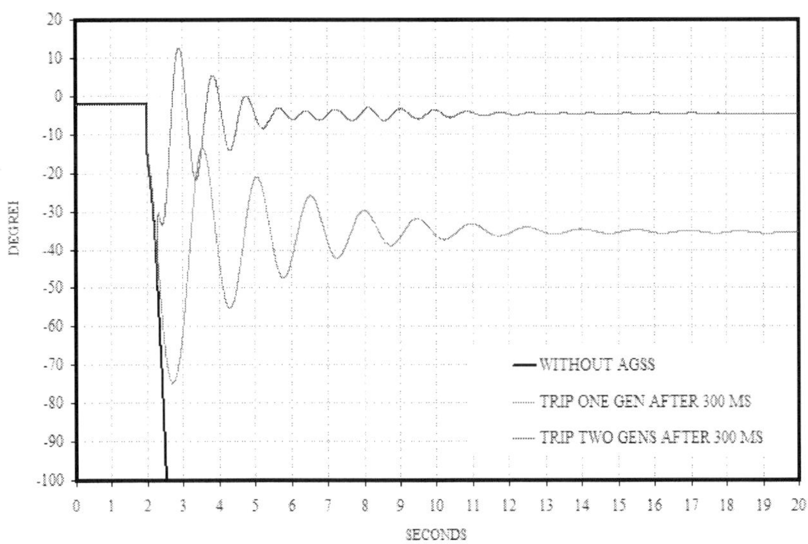

Fig. 19b Angle difference between same buses without AGSS and when the AGSS trips one and two generators 300 ms after the double contingency occurs.

7.3. Angular Differential Protection Scheme Using Synchronized Phasor

A new proposed AGSS could use the angle difference information to make trip decisions or be used to supervise existing schemes. For this application we take into account that the Chicoasen, Sabino, and Angostura substations are directly interconnected through 400 kV transmission lines. For this reason, the angle difference changes instantaneously at these buses when one of the 400 kV links is lost. Fig. 20 shows the logic of the improved angle-difference-based AGSS. With the added angle difference information, the logic of the scheme is simplified and depends only on one communications channel.

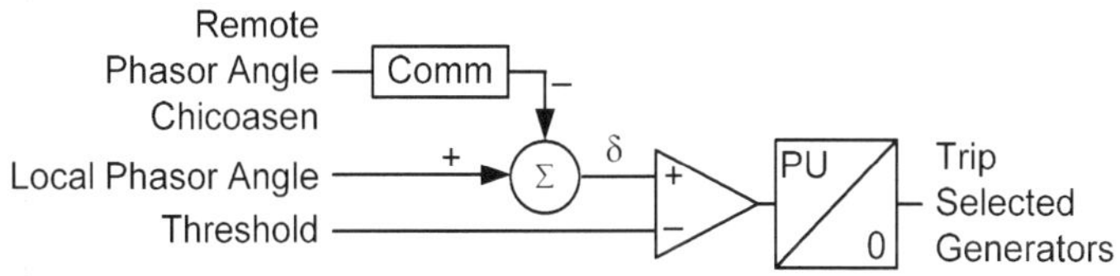

Fig. 20. Improved Angle-Based AGSS Logic

The angle difference must be compared against a threshold. If the angle difference indicates that the 400 kV link between Chicoasen and Angostura was lost because of a double contingency condition, the scheme sheds generation. An intentional time delay may be included in some applications to avoid tripping generation or arming the AGSS during transient or fault conditions. This application does not require such delay. Appendix A has a detailed implementation description of a synchronized real-time control network.

With load flow and stability studies (Table 1), the following was determined:

- Maximum angle difference for conditions where there is no need to shed generation. Contingencies on other links, such as the 115 kV parallel network, should also be considered to ensure that maximum power transfer is achieved between these two hydroelectric plants.
- Minimum angle difference for conditions where the system requires generation shedding. In this case, Chicoasen and Angostura are connected only through the 115 kV network.
- Contingencies at other power system locations that affect the bus voltage angles in the region of interest need to be considered.

CFE installed the proposed scheme with continuous remote monitoring to observe the performance of the AGSS and real-time angle difference measurement during different system operating conditions and different contingencies. The measurements were validated with an accurate power system model that includes the following: generator dynamics, power system stabilizers, automatic voltage regulators, governor dynamics, and system loads.

Two PMCUs were installed, one at Chicoasen and one at Angostura. Each of the PMCUs is connected to monitor its corresponding bus voltage and currents from two lines. The PMCUs are interconnected through a fiber-optic multiplexer with EIA-232 (V.24) asynchronous interface at 38,400 bauds.

We used only serial Fast Message protocol for this test. Another serial port is connected through a serial-to-ethernet converter and sends synchronized phasor data to remote monitoring systems located at CFE regional and national offices.

We captured synchronized phasor measurements, at a rate of 20 messages per second, during programmed line trip and close operations in the region under study with normal system loading conditions. The largest angular difference measurement between Chicoasen and Angostura, for a single contingency, occurred when Line A3030, A3130 or A3T60 tripped at Chicoasen (MMT) substation. Figs. 21a and 21b shows the network under study and the angular difference between Chicoasen and Angostura (ANG) for each condition. Fig. 21a shows the angular difference simulation results for three cases. Fig. 21b shows the same measurements angular difference. From Table 2, we can observe that these results match the measurements within a quarter of a degree. These results validate the model and the measurements.

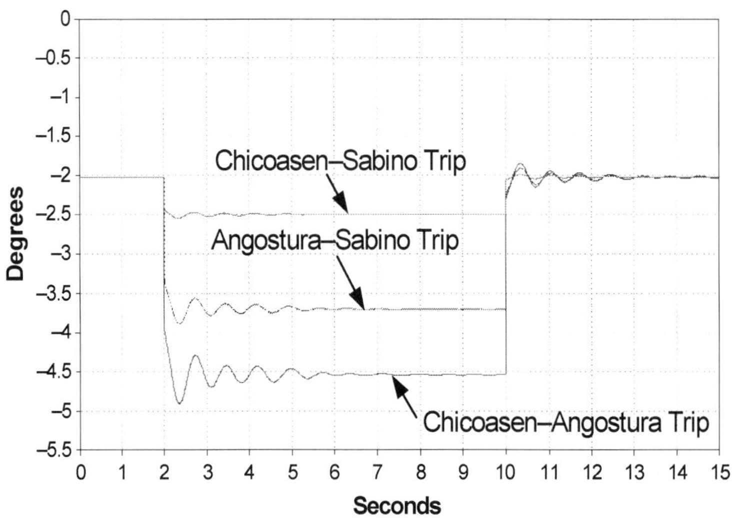

Fig. 21a Angular Difference, Between Chicoasen (MMT) and Angostura (ANG), simulation when the Line A3030 (Chicoasen-Angostura), A3031 (Angostura-Sabino) or A3T60 (Chicoasen-Sabino) Trips and Closes. Six minutes of data during the open condition are not show to shorten the graph.

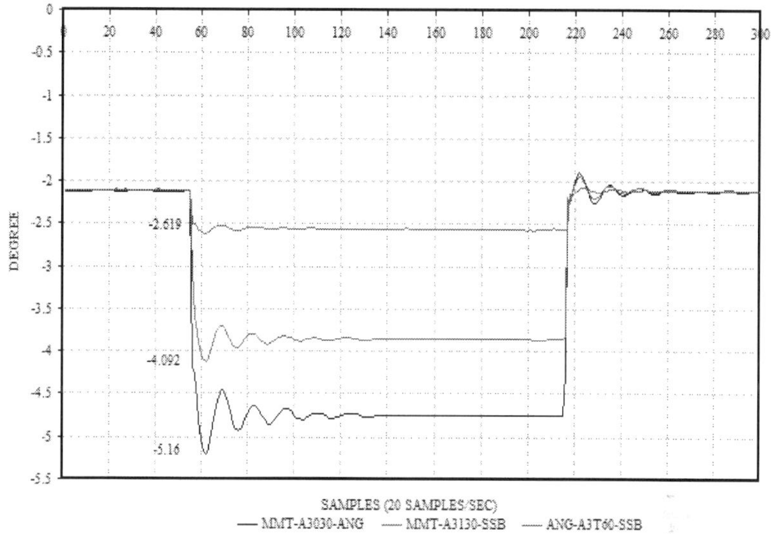

Fig. 21b Measurement values using PMUs in both substations for the same conditions.

	Steady-State Initial Angle	Maximum Angle During Oscillation
Simulation	-2.01°	-4.91°
Measurement	-2.13°	-5.20°

Table 2. Simulation Results and Measurements. Initial Conditions and Maximum Angle Difference When Line A3030 Opens and Closes

Below are additional objectives of performing field tests:

- Test communication channel performance and communication interfaces
- Test the logic that calculates angle difference and measures scheme operating times at different angle threshold levels

We programmed four angle difference elements to test angle difference element logic and measure scheme operating time. We set the angle difference to 3, 4, 5 and 10 degrees, respectively. The oscillographic record, was taken directly from the PMCU located at Chicoasen during the MMT-A3030-ANG line trip. The oscillogram showed the current at both lines and bus voltage at Chicoasen. They operated within 92 ms. After initial instantaneous angle change, Angostura machines accelerate, the angle difference increases, and the angle diffrence element operates after 292 ms.

8. Conclusions

- The architecture of SIMEFAS presents an alternative solution for the integration of different models of PMUs and different manufacturers under the IEEE 37.118 protocol. SIMEFAS architecture is based on PDCs that synchronizes and integrate phasor measurements through software providing a low cost alternative without limiting the integration of additional PMUs.
- CFE has also considered the application of SIMEFAS in monitoring of wind energy farms which are being introduced in Mexico. Continuous monitoring will improve CFE's knowledge of wind farm dynamics and will aid in the elaboration of the Network Code of the Electrical System
- CFE decided to evaluate the use of angle difference on this specific AGSS for three main reasons: It is one of the simplest AGSSs in the network, availability of fast communications channel and the need to accommodate future network changes in the region such as the interconnection to the Guatemala and Central America network.
- Use of PMCUs will reduce operating time and improve reliability if compared with traditional AGSSs based on traditional measurement, separate PLCs, and several remote communication channels.
- Fast communications channels and available PMCUs allow the angle-difference-based AGSS to operate in less than 200 ms.
- Present PMCUs are able to send up to 60 voltage and current synchrophasors per second. This message rate requires a communication channel bandwidth that is not available at these substations at the moment. For this reason, CFE decided to use only voltages at 20 samples per second (one phasor every 50 ms) to limit record size and bandwidth requirements. CFE would like to send voltage and currents to calculate power from synchronized phasor measurements and use it as a permissive signal, but multiplexer card bandwidths need to be changed.
- To safeguard information, CFE has developed a communications project using a Virtual Local Area Network (ViLAN) with fiber optics. It has a high bandwidth capability and TCP/IP access guaranteeing reliability, speed, and security in data transmission between PMUs.
- Records of angle difference measurements for single line contingencies validate measurements and simulation models. AGSS must operate only when two parallel lines are lost and studies should consider sequential or simultaneous double contingencies.
- CFE has not yet worked with special protection schemes that use digital relays simultaneously providing PMU functionalities and protective actions. The drawbacks are limitations on logical variable assignments in the prototype PMCUs and strong dependency of the GPS signal of each device that can affect the calculation of angular differences.
- At the moment the functions of a PMU in some relays of a fasor have been implemented, nevertheless, to implement the functions of control in a PMU, they have majors applications in schemes of protection and control of systems, since a PMU has more fasores and minor dependency of the GPS to maintain the synchronization.

9. References

Daniel Karlsson and Xavier Waymel "System Protection Schemes in Power Networks" Task Force 38.02.19 CIGRE, June 2001.

E. Martínez and Juan J. Márquez "Application of Phasor Measurement Units in the Adaptive Protection of Interconnected Systems" Power Systems and Communication Systems Infrastructures for the Future. Beijing, Peoples Republic of China, September 23-27, 2002

E. Martínez, Nicolás Juárez, Armando Guzmán, Greg Zweigle, and Jean León, "Using Synchronized Phasor Angle Difference for Wide-Area Protection and Control" WPRC, Spokane, WA, USA 2006.

E. Martínez "Analysis of Contingencies with PMUs, Causes and Effects in Power Systems and Their Components" CRIS, Third International Conference on Critical Infrastructures, Alexandria, VA, September 2006.

E. Martínez "SIMEFAS: A Phasor Measurement System for the Security and Integrity of Mexico's Electric Power System" IEEE power & Energy Society 2008 General Meeting, Pittsburgh, Pennsylvania USA 20-24 July 2008.

IEEE Synchrophasors for Power Systems, IEEE Standard C37.118-2005.

Joana Rasmussen and Preben Jorgensen "Synchronized Phasor Measurements of Power System Event in Eastern Denmark" IEEE Transactions on Power Systems, Vol. 21, No.1, February 2006

S. H. Horowitz, A. G. Phadke, and J. S. Thorp, "Adaptive Transmission System Relaying" Paper N° 87 SM 625-77, in Proc. IEEE PES Summer Meeting, San Francisco, CA, July 1987.

V. Centeno, J. De La Ree, A. G. Phadke, G. Michel, R. J. Murphy, and R. Burnett, "Adaptive out-of-step relaying using phasor measurement techniques" Memoria Técnica del Segundo Simposio Iberoamericano sobre Protección de Sistemas Eléctricos de Potencia, Monterrey, N. L., México, 14 al 19 de Noviembre de 1993.

22

Multi-wave differential optical absorption spectroscopy surface ozone measurement with open path meters

Leonid Balatsko, Victar Dziomin, Aliaksandr Krasouski, Alexander Liudchik and Victar Pakatashkin
National ozone monitoring research and education centre, Belarus

1. Introduction

Most of the scientific research activity has focused on the atmospheric boundary layer, where the photochemical smog constituents reach to high concentrations, resulting thus to the often occurrence of air pollution episodes, especially over mega-cities (Jacovides *al.*, 1994; Kondratyev & Varotsos, 1996, 2002; Ozone monitoring, mapping, and public outreach, 1999; Varotsos *et al.*, 2001, 2005). Surface ozone is known as one of the most important air pollutants, therefore, much attention is paid to ozone concentration monitoring.

Nowadays photometric surface ozone concentration meters are being widely exploited. These instruments employ a principle of taking for the analysis air samples at a local point of space (List of designated reference and equivalent methods, 2000; Girgzdiene & Girgzdys, 2003; Гиргждене *et al.*, 2008). To provide high quality measurements (Klausen *et al.*, 2003) with the instruments of such type their periodic calibration against the travelling (mobile) standard is used everywhere. The results of calibration allow to introduce correction into measurements for the given ozone analyzer.

Besides the air sampling technique a method for measuring surface ozone and other small atmosphere constituents by sounding an open lengthy path with radiation becomes quite popular (Bolot'ko *et al.*, 2004, 2005, 2007, 2008a, 2008b; Edner, 1993; Krasouski *et al.*, 2006; Opto-analyser AR-500. User's manual, 1993; Platt *et all.*, 1979; Transceiver-130. User's manual, 1994). The main advantage of such an approach is a possibility to extend results of measurements over a rather large area, since the method gives concentrations averaged over a long path instead of ones measured at a local point. It is clear that the averaged in space concentrations are less sensitive to the influence of such (hardly estimating and depending on sampling place and time) factors as local air microcirculation features and their variability under changing meteorological conditions (Bolot'ko *et al.*, 2007). The advantages of the open path measurements are displayed most effectively in the multipath version of the instrument presented by the OPSIS firm (Transceiver ER-130. User's manual, 1994). A source of radiation and an instrument receiver are placed in one module (to return radiation from a path corner (retro) reflectors are used). This modification provides measurements of the gas average concentration to be analysed in different directions from the radiator-receiver over a path about 250 m length. Such advancing of the idea of the open path measurements makes them similar to lidar systems but provides in contrast to the latters a metrological accuracy of measurements. The authors of this chapter prefer open path measurements namely for this reason and have been intensively working for several years to design construction of the instrument and develop methods of measurements (Bolot'ko *et al.*, 2004, 2005, 2007, 2008a, 2008b). To specify further discussion, theoretical results will be applied mainly for the TrIO-1 device as well as a procedure used to measure ozone concentration.

Quite evident, that checking and calibration of open path meters can not be performed by means of the travelling standard, since there is no possibility to fill the path of sondage with calibrated quantity of the gas to be analysed. The idea of calibration by measurements on parallel paths also raises valid objection. That's why, a fundamentally different approach based on the principle of measurements absoluteness is required to solve the problem of quality of open path measurements.

Formally, a photometric method (like in meters with air sampling) and a spectrophotometric method of differential optical absorption spectroscopy (like in the case of measurements on the open path) are both absolute ones, because they give a possibility to measure target concentrations without any prior calibration against a standard instrument. One has to satisfy only definite requirements on the instrument alignment and to determine contributions to a resulting uncertainty of ozone concentration from main sources of this uncertainty.

This study aims at presenting formulation of the multi-wave method for open path surface ozone concentration meters and assessing the effect of several typical sources of the uncertainty in the surface ozone concentration which are common with open path meters.

The mathematical formalism of the multi-wave method provides a possibility to link directly resulting ozone concentration with measured optical signals. Proposed mathematical formalism is universal and can form a basis for the analysis of various approaches towards measuring concentrations of small gas components of the atmosphere by the DOAS-method. A method applied for the OPSIS firm instruments (Opto-analyser AR-500. User's manual, 1993) is discussed as an example of such analysis in section 5. The formalism also enables to evaluate influence of different sources of errors on the results of ozone concentration measurements in terms of general approach. To realize this, random and systematic errors in the instrument parameters and the calculation method as well as the uncertainties of signal measurements are interpreted as the distinction of sounding radiation spectra on working and reference paths. Estimating further the effect of such distinction on the results of calculation, it is possible to determine the role of each error source in the formation of resultant uncertainty of the measured ozone concentration. We have derived analytical expressions to assess the influence of error sources of different nature on the results of ozone concentration definition by means of the open path technique. Numerical values of the contribution to the resultant ozone uncertainty from several error sources are given as an example for the multi-wave method implemented for the TrIO-1 instrument.

Some results of surface ozone concentration measurements with the TrIO-1 device at the Minsk ozone station of the National ozone monitoring research and education centre (NOMREC) are presented at the end of the chapter.

2. An open path ozone concentration meter TrIO-1

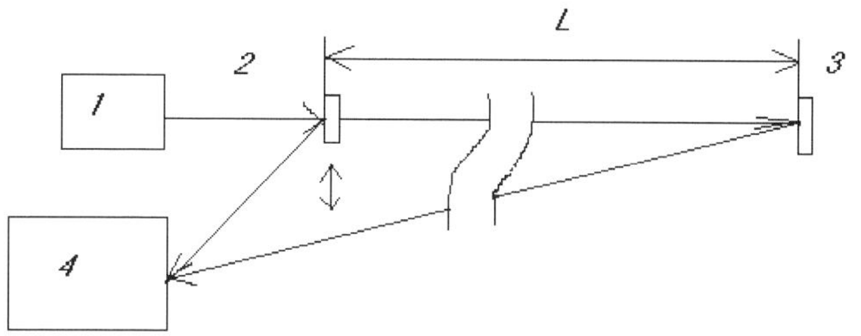

Fig. 1. A simplified optical scheme of the TrIO-1 meter. 1 - the source of radiation, 2 - the moveable mirror to create the reference path, 3 - the reflecting mirror of the open path, 4 - the spectrometer with the fore optics and the unit of registration

The device TrIO-1 is designed in NOMREC to measure average ozone concentration along the open path of sondage with the ultraviolet radiation (Bolot'ko *et al.*, 2004, 2008a). The instrument consists of 4 main modules (see figure 1):

1. The source of ultraviolet radiation (a halogen lamp; the unit 1 in the figure 1);
2. The mirror to reflect radiation on the working path to the receiver (the unit 3 in the figure 1);
3. The moveable mirror to form by a relocation of the mirror the reference path of a short length without absorbing substances (the unit 2 in the figure 1);
4. The spectrometer including a photoreceiver and a corresponding electronics to register optical signals on the working and reference paths at the selected working wavelengths (the unit 4 in the figure 1). A special fore optics including an interference filter set at the entrance of the spectrometer to plain a spectral intensity of the sounding radiation.

The TrIO-1 instrument has some characteristic properties being different from known open path gas analyzers.

1. The TrIO-1 employs the spectral interval of 260-295 nanometers assigned to a long-wave region of the ozone absorption band for measurements of ozone concentration. An appearance and a significant growth of the scattered sun radiation intensity on the wavelengths exceeding 295 nanometers stipulate this choice. Correcting for such radiation is impossible.

2. Unlike the techniques usually used in DOAS-analyzers, a measurement of concentration in TrIO-1 is accomplished at a smooth region of the ozone absorption band.

3. A short reference path is applied in the device to maintain the regular control of intensity of the sounding radiation flow directed to a working path. An optical thickness of the reference path practically equals zero.

The construction of the device differs from the majority of the OPSIS open path instruments since the latter does not use a reference path, and the source and the receiver are placed at the different ends of the working path. Meanwhile, the OPSIS model Transceiver ER-130

(Transceiver-130. User's manual, 1994) combines the emitter and the receiver in the one module. Thus, the TrIO-1 instrument is closer by a construction to the latter.
Two optical path surface ozone concentration meters TrIO-1 were made and put into operation in the middle of 2004. One instrument is located at the Station of a combined background monitoring of the Berezina biosphere reserve (54° 44′ N, 28° 21′ E) which is considered as a vast ecologically clean zone. The second instrument is run at the Minsk ozone station of NOMREC (53° 50′ N, 27° 28′ E). At the ozone station we have in operation a certified ozone analyzer TEI 49C to conduct a scientific research and maintain the quality control of the measurements received from the TrIO-1 meter.

3. Multi-wave technique of open path measurements of the surface ozone concentration

Differential optical absorption spectroscopy is based upon the Bouguer-Lambert-Beer law. A spectrophotometric method for measuring concentrations of atmosphere components can be formally considered as an absolute one, since it does not require any preliminary calibration against the reference instrument. However, the absoluteness of a method for open path measurements cannot be provided without taking into account imperfection of radiation sources and optical systems for measurement.
Unlike the devices dealing with air samples pumped into a photometric cell, open path meters perform measurements of gas concentrations along a lengthy path directly in the atmosphere. In that case it is impossible to remove the analysed gas from the path of sondage to estimate the intensity of the sounding radiation emitted on the path. That's why, open path measurements need more than one wavelength of the sounding radiation per one gas to be analysed. Formally, two wavelengths are sufficient for one atmosphere constituent to be defined (Bolot'ko *et al.*, 2005). Moreover, an additional reference path of a small length excluding influence of radiation absorbers is used to determine spectral distribution of sounding radiation. In some cases (for example for the device TrIO-1) reflecting mirrors returning sounding radiation are installed at the ends of both paths. The reflection spectral factors coincide for both mirrors. In other modifications a source of radiation and a receiver can be placed at the different ends of a path; it is not essential for a developed technique. In all cases the length of the open path exceeds 100 m. For TrIO-1 meters the spectral interval of measurements is equal to 260-300 nm.
The signal measured at a wavelength λ can be presented in the form of

$$S_\lambda = g\gamma_\lambda I_\lambda \exp(-\tau_\lambda) \qquad (1)$$

where:
g - is the geometrical factor, describing radiation attenuation on the path due to the divergence of the radiation beam, limited sizes of the mirrors, nonselective attenuation of the radiation on the path, etc. The geometrical factor, by definition, does not depend on the wavelength, but, it differs for different paths;
γ_λ stands for the absolute spectral sensitivity of the device. To simplify the notation, we have included spectral dependence of the mirrors reflection in γ_λ;
I_λ - is spectral density of the radiation intensity emitting to the path. In general, it depends on the path type, i.e. working (open, sounding, etc.) or reference;

τ_λ - is the optical thickness of the path. In case of the reference path $\tau_{\lambda 0} \equiv 0$ (further the reference path (0) and the working path (l) will be marked with the second index, where it is necessary).
Optical thickness of the working path is the sum of optical thickness of ozone absorption $\tau_{O3\lambda}$, optical thickness of molecular scattering $\tau_{mol\lambda}$ and optical thickness of radiation attenuation by aerosols and precipitations τ_{aer}. The latter can be viewed as a wavelength independent in a rather narrow working spectral interval. Thus,

$$\tau_{\lambda l} = \tau_{O3\lambda} + \tau_{mol\lambda} + \tau_{aer}. \quad (2)$$

Assuming spectral nonselectivity of τ_{aer}, it can be included in the geometrical factor of the working path. Thus, the geometrical factor ceases to be a device constant and starts to depend on the measurements conditions (τ_{aer} value), but, that is allowed by the proposed technique. A contribution to the optical thickness $\tau_{mol\lambda}$ is calculated by means of known formulas (Frohlich & Shaw, 1980; Bolot'ko *et al.*, 2005). A contribution due to the ozone absorption is proportional to the average ozone concentration along the sounding path n and the path length l:

$$\tau_{O3\lambda} = \sigma_\lambda n l, \quad (3)$$

where σ_λ - is a cross-section of radiation absorption at a wavelength λ by ozone.
The negligible contributions due to radiation absorption by other small gas components (for example, SO_2 and NO_2) are not accounted in equation (2). Estimations show that these contributions in a working spectral range are inappreciable comparing to the ozone absorption.
Let us introduce designations:

$$z_\lambda = \ln(S_{\lambda 0}) - \ln(S_{\lambda l}) - \tau_{mol\lambda}, \quad (4)$$

$$G = \ln(g_0 / g_l) + \tau_{aer}, \quad (5)$$

$$X = nl. \quad (6)$$

Then, combining the expressions (1) - (6), yields

$$z_\lambda = G + X\sigma_\lambda. \quad (7)$$

According to (4), values z_λ can be considered as transformed registered signals from the reference and working paths, including errors of various origin. G and X parameters are subject to determination. Equation (7) due to errors in measured signals and approximations used can not be satisfied with fixed values of G and X for all z_λ if a number of working wavelengths exceeds 2. Still, this relation can be viewed as the equation for the straight line drawn by means of the least squares method through the range of experimentally measured values z_λ that are related to the set of selected values of σ_λ. Thus, minimizing the sum of squared deviations of the equation (7) from really measured values z_λ (Bolot'ko *et al.*, 2008b), yields

$$X = D_{z\sigma} / D_{\sigma\sigma}, \tag{8}$$

$$G = z_{\text{av}} - \sigma_{\text{av}} X \tag{9}$$

where $D_{ab} = \sum_1^N (a_i - a_{\text{av}})(b_i - b_{\text{av}})$, $a_{\text{av}} = \frac{1}{N}\sum_1^N a_i$ - is the average value of the set a, N - is the total number of working wavelengths. In the case of $a = b$ the covariance D_{ab} reduces to a dispersion of a.

4. Influence of sounding radiation spectra discrepancy over the reference and working paths

Consider a case when a spectrum of sounding radiation on the reference path doesn't coincide with a spectrum on the working path. The distinction in spectra can be referred to the light beams of different temperatures irradiated from different segments of a lengthy emitter to the reference and working paths. They are also attributed to the distinction in spectral characteristics of the mirrors used for paths formation, change of a source radiation spectrum in a time interval between spectrometer scans on the reference and working paths, etc.

Let the difference of logarithms of sounding radiation spectra intensity on the reference and working paths be described by the function $\varphi(\lambda)$:

$$\ln I_{\lambda 0} = \ln I_{\lambda l} + \varphi_\lambda. \tag{10}$$

Then, instead of the result of measurements (4), we receive

$$z'_\lambda = z_\lambda + \varphi_\lambda. \tag{11}$$

Using these data to calculate X and G (equations (8), (9)), the following is obtained:

$$X' = X + D_{\sigma\varphi} / D_{\sigma\sigma}, \tag{12}$$

$$G' = G + \varphi_{\text{av}} - \sigma_{\text{av}} D_{\sigma\varphi} / D_{\sigma\sigma}. \tag{13}$$

Here X and G are realised with a case of $\varphi(\lambda) \equiv 0$. Thus, *if sounding spectra differ, the calculation leads to a constant shift of the resulting concentration and the ratio of the geometrical factors. The shift is proportional to the covariance $D_{\sigma\varphi}$ and doesn't depend on the real concentration of ozone*. Obviously, that constant systematic error is subject to definition for providing the absoluteness of measurements. Note that the shift appears only if there is a distinction in relative spectra; constancy of $\varphi(\lambda)$ (proportionality of spectra) leads to zero shift because in that case $D_{\sigma\varphi} \equiv 0$.

The set of the values φ_λ can be written as the sum of two items:

$$\varphi_\lambda = \frac{D_{\sigma\varphi}}{D_{\sigma\sigma}}\sigma_\lambda + q_\lambda \text{, where } D_{\sigma q} = 0 . \tag{14}$$

Such representation is a discrete analogue of the selection of contributions to $\varphi(\lambda)$ from the function $\sigma(\lambda)$ and the orthogonal to it function $q(\lambda)$. The equality $D_{\sigma q} = 0$ serves here as analogue of an orthogonality condition. In statistics it implies lack of correlation of two sets of random variables. In this particular case there are no random variables, especially if random errors in z_λ are not accounted. Therefore, the term 'orthogonality', despite its obvious inaccuracy, appears to be more suitable.
Let us introduce the notation

$$\Delta z_\lambda = z'_\lambda - G' - X'\sigma_\lambda ,$$

describing a deviation of the experimentally measured values z'_λ from those calculated on the basis of determined values X' and G'. In the assumption that $z_\lambda - G - X\sigma_\lambda \equiv 0$ (measurement errors are absent), we have

$$\Delta z_\lambda = q_\lambda - q_{\text{av}} . \tag{15}$$

The visible difference Δz_λ between the measured ozone absorption optical thickness and its theoretical estimation is completely determined by function $q(\lambda)$ - an orthogonal to $\sigma(\lambda)$ part of $\varphi(\lambda)$. From the expression (15) we derive quite obvious equalities $(\Delta z)_{\text{av}} \equiv 0$ and $D_{\sigma\Delta z} \equiv 0$. Values Δz_λ do not depend on the real ozone concentration. Naturally, the random errors of signals measurement will affect the values. Still, the effect of those errors can be easily minimized by averaging the results of the ozone concentration time series (for details see section 7.1). Thus, values Δz_λ show an orthogonal to σ_λ part of spectra difference of sounding radiation on both paths, and the dispersion $D_{\Delta z \Delta z}$ serves as a quantitative estimation of this part of the difference. The difference can be revealed when the function $\varphi(\lambda) - \varphi_{\text{av}}$ *is not strictly proportional to the dependence* $\sigma(\lambda) - \sigma_{\text{av}}$.
If only two wavelengths are used for measurements, such a proportionality is always provided, and distinction in spectra of sounding radiation cannot be revealed. Hence, systematic errors cannot be detected, while calculating ozone concentration. To somehow avoid such a situation, *a number of the wavelengths should be taken.* One can expect the $\varphi(\lambda)$ to include always a significant component of $q(\lambda)$, whose 'removal' by means of optical system alignment will cause a simultaneous 'removal' of a component proportional to the $\sigma(\lambda)$. *Nevertheless, the method loses its absoluteness, if coincidence of sounding radiation spectra on the reference and working paths is not provided by a technique and a scheme of instrumentation alignment.*
The discussed divergence in spectra of sounding radiation is expected to be a rather smooth function. That's why, the more complex trend is revealed by the ozone absorption for the set of selected wavelengths, the lesser effect will be produced by the distinction in spectra of sounding radiation.
Two simple models have been studied to assess the effect of quantity of used wavelengths

on the uncertainty of the measured ozone concentration:

1. $\varphi_\lambda = \lambda - 265$

2. $\varphi_\lambda = (\lambda - 265)^2$, where a wavelength is set in nm.

Various sets of working wavelengths from the list of 245, 250, 254, 258, 263, 280, 285, 295 nm have been used. Data on ozone absorption cross-sections for the chosen wavelengths allow us to calculate the systematic error of ozone concentration determination and a shape of a 'visible' part of spectra distinction $q_\lambda = \varphi_\lambda - (D_{\sigma\varphi} / D_{\sigma\sigma})\sigma_\lambda$. The absorption cross-sections and both functions φ_λ (scales for presentation are changed) have been given in figure 2.

A relative change in the systematic error of the ozone concentration using various sets of wavelengths, is shown in the table 1. The error obtained using a pair of wavelengths 295, 285 nm is assumed as the measurement unit. In the first case the increase in the number of working wavelengths reduces the systematic error no more than twice. In the second case the similar increase makes up almost the order in reducing the error.

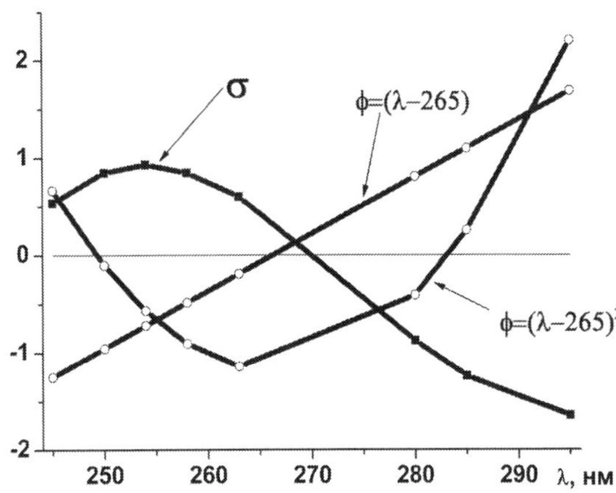

Fig. 2. Functions σ_λ and φ_λ for the chosen working wavelengths. The scale is changed for presentation and a shift of the ordinate axis is made.

Number of wavelengths	The contribution to a systematic error (model 1)	The contribution to a systematic error (model 2)
2	1.00	1.00
3	0.79	0.71
5	0.55	0.21
8	0.65	0.14

Table 1. Change of the contribution to a systematic error of ozone concentration calculation due to distinction in spectra of sounding radiation

Graphs of the normalized (divided by $\sqrt{D_{qq}}$) functions q_λ for sets of three (295, 285, 280 nm), five (295, 285, 280, 263, 258 nm) and all eight wavelengths as well as both variants of the distinction in spectra of the sounding radiation are given in figures 3a, 3b.

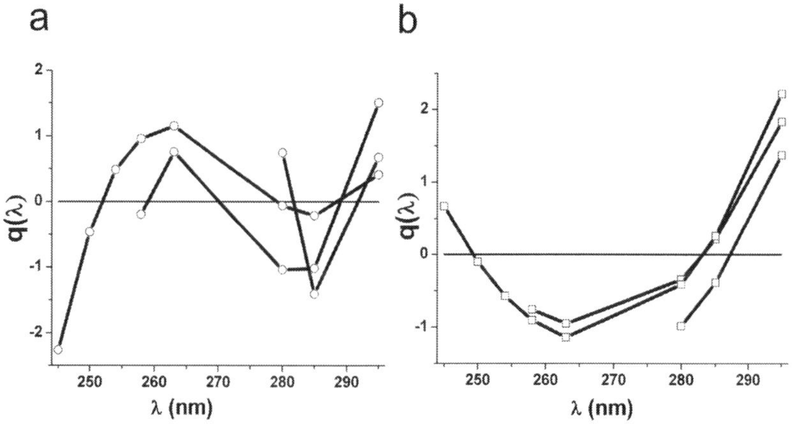

Fig. 3. Normalized 'visible part' in spectra distinction for cases of three, five and eight working wavelengths. a - distinction in spectra is described by a linear dependence, b - by a square-law dependence.

To assess quality and stability of the alignment, one can use the value $D_{\Delta z \Delta z} = D_{qq}$ showing an estimation of the difference of a values set $\{\Delta z_\lambda\}$ from zero. The condition $D_{\Delta z \Delta z} = 0$ is necessary but insufficient for the coincidence of spectra of sounding radiation on both paths. It is clear that eliminating only the visible part of the distinction in spectra does not influence the measured ozone concentration. The most essential segment of the distinction remains 'invisible', and it is impossible to estimate its impact on the calculated ozone error and to introduce the corresponding correction. This correction can be only determined if the possibility of the measurement and the comparison of source spectra on the reference and working paths are provided. On the other hand, the value of $D_{\Delta z \Delta z}$ is a good criterion to assess a device alignment and can be used to maintain the quality control. If condition of spectra coincidence is realized, the system becomes an absolute measuring instrument of ozone concentration and does not require prior calibration.

5. The OPSIS method

There is a reason to discuss here a particular case with measurements being performed in a rather broad spectral interval, where detailed spectrum is registered, and the original results of measurements being further processed. The transformed data are used for the final calculation of concentrations of the atmosphere components absorbing radiation in a working spectral region. These data are referred to a rather narrow spectral intervals, where fine structure details of radiation absorption by atmosphere components are well pronounced.

A method used to measure concentrations of small gas atmosphere components with an Opto-analyser AR 500 open path meter of the OPSIS firm is actually reduced to that case (Opto-analyser AR-500. User's manual, 1993). The method is based on the assumption that gases to be analyzed have narrow absorption spectral bands, and radiation source spectrum does not change in the first approximation within width of these bands. In fact, the spectral distribution of the source radiation is accounted in calculations, but its registration is carried out extremely rare. The sense of the previously mentioned preliminary data processing lies, in that case, in approximating the spectrum registered within the whole working spectral region by the 5th order polynomial and subtracting the result from the original data. To obtain a more consistent substantiation of the method, we shall slightly extend an approach, replacing that approximation by smoothing, which will be interpreted as a convolution of original data with a nonnegative curve of a finite width. This is a version of standard signal filtration procedure (Hamming, 1989). An important requirement for the method applicability is the absence of structural details of the sounding radiation spectrum in the working spectral region which have a comparable width with the absorption bands of gases to be determined.

Elaborating the OPSIS method allows us to make smoothing of z'_λ being measured with short wavelength step (this is a mandatory requirement of the OPSIS method application) and then subtract the result from the original data. Smoothed data will be labelled with a tilde symbol. Considering equality $\tilde{\varphi}_\lambda = \varphi_\lambda$, which is satisfied with high accuracy due to assumed by the OPSIS method smoothness of the radiation spectrum and subsequent smoothness of the function φ_λ within the working spectral region, we derive

$$z'_\lambda - \tilde{z}'_\lambda = z_\lambda - \tilde{z}_\lambda = X(\sigma_\lambda - \tilde{\sigma}_\lambda)\,.$$

Thus, the spectrum of sounding radiation and the parameter connected with the g-factors are eliminated from data used to calculate gas concentration. Clearly, the method application requires two conditions be satisfied: $\tilde{\varphi}_\lambda = \varphi_\lambda$, $\sigma_\lambda \neq \tilde{\sigma}_\lambda$ in the wavelength region, employed to measure concentrations. In other words, halfwidth of the smoothing filter should be much greater than halfwidth of absorption lines, and there must be absent structural elements in the spectrum of sounding radiation with width comparable to that of absorption spectral lines of gas components to be determined. More specifically, these elements must be absent in the difference of logarithms of sounding radiation source spectrum and source spectrum registered earlier (may be under other conditions). Two variants of the method are applicable for the ozone case: a) using existing low amplitude structural features in ozone absorption spectrum near its maximum, and, b) using practically whole wavelength region captured by the Hartley band. It is evident, that in the first case the method will be less sensitive to ozone due to low values of differences $\sigma_\lambda - \tilde{\sigma}_\lambda$, in the second case a problem arises with measuring signals in a rather broad spectral interval with a small step of a wavelength scanning. Basically, these difficulties reduce accuracy of ozone concentration measurement within the OPSIS method compared to other gases with a greater structural absorption spectrum (Opto-analyser AR-500. User's manual, 1993).

6. Sources of errors

Sources of errors discussed below are inherent to any spectrometric instrument and, hence, may be viewed as of common character. It is only their numerical contribution to the uncertainty of the measurement result varies, depending on a construction of the instrument and a measurement technique.

Random errors of signals measurement arise because of imperfection of the registration system and mechanical units of the spectrometer performing wavelength setting, and also due to atmosphere transparency fluctuations generated by turbulent pulses along the path of sondage.

Each spectrometer has a systematic error of wavelength setting, following practical impossibility to define precisely the origin of the working spectral region. This kind of error may be interpreted as a systematic error in signal measurement or as a systematic error in defining ozone absorption cross-sections. Clearly, both interpretations must lead to the same numerical estimation of resulting errors for the given instrument, method of measurement and method of ozone concentration calculation.

Systematic errors of signal measurement also result from a nonlinearity of the registration system and inaccurate account of the 'electronic pedestal' being formed by dark current of a photoreceiver and by a preamplifier of signals. The construction of the TrIO-1 device minimizes errors of such type, using neutral gauzes to attenuate radiation on the reference path, and an interference filter to align radiation source spectrum in the working spectral region. Moreover, the level of the electronic pedestal is almost two orders lower than signals being measured. Therefore, we will not consider errors of such type.

A significant role belongs to the errors generated by the instability of the radiation source and distinction of reflecting spectral coefficients of mirrors used to form the working and reference paths.

Errors in defining absorption cross-sections and cross-sections of molecular scattering for the working wavelengths as well as errors in determination of the working path length are specified within one class of errors. They are believed to have direct link with the sought ozone concentration, practically, in all existing methods for the open path meters.

Quite essential, the role of the majority of random and systematic errors may be interpreted as the distinction in sounding spectra on the working and reference paths generated by these errors. As follows, it is easy to estimate through the expression (12) contribution of each error source into the uncertainty of the calculated ozone concentration.

7. Mathematical representation of different error sources and estimation of their influence on the error of ozone concentration determination

Influence of the given error source on the result of ozone concentration measurement is viewed under the assumption that all other error sources are excluded. The combined effect of simultaneously acting several uncorrelated sources of errors is expressed according to well-known formula: systematic errors from different sources are summed up; dispersions of random errors from different sources are summed up.

7.1. Random errors of signal measurements

Let us accept a rather natural assumption that errors δS_λ of signal measurement at each working wavelength are of random character, and results of measurement at different wavelengths are uncorrelated with each other. This means that following equalities are realized: $\overline{\delta S_\lambda} = 0$, $\overline{\delta S_\lambda \delta S_{\lambda'}} = 0$, if $\lambda \neq \lambda'$. Here an upper bar designates averaging over a set of repeated measurements of signals at the wavelength λ, or over a set of repeated measurements of pairs of signals at the wavelengths λ and λ'. Considering expression (4), it may be concluded that analogical equalities are fulfilled in relation to variables z_λ: $\overline{\delta z_\lambda} = 0$, $\overline{\delta z_\lambda \delta z_{\lambda'}} = 0$, if $\lambda \neq \lambda'$. Let us designate the dispersion of the random variable δz_λ as $d_\lambda = \overline{\delta z_\lambda \delta z_\lambda}$.

One may assume the values of δz_λ (deviations of measured values z_λ from the exact ones) as the difference of logarithms of sounding radiation spectra on the both paths appearing in the given measurement of the ozone concentration. Then, a set of $\delta z_\lambda \neq 0$ yields an absolute error in determination X according to (12) as

$$\delta X = D_{\sigma\,\delta z} / D_{\sigma\sigma},$$

and the error dispersion (while deriving the final formula, an absence of measurement errors correlation at different wavelengths and zero equality of their mean values for a set of repeated measurements are essential) as

$$\sigma_X = \left[\sum_\lambda (\sigma_\lambda - \sigma_{\text{av}})^2 d_\lambda\right] / \left(N^2 D_{\sigma\sigma}^2\right). \tag{16}$$

If to define now the averaged over working wavelengths dispersion of δz_λ according to

$$d = \left[\sum_\lambda (\sigma_\lambda - \sigma_{\text{av}})^2 d_\lambda\right] / \left(N D_{\sigma\sigma}\right),$$

Then

$$\sigma_X = d / (N D_{\sigma\sigma}).$$

The same result is received if $d_\lambda = d$ (dispersions of measured z_λ at all wavelengths are equal). Let's express now the random error in z_λ through relative errors of signal measurements $\varepsilon_{\lambda 0} = \delta S_{\lambda 0} / S_{\lambda 0}$ and $\varepsilon_{\lambda l} = \delta S_{\lambda l} / S_{\lambda l}$ and link the dispersion d_λ with the relative dispersion of signals. Assuming (1) and smallness of signal errors, one may see that the error in the value z_λ is equal to the difference of relative errors of signals measurements on the both paths:

$$\delta z_\lambda = \delta S_{\lambda 0} / S_{\lambda 0} - \delta S_{\lambda l} / S_{\lambda l} = \varepsilon_{\lambda 0} - \varepsilon_{\lambda l}.$$

Under a rather natural assumption that errors $\varepsilon_{\lambda 0}$ and $\varepsilon_{\lambda l}$ do not correlate, we obtain

$$d_\lambda = s_{\lambda 0} + s_{\lambda l},$$

where $s_{\lambda 0} = \overline{\varepsilon_{\lambda 0}^2}$, $s_{\lambda l} = \overline{\varepsilon_{\lambda l}^2}$ - are relative signal dispersions, easily being assessed during direct signal measurements. Thus,

$$d = \left[\sum_\lambda (\sigma_\lambda - \sigma_{\text{av}})^2 (s_{\lambda 0} + s_{\lambda l})\right] / (N D_{\sigma\sigma}).$$

The presented expression link relative signal dispersions with the dispersion of the calculated ozone concentration. Consider an example of the currently employed method of the ozone concentration measurement for the TrIO-1 instrument. For measurement 5 wavelengths are used: 264, 268, 280, 285, 295 nm. Relative dispersions of signals for all working wavelengths on the reference and working paths (under good weather conditions) are approximately equal and do not exceed $(0.004)^2$. Table values of ozone absorption cross-sections give $D_{\sigma\sigma} = 1.3 \cdot 10^{-35}\,\text{cm}^4$. As a consequence, the ozone concentration dispersion generated by random errors of signal measurement is about $(3\text{ ppb})^2$.

Let us return now to the problem of a possible stable discrepancy of sounding radiation spectra on the working and reference paths. We have noted earlier that the nonzero value $D_{\Delta z \Delta z}$ may serve as an indicator of the existence of such distinction, and, hence, be used to monitor the quality of instrument adjustment and its stability in time. Actually, the dispersion $D_{\Delta z \Delta z}$ can be never reduced to zero due to random errors of signal measurements. In particular, if a systematic distinction of spectra is absent, averaging the dispersion $D_{\Delta z \Delta z}$ over a series of ozone concentration measurements yields

$$\overline{D_{\Delta z \Delta z}} = \left[(N-1)/N^2\right] \sum_\lambda d_\lambda \geq 0.$$

Random errors of signal measurement contribute to the value $D_{\Delta z \Delta z}$ and make it positive even when sounding spectra on the both paths coincide (in average). Therefore, even at the ideal adjustment of the instrument the dispersion $D_{\Delta z \Delta z}$ will be greater than zero.

On the other hand, values Δz_λ are independent of the actual ozone concentration along the path. They can change only due to random errors of signal measurements. Hence, a procedure of long averaging Δz_λ over a series of measurements is possible aiming to reduce influence of such errors. In this case, as number of measurements increases, the dispersion $D_{\overline{\Delta z}\,\overline{\Delta z}}$ of the averaged values $\overline{\Delta z_\lambda}$ tends to zero provided the systematic distinction absence of sounding radiation spectra.

7.2. Errors of wavelength setting

Generally, to measure signals on the working and reference paths, each wavelength is being set independently for each path. Then, $\delta\lambda_0 \neq \delta\lambda_l$, i.e. an error of the wavelength setting has a random character and a specific random value for each path. Saving terms of the first order of smallness yields

$$\delta z_\lambda = \frac{\partial \ln S_{\lambda 0}}{\partial \lambda}(\delta\lambda_0 - \delta\lambda_l) + X\frac{\partial \sigma_\lambda}{\partial \lambda}\delta\lambda_l .$$

If measurements of signals on the working and reference paths at each wavelength are conducted without wavelength resetting (for example, by exchanging of paths using a moveable mirror (Bolot'ko *et al.*, 2004, 2005), then, $\delta\lambda_0 = \delta\lambda_l = \delta\lambda$ (errors of wavelength setting are equal for the working and reference paths), and

$$\delta z_\lambda = X(\partial\sigma_\lambda / \partial\lambda)\delta\lambda .$$

In both cases deviations $\delta\lambda_0$, $\delta\lambda_l$ are random variables with zero mean value and they are not correlated for different wavelengths. Hence, the same is true in relation to variables δz_λ. Thus, the dispersion of the calculated ozone concentration is defined by the expression (16) and in the final form for the second case becomes

$$\sigma_X = X^2 d_{\delta\lambda}\left[\sum_\lambda (\sigma_\lambda - \sigma_{av})^2 (\partial\sigma_\lambda / \partial\lambda)^2\right] / \left(N^2 D_{\sigma\sigma}^2\right), \tag{17}$$

where a natural assumption of the dispersion $d_{\delta\lambda}$ independence of the wavelength setting error from the actual value of the wavelength is employed. For the first case one can obtain a rather complicated formula, which we do not present, because the method of measurements for the TrIO-1 instrument applies, namely, to the second case. In particular, for a given set of working wavelengths the sum in the right side of (17) is approximately equal to $5.5 \cdot 10^{-72}$ cm^8/nm^2. As follows, the relative error in the ozone concentration definition caused by a random error of wavelength setting in limits ±0.5 nm does not exceed 1.5%.

Finally, if we speak about systematic error in determination of the working spectral range origin (i.e. wavelength scale shift), the error of wavelength setting is constant for all working wavelengths and does not depend on a path selection: $\delta\lambda_0 = \delta\lambda_l = \delta\Lambda = const$. In this case we have

$$\delta z_\lambda = X\delta\Lambda\partial\sigma_\lambda / \partial\lambda ,$$

and a generated systematic error in the ozone concentration is

$$\delta X = X\delta\Lambda D_{\sigma\,\partial\sigma/\partial\lambda} / D_{\sigma\sigma}$$

and has zero dispersion. For the TrIO-1 instrument $D_{\sigma\,\partial\sigma/\partial\lambda} \approx -1.7 \cdot 10^{-37}\ \mathrm{cm}^4/\mathrm{nm}$, and a relative error in the ozone concentration due to wavelength scale shift of about 0.5 nm accounts for 0.7%. Such a rather low sensitivity of the ozone error to a significant wavelength scale shift seems to be improbable. To verify that conclusion, we have made calculations with actually measured signals for the case of 0.5 nm wavelength shift. The resulting relative error has been found in the limits of 0.4 - 0.8%.

7.3. Errors of ozone concentration definition resulted from instability of the radiation source

Errors of such type may appear when a signal measurement at each working wavelength on the working path and reference path is separated by a considerable time interval. This may occur, if within the selected method signals on the reference path are measured rather rare, and, then, for a long time interval Δt only signals on the working path are measured to receive a series of ozone concentrations. If one further defines difference in spectra of sounding radiation as $\varphi_\lambda(\Delta t) = \ln I_{\lambda 0}(t) - \ln I_{\lambda 0}(t+\Delta t)$, an originating error in the ozone concentration is derived from the expression (12). Now, expecting a linear dependence of φ_λ on Δt, it is possible to estimate maximal time interval between signal measurements on the reference path provided forecast error does not exceed the given level. After the interval is finished, measurements have to be repeated on the reference path. Importantly enough, that changing in time of the spectral composition of the sounding radiation is much more essential than changing only its integral amplitude. Thus, in a particular case of $\varphi_\lambda = \mathrm{const}(\lambda)$ (a proportional variation of the source spectrum has occurred at all working wavelengths), the error in the ozone concentration generated by such instability is equal to zero.
If measurements at every wavelength on the working and reference paths follow each other, the role of errors originated by a source instability is minimized. In this case, as we have noticed earlier, an influence of random errors of wavelength setting also reduces. Thus, such variant of the method looks as more preferable.

7.4. Uncertainties resulted from the errors in the method parameters

Let $\delta\sigma_\lambda$ be a deviation of the used value of ozone absorption cross-section from the accurate one. Then, according to (8), the error of the ozone concentration calculation is

$$X + \delta X = D_{z(\sigma+\delta\sigma)} / D_{(\sigma+\delta\sigma)(\sigma+\delta\sigma)} \,.$$

From this to the first order of smallness on the value of deviation $\delta\sigma_\lambda$ we find

$$\delta X \approx (D_{z\delta\sigma} - 2XD_{\sigma\,\delta\sigma}) / D_{\sigma\sigma} \,.$$

But from relation (7) $D_{z\delta\sigma} = XD_{\sigma\,\delta\sigma}$. Hence,

$$\delta X / X \approx -D_{\sigma\,\delta\sigma} / D_{\sigma\sigma} \,.$$

A maximal value of the covariation $D_{\sigma\,\delta\sigma}$ at the constant $D_{\sigma\sigma}$ is achieved when deviations are proportional to absorption cross-sections: $\delta\sigma_\lambda = \alpha\sigma_\lambda$. In this case the relative error of the ozone concentration determination equals to the relative error of absorption cross-sections definition.
Influence of the error of determination of the working path length on the results of the ozone concentration calculation is evaluated on the basis of relation (6). To the first order approximation (dependence of the optical thickness of molecular scattering on the length of the working path introduces an error into calculated ozone concentration of the second order of smallness) the relative error in ozone concentration $\delta n / n = -\delta l / l$ - equals to the relative error of the path length definition. As soon as the length of the path can be measured with a high accuracy, we may ignore influence of such type errors.

Errors in the definition of the optical thickness of molecular scattering on the working path $\tau_{\text{mol}\,\lambda} = l\beta_\lambda$ (Bolot'ko *et al.*, 2005) may appear as a result of the incorrect definition of the path length or incorrect definition of the scattering coefficient β_λ. In the latter case the error arising in the definition of variable z_λ is equal to $\delta z_\lambda = -l\delta\beta_\lambda$ and, considering relation (12), we have $\delta X = -lD_{\sigma\,\delta\beta} / D_{\sigma\sigma}$. The scattering coefficient is a smooth monotonically decreasing with wavelength function. Consequently, the most possible errors are of the type $\delta\beta_\lambda = \alpha\beta_\lambda$. For the given example $D_{\sigma\beta} = 1.3 \cdot 10^{-24}$ cm, and $\delta n \approx 4\alpha$ in units of ppb. If $\alpha = 0.02$ (2% increase of the coefficient), an originating error in ozone concentration does not exceed 0.1 ppb.

8. Some results obtained with the open path instrument and multi-wave technique of measurement

The need for regular measurement of surface ozone concentration has arisen with understanding of its active influence both over a climate and ecological conditions of industrial regions. In Belarus the network of stations of atmospheric air quality control has been steadily progressing to monitor such pollutants of the atmosphere as nitrogen and sulfur oxides, volatile hydrocarbons, firm particles, etc. Despite that, until 2004 there were no measurements of the surface ozone concentration regarding surface ozone as a basic participant of chemical and photochemical reactions in a boundary layer of the atmosphere as well as a greenhouse gas and the indicator of some other gases-pollutants. A significant region of Eastern Europe missed analysis of influence over an ecological condition of the atmosphere by such an active component as ozone. The impact of meteorological factors (mainly, a western air masses transfer and presence of local sources of anthropogenic photoactive atmosphere pollutants) on formation of the ozone concentration in a surface layer was not studied.

The TrIO-1 instrument designed in NOMREC has been the first instrument since 2004 to monitor surface ozone in Belarus. As was expected, five years of operation at the Minsk ozone station of NOMREC as well as reference data (Звягинцев, 2003, 2004; Звягинцев & Кузнецова, 2002) have shown significant daily and seasonal fluctuations of the surface ozone concentration depending on meteorological and anthropogenic factors. In figure 4 day time data (ozone concentration at local noon, near 10 a.m. Greenwich Mean Time) are presented for the whole period of observations.

Clearly enough, despite some annual distinctions, seasonal changes of the surface ozone concentration have been repeated in the main features year after year. By the way, the monthly averaged values of the ozone concentration have appeared to be more informative and revealed a specific character of the seasonal course for different years of observation. In fig. 5a annual seasonal peaks (spring and summer) are visible; in autumn and winter the concentration of surface ozone is minimal. However, usually distinct spring maximum occurs not every year. In Звягинцев, 2003 one may find some valid ideas concerning a dynamic nature of the spring maximum. The point is stressed that the maximum is generated by an activation of the atmosphere vertical mixing. As a result, the air with increased ozone amount floats down from a free troposphere to a surface of the Earth. In the opposite, the summer maximum is supposed to have mainly a photochemical origin and connected with intensification of the ozone formation due to UV radiation in the polluted air. From our point of view, while considering a diurnal course of the ozone concentration in summertime (see figure 6), both mentioned factors contribute to a formation of the observed ozone concentration. Actually, the ozone concentration peaks in the afternoon and becomes minimal at night. Such behaviour conforms well with a daily course of activity for both basic mechanisms of formation of the surface ozone concentration: the vertical mixing in the troposphere and the photochemical ozone generation in a polluted air of the surface layer of the atmosphere. Both processes reach a maximum namely in the afternoon. There are sharp short-term fluctuations revealed in a diurnal course of ozone concentration against a background of a general smooth course. This phenomenon is typical for surface ozone and observed in all seasons and at various time of a day. The obvious reason for the registered fluctuations is a thermal and dynamic turbulence in a combination with a significant vertical gradient of the ozone concentration.

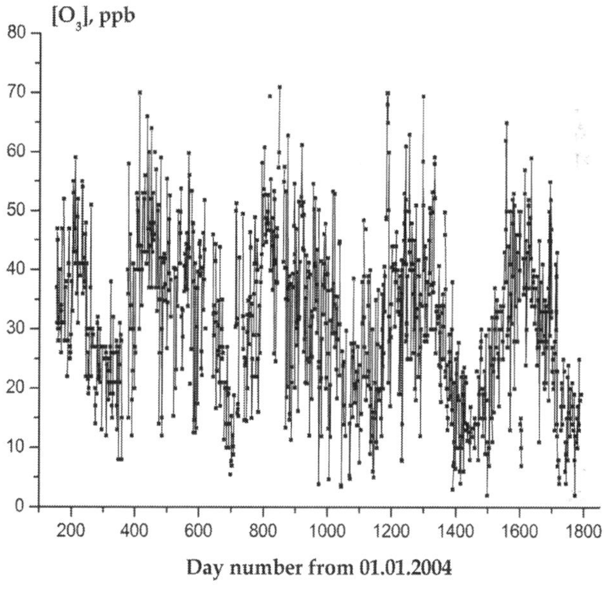

Fig. 4. Five years ozone concentration data measured at Minsk ozonometric station of NOMREC

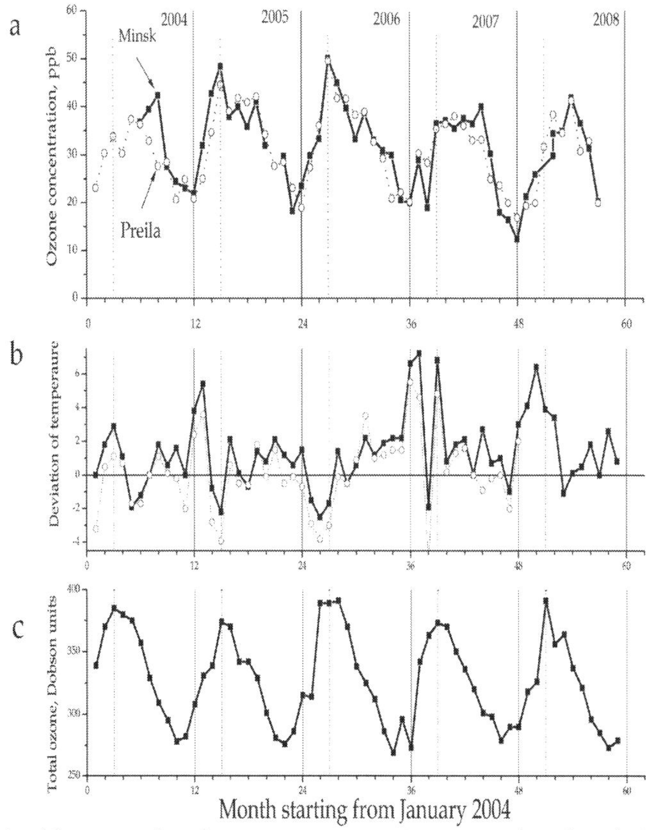

Fig. 5. Monthly averaged surface ozone concentrations in Minsk and Preila (a), deviation of monthly averaged temperature from the climatic norm in Minsk and Preila (b), monthly averaged total column ozone over Minsk (c).

One should mind, the daytime data at the NOMREC station correlate in general with data of the Preila station. As an illustration, the results of measurements for the first half of 2008 are presented in figure 7. Despite some distinctions in everyday results, there is on the whole a good correspondence in the measurements performed. The similar conclusion can be drawn, if analyzing data for different years. A quality of measurements at both stations is undoubtedly maintained at a high level, and observed discrepancies have been verified by independent measurements. In particular, very high values of the ozone concentration in the beginning of April and very low ones in May, obtained in Minsk in 2008 prove to be true as they are confirmed by parallel measurements of devices TrIO-1 and TEI-49C. The comparative analysis of a daily course of the ozone concentration for both stations shows (figure 8) the amplitude of a daily course in Minsk rather exceeding, than that of the Preila station. This can be explained through a close location of the Preila station to the Baltic Sea

(Звягинцев, 2004). Quite visible also, that a trend of concentration variation does not always completely correlate for both stations.

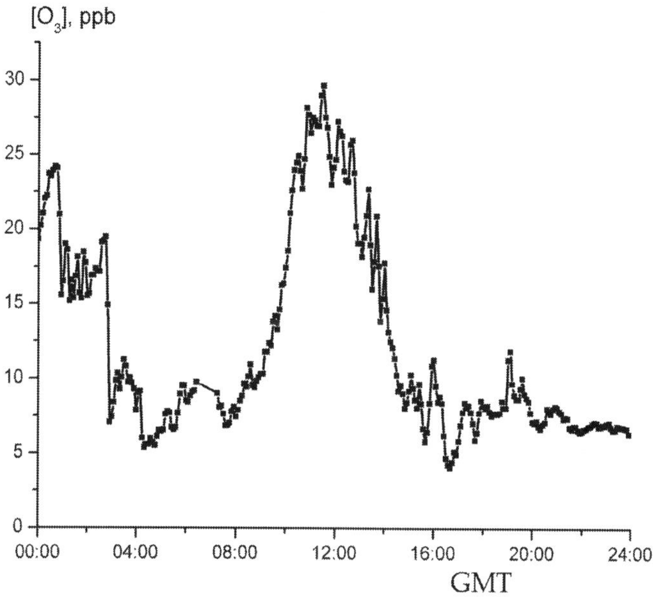

Fig. 6. A typical diurnal cause of the surface ozone concentration in Belarus. Measured on 28.08.2008. Local time is approximately 2 hours greater than Greenwich Mean Time. Short time fluctuations of the ozone concentration are clearly visible.

However, while considering monthly average values of ozone concentration, a seasonal course for both stations (figure 5a) is observed at a good match. Only for a few months for the whole period we have distinctions in figures reaching 10 ppb.

Belarus and Lithuania belong to the same climatic zone. The main differences are stipulated by the Lithuania proximity to the sea. The softer climate of Lithuania is characterized with a lower difference of climate temperature norms. The sea proximity also results in a moderate amplitude of a daily course of ozone concentration (Звягинцев, 2003).

In figure 5b deviations of a monthly average temperature from a climatic norm for Minsk and similar deviations for the Preila station are presented. The interrelation between fluctuations of temperature in March and absence or presence of a spring maximum of surface ozone concentration are well exposed for both sites of the observation. For both sites of the observation amplitude of the maximum is inversely proportional to the temperature of air: the higher monthly average temperature of March, the lower amplitude of the spring maximum. In particular, in figure 5b one can see the significant positive anomalies of the temperature in March to have been occurred in 2004, 2007, 2008. In these very years the spring maximum of the ozone concentration was, practically, missed (figure 5a). On the contrary, in 2005 and 2006 (and in 2009 too; data are not presented here) the monthly average temperature at both points was below the norm and it was accompanied with the strongly pronounced spring maximum of the surface ozone concentration.

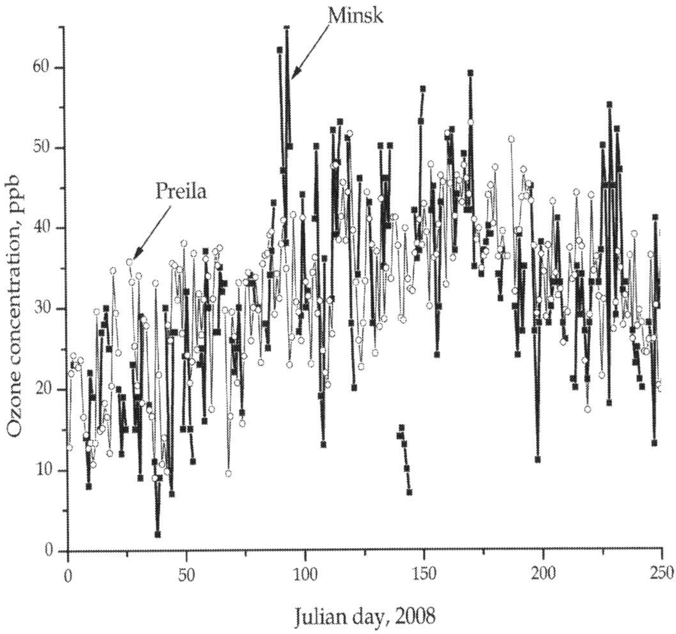

Fig. 7. Comparison of day time ozone concentrations in Minsk and Preila in 2008

Positive deviations of temperature for the Preila station are a bit lower in the most cases than on the average for Belarus, and negative deviations are a bit higher on the amplitude. However, to directly link this distinction with the difference in monthly average values of the ozone concentration does not seem to be possible. It is also difficult to analyze relation of temperature fluctuations and the surface ozone concentration in winter months, when concentration of ozone is minimal, and an influence of measurement errors can be quite essential.

An actual similarity of seasonal changes of the surface ozone concentration in Minsk and Preila, their identical dependence on the temperature (for the spring maximum) indicate that both stations most of the time are affected by the same field of the tropospheric ozone, originated from a global circulation of air masses in the Northern hemisphere. Only local factors specifically 'modulate' this field, resulting in differences of daytime results of measurements and a diurnal course. These distinctions disappear after averaging over the time interval, exceeding the period of the synoptic phenomena typical for territories of both countries (a few days and more).

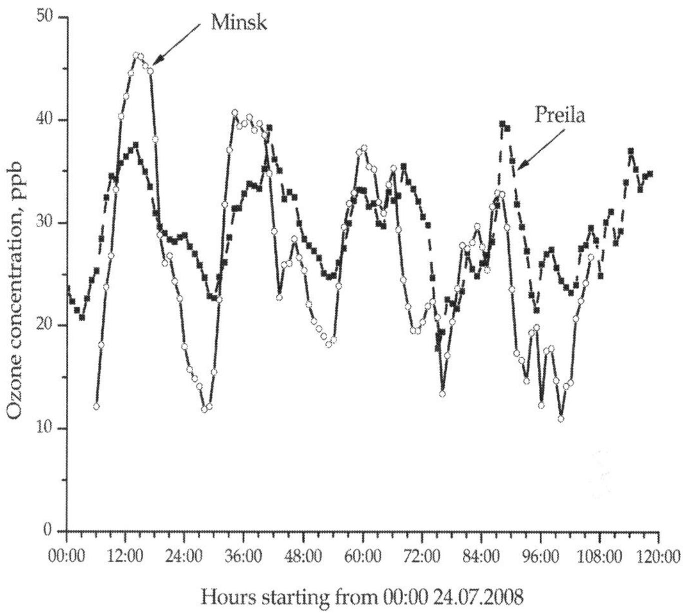

Fig. 8. Diurnal course of the surface ozone concentration (average hour data) in Minsk and in Preila during 24-28 July 2008.

One can view intensification of exchange with the upper troposphere under the exposition of the growing solar radiation as a main reason for the origination of the spring maximum of the surface ozone concentration. This suggestion is repeatedly expressed in Звягинцев, 2003, 2004; Звягинцев & Кузнецова, 2002 and other papers, though, the argumentation brought herein, has an indirect character. Quite evident, that the revealed anticorrelation of surface ozone concentration with the air temperature at a time of the spring maximum is not in any agreement with the mechanism of a photochemical generation of ozone in the polluted air.

The similar situation has evolved with such a parameter of the atmosphere, as the total ozone amount (TOA). Concentration of the surface ozone is a TOA component and according to Белан *et al.*, 2000 gain in the TOA leads to a reduction of the surface ozone concentration because of easing sun UV radiation initiating a surface photochemistry. We have compared a course of the total ozone amount over Belarus and the surface ozone concentration for the whole period of observation. As a result, any correlation appears not to be revealed (compare figures 5a, 5c).

9. Conclusion

The proposed multi-wave technique for determining surface ozone concentration by means of an optical open path meter is, in principle, of general type and may be modified to consider any specific realization. The given mathematical formalism allows analysing the influence of spectra discrepancy of the sounding radiation that emitted to the working and reference paths. It is shown that this discrepancy results in a constant systematic error of the ozone concentration.

The given mathematical formalism is employed to assess the role of different errors in the formation of the resulting uncertainty in the calculated ozone concentration. An idea lies in finding the link between errors in signal measurements, errors in parameters of the instrument and parameters of the calculation method and generated by those errors distinction in sounding radiation spectra on the working and reference paths. Such a distinction is simply referred to an error in a calculated ozone concentration. Analytical expressions are given to estimate role of errors of different nature. For example, for the optical open path surface ozone meter TrIO-1 it is shown that the errors in signal measurements and errors in definition of absorption cross-sections have the dominant role in the formation of the resulting error in ozone concentration measurement.

Some results received with the optical open path ozone meter TrIO-1 within the multi-wave method are presented. A revealed link of the spring ozone maximum and average temperature is of special interest. In particular, it is obviously conflicts with the hypothesis of the photochemical origin of the spring ozone maximum.

10. References

Bolot'ko, L.M.; Pokatashkin, V.I.; Krasovskii, A.N. & Tavgin, V.L. (2004). Optical trace measurer of surface ozone concentration. *Journal of Optical Technology*, Vol. 71, No. 1, pp. 48-50, ISSN 1070-9762

Bolot'ko, L. M.; Krasovskii, A. N.; Lyudchik, A. M. & Pokatashkin V.I. (2005). Measurement of the terrestrial ozone concentration by absorption UV spectroscopy. *Journal of Applied Spectroscopy*, Vol. 72, No. 6, pp. 911-916, ISSN 0021-9037 (Print) 1573-8647 (Online)

Bolot'ko, L.M.; Krasovskii, A.N., Lyudchik, A.M. & Pokatashkin, V.I. (2007). Optical track measurements of the near-earth ozone concentration using a multiwave technique under conditions of local thermal convection. *Journal of Optical Technology*, Vol. 74, No. 4, pp. 270-273, ISSN 1070-9762

Bolot'ko, L. M.; Krasovskii, A. N.; Lyudchik, A. M. & Pokatashkin, V. I. (2008a). Ozone optical track analyzer TrIO-1 and its metrological certification features. *Measurement Techniques*, Vol. 51, No 10, pp. 1139-1142, ISSN 0543-1972 (Print) 1573-8906 (Online)

Bolot'ko, L. M.; Demin, V.S.; Krasovskii, A. N.; Lyudchik, A. M. & Pokatashkin V.I. (2008b). Multiwavelength procedure for determining the concentration of near-ground ozone by an optical open-path measurement system. *Journal of Applied Spectroscopy*, Vol. 75, No 2, pp. 268-274, ISSN 0021-9037 (Print) 1573-8647 (Online)

Edner, H.; Ragnarson, P.; Spännare, St. & Svanberg, S. (1993). Differential optical absorption spectroscopy (DOAS) system for urban atmospheric pollution monitoring. *Appl.Opt.*, Vol. 32, No. 3, pp. 327-333, ISSN 0003-6935 (Print) 1593-4522 (Online)

Frohlich, C. & Shaw, G.E. (1980). New determination of Rayleigh scattering in the terrestrial atmosphere. *Appl. Opt.*, Vol. 19, No. 11, pp. 1773-1775, ISSN 0003-6935 (Print) 1593-4522 (Online)

Girgzdiene, R. & Girgzdys, A. (2003). Variations of the seasonal ozone cycles in the Preila station over the 1980-2001 period. *Environmental and chemical physics*, Vol. 25, No. 1, pp. 11-15, ISSN 1392-740X

Hamming, R.W. (1989). *Digital filters*, Prentice-Hall, ISBN 0-486-65088-X (pbk.), Englewood Cliffs, New Jersey

Jacovides, C.; Varotsos, C.; Kaltsounides, N.; Petrakis, M. and Lalas, D. (1994). Atmospheric turbidity parameters in the highly polluted site of Athens basin. *Renewable Energy*, Vol. 4, pp. 465–470, ISSN 0960-1481

Klausen, J.; Zellweger, Ch.; Buchmann, Br. & Hofer, P. (2003). Uncertainty and bias of surface ozone measurements at selected Global Atmosphere Watch sites. *J. Geophys. Res.*, Vol. 108, No. D19, pp. 4622-4632, ISSN 0148-0227

Kondratyev, K.Y. & Varotsos, C.A. (1996). Global total ozone dynamics - Impact on surface solar ultraviolet radiation variability and ecosystems. 1. Global ozone dynamics and environmental safety. *Environmental Science and Pollution Research*, No. 3, pp. 153-157, ISSN 0944-1344(Print) 1614-7499 (Online)

Kondratyev, K.Y. & Varotsos, C. (2002). Remote sensing and global tropospheric ozone observed dynamics. *International Journal of Remote Sensing*, Vol. 23, pp. 159-178, ISSN 0143-1161

Krasouski, A.N.; Balatsko, L.M. and Pakatashkin V.I. (2004). Surface ozone DOAS instrument with zero path, *Proceedings of Quadrennial Ozone Symposium, 1-8 June 2004.* Kos, Greece, Vol. 2, pp. 893-894, ISBN 960-630-103-6 960-630-105-2

Krasouski, A.; Balatsko, L.; Liudchik, A. & Pakatashkin, V. (2006). Surface ozone DOAS measurements comparison with the instrument measuring local ozone concentrations, *TECO-2006-WMO technical conference on meteorological and environmental instruments and methods of observation. Geneva, Switzerland, 4-6 December 2006.* www.wmo.int/pages/prog/www/IMOP/publications/IOM-94-TECO2006/P1(01)_Krasouski_Belarus.pdf

List of designated reference and equivalent methods. (2000). United States inveronmental protection agency. National exposure research laboratory. Human exposure & atmospheric sciences division (MD-46). Research triangle park, NC 27711, 919-541-2622. www.epa.gov/ttn/amtic/criteria.html

Opto-analyser AR-500. User's manual. (1993). Opsis AB - Furulund

Ozone monitoring, mapping, and public outreach. Delivering real-time ozone information to your community. (1999). United states environmental protection agency, National risk management laboratory office of research and development Cincinnati, Ohio 45268. EPA/625/R-99/007

Platt, U.; Perner, D. & Patz, H.W. (1979). Simultaneous measurement of atmospheric CH_2O, O_3, and NO_2 by differential optical absorption. *J. Geophys. Res.*, Vol. 84, No. C10, pp. 6329-6334, ISSN 0148-0227

Transceiver ER-130. User's manual. (1994). Opsis AB - Furulund

Varotsos, C.; Kondratyev, K.Y. and Efstathiou, M. (2001). On the seasonal variation of the surface ozone in Athense, Greece. *Atmospheric Environment*, Vol. 35, No. 2, pp. 315-320, ISSN 1352-2310

Varotsos, C.; Ondov, J. & Efstathiou, M. (2005). Scaling properties of air pollution in Athens, Greece and Baltimore, Maryland. *Atmospheric Environment,* Vol. 39, pp. 4041-4047, ISSN 1352-2310

World Meteorological Organization (WMO). Global Atmosphere Watch measurements guide, Tech. Doc. 1073. GAW Rep. 143. (2002). Geneva, Switherland

Белан, Б.Д.; Зуев, В.В.; Скляднева, Т.К.; Смирнов, С.В. и Толмачев, Г.Н. О роли суммарного озона в фотохимическом образовании его тропосферной части. (2000). *Оптика атмосферы и океана,* Т. 13, № 10, с. 928-932, ISSN 0235-6880

Гиргждене, Р.; Болотько, Л.М.; Демин, В.С.; Красовский, А.Р.; Людчик, А.М. и Покаташкин, В.И. Мониторинг приземного озона в Беларуси и Литве. (2008). *Природные ресурсы,* No. 1, с. 60-66, ISSN 1810-9810

Звягинцев, А.М. О сходстве долговременных рядов наблюдений приземного озона на станциях Долгопрудный Московской области и Бельск, Польша. (2003). *Известия РАН. Физика атмосферы и океана,* Т. 39, № 4, с. 510-514, ISSN 0869-5695

Звягинцев, А. М. Основные характеристики изменчивости содержания озона в нижней тропосфере над Европой. (2004). *Метеорология и гидрология,* № 10, с. 46-55. ISSN 0130-2906

Звягинцев, А.М. и Кузнецова, И.Н. Изменчивость приземного озона в окрестностях Москвы: результаты десятилетних регулярных наблюдений. (2002). *Известия РАН. Физика атмосферы и океана,* Т. 38, № 4, с. 486-495, ISSN 0869-5695

Ультрафиолетовый трассовый газоанализатор ДОАС М1. Назначение и краткое техническое описание. (2002). ООО «Обнинская фотоника», Обнинск

23

Intermediate measures consideration for a value chain or multistage system: an efficiency analysis using DEA approach

Wai Peng Wong[1] and Kuan Yew Wong[2]
[1]*School of Management, Universiti Sains Malaysia, Malaysia.*
[2]*Faculty of Mechanical Engineering, Universiti Teknologi Malaysia, Malaysia.*

1. Introduction

It has been recognized that performance evaluation is extremely important as the old adage says "you can't improve what you don't measure". Companies using performance measurement were more likely to achieve leadership positions in their industry and were almost twice as likely to handle a major change successfully (Wisner et al., 2004). Today, business performance is evaluated not only in terms of a single business unit but rather the entire value chain. Performance measurement of the entire value chain is a lot more difficult and complex compared to the performance measurement of a single business unit. When managing a value chain, apart from the formidable multiple performance measures problem, assessing the performance of several tiers, e.g., suppliers, manufacturers, retailers and distributors further complicates the matter. Basically, there are two main problems in value chain performance measurement, which are a) existence of multiple measures that characterize the performance of each member, for which the data must be acquired, b) existence of intermediate measures between them, e.g., the output from the upstream can become the input to the downstream which further complicates the performance assessment.

As noted in Wong and Wong (2007 and 2008), DEA is a powerful tool for measuring value chain efficiency. DEA, developed by Charnes et al. (1978), is a well-established methodology used to evaluate the relative efficiency of a set of comparable entities called Decision Making Units (DMUs) with multiple inputs and outputs by some specific mathematical programming models. DEA can handle multiple inputs and outputs and it does not require prior unrealistic assumptions on the variables which are inherent in typical supply chain optimization models (i.e. known demand rate, lead time etc) (Cooper et al. 2006). These advantages of DEA enable managers to evaluate any measure efficiently as managers do not need to find any relationship that relates the measures.

We point out that DEA's vitality, real-world relevance, diffusion and global acceptance are clearly evident, as supported from such literature studies as Seiford (1996) and Gattoufi et al. (2004a and 2004b). There are a number of DEA studies on value chain efficiency; yet, most of them tend to focus on a single chain member. This can be partly due to the lack of DEA models for the entire value chain or multi-stage systems. Note that DEA cannot be directly applied to the problem of evaluating the entire value chain efficiency because the value chain cannot be simply viewed as a simple input-output system as conceptualized in DEA.

Within the context of DEA, there are some recent models, e.g., by Fare and Grosskopf (2000) and Golany et al. (2006), which have the potential to address a value chain in a powerful way. Recently, Liang et al. (2006) developed two classes of DEA-based models for supply chain efficiency using a seller-buyer supply chain setting. They used the game theory approach to analyze the effect of one member having on another. One similarity of the recent models for addressing the chain effect or multilayer system is that they take into consideration the presence of intermediate measures; their differences lie in their mechanic system design. The issue of intermediate measures was initially addressed by Banker and Morey (1986) in a service industry which operates in a single layer. The model separates the inputs/outputs into two groups, i.e., discretionary and non-discretionary; non-discretionary inputs/outputs are exogenously fixed inputs/outputs that are not controllable and their values are predetermined.

The current chapter provides an alternative way to measure value chain or multistage efficiency which is by taking into consideration the effect of the intermediate measures in the system. We draw on previous Banker's model and extend the model construction for value chain. We analyze its dual formulation and explain how it suits the value chain setting. This chapter contributes to the existing value chain or multistage system literature by providing an alternative model to measure value chain or multistage efficiency. This model is simple and easy to understand. Though, this model may not have addressed all the concerns in value chain or multistage system, it can still serve as a tentative solution for measuring the efficiency of these systems.

In the following section, we will review Banker's model by analyzing its dual formulation and then provide the insight on how it can address the value chain or multistage efficiency. Then we present an application study to show the usefulness of the model.

2. Theoretical foundations

In this section, we first discuss the foundations of DEA. Then, we show the dual formulation of the Banker's model and explain how it can better characterize the value chain or multistage system.

2.1 Basic DEA methodology

Build upon the earlier work of Farrell (1957), data envelopment analysis (DEA) is a mathematical programming technique that calculates the relative efficiencies of multiple decision-making units (DMUs) based on multiple inputs and outputs.

Assume S to be the set of inputs and R the set of outputs. J is the set of DMUs. Further assume that DMU$_j$ consumes $x_{sj} \geq 0$ of input s to produce $y_{rj} \geq 0$ of output r and each DMU has at least one positive input and one positive output (Fare et al., 1994). Based on the efficiency concept in engineering, the efficiency of a DMU, says DMU j_0 ($j_0 \in J$), can be estimated by the ratio of its virtual output (weighted combination of outputs) to its virtual input (weighted combination of inputs).

To avoid the arbitrariness in assigning the weights for inputs and outputs, Charnes et al. (1978) developed an optimization model known as the CCR model in ratio form to determine the optimal weight for DMUj_0 by maximizing its ratio of virtual output to virtual input while keeping the ratios for all the DMUs not more than one. The fractional form of a DEA mathematical programming model is given as follows:

$$
\begin{aligned}
&\max \frac{\sum_{r\in R} u_r y_{rj_0}}{\sum_{s\in S} v_s x_{sj_0}} \\
&\text{s.t.} \quad \frac{\sum_{r\in R} u_r y_{rj}}{\sum_{s\in S} v_s x_{sj}} \le 1 \\
&\quad u_r, v_s > 0 \qquad s\in S,\ r\in R
\end{aligned}
\tag{1}
$$

where u_r and v_s are the weights for the output r and input s respectively.
The objective function of Model (1) seeks to maximize the efficiency score of a DMUj_0 by choosing a set of weights for all inputs and outputs. The first constraint ensures that, under the set of chosen weights, the efficiency score of the observed DMU is not greater than 1. The last constraint ensures that the weights are greater than 0 in order to consider all inputs and outputs in the model. A DMUj_0 is considered efficient if the objective function of the associated Model (1) results in an efficiency score of 1, otherwise it is considered inefficient.
Using the Charnes-Cooper transformation, this problem can be further transformed into an equivalent "output maximization" linear programming problem as follows:

$$
\begin{aligned}
&\max \sum_{r\in R} u_r y_{rj_0} \\
&\text{s.t.} \quad \sum_{r\in R} u_r y_{rj} - \sum_{s\in S} v_s x_{sj} \le 0, \quad j\in J \\
&\quad \sum_{s\in S} v_s x_{sj_0} = 1 \\
&\quad u_r, v_s > 0 \qquad s\in S,\ r\in R
\end{aligned}
\tag{2}
$$

Model (2) is known as the CCR model in multiplier form. If the objective function value of (2) is equal to 1, it implies that the DMU concerned is relatively efficient since we can find a weight combination to make its efficiency score to be equal to one. Despite the linear form of (2), efficiency score is usually calculated based on its dual problem:

$$\begin{aligned} & \min \ \theta \\ & \text{s.t.} \ \sum_{j \in J} x_{sj}\lambda_j \le \theta x_{sj_o}, \quad s \in S \\ & \sum_{j \in J} y_{rj}\lambda_j \ge y_{rj_o}, \quad r \in R \\ & \lambda_j \ge 0, \qquad j \in J \end{aligned} \tag{3}$$

Model (3) is known as the input-oriented CCR in envelopment form or the Farrell model, which attempts to proportionally contract DMUj_0's inputs as much as possible while not decreasing its current level of outputs. The λ_j values are the weights (decision variables) of the inputs/outputs that optimize the efficiency score of DMU j_0. These weights provide measures of the relative contributions of the inputs/outputs to the overall value of the efficiency score. The efficiency score will be equal to one if a DMU is efficient and less than one if a DMU is inefficient. The efficiency score also represents the proportion by which all inputs must be reduced in order to become efficient. In a similar way, we can also derive the output-oriented CCR in envelopment form if efficiency is initially specified as the ratio of virtual input to virtual output. A large number of extensions to the basic DEA model have appeared in the literature as described by Ramanathan (2003) and Cooper et al. (2006). We shall limit our discussion to this basic model as this is sufficient to lead us to the explanation of the following model to address a value chain or multistage system.

2.2 The DEA analysis of value chain efficiency.

Consider a value chain relationship as follows, e.g., supplier - manufacturer with inputs and outputs as described in Figure 1. This may also be viewed in terms of a multistage process, e.g., a product has to go through two stages of a manufacturing process: assembly (stage 1) and testing (stage 2). We may further categorize the inputs and outputs into two types, i.e., direct and indirect or intermediate. *Direct* inputs/outputs are associated with a single stage or member only and they do not affect the performance of other stages / members. For example, supplier cost and supplier revenue are direct inputs and outputs for the supplier only, they have no impact on the manufacturer. *Intermediates* are those inputs/outputs that are associated with two or more stages/members. For instances, ontime delivery is the performance of the supplier in delivering its products; it is also a cost measure to the manufacturer which relates to inventory holding cost.

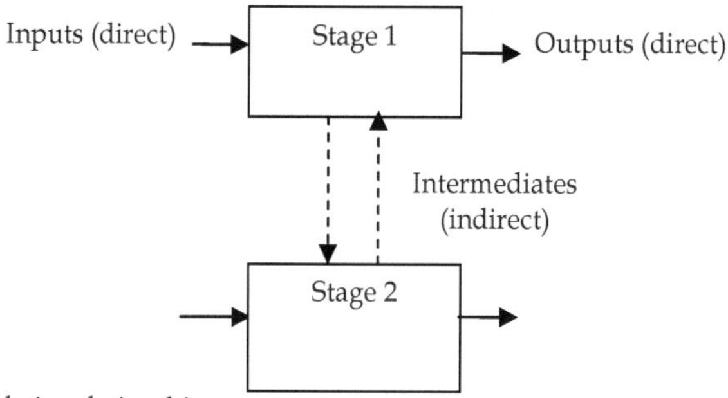

Fig. 1. A simple chain relationship

Note that if the intermediate measures are treated as both inputs and outputs in the model, all the DMUs (decision making units) will become efficient. This does not necessarily indicate efficient performance in an individual chain member. Due to the presence of intermediate measures in the value chain or multistage system, the performance of one member will affect the performance or efficiency status of the other members.

Alternatively, we may consider the effect of the intermediate measures using Banker's model. We will now elaborate how the value chain or multistage efficiency will be characterized if we take into consideration the intermediate (indirect) measures.

Let's use a simple scenario; for example, there are two value chains, i.e. DMU A and DMU B, and each of them is a dual-channel (supplier-manufacturer) system. Let's say that the manufacturers of A and B are the same. Also, let's assume that supplier A is very efficient while supplier B is less efficient compared to A. Note that the efficiency of the individual supply chain member can be obtained using the DEA CCR model as explained earlier. Recall that the best practice of one channel does not mean that it fits the other channel. In this case, the impact from the performance of the supplier may affect the efficiency status of the manufacturer in such a way that the manufacturer A may seem to be less efficient compared to the manufacturer B; by right, they should be equally good because they are the same manufacturer. This shows that a member's inefficiency may be caused by another's efficient operations.

In order to better characterize the value chain, we have to 'discount' or remove the impact of the performance improvement of one supply chain member that affects the efficiency status of the other. We will illustrate how this discounting concept can be realized using the intermediate (indirect) measures in Banker's model. From the basic DEA model in fractional (ratio) form, let's denote *IS* as the set of intermediate inputs, *DS* as the set of direct inputs, x_{tj} as the tth intermediate input of DMU j and x_{tj_0} as the tth intermediate input for the observed DMU j_0. Note that $DS \cup IS = S$.

$$\max \frac{\sum_{r\in R} u_r y_{rj0} - \sum_{t\in IS} v_t x_{tj0}}{\sum_{s\in DS} v_s x_{sj0}}$$

$$\text{s.t.} \quad \frac{\sum_{r\in R} u_r y_{rj} - \sum_{t\in IS} v_t x_{tj}}{\sum_{s\in DS} v_s x_{sj}} \le 1$$

$$u_r > 0, \quad r \in R$$

$$v_s > 0, \quad s \in DS \tag{4}$$

$$v_t \ge 0, \quad t \in IS$$

where v_t is the weight for the intermediate variables.

All the other notations used have been previously defined in Section 2. Note that the weights for the intermediate variables may be zero, but for the direct variables, the weights must always be positive. Note also that the difference between (4) and (1) is the subtraction of the intermediate term. This term represents the performance of one chain member (e.g. the upstream channel) that feeds into the other chain member (e.g. the downstream channel). By subtracting the intermediate term in such a way is analogous to 'discounting' the impact of one's performance that affects the other. From Model (4), it is obvious that the impact of the indirect factor is removed; and the efficiency obtained in this model will be the best case efficiency. Though the 'discounting' concept may not have fully addressed all the issues in a value chain or multi stage system, it can serve as a tentative solution to measure the chain or multistage efficiency.

Model (4) can be further transformed into its equivalent linear form as shown in Model (5) (the primal model) and Model (6) (the dual model).

CCR multiplier model

$$\max \sum_{r\in R} u_r y_{rj0} - \sum_{t\in IS} v_t x_{tj0}$$

$$\text{s.t.} \quad \sum_{r\in R} u_r y_{rj} - \sum_{t\in IS} v_t x_{tj} - \sum_{s\in DS} v_s x_{sj} \le 0, \qquad j \in J$$

$$\sum_{s\in DS} v_s x_{sj0} = 1 \tag{5}$$

$$u_r > 0, \quad r \in R$$

$$v_s > 0, \quad s \in DS$$

$$v_t \ge 0, \quad t \in IS$$

CCR envelopment model

$$
\begin{aligned}
&\min \Omega \\
&\text{s.t.} \sum_{j \in J} \lambda_j x_{sj} \leq \Omega \, x_{sj_o}, && s \in DS \\
&\sum_{j \in J} \lambda_j x_{tj} \leq x_{tj_o}, && t \in IS \\
&\sum_{j \in J} \lambda_j y_{rj} \geq y_{rj_o}, && r \in R \\
&\lambda_j \geq 0, && j \in J
\end{aligned}
\tag{6}
$$

Note that all the notations used have been previously defined in the above section. Note that Model (6) is an input oriented model whereby it aims to reduce the inputs as much as possible while not decreasing the level of the outputs. Note that the third constraint (i.e. for the outputs) can actually be separated into two constraints, i.e., one for direct and another for indirect terms. Since the indirect term for the output will not affect the objective function, we did not explicitly write it into two separate constraints; for conciseness purpose of the model, we combined them into one constraint.

Given Model (6), one way to evaluate the entire value chain or multistage efficiency, is to estimate the efficiency, Ω as the normalized (weighted) efficiency of all the members or stages. That is,

$$
\Omega^* = \frac{\sum_{i \in I} w_i \Omega_i^*}{\sum_{i \in I} w_i} \tag{7}
$$

where Ω^* is the optimal efficiency score of the value chain or multistage system, I is the set of members or stages in the system, Ω_i^*, $i \in I$, is the optimal efficiency score for a specific chain member (channel) or stage and w_i is the weight reflecting the extent of each channel or stage contributing to the evaluation of the entire value chain or multistage efficiency. These weights can be estimated using various methods such as AHP (Analytic Hierarchical Process), Delphi method and Pareto analysis (Clemen and Reilly, 2001; Kirkwood, 1997). In this research, we consider all channels (stages) have equal contribution to the value chain (multistage) system performance. As the indirect effect, i.e., the performance improvement of one channel affecting another channel has already been removed or discounted from Model (6), the weight measures proposed in such a way would be reasonable and the 'double counting' effect on the performance of the entire value chain will not be very significant. Note that in this study we set $w = 1$.

From Model (6), a chain or multistage system is efficient if $\Omega^* = 1$. Note that it is possible among all DMUs, the highest value of Ω^* is < 1. In this case, it means that none of the DMUs is efficient. Comparing Model (6) to (3), as the values of Ω^* and θ^* have to be greater than 0 and less than or equal to 1, and as Model (6) has less restriction on the intermediate inputs, the value of Ω^* from Model (6) will always be less than or equal to the value of θ^* from Model (1) i.e. $\Omega^* \leq \theta^*$.

Proposition 1. *The efficiency score, Ω^* of (6) for any DMU j_0 is less than or equal to the corresponding efficiency score from θ^* (3).*
To prove this proposition, we first note that $\theta^* \leq 1$ in the optimal solution of (1) because DMU j_0 is itself one of the $j_0 \in J$ referent observations. By comparing the constraint sets in the two linear programs, we see that any optimal solution to (3) is a feasible solution for (6); hence, $\Omega^* \leq \theta^*$.

Model (6) yields the target values on the performance measures for an inefficient supply chain to reach the best practice by using its slack information. The model assumes that the inputs could be reduced while maintaining all the outputs at the same level. The target values are obtained as follows. We denote $x^*_{sj_o}$ and $x^*_{tj_o}$ as the direct and indirect input targets i.e., $x^*_{sj_o} = \Omega^{a*} x_{sj_o} - s^-_{sj_o}$ and $x^*_{tj_o} = x_{tj_o} - e^-_{tj_o}$
where $s^-_{sj_o}$ and $e^-_{tj_o}$ are the direct and indirect input slacks respectively.

3. An application study

In this section, we discuss an application study on the proposed approach. First, we explain the conceptual configuration of the value chain or multistage setting which comprises the variables or metrics used in the study. Then, we provide some descriptions of the data, followed by results discussion of the empirical analysis.

3.1 Configuration

To illustrate our proposed approach, we model a value chain setting based on the global value chain system of multinational semiconductor corporations. There are three levels in the proposed setting, e.g., supplier, manufacturer and retailer. This can also be viewed in terms of a multistage process, e.g., first stage (assembly), second stage (testing) and third stage (final inspection/packaging). We use the supply chain operations reference (SCOR) to determine the value chain performance metrics. The metrics used are the financial and operational measures. Table 1 shows the categorization of the metrics.

Direct Inputs	**Cost (including labor , variable, and capital components) ($)**
Intermediates	**Fill rate (%), on-time delivery (%), cycle time (days)**
Direct output	**Revenue ($)**

Table 1. Metrics categorization

A brief definition for each measure is given below. For financial measures, the elements are cost and revenue; for operational measures, the elements include fill rate, on-time delivery and cycle time.

Financial measures:

a. Revenue - This is a common measure of efficiency in various profit-oriented organizations. In value chain studies, emphasis is often placed on the final revenue, i.e., revenue of the final product. In our experiment we consider the effect of revenue of one member affecting another's performance is minimal.

b. Cost - This is the performance attribute for value chain costs, i.e. the total cost associated with operating the value chain. The total cost comprises labor, variable and capital components. We consider the total cost of each member separately.

Operational measures:

a. Fill rate - This is a performance attribute for value chain reliability. In the broadest sense, fill rate refers to the service level between two parties. It is usually a measure of shipping performance expressed in percentage. Being an output to the upstream channel, the upstream channel will always desire to have a high fill rate so that it is able to satisfy customer demand. However, for the downstream channel, a high fill rate means additional storage and holding cost. Therefore, fill rate affects two parties; hence, it is generally viewed as an intermediate measure. An optimal level of fill rate is usually determined from the tradeoff between the rate of customer order fulfillment and inventory level.

b. On-time delivery rate - This is another common performance attribute for value chain delivery reliability. It is usually expressed in percentage; it refers to the performance of the value chain in delivering the correct product, to the correct place, at the correct time, in the correct condition and packaging, in the correct quantity, and with the correct documentation to the correct customer. It affects two members; as an output, the member will want this to be as high as possible; alternatively, as an input, it can be viewed as a cost to the associated member.

c. Cycle time - This is the performance attribute for 'production flexibility'. It refers to the agility of a value chain in responding to marketplace changes to gain or maintain competitive advantage. It is also known as the 'upside production flexibility'. It refers to the number of days required to achieve an unplanned, sustainable, a certain percent increase in production. One of the common constraints to cycle time is material availability. Cycle time affects the performance of one member with another, e.g., if the supplier cycle time is high, then, the manufacturer will not be able to meet its production and will be seen as inefficient and vice versa.

In short, we consider all the operational measures as intermediate measures because they affect the performance of one member with another. Note that, the particular setting may not be applicable to all types of industries or systems, as different industries or multistage systems may have different types of configuration. In our case, the value chain configured is sufficient to evaluate the model.

3.2 Data descriptions

The analysis uses observational data from the semiconductor companies based in Malaysia. As one of the requirements of DEA is to have a homogenous set of DMUs for fair comparison, the companies selected are based on a similar logistic distribution network and business. The data required, e.g., the total costs and revenues, are obtained from the companies' annual reports. Total cost comprises three components, i.e., labor, variable and capital components. Specifically, labor relates to the number of employees working directly for a particular channel or stage and the price of labor is measured using the average wages of the employees. As capital comprises buildings, facilities and other peripheral equipment, it is impossible to allocate the capital costs to individual components. According to Arnold (2004), stocks (inventory) play an important role in an operation activity and the costs associated with them are related with each individual capital component. We thus used the value of capital (or capital stock) as a proxy to the amount of physical capital used in the value chain. The value of capital can be obtained through the division of net operating income by the return on capital asset (ROA). On the other hand, the revenue figures may include revenue generated from other businesses; however, as the selected companies for the study have been filtered and ensured of having a similar logistic distribution network, the impact of revenue generated from other businesses would be minimal.

It is also difficult to obtain a full set of data due to some data, e.g., fill rate and cycle time, are considered confidential by most of the companies. To overcome this hurdle, we gathered the required data via several methods; site interviews with managers were conducted and experts' judgments were collected to gauge the mean value of these uncertain variables. Note that Wong et al. (2008) developed a method called the Monte Carlo DEA to measure the efficiency when there are uncertainties in the data. This method is based on the Bayesian framework where they used the distribution of the inputs/outputs to estimate the distribution of the efficiencies. In this chapter, we will apply the DEA Model (6) in a deterministic setting, which is, we assume that all data are finally available for the experiment. Table 2 shows the data used for the numerical experiment. All the monetary values are denominated in current US dollars.

DMU	Cost (stage 1) *$Million* USD	Cost (stage 2) *$Million* USD	Cost (stage 3) *$Million* USD	Fill rate (stage 1) %	On time delivery (stage 1) %	Cycletime (stage 1) days	Fill rate (stage 2) %	On time delivery (stage 2) %	Cycletime (stage 2) days	Revenue *$Million* USD
1	115	195	154	95	84	11	80	83	6	499
2	105	170	211	86	79	3	77	95	5	504
3	131	148	132	95	88	4	78	87	4	423
4	170	140	175	82	74	3	81	80	3	497
5	125	63	113	85	83	3	72	84	5	359
6	196	121	164	83	89	2	79	82	7	528
7	177	165	124	77	89	3	93	89	8	526
8	151	164	112	84	79	8	79	90	5	455
9	149	108	154	92	87	11	86	94	7	428
10	101	183	178	95	93	8	81	74	6	526
11	169	194	146	93	93	7	88	83	5	558
12	180	131	111	85	98	8	98	84	4	457
13	150	174	161	81	91	4	82	79	8	514
14	143	133	102	88	90	8	98	98	5	404
15	166	121	147	77	83	3	83	89	6	503
16	182	165	107	90	86	10	89	82	6	492
17	189	112	156	93	95	5	96	86	6	478
18	84	125	104	77	74	9	89	84	10	300
19	167	126	196	96	90	1	92	83	7	537
20	179	106	115	82	72	5	79	81	10	414

Table 2. Data

Note that for the semiconductor industry, the gross profit margin is approximately 8% and the net profit margin is approximately 4% of the total revenue. We evaluate a total of 20 value chains (multistage). For confidential purposes, the names of the companies have been omitted and some of the information has been disguised. The results are analyzed using MS Excel and its linear optimization solver.

3.3 Empirical analysis

The ultimate purpose of the experiment is to provide an insight on the importance of characterizing a value chain or multistage system. We will show from the results that, by considering the presence of intermediate measures in the value chain of multistage system, there are potential input savings in the system. We present the DEA efficiency results in Table 3.

DMU	CCR model (Model 3)				Banker's model (Model 6)			
	Stage1 (θ_1^*)	Stage 2 (θ_2^*)	Stage 3 (θ_3^*)	Average (θ^*)	Stage1 (Ω_1^*)	Stage 2 (Ω_2^*)	Stage 3 (Ω_3^*)	Chain (Ω^*)
1	1.000	0.865	0.988	0.951	0.944	0.428	0.839	0.737
2	0.883	1.000	1.000	0.961	0.874	1.000	1.000	0.958
3	0.875	0.880	1.000	0.918	0.771	0.491	0.811	0.691
4	0.524	1.000	1.000	0.841	0.516	1.000	1.000	0.839
5	0.743	1.000	1.000	0.914	0.729	1.000	0.688	0.806
6	0.501	1.000	1.000	0.834	0.497	1.000	1.000	0.832
7	0.554	1.000	1.000	0.851	0.551	1.000	0.953	0.835
8	0.600	0.974	1.000	0.858	0.597	0.815	0.985	0.799
9	1.000	1.000	0.876	0.959	0.722	0.978	0.603	0.768
10	1.000	0.822	1.000	0.941	1.000	0.410	1.000	0.803
11	0.661	0.839	1.000	0.833	0.602	0.477	1.000	0.693
12	1.000	1.000	1.000	1.000	0.595	1.000	1.000	0.865
13	0.666	0.951	0.986	0.868	0.665	0.652	0.823	0.713
14	0.690	1.000	1.000	0.897	0.690	1.000	0.877	0.856
15	0.560	1.000	0.965	0.841	0.549	1.000	0.851	0.800
16	0.588	0.871	1.000	0.820	0.539	0.648	1.000	0.729
17	1.000	1.000	0.922	0.974	0.547	1.000	0.695	0.747
18	1.000	1.000	1.000	1.000	1.000	1.000	0.624	0.875
19	1.000	1.000	0.932	0.977	0.610	1.000	0.680	0.764
20	0.495	1.000	1.000	0.832	0.487	1.000	0.779	0.755

Note: θ_i^* , i ={1, 2, 3} refers to the efficiency score obtained using CCR.
Table 3. Value chain efficiency

We compare Banker's Model (6) with the original CCR model; the original CCR model is applied separately on each stage and the value chain efficiency is obtained by taking the average. Note that the value chain efficiency from Model (6) (Banker's model) is always less than or equal to the value chain efficiency from the CCR model. The reduction of the value chain efficiency score in Model (6) is due to the removal of the impact from the indirect (intermediate) measures.

From the analysis, none of the value chain is efficient. We further interpret the target adjustments for the inefficient DMUs. As an example for discussion, we select DMU 3, which is the least efficient DMU.

DMU 3	Original value	Target value	% Change
Stage 1 cost	131	101.00	-22.9
Stage 2 cost	148	72.67	-50.9
Stage 3 cost	132	107.05	-18.9
Fill rate _stage 1	95	95	0
On time delivery _stage 1	88	92	4.5
Cycle time_stage 1	4	4	0
Fill rate _stage 2	78	78	0
On time delivery _stage 2	87	89	2.3
Cycle time_stage 2	4	4	0

Table 4. Target values for inputs, outputs and intermediate variables for DMU 3.

For example, for DMU 3, its average efficiency is 0.918 and the value chain efficiency (using Banker's model) is 0.691. The values of $\Omega_1^* = 0.771$, $\Omega_2^* = 0.491$ and $\Omega_3^* = 0.811$ for DMU 3 (from Table 4) indicate that all the three channels (stages) are inefficient. In order to reach the best practice, each channel, i.e., stage 1, 2 and 3 could reduce their inputs while maintaining the same level of outputs (based upon Ω_1^*, Ω_2^* and Ω_3^*, which are less than 1). In the case of DMU 3, all its direct input slacks have zeros values. Thus, the cost for stage 1 could be reduced to 101; while the cost for stage 2 and 3 could be reduced to 72.67 and 107.05 respectively. This is equivalent to a 22.9% reduction of cost for stage 1, 50.9% reduction of cost for stage 2 and 18.9% reduction of cost for stage 3. In addition, the on time delivery between stage 1 and 2 could be increased to 92% from the current rate of 88%; and the on time delivery between stage 2 and 3 could be increased to 89% from the current rate of 87%. These solutions indicate that based upon the best practice, the associated channels (stages) would be able to maintain the on time delivery rates, i.e., 92% and 89% respectively while cutting down costs. These are the potential savings that can be realized from the value chain or multistage system if it is characterized in a better way. Similarly, the adjustment for other DMUs and their system potential savings could be interpreted using the same way.

4. Conclusions

This chapter draws on previous DEA models and advances the construction of the models for measuring the entire value chain or multistage efficiency. The chapter contributes to the existing value chain (multistage system) literature by providing a simple alternative model to measure the efficiency of the system. This model removes the indirect effect of one's channel performance which affects the efficiency status of another channel. The results show if we characterize the value chain through consideration of the impact of intermediates measures, potential savings can be realized in the system. Though, this model may not have addressed all the concerns in value chains or multistage systems, it can serve as a tentative solution for measuring the efficiency of these systems. This model can be further enhanced by analyzing how different settings of weights affect the overall value chain or multistage performance. In addition, future research can also look into how to adapt the model in uncertain environments, e.g., by utilizing the Monte Carlo method. Lastly, this chapter serves as an exposition to the awareness on the potential of simple conventional models to address more complex problems.

5. References

Arnold, J.R., Chapman, S.N, Clive L.M (2008). *Introduction to Materials Management*. Upper Saddle River, Prentice Hall. ISBN-13: 978-0132425506, New Jersey.

Banker, R.D., Morey, R.C. (1986). Efficiency analysis for exogenously fixed inputs and outputs, *Operations Research*, Vol.34, No.4, pp. 513-521, ISSN 0030-364X.

Charnes, A., Cooper, W.W., Rhodes, E. (1978). Measuring the inefficiency of Decision Making Units. *European Journal of Operational Research*, Vol.2, No.6, pp.429-444, ISSN: 0377-2217.

Färe, R., Grosskopf, S., Lovell, C.A.K., 1994a. *Production frontiers*. Cambridge University Press, ISBN-13: 978-0521420334, Cambridge.

Färe, R., Grosskopf, S. (2000). Network DEA. *Socio-Economic Planning Sciences*, Vol.34, No.1, pp.35-49, ISSN: 0038-0121.

Cooper, W.W., Seiford, L.M., Tone, K. (2006). Introduction *to Data Envelopment Analysis and its Uses: with DEA-Solver Software and Reference*, Springer, ISBN-13: 978-0387285801,New York.

Gattoufi, S., Oral, M., Kumar, A., Reisman, A. (2004). Epistemology of data envelopment analysis and comparison with other fields of OR/MS for relevance to applications. *Socio-Economic Planning Sciences*, Vol.38, No.2/3, pp.123-140, ISSN: 0038-0121.

Gattoufi, S., Oral, M., Reisman, A. (2004). Data Envelopment Analysis literature: a bibliography update (1996-2001). *Socio-Economic Planning Sciences*, Vol.38, No.2/3, pp.122-159, ISSN: 0038-0121.

Golany, B., Hackman, S.T., Passy, U. (2006). An Efficiency Measurement Framework for Multi-Stage Production Systems. *Annals of Operations Research*, Vol.145, No.1, pp.51-68, ISSN: 0254-5330.

Kirkwood, C.W. (1997). *Strategic Decision Making: multiobjective decsion analysis with spreadsheets*, Duxbury, ISBN-13: 978-0534516925, California.

Clemen, R.T., Reilly, T. (2001). *Making hard decisions with DecisionTools*. Duxbury Thomson Learning, ISBN 0-534-36597-3, U.S.

Liang, L.F. Yang, Cook, W.D., Zhu, J. (2006). DEA models for supply chain efficiency evaluation, *Annals of Operations Research*, Vol.145, No.1, pp.35-49, ISSN: 0254-5330.

Ramanathan, R., 2003. *An Introduction to Data Envelopment Analysis: a Tool for Performance Measurement*. Sage Publications, ISBN-13: 978-0761997603, New Delhi.

Seiford, L.M. (1996). Data envelopment analysis: the evolution of the state of the art (1978-1995). *Journal of Productivity Analysis*, Vol.7, No.2/3, pp.99-137, ISSN: 0895-562X.

Wisner, J.D., Tan, K.C., Leong, G.K. (2009). *Principles of Supply Chain Management: a Balanced Approach*. South-Western Cengage Learning, ISBN: ISBN13: 9780324657913, U.S.

Wong, W.P., Wong, K.Y. (2007). Supply chain performance measurement system using DEA modeling. *Industrial Management & Data Systems*, Vol. 107, No.3, pp.361-381, ISSN: 0263-5577.

Wong, W.P., Wong, K.Y. (2008). A review on benchmarking of supply chain performance measures. *Benchmarking: an International Journal*, Vol.15, No.1, pp.25-51, ISSN: 1463-5771.

Wong,W.P., Jaruphongsa, W., Lee, L.H. (2008). Supply Chain Measurement System - A Monte Carlo DEA based approach. *International Journal of Industrial and Systems Engineering*, Vol. 3, No.2, pp.162-188, ISSN:1748-5037.

24

Analog to Digital Conversion Methods for Smart Sensing Systems

José Miguel Pereira, Octavian Adrian Postolache and Pedro Silva Girão
Instituto de Telecomunicações
Portugal
ESTSetubal (LabIM), Polytechnic Institute of Setúbal
Portugal

Abstract

The new capabilities of smart sensing systems namely, adaptability, reconfiguration, low-energy consumption and cost, between others, require a wisely selection of the methods that are use to perform analog to digital conversion. It is very important to optimize the trade-offs between, resolution, accuracy, conversion rate, and energy consumption, between others, and above all to adapt dynamically the conversion parameters for different signals characteristics and applications' purposes. Establishing the best trade-offs are even more important when signals to be digitized have different signal-to-noise ratios (S/N) ratios, different requirements of measuring accuracy and acquisition rate, their characteristics are time-variant and above all if they are sharing the same digitalization device.

Very low resolution or conversion rate of data acquisition (DAQs) systems are generally not compliant with measurement systems' requirements since signal information is lost without any possible recovery procedure. Otherwise, if resolution or data acquisition rate are excessively high that means the sampling rate is much higher than its minimum value (Nyquist rate), the excessive amplitude and time resolutions provided by A/D conversion or frequency-to-digital conversion (FDC) does not improve measurements system's performance. Moreover, the excessive resolution or data acquisition rate implies an increase of hardware and software complexity, data processing load and a higher implementation cost, without any benefits. So, for any A/D or FDC conversion method the best trade-off between different conversion characteristics must be established considering applications' purposes. For example, in wireless sensing and actuating networks (WSAN) energy wastes are particularly important because a wrong choice of conversion method can affect deeply measurement system autonomy.

Whenever possible, classical A/D conversion methods are being replaced by discrete A/D conversion methods that are supported by low cost microcontroller (μC) (Microchip, 2010) connected to a few external resistive or capacitive components. This solution takes full advantage of μCs benefits, namely specific hardware and software capabilities and it provides a conversion rate that can be higher that several hundreds of kHz that is sufficient

for a large number of applications. These conversion also enables an easy implementation of trade-offs between resolution, conversion rate, and energy consumption (EC), errors' compensation, namely offset and gain errors, and linearization capabilities without need of any additional circuitry, look-tables or linearization algorithms. However, the main drawbacks of these conversion methods include a large sensitivity to noise and the need of frequent self-calibration routines.
Alternative digitalization methods that can be implemented without need of classical A/D converters are based on FDC methods.
Referring to the contents of the present chapter, the first session includes an introduction, state of the art and the main requirements of smart sensing systems (SSS) that must be considered in terms of signals' conditioning and digitalization.
The second session summarizes the main characteristics of two A/D digitalization methods that seem to be particularly suitable for smart sensing applications, namely, discrete A/D and FDC methods. Some implementation details, particularly those ones that take full advantage of low-cost microcontrollers and specific integrated circuits will be underlined. This section also includes the description of the working principle and main advantages of each conversions FDC method together with some simulation and experimental results. The last session is dedicated to conclusions.

Keywords: digitalization methods, smart sensing, error compensation, resolution, signal-to-noise ratio.

1. Introduction

Analogue to digital (A/D) and digital to analogue (D/A) conversions are subjects of paramount importance in every digital data acquisition system. In the last years, especial demands appear in a large number of applications namely the ones that are supported by smart sensing devices. These demands were mainly related with A/D conversion resolution, data sampling rate and energy consumption. Improving the performance is each parameter is obviously possible but in more than one conversion parameter needs to be improved simultaneously it is not an easy task because those are correlated. A clear correlation exists always between two main conversion parameters; if resolution is increased, the maximum A/D conversion's rate always decreases. Other compromise, particularly important for wireless measuring systems, exists between sampling and energy consumption. If low energy consumption is a main requirement for a given measuring system, its maximum A/D conversion rate must be reduced as much as possible taking account to the Nyquist[1] frequency required for the input signal's bandwidth.
Even when an A/D converter is used to convert a single signal with time variable characteristics, for example harmonic contents or voltage range, there is also the need do adapt its conversion parameters according to the dynamic characteristics of the signal.
Using multiple A/D converters in the same data acquisition unit and processing their outputs is a possible solution to surpass previous drawbacks. However, this is not a cost

1 Nyquist rate is two times the bandwidth of a band-limited signal or a band-limited channel.

effective solution and it is not acceptable solution for smart [2]sensing systems, especially when many sensing unit are remote nodes of a wireless sensing network. In these applications, a single A/D with a flexible configuration must be used to optimize A/D conversion performance according signals' characteristics. These A/D converters, denominated as flexible analogue-to-digital converters (Pereira J. et al, 2001; Pereira J. et al, 2007), throughout this chapter, are aimed to establish the best trade off between the main A/D conversion's parameters in order to improve measuring system's performance. It is also important to underline previous trade-off must also A/D conversion errors[3], namely, linearity errors, offset and gain errors, signal to noise and distortion rate, jitter and latency time, to name a few, that must be minimized.

Additional advantages that are associated with a flexible A/D converter include an easy implantation on non-uniform sampling, particularly important in which concerns minimization of measuring system's energy consumption, generation of non-linear digital transfer characteristics and capability to implement different conversion methods. Since conversion rate is not a crucial requirement for the applications that are focused in the present chapter, the hardware and software parts of the A/D converter are mainly implemented by low cost microcontrollers (μCs), digital signal processors (DSPs) or field programmable gate arrays (FPGAs), between others processing devices.

2. Suitable digitalization methods for smart sensing applications

There are a large number of non-traditional and low-cost A/D conversion methods very suitable for SSS applications. Between these methods it is possible to underline pulse-width modulation methods, potentiometric methods, voltage to frequency methods, RC charge time constant methods, single, dual and multislope methods, sigma-delta methods and frequency to digital (F/D) conversion methods. It is important to underline that any of these methods can provide low energy power consumption, measurements error's compensation, namely, offset, gain and non-linearity errors compensations, easy implementation of the trade-offs between A/D or D/A conversion parameters, namely accuracy and conversion rate, and capability to implement multiple variable measurements with a single virtual sensor[4]. Moreover, it is also important to remember that in SSS applications accuracy is not the main goal of several applications and usually a number of bits lower than ten can fulfill applications' requirements. Data processing techniques that are exploited in SSS enhance conversion accuracy using different signal processing techniques after digitalization (e.g. data averaging, correlation and prediction techniques, errors' compensation, between others) and can compensate errors caused by influence variables[5] such as temperature, humidity and mechanical vibrations.

---

[2] A smart sensors or smart transducer is a device with built-in intelligence, whether apparent to the user or not.

[3] Institute of Electrical and Electronics Enginneers, Inc., IEEE Std 1241 DRAFT: Standard for Terminology and Test Methods for Analog-to-Digital Converters.

[4] A virtual sensor is a sensor that can provide indirect measurement of several variables by using data processing algorithms that are applied to direct measurement's results provided by a single sensor.

[5] Influence variables are variables that are cross-correlated with measuring variables affecting measurement's accuracy.

It is beyond the scope of this chapter to describe a huge number of alternative conversion methods that can be used for SSS applications. Instead, two A/D conversion methods already utilized by the authors will be presented. One of these methods is based on an improved PWM A/D conversion method and the other one is based on a FDC converters. Some implementation details, particularly those ones that take full advantage of low-cost microcontrollers and specific integrated circuits are underlined together with some limitations of each method.

2.1 PWM based A/D conversion

Traditional PWM based A/D converters require a very high resistive-capacitive (RC) time constants and their sampling rate is very low. Typically, lower than a few tens of samples per second (S/s) for an eight bits' converter. This conversion method uses a PWM signal use mean value, obtained from the output of a low-pass-filter (LPF) is sequentially incremented by the quantization step amplitude and then compared with the input voltage. However, this A/D conversion method contains several disadvantages, namely, as previously referred, this method only supports very sampling rates, and it is very sensitive to noise and to components' deviations from their nominal values, usually resistors and capacitors tolerances.

A variant much more interesting of the traditional PWM based conversion method can be obtained if the input voltage conversion is performed using a single a square signal that contains a number (n) of high levels pulses and a number of low level pulses that depends on input voltage's amplitude to be converted. The resolution of this method is easily controlled by software. Each time the number of pulses contained in the conversion cycle is doubled an additional resolution bit is obtained. For each positive or negative conversion pulse, the voltage variation of LPF output is incremented or decremented by the quantization step amplitude of the converter, respectively. This A/D conversion method is especially suitable for SSS with a sampling rate requirements lower than a few hundreds of samples per second. It enables analogue-to-digital conversion without need of dedicated A/D integrated circuits, takes full advantage of microcontrollers' compare, capture and PWM (CCP) capabilities and supports an easily implementation of self-calibration routines to compensate gain, offset and nonlinearity errors without any requirements of additional circuitry, look-tables or linearization algorithms.

2.1.1 Improved PWM based A/D converter: working principle

Figure 1 represents the hardware block diagram of the converter. Its hardware part includes two digital ports of the μC, an analog comparator (Comp.) that can be inside the μC, a low-pass (RC) filter and an analogue switch (S).

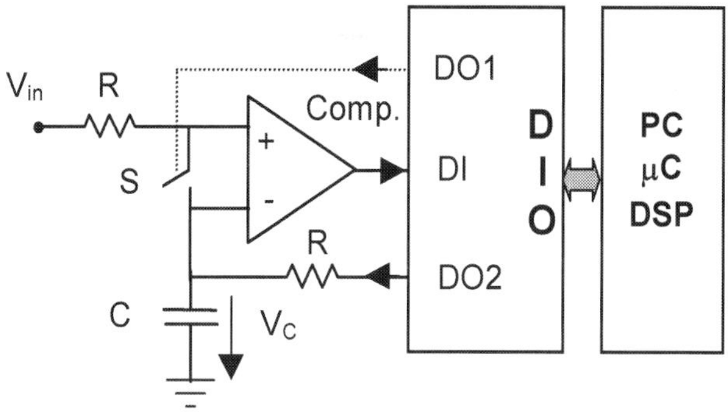

Fig. 1. Hardware block diagram (Comp.- comparator; DI- digital input; DO- digital output; DIO- digital input-output interface; PC- personal computer; μC- microcontroller; DSP- digital signal processor)

Before each A/D conversion, the analogue switch (S) is closed, and, the capacitor is charged with an initial voltage (V_{C0}) equal to the input voltage (V_{in}). It is assumed that the signal amplitude variation during acquisition time is lower than one LSB (Least Significant Bit). After this initialization cycle, the analogue switch (S) is opened, and the measurement cycle takes place. During this phase, the voltage delivered by the digital output (DO2) tends to maintain the capacitor charge by monitoring the output voltage of the comparator (Comp.). If the digital input (DI) is high, meaning that capacitor voltage is lower than V_{in}, digital output (DO2) is settled to the high level voltage (V_H), otherwise digital output (DO2) is settled to the low level voltage (V_L). Different values can be assigned for V_H and V_L and it is this way possible to reduce the voltage range of the A/D converter and to increase its resolution. Assuming that the R·C time constant is much higher than the pulse period (T_P) and lower than the capacitor's voltage variation in the order of few LSB's during T_P, it is possible to obtain the relationship between the input voltage amplitude (V_{in}) and the A/D converter's output code (n) (Pereira J. et al, 2001),

$$V_C(T_C) = V_{in} + n \cdot (V_H - V_{C0}) \cdot \left(1 - e^{-\frac{T_{P1}}{RC}}\right) + m \cdot (V_L - V_{C0}) \cdot \left(1 - e^{-\frac{T0}{RC}}\right) \quad (1)$$

where n and m represent the number of positive and negative pulses delivered by the μC during the conversion period (T_C), V_H represents the voltage amplitude of the positive pulses that have a time duration equal to (T_{P1}) and V_L represents the voltage amplitude of the negative pulse that have.

For a binary A/D converter coding $n+m=2^B$, where B represents the number of A/D converter's bits, from the previous relationship it is possible to obtain,

$$V_{in} = \frac{nk}{(2^b - n) + nk} V_H + \frac{2^B - n}{(2^B - n) + nk} V_L \quad (2)$$

where $k = \frac{1 - e^{-T_{P1}/RC}}{1 - e^{-T_{P2}/RC}}$ is a non-linear factor modulated by the relation between positive (T_{P1}) and negative (T_{P0}) pulse duration of the PWM signal, B is the number of bits of the ADC and V_H/V_L are the high and low digital voltage levels delivered by digital output (DO2), respectively.

Figure 2 represents the A/D converter's characteristic for five different values of k parameter. An eight's bits A/D converter with a ratio between pulse duration and time (τ=RC) is equal to 1/18. As expected for k=1, $T_H=T_L$, a linear characteristic is obtained.

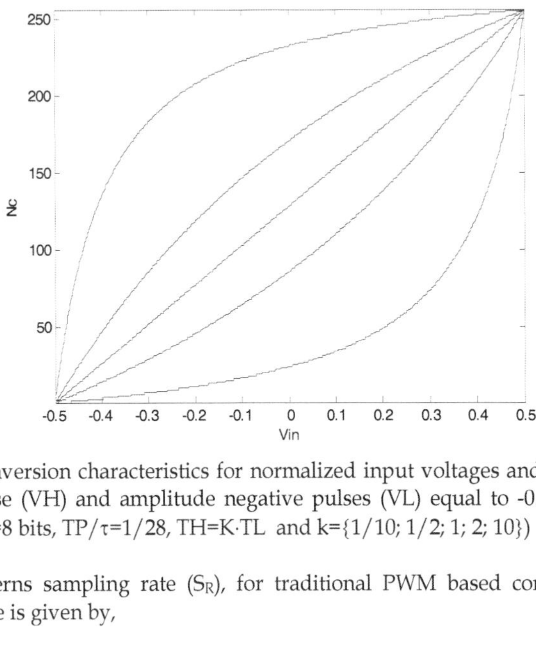

Fig. 2. A/D conversion characteristics for normalized input voltages and for an amplitude of positive pulse (VH) and amplitude negative pulses (VL) equal to -0.5 V and to 0.5 V, respectively (B=8 bits, TP/τ=1/28, TH=K·TL and k={1/10; 1/2; 1; 2; 10})

In which concerns sampling rate (S_R), for traditional PWM based converters, its value maximum value is given by,

$$(S_R)_{max} = \frac{(f_{\mu C})_{clk}}{4^B \cdot N} \tag{3}$$

where B represents the converter's number, $(f_{\mu C})_{clk}$ represents the converter's clock rate and N represents the number conversions that are performed during one period of the sinusoidal input signal (oversampling factor). Figure 3 represents the maximum frequency of a sinusoidal signal that can be converted by a traditional PWM based A/D converter as a function of the number of samples per second. As represented the maximum sampling rate is almost equal to 10 Hz S/s for the Nyquist rate (N=2).

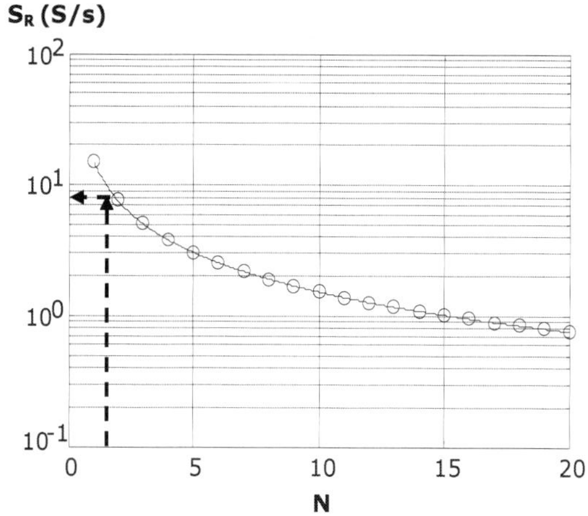

Fig. 3. A/D conversion rate for a converter with 8 bits resolution and running under a μC frequency clock equal to 1 MHz (traditional PWM based A/D converter)

If similar conditions are considered for the improved PWM based A/D converter the maximum sampling rate value is given by,

$$(S_R)_{max} = \frac{(f_{\mu C})_{clk}}{2^B \cdot N} \tag{4}$$

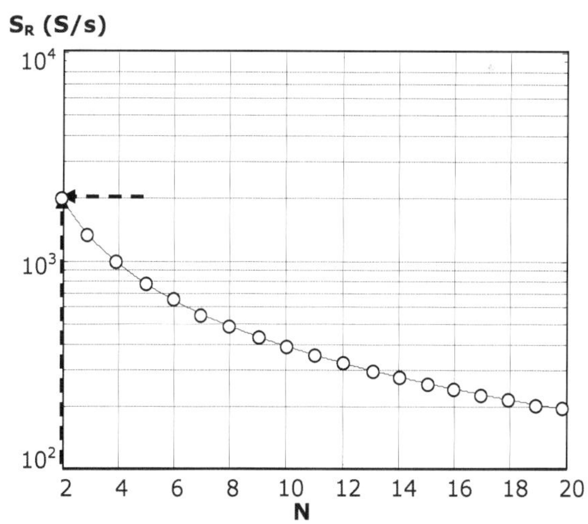

Fig. 4. A/D conversion rate for a converter with 8 bits resolution and running under a μC frequency clock equal to 1 MHz

Figure 4 represents the maximum frequency of a sinusoidal signal that can be converted by an improved version of the traditional PWM based A/D converter as a function of the oversampling factor. As represented in the figure signal's bandwidth is almost equal to 2 kS/s for the Nyquist rate (N=2).

2.1.2 Self-calibration capability

In order to improve the accuracy of the A/D converter, the measurement cycle can include an auto-calibration procedure (Goes F. & Gerard M., 1997; Goes F. & Gerard M., 1997), multiplexing three input signals: V_{in}, V_{REF} and V_{OFF}. The hardware block diagram of the PWM-A/D converter with continuous auto-calibration capability is represented in Figure 5.

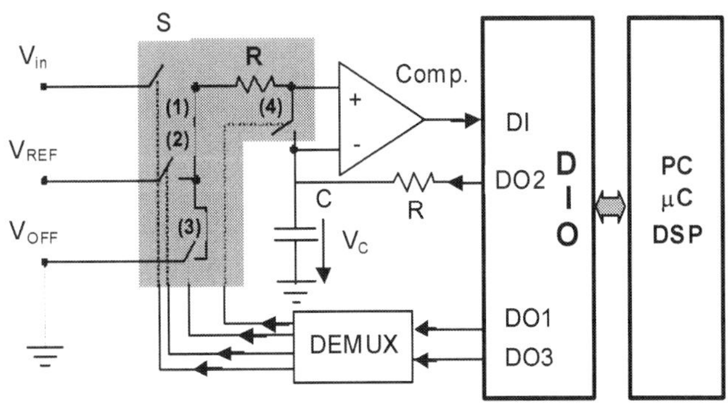

Fig. 5. Hardware block diagram of the PWM-A/D converter with continuous auto-calibration capability (Comp.- comparator; DI- μC's digital input port; DO- μC's digital output port; DIO- μC's digital input-output communication's interface; DEMUX-demultiplexer; PC- personal computer; μC- microcontroller; DSP- digital signal processor)

The reference voltage (V_{REF}) is usually equal to the ADC full-scale voltage (FS) and is used to correct the gain error. This reference voltage must have, at least, accuracy higher than ½ LSB of the A/D converter. For example, for an 8-bit converter, the reference voltage error must be lower than 0.195% of FS. The offset voltage (V_{OFF}) is usually equal to ground voltage level and is used to correct the offset error.

If the measurement results for V_{in}, V_{REF} and V_{OFF}, are equal to n, n_{REF} and n_{OFF}, respectively, the offset and gain errors can be continuously cancelled by using the following relation:

$$n_C = \frac{n - n_{OFF}}{n_{REF} - n_{OFF}} \cdot 2^B \tag{5}$$

where n_C represents the offset and gain compensated A/D converted code.

Another advantage of this measurement technique is that low-frequency noise (1/f) and temperature drift errors are almost cancelled because their bandwidth is generally much lower than the conversion period (T_C).

2.1.3 Experimental results

In this section we will considered the linearization characteristic of a circuit that includes a nonlinear thermistor (ON-400) (Omega, 1995) whose characteristic is represented in Figure 6. Linearity errors are represented for three different cases: without any correction of linearization errors for the temperature range [0, 50] (figure 7); with a first order least mean square (LMS) curve fitting (figure 8) and with the proposed PWM based A/D converter (figure 9). The advantage of the proposed solution for linearization purposes is clearly shown in Figure 9. It enables a non-linearity error reduction of an almost equal to 10 and 5 when compared with the terminal based linearity error and the first order LMS curve fitting, respectively, without the need of look-up tables or mathematical routines for linearization.

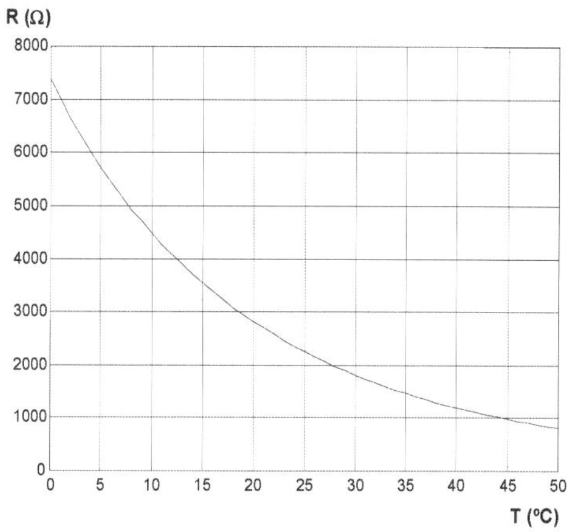

Fig. 6. Resistance-temperature characteristic of an ON-400 thermistor probe with standard 10″ leads

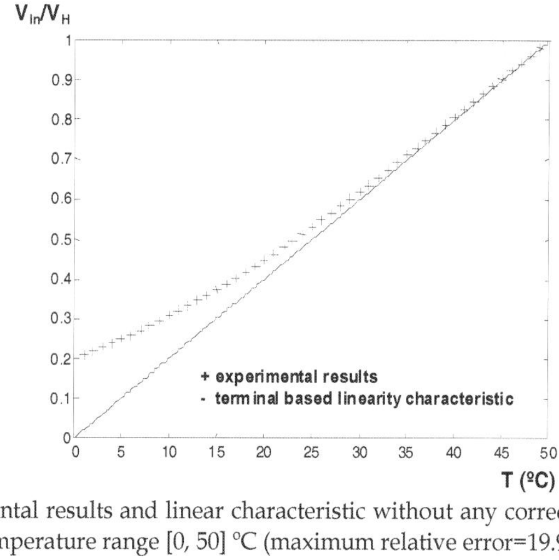

Fig. 7. Experimental results and linear characteristic without any correction of linearization errors for the temperature range [0, 50] °C (maximum relative error=19.9 % of FS)

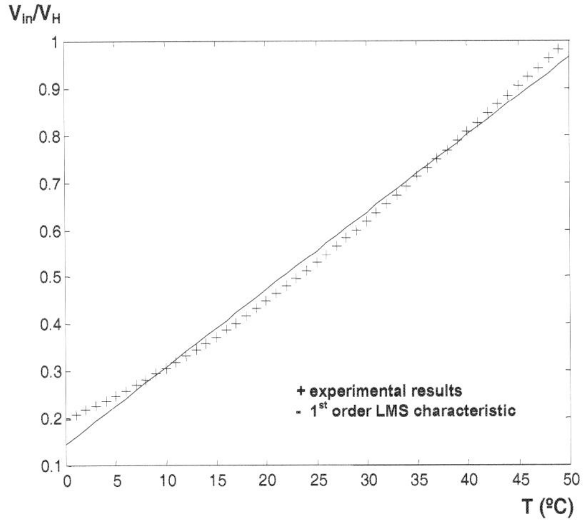

Fig. 8. Experimental results and 1st order LMS for the temperature range [0, 50] °C (maximum relative error=5.4 % of FS)

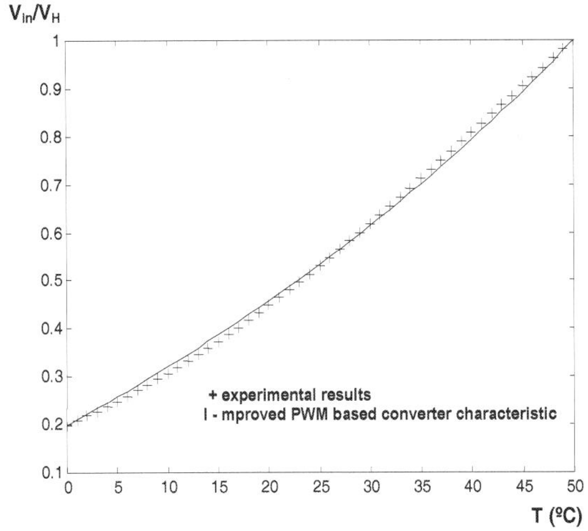

Fig. 9. Experimental results and curve fitting characteristic obtained with the improved PWM based A/D converter for the temperature range [0, 50] °C (maximum relative error=1.4 % of FS, $k=k_{OPT}=0.72$)

2.2 Frequency to digital conversion methods

Nowadays the number of transducers that use frequency-time domain parameters modulated by a physical, chemical or biological quantity is increasing and they are a promising alternative that must be considered in the design and implementation of any measurement system. Several advantages are associated with FDC methods. Based on

calibration data or on-line historical measurement data, an appropriate accuracy of the FDC device can be settled avoiding excessive conversion duration (T_C) that restricts, without any benefit, the maximum sampling rate of the measurement system. An intrinsic advantage of FDC conversion methods is that they eliminate the need of traditional A/D and D/A converters and they can compensate a large number of, internal and external, error sources that are associated with any measurement system. Additional advantages include higher accuracy, resolution, dynamic range, noise immunity, simplicity of implementation and interfacing, and a lower cost compared with equivalent digitalization methods with identical performances. Moreover, the advantages, previously referred, the high accuracy of this conversion method is related with accuracy. It can achieve a relative errors (δ) lower than 0.001 % FS (full-scale), being this error negligible for almost any measuring system.
The next paragraphs will underline main characteristics of FDC conversion methods presenting their working principle and their self-adaptive characteristics that make them a suitable for SSS applications.
In which concerns implementation, any microcontroller or a set of commercial off-the-shelf (COTS) integrated circuits, without any complex circuitry, can support the hardware that is required for the implementation of a FDC.

2.2.1 Frequency to digital conversion methods: working principle

The working principle of FDC methods is based on different techniques that can perform time and frequency measurements (IFSA, 2004, Kirianaki et al, 1998; Kirianaki et al, 2002). Between these techniques, we can underline the method of dependent count (MDC) due to its advantages and easy implementation.
Using this method a fixed number of pulses is counted according to the required measurement accuracy. Figure 10 represents the timing diagram of the MDC during a single measurement cycle for accuracy better than δ.

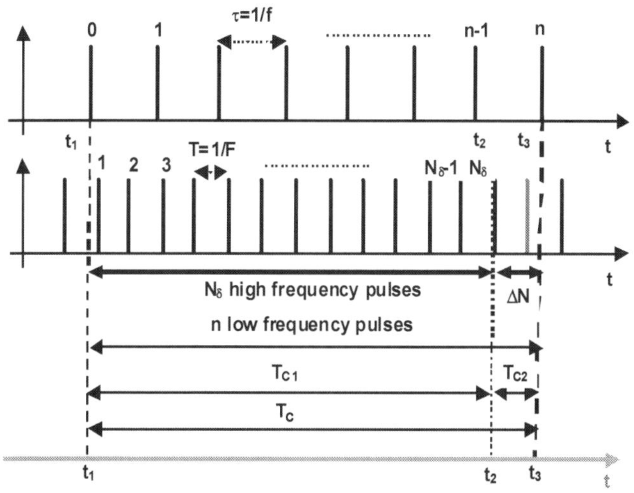

Fig. 10. Timing diagram of the MDC during a single measurement cycle (accuracy better than δ)

To obtain a relative measurement error lower than δ we must count at least $N_\delta=1/\delta$ pulses of the signal with higher frequency (F). If F is higher frequency signal's value and f the lower frequency signal's value and the unknown frequency (f_x) is given by,

$$\begin{cases} f_x = f_0 \cdot \dfrac{n}{N_\delta} & \text{if } f_x \leq f_0 \\ f_x = f_0 \cdot \dfrac{N_\delta}{n} & \text{if } f_x > f_0 \end{cases} \tag{6}$$

where f_x represents the unknown frequency, f0 represents the reference frequency, N_δ and n represents the number of high frequency and low frequency pulses counted during conversion period (T_C), respectively. As represented in figure 10, T_C is an integer number (n) of the low frequency signal period (τ).
The conversion period contains a number of high frequency pulses equal to N_δ plus a remaining number of pulses ΔN contained in time interval T_{C2}. Minimum and maximum values of quantization error are associated with the minimum and maximum number remaining pulses counted during time interval [t_2, t_3], respectively. These values are given by,

$$\begin{cases} \Delta N_{min} = 0 \\ \Delta N_{max} = \dfrac{\tau}{T} = \dfrac{F}{f} \end{cases} \tag{7}$$

From the previous expression, taking account the real number of high frequency pulses that are counted, the minimum and maximum values of quantization errors are given by,

$$\begin{cases} \delta_{max} = \dfrac{1}{N_\delta + \Delta N_{min}} = \dfrac{1}{N_\delta} \\ \delta_{min} = \dfrac{1}{N_\delta + \Delta N_{max}} = \dfrac{1}{N_\delta + F/f} \end{cases} \tag{8}$$

We can conclude that MDC is particularly suitable for time and frequency measurements. This method enables an easy adjustment of measurement's accuracy by selecting an appropriate value of N_δ for the requirements of each application. It is important to underline that it does not make any sense to define a quantization error much lower than the required accuracy since sampling rate and processing load are unnecessary affected. If the reference frequency (f_0) is selected to minimize the ratio F/f from expression (8) we can conclude that the measurement error is almost constant an equal to $1/N_\delta$. In which concerns conversion time (T_C) its minimum and maximum values are given by,

$$\begin{cases} (T_C)_{min} = N_\delta \cdot T \\ (T_C)_{max} = (N_\delta + F/f) \cdot T \end{cases} \tag{9}$$

From the previous relationships, we can conclude that as long as N_δ is much higher that F/f ratio the conversion time is almost constant without any waste overtime.

As expected MDC is not an exception, in which concerns accuracy and conversion rate, the conversion rate is inversely proportional to the requested accuracy.

2.2.2 Experimental results

In this section, we will consider a relative humidity measurement system with temperature compensation.

A smart sensing temperature kit (Smartec, 2010) containing four temperature sensors (SMT 160-30) was used to perform temperature measurements. The electrical schematic diagram and view are represented in figures 11 and 12, respectively. Each temperature sensor delivers a duty-cycle modulated signal.

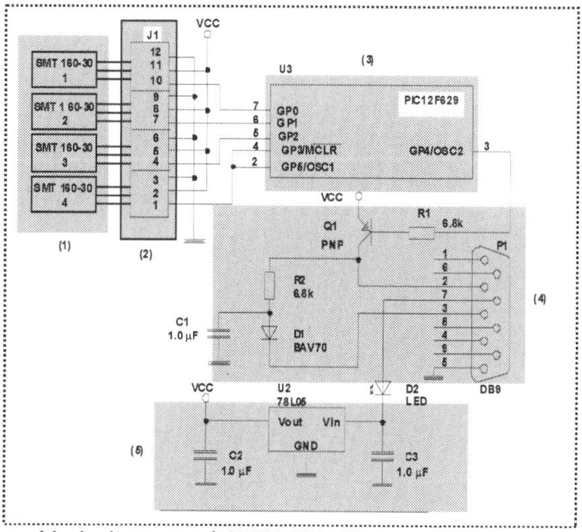

Fig. 11. Hardware block diagram of SMTAS04: (1) temperature sensors; (2) connecting terminals; (3) µC- PIC12F629; (4) DB9- RS232 interface plug; (5) voltage regulator

Fig. 12. View of the Evaluation Board SMTAS04 (with permission from SMARTEC Inc.)

The output signal for each sensor is a square wave with a well-defined temperature-dependent duty cycle. The temperature is linearly related with the DC according to the following relationship,

$$T = \frac{DC - 0.320}{0.00470} \tag{10}$$

where DC represents the duty-cycle and T the temperature in °C.

Since the DC modulated signal delivered by SMT 160-30 is TTL/CMOS compatible it can be directly connected to the digital input of a microcontroller device, for example, without need of any interface circuits. Measurement's accuracy can be increased by sampling the DC signal over a larger number of periods, however this cause a lower conversion rate and a higher value of power consumption. According to applications' requirements it is possible to optimize the trade-offs between accuracy, conversion rate and power consumption that can eventually change during real time operation. Statistical parameters of measurement or self-calibration data can be used to improve measurement's system performance (Pereira J. et al, 2009).

The relative humidity transducer (HF3223) (Humirel Inc., 2001) delivers a square wave signal whose frequency is given by,

$$F_{out} = 9740\text{-}18 \cdot RH \quad (Hz) \tag{11}$$

where RH represents relative humidity in percentage and F_{out} is the frequency in kHz. The frequency varies between 9560 Hz and 8030 Hz when RH varies between its minimum and maximum values, 10 % and 95 %, respectively.

Time and frequency measurements were performed using the universal frequency to digital converter integrated circuit UFDC (IFSA, 2010) whose block diagram and view are represented in figures 13 and 14, respectively.

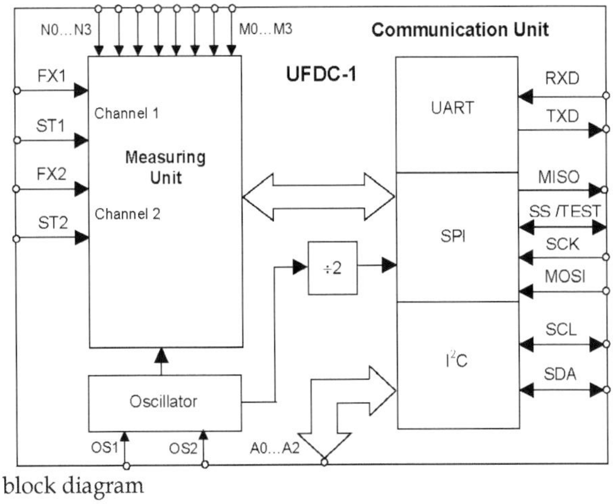

Fig. 13. UFDC block diagram

Fig. 14. View of the Evaluation Board SMTAS04 (with permission of Sergey Yurish)

The experimental tests of the temperature compensated humidity measurement system were performed using the UFDC that performs frequency, period, its ratio, duty-cycle and phase-shift measurements using a set of programmable accuracies levels. In the following, we propose a method that can be used to configure the FDC. The FDC parameters were configured to improve measurement's system performance, namely, to prevent FDC waste overtime and superfluous processing load (Pereira J. et al, 2005).

The method that was used to reach those objectives is based on the statistical parameters of measurement and self calibration data in order to configure properly the FDC device. The main tasks of the software programs that were include the uncertainty evaluation of the measurement data, the accuracy selection of the UFDC device, the detection of abnormal and faulty working conditions and the compensation of RH measurements caused by temperature variations.

The criteria used for UFDC accuracy selection is based on a Gaussian curve fitting histogram of the calibration data. Being the standard deviation (σ_c) and the mean value of the calibration (X_c) data defined by:

$$\sigma_c = \sqrt{\frac{\sum_{i=1}^{N}\left(X_c(i)-\overline{X}_c\right)^2}{N-1}} \qquad \overline{X}_c = \frac{\sum_{i=1}^{N} X_c(i)}{N} \tag{12}$$

where $X_c(i)$ represents the calibration measurements of a set of N points. The standard deviation, or a multiple of its value, is generally used to quantify measurement uncertainty. An uncertainty value equal to $\pm 3\cdot\sigma_c$ of the calibration data set assures, for a statistical

Gaussian distribution, that 99.7% of the measurements, of a given quantity (Xm), are within the interval $[X_m-3\cdot\sigma_c; X_m+3\cdot\sigma_c]$.
Based on these premises, the UFDC selected accuracy (δ_{max}) is the higher one that verifies the following condition:

$$\delta \leq 100 \cdot \min\left(\frac{3\cdot\sigma_C}{\overline{X}_C}\right) \quad (\%) \tag{13}$$

where σ_c represents the standard deviation obtained from Gaussian curve fitting histogram, $\overline{X}_C$ represents the average value of measurement results, obtained during calibration phase, and min() represents the minimum value of the argument evaluated for the set of N calibration points.
The above criteria assures that selected UFDC accuracy is adjusted to 99.7 % of the calibration data being the major part of measurement noises' automatically cancelled without need of superfluous computing procedures (e.g. averaging). This result optimizes measurement's system performance especially when the UFDC device converter multiple signals with different accuracies and sampling rates requirements. Figures 15 and 16 represent the Gaussian curve fitting for a set calibration points (N=1024) and the front panel of the virtual instruments (VI) (NI, 2010) used to display temperature and humidity measurement results, respectively.

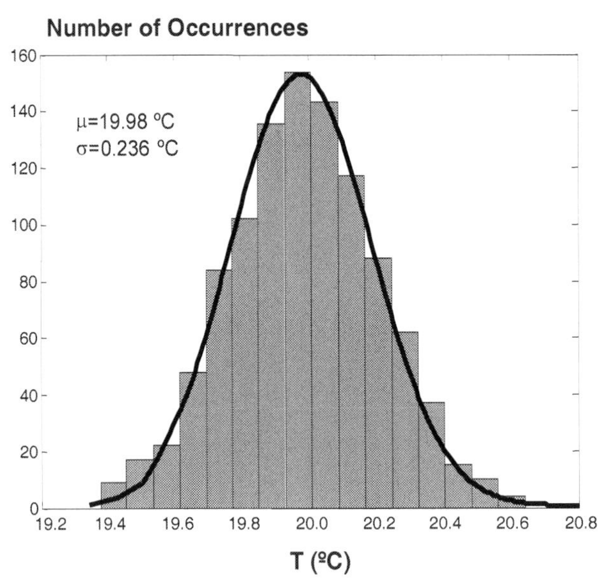

Fig. 15. Gaussian curve fitting of measurement data distribution

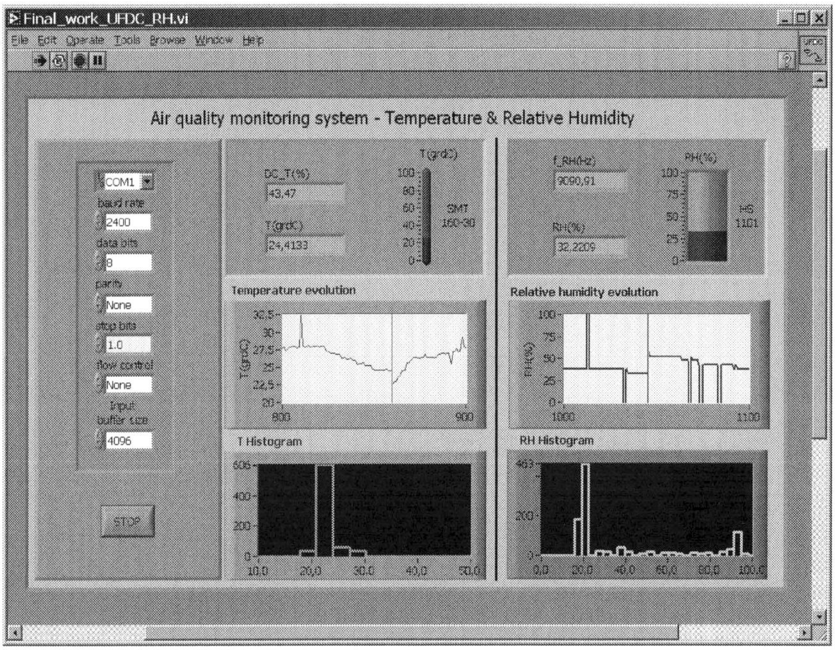

Fig. 16 VI front panel of the VI used to display temperature and relative humidity measurement results

In this example, the normalized standard deviation of measurement data is approximately equal to 3.5 %. Therefore, since the UFDC integrated circuit is used to perform FDC is implanted by the UFDC, accuracy equal to 0.5% is enough to fulfill application's requirements. If the maximum measurement accuracy of the UFDC was selected (δ=0.001%) the conversion time would be 500 times higher without any performance gain.

3. Conclusions

This chapter presents two A/D conversion methods that are particularly suitable for SSS applications. Both methods exhibit a set of characteristics that ensures an easy establishment of trade-offs between, resolution, accuracy, conversion rate, dynamic range and energy consumption. Moreover, both methods enable the establishment of flexible choice of conversion parameters that can be dynamically adjusted to optimize measurement's systems performance for different input signals characteristics and applications' purposes.
Finally, it is important to underline that exists many others A/D conversion solutions, not based on conventional A/D, that are also suitable for SSS A/D conversion. However, as previously referred, it is beyond the scope of this chapter to describe a huge number of alternative conversion methods that can be used for SSS applications.

4. References

Goes F & Meijer G. (1997). A Universal Transducer Interface for Capacitive and Resistive Sensor Elements, Analog Integrated Circuits and Signal Processing, 14, pp. 249-260, Kluwer Academic Publishers, Boston, 1997

Goes F. & Meijer G. (1997). A Simple Accurate Bridge-Transducer Interface with Continuous Autocalibration, IEEE Trans. Instrum. Meas., Vol. 46, No.3, pp. 704-710, June 1997

Humirel Inc. (2010). Humidity Module-HF3223, available: http://www.datasheetarchive.com/HF3223-datasheet.htm, Jan. 2010

IFSA, International Frequency Sensors Association (2004). Universal frequency-to-Digital Converter (UFDC-1) - Specification and Application Note, 2004

IFSA (2010), International Frequency Sensors Association. Universal Frequency to Digital Converter, available: http://www.sensorsportal.com/HTML/E-SHOP/PRODUCTS_4/UFDC_1.htm, Jan. 2010

Kirianaki N; Yurish S. & Shpak N. (1998). New Processing Methods for Microcontrollers Compatible Sensors with Frequency Output, Proceedings of the 12th European Conference on Solid-State Transducers and the 9th UK Conference on Sensors and their Applications, Southampton, UK, 13-16 September 1998

Kirianaki N.; Yurish S.; Shpak N. & Deynega V. (2002), Data Acquisition and Signal Processing for Smart Sensors, John Wiley and Sons LTD, 2002

Microchip (2010). Microchip COM, Microcontroler Data Book, available: http://www.microchip.com, Jan. 2010

OMEGA Technology Company (1995). ON-400 Series Thermistor Probes, Manual M0294/0292, 1995

NI (2010), National Instruments. LabVIEW 8.2, available: http://www.ni.com/labview, Jan. 2010

Pereira J.; Postolache O. & Girão P. (2001). Extending Digital Input/Output Capabilities to Low-Cost and Non-Linear A/D Conversion, Proceedings of the 11th Symposium on Trends in Electrical Measurement and Instrumentation - IMEKO TC4, Vol.I, pp. 101-105, Lisbon, Portugal, Sep. 2001

Pereira J.; Postolache O. & Girão P. (2005). Using a Dual-Channel FDC Device and ANN Techniques to Improve Measurements Accuracy, Sensors & Transducers Journal, Vol.62, Issue 12, pp. 462-472, Dec. 2005

Pereira J.; Postolache O. & Girão P. (2001). A Digitally Programmable A/D Converter for Smart Sensors Applications, IEEE Transactions on Instrumentation and Measurement, Vol.56, No.1, pp. 158-163, Feb. 2007

Pereira J.; Postolache O. & Girão P. (2009). PDF-based Progressive Polynomial Calibration Method for Smart Sensors Linearization, IEEE Transactions on Instrumentation and Measurement, Vol. 58, No. 9, pp. 3245-32-52, Sep. 2009

Smartec (2010). Smartec- Sensors Catalog,available: http://www.mmselectronics.co.uk/smt160-30.htm, Jan. 2010